Electronic Devices and Circuits

Second Edition

Dharma Raj Cheruku
Director of Academic Affairs
Professor of Electronics and Communication Engineering,
College of Engineering, GITAM University,
Visakhapatnam, India.

Battula Tirumala Krishna
Associate Professor
Electronics and Communication Engineering,
College of Engineering, GITAM University,
Visakhapatnam, India.

Pearson

ISBN 978-81-317-0098-3

First Impression

Published by Pearson India Education Services Pvt. Ltd, CIN: U72200TN2005PTC057128

Head Office: 1st Floor, Berger Tower, Plot No. C-001A/2, Sector 16B, Noida – 201 301,Uttar Pradesh, India.

Registered Office: Featherlite, 'The Address' 5th Floor, Survey No 203/10B, 200 Ft MMRD Road, Zamin Pallavaram, Chennai - 600044
Website: in.pearson.com; Email:companysecretary.india@pearson.com

Digitally printed in India by Trinity Academy for Corporate Training Ltd, New Delhi in the year of 2025.

To my beloved father

Late Sri C V V R Krishna Murty

– *Dharma Raj Cheruku*

To my mother

Smt B Alamelu Mangamma

– *Battula Tirumala Krishna*

Contents

Preface

In the modern-day world, our lives are influenced beyond measure by electronics and its applications. It has stolen its way into our day-to-day activities, and even a brief disruption in its comforts puts us to great inconvenience.

In this scenario, an engineering undergraduate is expected to be familiar with the basics of electronics. This book, the second edition of *Electronic Devices and Circuits*, aims to provide the student with information on the core concepts governing the subject, with emphasis laid on the principles of electronic engineering. It will serve as an ideal text for students of electronics and its allied fields, and a good source of reference to those in the electronics industry as well as students of associate member certificate and diploma courses. Students preparing for competitive examinations like GATE and IES also stand to be benefited by this book.

The need for a comprehensive book on electronic devices and circuits has been felt for long, with a wide range of topics being included in the syllabi and courses of study of various universities across the country. We present a concise text that can be understood by an average student, and that which meets his requirements. We have dealt with the subject in a lucid language and introduced the concepts in simple logical sequence. A new chapter (Chapter 13) on operational amplifiers has been added in this edition to help the learner gain an insight into this widely used electronic device that has a myraid of consumer, industrial, and scientific applications. Complex problems are explained with a clarity that is sometimes found lacking in other contemporary works on the subject. This will enable students to obtain the information they seek effortlessly. The narrative style of the book, its clear illustrations, and ample number of strategically placed worked-out problems and review questions will aid the student in quick recapitulation and a better understanding.

Chapter 1 deals with *'what is electronics?'* and articulates upon the immense range of applications of electronic devices and circuits. This makes the student feel comfortable with the subject and proceed with confidence. Chapter 2 deals with the physics of electronic devices and discusses their operations to a minute level. Chapter 3 explains at length the principle, features, types and applications of the P-N junction. Chapter 4 analyses the circuits using diodes, while Chapter 5 discusses about a popular electronic component—the transistor—the invention of which revolutionized the world of electronics.

If a transistor is to be used for a given application, it is to be biased properly. Various types of biasing circuits and the thermal stability of the transistor circuits are discussed in Chapter 6. A major application of the transistor is in amplification. The design, analysis and applications of transistor amplifier circuits are presented in Chapter 7. Field Effect Transistors are introduced in Chapter 8, while power electronic devices like TRIAC, DIAC and UJT are dealt with in detail in Chapter 9. Feedback amplifiers and sinusoidal oscillators which form an essential part of most electronic gadgets are examined in Chapters 10 and 11, respectively.

Chapter 12 looks at amplifiers that can handle large signals. The concept of narrowband tuned amplifiers is also covered in this chapter. Chapter 13, as mentioned earlier, is on operational amplifiers. As the name suggests, this device can perform all possible operations in the field of electronics and any electronic circuit can be built using it. We take a close look at the principles and operations that work behind this component. Chapter 14, the concluding chapter of the book, is devoted to the CRO, which displays any voltage or current visually on a screen. The electronics relevant to the understanding of the oscilloscope is elaborately explained here.

The solved examples in this edition have been revised to expose the student to an array of problems of varying complexities. Further, solutions to all exercise problems can be accessed at *www.pearsoned.co.in/dharmarajcheruku*. Comments and feedback on the contents of the text are also welcome.

Acknowledgements

We are very grateful to the management of GITAM University and to its President, Dr M. V. V. S. Murthy, in particular, for their continuous encouragement and support in bringing out this edition of the book. It might not have been possible to realize this venture if it had not been for their enduring support and assistance.

We thank Suneetha Cheruku and Krishna Murthy Cheruku for their patience and support at the time of preparing the manuscript. We are grateful to Dr Ramana Rao, D. Rukmini, D. Rajasekhar and Surya Kumari for their encouragement and inputs to enhance the quality of the book. Special thanks are due to K. A. Mangamma, R. Madhavi and B. Vijaya Saradhi.

We extend our regards to all our teachers and well wishers, and are obliged to our colleagues and supporting staff at the Department of ECE, GITAM University, for their cooperation in bringing this project to fruition.

Dharma Raj Cheruku

Battula Tirumala Krishna

About the Authors

Dharma Raj Cheruku started his career as a scientist at LRDE, Bangalore, and later switched to MACE, Visakhapatnam. A postgraduate in the field of microwave and radar engineering from Osmania University, he was also a gold medalist at the undergraduate level. He joined the College of Engineering, GITAM University, Visakhapatnam, in 1990 as Assistant Professor, and was successively elevated as Professor, Head of the Department, and then Director of Academic Affairs, a position which he still holds. He is in addition, the Chairman of the Board of Studies of BE (ECE) at GITAM University, and a member of the Board of Studies in Electronics at Nagpur University. An ardent researcher in the field of antennas and biomedical engineering and an active member of numerous professional societies, Professor Cheruku is also the treasurer of Biomedical Engineering Society of India.

Battulu Tirumala Krishna is Associate Professor, Department of Electronics and Communication Engineering, GITAM University, Visakhapatnam. A postgraduate from Andhra University, he is currently pursuing his Ph.D. in digital signal processing. He has rendered several papers at national and international conferences, the significant ones being *First order and second order IIR type digital differentiator for real time applications* presented at Guru Nanak Dev Engineering College, Ludhiana, in November 2006, and *A new procedure for the design of fractional order digital differentiators and integrators* presented at Andhra University Autonomous College of Engineering, Visakhapatnam, in February 2007.

Chapter 1

Introduction

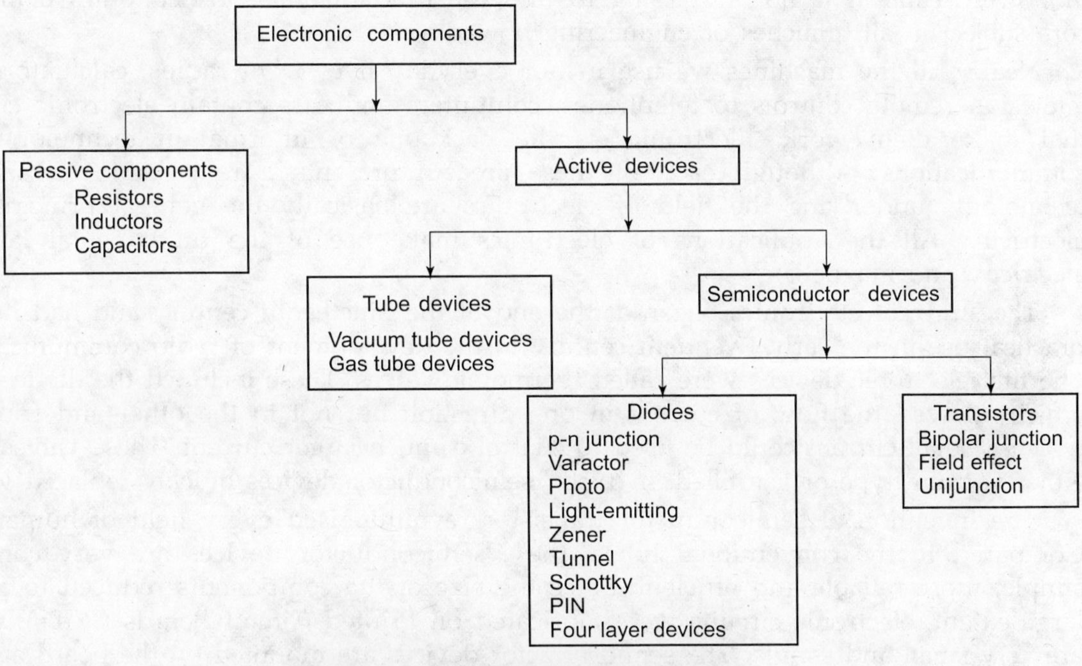

Classification of electronic components

Chapter Outline

This introductory chapter outlines the fundamental terms related to electronics and its applications while delving briefly into the concepts of electronic components. These are essentially awareness stimulation techniques before going into the serious thoughtful topics. One may call this the appetizer before the feast—'Electronic Warming Up'.

1.1 WHAT IS ELECTRONICS?

The word *electronics* is derived from *electron mechanics*, which means the study of the behaviour of an electron under different conditions of externally applied fields. Electronics can also be defined as physics of electrons and their utilisation. It is the study of how electrons behave and how they can be controlled so that they can do useful jobs. Electronics can also be referred to as *'the field of science and engineering, which deals with electronic devices and their utilisation'*. An ***electronic device*** is a device in which conduction takes place by the movement of electrons through vacuum, gas or semiconductor.

A tiny electronic component requires very minute power in the range of milliwatts or microwatts to function, but it can control or do works involving kilowatts or megawatts of power. Thus, it is not surprising to find the fundamentals of electronics being the core subject in all branches of engineering nowadays.

Nearly all the machines we use in our everyday lives from radios, calculators and television remote controls to telephones, computers and cars contain electronic circuits that make them work. Electronics is the backbone of information technology and communications. Although each of these areas represents a remarkable advance in technology, underlying the field of electronics are basically the general principles of electricity. All the applications of electronics make use of the fundamental laws of electricity and magnetism.

The study of electronics began at the end of the nineteenth century and had its first practical use in the early twentieth century in the development of radio communications. The first electronic devices were called thermoinic valves. These included the diode valve, which allowed the flow of current in one direction but not in the other and triode in which a small current could be used to control a much larger current. These valves were either vacuum type or gas filled. In 1950s, semiconductor devices quickly replaced valves.

The invention of semiconductor transistor revolutionised every field of human life. Compared to the conventional tubes, these semiconductor devices are very compact, simple, more reliable and efficient. Since the size of the components reduced to a very large extent, electronic circuits were fabricated on Printed Circuit Boards (PCBs), which are very small and simple. The semiconductor devices are mounted on the board and the interconnections between the components are established easily by etching the conducting tracks on the board. These electronic circuits are very compact and are capable of doing almost any job.

The development of the Integrated Circuits (ICs) made electronics occupy all possible fields in domestic, industrial, medical and defence applications. An IC is an electronic circuit that is much more compact compared to PCB circuits in that the components as well as the total circuits are fabricated on a single substrate. Very complex circuits could be fabricated on a single simple substrate, reducing the size of the electronic circuits to microscopic level. The growth of the IC fabrication techniques is tremendously high—these days millions of transistors are fabricated on to a single microscopic-sized IC. With these drastic developmental changes in IC fabrication, the growth of IC applications is extremely rapid and has occupied all walks of life. Ultra-high-speed electronic devices were possible due to miniature-sized components. The speed with which an electronic gadget can operate is of the order of gigacycles per second. This tremendous speed of operation has enabled the electronic engineer to devise new gadgets to meet all possible human requirements. These minute electronic components can control extremely large cargo ships at sea, gigantic steel mills, small robots and so on. It is no wonder today that these tiny electronic components do the most wondrous things in this wonderful world of ours.

The evolution of digital logic and digital electronics is the basic boost to the developments in the fields of computer science, information technology, communications and allied fields.

1.2 APPLICATIONS OF ELECTRONICS

The field of electronics is so broad that all human endeavours involve some form of electronics application. The following are a few of the innumerable areas that are dependent on electronics.

Home: Every nook and corner of a modern house is stacked with electronic gadgets. Electronic devices have simplified and speeded up mundane tasks of cleaning, cooking, lighting and security. Recreation at home is essentially the electronic way. Microwave oven, telephone, answering machine, home security, personal computer, light dimmers, appliances, toys, air conditioners, refrigerators, watches, clocks, washing machines, grinders, vacuum cleaners, electronic regulators, digital cameras, movie and video cameras and so on are few of the electronic devices used in the present day homes.

Information, communications and entertainment: The progress of a country depends upon the availability of economical and rapid means of communication. During the early part of twentieth century, the main application of electronics was in the field of telephony and telegraphy. This utilises a pair of wires. However, mobile communication using wireless communication systems gained wide popularity due to ease of communication between any two remote points on the globe. The wireless communication took shape and it has rapidly advanced in the last two or three years so much so that mobile communication is the buzzword of the day. Communication between any two remote places on the globe is possible, thereby shrinking the size of the globe. The other communication media these days are fax, Internet and so on, where any form of information (data, audio and video) travels with the speed of light. Entertainment has also drastically changed its face after the tremendous advancement of electronics. TV, radio, VCD, tape recorders, movie and so forth are part and parcel of domestic life these days. Electronic toys and games have revolutionised the field of entertainment to a very large extent. Each and every stage in the big screen and small screen makes endless use of electronics. Music synthesising, music recording, music enhancement, amplifying, special effects, stage special effects, public address system, digital images, graphics and so on are few stages involved in this field.

Transportation: In this jet age of travel, electronics is used in supersonic aircrafts to bullet trains to even a musical horn in a bicycle. Navigation of ships, space shuttles, air and auto traffic control, global positioning systems and so on are all electronic.

Business: Business transaction across the seas, buying and selling of not only commodities but also shares and stocks are all electronic these days.

Medicine: Bereft of electronics, medical sciences are severely handicapped. It is not very surprising to note that the functions of the medical field would be severely hampered these days without electronics. One cannot think of diagnostics, surgery or treatment these days without the use of electronics in the medical field. Really, one has to think how these might have been possible without electronics in the good old days. Electronics and biomedical instrumentation have a huge role in this key field of medicine. Amazing advances in diagnosis, surgery, and pharmacy have been possible due to aid of electronics. Medical transcription and coding have made it possible to share medical information the electronic way. Diagnostics such as ECG, MRI and X-ray; electronic microscope, scanning, cancer treatment, surgery, patient monitoring, treatment; therapies such as laser, ultrasound and electronic puncture; rehabilitation aids and technique such as pacemakers, drug delivery, physical therapy and biofeedback; patient aids such as artificial limbs and hearing aids; biomedical instrumentation, telemedicine and many more are all electronic.

Instrumentation: Automation in industries is the order of the day. Quality control, standardisation of techniques, time and cost saving methods are possible only due to the application of electronics. Electronic gadgets control the process of production in an industry, whatever may be the type of product. For example, control of thickness, quality, weight, moisture, temperature and so on. Automatic door openers, lighting systems, power systems and safety devices, are used in establishments and shops. Industrial heating, welding, power device controls, and so on are also electronic. Cathode-ray oscilloscopes, ammeters, voltmeters, digital and analog multimeters, flux meters, pH meters, strain gauge, and precision measuring equipment are a few more examples of electronic instrumentation.

Scientific research: Electronics make scientific research simple by means of simulations, modelling and so on.

Environmental protection: Electronics find application in the protection of environment also. Electronic alarms can be employed which alert when a specified level of pollution is exceeded. Electric car, the recent invention, emits minimum or no pollutants into air. Waste treatment or waste water recycling are all controlled by electronic gadgets.

Athletics and sports: Optimisation of performance through electronic analysis of motion, control exercise equipment and so on depend on electronics. Live coverage, analysis and umpiring of any type of sporting action are with the help of electronic equipment.

Military: Defence operations currently are largely electronic. The radar gives the exact coordinates and tracks the target with extreme precision. The missiles are guided by electronic equipment. Global positioning systems give exact enemy location. Secrecy of information is achieved these days with the help of digital communications. To maintain the secrecy of base military camps and stations, electronic countermeasures, electronic jammers and so on are employed.

Computers: All hardware circuits and modules of digital computers such as CPU, monitors, keyboard, printers, scanners, CD writers, players, floppy drives, memory—RAM or ROM, uninterrupted power supplies and web cams are all electronic. Hardware components of LAN, MAN, WAN of various computing systems, e-mail, e-chat, all networking modems, modules, robots, computer communications, data communications, e-commerce, e-business, e-transcription, e-manufacturing and e-banking (for example, credit cards and ATMs) make use of the principles of electronics.

Education: Education at present employs large amount of electronics. E-learning and satellite TV form an integral part of distance-mode education system. Even for classroom teaching one requires LCD projectors, slide projectors, PA systems and so on, which are all electronic.

Astronomy: The space shuttle and its control, the communication gadgets employed in the study of extra terrestrial features of space, other planets, moon and so on are all electronic.

1.3 ELECTRONIC COMPONENTS

An electronic circuit may appear to be complex but is capable of performing innumerable functions. The electronic circuit is built with a few basic components of electronics.

The *electronic components* may be classified broadly as *passive* and *active* as shown in Fig. 1.1. Passive components are those components that cannot by themselves amplify or generate electric signals. They need no external biasing for their operation. Under this category, *resistors, inductors* and *capacitors* can be listed. Resistors are dissipative elements and they offer resistance to the flow of electric current through them. They dissipate energy in the form of heat. There are many types of resistors. They may be classified based on the applications, material composition, power rating and so on. Inductors and capacitors are energy storage device. They store electric energy and release as and when required. An inductor is a current-controlled device whereas a capacitor is a voltage-controlled device. These components can be classified based on the make, material used in

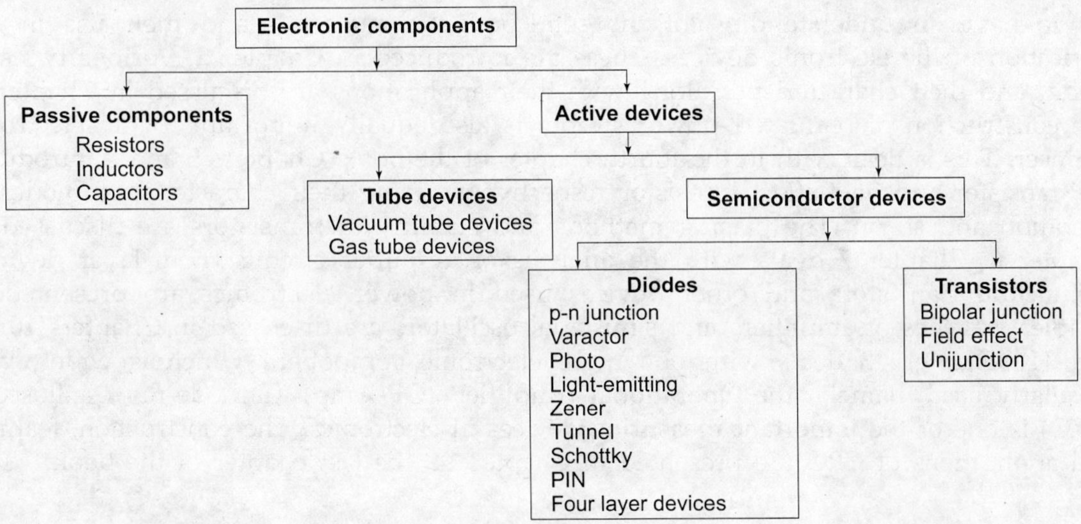

Fig. 1.1 *Classification of electronic components*

fabrication, power rating and so on. Variable resistors, capacitors or inductors are also possible.

Active devices are the subject of this present work. Active devices are classified into *tube devices* and *semiconductor devices.* Tube devices are not dealt with here since they are almost obsolete. Tube devices are classified once again into *vacuum tubes* and *gas tubes*. The examples of vacuum tubes are vacuum diodes, vacuum triodes, vacuum pentodes and so on. The examples of gas tubes are gas diodes, thyratrons and so on. With the invention of compact and more efficient semiconductor devices, these tube devices find no place these days in electronics.

Semiconductor devices are the subject of treatment in the present work. There are various components in the field of electronics. The first and foremost is the *p-n semiconductor junction.* All subsequent developments in the electronic components are based on the principle of this semiconductor junction. The *junction diode* and the *transistor* are the broad classification of these devices. There are large numbers of diodes possible, depending on the application. They can be *p-n junction diode, varactor diode, photodiode, light-emitting diode, Zener diode, tunnel diode, Schottky diode, PIN diode* and so on. These diodes are tailored for a given application. These diodes can rectify, detect, switch, regulate, amplify or oscillate. Under the transistor classification, there are *bipolar junction transistors, field-effect transistors, unijunction transistors,* and so on. These devices can amplify, oscillate, control power, switch and so on.

The construction, structure, characteristics, equivalent circuits and applications of these semiconductor devices are presented in total in this book. Few important circuits built with these devices, namely, rectifiers, amplifiers and oscillators are also presented and analysed in this work. Before leaping into the details of the electronic components, one has to have an understanding of the semiconductor materials and their use in the fabrication of the electronic devices. These are introduced in Chapter 2. Various types of diodes and their characteristics along with their applications are discussed in Chapter 3. The construction of regulated power supply is also equally important for an electronic engineer. This is dealt with in the fourth chapter of the book. Chapters 5 and 8 introduces the transistor and field-effect transistor respectively, giving their characteristics, modes of operation and so on. The biasing methods and stability of transistors are discussed in Chapter 6. Chapter 7 deals with the analysis of amplifier circuits. Four-layer devices, unijunction transistors and other devices used in power electronics are presented in Chapter 9. Types of amplifiers and sinusoidal oscillators are discussed in Chapters 10, 11 and 12. Chapter 13 deals with one more electronic component which is widely and popularly used, namely, the Operational Amplifier, (OP-Amp). Cathode-ray oscilloscope (CRO) is one of the important measuring devices in electronics. The construction, features and applications of CRO are presented in Chapter 14, the last chapter of the book.

CHAPTER

2

Semiconductors

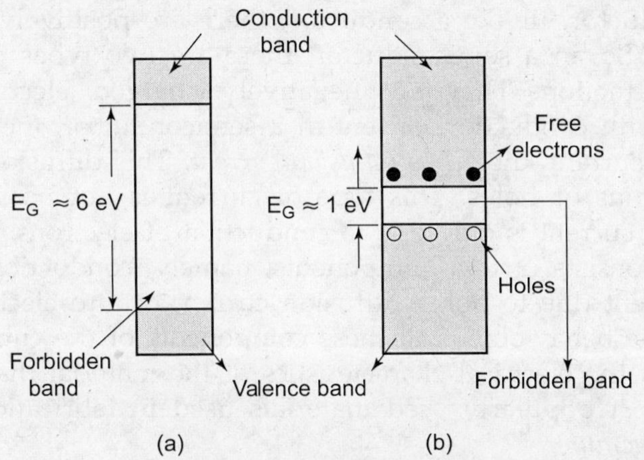

Energy-band structure of (a) insulator, (b) semiconductor, and (c) metal

Chapter Outline

The concepts introduced in this chapter are:

- *Conductors and semiconductors*
- *Different types of semiconductors*
- *Current components in a semiconductor*
- *The behaviour of minority carriers in semiconductors*
- *Concept of semiconductor junction*

2.1 INTRODUCTION

The physics of the materials used in the construction of various electronic devices is very important to understand the basic operation of the device. The materials used in the electronic devices are *semiconductors.* These are the materials which neither fall in the category of conductors nor insulators. Unlike the dielectric materials, they have some conductivity, which can be controlled by external agents. This advantage of controlling the conductivity of the material at designers' requirements makes the material most suitable for electronic devices. In other words, the electron, the basic entity, is to be made to do work as per the requirement. The tiny energy associated with the tiny particle is contributed in the direction that is useful for a given application. The electrons, which are free and are not bound to any particular atom or molecule, are the charges that contribute to the conduction. In a semiconductor, unlike a conductor, there are positively charged holes that contribute conduction. So, in a semiconductor, there are two types of charge carriers, that contribute to conduction. They are negatively charged electrons and positively charged holes. Along with conduction current in a semiconductor, there is also a possibility of one more type of current known as *diffusion current*. This diffusion current may be once again due to electrons or holes. This type of current is not possible in a conductor. So, in a conductor, the current is only due to conduction of electrons. Whereas, in a semiconductor, the current consists of four components, namely, conduction current due to electrons, conduction current due to holes, diffusion current due to electrons and diffusion current due to holes. The behaviour of all these components of the current is to be known in detail to understand the electrical characteristics of the material that is used in the electronic devices. The most popularly used materials used in fabrication of the active devices are *silicon* and *germanium.*

The charge of negative electricity and the mass of the *electron* are known to be 1.6×10^{-19} C and 9.11×10^{-31} kg, respectively. Since 1 A of current is flow of 1 C/s, 1×10^{-12} A (1 pA) represents the motion of approximately 6000000 electrons per second. But, 1 pA of current is in general, very minute. The radius of the electron is estimated to be 10^{-15} m. This being a very small value, the electron is assumed to be point in all the discussion to follow.

A semiconductor material has two types of charge carriers. They are electrons and holes. Holes are formed when an electron is missing from the covalent bond that is formed between neighbouring atoms of the semiconductor material. Since an electron is missing from the stable covalent bond, the empty space in the bond structure is referred to as *hole.* The electron is negatively charged and so, the hole is positively charged. The magnitude of the charge of a hole is same as that of the electron. These holes also contribute to conduction and the details of all of the current components will be taken up in detail later in the chapter.

The unit of energy, Joule, is in general used in mks system. Depending on the magnitude of the energy, the joule can be expressed in kilo, mega, micro, or pico in the discussion of basic electronic theory; since the energies associated with the electron are very small, expressing them in joules is difficult. Thus, a new unit of energy is used for such tiny magnitudes of energies.

One *electron volt* (eV) can be defined as

$$1 \text{ eV} = 1.6 \times 10^{-19} \text{ J} \quad \text{...(2.1)}$$

If an electron falls through a potential of 1 V, its kinetic energy will increase with decrease in potential energy. Thus,

$$qV = (1.6 \times 10^{-19} \text{ C}) \times (1 \text{ V}) = 1.6 \times 10^{-19} \text{ J} = 1 \text{ eV} \quad \text{...(2.2)}$$

This unit is unit of energy and can be used for any type of energy.

2.2 POTENTIAL BARRIER

The concept of *potential barrier* finds a very interesting and most useful application in semiconductor junctions. To understand this concept, consider two parallel electrodes *A* and *B* separated by distance *d*, with *B* at a negative potential V_d with respect to *A* as shown in Fig. 2.1.

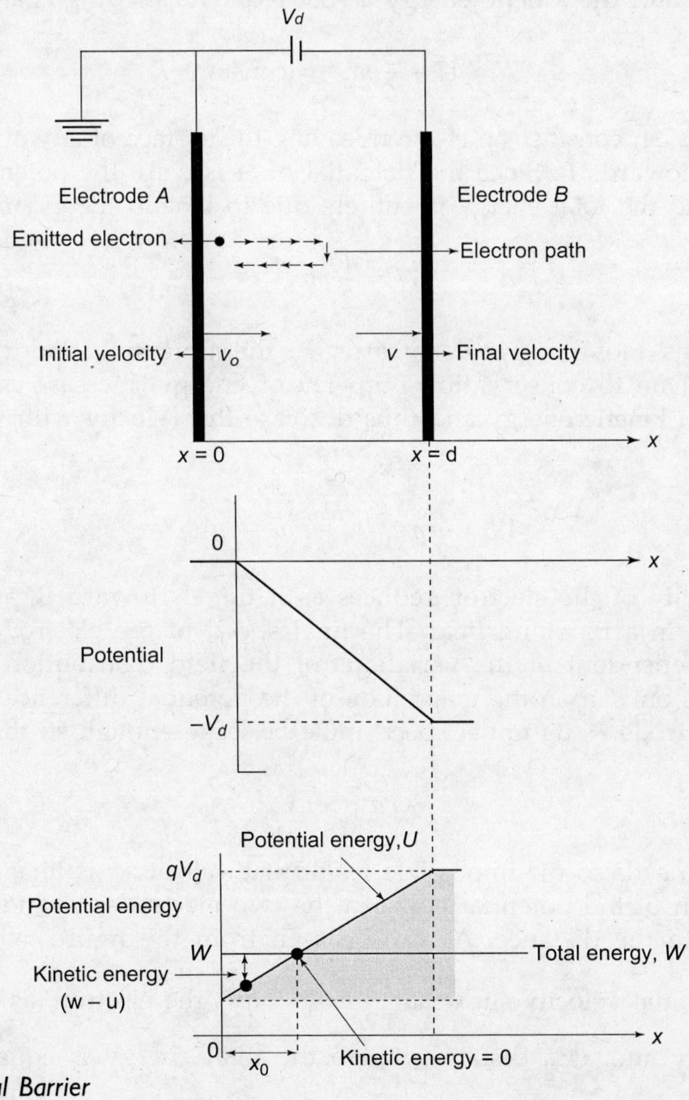

Fig. 2.1 *Potential Barrier*

The **force f** on charge q in presence of electric field is defined as charge times the electric field intensity ε. The same force may be expressed in terms of Newton's second law of motion as its mass m times acceleration if the charge moves with a velocity v.

$$\mathbf{f} = q\varepsilon = m(dv/dt) \quad \text{...(2.3)}$$

The work done per unit charge in moving a point charge from point A to point B is known as the *potential* V of the point B.

$$\varepsilon = -dV/dx \quad \text{...(2.4)}$$

And, the *potential energy U is charge times the potential,*

$$U = qV \quad \text{...(2.5)}$$

According to the law of conservation of energy the total energy in a system remains constant. Thus, the two components of energy available in this system are potential energy U [Eq.(2.5)] and the kinetic energy associated with moving charge (½ mv^2).

$$W = U + \frac{1}{2}mv^2 = \text{constant} \quad \text{...(2.6)}$$

Referring to Fig. 2.1, consider an electron leaving the surface of A with an initial velocity v_0 in the direction towards B. Since the potential of A is zero, the potential energy at this electrode is zero and the total energy is entirely due to kinetic energy and is equal to

$$W = \frac{1}{2}mv_0^2 \quad \text{...(2.7)}$$

As the electrons move forward towards B, the potential, and thus the potential energy, increases. Due to conservation property of energy, increase in potential energy leads to decrease in kinetic energy and thus decrease the velocity with which the electron travels. Thus,

$$W = \frac{1}{2}mv_0^2 = \frac{1}{2}mv^2 + qV_d \quad \text{...(2.8)}$$

Thus, the velocity of the electron reduces as it travels toward B, as is true since the electron is moving in a repelling field. The final speed of the electron, as it reaches the electrode B, is independent of the variations of the field distribution between the two plates and depends only upon the magnitude of the potential difference V_d. If the electron is to reach the electrode B, its initial speed must be large enough so that

$$\frac{1}{2}mv_0^2 > qV_d \quad \text{...(2.9)}$$

If not, Eq. (2.8) leads to the impossible result that velocity v is imaginary. Fig. 2.1 also shows the variation of the potential between the two electrodes A and B. It also gives a plot of energy with the distance. As can be seen from the figure, when an electron is released with an initial velocity such that $\frac{1}{2}mv_0^2 < qV_d$ the electron, as it travels forward, reduces its velocity and at a distance where the total energy is equal to the potential energy, the speed of the electron becomes zero and the electron will have to halt. Which means at this distance, there is no kinetic energy and the electron is at rest. Further, the electron cannot move in the forward direction to reach the electrode *B* and will have to reverse its direction toward electrode *A*. The distance travelled by the electron between the electrodes, x_0, depends on the total energy of the system or, in other words, on the initial kinetic energy of the electron. The higher the initial velocity, larger is the value of x_0. If the electron were to reach the other electrode, the initial energy should be more than the potential energy at the electrode *B*. This is as shown in Fig. 2.1.

From the above discussion it can be concluded that the electron never reaches the shaded portion of the energy diagram as shown in the figure. The electron travels forward with an initial velocity toward *B* until the electron hits the potential *hill*. On hitting this hill, it reverses its direction and travels in the opposite direction but can never reach the hill. No electron can thus exist in the shaded portion of the plot since if it exists, it is supposed to have negative kinetic energy, that is, negative velocity, which is an impossible condition practically. This *potential hill* is a fictitious barrier, which obstructs the electron from entering the other side of the barrier. Physically, there is nothing like this potential hill but the electron behaves as if it is bouncing back after colliding with the potential hill. This is the *potential barrier*, which is important in understanding the behaviour of the semiconductor junction.

2.3 ENERGY-BAND THEORY

The general structure of metals and semiconductors is crystalline in nature, that is, the atoms are arranged in arrays in the form of a definite fundamental pattern. The fundamental pattern may either consist of atoms or molecules depending up on the material under consideration. The energy associated with electrons of a single atom is not applicable to these crystalline structures. The atoms are not free as in a gas, but are shared by the neighbouring atoms. So, it is not possible to consider a single atom separately and discuss about the discrete energy levels of the electrons in the atom. The energy associated with electrons in a crystal is not the same as that when the atom is singled out. This is so because the potential characterising the crystalline structure is now a periodic function in space whose value at any point is appreciably by the presence of the neighbouring atoms. The levels of the outer or valence shell are affected and the inner shell electrons are not much disturbed since the neighbouring atoms share the electrons from the outermost shell. Or, in other words, the valence shell electrons are no more an inherent property of the base atom but are shared by more than one atom. The new energy levels are determined by quantum mechanics and sharing the valence shell electrons results in an *energy-band* of closely spaced energy states instead of widely separated energy levels of the isolated atom. That is, the energy levels are no more discrete and separated by large amounts of levels as in the case of an isolated atom, but are in very closely packed energy states such that it seems to be a continuous energy-band where electrons occupy different levels in the band.

For a better understanding of the concept of energy-band theory, consider group IVA elements [carbon (C), silicon (Si), germanium (Ge) and tin (Sn)] with four valence-shell electrons. There are *N* atoms in a considered crystal and so, there will be 4*N* electrons in the outer subshell of the system. Of these, 2*N* electrons are s type and remaining 2*N* are *p* type. Assume a situation where it is possible to vary the spacing between the neighbouring atoms without changing the basic structure of the crystal. When the interatomic distance is very large, the atoms can be considered to be isolated and the energy levels discrete, as in the case of a gas. The electrons in the inner shells are not considered since they are not affected. In the crystalline system thus formed, there are 2*N* *s* electrons occupying all the possible 2*N* levels. Of the possible 6*N* *p* states in the valence shell, only 2*N* levels are occupied and the remaining 4*N* levels of *p* electrons are vacant.

Now, if the interatomic spacing in the crystal is decreased, an atom will exert an electric force on its neighbours. A coupling between atoms is formed and the wave functions overlap, and the crystal becomes an *electronic system* that must follow *Pauli exclusion principle.* Pauli exclusion principle states that *in an electronic system, no two electrons can have same four quantum numbers, n, l,* m_l and m_s. Or, in other words, no two electrons can occupy the same quantum state in an electronic system. So, when an electronic system is formed by bringing the atoms close to each other in a crystal, the discrete energy levels of the electrons of single atom spread out into an *energy-band,* which is nothing but a large number of closely packed energy levels. As the interatomic spacing is decreased, since *N* is a very large number, and the separation between levels is small, the total spread between the minimum and maximum energy may be several electron volts. As the spacing is further decreased, the energy bands of both the *s* electrons and *p* electrons spread out as shown in Fig. 2.2. It may be noted that the *s* electron-band and the *p* electron-band are separated by an energy gap (forbidden energy gap), which decreases as the interatomic spacing is decreased. Thus, as can be seen from the figure, the 2*N* electrons occupy all the *s* states, while only 2*N* states of the possible 6*N* states of the *p* type energy-band are occupied.

As the spacing between the atoms is further reduced, the energy gap between the two energy levels decreases and tends to combine at a particular spacing value. That is, at this value of spacing, the two bands overlap and the energy gap vanishes. Now, there are 8*N* possible states in the energy-band, of which 4 *N* states are occupied and 4*N* are vacant. It

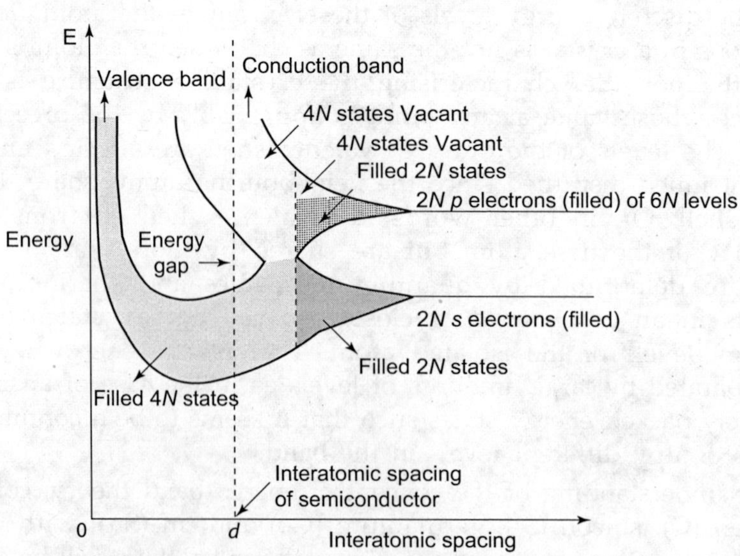

Fig. 2.2 *Energy bands*

may be noted at this juncture that the identity of the electron is lost—the *s* and *p* electrons are no more different and these electrons belong to the crystal as such. Simply, there are 4*N* electrons in the possible 8*N* states. Since the atom contributes four electrons to the crystal from its outermost shell, it is referred to as *tetravalent* and the energy band as *valence band.*

Reducing the spacing further is a very complex situation and it may be represented as shown in Fig. 2.2. As can be seen from the figure, the two bands, which merged with each other, will separate once again with a forbidden energy gap. Now, there are two energy bands, the lower one is referred to as *valence band* and the upper one as *conduction band,* separated by an *energy gap.* The valence band consisting of 4*N* states is totally occupied by the 4*N* electrons and the conduction band having 4*N* states is totally vacant. This is where the interatomic spacing of a tetravalent semiconductor material is found in nature. The energy gap between the energy-bands depends on the type of material and is very large for an insulator, small for a semiconductor and in case of a metal there is no energy gap and the two bands overlap.

2.4 CLASSIFICATION OF MATERIALS

A material may be classified as *metal, insulator* or *semiconductor* based on the energy-band structure. The conductivity of the material depends on the electrons available in the *conduction band.* The more the electrons are in this band, more is the conductivity of the material. The electrons available in this energy-band are not bound to the crystal structure but are *free* to move about randomly throughout the material. In all the discussion, the electrons referred to are these electrons, which are available in the conduction band, and the electrons in the lower energy-bands are of no use as far as conduction is concerned. The energy-band structure for all three types of materials is as shown in Fig. 2.3.

In an *insulator,* as shown in Fig. 2.3, the forbidden energy gap between the conduction band and valence band is very large, of the order of 6 eV (diamond, for example). The

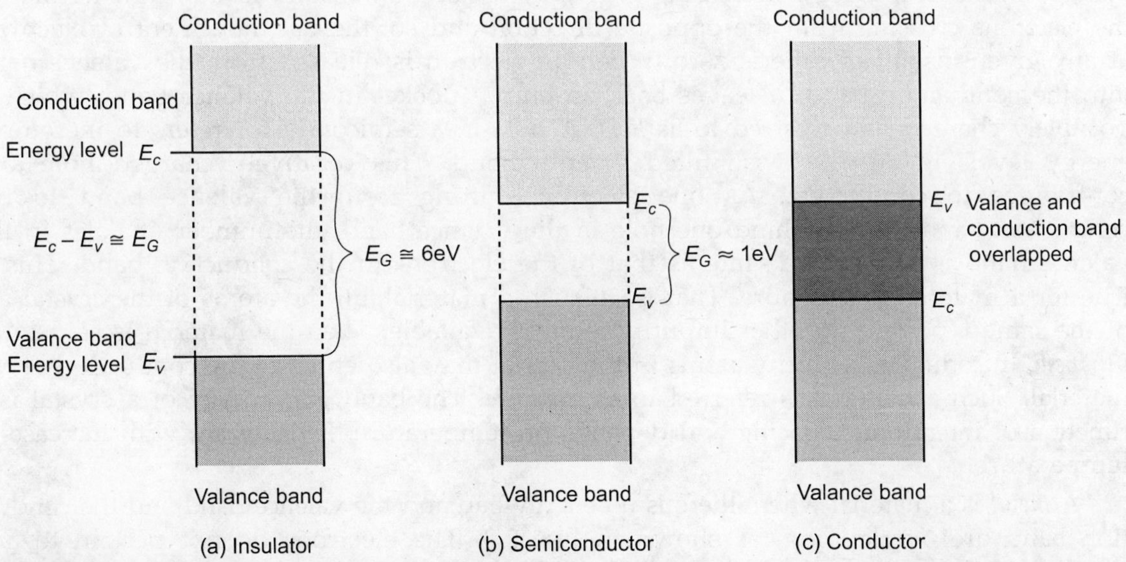

Fig. 2.3 *Energy-band structure*

valence band is totally filled with electrons and there are no electrons available in the conduction band. That is, all the electrons are bound to the crystalline structure and there are no electrons that can contribute to conduction. If one requires conduction in a material, the electrons from the valence band are to be shifted to conduction band. In other words, an electron that is bound in the covalent bond structure of the material is to be given a suitable energy such that the bond is broken and the excited electron reaches the conduction band level. It may be noted that the electron can only exist in the valence band or the conduction band, but if an electron in the valence band receives energy which will launch it into the forbidden energy gap, it will exist only in the valence band. Until the electron receives sufficient energy to overcome the forbidden energy gap, and thus to jump from the valence band to the conduction band, it remains in the valence band. In case of an insulator, this energy gap is so large that shifting electrons from the lower band to the conduction band is not possible and so there are no electrons in the conduction band. Thus, there is no conduction possible in these materials.

A material that has a relatively small energy gap, of the order of 1 eV, is referred to as a *semiconductor*. The energy gap E_G is 0.785 and 1.21 eV for germanium and silicon, respectively, at 0 K. At this absolute temperature (0 K), these materials are also insulators since the valence band is full and conduction band is totally empty. There is no conduction possible with such materials even at extremely small temperatures. But, as the temperature increases, the electrons in the valence band receive energy in the form of thermal energy and since E_G is small, small energy increments are sufficient to shift the electrons from the lower to upper bands. As the temperature increases, the system receives more and more energy from thermal excitations and more and more electrons are displaced from the valence band to the conduction band. Thus, the conductivity of the material improves. In other words, the covalent bond between the atoms in the crystal is broken and the electrons are released. These electrons are not bound to any structure now and are free to wander anywhere in the material. Their motion is very random, but the average number of electrons crossing a given cross-section would be the same as that of the electrons crossing from the opposite direction and so, the net movement of electrons at any given instant of time is zero. When an electron is released from the valence band into the conduction band, it leaves back an empty pocket in the valence band, which is positively charged and referred to as *hole*. A hole in a semiconductor refers to the empty energy levels in an otherwise filled valence band. This positively charged hole also contributes to conduction. As one electron shifting from the valence band to the conduction band leaves behind one hole in the valance band, the number of holes in the valence band is exactly the same as that of the electrons in the conduction band. This is true for a material that is pure. That is, in such a material, all the atoms of the crystal are of the same type and no other impurity atom is available. Then the material is said to be *intrinsic*. Introducing impurity atoms in the crystal may also enhance the conduction of the material. Such a material is referred to as *extrinsic*. The band gap energy of a crystal is a function of interatomic spacing and depends on temperature. It decreases with increase in temperature.

A *metal* is a material where there is no energy gap and the valence band and the conduction band are overlapping, as shown in Fig. 2.3. The electrons do not pertain to any definite band and so applying a small electric field to the material excites the electrons to a higher energy level and these electrons move under the influence of the applied field to constitute conduction current. There is no possibility of a hole in a metal or a conductor since there is no differentiation of valence and conduction bands and the energy gap is absent. Electrons can exist in any level of energy in the band.

2.5 INTRINSIC AND EXTRINSIC SEMICONDUCTORS

An intrinsic semiconductor is the one that is pure and has no impurity atoms in its crystalline structure. An *extrinsic semiconductor* is the one in which impurities are added purposefully to increase the conductivity of the material. For example, either trivalent or pentavalent atoms are added to tetravalent atoms. Adding the impurities, the conductivity of the material increases to a great extent.

The conductivity of the material depends on the number of atoms in the conduction band of the material. For a good conductor the electron concentration n (number of electrons per unit volume) is very large ($\approx 10^{23}$ electrons per m^3); for an insulator it is very small ($\approx 10^7$) and for a semiconductor its value lies between these two values. The valence electrons in a semiconductor are not free to wander about as they are in a metal, but rather are trapped in a bond between two adjacent ions.

All the Group IVA elements (C, Si, Ge and Sn) have four electrons in the valence shell and exhibit properties of semiconductors, of which silicon and germanium are widely used in electronic devices. The crystalline structure is a three dimensional tetrahedron with one atom affixed at each vertex. The basic unit repeats in all the three directions in the structure. The structure is as shown in Fig. 2.4. The silicon or germanium material has four electrons in the valence shell and these electrons of each atom are shared by the neighbouring four atoms to form covalent bonds. Thus, an *electron pair* is formed, which is the covalent bond. Sharing four electrons from the neighbouring atoms along with the four atoms of its valence shell constitutes eight electrons in the system and thus stability is

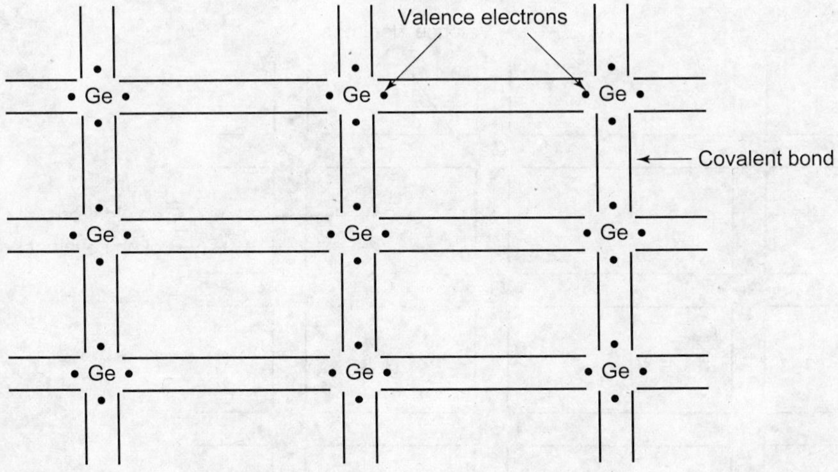

Fig. 2.4 *Crystal structure of germanium at 0° K*

achieved in the structure. Since four electrons are donated to the neighbouring atoms to form covalent bonds, the left-over part of the atom is an ion positively charged four times to that of the magnitude of the charge carried by each electron. There is a binding force between the adjacent atoms due to this sharing of electrons forming the covalent bonds. These valence electrons are tightly bound to the nucleus and so they are not free to wander about in the material. Even though there are four electrons available in the valence band of the atom, the crystal conductivity is very low since these electrons are bound.

The above discussion refers to the case when the material is ideal at 0° K. In this condition, the material behaves as an insulator since there are no free electrons. At room temperature, the conduction of the material increases since some of the electrons gain energy from the thermal agitation and acquire sufficient energy to get transferred from valence to conduction band. The thermal energy is sufficient to break the covalent bond and the electron that is bound previously gets released and does not belong to any particular atom. This electron is free and can go anywhere in the material. While a negatively charged electron is released to be free, it leaves behind an empty space, that is positively charged. This empty location in the valence band of the atom is referred to as hole. This positively charged hole also contributes to conduction like the free electrons in a semiconductor. Thus, breaking a covalent bond in the structure creates two carriers of current—charge carriers—one electron and one hole. The concept of electron and hole charge carriers is as shown in Fig. 2.5.

The positively charged hole contributes to conduction. The electrons in the neighbouring covalent atoms move to occupy the empty spaces in the valence band, that is, the hole. The electron moving from a bond to fill a hole leaves a hole in its initial position. Hence, the hole effectively moves in the direction opposite to that of the electron. An electron from another covalent bond may now fill this hole, in its new position, and the hole will correspondingly move one more step in the direction opposite to the motion of the electron. This conduction of the electrons from one hole to the other is equal to holes moving in the opposite direction. This mechanism of hole conduction does not involve any

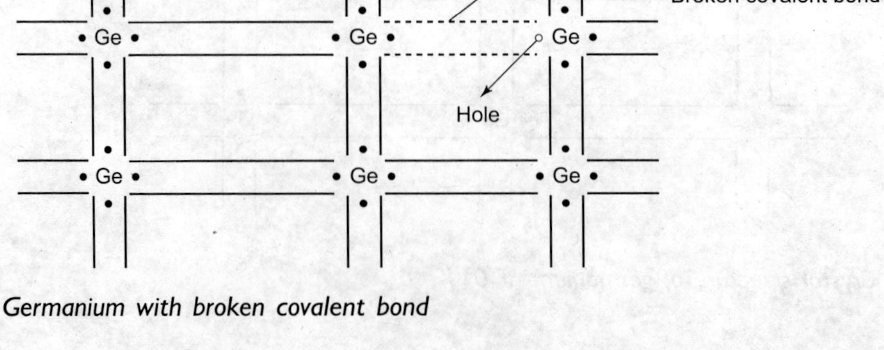

Fig. 2.5 *Germanium with broken covalent bond*

free electrons. So, the conduction of electricity is entirely due to holes moving in the opposite direction to that of the motion of bound electrons. This motion of positively charged holes in the forward direction is also conduction and since it does not involve free electrons, it is the conduction current or drift current due to holes in a semiconductor. Since existence of holes in a metal is not possible, such conduction current is not possible in a conductor.

The concept of hole current can be explained more elaborately with the help of Fig. 2.6. There are large numbers of filled covalent bonds in the crystalline structure of the material, as shown. They are assumed to be in straight line with a filled position at 6, as in Fig. 2.6(a). This represents a hole. Now, if the electron from position 7 leaves to occupy the vacant position 6, position 6 is occupied and an unfilled hole is available at 7. Effectively, even though the electron has moved from 7 to 6, one feels that the hole has shifted from 6 to 7. This continues and the hole can move in the forward direction when the bound electron is moving in the reverse direction. Since conventional current direction is reverse to that of the electron movement, the direction of movement of hole and the conventional current flow would be the same.

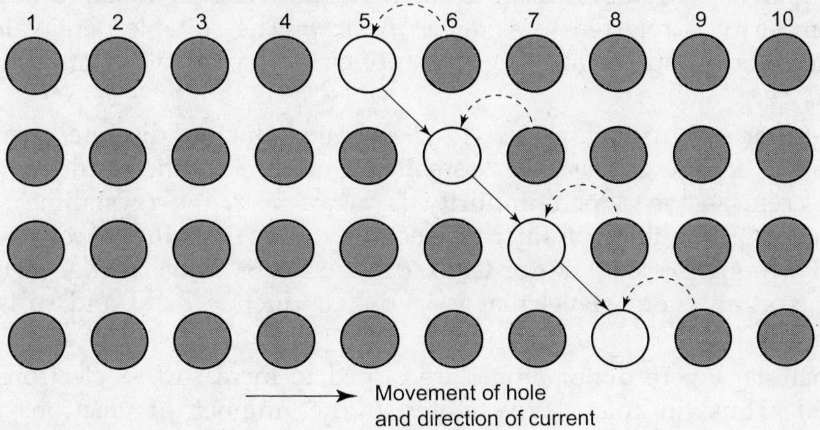

Fig. 2.6 *The hole current*

In a pure *intrinsic semiconductor*, the number of electrons in the conduction band is same as the number of holes in the valence band. Thus, n is the electron concentration, number of electrons per unit volume, and p is the hole concentration, number of holes per unit volume, then,

$$n = p = n_i \qquad ...(2.10)$$

where n_i is called the *intrinsic concentration* of the semiconductor material. This value of intrinsic concentration is a function of temperature, since the number of holes and electrons depends on the temperature. The higher the temperature, the greater is the value of n_i. To increase the conductivity of the material, one requires large number of charge carriers (free electrons or holes) in the material. So, when an intrinsic material is considered, its conductivity can be improved by increasing the temperature. But, increasing the temperature beyond a limit is not possible practically. Thus, increasing the charge carrier concentration by increasing the temperature is not possible. An alternative is to increase the free electrons in the conduction band or to increase the holes in the conduction band by some artificial means. This increases the conductivity of the material even at room temperature. One way out is to add impurity atoms to the crystal structure and make the material *extrinsic* or impure by adding atoms other than the parent atoms in the structure purposefully. A small percentage of trivalent (B, Ga, In) or pentavalent (P, As, Sb) atoms is *doped* to the *intrinsic* silicon or germanium to make it extrinsic and improve conductivity.

2.6 *p*- AND *n*-TYPE SEMICONDUCTOR MATERIALS

Donor impurities: A pentavalent atom has five electrons in the valence shell. When these atoms are doped with either silicon or germanium, they displace some of the parent atoms. Four electrons of the five in the valence shell form covalent bonds with the neighbouring atoms as in the case of a parent atom and one electron is left over. This left over electron, when in the conduction band, can contribute to conduction. The amount of energy required to free the excess electron is very small, of the order of 0.01 eV for Ge or 0.05 eV for Si. Thus, there are new electrons that are free and with energy levels in the range of conduction band. These electrons are very useful for conduction. In other words, the impurity atom *donates* one electron to the material, which is a charge carrier. Thus, the pentavalent impurity is referred to as *donor impurity.* The suitable pentavalent atoms are antimony, phosphorus and arsenic. The structure of the crystal lattice of such a donor is as shown in Fig. 2.7.

When a donor impurity is added to the semiconductor, discrete energy levels are introduced in the energy gap at a very small distance below the conduction band. These levels are discrete as the doped impurity is away from the crystalline structure. The energy difference of this impurity atom's electron and the conduction band lower edge is very small, of the order of 0.01 eV for Ge or 0.05 eV for Si. Thus, at room temperature, the fifth electron is at an energy level that is in the conduction band and so is a very good carrier of current.

Such a material where donor atoms are doped to increase free electrons is known as *n-type material.* Thus, in this *n*-type material, the number of electrons is very large compared to holes. The initial concentrations of the electrons and holes are equal to the

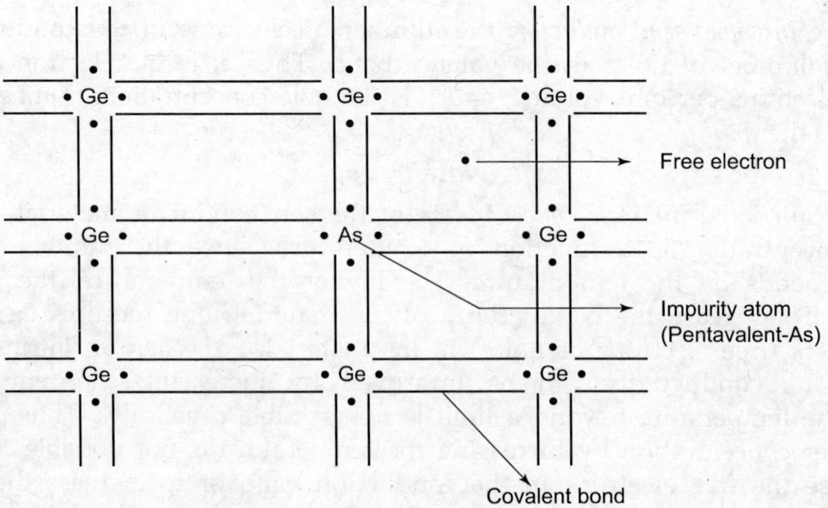

Fig. 2.7 *Donor impurity*

intrinsic concentration. After doping, the electron concentration increases to the extent of doping concentration, but the hole concentration would be low compared to the intrinsic concentration since the excess electrons from the impurity atoms can recombine with the holes. As a whole, the electron concentration is very large compared to hole concentration in an n-type material. Thus,

$$n >> p \quad ...(2.11)$$

Acceptor impurities: When a trivalent atom having three electrons in the outermost shell is doped with a semiconductor material, a *p-type material* is formed. The three electrons available in the valence shell are shared by the neighbouring Si or Ge atoms to form covalent bonds. The fourth bond is not formed and there is a vacancy in the bond, which leads to the creation of a hole, as shown in Fig. 2.8. These vacant spaces, holes, are positive charge carriers, which can conduct current. These holes *accept* electrons for conduction to take place and so, the trivalent atoms, boron, gallium and indium are known as *acceptor impurities.*

Adding acceptor impurity atoms to the semiconductor leads to insertion of discrete energy levels in the forbidden energy gap of the semiconductor. This level is nearer to the upper edge of the valence band. Since very minute energy is sufficient for the electron in the valence band to jump into this discrete energy level, large numbers of electrons contribute for the conduction through the holes created by the impurity doping. The conductivity of the material is multiplied by adding impurity atoms to the semiconductor material.

The p-type material thus has a large number of holes compared to that of electrons in the material. Holes are referred to as *majority carriers* in a p-type material whereas electrons in the p-type material are known as *minority carriers.* In a p-type material, hole concentration is very large compared to electron concentration and thus,

$$p >> n \quad ...(2.12)$$

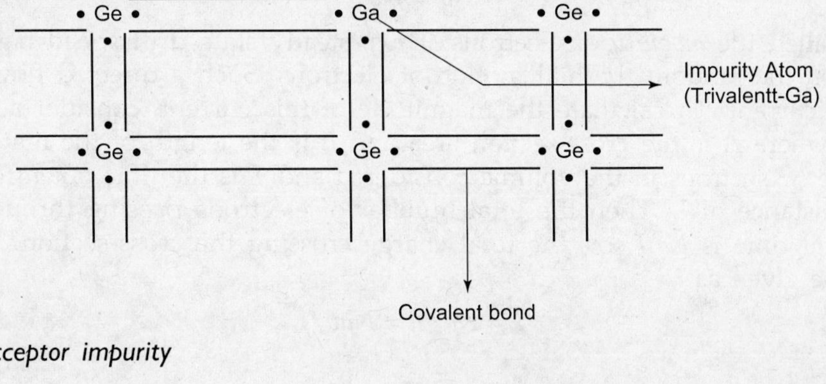

Fig. 2.8 *Acceptor impurity*

Mass action law: By adding acceptor atoms the hole concentration is increased and by adding donor atoms the electron concentration is increased in a semiconductor material. At thermal equilibrium, if n is electron concentration and p is the hole concentration in a given type of material, whatever may be the concentration of the impurities added, the *Mass Action Law* states that

$$np = n_i^2 \qquad ...(2.13)$$

where n_i is the intrinsic concentration and is the function of temperature.

2.7 MOBILITY AND CONDUCTIVITY

Considering a metal, there are a large number of electrons that are free and are not bound to an atom or molecule. Since there is no energy gap in metals, each atom contributes an electron or more for conduction. When an electron is removed from an atom, the remaining portion of the atom is a positively charged ion. So, one can imagine the situation as large number of electrons roaming around in the material where the positive heavy masses of ions are stationary. This is known as *electron gas theory* of a metal. So, a cloud of electrons wandering around in the material in the vicinity of fixed positive ions represents the metal.

These free electrons in the metal are not stationary but are moving randomly in the material. The electrons moving in a given direction collide with the stationary ions and change their direction and proceed further. The distance between two collisions is very random and the average value of such distances is called as *mean free path.* Since the direction is random, the number of electrons crossing a given cross-section at a given time would be same as that of the number of electrons moving in the opposite direction and hence the net current flowing would be zero.

When an electric field ε is now applied to a metal, the electrons would be accelerated in the direction opposite to that of the applied field and velocity would increase infinitely with time when there are no collisions with the static positive ions. However, due to the presence of these static ions, the collisions take place and the electron loses energy. When steady state is achieved, the velocity of the electrons becomes finite. This velocity, referred to as *drift velocity,* is in opposite direction to that of the applied field. The speed at time t between collision is at, where a is the acceleration and can be given as $q\varepsilon/m$. Thus, the average speed is proportional to the applied field and so,

$$v = \mu\varepsilon \qquad ...(2.14)$$

where μ is called the *mobility* of electrons. Thus, steady state drift speed has been superimposed upon the random thermal motion of electrons. Such a directed flow of electrons constitutes a current. To calculate the magnitude of this current, consider a metal of volume $A \times L$, where A is the cross-section area and L is the length of the material. Assume that there are N electrons in the volume considered and T is the time taken for an electron to travel a distance of L. Then the total number of electrons passing through any cross-section in unit time is N/T, or, the total charge crossing the cross-section, by definition, *current* can be given as

$$I = Nq/T = Nqv/L \qquad ..(2.15)$$

since L/T is the average drift speed v of the electrons. Current density is current per unit area of cross-section and so,

$$J = I/A \quad \text{...(2.16)}$$

Thus,

$$J = Nqv/(LA) \quad \text{...(2.17)}$$

Since N is total number of electrons in a volume LA of material considered, electron concentration n is

$$n = N/LA \quad \text{...(2.18)}$$

Therefore,

$$J = nqv = \rho v \quad \text{...(2.19)}$$

where $\rho = nq$ is the charge density and v is the drift velocity.

From Eq. (2.14),

$$J = nqv = nq\mu\varepsilon = \sigma\varepsilon \quad \text{...(2.20)}$$

where $\sigma = nq\mu$ is the *conductivity* of the metal. The above equation is nothing but Ohm's Law.

If now a semiconductor is considered, there are two types of charge carriers. The first type are free electrons, which are similar to those present in a conductor and, in addition to this, positively charged holes also contribute to current as discussed in the previous sections. So, now in a semiconductor, the total current would be superposition of the two independent components of currents due to electrons and holes. If J_n is electron current density in a semiconductor, it can be given as

$$J_n = nqv = nq\mu_n\varepsilon \quad \text{...(2.21)}$$

where μ_n is the mobility of the electrons in the semiconductor.

Similarly, the hole current density J_p can be expressed as

$$J_p = pqv = pq\mu_p\varepsilon \quad \text{...(2.22)}$$

where μ_p is the mobility of holes in the semiconductor material and p is the hole concentration.

Thus, the total current density can be given as

$$J = (n\mu_n + p\mu_p)\, q\varepsilon = \sigma\varepsilon \quad \text{...(2.23)}$$

where $\sigma = (n\mu_n + p\mu_p)q$ is the conductivity of the semiconductor.

2.8 PROPERTIES OF SEMICONDUCTOR MATERIALS

Charge densities in a semiconductor: Consider a semiconductor that is doped with donor atoms of concentration N_D and acceptor atoms of concentration N_A. n is the electron concentration after doping and p is hole concentration. Now, since N_D number of donor atoms are added per unit cubic metre of the material, it releases N_D number of free electrons per unit volume. This means that it is leaving behind N_D positive ions in the system. Now, the total positive charge available in the system would be the number of holes p plus the number of positive ions added by doping of donor atoms, N_D, per unit volume. Similarly, if N_A acceptor atoms are added per unit volume, N_A holes and N_A negative ions per unit volume are released into the material. Thus, there are n plus N_A number of negative charges in the material per unit volume. To maintain the neutrality in the system, the number of positive charges should be the same as that of negative charges and thus,

$$N_D + p = N_A + n \qquad ...(2.24)$$

To make an n-type material, one adds only donor atoms and not acceptor atoms into the semiconductor. In that case,

$$N_D + p = n \qquad ...(2.24a)$$

In an n-type material, it is well understood that the number of electrons is very large compared to the number of holes and so, p is very small compared to n and can be neglected. Thus,

$$N_D \approx n \qquad ...(2.25)$$

That is, the *electron concentration in an n-type material is approximately equal to the added impurity donor concentration.*

On similar argument, in a p-type material,

$$N_A \approx p \qquad ...(2.26)$$

That is, the *hole concentration in a p-type material is approximately equal to the added impurity acceptor concentration.*

Applying mass action law in an n-type material,

$$n_n p_n = n_i^2 \qquad ...(2.27)$$

where n_n is the electron concentration and p_n is hole concentration in an n-type material. But since in an n-type material, $N_D \approx n_n$

$$N_D p_n = n_i^2$$

Or,

$$p_n = n_i^2 / N_D \qquad ...(2.28)$$

And similarly, in a p-type material,

$$n_p p_p = n_i^2 \text{ and } n_p = n_i^2 / N_A \qquad ...(2.29)$$

Eqs. (2.28) and (2.29) give the minority carrier concentration in n-type and p-type materials, respectively.

It may be also observed that if in a material both donor and acceptor atoms are added, then the material would be intrinsic if the donor and acceptor atom concentration is same. The material can be recognised as n-type if in the material $N_D >> N_A$ and as p-type if $N_A >> N_D$. If one wants to convert a p-type material to n-type, one must add donor atoms in large concentration such that $N_D >> N_A$ is satisfied.

Intrinsic concentration: As already discussed, the intrinsic concentration of the semiconductor is a function of temperature and increases with temperature. Thus, increasing the temperature increases the new electron hole pairs in the material and thus the conductivity of the material improves. The definite relation of the intrinsic concentration with temperature is

$$n_i^2 = A_o T^3 e^{-E_{G0}/kT} \quad \text{...(2.30)}$$

where E_{G0} is the energy gap at 0 K in eV, k is the Boltzmann's constant in eV/K and A_0 is a constant independent of temperature.

The constants E_{G0}, μ_n , μ_p and so on for Si and Ge materials are as listed in Table 2.1. From the table, it can be observed that only 1 atom in about 10^9 atoms contributes free

Table 2.1 ***Electrical Properties of Germanium and Silicon***

Property	*Ge*	*Si*
Atomic number	32	14
Atomic weight	72.6	28.1
Density g/cm^3	5.32	2.33
Dielectric constant (relative)	16	12
Atoms/cm^3	4.4×10^{22}	5.0×10^{22}
E_{G0}, eV at 0 K	0.785	1.21
E_G, eV at 300 K	0.72	1.1
n_i at 300 K, cm^{-3}	2.5×10^{13}	1.5×10^{10}
P_i at 300 K, Ωcm^{-3}	45	230000
μ_n cm^2/v-s	3,800	1300
μ_p cm^2/v-s	1800	500
D_n, cm^2/s = $\mu_n V_t$	99	34
D_p, cm^2/s = $\mu_p V_t$	47	13
σ_i at 300°K (Ω-cm)$^{-1}$	0.0224	4.35×10^{-6}
Percentage increase in conductivity per degree rise in temperature	6	8
D_n / D_p (approximately)	2	3
Cut in Voltage V_r, V [Diodes]	0.2	0.6
Variation of I_0 with temperature per °C	11	8
Permissible Temperature of Operation (Transistors)	100°C	200°C

electrons and holes to the crystal because of broken bonds in germanium. For silicon, the ratio is further less, 1 in 10^{12}.

The energy gap: The forbidden energy gap between the valence and conduction bands is not fixed for a given semiconductor and is a function of temperature. The dependence of energy gap on temperature is as follows:

$$E_G(T) = 1.21 - 3.6 \times 10^{-4} T \quad \text{...(2.31)}$$

for silicon and that for germanium is

$$E_G(T) = 0.785 - 2.23 \times 10^{-4}T \qquad ...(2.32)$$

The mobility: The mobility of electrons and holes in a semiconductor depends not only on temperature but also on the applied electric field as follows:

$$\mu \propto T^{-m} \qquad ...(2.33)$$

in the temperature range of 100 – 400 K. The value of m is as follows:

$$\begin{aligned} m &= 2.5 \text{ for electrons in Si} \\ &= 2.7 \text{ for holes in Si} \\ &= 1.66 \text{ for electrons in Ge} \\ &= 2.33 \text{ for holes in Ge} \end{aligned} \qquad ...(2.34)$$

The mobility is independent of electric field ε when $\varepsilon < 10^3$ V/cm in n-type silicon. When $10^3 < \varepsilon < 10^4$ V/cm, μ_n varies approximately as $\varepsilon^{-1/2}$. For higher fields, μ_n is inversely proportional to ε.

2.9 HALL EFFECT AND ITS APPLICATIONS

The type of extrinsic semiconductor and the mobility of the material can be estimated with the help of *Hall effect*. The Hall effect can be explained as follows.

Consider a material, either a semiconductor or a metal, as shown in Fig. 2.9. Assume that the specimen is carrying a current I and is subjected to a transverse magnetic field B. Then, an electric field ε is induced in the direction perpendicular to both B and I. This phenomenon is known as *Hall effect*. This is very useful in semiconductors to determine the type of material, carrier concentration and mobility of the material.

As shown in Fig. 2.9, if current I is assumed to be flowing in positive x direction, and the applied magnetic field B is transverse to x axis, that is, z direction, then the induced electric field will have a force in the negative y direction by the Hall effect. This induced electric field in the specimen exerts force downwards on the charge carriers flowing in the material. If the material is a metal, the charge carriers are electrons flowing in the negative

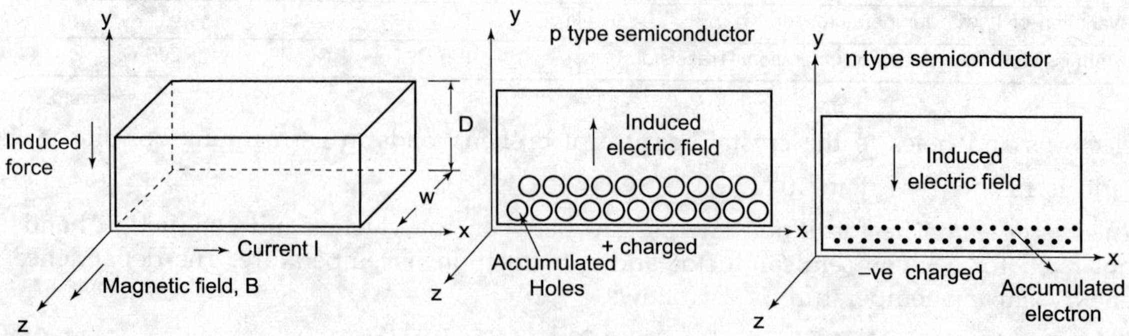

Fig. 2.9 *The Hall effect*

y-direction since the current is assumed to be flowing in the positive x direction. So, by the induced field these charge carriers will experience a force downwards in the negative y direction and one expect accumulation of electrons on the lower edge of the sample. Since the electrons assemble on the bottom of the specimen, equal amount of positive charge has to be deposited on top of the specimen. Thus, the electric field induced by the virtue of Hall effect will have positive field on the top whereas it will be negative at the bottom. This induces a field in the negative y direction.

When a semiconductor sample is subjected to Hall effect, the induced force due to the field will be on the charge carriers as in the case of a metal. If the semiconductor is n-type, the induced force will be exerted on both positive holes and negative electrons in the material. Since the electrons dominate the holes in an n–type semiconductor, the induced field would be same as in the case of a metal and will be in the negative y-direction. But if the specimen is a p-type semiconductor, since there are large number of positive holes, the induced field will force these holes towards the lower edge of the sample and so positive charge is accumulated on the bottom with an equal amount of negative charge on the top. This leads to the fact that the induced electric field is having a direction in the positive y-direction.

If an unknown semiconductor is subjected to the Hall effect, as explained in the earlier paragraphs, the type of material could be determined. If the induced field is in the positive direction of y axis, then the material is p type and if the induced field is in negative y-direction then the sample is n-type.

In an equilibrium state, the induced electric field ε due to Hall effect must exert a force on the carrier, which must balance the magnetic force. Thus,

$$q\varepsilon = Bqv \qquad \text{...(2.35)}$$

where q is the magnitude of the charge on the carrier and v is the drift velocity with which the carrier is moving.

The electric field

$$\varepsilon = V_H/d \qquad \text{...(2.36)}$$

where V_H is Hall voltage and d is the distance between the faces 1 and 2 of the specimen. The current density J can be given as

$$J = \rho v = I/wd \qquad \text{...(2.39)}$$

Then,

$$V_H = \varepsilon d = Bvd = BJd/\rho = BI/(\rho w) \qquad \text{...(2.40)}$$

From the above equation, if V_H, B and I are measured, ρ can be determined. The *Hall coefficient* R_H can be defined as

$$R_H = 1/\rho \qquad \text{...(2.41)}$$

Thus,

$$R_H = V_H w/(BI) \qquad \text{...(2.42)}$$

The conductivity σ can be given as

$$\sigma = \rho\mu \qquad ...(2.43)$$

or,

$$\mu = \sigma R_H \qquad ...(2.44)$$

From Eq. (2.44), if R_H is determined from Hall effect and conductivity measured for a given semiconductor, the mobility can be easily determined.

2.10 THERMISTOR AND SENSISTOR

The conductivity of the semiconductor depends on the number of free electrons and holes available for conduction in a material. The number of the charge carriers depends on temperature in an intrinsic material and so the conductivity of the semiconductor depends on temperature. As the temperature increases, the electron-hole pairs increase, thus increasing the conductivity of the semiconductor. The conductivity of the material can also be improved by doping the intrinsic material with either trivalent of pentavalent atoms to form *p*-type or *n*-type semiconductor. The variation of conductivity of a material with respect to a given parameter like temperature or light intensity is known as *conductivity modulation.*

From Eq. (2.30), it is seen that the conductivity increases approximately by 6% per degree increase in temperature for germanium and the rate is 8% for silicon. In fact, such a large variation of the conductivity with temperature is a drawback in some circuits. But, this phenomenon is very useful in some applications. A semiconductor that is designed for such an application is called a *thermistor.* The thermistor is a resistor made up of semiconductor instead of conventional carbon or other materials. An ordinary resistor exhibits resistance constant value irrespective of temperature. But a thermistor, being a temperature sensitive resistor, exhibits a resistance that decreases with increase in temperature. This thermistor finds applications in thermometry, microwave power measurement, thermal relay and other thermally controlled devices. Since silicon and germanium are too sensitive to impurities, they are not used as thermistor materials, but sintered mixtures of oxides such as NiO, Mn_2O_3 and Co_2O_3 are used.

When a metal is considered, its conductivity is also not a constant value with respect to temperature. It varies with temperature. The conductivity decreases with increase in temperature in a metal. Increase of temperature in a metal results in greater thermal agitation of the ions and this results in decrease of mean free path of the free electrons. Since mean free path depends on the mobility of the material, the mobility decreases. The conductivity of the metal is directly proportional to the mobility and thus the conductivity decreases. Thus, the resistance of the metal increases with increase in temperature. This rate of increase is of the order of 0.4% per degree rise in temperature. Thus, a metal has a positive coefficient of resistance while a semiconductor has a negative coefficient of resistance.

A very heavily doped semiconductor behaves in a manner similar to that of a metal. Thus, this semiconductor also exhibits positive coefficient of resistance like that of a metal due to decrease in carrier mobility with increase in temperature. Such a device is known as *sensistor.* A sensistor is a resistor made up of a heavily doped semiconductor having positive coefficient of resistance. This can have a temperature coefficient of resistance of the order of 0.7% per degree rise in temperature over the range of – 60 to + 150 °C.

The conductivity of a semiconductor can be improved by creating new electron-hole pairs in the system. Various possible schemes have been discussed in the above sections. Another scheme is to achieve *conductivity modulation* by light intensity variations. Such a material whose conductivity can be modulated by light can be referred to as a *photoconductor*. When the radiated light is exposed to the covalent bonds of the semiconductor, the photons strike the bonds to break them. Breaking one covalent bond releases two charge carriers, one electron and one hole. The number of covalent bonds broken would depend on the number of photons incident on the material. Thus, larger the intensity of the light, larger is the conductivity of the material. Such a material exhibiting photoconductivity modulation can be either referred to as a *photoconductor* or *a photoresistor*. Such a device finds large applications in the measurement of illumination (light meter), recording modulation intensity (sound track recording), on-off relay (digital control systems) and so on. The photoconducting device with the widest application is the cadmium sulfide cell.

2.11 TIME VARIATION OF EXCITED MINORITY CARRIERS

In equilibrium, at room temperature, in a semiconductor, the number of free electrons and holes remains constant. This does not mean that the given electron is always present and will contribute to conduction. The free electrons and holes are not stationary but wander throughout the material randomly due to thermal energy available at room temperature. In this process, recombination and generation of electrons take place continuously, but at any given time, the number of free electrons or holes is unaltered. To create charge carriers, one has to break a covalent bond. To break a covalent bond, energy from an external agency is expended and so, one loses energy in creating an electron-hole pair. On the other hand, when an electron recombines with a hole, one loses two charge carriers (one electron and one hole) and energy is gained by the external agency. In a system, generation and recombination take place continuously and the *lifetime* of a charge carrier is the time between generation and recombination. The term *mean lifetime* of a charge carrier is the average of the lifetimes of these carriers. τ_n and τ_p are the mean lifetimes of an electron and a hole, respectively.

Consider an *n*-type semiconductor bar. Due to thermal agitation, electron-hole pairs are generated—let the rate of generation be g. On the other hand, equal number of electron-hole pairs disappear due to recombination after thermal equilibrium is achieved. That is, the rate of generation is same as the rate of recombination at a given temperature at equilibrium. The recombination is due to electron falling into empty valence band, resulting in loss of a pair of charge carriers. Let p_0 and n_0 be the concentrations of holes and electrons at room temperature at equilibrium. At a time $t = t'$ (as shown in Fig. 2.10) the bar is subjected to illumination, which can break the covalent bonds. Thus, new electron-hole pairs are generated and in this condition the rate of generation is more than recombination and thus the electron and hole concentration increases until new equilibrium is achieved. The additional electrons and holes generated would be same in number. Let $\overline{p}$ and $\overline{n}$ be the new concentrations at equilibrium after excitation. Thus,

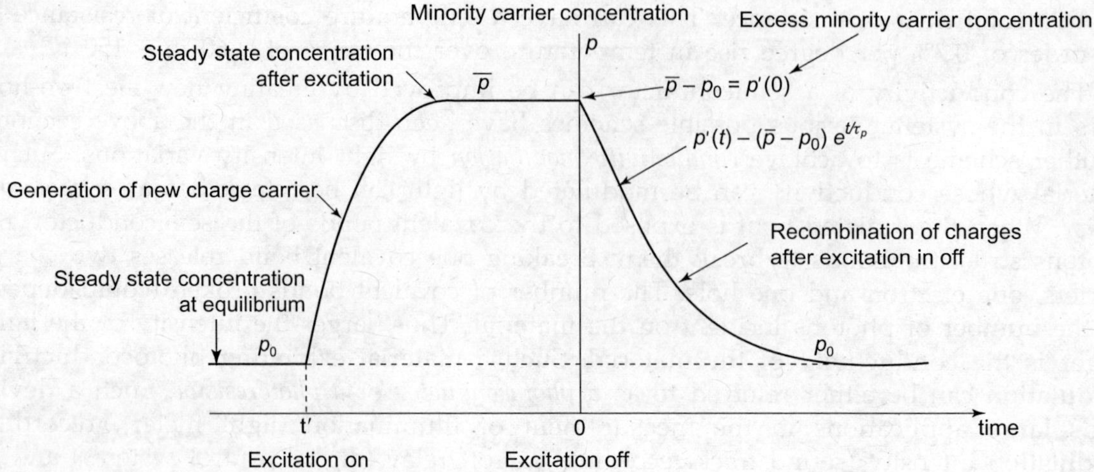

Fig. 2.10 *Minority carrier recombination*

$$\overline{p} - p_0 = \overline{n} - n_0 \qquad \text{...(2.45)}$$

Although the increase in hole concentration is same as that in electron concentration, since *n*-type bar is considered, there are large numbers of electrons and adding a few more electrons leads to very small rate of increase in electrons. But, when holes, the minority carriers, are considered—since they are very small in number—adding a few more leads to a large rate of change in concentration. So, the increase in the electrons can be neglected and the minority carrier behaviour is studied.

After a steady state is reached, the excitation is turned off, say, at $t = 0$. When the excitation is switched off, the rate of generation decreases and the rate of recombination would be more than that of generation, thus leading to loss of charge carriers. To understand how the rate of recombination affects the charge concentration, the following analysis is done.

Assuming that the mean lifetime τ_p is independent of hole concentration,

p/τ_p = decrease in hole concentration per second due to recombination, ...(2.46)

g = increase in hole concentration per second due to thermal agitation. ...(2.47)

At equilibrium, since no charge can be created or destroyed, the above two rates are equal. Or, the rate of change of hole concentration would be zero,

$$dp/dt = 0. \qquad \text{...(2.48)}$$

But after the excitation is switched off at $t = 0$, the rate of change of hole concentration would be the difference between the generation rate and recombination rate. Thus,

$$dp/dt = g - p/\tau_p \qquad \text{...(2.49)}$$

When no radiation is falling on the sample, the generation rate would be p_0/τ_p and therefore,

$$dp/dt = (p_0 - p)/\tau_p \qquad \text{...(2.50)}$$

If p' denotes the *excess* or *injected* carrier density, defined as increase in minority carrier concentration above equilibrium level, then, p' would be a function of time and

$$p' = p - p_0 = p'(t) \qquad ...(2.51)$$

Since p_0 is constant, substituting the above equation in Eq. (2.50),

$$dp'/dt = -\, p'/\tau_p \qquad ...(2.52)$$

The minus sign in the above equation indicates that there is a decrease in the hole concentration. Solving the above differential equation for $t \geq 0$,

$$p'(t) = p'(0)\, e^{-t/\tau_p} = (\bar{p} - p_0)\, e^{-t/\tau_p} = p - p_0 \qquad ...(2.53)$$

Thus, *the excess minority carrier concentration* after the external excitation is switched off *decays towards the equilibrium value exponentially with the time constant as the mean lifetime of the carriers*.

It is important to note here that all the electrons encountering a hole will not recombine. In fact, recombination results in release of energy and loss of two charge carriers. In order to recombine, when an electron collides with a hole both should have the same momentum and they must travel in the opposite directions. This is a very stringent condition and very rare. So, recombination by such direct encounters is very small.

In practice, recombination is not by direct collision as explained above, but the mechanism involves *recombination centres* or *traps*. These traps contribute electronic states in the energy gap of the semiconductor. Such a location acts effectively as a third body that can satisfy the conservation of momentum requirement. These new states are due to imperfections in the crystal. Metallic impurities in the semiconductor are capable of introducing energy states in the forbidden gap. Not only the volume imperfections, but also surface imperfection leads to the traps in the crystal. So, purposefully, gold is introduced in a controlled fashion into silicon to achieve the desired traps in the crystal.

2.12 CONCEPT OF DIFFUSION CURRENT

In a semiconductor, in addition to the conduction current due to electrons and holes, there are two more components of current due to *diffusion* of holes and electrons. This diffusion is not observed in metals. The concept of diffusion is as explained below.

Consider a p-type semiconductor bar where the doping is not uniform throughout the material, but is very large towards one end of the bar and the concentration decreases as one moves towards the opposite end. That is, the doping of acceptor atoms is controlled and is non-uniform. Assume that the bar is oriented in x direction with maximum doping end at $x = 0$. As seen from the figure, the concentration of the impurity in the semiconductor is a function of x and decreases with increase in x. So, there is a concentration gradient dp/dx and thus the density of carriers is also not constant. Now, consider an imaginary cross-section of the bar at some x where $x > 0$ as shown in the figure. To the left of this imaginary cross-section the concentration of the charge carriers, holes, is large compared to that to the right. This is because the concentration is maximum at $x = 0$ and decreases with increase in x. These holes are not stationary as seen previously

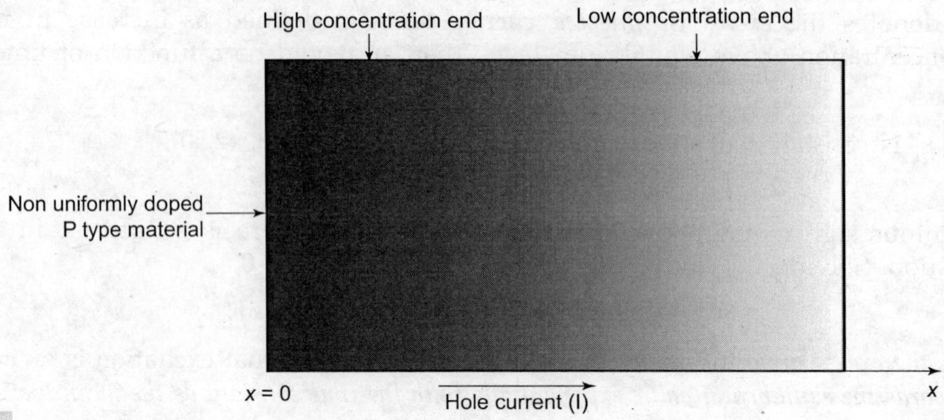

Fig. 2.11 *Diffusion*

and are in random motion due to thermal agitation. Because of this random motion, the holes cross the imaginary cross-section in both directions. In a given time interval, there is a probability that the number of holes crossing from left to right would be more than that of holes crossing from right to left. This is a statistical phenomenon since the concentration gradient occurs in the semiconductor. Since the number of charge carriers crossing the cross-section is not the same in both the directions, there is a net transport of charge carriers in the direction of positive x direction. This leads to a current flow in the positive x direction as shown in Fig. 2.11. It should be noted that this current is not due to the repulsions of the charge carriers, but is purely a statistical phenomenon. This current is referred to as *diffusion current*.

As one expects, this diffusion current is proportional to the concentration gradient in the semiconductor. Thus, *diffusion current density is proportional to the concentration gradient.* The proportionality constant is split into components, as seen below, for convenience,

$$J_p \equiv -qD_p(dp/dx) \qquad ...(2.54)$$

where q is the charge associated with each charge carrier and D_p is the d*iffusion constant* for holes measured in square metres per second. Since p decrease with increase in x, the concentration gradient dp/dx is negative. Since the hole current is in positive direction, the negative sign in the equation is inserted.

Similarly, if pentavalent atoms non-uniformly dope a semiconductor, a concentration gradient of n-type material can be achieved. In such a semiconductor, diffusion current due to electrons can be achieved. The constituent diffusion current density of electrons can be given as

$$J_n \equiv qD_n(dn/dx) \qquad ...(2.55)$$

where D_n is the diffusion constant of electrons. Minus sign is not required in this equation since the conventional current and the concentration gradient coincide.

Thus, in a semiconductor various components of current are possible. The electron current density due to conduction and diffusion can be given as

$$J_n = q\mu_n n\varepsilon + qD_n(dn/dx) \qquad ...(2.56)$$

The hole current density due to conduction and diffusion is

$$J_p = q\mu_p p\varepsilon - qD_p(dp/dx) \qquad ...(2.57)$$

Einstein relationship: Both the mobility and the diffusion are statistical phenomena. So, D and μ are not independent. The relationship between them is given by Einstein equation:

$$D_p/\mu_p = D_n/\mu_n = V_T \qquad ...(2.58)$$

where V_T is the *volt equivalent of temperature* and is defined as

$$V_T = \bar{k}T/q = T/11600 \qquad ...(2.59)$$

where $\bar{k}$ is the Boltzmann's constant in joules per kelvin. At room temperature, T = 300 K and V_T = 26 mV.

2.13 CONTINUITY EQUATION IN SEMICONDUCTORS

As discussed in Sec. 2.11, if the semiconductor material is disturbed from equilibrium, the concentrations of holes and electrons vary with time. However, the carrier concentration in the body of a semiconductor is a function of distance, too. The *continuity equation* is a differential equation that governs the relationship of the carrier concentration both with time and distance. This equation is derived on the basis of conservation of charges property.

Consider an infinitesimal element of p-type semiconductor having a volume $L \times dx$. Let the average hole concentration in the volume be p. The problem is assumed to be one

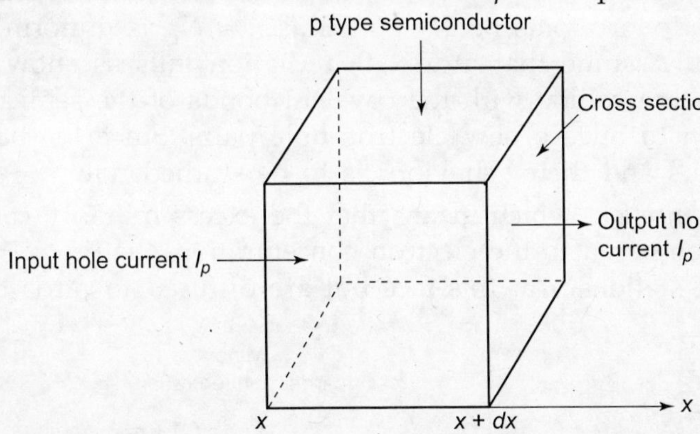

Fig. 2.12 *Continuity equation*

dimensional and the current I_p in the element is flowing in x direction. If I_p is the current entering into the element at x and $I_p + dI_p$ at $x + dx$ at the same time, dI_p coulombs of charge per second are leaving in addition to the input. Hence, the charge carriers in the element decrease within the volume. Since magnitude of the charges is q, dI_p/q is the decrease in the number of holes per second within the element. If the volume is Adx, then the *decrease in hole concentration per second* can be given as

$$\frac{dI_p}{qAdx} = \frac{dJ_p}{qdx} \qquad ...(2.60)$$

If $g = p_0/\tau_p$ is the increase of holes per unit volume per second due to thermal agitation, and p/τ_p is the decrease of holes per unit volume per second due to recombination, as we have seen in Sec. 2.11, the rate of hole concentration change due to thermal agitation and recombination can be given as $(p_0 - p)/\tau_p$. Thus, taking into account the conservation of charges from the above two concepts, the rate of change of hole concentration can be stated as

$$\frac{\partial p}{\partial t} = \frac{p_0 - p}{\tau_p} - \frac{1}{q}\frac{\partial J_p}{\partial x} \qquad \text{...(2.61)}$$

The above equation is called the *continuity equation*. This can also be referred to as *law of conservation of charge*. This law can also be applied to electrons and n is substituted for p in the above equation.

2.14 INJECTED MINORITY CARRIERS

The minority carrier behaviour is seen to be exponentially decaying with time when an external excitation is removed and the time constant is the lifetime of the carrier. Now, the behaviour of the minority carrier with distance is to be considered when introduced at one end of a semiconductor. This also decays exponentially with distance, the time constant being *diffusion length* this time.

Consider a uniformly doped n-type bar of semiconductor as shown in Fig. 2.13. Let the concentration of the donor atoms be N_D and thus $n \approx N_D$ is uniform throughout the bar under consideration. Assume that at $x = 0$, radiation falls as shown in the figure. The photons of the radiation collide with the covalent bonds of the semiconductor bar on this face and break them to release new electron-hole pairs. Since the bar is n-type, the holes are minority carriers and their behaviour is to be studied now.

Consider a *low-level injection*, which means that the excess minority carriers created in the bar are very small compared to the electron concentration, that is, $p' \ll n$. The excess minority carriers are not stationary at this face but are diffused towards the other end of

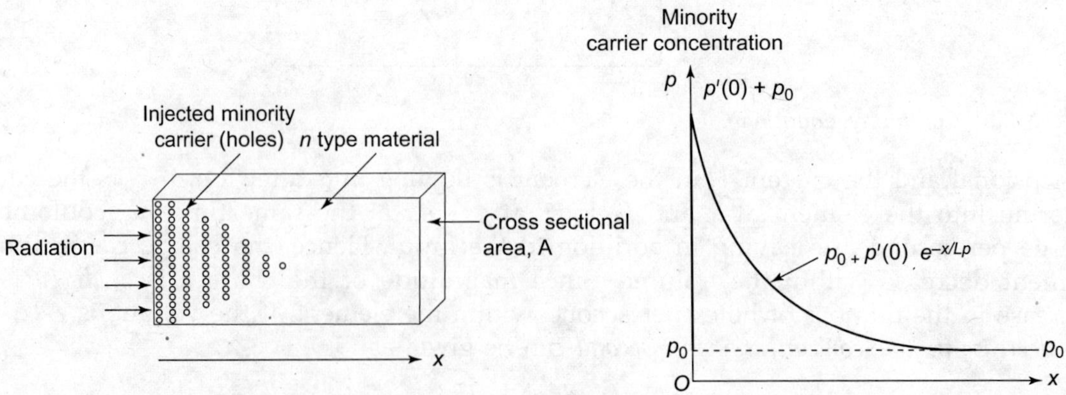

Fig. 2.13 *Injected minority carriers*

the bar and as they diffuse, recombination takes place with the free electrons and the minority carrier concentration decreases. Since the drift current is proportional to the charge carrier concentration, the drift current due to holes can be neglected as the amount of holes in the bar is very small compared to electron concentration, that is, $p_0 + p' << n$. So, the hole current is entirely due to diffusion. The other components of electron current are present in the bar. Then,

$$J_p = -qD_p(dp/dx) \quad \text{...(2.62)}$$

From the continuity equation (2.61), at steady state, rate of change of hole concentration is zero, $\partial p/\partial t = 0$, which leads to

$$\frac{d^2p}{dx^2} = \frac{p - p_0}{D_p \tau_p} \quad \text{...(2.63)}$$

Diffusion length for holes is defined as

$$L_p = (D_p \tau_p)^{1/2} \quad \text{...(2.64)}$$

Thus, the above differential equation can be rewritten as

$$\frac{d^2p'}{dx^2} = \frac{p'}{L_p^2} \quad \text{...(2.65)}$$

The general solution to such a second order differential equation can be written as

$$p'(x) = K_1 e^{-x/L_p} + K^2 e^{+x/L_p} \quad \text{...(2.66)}$$

where K_1 and K_2 are constants. When a very long bar is considered, as x increases, the second term increases if K_2 were to be a given value other than zero. That means as x tends to infinity, the hole concentration increases with distance, which is an impossible condition. Thus, $K_2 = 0$. If at $x = 0$, the excess concentration of holes is assumed to be $p'(0)$, then $K_1 = p'(0)$. Then,

$$p'(x) = p'(0)e^{-x/L_p} = p(x) - p_0 \quad \text{...(2.67)}$$

Thus, the minority carrier concentration decreases exponentially with L_p as constant. The *diffusion length* can be redefined as *the length at which the concentration falls to 1/e times that of the initial value* at $x = 0$. It can also be seen as the average distance that an injected hole travels before recombining with an electron.

Diffusion current: From Eq. (2.54) for current density of diffusion current due to holes and remembering $J_p = I_p/A$,

$$I_p(x) = \frac{AqD_p p'(0)}{L_p} e^{-x/L_p} = \frac{AqD_p}{L_p}[p(0) - P_0]e^{-x/L_p} \quad \text{...(2.68)}$$

Thus, the diffusion current falls exponentially, as is the case with hole concentration. Now, the majority carrier electron diffusion current is

$$I_n(x) = AqD_n(dn/dx) \quad \text{...(2.69)}$$

The excess minority and majority carriers would be the same since breaking a bond creates a electron-hole pair. Thus, $p' = n'$ or,

$$p - p_0 = n - n_0 \qquad ...(2.70)$$

Since p_0 and n_0 are independent of x,

$$dp/dx = dn/dx \qquad ...(2.71)$$

Thus, the electron-hole current is

$$I_n(x) = AqD_n(dn/dx) = AqD_n(dp/dx) = -I_p(D_n/D_p) \qquad ...(2.72)$$

Drift current: Since the semiconductor bar is open circuited, the net current flowing in the bar, that is, the sum of drift current due to electrons and diffusion current due to holes and electrons, would be zero. Thus, if I_{nd} is the drift current due to electrons,

$$I_p + \left(I_{nd} - \frac{D_n}{D_p} I_p\right) = 0 \qquad ...(2.73)$$

or

$$I_{nd} = \left(\frac{D_n}{D_p} - 1\right) I_p \qquad ...(2.74)$$

Thus, the electron drift current also decreases exponentially with distance.

As seen, if a drift current exists in the bar, this current would be due to a potential drop, that is, there should be some electric field existing in the bar. This can thus be calculated as

$$\varepsilon = \frac{1}{Aqn\mu_n}\left(\frac{D_n}{D_p} - 1\right) I_p \qquad ...(2.75)$$

When an electric field is present in the bar, it should also cause drift current in the bar due to holes. Hence, the drift current due to holes, $I_{pd,}$ can be given as

$$I_{pd} = Aqp\mu_p\varepsilon = \frac{p\mu_p}{n\mu_n}\left(\frac{D_n}{D_p} - 1\right) I_p \qquad ...(2.76)$$

Since $p << n$ in a n-type bar, $I_{pd} << I_p$. Thus, the hole drift current is very small compared to the hole diffusion current, justifying the assumption made in the beginning of this discussion under low-level injection.

2.15 GRADED SEMICONDUCTOR JUNCTION

Consider a semiconductor in which the doping of the impurity atoms is non-uniform. Such a semiconductor can be referred to as *graded* semiconductor. Assume that the considered bar is doped with trivalent atoms to form the p-type semiconductor. Assume that there is no external excitation given to the bar and the bar is at steady state. Even though the net current in the bar is zero, there is movement of the charge carriers, holes, randomly in the bar due to thermal agitation. Since p is not uniform in the bar, one can expect diffusion current due to holes to exist in the bar. In order to make the total current due to holes zero at steady state, there should be equal and opposite drift current due to holes in the bar (same is the case with electrons—total electron current should be zero). If a drift current is to be flowing in a material, there should be potential drop across the bar establishing an electric field. So, by virtue of non-uniform doping of impurities in the semiconductor, a potential drop is induced into the bar. This potential is investigated now.

Setting the total current density due to holes, J_p equal to zero, from Eq. (2.57),

$$q\mu_p p\,\varepsilon = qD_p(dp/dx) \qquad \text{...(2.77)}$$

Since $V_T = D_p/\mu_p$,

$$\varepsilon = \frac{V_T\,dp}{pdx} \qquad \text{...(2.78)}$$

The relation between the electric field and potential is $\varepsilon = -\,dV/dx$. Thus,

$$dV = -\,V_T(dp/p) \qquad \text{...(2.79)}$$

To find out the potential drop between two points 1 and 2, the above equation is to be integrated within the limits x_1 and x_2. If p_1 is the concentration at x_1 and p_2 is the concentration of holes at x_2,

$$V_{21} = V_2 - V_1 = V_T ln\,(p_1/p_2) \qquad \text{...(2.80)}$$

V_{21} is the potential drop between 2 and 1 as shown in Fig. 2.14. Thus, the potential drop between the two points depends only on the concentrations of holes at the two points, but does not depend on the position of the two points under consideration. The

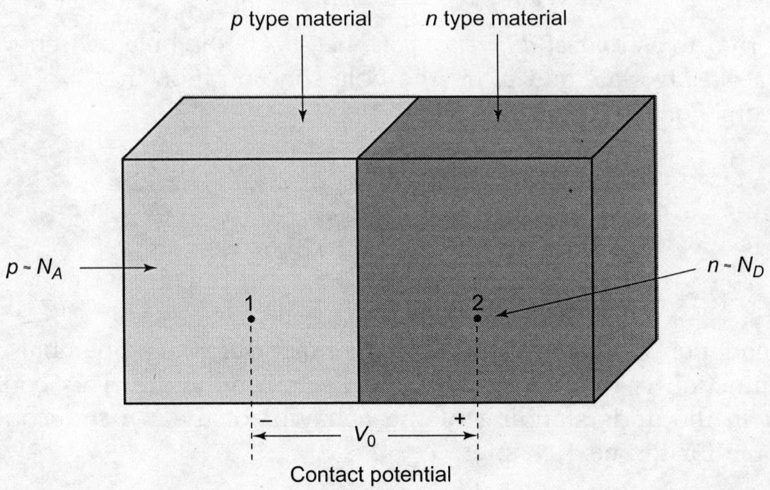

Fig. 2.14 *Contact potential*

potential is independent of the distance between the two points. Equation (2.80) can be recast as follows to estimate the concentration at a given point to establish the defined potential:

$$p_1 = p_2 e^{V_{21}/V_T} \quad \text{...(2.81)}$$

This is the Boltzmann's relationship of kinetic gas theory.

Carrying out the discussion for electrons as is done in the above for holes, where the electron current density is zero, leads to derivation of the following equation:

$$n_1 = n_2 e^{-V_{21}/V_T} \quad \text{...(2.82)}$$

From Eqs. (2.81) and (2.82),

$$n_1 p_1 = n_2 p_2. \quad \text{...(2.83)}$$

This equation states that the product np is a constant independent of x, and hence of the amount of doping, under thermal equilibrium. For an intrinsic semiconductor, $n = p = n_i$ and hence,

$$np = n_i^2 \quad \text{...(2.84)}$$

which is the mass action law.

Step-graded junction: Consider a junction formed between two semiconductors as shown in Fig. 2.14. The junction is formed between p-type and n-type semiconductors and the doping in the material is assumed to be uniform. Such a junction is referred to as *step-graded junction* since the concentration of impurities is suddenly changed from p to n type in a step. Let the concentration of impurities towards the left of the junction be N_A and to the right the concentration of n-type material be N_D. The above discussion shows that the potential developed due to concentration gradient depends on the concentration but is independent of the distance. The potential drop between any point on the left of the junction and any point to the right of the junction can be given as

$$V_0 = V_{21} = V_T ln\,(p_{p0}/p_{n0}), \quad \text{...(2.85)}$$

where V_0 is referred to as *contact difference potential,* p_{p0} is the hole concentration in a p-type semiconductor at steady state and p_{n0} is the hole concentration in n-type semiconductor at steady state. From Eq. (2.29),

$$p_{p0} = N_A \quad \text{and} \quad p_{n0} = n_i^2/N_D \quad \text{...(2.86)}$$

Thus,

$$V_0 = V_T\, ln \frac{N_A N_D}{n_i^2} \quad \text{...(2.87)}$$

The above equation gives the value of the contact difference potential developed when a step-graded junction is formed between two semiconductors. This contact potential is very important in the understanding of the behavior of the p-n semiconductor junction, which is the topic for the next chapter.

SUMMARY

- Energy-band theory of a crystal suggests the existence of a semiconductor whose conductivity can be increased by varying the temperature or adding impurities.
- Tetravalent atoms (Si and Ge) are basic materials used in electronic devices. Doping with trivalent or pentavalent impurities increases the conductivity.
- Trivalent atoms introduce a large number of holes, while pentavalent atoms introduce electrons. The materials are then referred to as *p* and *n* type, respectively.
- In an intrinsic pure material at a given temperature, the number of holes is same as that of electrons, while mass action law states that product of two concentrations is equal to square of intrinsic concentration in an extrinsic material.
- Holes constitute positively charged carriers and so due to drift there are two components of current in a semiconductor—current due to holes and electrons.
- Another source of current in a semiconductor is diffusion, which is not possible in a metal. This is due to concentration gradient in a semiconductor.
- The two components of current due to diffusion are due to holes and electrons. Thus, there are four different current components in a semiconductor.
- Hall effect helps one to determine the type of unknown semiconductor and it also gives its mobility.
- Thermistor is a resistor whose resistance decreases with increase in temperature, while sensistor is a resistor whose resistance increases with increase in temperature.
- Photoconductor is a device whose resistance decreases with light intensity.
- Breaking of one covalent bond requires external energy and release two charge carriers—an electron and a hole.
- Recombination of an electron and a hole in a trap in semiconductor results loss of a pair of charge carriers but it releases energy.
- Carriers are continuously being generated due to thermal creation and are simultaneously disappearing due to recombination at equlibrium.
- The newly generated minority carriers in a semiconductor due to external excitation recombine with the existing majority carriers after the excitation is removed in an exponential decay with time.
- Continuity equation governs the flow of charge carriers. It is based on the principle of charge conservation.
- The injected minority carriers decay exponentially with distance due to recombination.

- The total current in an unbiased graded semiconductor is zero. So, equal and opposite drift current should flow due to drift of charge carriers. This leads to the concept of contact potential in a graded semiconductor.
- The contact potential depends on the concentration between the two points but not on the distance between them. The higher the difference in concentrations, higher is the potential.

SOLVED PROBLEMS

2.1 Prove that the concentration n of free electrons per cubic meter of a metal is given by

$$n = \frac{dv}{AM} = \frac{A_0 dv \times 10^3}{A}, \text{ where}$$

d = density, kg/m^3

v = valence free electrons per atom

A = atomic weight

m = weight of atom of unit atomic weight, kg

A_0 = Avogadros' number, molecules/mole.

Solution The weight of an atom is considered to be equal to the atomic mass times the mass of an atom of unit atomic weight.

$$n = \left[\frac{\text{atoms}}{Am\,\text{kg}}\right]\left[v\frac{\text{electrons}}{\text{atom}}\right]\left[d\frac{\text{kg}}{\text{m}^3}\right] = \frac{dv}{AM}\text{ electrons/m}^3.$$

If n = no. of atoms per molecule, then AM = Molecular weight of one mole in grams, and

$$n = \left[\frac{1}{An}\frac{\text{mole}}{g}\right]\left[A_0\frac{\text{molecules}}{\text{mole}}\right]\left[n\frac{\text{atoms}}{\text{molecule}}\right]\cdot\left[10^3\frac{\text{g}}{\text{kg}}\right]\left[d\frac{\text{kg}}{\text{m}^3}\right]\left[v\frac{\text{electrons}}{\text{atom}}\right]$$

$$= \frac{A_0 dv}{A} \times 10^3 \text{ electrons/m}^3.$$

2.2 If a donor type impurity is added to the extent of 1 atom per 10^8 Ge atoms, calculate the value of resistivity.

Solution If there is 1 donor atom per 10^8 Ge atoms, then,

$$N_D = \frac{\text{atoms/cm}^3}{10^8} = \frac{4.4 \times 10^{22}}{10^8} = 4.4 \times 10^{14} \text{ atoms/cm}^3$$

Hence, $n = N_D = 4.4 \times 10^{14}$

$$\therefore p = \frac{n_i^2}{N_D} = \frac{(2.5\times10^{13})^2}{4.4\times10^{14}} = 1.42 \times 10^{12}$$

Therefore, $p << n$, neglecting p.

$$\sigma = nq\mu_n = 4.4 \times 10^{14} \times 1.6 \times 10^{-19} \times 3800$$
$$= 0.1268 \ (\Omega\text{-cm})^{-1}$$

$$\therefore \text{Resistivity } \rho = \frac{1}{\sigma} = 3.728(\Omega\text{-cm})$$

2.3 (a) Find the resistivity of intrinsic silicon at 300 K

(b) If a donor type impurity is added to the extent of 2 atoms per 10^8 Si atoms, find the resistivity

Solution (a) Conductivity $s_i = n_i q\ (\mu_n + \mu_p)$ for an intrinsic semiconductor.

$$\therefore \sigma_i = 1.6 \times 10^{-19} \times 1.5 \times 10^{10}\ (1300 + 500)$$
$$= 4.32 \times 10^{-6}\ (\Omega\text{-cm})^{-1}$$

$$\text{Hence, resistivity } \rho_i = \frac{1}{\sigma_i} = 2.3 \times 10^5(\Omega\text{-cm})$$

$$\text{(b) } N_D = \frac{5\times10^{22}}{2\times10^8} = 2.5 \times 10^{14} \text{ atoms/cm}^3$$

Hence, $n = N_D = 2.5 \times 10^{14}$ atoms/cm^3

$$p = \frac{ni^2}{N_D} = \frac{(1.5\times10)^2}{2.5\times10^{14}} = 0.9 \times 10^6$$

Hence, $p << n$, so p can be neglected.

$$\therefore \sigma = nq\mu_n = 1.6 \times 10^{-19} \times 2.5 \times 10^{14} \times 1300$$
$$= 0.052\ (\Omega\text{-cm})^{-1}$$

$$\text{Hence, resistivity } \rho = \frac{1}{\sigma} = 19.23\ \Omega\text{-cm.}$$

2.4 (a) Find the concentration of holes and electrons in p-type Ge at 300 K if the conductivity is 60 $(\Omega\text{-cm})^{-1}$.

(b) Repeat part (a) for n-type silicon if the conductivity is 0.13 $(\Omega\text{-cm})^{-1}$.

Solution (a) We know that $s = nqm_n + pqm_p$

for p-type semiconductor $p >> n$

$$\therefore \sigma_p \approx pq\mu_p$$

$$\Rightarrow \quad p = \frac{\sigma_p}{q\mu_p} = \frac{60}{1.6\times10^{-19}\times1800} = 2.083 \times 10^{17} \text{ holes/cm}^3.$$

$$\text{Hence, } n = \frac{n_i^2}{p} \text{ from mass action law.}$$

$$= \frac{(2.5\times10^{13})^2}{2.083\times10^{17}} = 3\times 10^9 \text{ electrons/cm}^3.$$

(b) For an n-type semiconductor $n >> p$

$$\therefore \sigma_n = nq\mu_n + pq\mu_p \approx nq\mu_n$$

$$\Rightarrow \quad n = \frac{\sigma_n}{q\mu_n} = \frac{0.13}{1.6\times10^{-19}\times1300}$$

$$= 6.25 \times 10^{14} \text{ electrons/cm}^3$$

$$\therefore p = \frac{n_i^2}{n} = \frac{(1.5\times10^{10})^2}{6.25\times10^{14}} = 36 \times 10^4 \text{ holes/cm}^3$$

2.5 Show that for an intrinsic Ge the conductivity increases by 5.5% per °C.

Solution For an intrinsic Ge $\sigma_i = n_i q(\mu_n + \mu_p)$.
Assume that μ is independent of temperature.

$$\text{and,} \quad n_i^2 = A_0 T^3 e^{-E_{G0}/KT}$$

$$\Rightarrow \quad n_i = A_0^{1/2} T^{3/2} e^{-E_{G0}/2KT}$$

Taking natural logarithm

$$\ln n_i = \ln A_0^{1/2} + \ln T^{3/2} + \ln e^{-E_{G0}/2KT}$$

$$= \ln A_0^{1/2} + \ln T^{3/2} - \frac{E_{G0}}{2KT}$$

Differentiating this equation

$$\frac{dn_i}{n_i} = \frac{3}{2}\frac{dT}{T} + \frac{E_{G0}}{2KT^2}dT$$

$$= \left[\frac{3}{2} + \frac{E_{G0}}{2KT}\right]\frac{dT}{T}$$

$$\text{For Ge,} \quad \frac{dn_i}{n_i} = \left[1.5 + \frac{0.785}{2\times0.026}\right]\frac{100}{300}$$

$$= 5.5\% \text{ per degree.}$$

$\therefore$ Change in conductivity is 5.5% per °C.

2.6 (a) Show that the conductivity of a semiconductor is minimum when it is lightly doped with *p*-type impurity such that $p = n_i \left[\frac{\mu_n}{\mu_p}\right]^{1/2}$

(b) Show that the minimum conductivity is $2n_i q\sqrt{\mu_n \mu_p}$

(c) Use the above result to find the minimum conductivity of silicon.

(d) Find the corresponding doping level.

Solution (a) Conductivity $\sigma = nq\mu_n + pq\mu_p$

$$= q\left(p\mu_p + \frac{n_i^2}{p}\mu_n\right) \qquad \text{...(1)}$$

To find minima, find $\frac{d\sigma}{dp}$ and equate it to zero.

$$\therefore \quad \frac{d\sigma}{dp} = \left(-\frac{n_i^2}{p^2}\mu_n + \mu_p\right)q = 0$$

$$\Rightarrow \quad \mu_p = \frac{n_i^2}{p^2}\mu_n$$

$$\therefore \quad p = n_i\sqrt{\frac{\mu_n}{\mu_p}} \qquad \text{...(2)}$$

(b) Substituting (2) in (1)

$$\sigma_{\min} = \left[\frac{n_i^2 \mu_n}{n_i\left(\frac{\mu_n}{\mu_p}\right)^{1/2}} + n_i\left(\frac{\mu_n}{\mu_p}\right)^{1/2}\mu_p\right]q$$

$$= \left(n_i\sqrt{\mu_n\mu_p} + n_i\sqrt{\mu_n\mu_p}\right)q$$

$$= 2n_i q\sqrt{\mu_n \mu_p}$$

(c) For silicon,

$$\sigma_{\min} = 2 \times 1.5 \times 10^{10} \times 1.6 \times 10^{-19}\sqrt{1300 \times 500}$$
$$= 3.87 \times 10^{-6} \ (\Omega\text{-cm})^{-1}$$

(d) We have charge balance equation

$$N_D + p = N_A + n$$

Since this is a *p*-type semiconductor,

$$N_A = p - n$$

$$= n_i\sqrt{\frac{\mu_n}{\mu_p}} - \frac{n_i^2}{n_i\sqrt{\frac{\mu_n}{\mu_p}}}$$

$$= n_i\sqrt{\frac{\mu_n}{\mu_p}} - n_i\sqrt{\frac{\mu_p}{\mu_n}}$$

On substituting the values

$$N_A = 1.48 \times 10^{10} \text{ atoms/cm}^3$$

2.7 A sample of Ge has *n*-type impurity concentration of 3×10^{14} donors/cm^3 and *p*-type impurity concentration of 4×10^{14} acceptors/cm^3. Find n and p at room temperature.

Solution $N_A = 4 \times 10^{14},\ N_D = 3 \times 10^{14}$

$$\therefore \quad N_D + p = N_A + n \qquad \text{...(1)}$$

$$\Rightarrow \quad p - n = N_A - N_D = 4 \times 10^{14} - 3 \times 10^{14}$$

$$= 1 \times 10^{14} \qquad \text{...(2)}$$

From mass action law $np = n_i^2$

$$= (2.5 \times 10^{13})^2 \qquad \text{...(3)}$$

From (2) and (3), solving for p and n,

$$p = 0.933 \times 10^{14} \text{ holes/cm}^3.$$

$$n = 6.698 \times 10^{12} \text{ Electrons/cm}^3.$$

2.8 Find the factor by which conductivity of pure Ge will change when the temperature is rised from 300 K to 600 K. $\mu_p(300) = 1800$, $\mu_p(600) = 300$, $\mu_n(300) = 3900$, $\mu_n(600) = 1300$.

Solution We know that $\sigma_i = n_i q(m_n + m_p)$

$$= q(m_n + m_p)\left(A_0 T^3 e^{-E_{G0}/KT}\right)^{1/2}$$

$$\therefore \quad \frac{\sigma(600)}{\sigma(300)} = \frac{(1300 + 300)}{(3900 + 1800)}\left(\frac{600}{300}\right)^{3/2}\left[\frac{e^{-0.785/8.6\times10^{-5}\times600}}{e^{-0.785/8.6\times10^{-5}\times300}}\right]^{1/2}$$

$$= \frac{1800}{5700} \times (2)^{3/2} \times \left[e^{\left(\frac{-0.785\times10^5}{5160} + \frac{0.785\times10^5}{2580}\right)}\right]^{1/2}$$

$$= 0.315 \times 2.828 \times \left(e^{\frac{0.785 \times 10^5}{5160}} \right)^{1/2}$$

$$= 1792$$

2.9 Show that the p-type semiconductor has same conductivity as that of intrinsic semiconductor if $p = n_i \left(\frac{\mu_n}{\mu_p} \right)$

Solution For a p-type semiconductor

$$\sigma_p = nq\mu_n + pq\mu_p$$
$$= q(n\mu_n + p\mu_p)$$

Using $p = n_i \left(\frac{\mu_n}{\mu_p} \right)$ in the above equation,

$$\sigma_p = q(n\mu_n + p\mu_p)$$

$$= q \left(\frac{n_i^2 \mu_n}{n_i \frac{\mu_n}{\mu_p}} + n_i \frac{\mu_n}{\mu_p} \mu_p \right)$$

$$= q(n_i\mu_p + n_i\mu_n) = n_i q(\mu_n + \mu_p)$$
$$= \sigma_i = \text{conductivity of intrinsic semiconductor.}$$

2.10 (a) How much donor impurity should be added to pure Ge, so that its resistivity drop to 10% of its original value.

(b) Find p and n of p-type Ge if its resistivity is 0.01 Ω-cm.

(c) Find n and p of a n-type Si if its resistivity is 10 Ω-cm.

Solution (a) 10% of 44.6 is 4.46, Since for Ge, $p_i = 44.6$.

$$\text{Hence, conductivity} = \frac{1}{4.46} = 0.224$$

$$\therefore 0.224 = N_D \, q\mu_n$$

$$\therefore N_D = \frac{0.224}{1.6 \times 10^{-19} \times 3800} = 3.687 \times 10^{14} \text{ atoms/cm}^3$$

(b)

$$\sigma p = N_A q \mu_n$$

$$\frac{1}{0.01} = N_A \times 1.6 \times 10^{-19} \times 1800$$

$$N_A = \frac{1}{0.01 \times 1.6 \times 10^{-19} \times 1800} = 3.47 \times 10^{17} \text{ atoms/cm}^3$$

Since $N_A >> n_i$, so $p = N_A = 3.47 \times 10^{17}$

$$n = \frac{n_i^2}{p} = 1.801 \times 10^9 \text{ electrons/cm}^3$$

(c)
$$\sigma_n = N_D q \mu_n = \frac{1}{\rho} = \frac{1}{10}$$

$$N_D = \frac{1}{1.6 \times 10^{-19} \times 10 \times 1300}$$

$$= 4.81 \times 10^{14} \text{ atoms/cm}^3$$

$$\therefore \quad n \approx N_D = 4.81 \times 10^{14} \text{ electrons/cm}^3$$

$$p = \frac{n_i^2}{n} = 4.68 \times 10^5 \text{ holes/cm}^3$$

2.11 Find the conductivity of germanium with:
(a) donor impurity of 1 part in 10^6.
(b) acceptor impurity of 1 part in 10^7.
(c) when both the above impurities are present simultaneously, given that n_i for Ge at 300 K is $2.5 \times 10^{13}/\text{cm}^3$, μ_n and μ_p for Ge are 3800 and 1800 $\text{cm}^2/\text{V-s}$ and the number of Ge atoms/cm^3 = 4.4×10^{22}.

Solution (a)
$$N_D = \frac{4.4 \times 10^{22}}{10^6} = 4.4 \times 10^{16}/\text{cm}^3$$

$$\therefore \quad n \approx N_D = 4.4 \times 10^{16}/\text{cm}^3$$

$$p = \frac{n_i^2}{n} = \frac{(2.5 \times 10^{13})^2}{4.4 \times 10^{16}} = 1.42 \times 10^{10} \text{ holes/cm}^3$$

$\therefore$ $n >> p$ Hence, we can neglect p in calculating the conductivity.

$$\sigma_n = nq\mu_n = 4.4 \times 10^{16} \times 1.6 \times 10^{-19} \times 3800$$

$$= 26.752 \ (\Omega\text{-cm})^{-1}$$

(b)
$$N_A = \frac{4.4 \times 10^{22}}{10^7} = 4.4 \times 10^{15}/\text{cm}^3$$

$$p \approx N_A \approx 4.4 \times 10^{15}$$

Hence,
$$n \approx \frac{n_i^2}{p} = \frac{(2.5 \times 10^{13})^2}{4.4 \times 10^{15}} = 1.42 \times 10^{11} \text{ electrons/cm}^3$$

$\therefore$ $p >> n$, hence we can neglect n while calculating the conductivity

$$\therefore \quad \sigma_p = pq\mu_p = 4.4 \times 10^{15} \times 1.6 \times 10^{-19} \times 1800$$
$$= 1.267\ (\Omega\text{-cm})^{-1}$$

(c) With both donor and acceptor impurities present,

$N_D = 4.4 \times 10^{16}/\text{cm}^3$, $N_A = 4.4 \times 10^{15}/\text{cm}^3$

$$\text{Since } N_D + p = N_A + n$$
$$n - p = N_D - N_A = 4.4 \times 10^{16} - 0.44 \times 10^{16}$$
$$= 3.96 \times 10^{16} \qquad ...(1)$$

From mass action law $np = n_i^2 = (2.5 \times 10^{13})^2 = 6.25 \times 10^{26}$ ...(2)

Solving (1) and (2),

we will get n and p. But in this particular problem, N_D and N_A are comparable, N_A holes combine with equal number of free electrons leaving the material n-type with net concentration $N_D - N_A = 3.96 \times 10^{16}/\text{cm}^3 = N'_D$

Hence, $n = 3.96 \times 10^{16}/\text{cm}^3$

$$\sigma = N'_D q \mu_n = 3.96 \times 10^{16} \times 1.6 \times 10^{-19} \times 3800$$
$$= 24.08\ (\Omega\text{-cm})^{-1}$$

2.12 An n-type Si bar is 2 cm long and has a cross section of 2 mm × 2 mm. When a 1 V battery is connected across it, a current of 8 mA flows. Find (a) doping level and (b) drift velocity.

Solution (a)

$$R = \frac{1\text{ V}}{8\text{ mA}} = 125\ \Omega$$

$$R = \frac{\rho L}{A} = \frac{L}{\sigma A} \Rightarrow \sigma = \frac{L}{RA}$$

$$\therefore \quad \sigma = \frac{2}{125 \times 0.02 \times 0.02} = 40\ (\Omega\text{-cm})^{-1}$$

Since this is an n-type Si bar, we can use approximate formula. So, $n = N_D$.

$$\sigma_n = nq\mu_n$$
$$40 = N_D \times 1.6 \times 10^{-19} \times 1300$$
$$N_D = \frac{40}{1.6 \times 10^{-19} \times 1300} = 192.30 \times 10^{15}/\text{cm}^3$$

(b) Drift velocity $v = \mu\varepsilon$

$$= 1300 \times \frac{1\text{ V}}{2\text{ cm}}$$

$$= 650\text{ cm/sec.}$$

2.13 A sample of n-type semiconductor has conductivity of 100 mho/cm at 27°C. Find the conductivity at 127°C. n_i at 27°C is 2×10^{10} atoms/cm^3 and the mobility of free electrons varies with absolute temperature. According to the equation $\mu_n = 2.1 \times 10^9 \times T^{-2.5}$ and $E_{G0} = 1.1$ eV.

Solution

$$\mu_n(27) = 2.1 \times 10^9 \times (300)^{-2.5} = 1347 \text{ cm}^2/\text{V-s}$$

$$\mu_n(127) = 2.1 \times 10^9 \times (400)^{-2.5} = 656 \text{ cm}^2/\text{V-s}$$

$$\frac{n_i^2(127)}{n_i^2(27)} = \left(\frac{400}{300}\right)^3 \frac{e^{-\frac{1.1}{8.6\times10^{-5}\times400}}}{e^{-\frac{1.1}{8.6\times10^{-5}\times300}}} = 100891$$

$$\therefore\ n_i(127) = 317.64\, n_i(27)$$

$$= 317.64 \times 2 \times 10^{10}$$

$$= 6.34 \times 10^{12}$$

$$\text{at } 27°\text{C},\ \sigma = N_D q \mu_n$$

$$100 = N_D \times 1.6 \times 10^{-19} \times 1347$$

$$N_D = 4.64 \times 10^{17} \text{ atoms/cm}^3$$

At 127°C, n_i increases to 6.34×10^{12} but still $N_D >> n_i$, so $n = N_D$

$$\sigma = N_D q \mu_n = 4.64 \times 10^{17} \times 1.6 \times 10^{-19} \times 656$$

$$= 48.70\ (\Omega\text{-cm})^{-1}$$

2.14 (a) Find the magnitude of the Hall voltage V_H in an n-type Ge bar, having majority carrier concentration $N_D = 10^{17}/\text{cm}^3$, assume $B_z = 0.3$ Wb/m^2, $d = 3$ mm, and $\varepsilon_x = 6$ V/cm.

(b) What happens to V_H if an identical p-type germanium bar having $N_A = 10^{17}/\text{cm}^3$ is used in part (a).

Solution (a) Hall voltage $V_H = Bvd = B\mu\varepsilon_x d$

$$= 0.3 \times 3800 \times 6 \times 10^2 \times 3 \times 10^{-3} \times 10^{-4}$$

$$= 205.2 \text{ mV}.$$

(b) If accepted atoms are added, then the polarity of V_H going to be changed, but the value remains the same to 205.2 mV.

2.15 The Hall-effect is used to determine the mobility of holes in a p-type silicon bar. Assume the bar resistivity is 200000 Ω-cm, the magnitude of B_z is 0.4 Wb/m^2, and $d = w = 3$ mm. The measured values of the current and Hall voltage are 100 μA and 100 mV, respectively. Find μ_p.

Solution We have

$$R_H = \frac{V_H w}{BI} = \frac{100\times10^{-3}\times3\times10^{-3}}{0.1\times100\times10^{-6}}$$

$$= 30$$

$$\mu_p = \sigma R_H = \frac{R_H}{\rho} = \frac{30}{2000} = 15 \times 10^{-3} \text{ m}^2/\text{V-sec}$$

$$= 150 \text{ cm}^2/\text{V-s}$$

2.16 A sample of Ge is doped to the extent of 10^{14} donor atoms/cm^3 and 7×10^{13} acceptor atoms/cm^3. At the temperature of the sample the resistivity of pure Ge is 60 Ω-cm. If the applied electric field is 2 V/cm, find the total conduction current density.

Solution We know that for a pure Ge

$$\sigma_i = n_i q(\mu_n + \mu_p)$$

$$\Rightarrow \quad n_i = \frac{\sigma_i}{q(\mu_n + \mu_p)} = \frac{1}{60 \times 1.6 \times 10^{-19} \times (3800 + 1800)}$$

$$= 1.86 \times 10^{13} \text{ atoms/cm}^3$$

Given that $N_D = 10^{14}/\text{cm}^3$ and $N_A = 7 \times 10^{13}$ atoms/cm^3

$$\therefore \quad N_D - N_A = 10 \times 10^{13} - 7 \times 10^{13} = 3 \times 10^{13} = p - n$$

$$np = n_i^2 = (1.86 \times 10^{13})^2$$

$$\therefore \quad n = \frac{N_D}{2} \pm \sqrt{\left(\frac{N_D}{2}\right)^2 + n_i^2}$$

$$= \frac{3 \times 10^{13}}{2} \pm \sqrt{\left(\frac{3 \times 10^{13}}{2}\right)^2 + \left(1.86 \times 10^{13}\right)^2}$$

$$= 3.88 \times 10^{13} \text{ electrons/cm}^3$$

$$p = \frac{n_i^2}{n} = \frac{(1.86 \times 10^{13})^2}{3.88 \times 10^{13}} = 0.89 \times 10^{13} \text{ holes/cm}^3$$

$$\therefore \quad \sigma = nq\mu_n + pq\mu_p$$

$$= 1.6 \times 10^{-19} (3.88 \times 10^{13} \times 3800 + 0.89 \times 10^{13} \times 1800)$$

$$= 0.02615 \ (\Omega\text{-cm})^{-1}$$

$\therefore$ Current density $J = \sigma \varepsilon$

$$= 0.02615 \times 2$$

$$= 52.307 \text{ mA/cm}^2$$

2.17 Given a 20 Ω-cm n-type Ge bar with material lifetime of 100 μ sec, cross-section of 1 mm^2, and length of 1 cm. One side of the bar is illuminated with 10^{15} photons/sec. Assume that each incident photon generated one electron-hole pair and these are distributed uniformly throughout the bar. Find the bar resistance under continuous light excitation at room temperature.

Solution When there is no radiation, ρ = 20 Ω-cm

Bar resistance $R = \dfrac{\rho L}{A} = \dfrac{20 \times 1}{0.1 \times 0.1} = 2000\ \Omega$

Conductivity $\sigma = \dfrac{1}{\rho} = \dfrac{1}{20} = (n\mu_n + p\mu_p)q$

$$\frac{1}{20} = 1.6 \times 10^{-19}\left(3800n + 1800 \times \frac{6.25 \times 10^{26}}{n}\right)$$

$$\Rightarrow n^2 - 8.22 \times 10^{13}\, n + 2.96 \times 10^{26} = 0$$

$$\therefore \quad n = \frac{8.2 \times 10^{13} \pm \sqrt{67.57 \times 10^{26} - 4 \times 2.96 \times 10^{26}}}{2}$$

$$= 7.84 \times 10^{13} \text{ electrons/cm}^3 \text{ (neglecting negative sign)}$$

$$\therefore \quad p = \frac{n_i^2}{n} = \frac{6.25 \times 10^{26}}{7.84 \times 10^{13}} = 0.8 \times 10^{13} \text{ holes/cm}^3$$

When the bar is illuminated, 10^{15} electron-hole pairs are created every second which last for 100 μ sec on average, that is, in 100 μ sec, 10^{11} electron-hole pairs are created and destroyed. Thus, the excess number of holes and electrons will stabilise to 10^{11} in the sample. But the sample volume is 1 cm × 1 mm × 1 mm = 0.01 cm^3

$$p' = \text{excess hole concentration} = \frac{10^{11}}{0.01} = 10^{13} \text{ holes/cm}^3$$

Similarly, n' = excess electron concentration = 10^{13} electrons/cm^3

$\therefore$ New electron concentration $= n = 0.8 \times 10^{13} + 10^{13}$
$= 1.8 \times 10^{13}$

New hole concentration $= p = 7.84 \times 10^{13} + 10^{13}$
$= 8.84 \times 10^{13}$

$$\therefore \quad \sigma = (n\mu_n + p\mu_p)q$$
$$= (8.84 \times 10^{13} \times 3800 + 1.8 \times 10^{13} \times 1800)\ 1.6 \times 10^{-19}$$
$$= 0.05894\ (\Omega\text{-cm})^{-1}$$

$$\therefore \quad R = \frac{L}{\sigma A} = \frac{1}{0.05894 \times 0.1 \times 0.1} = 1696.64\ \Omega$$

2.18 (a) Radiation is incident on one end of n type molecule which causes minority carrier to be injected as shown in fig 2.13. Show that the equation of conservation of charge is

$$\frac{dp}{dt} + \frac{p}{\tau} = \frac{\overline{p}}{\tau}$$

(b) Verify that the concentration is given by the Equation

$$p = \bar{p} + (p_o - \bar{p})e^{-t/\tau}$$

Solution (a) $g = \dfrac{p_o}{\tau}$ = No. of holes per second generated thermally (increase).

$\dfrac{p}{\tau}$ = No. of holes per second recombining (decrease).

$\dfrac{\bar{p} - p_o}{\tau}$ = No. of holes per second injected due to light (increase).

Hence, $$\frac{dp}{dt} = \frac{p_o}{\tau} + \frac{\bar{p} - p_o}{\tau} = \frac{\bar{p} - p}{\tau}$$

$$\therefore \quad \frac{dp}{dt} + \frac{p}{\tau} = \frac{\bar{p}}{\tau}$$

(b) The solution of the differential equation is

$$p = \bar{p} + Ae^{-t/\tau}$$

$$\text{at } t = 0,\ p = \bar{p} + A = p_o$$

$$\Rightarrow A = p_o - \bar{p}$$

$$\therefore \quad P = \bar{p} + (p_o - \bar{p})e^{-t/\tau}$$

2.19 The hole concentration is a semiconductor specimen is shown.

(a) Find an expression for and sketch the hole current density $J_p(x)$ for the case in which there is no externally applied electric field.

(b) Find an expression for and sketch the built-in electronic field that must exist if there is no net hole. Current associated with the distribution shown.

(c) Find the value of the potential between the points $x = 0$ and $x = w$ if $\dfrac{P(0)}{P_o} = 10^3$.

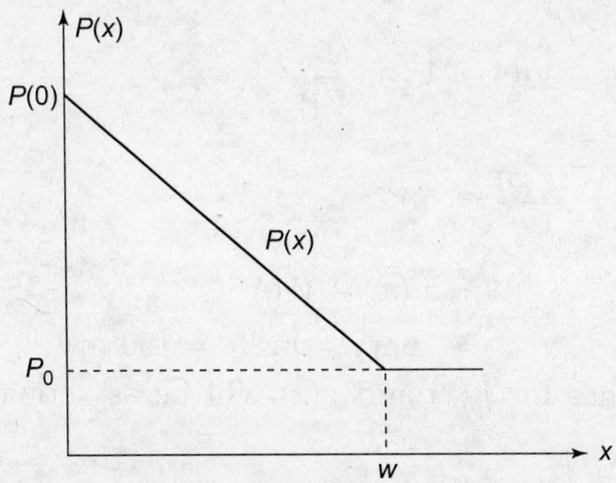

Solution (a) $$J_p(x) = -qD_p\frac{dp}{dx}$$

$$= qD_p\frac{P(o) - P_o}{w} \text{ for } x < w$$

$$J_p(x) = 0 \text{ for } x > w$$

(b) for $x > w$, $$J_p(x) = pq\mu_p\varepsilon - qD_p\frac{dp}{dx} = 0$$

$$p\mu_p\varepsilon(x) = D_p\frac{dp}{dx}$$

$$\varepsilon(x) = \frac{1}{p(x)}\frac{D_p}{\mu_p}\frac{dp}{dx}$$

$$= \frac{V_T}{p(x)}\cdot\frac{dp}{dx} \quad (\text{Since } D_p = \mu_p V_T)$$

$$\therefore \quad \varepsilon(x) = \frac{V_T}{p(x)}\cdot\frac{P_o - P(o)}{w}$$

$$\varepsilon(x) = -\frac{dV}{dx} = \frac{V_T}{P(x)}\cdot\frac{dp}{dx}$$

$$\Rightarrow \quad \frac{dV}{V_T} = -\frac{dp}{P(x)}$$

$\therefore$ Integrating within the limits from $x = 0$ to $x = w$.

$$\int_{V(o)}^{V(w)}\frac{dV}{V_T} = -\int_{P(o)}^{P_o}\frac{dP}{P(x)}$$

$$\Rightarrow \quad V(w) - V(0) = V_T \ln\frac{P(o)}{P_o}$$

(c) Since $$\frac{P(o)}{P_o} = 10^3$$

$$\therefore \quad V = V(w) - V(0)$$

$$= 26\text{mV} \times \ln 10^3 = 180 \text{ mV}$$

The sketches for $J_p(x)$ and $\varepsilon(x)$ will be as shown.

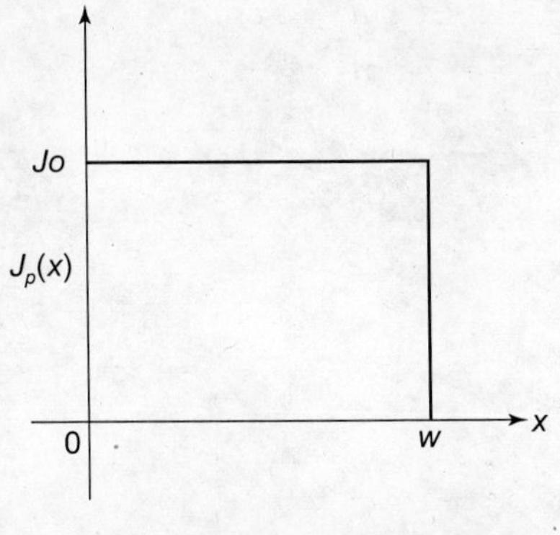

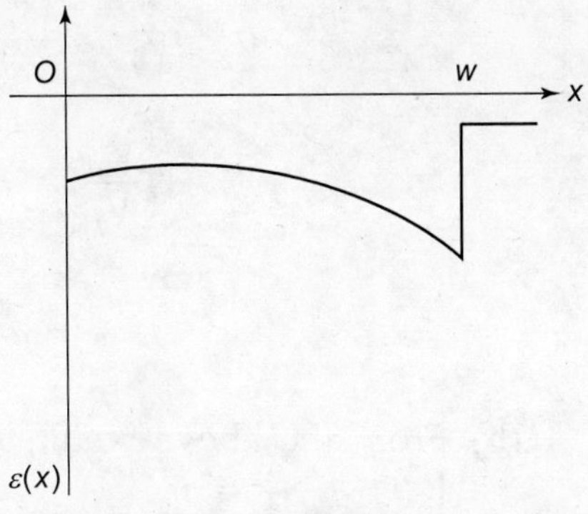

2.20 (a) For an open-circuited graded semiconductor verify the Boltzmann equation for electrons.

(b) For the step graded semiconductor derive the expression for contact potential V_0, starting with $J_n = 0$.

Solution (a)
$$J_n = qD_n\frac{dn}{dx} + nq\mu_n\varepsilon = 0$$

$$\therefore \qquad \varepsilon = -\frac{D_n}{\mu_n}\frac{1}{n}\frac{dn}{dx}$$

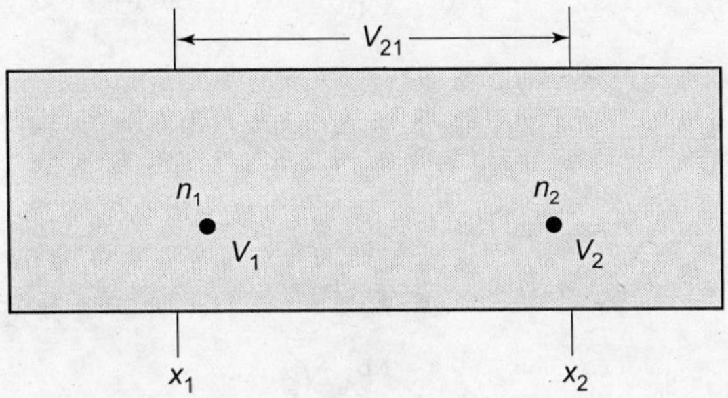

We know that $\qquad D_n = \mu_n V_T$

Using this in the above equation,

$$\varepsilon = -V_T\frac{1}{n}\frac{dn}{dx} = -\frac{dV}{dx}$$

$$\Rightarrow \qquad dV = V_T\frac{dn}{n}$$

Integrating within the limits from x_1 to x_2,

$$\int_{V_1}^{V_2} dV = V_T\int_{n_1}^{n_2}\frac{dn}{n}$$

$$V_2 - V_1 = V_{21} = V_T \ln \frac{n_2}{n_1} = -V_T \ln \frac{n_1}{n_2}$$

$$\therefore \quad \frac{V_{21}}{V_T} = -\ln \frac{n_1}{n_2}$$

$$\Rightarrow \quad n_1 = n_2 e^{\frac{-V_{21}}{V_T}}$$

(b) From the above result, $n_1 = n_2 e^{\frac{-V_{21}}{V_T}}$

Let $\quad n_1 = n_{Po}$

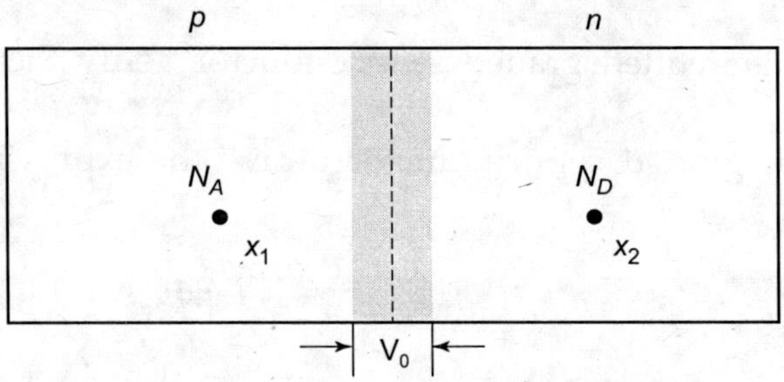

= thermal equilibrium concentration in p-type region of electrons.

$n_2 = n_{no}$ and, = thermal equilibrium electron concentration in n-type region.

$= N_D$.

$$n_1 = \frac{n_i^2}{N_A}$$

$$SO, \frac{n_i^2}{N_A} = N_D \frac{-V_o}{eV_T}$$

$$\therefore \quad V_o = V_T \ln \frac{N_A N_D}{n_i^2}$$

2.21 Consider a step graded Ge semiconductor with $N_D = 2000\, N_A$ and with N_A corresponding to 1 acceptor atom per 10^8 Ge atoms. Find the value of contact potential V_0 at room temperature.

Solution

$$N_A = \frac{4.4 \times 10^{22}}{10^8} = 4.4 \times 10^{14}/\text{cm}^3$$

$$N_D = 2000\, N_A = 8.8 \times 10^{17}/\text{cm}^3$$

$$n_i = 2.5 \times 10^{13}/\text{cm}^3$$

$$V_0 = V_T \ln \frac{N_A N_D}{n_i^2} = 26 \times 10^{-3} \ln \frac{4.4 \times 10^{14} \times 8.8 \times 10^{17}}{(2.5 \times 10^{13})^2}$$

$$= 346.754 \text{ mV.}$$

EXERCISE PROBLEMS

2.1 If we were living at an ambient temperature of 1000 K in a world, would silicon and germanium still be conductors.

2.2 In a semiconductor at room temperature the intrinsic carrier concentration and resistivity are $1.5 \times 10^{16}/m^2$ and 2×10^3 σ-m, respectively. It is converted into an extrinsic semiconductor with a doping concentration of $10^{20}/m$. Calculate

(a) minority carrier concentration

(b) resistivity

2.3 A specimen of metal has 7.87×10^{28} free electrons/m^3. The mobility of electrons is 34.8 cm^2/v-s. Compute

(a) conductivity of the metal

(b) If an electric field of 30 V/cm is applied across the specimen, calculate the drift velocity of electron and the current density.

2.4 The resistivity of *p*-type Si specimen is 0.12 Ω-m. Find electron and hole concentration.

2.5 1 kΩ resistor is to be fabricated as a narrow strip of *p*-type Si 4mm thick. If the strip is 20 μm wide and 400 μm long, what concentration of acceptor atoms is required?

2.6 A *p*-type material has an acceptor ion concentration of 1×10^{16} per cm^3. Its intrinsic carrier concentration is $1.48 \times 10^{10}/cm^3$. The hole and electron mobilities are 0.05 m^2/V-s and 0.13 m^2/V-s. Calculate the conductivity of the material.

2.7 The Hall effect is used to determine the mobility of holes in a *p*-type Si bar. Assume the bar resistivity is 300,000 Ω-cm. The magnetic field B_z = 0.1 Wb/m^2 and $d = \omega$ = 6 mm. The measured value of the current and voltage are 10 μA and 60 mV. Find μ_p.

2.8 (a) Find the resistivity of Ge and Si at 600 K.

(b) If donor impurity is added to an extent of impurity atom per 5×10^7 Ge atoms, at room temp, find the resistivity of Ge.

2.9 A Si bar 0.1 cm long and 100 μm^2 in cross-sectional area is doped with $10^{17}/cm^3$ antimony. Find the current at 300 K with applied voltage of 10 V.

2.10 Find the concentration of holes and electrons in *p*-type Ge at 300 K if the conductivity is 300 $(\Omega\text{-cm})^{-1}$.

REVIEW QUESTIONS

1. What is the distinction between an intrinsic and an extrinsic semiconductor? Give two examples of elements used for *p*-type doping and *n*-type doping.
2. Explain the phenomena of drift and diffusion. Describe the recombination process in the semiconductor. How are these parameters related in the continuity equation?
3. How do you explain the fact that the semiconductors have negative temperature coefficient of resistance.
4. Sketch the energy band diagrams for intrinsic, *p*-type and *n*-type semiconductors.
5. What properties of semiconductor are determined from Hall effect?
6. Explain why a constant difference of potential must develop across a *p-n* junction.
7. What is conductivity modulation? How are thermistors and sensitors different from resistors? Explain.
8. What is a photo resistor? Explain how conductivity modulation achieved with such a resistor.
9. Bring out the effect of temperature on semiconductors and explain the concept of movement of a hole.
10. Describe the Hall effect phenomenon in semiconductors.
11. What is potential barrier? Explain its significance in a *p-n* junction.
12. Explain the effect of temperature on an intrinsic semiconductor.
13. Give the equation for conductivity in an extrinsic semiconductor. Discuss the effect of temperature on the energy gap and mobility of a semiconductor.
14. Derive an equation for minority carrier concentration behaviour after the external excitation is removed.
15. Discuss in detail about the low-level injection of minority carriers into a semiconductor.
16. Derive the equation for the contact potential of a step-graded semiconductor junction.
17. Define mobility and conductivity and give expressions for both in a semiconductor.
18. Define intrinsic concentrations of holes. What is the relationship between this density and the intrinsic concentration for electrons? What do these equal at 0 K?
19. State mass action law for extrinsic semiconductor. Give the expressions for minority carrier concentrations in extrinsic semiconductors of *p* and *n* type?
20. What is the relationship between mobility and diffusion constant? Give its equation and discuss.

21. Define mean lifetime and mean drift velocity and bring out their significance in a semiconductor.
22. What is Hall coefficient? Give its equation. How is it useful in determination of mobility of a semiconductor?
23. What are recombination centres? Why are they required in a semiconductor? How are they introduced into a semiconductor?
24. Explain qualitatively why a contact potential is developed in a graded semiconductor. Derive its equation for step-graded junction.
25. What is a photoconductor? Explain the process of photogeneration in detail.

CHAPTER

Semiconductor Diodes

3

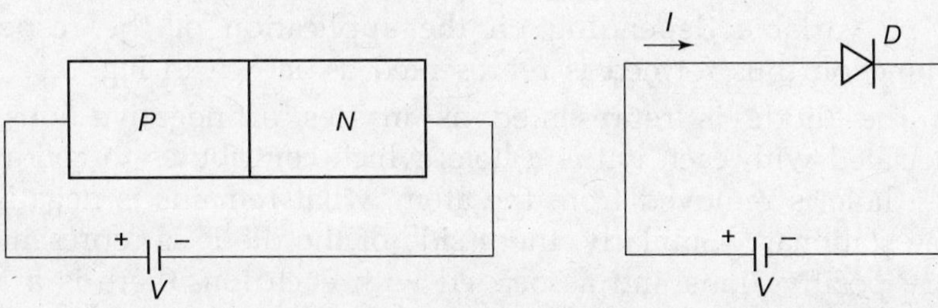

Forward-biased p-n junction diode

Chapter Outline

The concepts introduced in this chapter are:

- *The p-n junction diode*
- *Current components, characteristics, biasing, equivalent circuit and applications of p-n junction diode*
- *Characteristics and applications of special diodes like varactor, Zener, tunnel, photodiode, and light emitting diode*

The junction formed between the *p*-type and *n*-type materials, called, *p-n junction diode,* is very useful in the world of electronics. It is the first and foremost solid-state electronic device. There are various types of diodes with fine variations for a wide range of applications. Basically this device is a switch, which can be controlled with the help of applied voltage. It allows current to pass through it in one direction and restricts its flow in the opposite direction. This makes the device a very good rectifier, which converts the alternating current to direct current. Apart from these two applications, there are varied applications of the device which transformed the world of electronics. All other components of electronics are based on the principle of the diode. Many types of semiconductor diodes and their variants are discussed at length in this chapter.

3.1 *p-n* JUNCTION DIODE

"A junction is formed between two semiconductors of *p* and *n* types, respectively. This junction is referred to as *p-n junction diode*." The doping is assumed to be uniform in the two layers on both sides of the junction. The doping concentration of both sides need not be same and can be variable depending on the application of the respective diode. Schematically, the junction thus formed is represented as shown in Fig. 3.1.

The *p* side of the diode is represented as masses of negative ions, which are stationary, and associated with each ion is a hole, which contributes to conduction. When a positively charged hole is removed from the atom, what remains is negatively charged mass assumed to be stationary. Similarly, the *n* side of the diode is represented by heavy masses of stationary positive ions and associated with each ion, there is a free electron, which helps in conduction. When an electron is pulled into the conduction band from valence band, it leaves behind a heavy mass of positively charged ions. Thus, there are holes towards the left of the junction and electrons towards the right of the junction as carriers of charge. These are the majority carriers whose concentrations are approximately equal to the impurity doping concentrations. In addition to these charge carriers on both sides of the junction, there are minority carriers which are very few in number.

Consider an open-circuited junction diode as shown in Fig. 3.1. Since there is a step-graded concentration at the junction of the device, a contact potential is developed and due to this potential, the free electrons from *n* side diffuse to *p* side to recombine with the holes present in the *p* side in the vicinity of the junction. So, in the vicinity of the junction, all the active charge carriers are now absent. The available carriers are been *depleted* by the carriers diffusing from the other side. There are unneutralised ions present on both sides of the junction and these are referred to as *uncovered charges.* There are no charge carriers in this region on both sides of the junction. The distribution of charge density is as shown in Fig. 3.1(b). The charge is neutral up to the region of depletion and from this point onwards, it increases negatively and is zero once again at the junction, going positive in the *n* region and back to zero at the end of depletion region. "*Depletion region* is the region where the charge carriers are depleted and there are no free electrons or holes for conduction. This is also referred to as *space charge region* or *transition region*." The thickness of this region is of the order of 0.5 μm and towards left of this, $p \approx N_A$ and towards right, $n \approx N_D$.

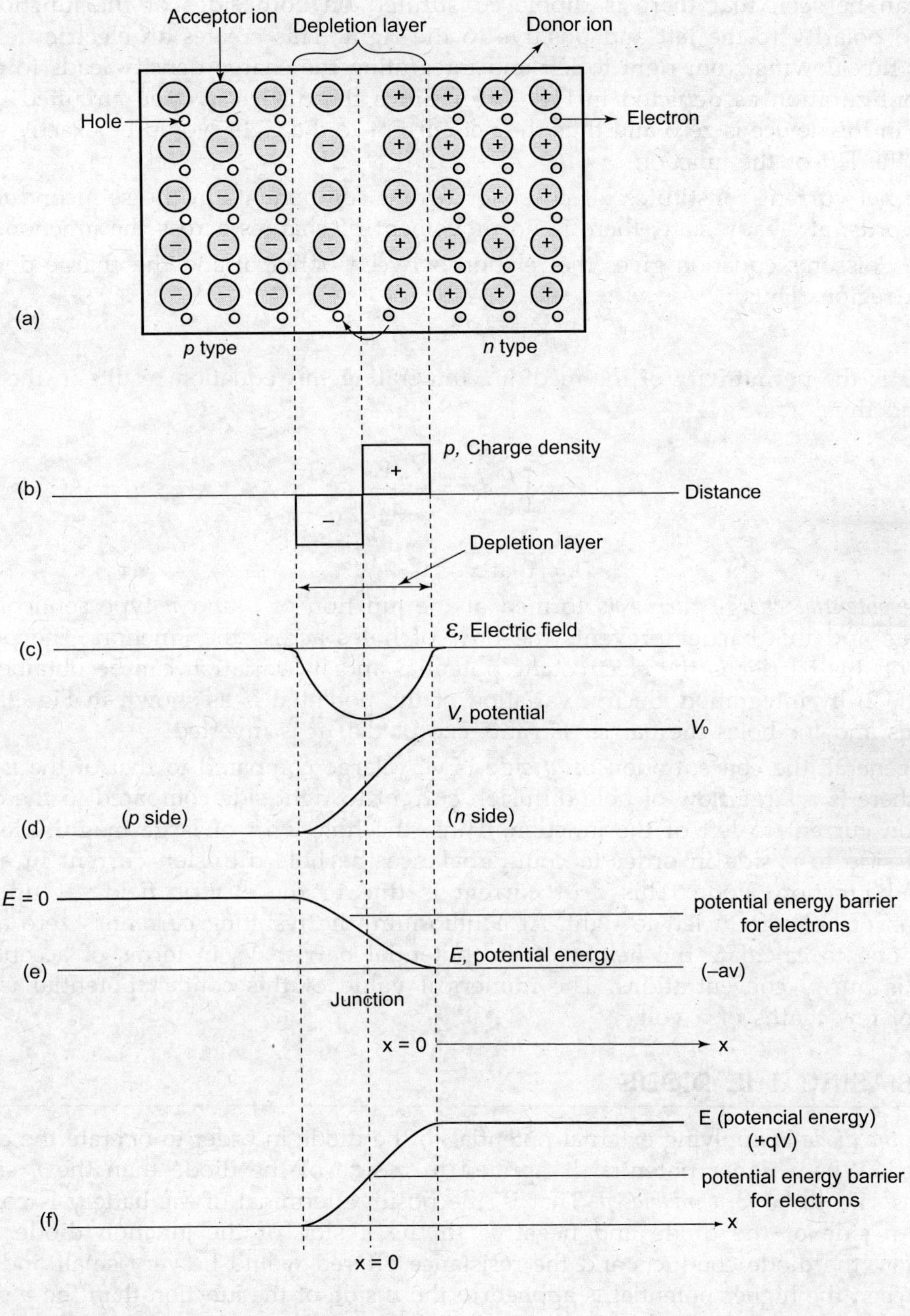

Fig. 3.1 *The open-circuited p-n junction diode*

It can be seen that there is dipole constituted on both sides of the junction with negative polarity to the left and positive to the right. This creates an electric field with lines of flux flowing from right to left and integrating the charge density leads to electric-field configuration as depicted in Fig. 3.1(c). Since the device is open circuited, the net current in the device is zero and thus the net current to the right would be exactly same as that to the left of the junction.

The net current constitutes all possible current components in the semiconductor. In other words, at steady state, there is no movement of charges across the junction.

The Poisson's equation gives the relation between potential and the charge density in a given region. Thus,

$$d^2V/dx^2 = -\rho/\varepsilon \qquad ...(3.1)$$

where ε is the permittivity of the medium. Integrating this equation results in the electric field and thus,

$$\mathcal{E} = \int_{x_0}^{x} \frac{\rho}{\varepsilon} dx = -\frac{dV}{dx} \qquad ...(3.2)$$

where $\mathcal{E} = 0$ at $x = x_0$

The *potential energy barrier* is formed at the junction of p and n type semiconductor materials and this barrier prevents the flow of holes across the junction. The potential height of this barrier is the electrostatic potential and its variation can be obtained from the Eq.(3.2) by integration. Such a variation of the potential is as shown in Fig. 3.1(e) for electrons and for holes the barrier is same except that it is inverted.

In general, the concentration on p side is very large compared to that of the n region. Thus, there is a large flow of hole diffusion current to right side compared to the electron diffusion current to left of the junction. Thus, the hole drift of large magnitude occurs from n side to p side in order to counterbalance the hole diffusion current in an open circuited junction diode. This drift current is due to the electric field set up by the diffusion of holes from left to right. At equilibrium, the resulting current is zero and this allows one to calculate the height of the potential barrier V_0 in terms of acceptor and donor impurity concentrations. The numerical value of this contact potential is of the order of few tenths of a volt.

3.2 BIASING THE DIODE

Biasing the diode is applying external potential to the diode in order to operate the diode as required. When higher potential is applied to p side of the diode than the n side, the diode is said to be *forward biased.* That is, the positive terminal of the battery is connected to the p side of the diode and negative to the n side of the junction diode. In this condition, the diode conducts and the resistance offered would be very small and ideally zero. When the higher potential is applied to the n side of the junction than the p side, the diode is *reverse biased.* In this biasing condition, the diode does not conduct and the current in the diode is very small and ideally zero. The diode is equivalent to an open circuited switch in the reverse bias and is equivalent to closed switch in the forward bias. The two conditions of biasing are considered more qualitatively in the following discussion.

Forward bias: An external voltage is applied to the diode as shown in Fig. 3.2 to *forward bias* the diode. By applying such a voltage externally to the diode, the contact potential is lowered by the applied voltage and the potential barrier height is decreased. Thus, more number of electrons and holes can cross the junction to reach the other side to complete the circuit. When the charge carriers cross the junction and reach the other side of the junction, they become minority carriers and so, diffuse recombining with the majority carriers in the region. Due to applied electric field, the holes in the *p* side are pushed into the *n* side and these are nothing but injected minority carriers. They recombine with the majority carriers and exponentially decay as they move away from the junction. Similarly, the electrons in the *n* side are attracted towards the positive edge of the applied electric field and thus, they cross the junction from *n* side to *p* side. When once they enter the *p* side they are minority carriers and thus decay exponentially as they travel away from the junction. Since the potential barrier is very small, the injected carriers are at a higher energy compared to the barrier height and thus conduction is possible. Thus, the total current in the diode is the sum of hole current at junction from left to right and the electron current from right to left of the junction. The symbol of the diode and direction of flow of conventional current are as shown in Fig. 3.2. The symbol of the diode is an arrowhead pointing the direction of conventional current when the diode is forward biased.

Reverse bias: The battery is connected to the diode to bias it in the *reverse bias* as shown in Fig. 3.3. In this condition, no current flows in the diode due to the majority carriers as they are drawn away from the junction. To be more specific, the potential barrier increases by the applied voltage not allowing any majority carrier to cross the junction. By virtue of applied field, the electrons are attracted towards the end of the diode away from the junction in side and same is the case with the holes in *p* side. The holes travel in the reverse direction and go towards left side in the *p* side. So, the thin depletion layer is going to spread throughout the diode layers, not allowing any hole from *p* side or electron from *n* side to cross the junction to constitute the current.

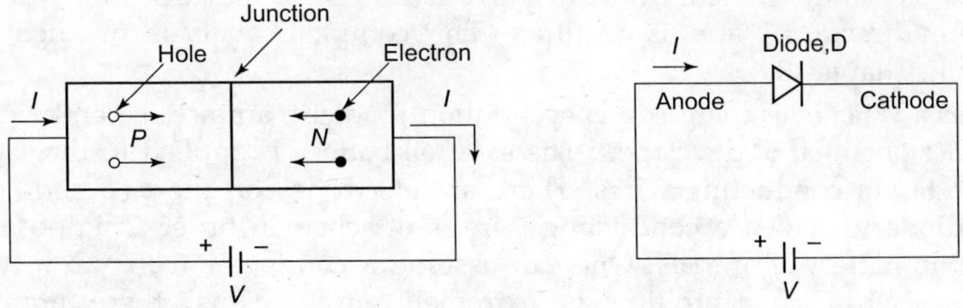

Fig. 3.2 *Forward-biased p-n junction diode*

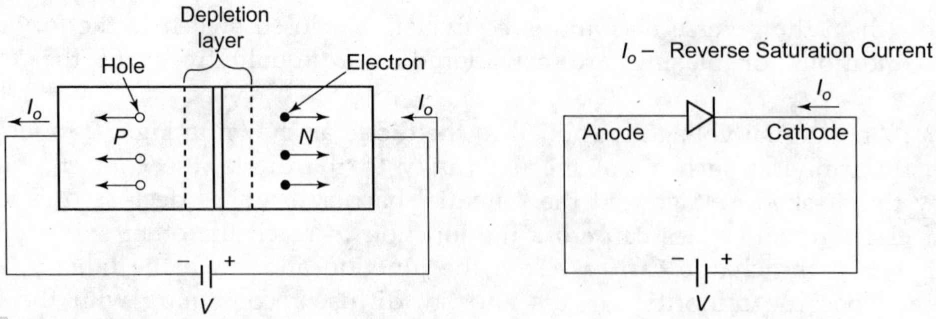

Fig. 3.3 *Reverse-biased p-n junction diode*

Ideally, the current in the reverse biased diode is zero and no charge crosses the junction as seen by the above discussion. But, along with the majority carriers, there are minority carriers present in the crystal. These minority carriers are due to thermal agitation and are present throughout the crystal. The applied external battery electric field is in such a direction that these minority carriers cross the junction. "The electrons in p side and holes in the n side cross the junction to formulate a small current. The magnitude of such a current is very small since the number of minority carriers is very small. This current is referred to as *reverse saturation current*." It is in saturation (maximum) since the number of minority carriers is limited and does not depend on the applied field. So, whatever may the magnitude of the applied reverse voltage, the magnitude of current is constant. This current is directed from n side to p side conventionally. The magnitude of this current depends on temperature since the number of minority carriers depends on the temperature. Higher the temperature, more are the minority carriers, and thus more is the reverse saturation current. In fact, this saturation current doubles for every ten degrees rise in temperature. The analytical treatment shall be considered in later sections. This is also known as *blocking bias*.

Thus, the diode allows current through it only in one direction but blocks the flow of current in the opposite direction. This principle finds a very good application in the field of electronics—*rectification*. "A device that rectifies is known as *rectifier*." A diode is a very good rectifier since it can convert the alternating current voltage to direct current voltage. When an ac signal is imposed on the diode, it allows the current to flow in forward direction but blocks the current in the negative half cycle of the ac signal. Thus, the output current is unidirectional and is rectified. The complete analysis of such rectifiers is considered in Chapter 3.

Ohmic contacts: Whenever there is a concentration gradient, a junction can be expected and thus a contact potential is developed. The external battery is applied to the semiconductor junction diode via conducting wires. There are electrodes on the two ends of the semi-conductor diode to which a conducting wire is attached in order to impose the electric field from the battery terminals. One can assume a conductor to be an n-type material where the electron concentration is extremely large. Thus, there are two metal-semiconductor junctions formed in a diode. A metal–p-type-semiconductor junction is at one electrode and the other junction is between the n-type semiconductor and conducting electrode at the other end of the diode. Thus, a contact potential is developed across the metal-semiconductor junction and the rectifying action can be expected if these junctions are left unchecked. This complicates the situation and the rectifying property of junction diode is lost. The applied external potential is not transferred to the semiconductor junction in such cases. To avoid such a complication, these metal-semiconductor junctions are fabricated so that they are nonrectifying. That is, they allow current through them with same magnitude in both the directions. "The contact potential developed would then be constant and so, the applied potential externally can thus be available at the semiconductor junction. Such contacts are known as *Ohmic contacts.*"

Short-circuited and open-circuited p-n junctions: By virtue of the contact potential developed between the two layers of the *p-n* junction, when the two terminals of the diode are short circuited, one can expect that a current flows in the conductor shorting the two ends of the bar. This is due to the electric field that exists between the two terminals. If such is the case, when a finite current flows in the connecting conducting wire, due to Ohmic losses, the wire gets heated and dissipates energy in the form of heat. Due to conservation property of the energy, since this wire dissipates energy and there is no source of energy in the system, the semiconductor junction should be cooled since it has to lose equal amount of energy. But, this is in contradiction with a given physical system, since cooling of one end and simultaneously heating of the other end in the same physical system without any energy generating sources is not possible. So, there should not be any current flowing in the short circuit. This can be explained with the help of Ohmic contact drops. In addition to the contact potential between the two semiconductor layers, there are two more Ohmic contacts in the system, which also develop potentials. Thus, these two Ohmic contact potentials are such that they are same in magnitude but opposing to the contact potential of semiconductor junction. Thus, the effective drop across the two ends of the junction is zero and no current can flow in the system when short circuited.

If one attempts to measure the contact potential of the semiconductor junction with a conventional meter when the diode is open circuited, the potential would read zero since it is not possible to measure the potential of the junction alone. The potential measured by a meter would be the effective drops of contact potential and the two Ohmic drops, which is exactly equal to zero since Ohmic potential compensates the semiconductor contact potential.

Large forward-bias voltage: When a forward bias is applied to the junction diode, the contact potential at the junction is lowered by the applied potential and the holes and electrons are allowed to cross the junction. As the bias voltage is increased, more number of carriers cross the junction and thus the current increases. But when the magnitude of voltage is very large, one can expect that the depletion layer at the junction vanishes and the current increases without limits. But, this is not possible practically and the depletion layer at the junction can never vanish in total and so, a minimum resistance is offered to the forward flow of the current at very high forward voltages. The resistance is the bulk resistance of the semi-conductor, which limits the current flow in the device. So, the bulk resistance and the Ohmic contact resistances govern the flow of current in a diode. Thus, the device almost acts as a linear device with a straight-line slope of volt-ampere characteristic.

3.3 THE CURRENT COMPONENTS IN A JUNCTION DIODE

When a diode is forward biased, the current in the diode is allowed to flow in the forward direction, as discussed in the previous sections. The current in the diode is not entirely due to a single component, but all possible four components of drift and diffusion currents are possible in the junction diode. The analytical treatment of these components of a *p-n* junction is presented in this section of the chapter.

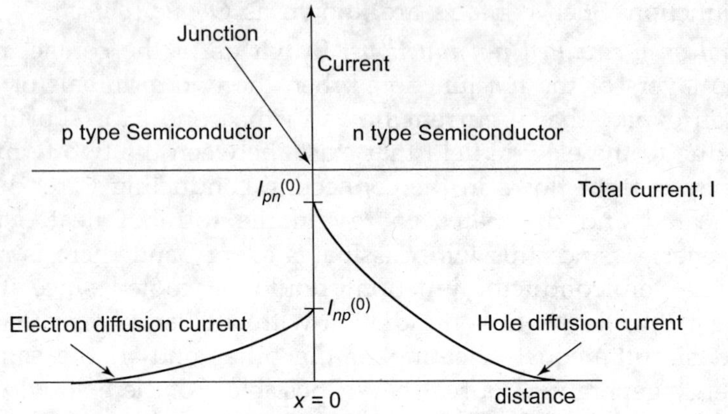

Fig. 3.4 *Current components of junction diode*

Consider a *p-n* junction diode forward biased as shown in Fig. 3.2. Holes are injected into *n* side and electrons into *p* side of the diode. Under low-level injection, the minority current is only due to diffusion and drift current due to minority carriers can be neglected as already justified in Sec. 2.14. It is also seen that the injected minority carrier diffusion current decays exponentially away from the junction with distance. Assume that the junction is placed at $x = 0$. Thus, the hole diffusion current, I_{pn} in the *n* side decreases exponentially with *x* in positive direction and at a length L_p, the value of the current is $1/e$ times that at the junction. Such a current is plotted in Fig. 3.4. Similarly, the electron diffusion current I_{np} from the *n* side into the *p* side decays exponentially and its plot is also shown in the figure. Since doping is not same on both sides of the junction and *p* side is large compared to *n* side, the electron diffusion current at the junction is very small compared to the hole diffusion current. Or,

$$I_{pn}(0) >> I_{np}(0) \qquad \text{...(3.3)}$$

In this discussion, the depletion layer is assumed to be very thin and is neglected. From Eq. (2.68), the value of I_{pn} at the junction can be given as

$$I_{pn}(0) = \frac{AqD_p}{L_p}[p_n(0) - p_{n0}] \qquad \text{...(3.4)}$$

The forward-bias voltage *V* lowers the barrier height and allows more carriers to cross the junction. Hence, the concentration of holes in *n* side at $x = 0$, $p_n(0)$ is a function of *V*. From the Boltzmann's relationship (2.81), $p_n(0)$ depends exponentially on *V*. Thus, it can be deuced that

$$p_n(0) = p_{n0}e^{V/V_T} \qquad ...(3.5)$$

This relationship gives the hole concentration at the edge of the n region just outside the transition region in terms of thermal equilibrium minority carrier concentration far away from the junction. This is called the *law of the junction*. With n and p suffixes interchanged, the above equation gives the electron concentration at the edge of p region in terms of potential V.

Substituting Eq. (3.5) into Eq. (3.4),

$$I_{pn}(0) = \frac{AqD_p P_{n0}}{L_p}[e^{V/V_T} - 1] \qquad ...(3.6)$$

The expression for the electron current $I_{np}(0)$ crossing the junction from n to p regions can be given as

$$I_{np}(0) = \frac{AqD_n n_{p0}}{L_n}[e^{V/V_T} - 1] \qquad ...(3.6a)$$

The total current in the diode, I, would be constant measured anywhere in diode. Thus, the sum of the two components of currents discussed above at x=0 gives the total current in the diode. The total current I is the sum of electron current crossing from right to left and the hole current crossing from left to right at the junction. Thus,

$$I = I_{pn}(0) + I_{np}(0) \qquad ...(3.7)$$

Since I is uniform throughout the diode, I is independent of x and so,

$$I = I_0[e^{V/V_T} - 1] \qquad ...(3.8)$$

where I_0 is the physical parameter of the diode and is equal to the magnitude of reverse saturation current in the diode. This equation is the equation that gives the behaviour of current with respect to voltage in a p-n junction diode.

When the junction is forward biased, the applied potential V would be positive. When V is a large value, the first term in the equation is very large compared to 1 and thus,

$$I = I_0[e^{V/V_T}] \qquad ...(3.9)$$

From the above equation, one can observe that the relation between the voltage and current in a diode is not linear but the current increases exponentially with voltage when forward biased. The current is assumed to be positive when the current flows from p to n side.

When the junction is reverse biased, the applied potential would be negative and then, the magnitude of the first term of Eq. (3.8) will be very small compared to unity and thus can be neglected. Thus,

$$I = -I_0 \qquad ...(3.10)$$

The magnitude of current when the diode is reverse biased is I_0 and since a negative sign is incorporated, the current flows from n to p side in the diode. So, this current I_0 is the *reverse saturation current*. Since p_{n0} and n_{p0} values depend on the temperature at steady state, the magnitude of I_0 also depends on temperature.

In the n region of the junction diode, the total current is constant independent of x. But it is observed that the hole diffusion current I_{pn} depends on distance and decreases with increasing x. Thus, the current due to majority carriers that is, electrons, $I_{nn,}$ should also be a function of x. This majority carrier current can be given as

$$I_{nn}(x) = I - I_{pn}(0) \qquad ...(3.11)$$

The variation of the currents is shown in Fig. 3.5. The majority current is shown in dotted lines. Thus, in the n region, as the minority diffusion current decreases, the majority current increases, making the total of these currents always constant, I. Similarly, the majority carrier current due to holes in the p region, I_{pp} is plotted as a function of distance. It may be recalled that the majority current consists of two components—one is diffusion due to minority carriers and the other is drift due to majority carriers.

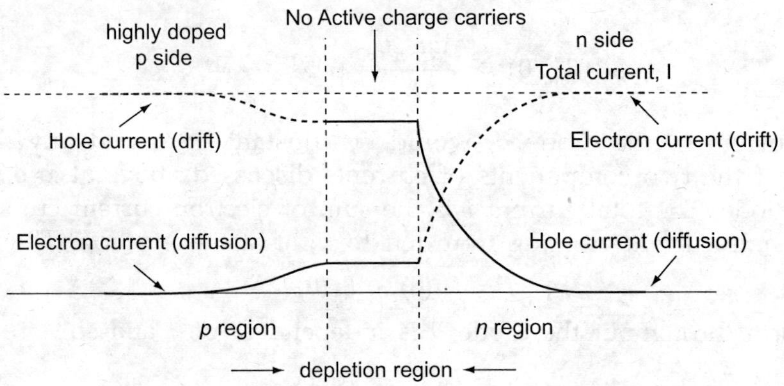

Fig. 3.5 *Detailed components of diode current*

The majority current is due to drift far away from the junction. I_{pp} the hole drift current, is sustained by a small electric field in the semiconductor. As the holes approach the junction, some of them recombine with the electrons, which are injected into the p side from the n side. The current I_{pp} thus decreases toward the junction. The remaining current at the junction enters the n side and becomes hole diffusion current I_{pn}. Similarly, the current components of the electrons, $I_{nn,}$ in n side can be described.

The current in the diode is due to movement of electrons from n to p region and holes from p to n region. Since there are two types of charge carriers, one positive and the other negative, the current in a diode is bipolar.

Generation or recombination of charge carriers is not possible in the transition region of the junction. This is due to the fact that all the mobile carriers in this region are depleted to either side of the junction. So, the net charge carriers entering the layer would be same as that leaving the layer. That is, the current magnitude will not get altered as it passes through the layer. This assumption is valid for large currents like in Ge diode but when the current is small, a constant η is to be appended so as to obtain the correct value of the current magnitude. Thus, η will take value of unity for Ge while it is equal to 2 for Si diode. Since the order of current is small in silicon η is taken to be 2 in this material, whereas the order of current in germanium would be large comparatively and thus $\eta = 1$ for this case. The diode current equation can be rewritten as

$$I = I_0[e^{V/\eta V_T} - 1] \quad \text{...(3.12)}$$

3.4 *V-I* CHARACTERISTICS OF A DIODE

The plot of the diode current equation [Eq.(3.12)] is as shown in Fig. 3.6. This equation is valid for the diodes which are either forward biased or reverse biased. The current and voltage are assumed to be positive when the diode is forward biased, that is, higher potential is connected to *p* side and lower potential to *n* side of the diode. Both the current and voltage take negative sign when the diode is reverse biased. As discussed in previous section, η is 2 for Si and unity for Ge diode, V_T is volt equivalent of temperature given by Eq.(2.59). At room temperature 300 K, its value is 26 mV.

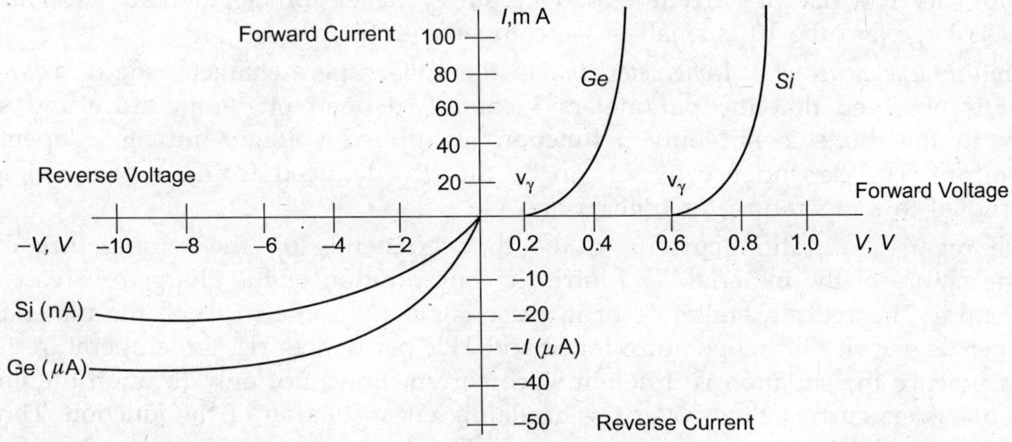

Fig. 3.6 *Volt–ampere characteristics of a p-n junction diode*

In the forward bias condition, the current increases exponentially for both Ge or Si diodes. The difference is at the initial point of raise in current for the two cases. The response of the current is different for the two diodes and is shown in Fig. 3.6. It can be observed that the current increases only beyond a *threshold voltage* V_γ. When the diode is reverse biased. When the diode is reverse biased, and the magnitude of applied voltage is large, the current tends to reverse saturation value of the diode and is independent of the applied voltage.

The magnitude of the reverse saturation current is very small, of the order of nanoamperes for Si and microamperes for Ge, whereas the order of current in the forward bias is of the order of milliamperes. Thus, the *volt–ampere* or *V-I characteristics* of the *p-n* junction are not drawn to the same scale in forward- and reverse-bias condition. The scale in the negative direction is generally enhanced to get a better understanding of the characteristic.

The cutin voltage V_γ: Both silicon as well as germanium diodes are available for commercial use in general. There are some differences between the two, which should be taken care of while in use. There exists a voltage known as *cutin* or *offset* or *break point* or *threshold voltage* in the diode. This is the voltage in the forward-bias condition of the diode until which the increase in the current is very small and may be neglected. That is, the current in the diode does not increase right from zero voltage and above, but the current responds significantly only after a certain voltage called cutin voltage. Below this voltage, the response of the current to voltage is so small that it can be neglected (it is less than 1% of the maximum rated value). Beyond this value, the current rises very rapidly, exponential with voltage, and the value of V_γ is not same for Si and Ge. Its value is

$$\begin{aligned} V_\gamma &\approx 0.6 \quad \text{for Si} \\ &\approx 0.2 \quad \text{for Ge} \end{aligned} \qquad ...(3.13)$$

The difference in the value of cutin voltage for Si and Ge is due to the fact that the magnitudes of the reverse saturation currents are not the same as already discussed. The magnitude of I_o in Ge is almost 1000 times that in Si. Also, the constant η is 2 for Si and unity for Ge. The rise of current would be surely faster for Ge than for Si. Thus, the current is large and the V_γ is small in Ge compared to Si.

Temperature dependence of V-I characteristic: In the volt–ampere characteristic of a *p-n* junction, it is observed that the parameters V_T and I_0 depend on temperature and so the current in the diode is not only a function of applied voltage, but also depends on temperature. The dependence of V_T can be directly deduced from its equation and is proportional to temperature for both Si and Ge.

The reverse saturation current in the diode depends on the temperature due to intrinsic nature of the material. The intrinsic concentration of the charges increases with temperature. Theoretically, this rate of increase is not the same for both the materials but is 8% per degree rise in temperature for Si and 11% per degree rise in temperature for Ge. But, in practice the situation is different since current flows not only through the junction but also leakage current flows over the insulating encapsulation of the junction. Thus, the rate of increase in reverse saturation current is 7% per degree rise in temperature for both the materials, or in other words, *the reverse saturation current doubles for every ten degree rise in temperature*. This can be expressed in analytical form as

$$I_0(T) = I_{01} \times 2^{(T-T_1)/10} \qquad ...(3.14)$$

In the above discussion, it is assumed that the voltage is constant and current is variable with temperature. But, if current were to be maintained constant and voltage being adjusted, it is found that either for silicon or germanium, the voltage would have to be decreased by 2.5 mV for every one degree rise in temperature to maintain current constant. Thus,

$$dV/dT = -2.5 \text{ mV/°C} \qquad ...(3.15)$$

3.5 PIECEWISE LINEAR MODEL

Ideally, the junction diode is a closed switch with resistance offered as zero in the forward bias and so it draws maximum current in this bias condition, whereas in the reverse bias condition the diode is equivalent to an open switch and will not draw any current, offering infinite resistance ideally. But, in contrast, in practice, the forward bias offers a very small resistance along with a small voltage drop of V_γ across the diode and in the reverse bias, it offers very large resistance. It is very advantageous to have an equivalent circuit of the device for analysis or design of a given circuit. In order to develop the equivalent circuit of the diode, the *V-I* characteristic is bifurcated into linear pieces and simplified for analysis. This model of the diode is known as *piecewise linear model of the diode.*

The ratio of voltage and current, V/I is defined as the resistance and is, in particular, referred to as *static resistance.* This concept of static resistance is of no great importance in diode circuits since the device is not linear, but nonlinear. The static resistance is dependent on the voltage and current relation and unlike a linear resistance, in a diode this relation is not constant. Thus, it is not a useful parameter.

When a diode uses ac signal, and when the input variation is very small, the signal is referred to as *small signal.* Within the small limit of use, in the small portion of the *V-I* characteristic, even though the total characteristic is nonlinear, it is assumed that the device is linear and analysis is performed. To determine the resistance offered within this small region of operation, static resistance is not going to give complete behaviour of the diode. Thus, a resistance known as *dynamic resistance* is useful. Dynamic resistance is the ratio of a small change in voltage to the corresponding change in the current. Thus,

$$r = dV/dI \qquad ...(3.16)$$

Even this dynamic resistance is not constant but depends on the operating region. Assuming the device to be linear in the small region of operation, the dynamic resistance gives the amount of resistance offered by the device within the region of interest. Since the equation governing the diode is current in terms of voltage, it is convenient to differentiate current with respect to voltage and thus this ratio of variation in current to corresponding voltage variation is the dynamic conductance, the reciprocal of which is the dynamic resistance. Thus, differentiating Eq. (3.12),

$$g = \frac{dI}{dV} = \frac{I_0 e^{V/\eta V_T}}{\eta V_T} = \frac{I + I_0}{\eta V_T} \qquad ...(3.17)$$

When the diode is forward biased, the magnitude of I is large compared to that of I_0 and so, I_0 can be neglected in the numerator. Thus, the dynamic resistance can be given as

$$r \approx \eta V_T / I \qquad ...(3.18)$$

Thus, it can be seen that the dynamic resistance is inversely proportional to the current and thus larger the current, the less is the resistance offered by the diode. The value of the dynamic resistance is very small, of the order of a few ohms, for example, at room temperature and at a current of 26 mA it is 1 Ω. Although the value of dynamic resistance is dependent on current and varies, for small signal operations, it is assumed to be constant.

So, when a diode is forward biased, it is equivalent to a resistance whose value can be determined as above. In addition to this resistance, there would be a voltage drop equal to the offset voltage V_γ across it. Thus the diode can be represented with a resistance in

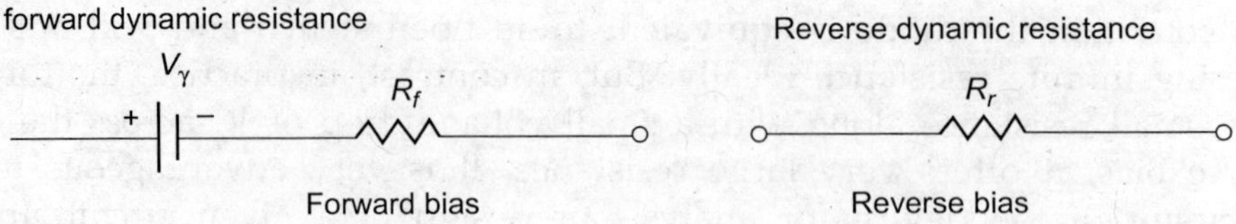

Fig. 3.7 *Equivalent circuit of a diode*

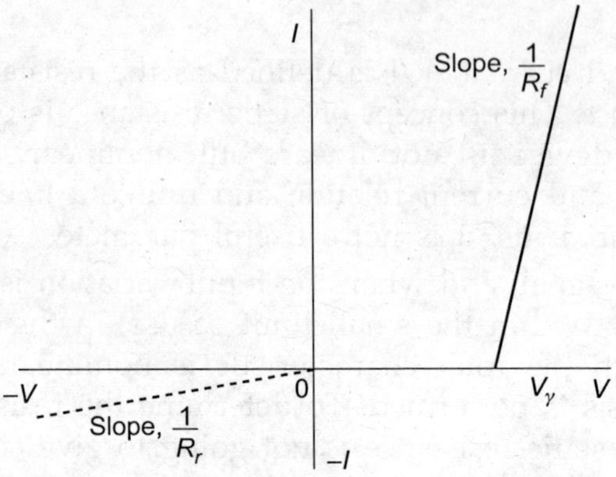

Fig. 3.8 *Piecewise linear characteristic of the diode*

series with a voltage supply of V_γ as shown in Fig. 3.7. Similarly, in the reverse bias, the diode offers a large dynamic resistance whose value can be determined similar to the above analysis. Substituting $I = -I_0$ in Eq. (3.17) leads to the discussion that the resistance is infinite, which is true ideally. But, in practice, the dynamic resistance of the device in reverse bias will not be infinite since the reverse current in the diode is not exactly constant with voltage. So, the value of this resistance can be determined from the characteristic itself. Thus, in the reverse bias, the diode is equal to a resistance whose value is very large. This is also shown in Fig. 3.7.

Fig. 3.8 thus shows the *V-I* characteristic, which includes V_γ and a straight line from that point whose slope is proportional to forward dynamic resistance of the diode. This indicates a linear resistance in the forward direction. In the negative region, the slope of the line gives the dynamic resistance of the diode when reverse biased. This is the *piecewise linear characteristic of the diode.*

3.6 TRANSITION CAPACITANCE

When the diode is reverse biased, all the mobile charge carriers are drifted away from the junction and there is a large immobile layer of charges, depletion layer, separating the accumulated charges at the edges of the diode. This leads to the action of a capacitance. This capacitance is known as *transition region* or *space charge* or *depletion-region capacitance.* This capacitance is similar to that of a parallel plate capacitor and the equation for such capacitance can be derived analytically. This capacitance depends on the magnitude of the applied reverse voltage and is inversely proportional to it.

In a reverse-biased diode, the holes and electrons drift away from junction due to the virtue of the applied voltage and thus there is no possibility of any majority carrier current crossing the junction. Now, the situation is such that the positively charged holes accumulate towards left end of the diode and negatively charged electrons at the other end of the diode. An uncovered charge region, the depletion layer, separates these two charge configurations. This depletion layer thickness increases with the applied reverse voltage. This can be considered to have capacitive effect. The incremental capacitance can now be determined.

The capacitance can be derived from the basic equation for current in a capacitance, which is

$$i = C_T(dV/dt) \quad \text{...(3.19)}$$

This can be rewritten in terms of increment in the charge accumulation as

$$C_T = |dQ/dV| \quad \text{...(3.20)}$$

where dQ is the differential increase in charge due to differential increase in voltage dV. Thus, as can be seen from the above equation, the capacitance is inversely proportional to the voltage between the two terminals of the diode.

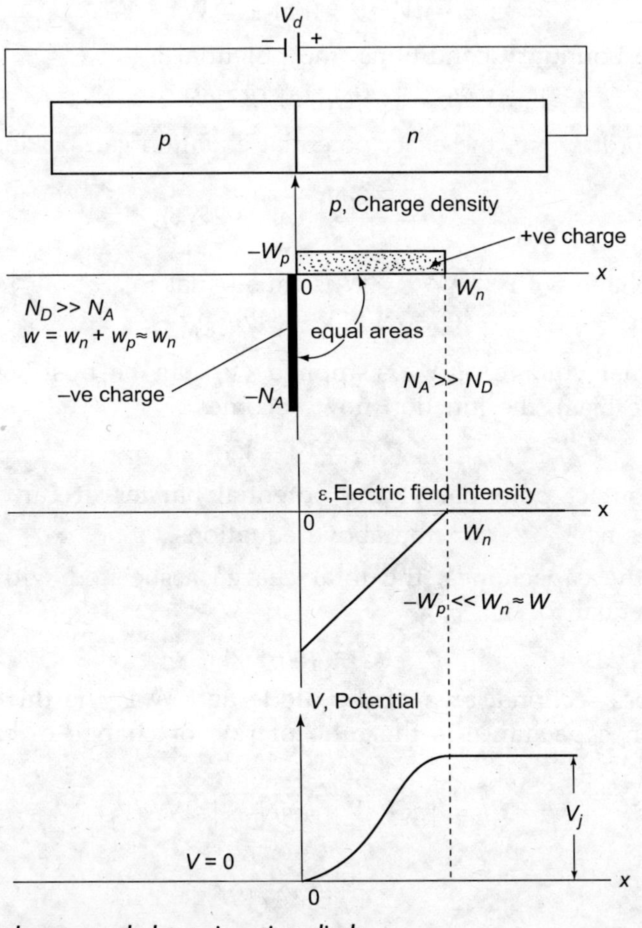

Fig. 3.9 *Reverse-biased step-graded p-n junction diode*

Consider a diode that is reverse biased, as shown in Fig. 3.9. Assume that the junction is step-graded, that is, concentrations in the p and n sides are uniform and the transition from p to n side is abrupt at the junction. If N_A is the acceptor concentration of p region and N_D is the concentration of donor atoms in n region, assume that $N_A >> N_D$. As seen from the figure, the charge distribution can be assumed of rectangular shape for simplifying the analysis. Since the concentration of p is large compared to n side, the penetration of the depletion layer into the n side is very large compared to that in p side. If W_p is the thickness of depletion layer in the p region and W_n is that in n region, then $W_p << W_n$. Thus, the total thickness of the depletion layer, W, can be

$$W \approx W_n \qquad ...(3.21)$$

The total charge should be zero and thus the areas under curve on both sides of the junction should be the same.

$$N_A W_p = N_D W_n \qquad ...(3.22)$$

Poisson's equation gives the relation between the potential and charge equation. Thus,

$$d^2V/dx^2 = -\,qN_D/\varepsilon \qquad ...(3.23)$$

Integrating Eq. (3.23), the electric field can be determined. The electric field is

$$\varepsilon = -\,dV/dx = 0 \text{ at } x = W_n \approx W \qquad ...(3.24)$$

Substituting the boundary conditions, the potential is

$$dV/dx = (-\,qN_D/\varepsilon)\,(x - W) = -\,\varepsilon \qquad ...(3.25)$$

Assuming $V = 0$ at $x = 0$, the exact potential equation can be derived as

$$V = \frac{-\,qN_D}{2\varepsilon}\,(x^2 - 2Wx) \qquad ...(3.26)$$

The junction voltage with $x = W_n = W$ is thus equal to

$$V_j = qN_D W^2/(2\varepsilon) \qquad ...(3.27)$$

When an external voltage of V_d is applied (V_d can be positive for forward bias or negative for reverse bias), the junction now becomes

$$V_j = V_o - V_d \qquad ...(3.28)$$

where V_o is the contact potential of the potential barrier. It can be observed that the barrier width varies as $V^{1/2}$ from the above equations.

To determine the capacitance, the total charge associated with the system is to be calculated. This is equal to

$$Q = qN_D WA \qquad ...(3.29)$$

where A is the cross sectional area of the diode and W is the thickness depletion layer. Thus, the transition capacitance is the ratio of rate of change of charge with respect to voltage

$$C_T = |dQ/dV| = qN_D A\,|dW/dV_j| \qquad ...(3.30)$$

or

$$C_T = \varepsilon A/W \qquad ...(3.31)$$

Eq.(3.31) is nothing but the equation of a parallel plate capacitor with distance between the plates as W.

Eq.(3.31) is derived for a step graded junction and with $W_p \approx 0$ and neglecting N_A. The equation will not change when the thickness of the depletion layer in p region is also considered. The derivation is slightly altered taking both concentrations of acceptor and donor also and width of depletion layer as sum of the layers in the p region and n region of the diode.

It is not that always the junction be step-graded. Some junctions may be *linearly graded*. That is, the transition from p to n is not abrupt but linearly varying with distance. The analysis for such a linearly graded junction is also similar to the above, except that the W varies as $V_j^{1/3}$ instead of $V_j^{1/2}$.

3.7 DIFFUSION CAPACITANCE

In the forward bias of a diode also there can be a capacitance. This capacitance is referred to as *diffusion capacitance*. This capacitance is due to the diffusion of minority carriers on both sides of the junctions. These charges get accumulated near the junction before they get diffused and recombine with the majority carriers. This accumulation of holes in the n region and electrons in the p region separated by a very thin depletion layer leads to the capacitance in the forward-bias condition. To derive the equation for such a capacitance, the total charge accumulation is to be determined. Fig. 3.10 gives the charge distribution for forward-biased and reverse-biased diode junctions.

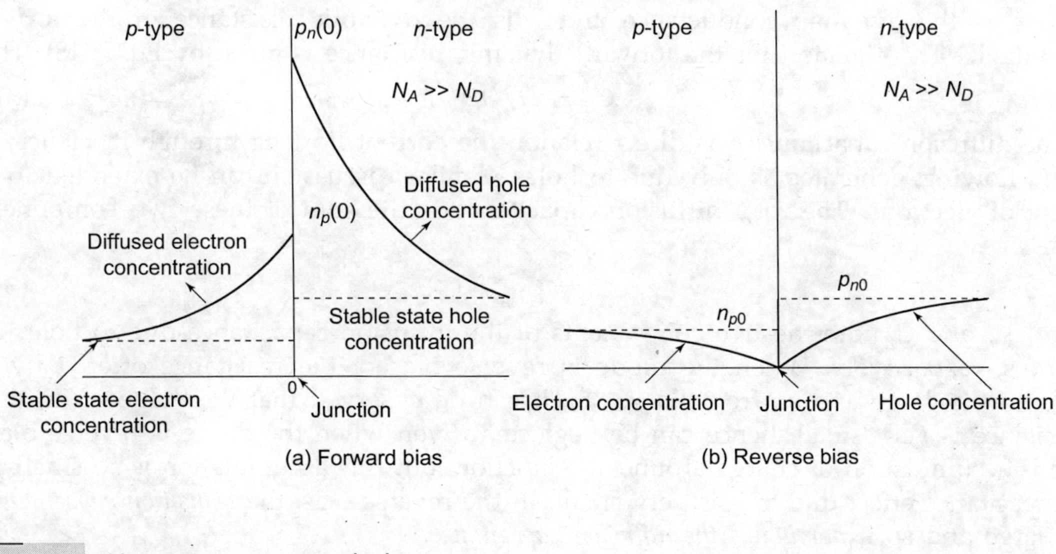

Fig. 3.10 *Minority carriers in a diode*

As seen from the figure, the minority carriers decrease exponentially away from the junction and the area under the shaded portion of the graph gives the total charge accumulated due to injected carriers. For simplicity in the analysis, it is assumed that p region is heavily doped compared to n region. Thus, the current can be assumed to be only due to holes from p to n side and thus, $I \approx I_{pn}(0)$. The excess minority charge Q will exist only on n side and is given by the shaded portion of the curve multiplied by the area. Thus,

$$Q = \int_0^\infty Aqp'(0)e^{-x/Lp} = A\ q\ L_p\ p'(0) \qquad ...(3.32)$$

At $x = 0$, the current is given by

$$I = \frac{AqD_p p'(0)}{L_p} \qquad ...(3.33)$$

From Eqs. (3.32) and (3.33),

$$I = Q/\tau \qquad ...(3.34)$$

where t is the mean lifetime of the holes $= L_p^2/D_p$. This equation is known as *charge control description of a diode.* It can be seen from Eq. (3.34) that the rate of disappearance of charges due to recombination is same as the rate at which they are generated. Hence, current I supplies charges that are lost due to recombination.

Thus, the *diffusion* or *storage capacitance* can be defined as the rate of change of injected charge with voltage. Thus,

$$C_D = dQ/dV = \tau dI/dV = \tau g = \tau/r \qquad ...(3.35)$$

where g is the dynamic conductance and r is the dynamic resistance of the forward biased-diode. The equation for the forward dynamic resistance is given by Eq. (3.18). Thus,

$$C_D = \tau I/(\eta V_T) \qquad ...(3.36)$$

The diffusion capacitance has direct relation the current flowing through it. Hence, the current flowing/generated is only due to holes. Similar discussion can be extended to the current of electrons. The total diffusion capacitance is the sum of these two components. Thus,

$$C_D = C_{Dp} + C_{De} \qquad ...(3.37)$$

where C_{Dp} and C_{De} are the two components of the diffusion capacitance due to holes and electrons, respectively. When the diode is reverse biased, the resistance offered by the diode is very large and so from Eq. (3.35), it can be observed that the value of diffusion capacitance is very small, hence can be neglected. Even when the diode is reverse biased there is accumulation of charge around the junction, but this accumulation is so small that the capacitance effect due to it is very small. In the reverse bias, the *transition capacitance* is very large and so it *dominates the diffusion capacitance.*

It can be anticipated that when the diode is forward biased, the minority carriers in the diode are drifted away and these may cause the transition capacitance. But the minority carriers being very small in number, the transition capacitance due to these carriers is very small and is neglected compared to diffusion capacitance. Thus, in *forward bias, the diffusion capacitance dominates the transition capacitance.*

3.8 VARACTOR DIODE

When a diode is reverse biased, a capacitance, which is based on the applied reverse bias, is observed as from the above discussions. This finds large applications when variable capacitors are required. The capacitance of the device can be varied not by physical parameter variations as in the case of parallel plate capacitor, but by the applied potential. Thus, a voltage variable capacitance can be achieved. Such a diode that is designed for voltage variable capacitance is known as *varactor* or *voltacap* or *varicap*.

The symbol and the equivalent circuit of such a *varactor* is as shown in Fig. 3.11. The transition capacitance, as seen from the above discussions, is inversely proportional to the applied potential. Thus, as the voltage is increasing, the transition capacitance comes down as shown in Fig. 3.12.

The series resistance R_s represents the body resistance of the diode. The typical values of C_T and R_s are 20 pF and 8.5 Ω, respectively. The shunt resistance shown in dotted line in the circuit is the reverse diode resistance R_r and since its value is very large, it can be neglected.

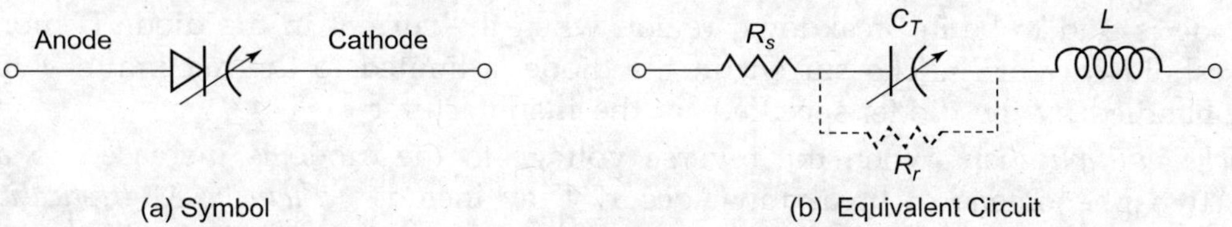

Fig. 3.11 *Varactor diode*

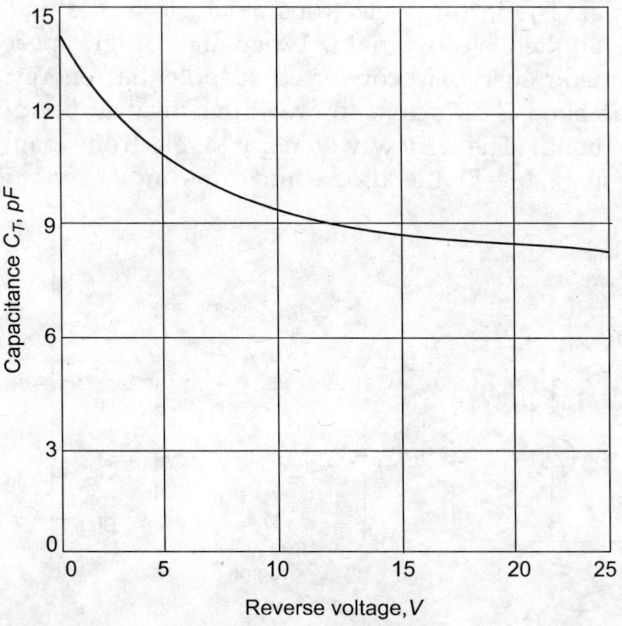

Fig. 3.12 *Variation of depletion capacitance with applied voltage*

The various applications of varactor are in *parametric amplifiers, FM modulators, voltage tuning of LC circuits, self-balancing bridge circuits* and so on.

3.9 ZENER DIODE

The diode, when reverse biased, allows reverse saturation current through it and the saturation current is constant with respect to applied voltage and depends on the temperature. It doubles for every degree rise in temperature. This is true until a voltage is applied. But as the reverse voltage to the diode increases, this breaks and the current in the diode increases tremendously at a particular voltage. This increase in current is due to either *Zener breakdown or Avalanche multiplication* in the diode when reverse biased. This feature of the diode, called b*reakdown*, finds large applications. It acts as a *voltage regulator.* A voltage regulator is a device that maintains the output voltage constant irrespective of the load or load current. The diodes that which are manufactured to be specifically operated in this breakdown region are referred to as *Zener* or *Avalanche* or *breakdown diodes.*

The *V-I* characteristics of Zener diode are as shown in Fig. 3.13. In the reverse bias the current is constant until a voltage is reached and at this voltage, referred to as *Zener voltage,* the current multiplies to a very large extent. This region of operation is known as *breakdown region.* The symbol and Zener diode as regulator is also as shown in Fig. 3.13. The diode is said to be in breakdown region when the current in the diode is more than the threshold current I_{zk}. The current in the diode is limited to the maximum value that can be handled by the diode, specified by the manufacturer.

Avalanche multiplication: When the reverse voltage to the diode is increased to a large extent, two phenomena of breakdown occur. They include *Avalanche Multiplication* and *Zener Breakdown.* In *Avalanche Multiplication,* at the high applied electric fields, the covalent bonds are broken by the virtue of collision of high speed electrons with the crystalline structure of the material. The intrinsic electrons available in the material get the required acceleration from the applied electric field. When these high speed electrons strike the covalent bond, the kinetic energy is converted to potential energy and if the potential energy imparted to the bond is sufficient to break the covalent bond, an electron hole-pair is released from the bond. These newly generated electrons gain acceleration by the electric field already available in the diode and thus move with very high velocity to

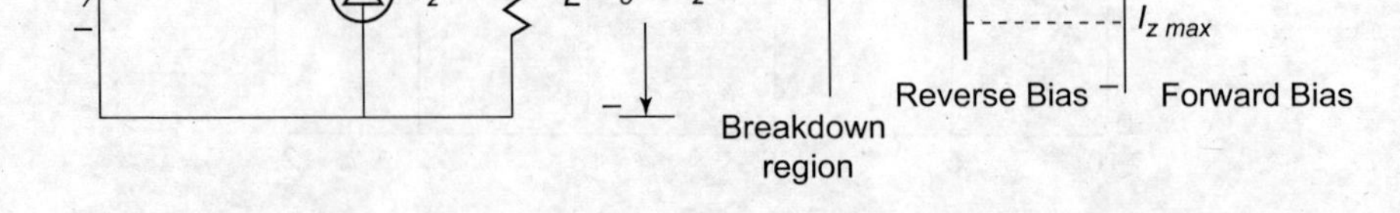

Fig. 3.13 *Zener diode*

strike and release a new pair of electron and hole. The process continues and large number of electron-hole pairs are generated which will drift towards the respective electrodes crossing the junction and get absorbed. This formulates very large currents at the applied reverse voltage. Thus, the current is only due to intrinsic charge carriers crossing the junction below a cutoff voltage, which is the reverse saturation current. But when the voltage applied is such that the electric field is sufficient to impart required energy to break the covalent bond by virtue of collision of high speed electrons, current in the diode increases sharply. Such a phenomenon of current multiplication is called as *Avalanche Multiplication*.

Zener breakdown: In addition to the above-discussed phenomenon, one more process of increasing the current in the breakdown is possible. This is *Zener breakdown.* When the applied voltage is sufficient to establish an electric field that can break the covalent bond, new electron-hole pairs are generated. In this process, no collision of the electrons with the bonds is encountered, but the electric field is itself sufficient to pull out an electron into the conduction band. Thus, a free electron and a hole in the valence band are created for each bond broken. At a given applied field, large number of bonds are broken and thus large number of charge carriers are produced. These charge carriers are accelerated and travel down to cross the junction to constitute current in the breakdown region.

Avalanche multiplication or Zener Breakdown can cause the current to breakdown when the diode is reverse biased. Whatever may be the phenomenon occurring in the diode, the diode is referred to as *Zener diode*. Depending upon the applied voltage and doping levels, either of the phenomenon is initiated first. When the current multiplication is initiated, it multiplies and there is no distinction between the two on going processes. The doping levels of the diode determine which process is initiated first. When the doping levels are high such that the electric field is of order of 2×10^7 V/m, Zener Breakdown occurs and this is possible upto breakdown voltages of 6V. When the doping is light, avalanche multiplication is initiated and this is possible with diodes having breakdown voltages above 6 V. Diodes can be fabricated with different breakdown voltage right from few volts to several hundred volts. The power handling capacity of such Zener diodes can be of the order of 50 W.

Zener breakdown or avalanche multiplication can even occur in ordinarily doped diodes. But this breakdown occurs at very large voltages. Since the voltage is very high and also the current held in the diode in the breakdown is very large, the power dissipation would be very large, beyond the power rating of the diode, such that the diode would be burnt out and no more useful. Thus, special highly doped diodes are manufactured for this application with rated *Zener voltage*. Reference voltages as low as 2 V are also available commercially.

The temperature also effects the breakdown voltage. The *Reference* or *Zener* or *Breakdown* voltage changes with temperature. The change may be having positive temperature coefficient of negative depending on what phenomenon is first set in. The temperature coefficient is negative, i.e., the breakdown voltage decreases with temperature in diodes with breakdown voltage less than 6 V. In diodes where avalanche multiplication is initiated first where the breakdown voltage is greater than 6 V, positive temperature coefficient is observed. In such diodes, the Zener voltage increases with increase in temperature. In both the cases, the change in the voltage is in the limit of ± 0.1% per degree of temperature. Avalanche multiplication predominates and the coefficient is positive.

Manufacturers specify the *knee current* I_{zk} below which the breakdown will not occur and so regulation would be poor. Ideally, the diode has zero dynamic resistance in the breakdown region if the curve is a vertical line. Some diodes exhibit very sharp knee even up to microampere region. Transition capacitance across the diode is of the order of 10 to 10000 pF. Since this capacitance depends on the cross-sectional area, the capacitance value would be high for high-power diodes.

3.10 TUNNEL DIODE

The *V-I* characteristics of a diode are totally changed when the impurity concentrations in the diode are increased to very large values. The characteristic exhibits a region known as *negative resistance*, which finds applications at microwave frequencies. In a general *p-n* junction diode, the order of doping would be around 1 in 10^8. If this is increased to 1 in 10^3, *tunneling phenomenon* takes place. The order of the thickness of the depletion layer for an ordinary diode would be around 1 μm and is reduced to the order of less than 10^{-6} cm. This is only about one-fiftieth of the wavelength of visible light. At this very thin depletion layer, tunneling of charge carriers takes place and the diode is called *tunnel* or *Esaki diode.*

A charge carrier must have energy at least equal to the potential barrier height in order to cross the junction. However, when the depletion layer is extremely thin, as indicted above, there is a probability that a carrier may penetrate through the barrier. This phenomenon of tunneling is the special character of these diodes, which alters the characteristics of an ordinarily doped diode. To understand this concept of tunneling, energy-band diagram is useful and the detailed discussion is as follows.

"*Fermi level* is defined as the energy state with 50% probability of being filled if no forbidden band exists." The Fermi level for an intrinsic semiconductor would be almost midway between the valence and conduction bands for an intrinsic semiconductor. When the impurities are added to the semiconductor to convert to extrinsic type, this Fermi level shifts up or down. If the acceptor atoms are added to make the semiconductor *p* type, the Fermi level is shifted towards the valence band and is just above the valence band. For an *n* type semiconductor, this Fermi level will shift up towards the conduction band. The complete energy-band diagram, including the Fermi level for an ordinary diode, is shown in Fig. 3.14. The Fermi level on both sides of the diode is seen to be at the same level and lies in the forbidden gap on both sides of the junction.

As seen from Fig. 3.15(a), the Fermi level is no more in the forbidden gap of the energy band for a highly doped tunnel diode. The Fermi level is pushed to valence band in the *p* region and is in the conduction band of the *n* region. The Fermi level is at the same energy level on both sides of the diode when the diode is open circuited. In the *p* region of a tunnel diode, there is an empty space above the Fermi level in the valence band where electrons can exist. But, in the *n* region of the diode, the valence band is totally full and electrons available up to the Fermi level in the conduction band as shown in the figure. There are electrons in the conduction band of *n* region and also electrons in the valence band of *p* region of same energy levels in this diode. Since the depletion layer

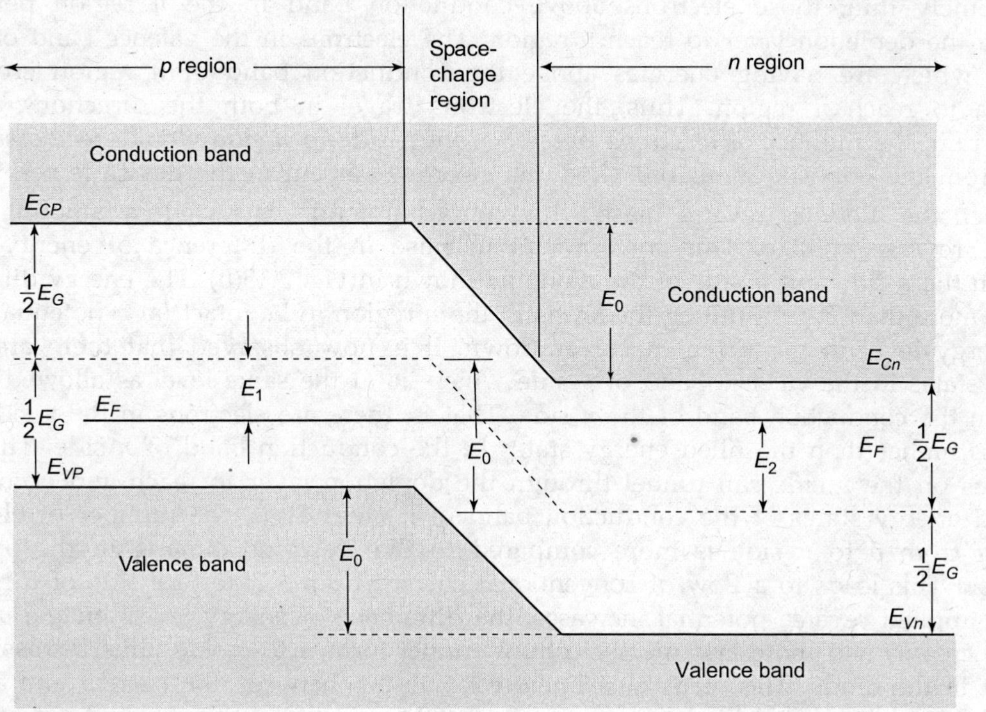

Fig. 3.14 *Complete energy-band diagram of p-n junction diode*

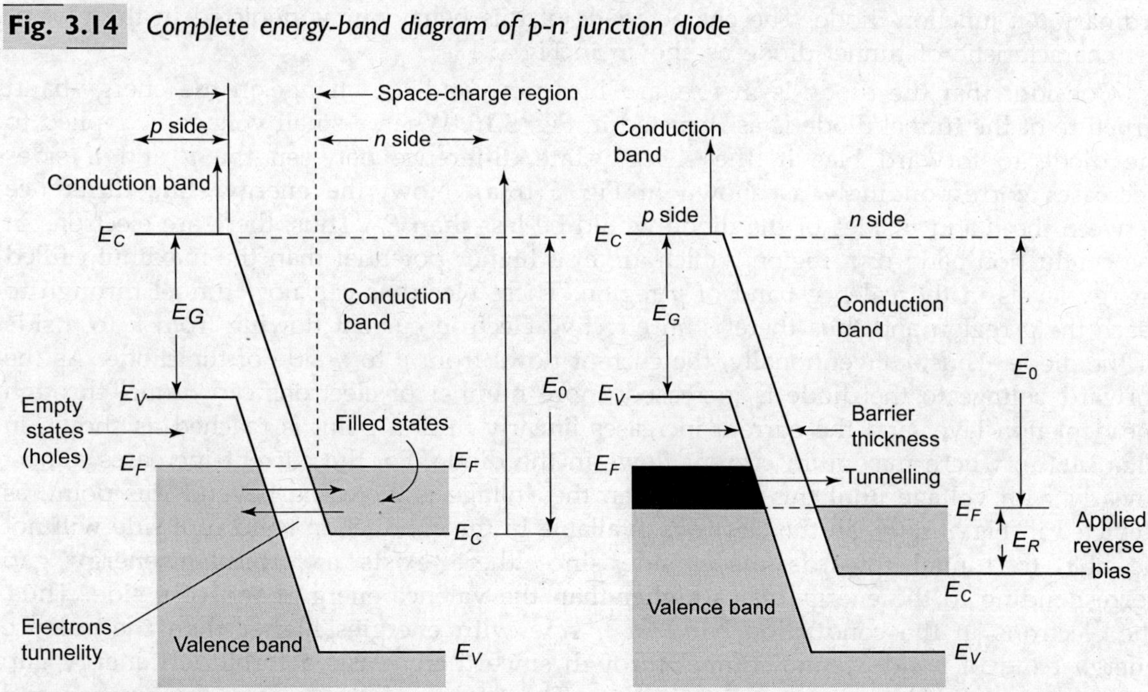

Fig. 3.15 *Energy diagram of a tunnel diode for open circuit and reverse bias*

is extremely thin, those electrons above conduction band in the *n* region penetrate through the depletion layer to reach *p* region. The electrons in the valence band of the *p* region, which are having energies above the conduction band of *n* region cross the junction to reach *n* region. Thus, the electrons travel in both the directions and at equilibrium, the number of electrons passing from *p* side to *n* side would be the same as those from the opposite direction. Thus, the effective current in the device is zero.

When the diode is reverse biased, the contact potential is raised by amount of the applied reverse voltage. This corresponds to raise in the difference of energy levels between the *p* side and *n* side of the diode as shown in Fig. 3.15(b). The energy difference will be more than E_G. Assuming the levels in the *p* region to be intact (at a potential zero), the energy level in the *n* region comes down. It is now observed that there are some energy states in the valence band of *p* side which lie at the same level as allowed empty states in the conduction band of the *n* side. That is, there are electrons in the *p* side with energies higher than the filled energy states in the conduction band of *n* side. Thus, the electrons on the *p* side can tunnel through the depletion layer to reach and occupy the unfilled energy states of the conduction band of *n* side. Thus, the number of electrons moving from *p* to *n* side is more compared to the electrons moving in the opposite direction. This leads to a flow of conventional current from *n* side to *p* side of the diode. As the applied reverse potential increases, the difference of energy levels on the *p* and *n* sides increases and more and more electrons tunnel form *n* to *p* side, thus increasing the current in the diode. This leads to a linear relationship between the current and applied potential in the reverse direction in contrast to the constant current, as observed in an ordinary *p-n* junction diode. The characteristic of this behaviour is depicted in the plot of *V-I* characteristic of tunnel diode as shown in Fig. 3.17.

Consider that the diode is in forward bias now and the change in the energy-band structure of the tunnel diode is as depicted in Fig. 3.16. When a small voltage is applied to the diode to forward bias it, the energy state difference between the *p* and *n* sides decreases correspondingly as shown in Fig. 3.16(a). Now, the energy state difference between the *p* and *n* sides of the diode would be less than E_G. Thus, there are electrons in the conduction band of *n* region, which are at a higher potential than the maximum filled energy levels in the valence band of *p* region. These electrons can now tunnel through to reach the *p* region and thus there is an effective electron current flowing from *n* to *p* side of the diode. Thus, conventionally, the current flows from *p* to *n* side of the diode. As the forward voltage to the diode is increased, more number of electrons can tunnel through the depletion layer and the current increases linearly until a point is reached as shown in Fig. 3.16(b), where maximum current flows in the diode. So, the current increases almost linearly with voltage until this point. When the voltage is increased beyond this point, as depicted in Fig. 3.16(c), all the electrons available in the conduction band of *n* side will not be able to tunnel towards the *p* side since there exists a forbidden energy gap corresponding to the energy states higher than the valence energy band of *p* side. Thus, the electrons in the conduction band of *n* side with energies higher than the valence energy band of *p* side cannot tunnel through since there exists a forbidden energy gap corresponding to these electron energy states. Thus, fewer electrons cross the junction and the current decreases. As the voltage is further increased, the electrons crossing the junction decrease further and the current is lowered with voltage. This is as represented in Fig. 3.17. In this region, as the voltage is increased, the current decreases. This is referred to as *negative resistance region.*

This decrease in current continues as the voltage is increased, until a point is reached where all the electrons in the conduction band of n side are at energy states corresponding to forbidden gap of the p side. At this juncture, as shown in Fig. 3.16(d), no current flows in the diode. Above this voltage, the band structure of an ordinary p-n junction diode, as shown in Fig. 3.14, is valid and the diode behaves normally from this point onwards. The total *V-I* characteristic of the *tunnel diode* is shown in Fig. 3.17(b).

As seen from the *V-I* characteristic of a tunnel diode, the diode conducts as a linear resistor when reverse biased and exhibits negative resistance when forward biased. The voltage at which the maximum current flows before the negative resistance region is known as *peak voltage* V_P and the corresponding current as *peak current* I_P. The voltage at which minimum current flows in the diode, when forward biased, is *valley voltage* V_V and

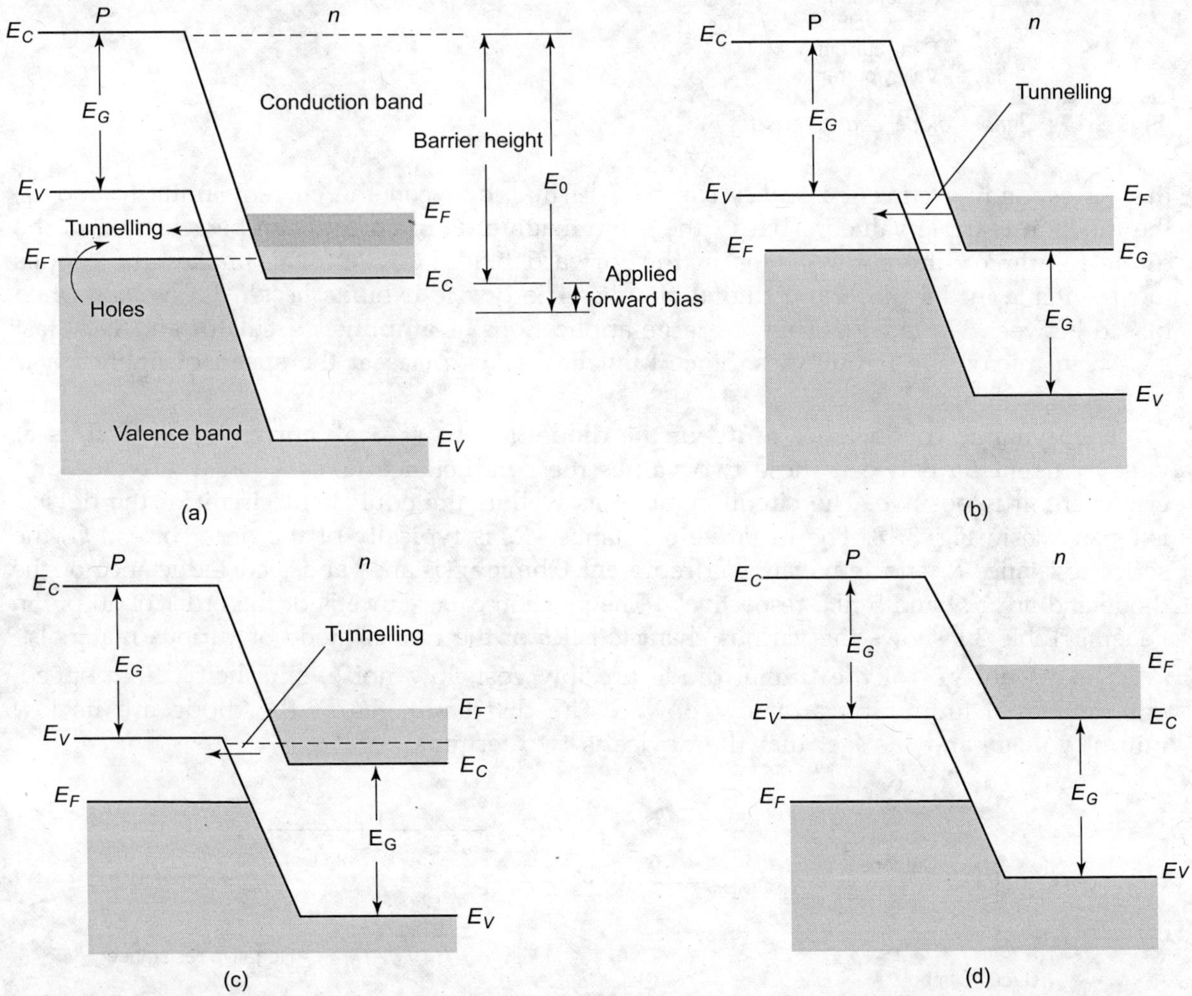

Fig. 3.16 *Energy-band diagram of a forward-biased tunnel diode*

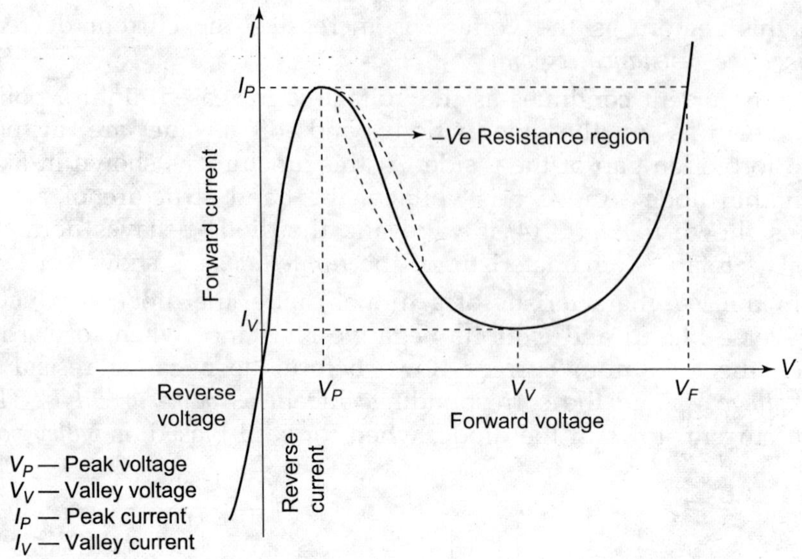

Fig. 3.17 *Tunnel diode characteristics*

the corresponding current is *valley current* I_V. Thus, between the current limits I_V and I_P, the function is triple valued. That is, the same magnitude of current can be achieved in the diode for three different voltages in the forward biased region. This makes the diodes most suitable in the pulse and digital circuits. The device exhibits *negative resistance* when biased between V_P and V_V. This has large applications as amplifier, oscillator and very fast switch in microwave frequencies. Since tunneling takes place at the speed of light, it acts as a very high speed switch.

The dynamic conductance dI/dV of the diode is zero at peak current I_P as well as at valley current I_V. Between these two values the conductance takes a negative value and elsewhere, it is positive. The circuit symbol, as well as the equivalent circuit of the device, is as shown in Fig. 3.18. The negative resistance $-R_n$ is typically of the order of –30 Ω, the series resistance R_s and inductance L_s represent Ohmic resistance and lead inductance of the diode and are 1 Ω and 5 nH, respectively. The junction capacitance is of the order of 20 pF, in general. Table 3.1 shows the various characteristics of the tunnel diode of various materials.

The advantages of the tunnel diode are low cost, low noise, simplicity, high speed, environmental immunity, and low power. The disadvantages of the diode are its low output voltage and the fact that the device is two terminal.

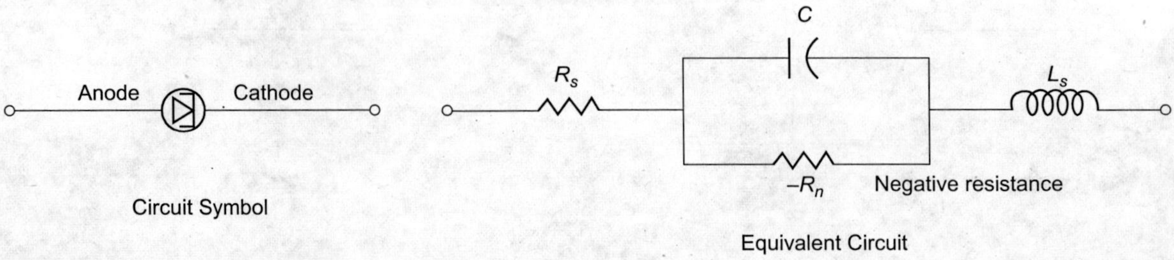

Fig. 3.18 *Circuit symbol and equivalent circuit of a tunnel diode*

Table 3.1 ***Typical Parameters of Tunnel Diode***

	Ge	*Si*	*GaAs*
I_p / I_V	8.0	4	15
V_P Volts	0.08	0.07	0.15
V_P Volts	0.5	0.45	0.5
V_F Volts	0.55	0.7	0.95

3.11 PHOTODIODE

When a reverse-biased *p-n* junction is exposed to light radiation, the photons in the light radiation strike the covalent bonds of the semiconductor to create charge carrier pairs. Under the influence of the applied voltage, the generated carriers cross the junction and constitute current. As the radiation intensity is increased, more photons are incident and consequently more bonds are broken, resulting in larger current in the diode. "Such a diode, which can convert light intensity variations to electric current variations, is known as *Photodiode.*"

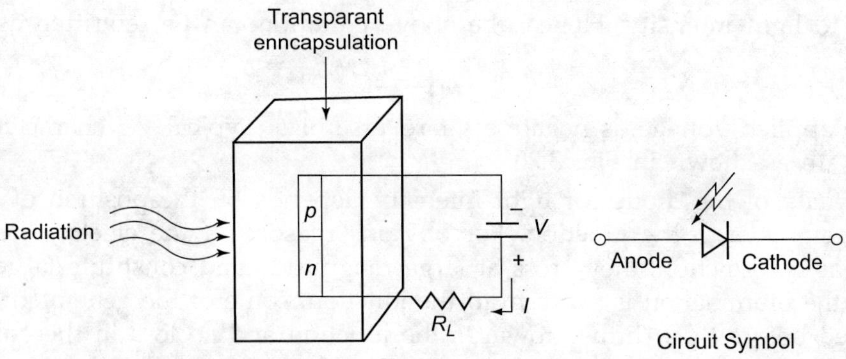

Fig. 3.19 *Photodiode and its circuit symbol*

Fig. 3.19 shows a transparent encapsulation of plastic, which houses a semiconductor *p-n* junction. When light radiation is incident on this reverse-biased junction, the current in the diode increases as discussed above. The current flowing in the diode is *dark current* when the diode is not subjected to light radiations. Thus, the current flowing in this reverse-biased diode is similar to that in the ordinary diode, the reverse saturation current due to thermally generated minority carriers. If light falls on the junction surface, additional electron-hole pairs are generated and the source of light is minority carrier injector. These injected minority carriers diffuse to the junction, cross it, and contribute to current.

The current in the reverse-biased diode is dependent on the minority carrier concentration. When the light falls on the junction, the newly generated carriers are proportional to the number of incident photons. Hence, the current in the reverse bias would be

$$I = I_0 + I_s \qquad \text{...(3.38)}$$

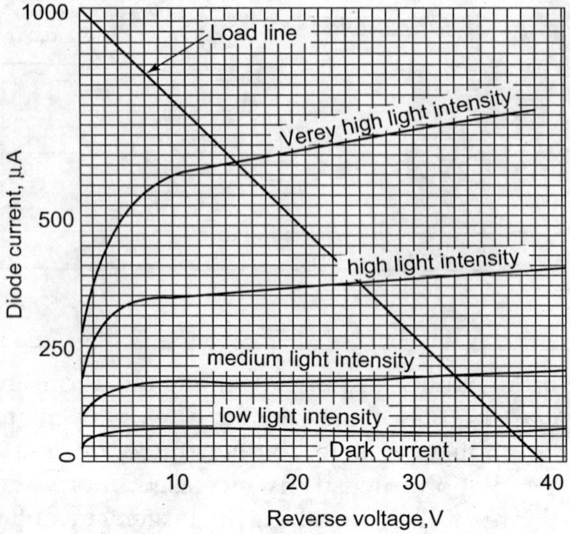

Fig. 3.20 *V-I characteristics of a photodiode*

where I_0 is reverse saturation current, the *dark current, and* I_s is the short-circuit current proportional to light intensity. Hence, the above equation can be rewritten as

$$I = I_s + I_0 (1 - e^{V/\eta V_T}) \qquad \text{...(3.39)}$$

where V, the applied voltage, is negative for reverse bias. Typical V-I characteristics of the photo diode are as shown in Fig. 3.20.

The *sensitivity* of the diode for light intensity depends on the position of illumination of the light reference to the junction. For obvious reasons, if the electron-hole pairs are generated near the junction, they cross at large magnitude and constitute large currents in the diode. If the illumination is away from the junction, some of the generated carriers may be recombined before they diffuse towards the junction and so loss in the charge carriers results in decrease in the magnitude of the current. Hence, as the illumination moves away from the junction, the current decreases. The result of this is shown in Fig. 3.21.

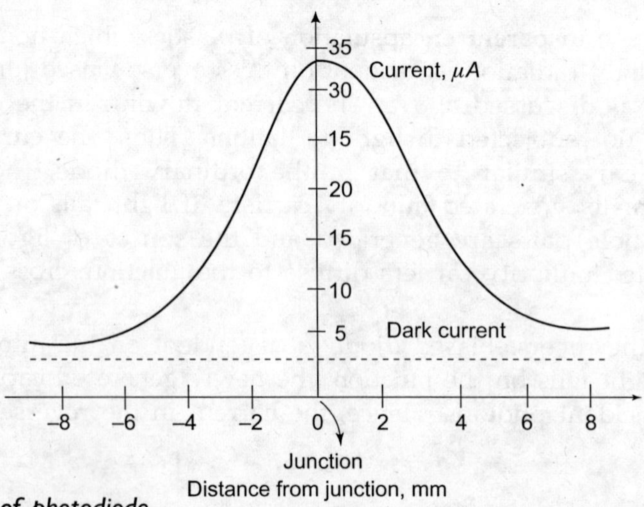

Fig. 3.21 *Sensitivity of photodiode*

The difference in the slope on either side of the junction in curve depicted in Fig. 3.21 is due to the difference in *diffusion length* for holes and electrons.

The barrier capacitance C_T of this photodiode is of the order of 10 pF, the dynamic resistance is approximately 50 M and Ohmic resistance is 100 Ω. An improved version of this is the *phototransistor*, which has better applications. Photodiode is widely used as a light-reading device, in light-detection circuits, reading of film sound tracks, light-operated switches, as a production line counting device and so on.

3.12 LIGHT-EMITTING DIODE

Light-emitting diode **(LED)** finds very large applications as output display unit. As the name indicates, the diode emits light whose intensity is proportional to that of the electric current through the diode. The phenomenon of emission of light is called *electroluminescence*. If one visualises the photodiode as a transducer which converts light energy to electric energy, LED is a transducer that converts electrical variations to light intensity variations. Since it emits light in proportion to the current through it, it finds large applications in the field of output displays.

When a *p-n* junction diode is forward biased, the holes from the *p* side diffuse to *n* side and get recombined with the electrons in the region there. While one requires energy to break a bond, recombination of an electron and a hole to form a bond releases energy. This energy, which is released on recombination of carriers, is dissipated in the form of heat. By Planck's theory, the released energy is, with certain frequency, multiplied by the Planck's constant. Thus, if the released energy is, with frequency, in the visible region of the frequency spectrum, the energy is released in the form of light instead of heat. Thus, if a material like gallium arsinide, GaAs, is used to construct the diode and the junction is covered with a transparent plastic encapsulation, the emitted energy can be *visible*. The light thus generated has an intensity proportional to that of the current flowing in the device. In the forward bias, the more the voltage, more is the current. Thus, more the forward voltage, more is the intensity of the emitted light energy.

The efficiency of generation of light increases with increase in injected current and with a decrease in temperature. The light is concentrated near the junction as the carriers are available within a diffusion length of the junction.

LEDs radiate different colours such as red, green, yellow, orange and white. Some LEDs even operate in the infrared light region. The wavelength of emitted light depends on the energy gap of the material. Hence, the colour of the emitted light depends on the energy gap of the material and the material used. For example, GaAs emits infrared radiations, GaP emits red or green and GaAsP emits red or yellow.

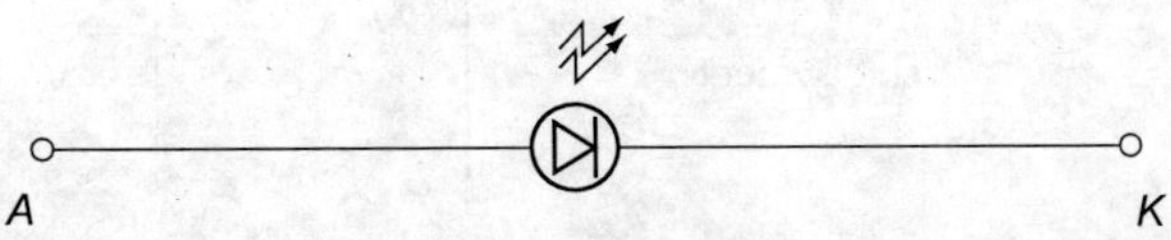

Fig. 3.22 *Circuit symbol of a LED*

LEDs do not emit light when reverse biased. They operate at voltages of the order of 1.5–3.3 V. The current is of the order of tens of milliamperes. The power requirement is 10–150 mW. The switching speed is very fast, of the order of 1 nS.

The typical applications of LEDs are antiburglar systems, picture phones, multimeters, calculators, digital meters, microprocessors, digital computers, electronic telephone exchanges, intercoms, electronic panels, digital watches, video displays, optical communications and so on. Also, there are two-lead LED lamps, which contain two LEDs so that a reversal in biasing will change the colour from green to red or vice versa.

3.13 SCHOTTKY DIODE

The germanium and standard silicon diodes are built with *p*- and *n*-type materials, but it can also be constructed by bonding a metal (aluminum or platinum) to *n*-type silicon. That is, the *p* region of the *p-n* junction diode is replaced with a metal. This type of diode is known as *Schottky diode.*

The circuit symbol and the *V-I* characteristics of the Schottky diode are as shown in Fig. 3.23. It can be seen that the characteristic of this diode is similar to that of a Ge diode. For a Schottky diode, as for a Ge diode, $V_\gamma \approx 0.3$ V. The breakdown voltages are also of the same range as that of Ge diode.

Schottky diode can switch On and Off much faster than the *p-n* junction diode. Also, the Schottky diode produces less unwanted noise than either Si or Ge diode. These two characteristics of the Schottky diode make it very useful in high-speed switching power circuits. The turn on voltage for the Schottky diode is same as that of a Ge diode; however, Ge diode is rarely used these days due to very high switching speeds of Schottky diodes.

Two types of metal-semiconductor junctions are possible—*Ohmic* and *rectifying.* The former is the type of contact desired when a lead is to be attached to a semiconductor. On the other hand, rectifying contact results in *Schottky barrier diode* with *V-I* characteristic very similar to those of *p-n* junction diode.

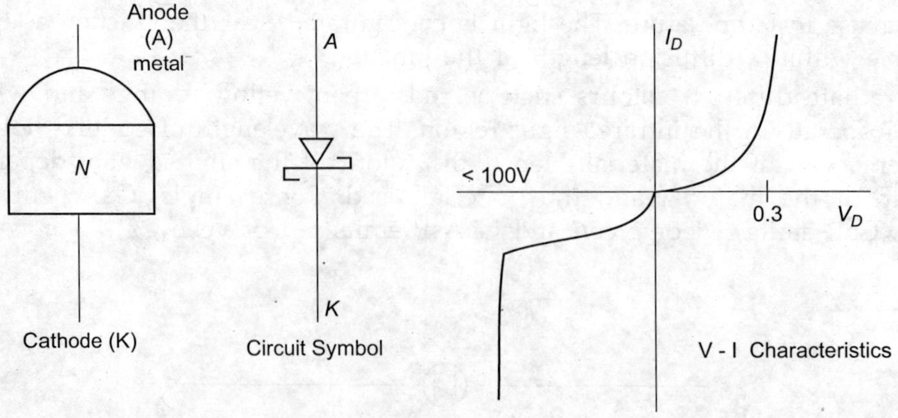

Fig. 3.23 *Circuit symbol and V-I characteristics of Schottky diode*

Aluminum acts as a *p*-type impurity when in contact with Si. If Al is to be attached as a lead to *n*-type Si, an Ohmic contact is desired and formation of *p-n* junction must be prevented.

The *V-I* characteristic of Schottky diode is essentially same as that of a *p-n* junction diode, but the physical mechanisms involved are more complicated. Note that in the forward direction, electrons from *n*-type Si cross the junction into metal, where electrons are plentiful. In this sense, this is a majority carrier device, whereas minority carriers account for a *p-n* junction diode characteristic. There is delay in switching a *p-n* diode from ON to OFF because the minority carriers stored at the junction must first be removed. Schottky diodes have a negligible storage time t_s because majority carriers carry the current predominantly. Electrons from *n* side enter the Al and become indistinguishable from electrons in the metal and hence are not stored near the junction.

There are no minority carriers in the reverse-bias condition, that is, no significant current from the metal to semiconductor with back bias. Thus, the delay present in junction diode, due to the hole-electron recombination time, is absent here. However, because of a larger contact area between metal and semiconductor than in the *point contact diode,* the forward resistance is lower and so is the noise.

The most commonly used semiconductors are Si and GaAs. As usual, GaAs has the lower noise and higher operating frequency limits, is easier to fabricate and consequently used at microwave *X*- band (8-12 GHz). Below these frequencies, *n*-type epitaxial materials are used and the metal is often a thin layer of titanium surrounded by gold for protection and low Ohmic resistance. The device is also known as *Esbar* or *hot electron diode.* Schottky diodes are available for microwave frequencies at least up to 100 GHz. They are used as detectors, mixers and so on. At frequency above X band, GaAs diodes are preferred since they are of lower noise. At the highest frequencies, point contact diodes are preferred since they have lower shunt capacitance.

3.14 POINT CONTACT DIODE

Point contact diode is also known as *crystal diode*. The noise figure of such a diode is very low at microwave frequencies, of the order of 6dB at 10 GHz. The construction of a typical point contact diode is as shown in Fig. 3.24.

An identical construction would be used for other semiconductor materials. It consists of a brass base on which a small pellet of Si, Ge, GaAs or indium phosphide is mounted. A fine gold-plated tungsten wire, with a diameter of 80-400 μm and a sharp point, makes contact with the polished top of the semiconductor pellet and presses down on it slightly for spring contact. This *cat's whisker,* as it is known, is connected to the top brass contact. The semiconductor and its cat's whisker are surrounded by wax to exclude moisture and are located in a metal-ceramic housing.

Such crystal diodes can be fitted into coaxial waveguide mounts and are available at frequencies in excess of 100 GHz, although they are then noisier than at *X* band. They are used as microwave mixers or detectors, there being some differences in diode

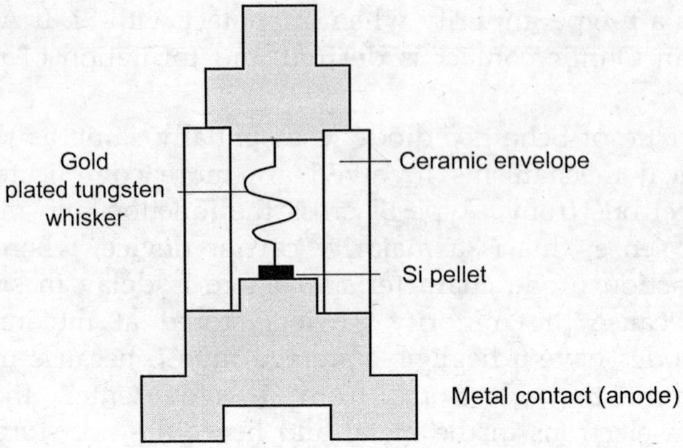

Fig. 3.24 *Point contact diode*

characteristics between the two applications. The diode must be mounted so that it provides a complete dc path for rectification, without unduly upsetting the RF field in the waveguide.

3.15 PIN DIODE

The *PIN diode* consists of a narrow layer of *p*-type semiconductor separated from an equally narrow layer of *n*-type material by a somewhat thicker region of *intrinsic* material. In fact, the intrinsic layer is a lightly doped *n*-type semiconductor. Although GaAs is used in the construction of PIN diodes, Si tends to be the main material. The reason for this is easier fabrication, higher powers handled and higher resistivity of intrinsic region. The PIN diode is used for microwave power switching, limiting and modulation.

The construction of the PIN diode is as shown in Fig. 3.25. The advantage of the planar construction is the lower series resistance while conducting. Encapsulation for such a chip takes any of the forms already shown for other microwave diodes. The in-line construction has a number of advantages, including reduced shunt capacitance. The construction shown in Fig. 3.25(c) is often preferred in practice, except perhaps for the highest powers. When fairly large dissipations are involved, the planar construction is better adopted to mounting on a heat sink.

The PIN diode acts more or less as an ordinary diode at frequencies up to about 100 MHz. However, above this frequency, it ceases to be a rectifier, because of the carrier storage in, and the transit time across, the intrinsic region. At microwave frequencies, the diode acts as variable resistor, with a simplified equivalent circuit as shown in Fig. 3.26.

When bias is varied on a PIN diode, its microwave resistance changes from a typical value of 5 kΩ to 10 kΩ under negative bias to the vicinity of 1 Ω to 10 Ω when the bias is positive. Thus, if the diode is mounted across a 50 Ω coaxial line, it will not significantly load the line when it is back biased, so that power flow will be unaffected. When the diode is forward biased, however, its resistance becomes very low, so that most of the power is reflected and hardly any is transmitted, the diode is acting as a switch. In a

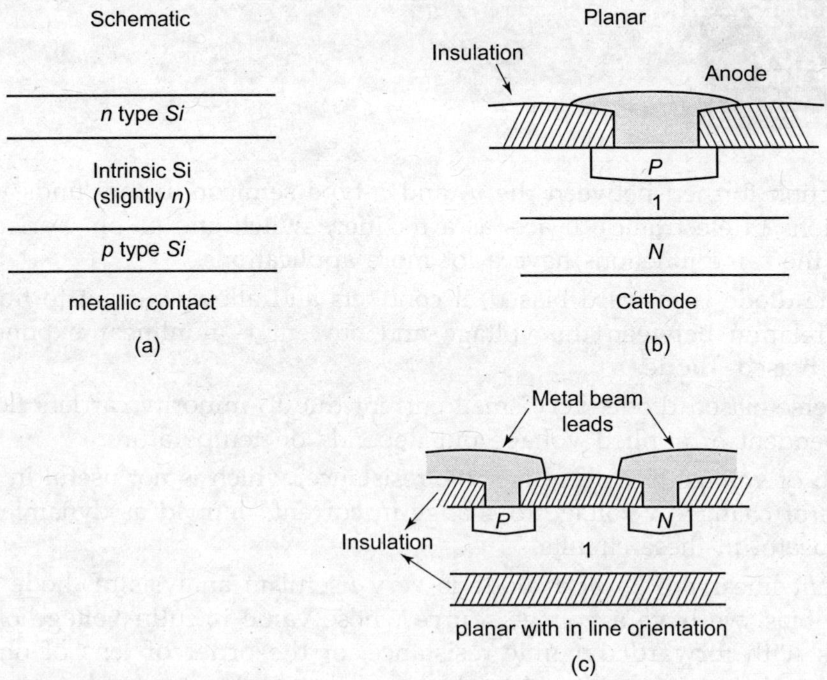

Fig. 3.25 *Construction of PIN diode*

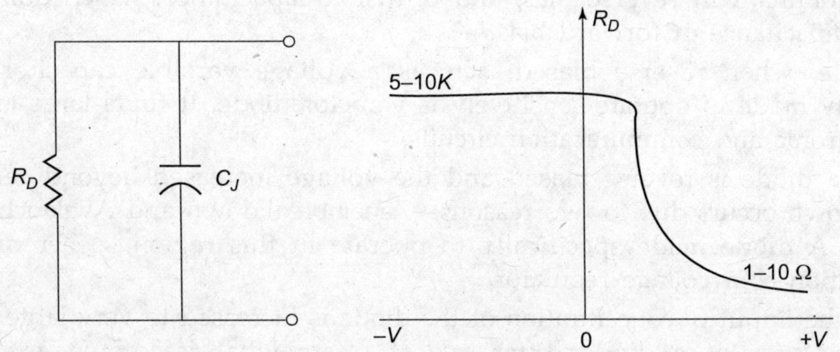

Fig. 3.26 *Equivalent circuit and V-I characteristics of PIN diode*

similar fashion, it may be used in series or in parallel in a waveguide or coaxial line, to increase the power handling or to reduce the transmitted power in Off condition.

The applications of PIN diode are as microwave diode switch. Diodes are available with resistance cutoff frequencies up to about 700 GHz. Individual diodes may handle powers up to 200 kW peak (200 W average). Several diodes can enhance this figure up to 1 MW peak. Actual switching times vary from 40 nS for higher power limiters to as little as 1 nS at lower powers.

SUMMARY

- The junction formed between the *p*- and *n*-type semiconductors finds very useful applications in electronics. It acts as a rectifier, switch and so on. Varied compositions of the same junctions have a lot more applications.
- When the diode is forward biased, it conducts and allows current to flow through it. The relation between the voltage and current is nonlinear exponential in a forward-biased diode.
- In a reverse-biased diode, very small current due to minority carriers flows, which is independent of applied voltage but depends on temperature.
- The ratio of voltage to current is static resistance, which is not useful in diode, and the ratio of change in voltage to change in current, defined as dynamic resistance, is very useful in these circuits.
- Equivalent linear model of a diode is very useful in analysis of diode circuits. In forward bias, we have a voltage source whose value is cutin voltage of the diode in series with forward dynamic resistance, of the order of tens of ohms, in the equivalent circuit. In reverse bias, the circuit is simply reverse dynamic resistance of the order of a few megaohms.
- There are two capacitances associated with diodes—depletion capacitance, which is predominant in reverse bias, and diffusion capacitance, which dominates the other capacitance in forward bias.
- A diode, when reverse biased, acts as a voltage variable capacitor. A diode specially made to operate sensitively is varactor diode. It finds large applications in electronic and communication circuits.
- When a diode is reverse biased and the voltage increased beyond certain limit, breakdown occurs due to two reasons—Zener breakdown and Avalanche multiplication. A diode made specifically to operate in this region, Zener diode, finds application as a voltage regulator.
- When the impurity concentration of the diode is increased to very large extent, the characteristics of a diode differ and it offers what is known as a negative resistance. Such diodes, tunnel diodes, find applications as oscillators, switches and so on at high frequencies.
- A Photodiode is a diode where the current is modulated in accordance with the light intensity. It is operated in reverse bias.
- A LED is the reverse of photodiode. It emits light, whose intensity is the function of forward current through the diode.
- Schottky and point contact diodes are metal-to-semiconductor junction diodes, which find large applications at microwave frequencies or above.
- A PIN diode has an intrinsic layer between the *p* and *n* layers of the diode. It acts as a very fast switch.

SOLVED PROBLEMS

3.1 (a) The p side and n side of a Ge diode have resistivities 1 Ω-cm and 2 Ω-cm, respectively. Find the height of the potential barrier V_0.

(b) What will be V_0 if the material is silicon instead of germanium?

Solution (a) $\sigma_p = pq\mu_p = N_A q\mu_p$

$$\Rightarrow \quad N_A = \frac{\sigma_p}{q\mu_p} = \frac{1}{\rho_p \cdot \mu_p \cdot q} = \frac{1}{1 \times 1800 \times 1.6 \times 10^{-19}}$$

$$= 3.472 \times 10^{15} \text{ atoms/cm}^3$$

$$\text{Similarly } N_D = \frac{1}{\rho_n \cdot \mu_n \cdot q} = \frac{1}{2 \times 3800 \times 1.6 \times 10^{-19}}$$

$$= 0.82 \times 10^{15} \text{ atoms/cm}^3$$

$$V_0 = V_T \, ln \frac{N_A N_D}{n_i^2} = 26 \times 10^{-3} \log_e \frac{3.4720.82 \times 10^{30}}{6.25 \times 10^{26}}$$

$$= 219.025 \text{ mV.}$$

(b) For silicon, $N_A = \dfrac{1}{1 \times 500 \times 1.6 \times 10^{-19}} = 12.5 \times 10^{15}$ atoms/cm^3

$$N_D = \frac{1}{2 \times 1300 \times 1.6 \times 10^{-19}} = 2.4 \times 10^{15} \text{ atoms/cm}^3$$

$$V_0 = 26 \times 10^{-3} \log_e \frac{12.5 \times 2.4 \times 10^{30}}{2.25 \times 10^{20}}$$

$$= 666.02 \text{ mV.}$$

3.2 Calculate the ratio of the hole diffusion current to the electron diffusion currents and express it in terms of σ_p, σ_n, L_n, L_p.

Solution We know that

$$I_{pn}(0) = \frac{AqD_p P_{no}}{L_p}\left(e^{\frac{V}{\eta V_T}} - 1\right) \qquad \text{...(1)}$$

$$I_{np}(0) = \frac{AqD_n n_{po}}{L_n}\left(e^{\frac{V}{\eta V_T}} - 1\right) \qquad \text{...(2)}$$

Dividing (1) and (2)

$$\frac{I_{pn}(0)}{I_{np}(0)} = \frac{D_p P_{no}}{L_p} \frac{L_n}{D_n n_{po}}$$

$$= \frac{D_p}{D_n} \cdot \frac{L_n}{L_p} \cdot \frac{P_{no}}{n_{po}}$$

$$= \frac{\mu_p \cdot V_T}{\mu_n \cdot V_T} \cdot \frac{L_n}{L_p} \cdot \frac{n_i^2/N_D}{n_i^2/N_A}$$

$$= \frac{\mu_p}{\mu_n} \cdot \frac{L_n}{L_p} \cdot \frac{N_A}{N_D}$$

$$= \frac{L_n}{L_p} \cdot \frac{N_A \cdot q \cdot \mu_p}{N_D \cdot q \cdot \mu_n} = \frac{L_n}{L_p} \cdot \frac{\sigma_p}{\sigma_n}$$

3.3 (a) Prove that the reverse saturation current I_0 is given by

$$I_0 = Aq\left(\frac{D_p}{L_p N_D} + \frac{D_n}{L_n \cdot N_A}\right) n_i^2$$

(b) Express the above equation in terms of conductivity as

$$I_0 = AV_T \frac{b\sigma_i^2}{(1+b)^2}\left[\frac{1}{L_p \sigma_n} + \frac{1}{L_n \sigma_p}\right], \text{ where}$$

$b = \mu_n / \mu_p$ and σ_i, σ_p and σ_n are conductivities of intrinsic, p and n type.

Solution (a) We have $I_0 = \dfrac{AqD_p P_{no}}{L_p} + \dfrac{AqD_n N_{po}}{L_n}$

$$= \frac{AqD_p \cdot \frac{n_i^2}{N_D}}{L_p} + \frac{AqD_n \cdot \frac{n_i^2}{N_A}}{L_n}$$

$$= Aq\left(\frac{D_p}{L_p \cdot N_D} + \frac{D_n}{L_n \cdot N_A}\right) n_i^2 \quad \text{...(1)}$$

(b) From (1) $I_0 = Aq\left(\dfrac{\mu_p \cdot V_T}{L_p \cdot N_D} + \dfrac{\mu_n \cdot V_T}{L_n \cdot N_A}\right) n_i^2$

$$\left(\text{Since } \frac{D_p}{\mu_p} = \frac{D_n}{\mu_n} = V_T\right)$$

$$\therefore \quad I_0 = Aq^2V_T\mu_n\mu_p\left(\frac{1}{L_pN_D\mu_nq} + \frac{1}{L_nN_A\mu_pq}\right)n_i^2$$

$$= AV_Tq^2n_i^2(\mu_n + \mu_p)^2\cdot\left(\frac{1}{\mu_n+\mu_p}\right)^2\cdot\mu_p\mu_n\cdot\left(\frac{1}{\sigma_pL_n} + \frac{1}{L_n\sigma_p}\right)$$

$$= AV_T\cdot\sigma_i^2\frac{\mu_p\mu_n}{(\mu_p+\mu_n)^2}\left(\frac{1}{\sigma_pL_n} + \frac{1}{L_p\sigma_n}\right)$$

$$= AV_T\sigma_i^2\frac{\mu_p\mu_n}{\mu_p{}^2\left(\dfrac{\mu_n}{\mu_p}+1\right)^2}\left(\frac{1}{L_n\sigma_p} + \frac{1}{L_p\sigma_n}\right)$$

$$= AV_T\sigma_i^2\frac{\mu_n/\mu_p}{\left(\dfrac{\mu_n}{\mu_p}+1\right)^2}\left(\frac{1}{L_p\sigma_n} + \frac{1}{L_n\sigma_p}\right)$$

$$= AV_T\sigma_i^2\frac{b}{(b+1)^2}\left(\frac{1}{L_p\sigma_n} + \frac{1}{L_n\sigma_p}\right)$$

3.4 (a) A Ge diode is having the following parameters. Determine the reverse saturation current I_o. Junction area is 4 mm^2 at room temperature, $\sigma_p = 1\ (\Omega\text{-cm})^{-1}$, $\sigma_n = 0.1\ (\Omega\text{-cm})^{-1}$, $L_n = L_p = 0.15$ cm.

(b) A Si diode is having the following parameters. Determine the reverse saturation current I_o. Junction area is 4mm^2, $\sigma_p = \sigma_n = 0.01\ (\Omega\text{-cm})^{-1}$, $L_n = L_p = 0.01$ cm.

Solution (a)

$$\sigma_i = n_i\cdot q(\mu_n + \mu_p)$$
$$= 2.5\times10^{13}\times1.6\times10^{-19}(3800 + 1800)$$
$$= 0.0224\ (\Omega\text{-cm})^{-1}$$

$$b = \frac{\mu_n}{\mu_p} = \frac{3800}{1800} = 2.11$$

$$A = 0.04\ \text{cm}^2$$

$$I_0 = AV_T\frac{b\sigma_i^2}{(b+1)^2}\left(\frac{1}{L_p\sigma_n} + \frac{1}{L_n\sigma_p}\right)$$

$$= 0.04 \times 26 \times 10^{-3} \times \frac{2.11 \times 0.0224}{(1+2.11)^2}\left(\frac{1}{0.15 \times 0.1} + \frac{1}{0.15 \times 1}\right)$$

$$= 8.34\ \mu\text{A}$$

(b) For silicon $\quad b = \dfrac{1300}{500} = 2.6$

$$\sigma_i = n_i q(\mu_n + \mu_p) = 1.5 \times 10^{10} \times 1.6 \times 10^{-19}(1300 + 500)$$
$$= 4.32 \times 10^{-6}$$

Using the formula $I_0 = AV_T \dfrac{b\sigma_i^2}{(b+1)^2}\left(\dfrac{1}{L_p\sigma_n} + \dfrac{1}{L_n\sigma_p}\right)$

$$= \frac{0.04 \times 26 \times 10^{-3} \times 2.6 \times (4.32 \times 10^{-6})^2}{(3.6)^2}\left(\frac{1}{0.01 \times 0.01} + \frac{1}{0.01 \times 0.01}\right)$$

$$= 77.9\ \text{pA}$$

3.5 Find the ratio of the reverse saturation current in germanium to that of silicon. Assume $L_n = L_p = 0.1$ cm and $\sigma_n = \sigma_p = 1.0\ (\Omega\text{-cm})^{-1}$ for Ge, whereas the corresponding values are 0.01 cm and 0.01 $(\Omega\text{-cm})^{-1}$ for silicon.

Solution We have

$$I_0 = AV_T \frac{b}{(b+1)^2}\sigma_i^2\left(\frac{1}{L_p\sigma_n} + \frac{1}{L_n\sigma_p}\right)$$

For silicon, $\quad b = \dfrac{\mu_n}{\mu_p} = 2.6$

$$\sigma_i = 0.432 \times 10^{-5}$$

$$\therefore \quad I_0 = AV_T \cdot \frac{2.6}{(3.6)^2}(0.432 \times 10^{-5})^2\left[\frac{1}{0.01 \times 0.01} + \frac{1}{0.01 \times 0.01}\right]$$

$$= AV_T\,(17.33 \times 10^{-8})$$

For germanium,

$$b = \frac{\mu_n}{\mu_p} = \frac{3800}{1800} = 2.11,$$

$$\sigma_i = n_i q(\mu_n + \mu_p) = 2.24 \times 10^{-2}$$

$$I_o = AV_T \cdot \frac{2.11}{(3.11)^2}(2.24 \times 10^{-2})\left[\frac{1}{0.1 \times 1} + \frac{1}{1 \times 0.1}\right]$$

$$= AV_T \cdot 21.89 \times 10^{-4}$$

$\therefore$ Ratio of reverse saturation current will be

$$\frac{AV_T \cdot 21.89 \times 10^{-4}}{AV_T \cdot 17.33 \times 10^{-8}} = 1.2632 \times 10^4$$

$$= 12632.5$$

3.6 (a) For what voltage will the reverse current in a *p-n* Junction Ge diode reach 90% of its saturation value at room temperature?

(b) What is the ratio of the current for a forward bias of 0.05 V to the current for the same magnitude of reverse bias?

(c) If the reverse saturation current is 10 μA, calculate the forward currents for voltages of 0.1, 0.2 and 0.3 V, respectively.

Solution (a) Diode-current equation is given by $I = I_o\left(e^{\frac{V}{\eta V_T}} - 1\right)$

90% of the reverse saturation value = – 0.9 I_o

For Germanium $\eta = 1$

$$\therefore \quad -0.9\, Io = I_o\left(e^{\frac{V}{V_T}} - 1\right)$$

$$\Rightarrow \quad -0.9 = e^{\frac{V}{V_T}} - 1$$

$$\Rightarrow \quad 0.1 = e^{\frac{V}{V_T}}$$

$$V = V_T \log_e (0.1) = -59.86 \text{ mV}$$

(b) $V = 0.05$ V for forward bias and is negative for reverse bias.

$\therefore$ Ratio of forward to reverse current will be

$$= \frac{I_o\left(e^{\frac{0.05}{26\times10^{-3}}} - 1\right)}{I_o\left(e^{\frac{-0.05}{26\times10^{-3}}} - 1\right)} = \frac{e^{1.923} - 1}{e^{-1.923} - 1}$$

$$= \frac{5.8414}{-0.8538} = -6.841$$

(c) *For V = 0.1 V*

$$I = I_o\left(e^{\frac{0.1}{26\times10^{-3}}} - 1\right) = 10 \times 10^{-6}(46.81 - 1)$$

$$= 458.126\ \mu\text{A}$$

For $V = 0.2$ V

$$I = 10 \times 10^{-6}\left(e^{\frac{0.2}{26\times10^{-3}}} - 1\right) = 10 \times 10^{-6}(2191.42 - 1)$$

$$= 21.90 \text{ mA}$$

For $V = 0.3$ V

$$I = 10 \times 10^{-6}\left(e^{\frac{0.3}{26\times10^{-3}}} - 1\right) = 10 \times 10^{-6}(102586.4912 - 1)$$

$$= 1.026 \text{ A}$$

3.7 (a) Calculate the anticipated factor by which the reverse saturation current of Ge diode is multiplied when the temperature is increased from 25°C to 75°C.

(b) Repeat for T = 25°C to 125°C.

Solution (a)

$$I_o(T) = I_o \cdot 2^{(T - T_1)/10}$$

$$I_0(75°) = I_o(25) \cdot 2^{(75-25)/10}$$

$$= I_o(25) \cdot 2^{50}$$

$$= 32 I_o(25)$$

$$\frac{I_o(75)}{I_o(25)} = 32 \text{ is the anticipated factor.}$$

(b)

$$I_o(125°) = I_o(+25)\, 2^{(125-25)/10}$$

$$= 2^{10}.I_o(25)$$

$$= 1024\ I_o(25)$$

$$\therefore \quad \frac{I_o(125)}{I_o(25)} = 1024$$

3.8 A silicon diode operates at a forward voltage of 0.4 V. Calculate the factor by which the current will be multiplied when the temperature is increased from 25°C to 125°C.

Solution The temperature equivalent of 125°C is 398 K.

$$V_T = \frac{398}{11,600} = 34.31 \text{ mV}$$

$$I_o(150) = I_o(25) \cdot 2^{(125-25)/10} = 1024\ I_o(25)$$

$$\frac{I(150)}{I(25)} = \frac{I_o(150)\left(e^{\frac{0.4}{2\times0.0343}} - 1\right)}{I_o(25)\left(e^{\frac{0.4}{2\times0.026}} - 1\right)}$$

$$\approx 1024 \cdot e^{\left(\frac{0.2}{0.0343} - \frac{0.2}{0.026}\right)}$$

$$\approx 159.19$$

3.9 An ideal germanium *p-n* Junction diode has at a temperature of 100°C and a reverse saturation current of 30 μA. At a temperature of 100°C find the dynamic resistance for a 0.2 V bias in (a) forward dirənction and (b) the reverse direction.

Solution

$$V_T(125) = \frac{100 + 273}{11{,}600} = 32.15 \text{ mV/°C}$$

For large forward bias $I = I_o e^{\frac{V}{V_T}}$

$$\frac{dI}{dV} = \frac{I_o}{V_T} \cdot e^{\frac{V}{V_T}}$$

and,
$$r = \frac{dV}{dI} = \frac{V_T}{I_o} \cdot e^{-V/V_T}$$

(a)
$$r(0.2) = \frac{0.03215}{30 \times 10^{-6}} \cdot e^{\frac{-0.2}{0.03215}} = 2.13 \ \Omega$$

(b)
$$r(-0.2) = \frac{0.03215}{30 \times 10^{-6}} \cdot e^{\frac{+0.2}{0.03215}} = 539.19 \text{ K}\Omega$$

3.10 A diode current is 0.6 mA when applied voltage is 400 mV, and 20 mA when applied voltage is 500 mV. Find η. Assume V_T = 26 mV.

Solution

$$I = I_0\left(e^{\frac{V}{\eta V_T}} - 1\right)$$

$$0.6 \times 10^{-3} \cong I_0 e^{\frac{400 \times 10^{-3}}{\eta \times 26 \times 10^{-3}}} \quad ...(1)$$

$$20 \times 10^{-3} = I_0 e^{\frac{500 \times 10^{-3}}{\eta \times 26 \times 10^{-3}}} \quad ...(2)$$

Dividing (1) and (2)

$$\frac{20 \times 10^{-3}}{0.6 \times 10^{-3}} = e^{\left(\frac{500}{26\eta} - \frac{400}{26\eta}\right)} = e^{\frac{100}{26\eta}}$$

$$33.33 = e^{\frac{100}{26\eta}}$$

Taking natural logarithm on both sides,

$$\log_e (33.33) = \frac{100}{26\eta}$$

$$\eta = \frac{100}{26 \log_e (33.33)} = 1.096$$

3.11 Find the *V-I* characteristics for a series combination of Ge and Si under large forward bias, as shown in the figure.

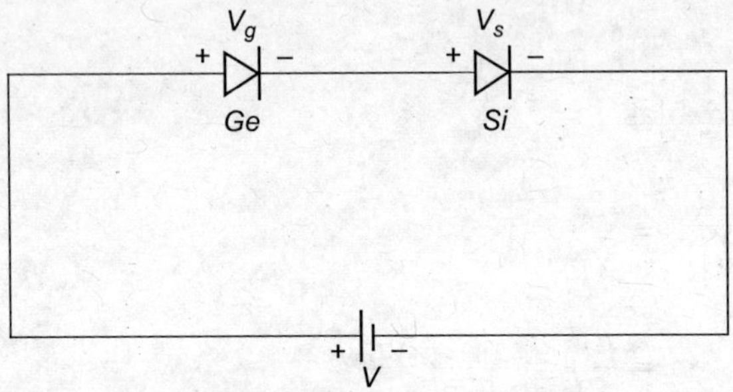

Solution Let I_g and I_s be the reverse saturation currents for Ge and Si, respectively. Under large forward bias

$$e^{\frac{V}{\eta V_T}} - 1 = e^{\frac{V}{\eta V_T}}$$

Applying KVL in the circuit

$$V_g + V_s = V$$

$$V_g = V_T \ln \frac{I}{I_g}, V_s = 2V_T \ln \frac{I}{I_s}$$

$$\therefore \quad V = V_T \ln \frac{I}{I_g} + 2V_T \ln \frac{I}{I_s}$$

$$V = V_T \ln \cdot \frac{I}{I_g} \cdot \left(\frac{I}{I_s}\right)^2 = V_T \ln \frac{I^3}{I_g I_s^2}$$

$$= V_T \ln \left[\frac{I}{(I_g I_s^2)^{1/3}}\right]^3$$

$$\therefore \quad I = \left(I_g I_s^2\right)^{1/3} \cdot e^{V/3V_T}$$

3.12 Calculate the current through 48 Ω resistor in the circuit. Assume the diode to be of silicon type and forward resistance of each diode is 1 Ω.

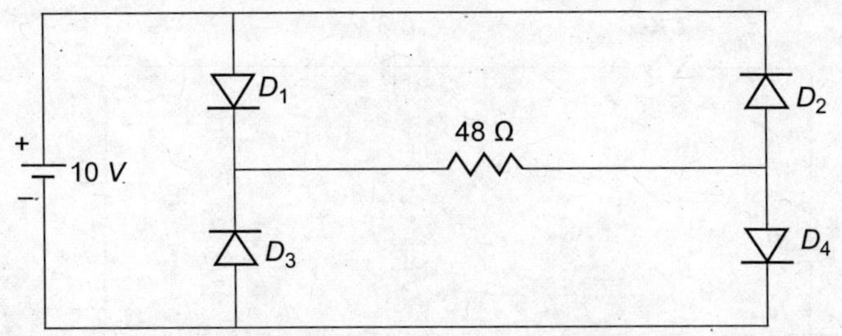

Solution D_1 and D_4 are forward biased and D_2 and D_3 are reverse biased. Replacing D_1 and D_4 by their equivalent circuits, that is, V_γ in series with R_f and D_2 and D_3 by their equivalent circuits, that is, by an open circuit, the circuit becomes as shown below.

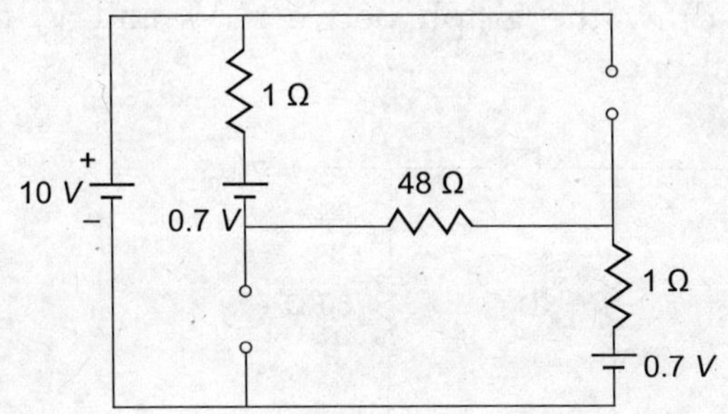

Current through 48 Ω resistor

$$= \frac{10 - 0.7 - 0.7}{50}$$

$$= 172 \text{ mA}$$

3.13 Find the current I in the circuit shown. Diodes are made of silicon, forward resistance of each diode is zero.

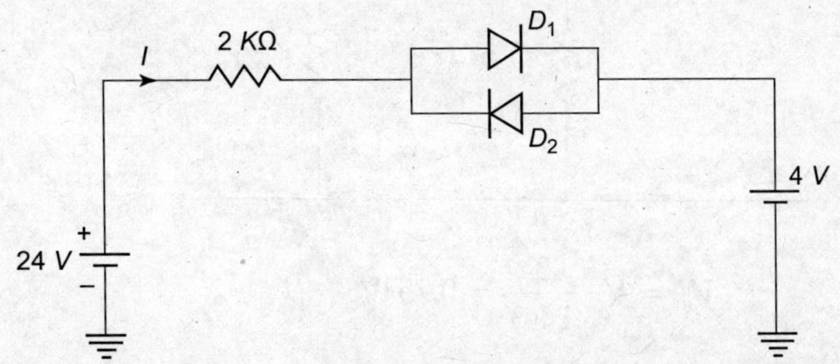

Solution Diode D_1 is forward biased, D_2 is reverse biased. Replacing D_1 with Cutin voltage and D_2 with an open circuit, the circuit will be as shown.

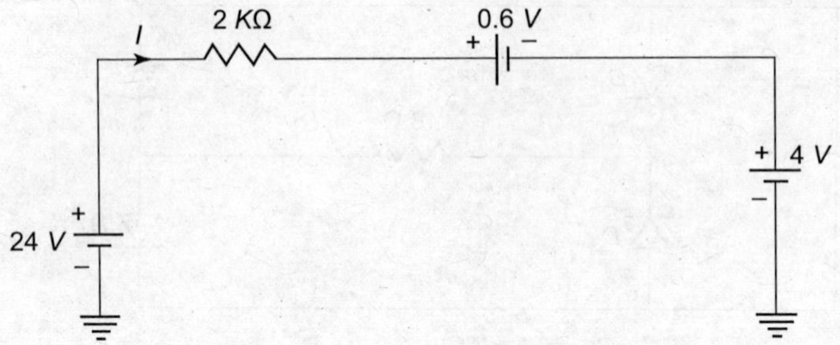

$$\therefore \quad I = \frac{24 - 4 - 0.6}{2k\Omega} = 9.7 \text{ mA}$$

3.14 For the following circuit draw the graph between V_0 and V_i for $-5 < V_i < 10$ assuming diode as an ideal one.

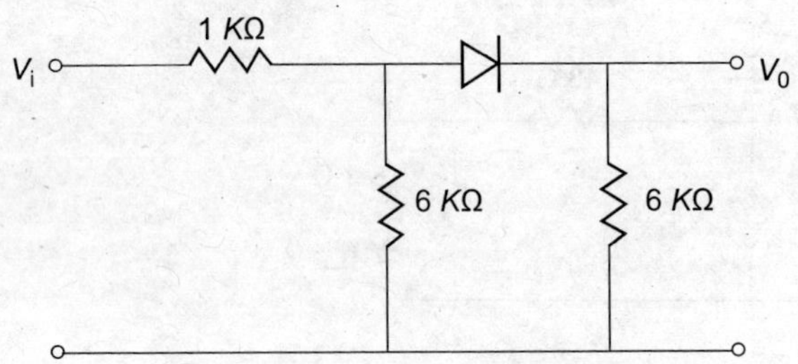

Solution When V_i is negative, the diode is reverse biased. So, $V_0 = 0$.
When V_i is positive, the diode is forward biased, and it acts as short circuit, then the circuit becomes.

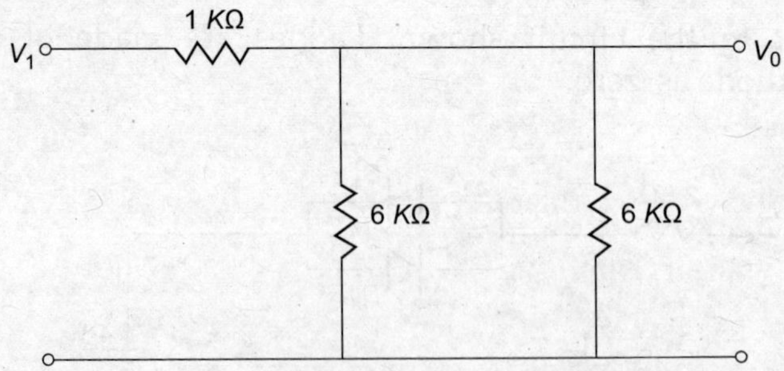

$$V_o = V_i \cdot \frac{3}{1+3} = 0.75\ V_i$$

(Since there is a parallel combination of two 6 kΩ resistors, this combination is replaced with 3 K). The graph between V_o and V_i as shown.

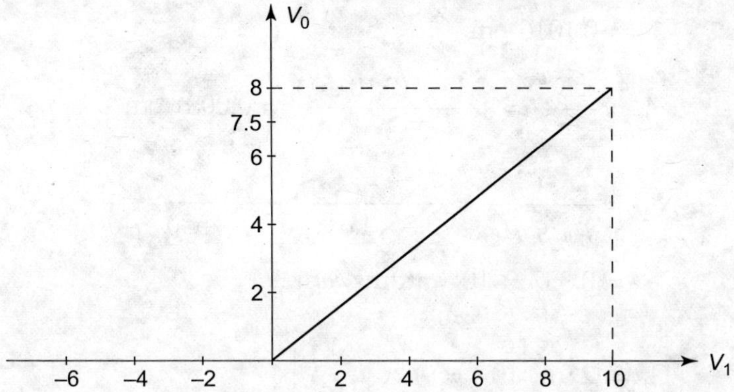

3.15 (a) Find the expression for transition capacitance of a Si diode in Pf/cm^2 in terms of the doping level (N_A) and junction voltage (V_J). Assume $N_D >> N_A$.

(b) A silicon diode has ρ_p= 3.5 (Ω-cm), V_o = 0.35 V, reverse bias of 5 V and cross sectional area is circular with dia = 40 mils. Find C_T.

Solution (a) We have $C_T = A\left(\dfrac{q\varepsilon N_D}{2V_j}\right)^{1/2}$

Since here $N_D >> N_A$, replace N_D with N_A in the above equation,

$$C_T = A\left(\frac{q\varepsilon N_A}{2V_j}\right)^{1/2}$$

Put $A = 10^{-4}$ m^2, $N_A = 10^6$ N_A/m^3, $\varepsilon = 8.849 \times 10^{-12} \times 12$F/m.

$$\therefore \quad C_T = 10^{-4}\, m^2\left[\frac{1.6\times10^{-19}\times 8.849\times10^{-12}\times N_A\times10^6}{2V_J}\right]^{1/2}$$

$$= 2.9 \times 10^{-4} \times 10^{-12}\left(\frac{N_A}{V_J}\right)^{1/2} \text{F/cm}^2$$

$$= 2.9 \times 10^{-4}\left(\frac{N_A}{V_J}\right)^{1/2} \text{pF/cm}^2$$

(b) $1 \text{ mil} = 0.001 \text{ in.}$

$$D = \text{diameter} = \frac{400}{1000} \text{ in.} = \frac{400}{1000} \times 2.54 \text{ cm}$$

$$= 0.1016 \text{ cm.}$$

$$A = \frac{\pi D^2}{4} = \frac{3.14 \times (0.1016)^2}{4} = 0.0081 \text{ cm}^2$$

$$N_A = \frac{1}{\rho_p \cdot q \cdot \mu_p} = \frac{1}{3.5 \times 1.6 \times 10^{-19} \times 500}$$

$$= 0.357 \times 10^{16} \text{ atoms/cm}^3.$$

$$C_T = 2.9 \times 10^{-4} \times A \times \left(\frac{N_A}{V_J}\right)^{½}$$

$$= 2.9 \times 10^{-4} \times 0.0081 \times \left(\frac{0.357 \times 10^{16}}{5 + 0.35}\right)^{½}$$

$$= 60.67 \text{ pF}$$

3.16 An abrupt junction diode has transition capacitance 20 Pf when reverse biased at 5V. By how much will the capacitance decrease if the voltage is increased by 3V?

Solution $C_T = \frac{K}{\sqrt{V}}$. Since for an abrupt junction diode $C_T \propto \frac{1}{\sqrt{V}}$

$$\text{So, } 20 = \frac{K}{\sqrt{5}} \Rightarrow K = 20\sqrt{5} = 44.72$$

$$\text{Capacitance for 3V increase} = \frac{K}{\sqrt{5+3}} = \frac{44.72}{\sqrt{8}}$$

$$= 15.81 \text{ PF}$$

$$\therefore \text{Decrease in capacitance} = 20 - 15.81$$

$$= 4.2 \text{ pF.}$$

3.17 Calculate the junction capacitance of a Ge diode whose thickness of Barrier layer is 4 microns and junction area is 1 mm × 1 mm, $\varepsilon_r = 16$.

Solution

$$C_T = \frac{\varepsilon A}{W} = \frac{\varepsilon_o \varepsilon_r A}{W}$$

$$= \frac{8.849 \times 10^{-12} \times 16 \times (10^{-3})^2}{4 \times 10^{-6}}$$

$$= 35.395 \text{ pF.}$$

3.18 Barrier voltage in an alloy junction Ge diode is 0.2 V. $N_A = 3 \times 10^{20}$ atoms/m^3 and $N_D >> N_A$. The junction area is 1 mm^2. Find the width of the depletion layer and transition capacitance for the following applied biases (a) 10 V reverse (b) 0.1 V reverse.

Solution We have depletion layer width $W = \left(\frac{2\varepsilon V_J}{qN_A}\right)^{½}$

$$= \left(\frac{2 \times 8.849 \times 10^{-12} \times 16\, V_J}{1.6 \times 10^{-19} \times 3 \times 10^{20}}\right)^{½}$$

$$= 2.42 \times 10^{-6} \sqrt{V_J} \text{ m}$$

$$C_T = \frac{\varepsilon A}{W} = \frac{(10^{-3})^2 \times 8.849 \times 10^{-12} \times 16}{W}$$

$$= \frac{141.584 \times 10^{-18}}{W}$$

If W is in microns, C_T is in Pf, then

$$C_T = \frac{141.584}{W} \text{ Pf.}$$

(a) $$W(-10) = 2.42\sqrt{10 + 0.2} = 7.73 \text{ microns}$$

$$C_T = \frac{141.584}{7.73} = 18.32 \text{ Pf}$$

(b) $$W(-0.1) = 2.42\sqrt{0.1 + 0.2} = 1.325 \text{ microns}$$

$$C_T = \frac{141.584}{1.325} = 106.81 \text{ pF}$$

3.19 (a) Find the barrier voltage in a linearly graded junction diode where the charge density varies as $\rho = ax$.

(b) Find the barrier capacitance.

Solution (a) The variation of charge density for a linearly graded junction diode will be as shown.

$$\frac{d^2V}{dx^2} = -\frac{\rho}{\varepsilon} = \frac{-ax}{\varepsilon}. \qquad ...(1)$$

Integrating equation (1)

$$\frac{dV}{dx} = \frac{-ax^2}{2\varepsilon} + C_1$$

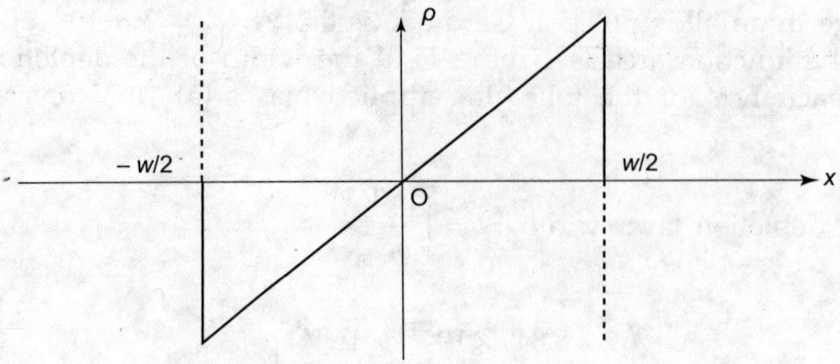

At $x = -W/2$, $\dfrac{dV}{dx} = 0$

$$\therefore \quad 0 = \frac{-aW^2}{8\varepsilon} + C_1 \Rightarrow C_1 = \frac{aW^2}{8\varepsilon}$$

$$\therefore \quad \frac{dV}{dx} = \frac{-ax^2}{2\varepsilon} + \frac{aW^2}{8\varepsilon}. \qquad ...(2)$$

Integrating equation (2),

$$V = \frac{-ax^3}{6\varepsilon} + \frac{aW^2x}{8\varepsilon} + C_2$$

At $\quad x = -W/2, V = 0$

$$\therefore \quad 0 = \frac{+aW^3}{48\varepsilon} - \frac{aW^3}{16\varepsilon} + C_2$$

$$C_2 = \frac{-aW^3}{48\varepsilon} + \frac{aW^3}{16\varepsilon} = \frac{aW^3}{24\varepsilon}$$

At $\quad x = +W/2, V = V_J$

$$\therefore \quad V_J = \frac{-aW^3}{48\varepsilon} + \frac{aW^3}{16\varepsilon} + \frac{aW^3}{24\varepsilon}$$

$$= \frac{aW^3}{12\varepsilon}$$

Hence proved.

(b) Charge $Q = \int_0^{W/2} A\rho dx = \int_0^{W/2} Aax dx$

$$= Aa \cdot \frac{x^2}{2}\bigg|_0^{W/2} = \frac{Aa}{2} \cdot \frac{W^2}{4} = \frac{AaW^2}{8}$$

$$C_T = \frac{dQ}{dV} = \frac{d}{dV}\left(\frac{AaW^2}{8}\right) = \frac{Aa \cdot 2W}{8}\frac{dW}{dV}$$

$$= \frac{AaW}{4}\frac{dW}{dV}$$

We have $V_J = \frac{aW^3}{12\varepsilon} = V$

$$dV = \frac{a3W^2}{12\varepsilon} \cdot dW = \frac{aW^2}{4\varepsilon} dW$$

$$\therefore \quad \frac{dW}{dV} = \frac{4\varepsilon}{aW^2}$$

Using this result in the expression for C_T

$$C_T = \frac{AaW}{4} \cdot \frac{dW}{dV} = \frac{AaW}{4} \cdot \frac{4\varepsilon}{aW^2}$$

$$= \frac{\varepsilon A}{W}$$

3.20 A silicon diode has diffusion capacitance of 10 μF when carrying a forward current of 10 mA. $N_A >> N_D$. Find the diffusion length of holes (L_p).

Solution We have an expression for $C_D = \frac{\tau_p I}{\eta V_T}$

$$\Rightarrow \quad \tau_p = \frac{C_D \eta V_T}{I}$$

Diffusion length $Lp = \sqrt{D_p \tau_p}$

$$Lp^2 = D_p \cdot \tau_p = D_p \cdot \frac{C_D \eta V_T}{I}$$

$$= \frac{13 \times 10^{-6} \times 10 \times 2 \times 26 \times 10^{-3}}{10 \times 10^{-3}}$$

$$= 676 \times 10^{-6}$$

$$Lp = 26 \times 10^{-3} \text{ cm} = 0.026 \text{ cm}$$

3.21 (a) Determine the current flowing through the Zener diode for the circuit shown in the figure. If R_L = 5000 Ω, input voltage = 50 V, R_S = 2000 Ω and output voltage = 30 V.

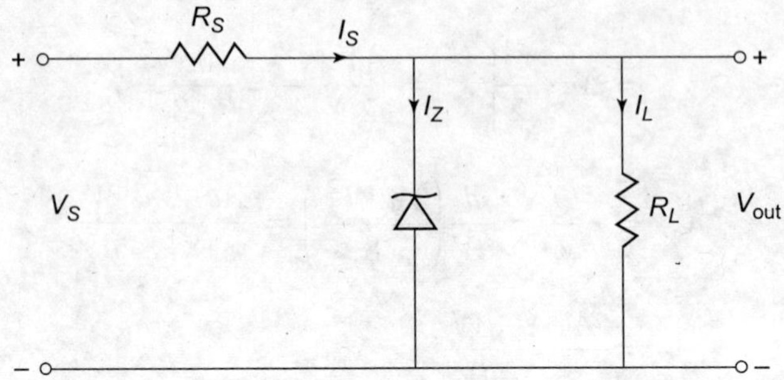

Solution Writing KVL equation in the circuit,

$$-V_S + I_S \cdot R_S + V_{out} = 0$$

$$I_S = \frac{V_S - V_{out}}{R_S} = \frac{50 - 30}{2000} = 10 \text{ mA}$$

$$\text{Load current } I_L = \frac{V_{out}}{R_L} = \frac{30}{5000} = 6 \text{ mA}$$

Current through Zener diode $I_Z = I_S - I_L$

$$= 10 - 6 = 4 \text{ mA}$$

(b) Find the maximum and minimum values of Zener Current if in the above circuit, R_L = 5000 Ω, R_S = 10000 Ω, V_{out} = 30 V and V_S varies between 100 and 120 V.

Solution Load current $I_L = \dfrac{V_{out}}{R_L} = \dfrac{30}{5000} = 6$ mA

The Zener Current is maximum when V_S is maximum, that is, 120 V. Then,

$$I_S = \frac{120 - 30}{10{,}000} = 9 \text{ mA. Then, } I_Z = 9 - 6 = 3 \text{ mA.}$$

The Zener Current is minimum, when V_S is minimum, that is, 100 V, then,

$$I_S = \frac{100 - 30}{10{,}000} = 7.0 \text{ mA}$$

Corresponding Zener Current = 7 – 6 = 1 mA

(c) Determine the temperature coefficient of a 5 V Zener diode at 25 °C, if the nominal voltage drops to 4.8 V at a temperature of 100 °C.

Solution Change in Zener voltage ΔV_Z = 4.8 – 5 = – 0.2 V

Change in temperature Δt = 100 – 25 = 75 °C

$$\text{Temperature coefficient of a Zener diode} = \frac{\Delta V_Z}{V_Z \Delta t} \times 100$$

$$= \frac{-0.2}{5 \times 75} \times 100$$

$$= -0.0533$$

3.22 The Avalanche diode shown in figure regulates at 50 V from 5 to 40 mA current. The supply voltage V is 100 V.

(a) Calculate R to allow voltage regulation from $I_L = 0$ to $I_L = I_{L\max}$. What is $I_{L\max}$?

(b) If R is set as above but I_L is fixed at 25 mA, then what is the permissible range of V?

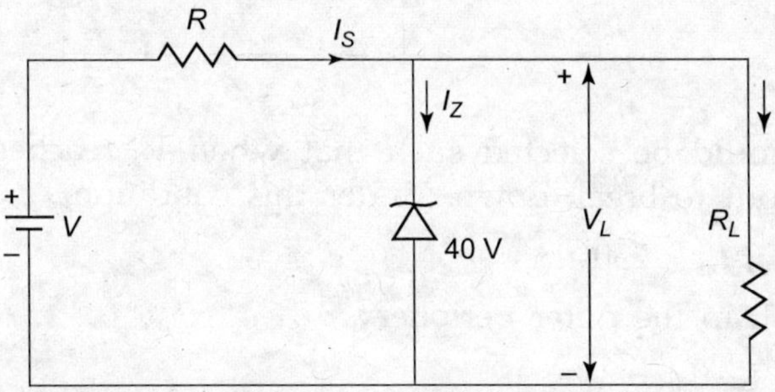

Solution (a) If Load Current is zero, then Maximum Current will flow through the Zener diode. This current must be limited to the maximum permissible value of **40 mA** by properly selecting R. In this condition $I_S = I_Z = 40$ mA.

$$R = \frac{V - V_Z}{I_s} = \frac{100 - 40}{40 \times 10^{-3}} = 1.5 \text{ k}\Omega.$$

When load is connected, some portion of I_S gets diverted to load and therefore, I_Z reduces. But I_Z should not be allowed to fall below the minimum permissible value of 5 mA.

$$I_{L\max} = I_S - I_{Z\min} = 40 - 5 = 35 \text{ mA}$$

(b) When I_L is fixed at 25 mA, then V can be increased upto a limit till I_Z is less than 40 mA.

$$V_{\max} = V_Z + I_S R = V_Z + (I_L + I_{Z\max})R$$

$$= 40 + (40 + 25) \times 1.5$$

$$= 137.5 \text{ V}.$$

$$V_{\min} = V_Z + (I_{Z\min} + I_L)R$$

$$= 40 + (5 + 25) \times 1.5$$

$$= 85 \text{ V}.$$

3.23 The Zener diode is used to protect the galvanometer which has 200 μA full scale rating and 560 Ω internal resistance, as shown in the figure. It should show full deflection when V_i = 20 V and if V_i is increased beyond 20 V, then the current in the galvanometer should remain constant at 250 μA. Find the values of R_1 and R_2.

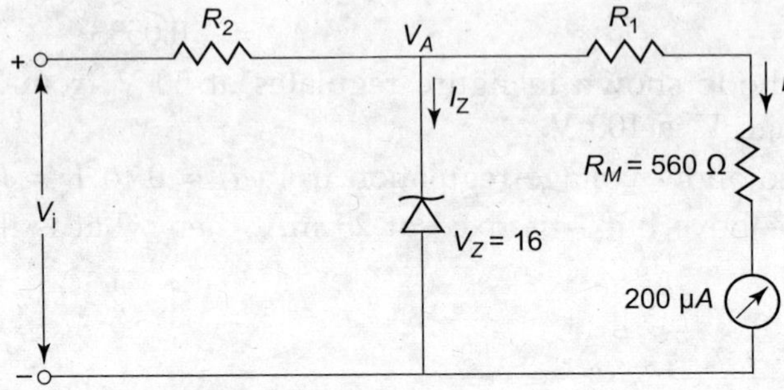

Solution R_1 and R_2 should be selected such that when V_i reaches 20 V, Zener diode should be about to break down. Under this condition,

$$V_A = 16 \text{ V}, I_m = 250 \ \mu\text{A}$$

Writing KVL in the outer periphery

$$(R_1 + R_2 + R_m)I_m = V_i$$

$$R_1 + R_2 + 560 = \frac{20}{250 \ \mu A}$$

$$R_1 + R_2 = 79.44 \text{ k}\Omega. \qquad ...(1)$$

Writing KVL in second mesh of the circuit,

$$V_A = I_m(R_1 + R_m)$$

$$R_1 + R_m = \frac{16}{250 \times 10^{-6}}$$

$$R_1 = \frac{16}{250 \times 10^{-6}} - 560 = 63.44 \text{ k}\Omega.$$

$$R_2 = 79.44 \text{ K} - 63.44 \text{ K} = 16 \text{ k}\Omega$$

3.24 The saturation currents of diodes are 1 *m*A and 2 *m*A. The breakdown voltage of diodes is 100 V.

(a) Find the current and voltage across each diode if V = 90 V and V = 110 V.

(b) Repeat part (a) if each diode is shunted by a 10 MΩ resistor.

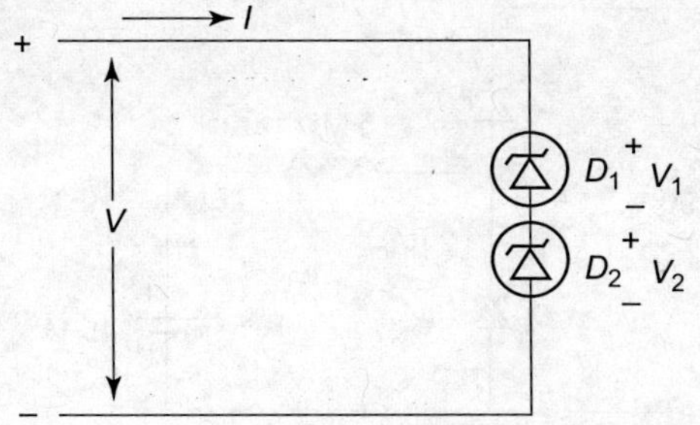

Solution Let I_{o1} and I_{o2} be the reverse saturation currents of diodes D_1 and D_2 respectively. Given that $V_{z1} = V_{z2} = 100$ V. Assume diodes are made up of silicon, so $\eta = 2$.

We have diode current equation

$$I = I_o\left(e^{\frac{V}{\eta V_T}} - 1\right)$$

$$\Rightarrow \quad V = \eta V_T \log_e\left(1 + \frac{I}{I_o}\right)$$

$$= 2V_T \log_e\left(1 + \frac{I}{I_o}\right)$$

(a) *For V = 90 V.* Since zener breakdown voltage is 100 V, none of D_1 and D_2 will breakdown and I is determined by the diode with smallest I_0 that is, D_1. Thus, $I = 1\ \mu$A and for D_2 $I = -I = -1\ \mu$A.

$$\therefore \quad V_2 = 2V_T \ln\left(1 - \frac{1}{2}\right) = 2 \times 26 \log_e (1/2)$$

$$= -36.04 \text{ mV}$$

$$V_1 = -89.964 \text{ V.}$$

When V = 110 V. D_1 is breakdown, D_2 is reverse biased. Thus, $I = I_{o2} = 2$ μA.

$V_1 = -100$ V and $V_2 = -10$ V.

(b) When 10 MΩ resistors shunt the diodes, the circuit will become as shown below.

Using Thevenin's theorm, with respect to Diode D_2, $R_{th} = 10\text{ M}||10\text{ M} = 5$ MΩ.

$$V_{th} = \frac{10\text{ M}}{10\text{ M} + 10\text{ M}} \times 90 \text{ V}$$

$$= 45 \text{ V.}$$

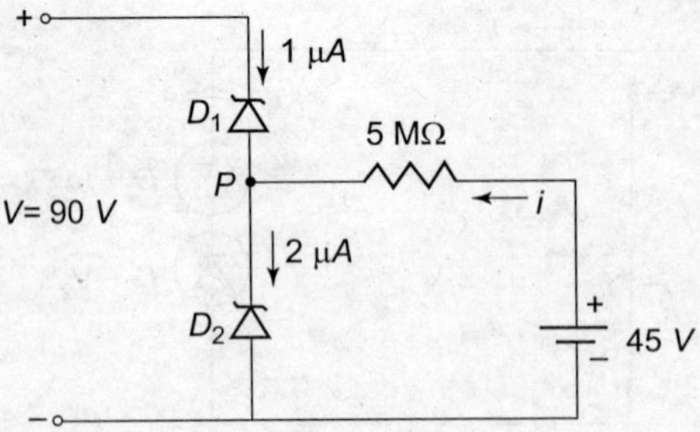

Then circuit becomes

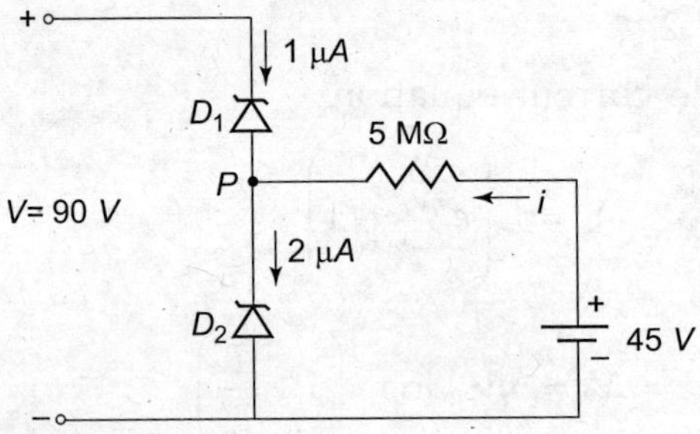

Writing KCL at point, P

$$1 + i = 2 \Rightarrow i = 1\ \mu\text{A}$$

$$V_2 = 45 - 5 \times 1 = 40 \text{ V}$$

$$V_1 = 90 - 40 = 50 \text{ V}$$

$$\text{for , } V = 110 \text{ V, } V_{th} = 110 \times \frac{10}{10+10} = 55 \text{ V}$$

$$V_2 = 55 - 5 \times 1 = 50 \text{ V}$$

$$V_1 = 110 - 50 = 60 \text{ V}$$

3.25 A diode whose V–I characteristics is shown in Figure (a) is connected as shown in (b). Find the current I.

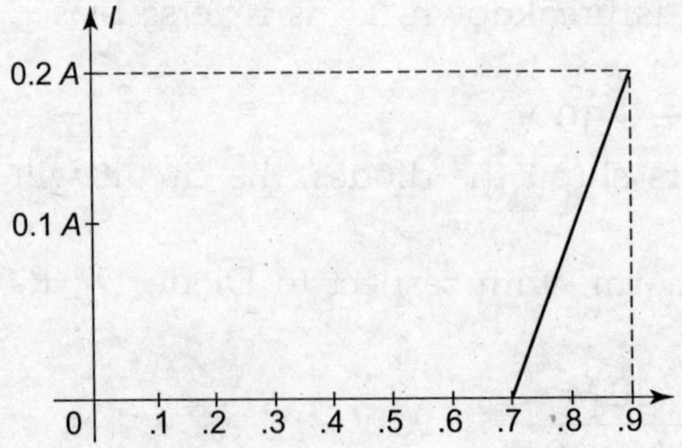

(a)

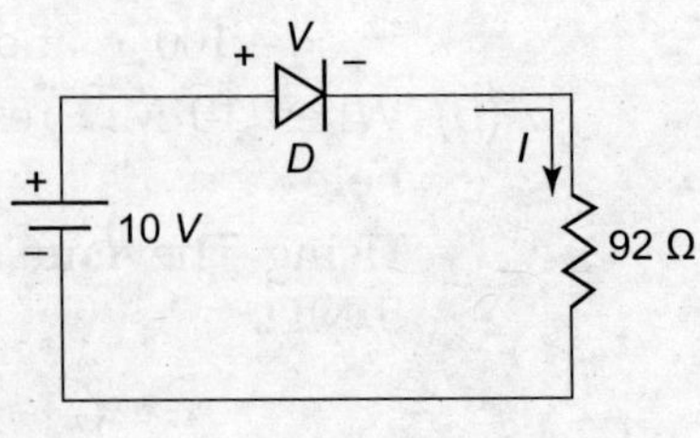

(b)

Solution Equation of *V–I* characteristics for $V > 0.7$ V is

$$I - I_1 = \frac{I_2 - I_1}{V_2 - V_1}(V - V_1)$$

Here, $(V_1, I_1) = (0.7, 0)$ and $(V_2, I_2) = (0.9, 0.2)$

$$I - 0 = \frac{0.2 - 0}{0.9 - 0.7}(V - 0.7)$$

$$\Rightarrow V = I + 0.7 \quad ...(1)$$

Applying KVL to the Fig. (b),

$$10 = (I + 0.7) + 92I$$

$$10 = 93I + 0.7$$

$$\Rightarrow \quad 93I = 9.3 \Rightarrow I = 9.3/93 = 0.1 \text{ A}$$

EXERCISE PROBLEMS

3.1 (a) When a reverse bias is applied to a Ge *p-n* junction diode, the I_0 at room temp is 0.3 μA. When 0.15 V forward bias is applied, find the current flowing.

(b) The reverse saturation current of a Si *p-n* junction diode is 10 μA find the diode current for a forward-bias voltage of 0.6 V at 25 °C.

3.2 Calculate the Barrier capacitance of a Ge p-n junction whose Area is 2 m × 2 mm and space-charge thickness is 2×10^{-4} cm. The dielectric constant of Ge is 16.

3.3 Calculate the anticipated factor by which the reverse saturation current of a Si diode is multiplied when temperature is increased from 25 °C to 80 °C.

3.4 A Ge diode operates at a forward voltage of 0.4 V, calculate the factor by which the current will be multiplied when the temperature is increased from 25 to 150 °C.

3.5 A diode current is 0.8 mA when the applied voltage is 400 mV and 1.0 mA when the applied voltage is 500 mV, find η. Assume $V_T = 26$ mV.

3.6 Find I_1, I_2 and I_{D2} for the network. (Si stands for Silicon).

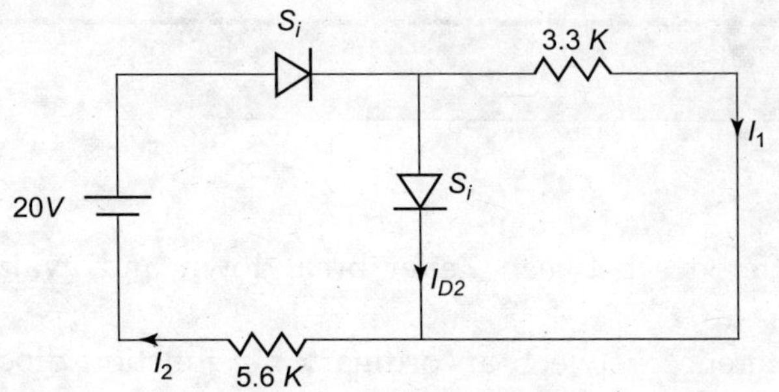

3.7. Determine V_o and I_d for the networks shown below. (Si stands for Silicon).

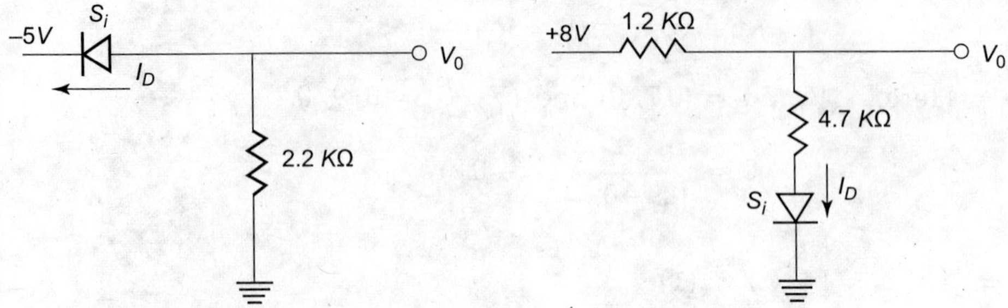

3.8. (a) Determine the currnet flowing through the zones diode for the circuit shown in figure
If R_L = 2 kΩ, V_S = 30 V, R_s = 1000 Ω, V_o = 10 V

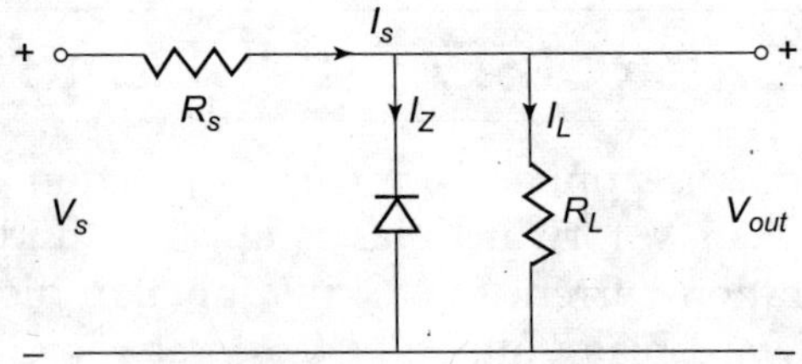

(b) Find the minimum and maximum values of zener current if R_L = 5000 Ω, R_s = 20 kΩ, V_{out} = 30 V and V_s varies between 200 to 220 V.

3.9. For the circuit of above figure, R_s = 2 kΩ, R_L = 5 kΩ, V_s = 60 V and V_{out} = 35 V. Find (a) voltage drop across the series resistance, (b) I_L, (c) I_s (d) I_z.

3.10. Find the temperature coefficient of a 7.5 V zener diode at 25 °C if the nominal voltage increases to 7.8 V at a temperature of 105 °C.

REVIEW QUESTIONS

1. Explain the difference between Zener breakdown and avalanche breakdown in a diode.
2. Explain the difference between an ordinary *p-n* junction diode and a Zener diode.
3. Distinguish between drift current and diffusion current. Discuss in detail the various current components in a forward-biased *p-n* junction diode.

4. What is a tunnel diode? Describe its construction and draw its characteristics.
5. Distinguish between depletion and diffusion capacitances of a *p-n* junction and derive an expression for diffusion capacitance. Discuss why the diffusio capacitance is negligible for a reverse-biased diode.
6. Explain the mechanisms involved in the breakdown of a reverse-biased *p-n* junction. Draw the *V-I* characteristics of a Zener diode and explain its operation.
7. Explain the effect of temperature on diode characteristics.
8. Discuss the switching characteristics of a diode. Derive an expression for the diffusion capacitance.
9. Explain the formation of potential barrier in a *p-n* junction diode. Derive an expression for contact potential. Describe qualitatively how the width of this layer varies with the applied voltage.
10. Explain how a *p-n* junction diode acts as a rectifier.
11. Derive the current equation of a *p-n* junction diode.
12. What are the possible capacitances associated with a *p-n* junction diode? What is the capacitance that is predominant when the diode is reverse biased? Derive its equation.
13. Derive an equation for the capacitance due to charge storage when the *p-n* junction diode is forward biased.
14. What is a varactor? Give its function and equivalent circuit.
15. What are the phenomena that lead to breakdown in a junction diode? Explain them in detail and give the application of such a diode in breakdown.
16. Which diode exhibits negative resistance? Discuss in detail about the function of such a diode with the help of energy-band diagrams.
17. Plot the *V-I* characteristics of a tunnel diode and explain each region of operation in detail. Give the equivalent circuit of the tunnel diode.
18. Explain the characteristics of a photodiode and draw the *V-I* characteristics of such a diode. Explain the sensitivity variation of the photodiode.
19. What is a LED? Give its operation and applications.
20. What is a Schottky diode? What is the basic difference between this diode and a *p-n* junction diode? Give the application of a Schottky diode.
21. What is the difference between a Schottky diode and a point contact diode? Draw the constructional diagram of a point contact diode and give its applications.
22. What are the applications of a PIN diode? Draw the equivalent circuit and *V-I* characteristics of such a diode.
23. How does the reverse saturation current vary with temperature? Give its equation. Give the equation for variation of voltage for fixed current when the temperature changes.
24. How does the dynamic resistance of a diode vary with current? Draw the piecewise linear characteristic of a diode. What is an ideal diode? Discuss its characteristics.
25. Write the equation for the voltage ampere characteristics of a photodiode. Give its circuit symbol and sketch its *V-I* characteristics and sensitivity graph.

CHAPTER 4

Diode Circuits

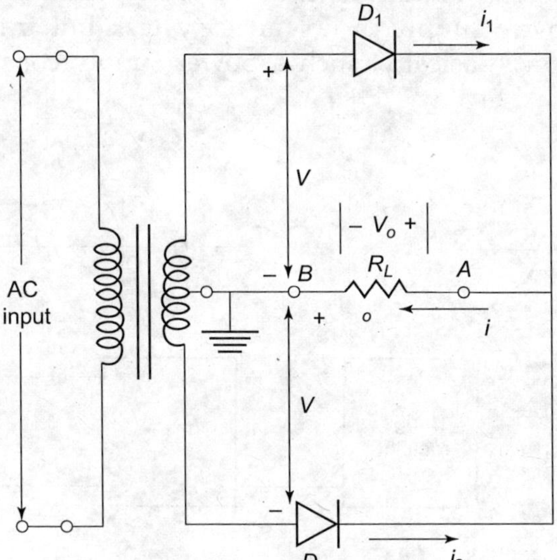

Full-wave rectifier

Chapter Outline

The concepts introduced in this chapter are:

- *Regulated power supply and its applications*
- *Different types of rectifiers and their performance*
- *Different types of filters and their characteristics*
- *Regulator circuits and their stability*

4.1 INTRODUCTION

Active devices used in electronic circuits require a dc power supply in order to be properly biased to operate in the required region. The basic source of dc power supply is the battery. But relying on the battery for the dc power is not possible as it would be very expensive and sometimes not even feasible to have batteries of required capacity for a given application. Commercially available supply is not dc, but ac. Thus, if the ac supply available commercially is converted to dc, it would be very useful to bias all the electronic circuits comfortably. So, one requires an electronic circuit that converts the available ac to dc power supply. It is not sufficient that ac is converted to dc, regulation is also of prime concern. That is, the output of the power supply should be able to maintain a standard rock stable voltage irrespective of variations in load, temperature or input fluctuations. So, the design of such a power supply is not an easy task but involves various stages of careful realisation of circuit. Basically, such a power supply consists of various blocks as shown in Fig. 4.1.

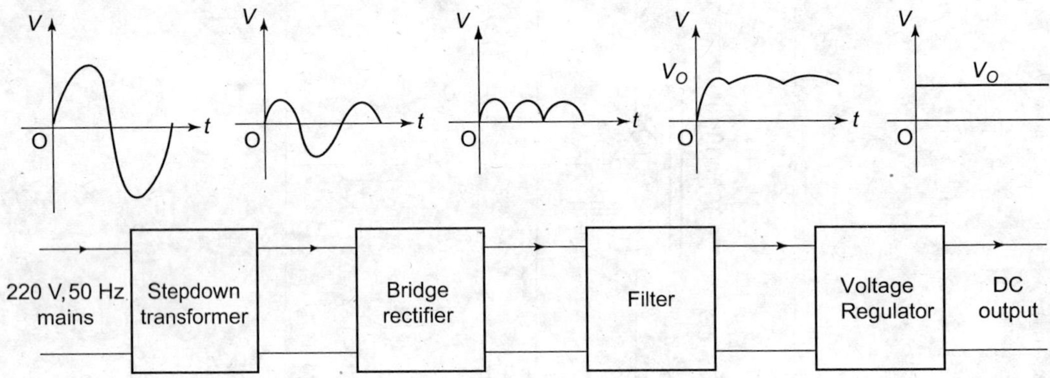

Fig. 4.1 *Block diagram of regulated power supply*

Since the power supply voltage required for biasing an electronic circuit is of the order of 15 V or less, the commercially available 220 V ac supply is to be first stepped down to a lower voltage. A step-down transformer converts the 50 Hz, 220 V (rms voltage) single-phase supply to a very low voltage of the order of 12 V (rms voltage), 50 Hz, single-phase supply. So, the first stage is the selection of appropriate transformer of required rating. The next stage of the supply sequence is the rectifier, which converts bidirectional voltage to unidirectional voltage. This is unidirectional voltage but is still pulsating. In other words, the rectifier converts a voltage of bidirectional nature to unidirectional voltage, but still the voltage is not dc. A *p-n* junction diode is employed in such rectifier circuits. There are various types of rectifier circuits that can be employed at this stage of the power supply. These will be discussed at length in the following sections of this chapter.

The pulsating nature of the unidirectional voltage is to be smoothened and made stable with respect to time to achieve a rock-stable voltage. This is achieved by the use of *filters*, which filter out the ripple in the voltage and make the voltage a stable one. Capacitors, inductors or a combination of these two elements are employed to achieve the required filtration of the ripples in the voltage. Since the circuits involved until this point are active electronic devices whose characteristics depend on temperature, there is a chance that the output voltage may vary with temperature. Also, the output voltage may be varying with respect to input ac supply voltage variations. Not only these, but also the load current may vary the output voltage of the power supply. But, one anticipates a voltage at the output of the power supply to be constant irrespective of any variations. This requires *regulation* of the power supply. A regulator is a device that regulates the voltage and maintains the output stable irrespective of any variations. Regulators can be made either with a transistor and Zener diode combination or *integrated circuit* (IC) *regulators* are directly available these days. A power supply that integrates all of these features is referred to as *regulated power supply.* The block diagram along with signal form at each stage of a *regulated power supply* is shown in Fig. 4.1.

4.2 RECTIFIERS

The *rectifier* is a device that converts a bidirectional power supply to a unidirectional one. *p-n* junction diodes are employed as circuit elements in these circuits. Either one, two or four diodes are employed in this rectification process. There are various types of rectifiers possible and their selection is based on the application. The rectifiers can be broadly classified into *half-wave rectifiers* and *full-wave rectifiers*. Full-wave rectifiers are once again of two types: one employing *centre-tapped transformer* and the other *bridge type.*

There are various parameters that can measure the performance of a rectifier. These are *ac voltage, dc voltage, ripple factor, transformer utilisation factor, peak inverse voltage, regulation, form factor, peak factor, conversion efficiency* and so on. The three types of rectifiers are discussed and they are compared based on the above-mentioned parameters for their performance in the following discussion.

4.2.1 Half-wave Rectifier

The classification of the rectifiers is mainly based on the conduction angle of the diode. If a single diode is employed in the circuit, it conducts in one direction and not in the opposite direction. Thus, for the positive half-cycle of the input it allows the current to flow, while it blocks the current when the input is in the negative half-cycle. Thus, the rectifier is referred to as *half-wave rectifier*. The circuit diagram of such a rectifier is as shown in Fig. 4.2.

As shown in the circuit of Fig. 4.2, the secondary of the step-down transformer is connected to a *p-n* junction diode in series to the load resistance R_L. The voltage developed across the load resistance is the load voltage and assuming the load to be purely resistive, the load voltage and current are linearly dependent. In the positive half-cycle of the input voltage, the potential on *p* side of the diode is higher compared to that on the *n* side, the diode conducts and a current flows in R_L, which develops a linear voltage across the load resistance which is sinusoid. Now, in the second half of the cycle, when *n* side of the diode is at higher potential compared to *p* side, the diode is reverse biased and so, only small negligible current of saturation flows to the load. Since the current is of the order of microamperes or nanoamperes, the corresponding voltage is very small, which can be neglected. Thus, the output voltage in the second half-cycle is

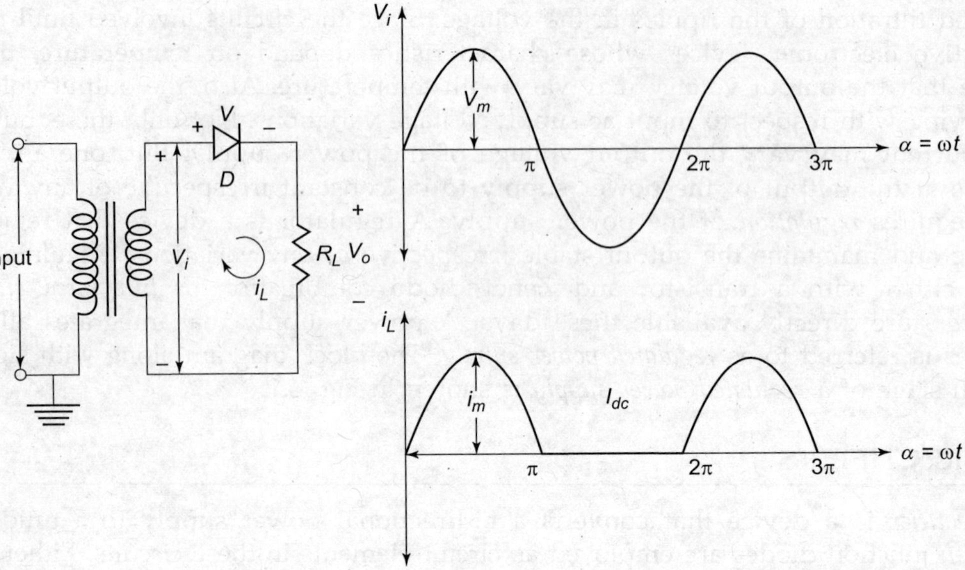

Fig. 4.2 *Half-wave rectifier and associated wave forms*

almost zero. If the diode is assumed to be ideal, it offers infinite resistance and the output voltage is zero and all the potential is dropped across the diode. When the diode is assumed to be ideal in forward bias in the positive half cycle, the drop across the diode would be zero and the total input voltage is transferred as it is to the output. Thus, as seen in the figure, the negative half-cycle voltage is made to be zero in such a rectifier.

The performance of the circuit can now be analysed by the parameters of the rectifier.

ac or rms output voltage: If the input voltage is a sinusoid, it can be represented mathematically as

$$v = V_m \sin(\omega t),\ 0 \le \omega t \le 2\pi \qquad ...(4.1)$$

where V_m is the maximum sinusoid voltage, $\omega = 2\pi f$ = angular frequency. Thus, the output voltage of the rectifier can be given as

$$\begin{aligned} v &= V_m \sin(\omega t), & 0 \le \omega t \le \pi \\ &= 0, & \pi \le \omega t \le 2\pi \end{aligned} \qquad ...(4.2)$$

The *root mean square* (rms) value of a current is defined as *the effective current that dissipates the same amount of energy when a dc current flows through a resistive circuit*. Thus, the rms current in the load is linearly related to rms voltage across the load as $V_m = I_m R_L$. The rms value of a voltage can be given as

$$V_{\text{rms}} = \left[\frac{1}{2\pi}\int_0^{2\pi} V_m^2 \sin^2(\omega t)d(\omega t)\right]^{\frac{1}{2}} \qquad ...(4.3)$$

$$= \left[\frac{1}{2\pi}\int_0^{\pi} V_m^2 \sin^2(\omega t) d(\omega t)\right]^{\frac{1}{2}}$$

$$= V_m/2. \qquad \text{...(4.4)}$$

dc or average output voltage: The *average or* dc content of the voltage across the load is given by

$$V_{dc} = \left[\frac{1}{2\pi}\int_0^{2\pi} V_m \sin(\omega t) d(\omega t)\right] \qquad \text{...(4.5)}$$

$$= \left[\frac{1}{2\pi}\int_0^{\pi} V_m \sin(\omega t) d(\omega t)\right]$$

$$= V_m / \pi \qquad \text{...(4.6)}$$

Ripple factor: The *ripple factor* gives the amount of ripple present in the output voltage wave form of the rectifier. Thus, the *ripple factor* gives the efficiency of the rectifier circuit to suppress the ripple or pulsating nature in the output wave form. It can be defined as the *ratio of ripple voltage to the dc voltage* or *ratio of rms value of ac component to dc value of the wave form.* It can be observed that the ripple factor should be as small as possible or ideally zero for a good design of the rectifier. Thus, ripple factor r can be given as

$$r = V_{r,\text{ rms}}/V_{dc} \qquad \text{...(4.7)}$$

where $V_{r,\text{ rms}}$ is the rms value of the ac component alone in the output wave form. Thus, if $(v - V_{dc})$ represents pure ac component in the wave,

$$V_{r,\text{ rms}} = \left[\frac{1}{2\pi}\int_0^{2\pi} [V_m \sin(\omega t) - V_{dc}]^2 d(\omega t)\right]^{\frac{1}{2}} \qquad \text{...(4.8)}$$

$$= \left[\frac{1}{2\pi}\int_0^{2\pi} [V_m^2 \sin^2(\omega t) + V_{dc}^2 - 2V_{dc}V_m \sin(\omega t)] d(\omega t)\right]^{\frac{1}{2}}$$

$$= [V_{rms}^2 - V_{dc}^2]^{1/2} \qquad \text{...(4.9)}$$

Thus, the ripple factor can be derived to be

$$r = \left[\left(\frac{V_{rms}}{V_{dc}}\right)^2 - 1\right]^{\frac{1}{2}} \qquad \text{...(4.10)}$$

and, for a half-wave rectifier,

$$r = \left[\left(\frac{V_m/2}{V_m/\pi}\right)^2 - 1\right]^{\frac{1}{2}}$$

$$= 1.21. \qquad (4.11)$$

Transformer utilisation factor: In the design of the power supply, it is necessary to determine the rating of the transformer. This is done with knowledge of the dc power to be delivered to the load and the type of rectifying circuit to be used. The *transformer utilisation factor* (TUF) can be defined *as the ratio of dc power delivered to the load to the ac rating of the transformer secondary*. For a half-wave rectifier,

$$\text{TUF} = P_{dc}/P_{ac\ rated} \qquad ...(4.12)$$

$$\text{TUF} = \frac{I_m^2 R_L/\pi^2}{V_m I_m/2\sqrt{2}}$$

where $V_m = I_m(R_f + R_L)$, R_f is the forward dynamic resistance of the diode. Therefore,

$$\text{TUF} = \frac{2\sqrt{2}R_L}{\pi^2(R_f + R_L)} \qquad ...(4.13)$$

$$= 2\sqrt{2}/\pi^2 = 0.287 \qquad ...(4.14)$$

when $R_L >> R_f$.

That is, if 1 kVA is the secondary rating of the transformer, only 287 W power is delivered to the load as dc power. If the output of the transformer is to be effectively converted to dc power, the TUF should be as large as possible, tending towards unity ideally.

Peak inverse voltage: *Peak inverse voltage* (PIV) is defined as *the maximum voltage available across the diode in a rectifier circuit*. It is a well-known fact that the voltage across the diode, when forward biased, is very small, of the order of V_γ, but when the diode is reverse biased, the total input voltage would occur across the diode and in the half-wave rectifier, the maximum voltage is V_m. Thus, *the PIV of a half-wave rectifier is* V_m. The significance of this factor is that it gives the maximum voltage that occurs across the diode in the circuit and thus the selection of diode depends on this factor. One has to choose such a diode in the circuit which can withstand the voltage equal to PIV and no breakdown occurs. That is, the diode's maximum rating should be more than that of the PIV, else the diode would breakdown or it may be totally damaged. Thus, the PIV gives the maximum voltage rating of a diode to be selected in the rectifier circuit.

Regulation: The variation of the dc output voltage as a function of dc load current is called *regulation*. The percentage regulation is defined as

$$\%\ \text{Regulation} = \frac{V_{no\ load} - V_{load}}{V_{load}} \times 100\ \% \qquad ...(4.15)$$

where $V_{\text{no load}}$ is the dc voltage at zero load current, that is, when $R_L = \infty$ and V_{load} is the dc voltage when normal load R_L is connected. For an ideal power supply, the output voltage is independent of the load and thus the % regulation is zero.

Form factor: *Form factor* is defined as the *ratio of rms value to the average value of the voltage.* Thus, for a half-wave rectifier,

$$\text{Form factor} = V_{\text{rms}}/V_{\text{dc}} = (V_m/2)/(V_m/\pi) = 1.57 \quad \text{...(4.16)}$$

Peak Factor: *Peak factor* is defined as the *ratio of peak value to the rms value of the voltage.* Thus, for a half-wave rectifier,

$$\text{Peak factor} = V_m/V_{\text{rms}} = (V_m)/(V_m/2) = 2 \quad \text{...(4.17)}$$

Conversion efficiency: *Conversion efficiency* or *rectification efficiency* or *ratio of rectification* is defined as *the ratio of the output dc power to input ac power.*

$$\text{Conversion efficiency} = \eta = \frac{\text{dc power delivered to load}}{\text{ac input power}} \quad \text{...(4.18)}$$

Thus,

$$\eta = P_{\text{dc}}/P_{\text{ac}} \quad \text{...(4.19)}$$

For a half-wave rectifier,

$$P_{\text{dc}} = I_{\text{dc}}^2 R_L = I_m^2 R_L/\pi^2$$

$$P_{\text{ac}} = I_{\text{rms}}^2 (R_f + R_L) = I_m^2 (R_f + R_L)/4$$

Therefore,

$$\eta = \frac{4}{\pi^2}\frac{R_L}{R_f + R_L}$$

$$= 0.406/(1 + R_f/R_L) \quad \text{...(4.20)}$$

When $R_f \ll R_L$, the maximum efficiency is 0.406. Or, in terms of percentage, the *maximum efficiency of a half-wave rectifier is 40.6%.*

4.2.2 Full-wave Rectifier

As already seen, the classification of rectifiers is based on the fire angle of conduction of the diode. In the above discussion only a single diode is considered and so, only in one half of the total cycle the diode is conducting and the circuit is referred to as half-wave rectifier. Half of the cycle, the output is zero and so, the performance of the circuit is very poor as seen by the parameter measured for half-wave rectifier from the above. To improve the performance of the rectifier circuits, rectification over the total cycle of the wave form is required. That is, the output should not be zero in one half-cycle, but if one is able to fold back the negative half-cycle of the wave to positive direction, the performance of the circuit would increase to a very large extent. This feat is achieved by the *full-wave rectifiers*. There are two types of full-wave rectifiers. They are classified as *full-wave rectifier with centre-tapped transformer* and *bridge rectifier*. The two have their own advantages and drawbacks and their analysis is taken up in detail in the following sections.

Centre-tappped Transformer Full-wave Rectifier

Figure 4.3 shows a *full wave rectifier* that employs a centre-tapped transformer along with the currents in the two diodes and also the output current in load.

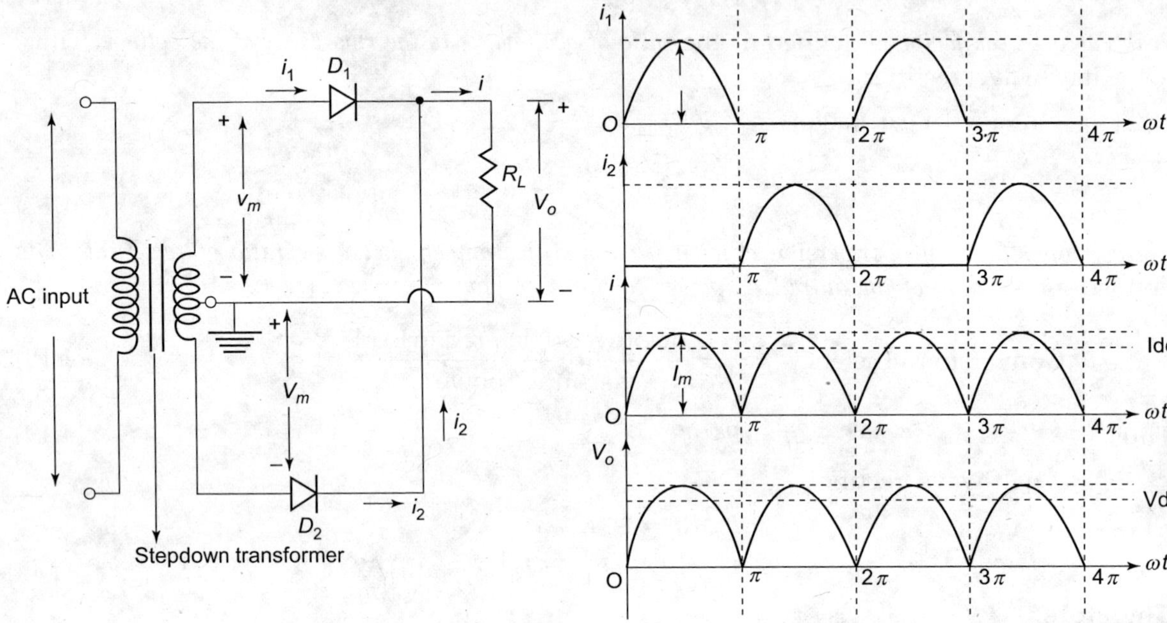

Fig. 4.3 *Full-wave rectifier and its wave forms*

The circuit consists of two identical diodes, and a load resistance. The ac wave form is delivered by a centre-tapped transformer whose windings are shown in the figure, with the centre tap being at ground potential. So, the rating of the transformer should be twice that of the secondary half-wave rectifier. The circuit seems to be two secondary half-wave rectifiers connected so that conduction takes place through one diode during one half of the cycle and through the other in the other half-cycle. In the positive half-cycle of the input wave form, p side of the diode D_1 is more positive than the n side and so, it is forward biased and conducts. At the same time, the diode D_2 is reverse biased since p side of the diode is at lower potential compared to n side. Thus, it is reverse biased and so no conduction takes place. So, the current i_1 in diode D_1 flows through the load resistance in this half cycle.

In the negative half-cycle, the condition of potentials is reversed to that of the above case and so the diode D_1 is reverse biased and the diode D_2 is forward biased. Now, a current i_2 flows through the diode D_2 that flows through the load resistance. So, whichever may be the conducting diode, the current in the load flows in the same direction. Thus, the current in the load has a form as shown in Fig. 4.3, which is as if the voltage in the negative half cycle of the input is folded to positive side at the output. Thus, the zero voltage region of half-wave rectifier is now filled with a voltage, which is also positive. It can be observed that the frequency of the output wave form is doubled compared to that of the input signal. It may be observed that in the half-wave rectifier the input as well as the output wave form has the same frequency. Since the load is linear, the voltage across the R_L would be of the same shape as that of the current through it. To estimate the performance of this circuit, the parameters are analysed as follows.

ac or rms output voltage: Thus, the output voltage of the full-wave rectifier repeats for every π radians and so, can be given as

$$\begin{aligned} V &= V_m \sin(\omega t), \qquad 0 \le \omega t \le \pi \\ &= V_m \sin(\omega t), \qquad \pi \le \omega t \le 2\pi \end{aligned} \quad \text{...(4.21)}$$

The rms value of a current is defined as *the effective current that dissipates the same amount of energy when an ac current flows through a resistive circuit*. Thus, the rms current in the load is linearly related to rms voltage across the load as $V_m = I_m R_L$. The rms value of a voltage can be given as

$$\begin{aligned} V_{\text{rms}} &= \left[\frac{1}{2\pi} \int_0^{2\pi} V_m^{\,2} \sin^2(\omega t)\, d(\omega t) \right]^{1/2} \\ &= \left[\frac{2}{2\pi} \int_0^{\pi} V_m^{\,2} \sin^2(\omega t)\, d(\omega t) \right]^{1/2} \\ &= V_m/\sqrt{2} \end{aligned} \quad \text{...(4.22)}$$

dc or average output voltage: The *average or dc* content of the voltage across the load for a full-wave rectifier is given by

$$\begin{aligned} V_{\text{dc}} &= \left[\frac{1}{2\pi} \int_0^{2\pi} V_m \sin(\omega t)\, d(\omega t) \right] \\ &= \left[\frac{2}{2\pi} \int_0^{\pi} V_m \sin(\omega t)\, d(\omega t) \right] \\ &= 2V_m/\pi \end{aligned} \quad \text{...(4.23)}$$

Ripple factor: From Eq. (4.10), the ripple factor for a full-wave rectifier can be derived to be

$$r = \left[\left(\frac{V_m/\sqrt{2}}{2V_m/\pi}\right) - 1\right]^{½}$$

$$= 0.482 \qquad \text{...(4.24)}$$

Transformer utilisation factor: The transformer utilisation factor for a full-wave rectifier using centre-tapped transformer is determined by considering the primary and secondary windings separately. Its value can be derived to be 0.693.

Peak inverse voltage: *Peak inverse voltage* (PIV) is defined as *the maximum voltage available across the diode in the rectifier circuit.* As can be seen from the circuit, applying KVL to the outer loop when the diode *D*1 is forward biased and diode *D*2 is reverse biased, the voltage across the diode *D*2 would be the sum of load voltage and the voltage across the second half of the primary. Thus, the voltage across this reverse biased diode would be double to that of the one section of secondary of the transformer. Thus, the maximum voltage across a reverse biased diode is not V_m, as in the case of a half-wave rectifier, but it is $2V_m$. *Thus, the PIV of a full-wave rectifier with centre-tapped transformer is* $2V_m$. One has to choose such a diode in the circuit which can withstand the voltage equal to $2V_m$. That is, the maximum voltage rating of the diodes should be more than $2V_m$.

Form factor: *Form Factor* is defined as the *ratio of rms value to the average value of the voltage.* Thus, for a full-wave rectifier,

$$\text{Form factor} = V_{\text{rms}}/V_{\text{dc}}$$

$$= (V_m/\sqrt{2})/(2V_m/\pi)$$

$$= 1.11. \qquad \text{...(4.25)}$$

Peak factor: *Peak factor* is defined as the *ratio of peak value to the rms value of the voltage.* Thus, for a full-wave rectifier,

$$\text{Peak factor} = V_m/V_{\text{rms}}$$

$$= (V_m)/(V_m/\sqrt{2})$$

$$= \sqrt{2} \qquad \text{...(4.26)}$$

Conversion efficiency: *Conversion efficiency* or *rectification efficiency* or *ratio of rectification* is defined as *the ratio of the output dc power to ac input power,*

$$\text{Conversion efficiency} = \eta = \frac{\text{dc power delivered to load}}{\text{ac input power}}$$

Thus,

$$\eta = P_{\text{dc}}/P_{\text{ac}}$$

For a full-wave rectifier,

$$P_{\text{dc}} = I_{\text{dc}}^2 R_L = 4I_m^2 R_L/\pi^2$$

$$P_{\text{ac}} = I_{\text{rms}}^2 (R_f + R_L) = I_m^2\,(R_f + R_L)/2$$

Therefore,

$$\eta = \frac{8}{\pi^2}\frac{R_L}{R_f + R_L}$$

$$= 0.812/(1 + R_f/R_L) \qquad \text{...(4.27)}$$

When $R_f \ll R_L$, the maximum efficiency is 0.812. Or, in terms of percentage, the *maximum efficiency of a full-wave rectifier is 81.2%.*

Bridge Rectifier

The requirement of centre-tapped transformer can be avoided when four diodes are used instead of two. The circuit employing four diodes and an ordinary transformer is similar to that of a bridge circuit and thus the name *bridge rectifier*. The output wave form of this circuit is same as that of full-wave rectifier with centre-tapped transformer. Thus, the performance of such a circuit is almost similar to that of the previous circuit. The circuit of *bridge rectifier* along with the wave forms is as shown in Fig. 4.4.

Four diodes are employed in this bridge circuit. Two diodes at a time are in conduction and the other two are in the reverse-bias condition. Thus, in the first half-cycle of the input signal, diodes *D*1 and *D*3 are in conduction while the diodes *D*2 and *D*4 are reverse biased. Thus, the current in the circuit will flow in the path shown in figure, through the diode *D*1, R_L and diode *D*3 back to secondary of the transformer. In the second half-cycle, the polarities of the voltages are exactly reversed and thus diodes *D*2 and *D*4 are in conduction now while the diodes *D*1 and *D*3 are back biased. The path of the current is through the diode *D*2, R_L, *D*4 and back to transformer. In this case also, the direction of current in the load is same as the previous case and thus the polarity of voltage in the load is not changed. Thus, the voltage developed across the load resistance is same as that of full-wave rectifier. To achieve the same output, the transformer secondary need not be centre-tapped and lower rating is sufficient. The winding of the

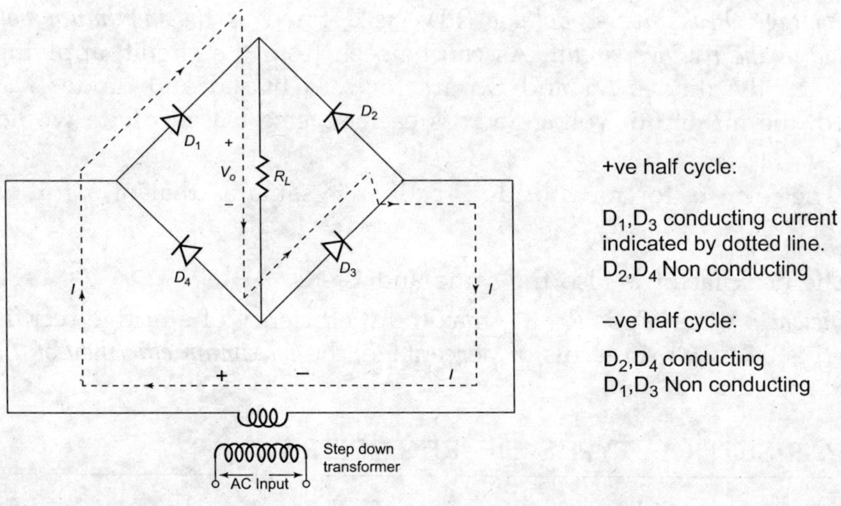

Fig. 4.4 *Bridge rectifier*

secondary should develop twice the required voltage in case of a full-wave rectifier but the rating of the secondary can be far reduced when a bridge rectifier is used, since rating of secondary is same as the voltage required.

The performance of the circuit is almost similar to the full-wave rectifier since the output wave form is same for both the circuits. Thus, the analysis can be presented as follows.

ac or rms output voltage: The rms voltage is same as that of the full-wave rectifier and thus,

$$V_{\text{rms}} = V_m/\sqrt{2} \qquad \text{...(4.28)}$$

dc or average output voltage: The dc voltage is also same as that of the full-wave rectifier and thus,

$$V_{\text{dc}} = 2V_m/\pi \qquad \text{...(4.29)}$$

Ripple factor: The ripple factor is the ratio of rms voltage of ac component to dc voltage and thus,

$$r = 0.482 \qquad \text{...(4.30)}$$

Transformer utilisation factor: The current flowing in the secondary of the transformer is pure sinusoid,

$$\text{TUF} = \frac{4I_m^2 R_L/\pi^2}{I_m^2(2R_f + R_L)/2}$$

R_f is the forward dynamic resistance of the diode. Therefore,

$$\text{TUF} = \frac{8}{\pi^2}\frac{R_L}{2R_f + R_L} \qquad \text{...(4.31)}$$

$$= 0.812 \qquad \text{...(4.32)}$$

when $R_L \gg R_f$.

Peak inverse voltage: *Peak inverse voltage* (PIV) is defined as *the maximum voltage available across the diode in the rectifier circuit.* As can be seen from the circuit, applying KVL to the outer loop when the diodes *D*1 and *D*3 are forward biased and diodes *D*2 and *D*4 are reverse biased, the maximum voltage across each reverse biased diode would be equal to V_m.

Form factor: The form factor for a bridge rectifier is same as that of a full-wave rectifier, 1.11.

Peak factor: The peak factor is also the same and is $\sqrt{2}$.

Conversion efficiency: When $R_f \ll R_L$, the maximum efficiency of a bridge rectifier, like a full-wave rectifier, is 0.812. Or, in terms of percentage, the *maximum efficiency of a bridge rectifier is 81.2%.*

4.3 COMPARISON OF TYPES OF RECTIFIERS

Table 4.1 gives the consolidated performance of all the three types of rectifiers discussed above.

Table 4.1 ***Comparison of Rectifiers***

Particulars	Type of rectifier		
	Half-wave	*Full-wave*	*Bridge*
No. of diodes	1	2	4
Max. efficiency	40.6%	81.2%	81.2%
V_{dc}	V_m/π	$2V_m/\pi$	$2V_m/\pi$
V_{rms}	$V_m/2$	$V_m/\sqrt{2}$	$V_m/\sqrt{2}$
Avg. current/diode	I_{dc}	I_{dc}	I_{dc}
Ripple factor	1.21	0.48	0.48
PIV	Vm	2Vm	Vm
Output frequency	f	2f	2f
TUF	0.287	0.693	0.812
From factor	1.57	1.11	1.11
Peak factor	2	$\sqrt{2}$	$\sqrt{2}$

As can be seen from Table 4.1, the performance of a half-wave rectifier is very inferior to that of a full-wave rectifier and so is not, in general, used in practice. So, the performance of two full-wave rectifiers is utilised in general.

The full-wave rectifier has an advantage over the bridge circuit in the sense that only two diodes are employed instead of four. Considering a diode in practice, it has a voltage drop of V_γ or is slightly more dropped across it when it is forward biased. In a full-wave rectifier, only one forward biased diode is in series with load and so the output voltage is decreased by V_γ. Whereas, in a bridge circuit, two forward biased diodes are in series with the load and so the voltage drop is $2V_\gamma$. Thus, in the bridge circuit, the output voltage is small compared to the full-wave circuit. One end of the load is grounded in a full-wave circuit, whereas no end of load is grounded in a bridge circuit. Thus, a bridge circuit cannot be employed when a grounded load is to be applied to the rectifier.

The bridge rectifier circuit has better advantages compared to the full-wave circuit and thus it is generally employed wherever possible. Centre tap of the transformer is not required. Having a centre tap in the transformer may be a great disadvantage because if the exact centre tap is not provided, the voltage in the two half-cycles of the output would not be same and so, the total wave form is disturbed. Also, the windings of the required secondary would be complex compared to a simple winding as required by bridge circuit. The secondary rating is small. Thus, the transformer is utilised to maximum extent. Diodes with lower maximum voltage rating can be employed in a bridge rectifier since PIV is almost half of that of a full-wave rectifier.

4.4 FILTERS

The output of the rectifying circuit consists of a dc component as well as an ac component of the voltage. This is due to the pulsating nature of the output of any rectifier circuit. But, for a dc power supply, one requires that the output of the supply should not have any ripple, that is, the ac component should be zero. This ripple is of the order of 48% in a full-wave rectifier and in a half-wave rectifier it is much higher, of the order of 121%. Before using this power supply, the ac component in the wave form should be filtered out totally. This can be achieved by employing appropriate filters in cascade with rectifying circuits. The basic components of filters are the *inductors* and *capacitors*. Various possible configurations of the filters are possible and in the following few of them are considered and analysed for their performance. Some of the important filters are the following:

- Capacitor filter
- Inductor filter
- LC or *L* section filter
- CLC or π filter

4.4.1 Capacitive Filters

The capacitor filter is suitable for light loads and is directly connected across the load, as shown in Fig. 4.5. The capacitor, which is uncharged initially, on application to the output of the rectifier gets charged as the input voltage rises at the rate of the input rise and so the capacitor voltage follows the input for the first quarter cycle of the wave until a maximum voltage V_m is reached. Beyond this point, the action of capacitance comes into play. The capacitor cannot discharge at the same rate of the input and it discharges exponentially at a time constant R_LC. If the value of C is chosen to be very high, the capacitor discharge takes a very long time and exponentially decays to zero in general. But, as seen from the graph of Fig. 4.5, as the capacitor voltage decreases, it encounters at a point the rising potential of the second half-cycle of the wave. The diode is forward biased from zero to V_m and when once the capacitor holds the voltage, the diode becomes reverse biased since the p side of the diode is at lower potential. So, the diode is reverse biased from the peak voltage to the point where the capacitor discharge voltage is equal to that of the second cycle voltage. At this juncture the diode is once again forward biased and capacitor gets charged to V_m following the input wave. Then once again the capacitor holds the voltage and the diode is reverse biased. Thus, as can be seen, the diode is in conduction only for a very small angle. The output voltage of the capacitor is as shown in Fig. 4.5.

When the capacitor value C is chosen to be high, the exponential decay takes a long time and thus the firing angle of the diode is reduced to a very large extent. If the exponential decay is assumed to be linear in this condition, the output wave form is triangular with a very small peak-to-peak voltage so that it can almost be approximated to a straight line. Thus, higher the time constant R_LC, the smaller the ripple in the wave. On the other hand, if the load is large with a smaller R_L value, the ripple in the wave form increases. The point at which the diode starts to conduct is called *cutin point* and that at which it stops conducting is called *cutout point*.

The capacitor voltage and the diode current are redrawn for analysis in Fig. 4.6. With reference to this figure, the cutout angle is determined. When the diode conducts, $v_0 = V_m \sin(\omega t)$ and I is the sum of the load resistor current I_L and the capacitor current I_C. Hence,

$$i = \frac{V_0}{R_L} + C\frac{dV_0}{dt} = \frac{V_m}{R_L}\sin\omega t + \omega C V_m \cos\omega t \qquad \text{...(4.33)}$$

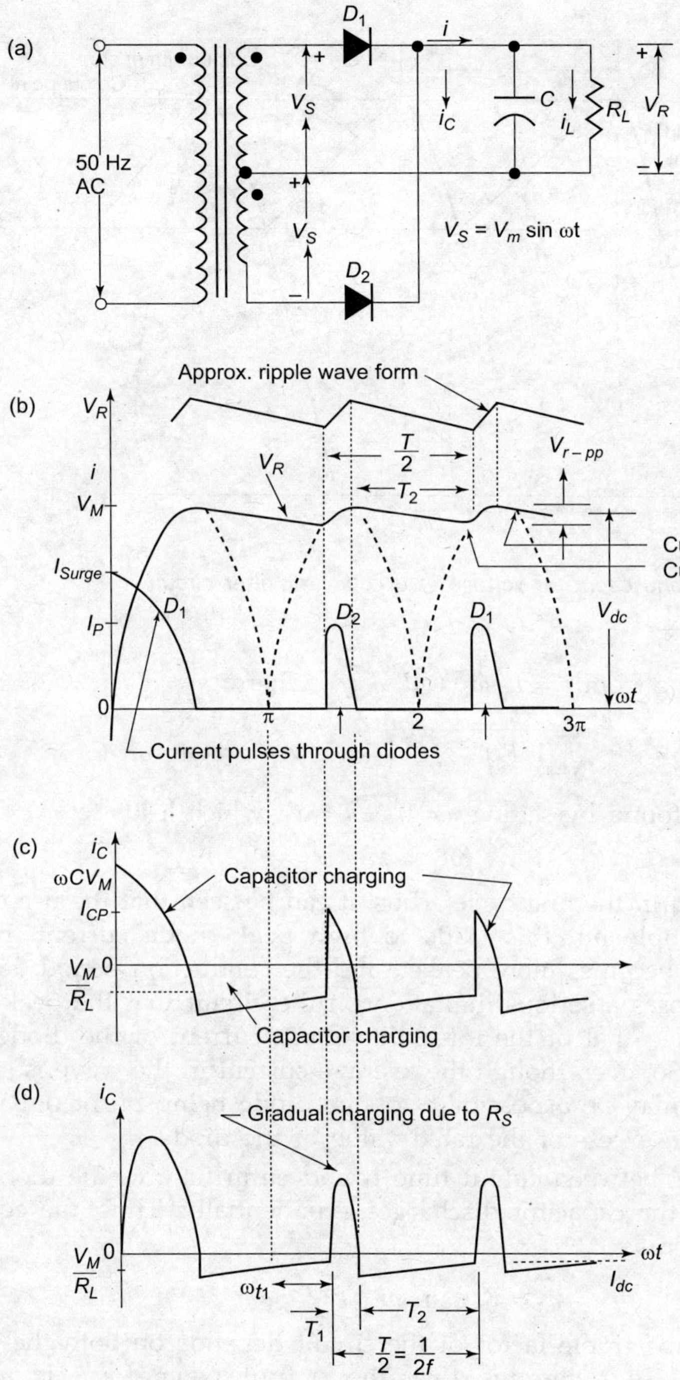

Fig. 4.5 *Capacitor filter and associated wave forms*

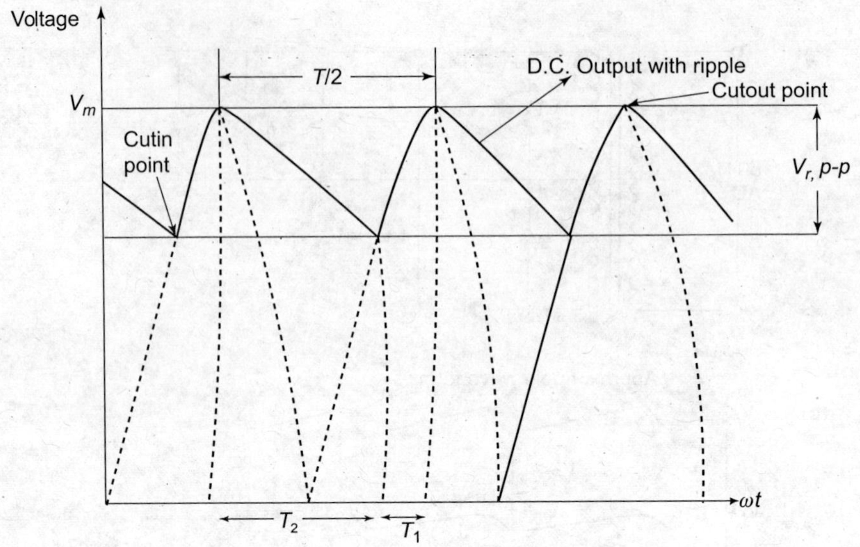

Fig. 4.6 *Diode current and capacitor voltage in a capacitor filter circuit*

This current is of the from $i = I_m \sin\,(\omega t + \psi)$, where

$$I_m = V_m \sqrt{[(1/R_L^2)+\omega^2C^2]} \text{ and } \psi = \arctan \omega R_L C \qquad \text{...(4.34)}$$

The cutout time is found by setting $i = 0$ at $t = t_1$, which leads to

$$\omega t_1 = \pi - \psi \qquad \text{...(4.35)}$$

for the cutout angle ωt_1 in the first cycle. Thus, it can be seen that the use of large capacitance decreases the ripple but this leads to high peak diode current. For a specified average load current, i becomes more peaked and the conduction period decreases as C is made larger. This imposes a serious limitation on the circuit that the peak current in the diode may be more than that of the rated maximum current of the diode when a large value of C is selected. So, even though the average current in the wave is small compared to diode rating, there may be a possibility of the diode being burnt out when the peak current is very large, in excess of the rated value of the diode.

In the time interval between cutout time t_1 and cutin time t_2, the diode is effectively out of the circuit and the capacitor discharges exponentially. Thus, the equation for this wave form can be given as

$$v_0 = (V_m \sin\,\omega t_1)\, e^{-(t-t_1)/CR_L} \qquad \text{...(4.36)}$$

It may observed that the ripple factor of the circuit depends on both the load resistance and capacitor values. To determine this, the output wave form is assumed to be triangular. From cutin point to cutout point, whatever the charge the capacitor acquires is equal to the charge the capacitor has lost during the period of nonconduction, that is, from the cutout point to the next cutin point.

$$\text{Charge acquired by the capacitor} = V_{r,\,p-p} \times C$$

$$\text{Charge lost by the capacitor} = I_{dc} \times t_2$$

Therefore,
$$V_{r,\,p-p} \times C = I_{dc} \times t_2 \qquad ...(4.37)$$

With an assumption of triangular wave form and that T_2 is half of the total period T,

$$T_2 = T/2 = 1/2f$$

And so,
$$V_{r,\,p-p} = I_{dc}/(2fC) \qquad ...(4.38)$$

For a triangular wave form,

$$V_{r,\,\text{rms}} = V_{r,\,p-p}/\left(2\sqrt{3}\right) \qquad ...(4.39)$$

Thus,
$$V_{r,\,\text{rms}} = I_{dc}/\left(4\sqrt{3}fC\right)$$

$$= V_{dc}/\left(4\sqrt{3}fCR_L\right) \qquad ...(4.40)$$

The ripple factor r can be given as

$$r = \frac{1}{4\sqrt{3}fCR_L} \qquad ...(4.41)$$

Thus, *the ripple factor may be decreased by increasing either R_L or C or both.*

As seen from the above discussion, selection of a large value of capacitance has many effects on the performance of the circuit. They can be consolidated as follows:

- dc load voltage approaches the limiting value of V_m.
- Peak-to-peak ripple voltage becomes less.
- The duration of the current pulses through the diode diminishes.
- The peak current in the diode increases.

4.4.2 Inductor Filters

The inductor filter is a choke in series with the load, as shown in Fig. 4.7. The characteristic of an inductor is to oppose any change in current through it. Or, it blocks the ac and allows only the dc component to reach the load. This inductor in series has a smoothening effect as shown in the figure. The voltage behaviour across a capacitor is same as current behaviour through an inductor. Thus, the capacitor is in shunt with the load whereas the inductor is in series with the load. These filters are employed when the load is heavy, that is, when the R_L is a small value.

As one can anticipate, the ripple factor in the output wave form depends on the R_L and L values. Ripple factor would decrease with decreasing R_L or increasing L value. The equation for the ripple factor is derived in the following.

For simplicity, the diode, choke and transformer resistances are neglected compared to R_L. Then, for dc component of current, $I_m = V_m/R_L$. But the impedance of L and R_L in series with ac component has a magnitude at 2ω of

$$Z = \sqrt{\left[R_L^2 + 4\omega^2L^2\right]} \qquad ...(4.42)$$

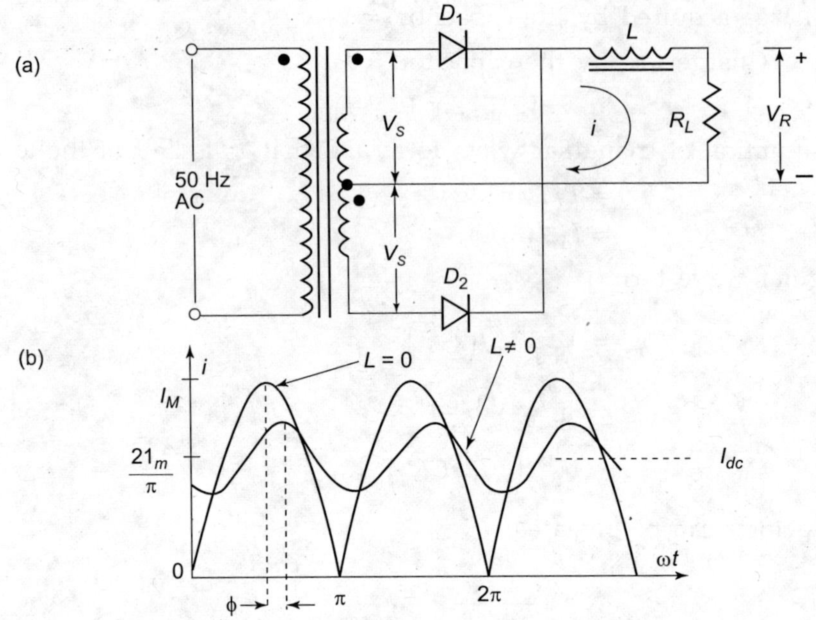

Fig. 4.7 *Inductor filter and associated wave forms*

Thus, for ac, $$I_m = V_m \Big/ \sqrt{\left[R_L^2 + 4\omega^2 L^2\right]} \qquad \text{...(4.43)}$$

And the resulting load current is given by

$$i \approx \frac{2V_m}{\pi R_L} - \frac{4V_m}{3\pi\sqrt{[R_L^2 + 4\omega^2 L^2]}} \cos(2\omega t - \phi) \qquad \text{...(4.44)}$$

where ϕ is the angle by which the load current lags behind the voltage and is given by

$$\phi = \tan^{-1}(2\omega L/R_L). \qquad \text{...(4.45)}$$

Using the definition of ripple factor as the ratio of the rms value of the ac component to dc value,

$$r = \frac{4V_m \Big/ \left\{3\pi\sqrt{2}\sqrt{R_L^2 + 4\omega^2 L^2}\right\}}{2V_m/[\pi R_L]} \qquad \text{...(4.46)}$$

$$= R_L/[\omega L 3\sqrt{2}] \qquad \text{...(4.47)}$$

if $4\ \omega^2 L^2/R_L{}^2 \gg 1$.

This equation shows that the ripple factor decreases as L is increased or R_L is decreased or both. As R_L tends to infinity, the value of r tends to $2/3\sqrt{2} = 0.471$. That is, at no load the ripple factor improvement is very small, from 0.482 to 0.471. The difference is due to the omission of higher harmonics. Thus, the inductor filter is of no significant use at large load resistance values. It is used when the load resistance is consistently small, that is, where the load current is high.

The advantages of the choke filter are the following:

- Current flows continuously in the inductor and so efficient use of transformer is achieved.
- Ripple content in the output is not only low but is also less dependent on the load current.

4.4.3 Other Types of Filters

There are two more important filters to be considered in this section. They are LC or *L section* or *choke input filter* and CLC or *pi* or *π filter*.

L section filter: In the previous discussions, we have observed that the ripple factor is dependent on the load resistance. The ripple factor is inversely proportional to load resistance in a capacitor filter, whereas in an inductor filter, the ripple factor is directly proportional to load resistance. So, if a filter with a combination of an inductor and a capacitor is realised such that the ripple factor does not depend on load and is independent of the R_L value, it is most convenient in design of power supplies. Such a filter is LC or *L section* or *choke input filter*.

The circuit of a L section filter is as shown in Fig. 4.8. If L is small, the capacitor will be allowed to charge up to V_m and cut off the diodes, allowing only short pulses of current. As L is increased, the pulses of current are smoothed and made to flow for a longer period but at a reduced amplitude until at some critical value of inductance, L_c, either one diode or the other is always conducting, with the result that the input voltage v and current i are full-wave rectified sine waves.

With suitable assumptions, the analysis can be simplified and the voltage across the load can be derived as

$$v = \frac{2V_m}{\pi} - \frac{4V_m}{3\pi}\cos 2\omega t \qquad \text{...(4.48)}$$

Thus, dc component can be given as

$$V_{dc} = (2V_m/\pi) - I_{dc}R$$

The ripple current that does pass through L is not allowed to develop much ripple voltage across R_L if the reactance of C at the ripple frequency is small compared with R_L. Thus, for a properly deigned LC filter,

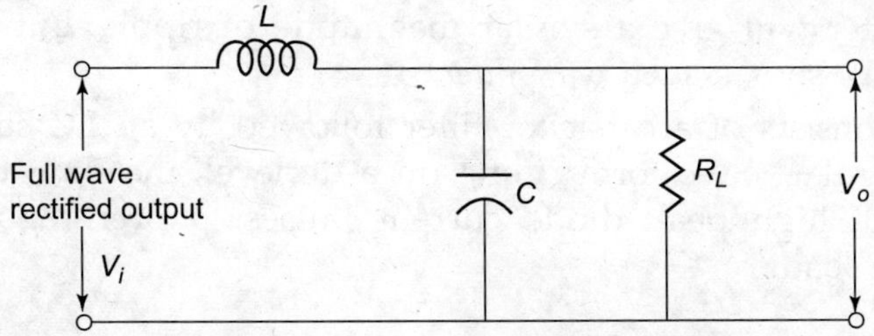

Fig. 4.8 *L section filter*

$$X_C \ll R_L \text{ and } X_L \gg X_C \text{ at } 2\omega = 4\pi \qquad \text{...(4.49)}$$

The rms value of ripple current is

$$I_{r,\,\text{rms}} = 4V_m/[3\pi X_L\sqrt{2}]$$
$$= V_{dc}\sqrt{2}/(3X_L) \qquad \text{...(4.50)}$$

The rms voltage of the ac component is

$$V_{r,\,\text{rms}} = \frac{X_C\sqrt{2}}{3X_L}V_{dc} \qquad \text{...(4.51)}$$

Therefore, the ripple factor is

$$r = \frac{\sqrt{2}}{3}\frac{1}{4\omega^2 CL} \qquad \text{...(4.52)}$$

Thus, *the ripple factor does not depend on the load.*

PI filter: Better filtering can be achieved using two or more L section filters, as shown in Fig. 4.9. These can be referred to as *multiple LC filters.*

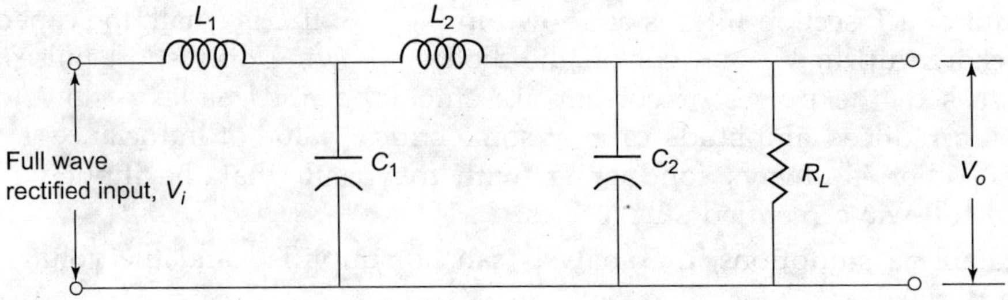

Fig. 4.9 *Multiple LC filter*

The ripple factor for such a filter can be given as

$$r = \frac{\sqrt{2}}{3}\frac{X_{C_2}}{X_{L_2}}\frac{X_{C_1}}{X_{L_1}} \qquad \text{...(4.53)}$$

The CLC or PI filters are used when the additional dc voltage drop across L cannot be tolerated. A CLC filter will give a similar magnitude of ripple, but with poorer voltage regulation. Figure 4.10 shows such a *PI filter*.

The CLC filter consists of a capacitor filter followed by an LC section. Consequently, v consists of a triangular wave form riding on a dc level, meaning that this filter brings back the problem of high peak diode current pulses, poorer regulation and a ripple dependent upon the load.

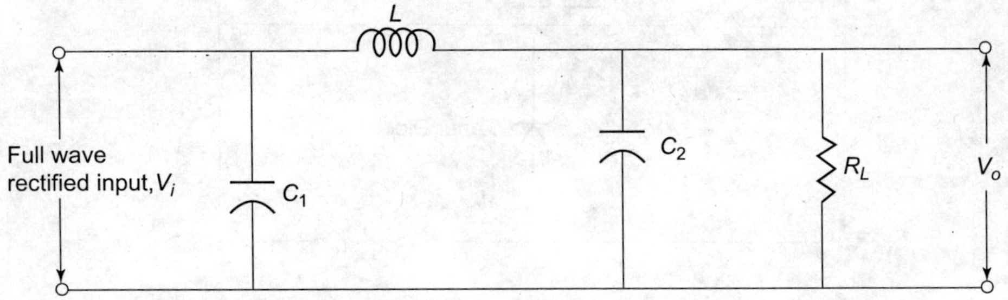

Fig. 4.10 *A PI filter*

$$r = \sqrt{2}\,\frac{X_C}{R_L}\,\frac{X_{C_1}}{X_{L_1}} \qquad ...(4.54)$$

Under some conditions of light load currents it is possible to replace L_1 with resistor R equal in value to the reactance of L_1 at the ripple frequency without changing the ripple factor. However, the increased resistance to dc current flow will make this impractical except where the load current is very small.

4.5 VOLTAGE REGULATORS

An *ideal regulated power supply* is an electronic circuit designed to provide a predetermined dc voltage V_0 that is independent of the current I_L drawn from V_0, of the temperature, and also of any variations in the ac line voltage. An unregulated power supply consists of a transformer, a rectifier and a filter.

The reasons for regulated power supply are the following:

- Poor regulation of unregulated power supply with load.
- dc output varies with ac input.
- dc output varies with temperature due to active devices used in the unregulated power supply.

There are various types of voltage regulators and of these some of them are considered here.

Zener diode shunt regulator: A Zener diode, under reverse bias breakdown condition, can be used to regulate the voltage across the load, irrespective of the supply voltage or load current variations. A simple Zener voltage regulator circuit is shown in Fig. 4.11. The Zener diode is selected with V_z equal to the voltage desired across the load. The Zener diode has a characteristic that under reverse bias condition, the voltage across it practically remains constant, even if the current through it changes by a large extent. Under normal conditions, the input current $I_i = I_L + I_z$ flows through the resistor R. The input voltage V_i can be written as $V_i = I_i R + V_z$.

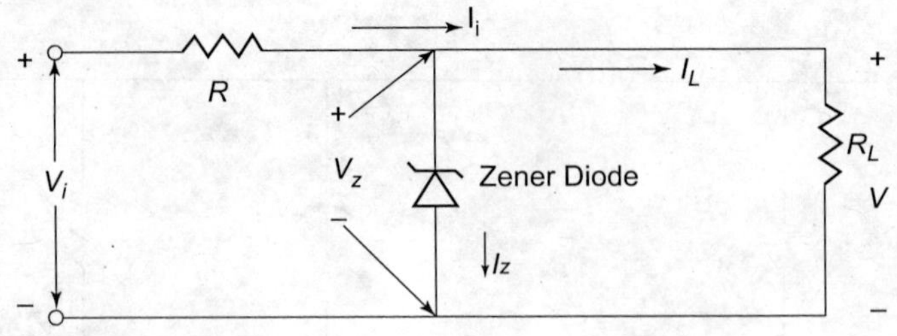

Fig. 4.11 *A Zener diode shunt regulator*

When the input voltage increases, as the voltage across the Zener diode remains constant, the drop across the resistor R will increase with a corresponding increase in $I_L + I_z$. As V_z is a constant, the voltage across the load will also remain constant and hence I_L will be a constant. Therefore, an increase in $I_L + I_z$ will result in an increase in I_z which will not alter the voltage across the load.

It must be ensured that the reverse voltage applied to the Zener diode never exceeds PIV of the diode and, at the same time, the applied input voltage must be greater than the breakdown voltage of the Zener diode for its operation. The Zener diodes can be used as "standalone" regulator circuits and also as reference voltage sources.

Series voltage regulator: Even though the Zener diode shunt regulator seems to be a good regulator, it has certain limitations. The output voltage is fixed and is equal to the Zener voltage. It cannot be changed and if one wants to do so, he has to change the Zener diode. Also, Zener diodes of standard Zener voltages are available and all possible voltages are not commercially available. The maximum current rating of the Zener diode limits the current drawn by the load. The diode may be damaged if the current drawn exceeds the rated value. This may happen when by mistake a short circuit is kept in place of the load. The drawbacks of the Zener diode regulator, namely, flexibility of the reference voltage, current handling capacity, short circuit protection and so on can be eliminated with this *series voltage regulator*. Not only this, the stabilisation of this circuit is very high compared to Zener circuit.

The circuit of series voltage regulator is as shown in Fig. 4.12. The transistor Q_1 is called the *pass transistor* since it allows all the current to flow through it. It is emitter follower configuration giving a voltage gain of unity but it can have good current handling capacity (the characteristics of the transistors are discussed in detail in Chapter 5). The emitter voltage is same as the base voltage since the voltage gain is unity. The base voltage is the sum of Zener voltage and the collector to emitter voltage of Q_2. Q_2 being in active region, the output voltage is in proportion to the base voltage. The base voltage is the fraction of output voltage selected by the voltage divider circuit R_1–R_2. Thus, adjusting the position of the potentiometer can alter the base voltage of Q_1. Thus, the flexibility of the output voltage over the reference Zener voltage can be achieved and any voltage can be set up at the output. Since a Zener voltage is employed, regulation is good. The current does not pass through the reference diode and is handled by transistor Q_1. Thus, the

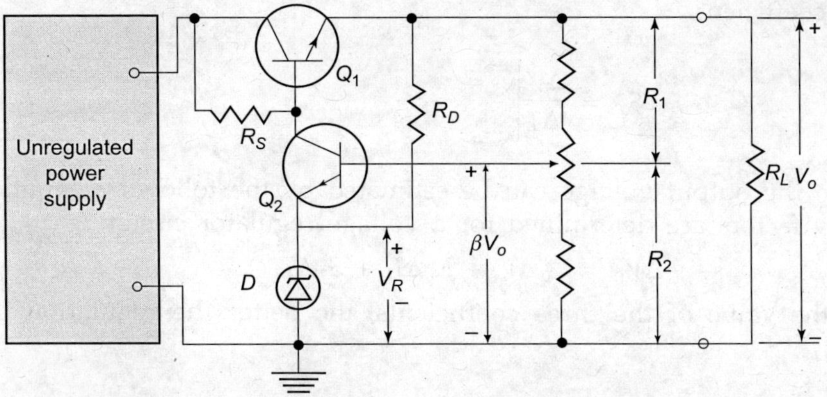

Fig. 4.12 *Series voltage regulator*

current handling capacity is very large compared to a Zener circuit. An ancillary transistor Q_3 is employed in Q_1 and Q_2 to have short-circuit protection in commercial circuits. An increase or decrease in voltage at the unregulated power supply point is served as the biasing to the transistor Q_1 and the differential voltage changes occur across R_3 and so, it has little effect on the output voltage of the regulator.

The drawback of this circuit is that the drop V_{CE} across the transistor Q_1 is seen to be affecting the output voltage. This can be avoided to a large extent when thyristors are used in the circuit. For better performance of the regulator circuits, operational amplifiers are employed.

4.6 PERFORMANCE OF REGULATORS

The performance of the voltage regulators can be estimated with the help of % regulation, as defined in Eq. (4.15). The % regulation should be as small as possible and ideally zero. This is also referred to as *load regulation.*

Line regulation is defined as the change in output voltage for a change in line supply voltage, keeping the load current and temperature constant.

In addition to the above two regulations, the stability of the circuit can be measured with three parameters. They are *input regulation factor, output resistance* and *temperature coefficient.* They are defined and related as follows.

Input regulation factor,

$$S_V = \left.\frac{\Delta V_0}{\Delta V_i}\right|_{\Delta T, \Delta I_L} \quad \text{...(4.55)}$$

Output impedance,

$$R_o = \left.\frac{\Delta V_0}{\Delta I_L}\right|_{\Delta T, \Delta V_i} \quad \text{...(4.56)}$$

Temperature coefficient,

$$S_T = \left.\frac{\Delta V_0}{\Delta T}\right|_{\Delta V_i, \Delta I_L} \qquad ...(4.57)$$

The change in the output voltage can be estimated by the following equation when the above stability factors are determined for a voltage regulator circuit:

$$\Delta V_0 = S_V \Delta V_i + R_o \Delta I_L + S_T \Delta T \qquad ...(4.58)$$

The smaller the value of the three coefficients, the better the regulation of the power supply.

4.7 IC REGULATORS

Integrated circuits (IC) are those circuits that are built on a single substrate. Regulators thus built are known as *IC regulators.* IC regulators are versatile, relatively inexpensive and are available with features such as programmable output, current-voltage boosting and floating operation for high voltage applications. Some IC regulators are the following:

- Fixed positive-negative output voltage regulators.
- Adjustable output voltage regulators.

Early IC regulators, μA 723, LM300 have Zener diode, high gain amplifier, current limiting and other useful features. 340 and 320 series of IC regulators are fixed type of positive and negative voltages, respectively. 317 series are adjustable type of regulators. The range of voltages is from 1.25 V to 37 V with a current rating of 1.5 A.

78 series IC regulators are three terminal, positive fixed-voltage regulators. The standard voltages available are 5, 6, 8, 12, 15, 18 and 24 V. The IC number is 78XX, where XX represent the voltage regulated by the specific IC.

79XX series regulators are negative fixed-voltage regulators, which complement the 78 series. In 79 series there are –2 and –5.2 V regulators available in addition to the above.

Fig. 4.13 shows the standard representation of a monolithic voltage regulator. The input capacitor C_i is used to cancel the inductive effects due to long distribution leads and the output capacitor C_o improves the transient response.

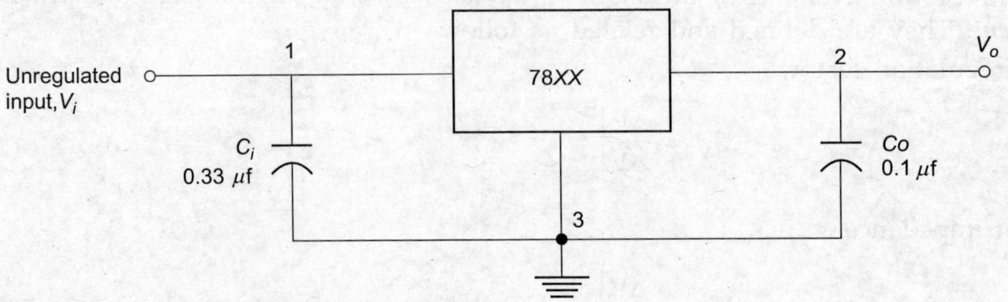

Fig. 4.13 *IC voltage regulator*

Dual tracking regulated power supplies can be obtained using 7814 and 7915 ICs. The advantage with such circuits is that they provide a wide range of voltages and higher current handling capacities.

LM723C is the general-purpose *adjustable voltage regulator*. The output can be adjusted from 2 to 37 V. The current rating is 150 mA and can be enhanced to 10 A by providing external pass transistor. It can be used as either a linear or switching voltage regulator. It can also be used as a negative voltage regulator.

SUMMARY

- Regulated power supply is required to bias electronic components and gadgets. It basically converts commercial supply to required dc power.
- Regulated power supply consists of a step-down transformer, a rectifier, a filter and a voltage regulator.
- A rectifier converts bidirectional current to unidirectional current. There are three types of rectifiers—half-wave, full-wave with centre-tapped transformer and bridge rectifiers.
- A rectifier employs diodes for rectification. There are one, two and four diodes employed in half-wave, full-wave and bridge rectifiers, respectively.
- The performance of the rectifiers is measured in terms of ac voltage, dc voltage, ripple factor, PIV, rectification efficiency, transformer utilisation factor and so on.
- A half-wave rectifier is least used since its performance is poor. Half-wave and bridge rectifiers find their own applications in practice.
- The ripple in the output of the rectifier is filtered by a filter. There are various types of filters—capacitor, inductor, LC, *pi* type and so on.
- A voltage regulator regulates voltage in the sense that voltage is maintained irrespective of variations in load, input voltage or temperature.
- The performance of a voltage regulator is measured by parameters like input regulation factor, temperature coefficient and output impedance.
- Zener diode is key element in a voltage regulator. It is biased in the breakdown region.
- There are various types of regulators—Zener, voltage series, fixed IC, variable IC regulator and so on.

SOLVED PROBLEMS

4.1 A sinusoidal voltage of amplitude 20 V and frequency 50 hz is applied to a half-wave rectifier shown in the figure. $R_L = 1000\ \Omega$, $V_\gamma = 0$, $R_f = 10\ \Omega$, $R_r = \infty$. Find (a) peak, average and rms values of load current; (b) dc voltage and dc power output; (c) ac input power; (d) η; (e) ripple factor; and (f) percent regulation.

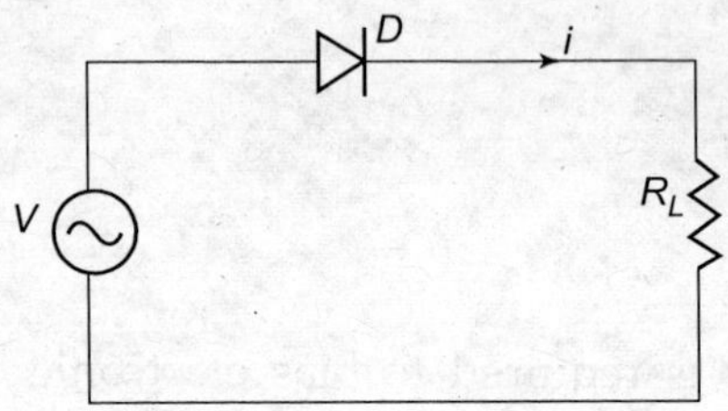

Solution (a) $I_m = \dfrac{V_m}{R_f + R_L}$

$$= \frac{20}{10 + 1000} = 19.8 \text{ mA}$$

$$I_{dc} = I_m/\pi = 6.303 \text{ mA}$$

$$I_{rms} = I_m/2 = 9.9 \text{ mA}$$

(b) $V_{dc} = I_{dc} \cdot R_L = 6.303$ V

$$P_{dc} = I_{dc}^2 R_L = 39.72 \text{ mW}$$

(c) $P_{in} = I_{rms}^2 (R_f + R_L) = 1010 \times (9.9)^2$

$$= 98.99 \text{ mW.}$$

(d) $\%\eta = \text{Efficiency} = \dfrac{P_{dc}}{P_{in}} \times 100$

$$= \frac{39.72}{98.99} \times 100 = 40.12\%$$

(e) Ripple factor $= \sqrt{\left(\dfrac{I_{rms}}{I_{dc}}\right)^2 - 1} = 1.21$

(f) Percentage regulation $= \dfrac{V_{NL} - V_L}{V_L} \times 100$

$$V_{NL} = V_{dc}\big|_{R_L \to \infty}$$

$$= \left.\frac{V_m R_L}{\pi(R_f + R_L)}\right|_{R_L \to \infty} = \frac{V_m}{\pi} = \frac{20}{\pi} = 6.366$$

$$V_L = V_{dc} = 6.3 \text{ V}$$

$$\% \text{ regulation} = \frac{V_{NL} - V_L}{V_L} \times 100$$

$$= \frac{6.366 - 6.3}{6.3} \times 100$$

$$= 1.05\%$$

4.2 In a centre-tapped full-wave rectifier two diodes with $R_f = 10\ \Omega$, $R_r = \infty$, $V_\gamma = 0$ are used. $R_L = 1000\ \Omega$. Find (a) peak, dc and rms load currents; (b) dc output voltage and dc power; (c) ac input power; and (d) $\%\eta$ if a 24 V peak sinusoidal signal is applied.

Solution (a)

$$I_m = \frac{V_m}{R_f + R_L} = \frac{24}{10 + 1000} = 23.76 \text{ mA}$$

$$I_{dc} = \frac{2I_m}{\pi} = \frac{2 \times 23.76}{\pi} = 15.12 \text{ mA}$$

$$I_{rms} = I_m/\sqrt{2} = 16.80 \text{ mA}$$

(b)

$$V_{dc} = I_{dc} \cdot R_L = 15 \cdot 12 \times 10^{-3} \times 1000 = 15 \cdot 12 \text{ V}$$

$$P_{dc} = I_{dc}^2 \cdot R_L = (15.12 \times 10^{-3})^2 \times 1000$$

$$= 228.26 \text{ mW}$$

(c)

$$\text{Power input} = I_{rms}^2 (R_f + R_L)$$

$$= (16.8)^2 (1010)$$

$$= 285.06 \text{ mW}$$

(d)

$$\text{Efficiency } \eta = \frac{228.26}{285.06} \times 100$$

$$= 80.07\%$$

4.3 A bridge rectifier uses $R_L = 2.4\ \text{k}\Omega$, each diode is idealised to have $R_f = 5\ \Omega$ and $R_r = \infty$, $V_\gamma = 0$. A sinusoidal voltage having an amplitude of 30 V is applied. Calculate (a) peak, dc and rms values of load current; (b) dc and rms output voltages, (c) dc output power (d) ac input power; (e) η; and (f) percentage regulation.

Solution (a)

$$I_m = \frac{V_m}{R_L + 2R_f} = \frac{30}{2.4 \text{ K} + 10} = 12.44 \text{ mA}$$

$$I_{dc} = \frac{2I_m}{\pi} = 7.924 \text{ mA}$$

$$I_{rms} = \frac{I_m}{\sqrt{2}} = 8.797 \text{ mA}$$

(b)
$$V_{dc} = I_{dc} \cdot R_L = 19.01 \text{ V}$$
$$V_{ac} = I_{rms} \cdot R_L = 21.11 \text{ V}$$

(c)
$$P_{dc} = I_{dc}^{\,2} \cdot R_L = (7.924)^2 \times 2.4 \times 10^3 \times 10^{-6}$$
$$= 150.69 \text{ mW}$$

(d)
$$P_{in} = I_{rms}^2 (R_L + 2R_f) = (8.797 \times 10^{-3})^2 \, (2.4 \times 10^3 + 10)$$
$$= 186.503 \text{ mW}$$

(e)
$$\%\eta = \text{Efficiency} = \frac{P_{dc}}{P_{in}} \times 100$$
$$= 80.79\%$$

(f) Percentage regulation $= \dfrac{V_{NL} - V_L}{V_L} \times 100$

$$V_{NL} = V_{dc}\Big|_{R_L \to \infty}$$
$$= \left.\frac{2I_m R_L}{\pi}\right|_{R_L \to \infty}$$
$$= \left.\frac{2V_m R_L}{\pi(2R_f + R_L)}\right|_{R_L \to \infty}$$
$$= \frac{2V_m}{\pi} = \frac{2 \times 30}{\pi} = 19.09$$
$$V_L = V_{dc} = 19.01 \text{ V}$$

$$\text{Percentage regulation} = \frac{19.09 - 19.01}{19.01} \times 100$$
$$= 0.466\%$$

4.4 Find the condition for the maximum power transfer in terms of R_f and R_L for a half-wave rectifier.

Solution We have
$$P_{dc} = I_{dc}^{\,2} \cdot R_L = \frac{V_m^2 R_L}{\pi^2 (R_f + R_L)^2}$$

For maximum value of P_{dc}, $\dfrac{dP_{dc}}{dR_L} = 0$

$$\Rightarrow \frac{V_m^2}{\pi^2}\left[\frac{(R_f+R_L)^2-2R_L(R_f+R_L)}{(R_f+R_L)^4}\right]=0$$

$$\Rightarrow (R_f + R_L)^2 - 2R_L(R_f + R_L) = 0$$

$$\Rightarrow R_f + R_L - 2R_L = 0$$

$\Rightarrow R_f = R_L$. Hence proved.

4.5 Express the regulation in terms of R_f and R_L for both half-wave and full-wave rectifiers.

Solution *For a half-wave rectifier*

$$V_{NL} = \frac{V_m}{\pi}$$

$$V_L = V_{dc} = \frac{V_m}{\pi}\frac{R_L}{R_f+R_L}$$

$$\%\ \text{regulation} = \frac{V_{NL}-V_L}{V_L}\times 100$$

$$= \frac{\dfrac{V_m}{\pi}-\dfrac{V_m}{\pi}\cdot\dfrac{R_L}{R_f+R_L}}{\dfrac{V_m}{\pi}\cdot\dfrac{R_L}{R_f+R_L}}\times 100$$

$$= \frac{\dfrac{V_m}{\pi}\left(\dfrac{R_f+R_L-R_L}{R_f+R_L}\right)}{\dfrac{V_m}{\pi}\cdot\dfrac{R_L}{R_f+R_L}}\times 100$$

$$= \frac{R_f}{R_L}\times 100$$

For full-wave rectifier, the V_{NL} and V_L are just doubled. Hence, the regulation remains the same as above.

4.6 A half-wave rectifier uses a R_L = 10 kΩ with a shunt filter capacitor of 10 μF. The sinusoidal voltage of amplitude 20 V, 50 Hz is applied. The diodes have $R_f = 0$, $R_r = \infty$ and $V_\gamma = 0$. The angle of conduction is 36°. Find (a) cutin and cutout angles; (b) I_{dc} and V_{dc}; (c) ripple voltage and ripple factor.

Solution (a) Cutout angle $\theta_2 = \pi - \tan^{-1}(WR_LC)$

$$= 180° - \tan^{-1}(2\pi \times 50 \times 10 \times 10^3 \times 10 \times 10^{-6})$$

$$= 180° - \tan^{-1}(10\pi)$$

$$= 91.83°$$

Cutin angle $\theta_1 = (\theta_2 - \theta_C) = (91.83 - 35) = 56.8°$.

(b)
$$I_{dc} = \frac{V_m}{2\pi R_L}\sqrt{1+(WR_LC)^2}\,(1 - \cos\,\theta_C)$$

$$= \frac{20}{2\pi \times 10^4}\sqrt{1+(10\pi)^2}\,(1 - \cos 35°)$$

$$= 1.81 \times 10^{-3}\text{ A} = 1.81\text{ mA}$$

$$V_{dc} = I_{dc} \cdot R_L = 18.1\text{ V}$$

(c) Ripple voltage $V_R = \dfrac{I_{dc}}{fC} = \dfrac{1.81 \times 10^{-3}}{50 \times 10 \times 10^{-6}}$

$$= 3.62\text{ V}$$

$$\text{Ripple factor} = \frac{1}{2\sqrt{3}fCR_L} = \frac{1}{2\sqrt{3} \times 50 \times 10 \times 10^{-6} \times 10 \times 10^3}$$

$$= \frac{1}{10\sqrt{3}} = 0.0577$$

4.7 The figure shows the output wave form of a full-wave rectifier with capacitor filter. The value of capacitor is 500 μF. Find (a) V_{dc}, (b) I_{dc}, (c) R_L and peak to peak ripple.

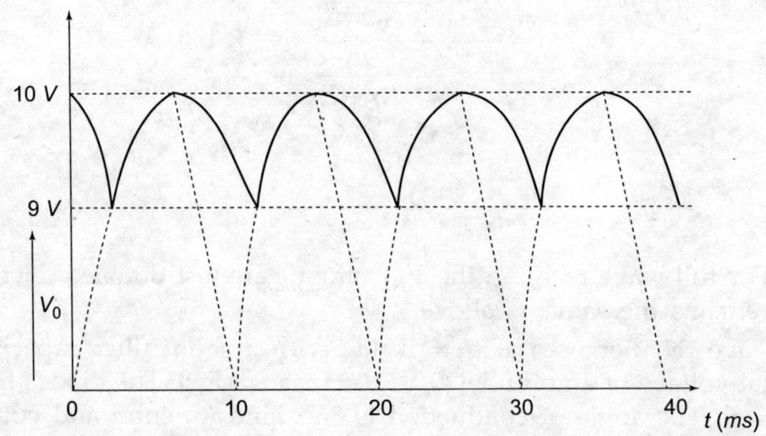

Solution

$$V_r = \text{peak to peak ripple} = 10 - 9 = 1 \text{ V}$$

$$V_{\text{rms ac}} = \frac{V_r}{2\sqrt{3}} = \frac{1}{2\times\sqrt{3}} = 0.2886$$

$$\text{ripple factor } r = \frac{V_{\text{rms ac}}}{V_{\text{dc}}} = \frac{0.2886}{9.5} = 0.03$$

To calculate I_{dc} we have $V_r = \dfrac{I_{\text{dc}}\cdot T/2}{C}$

$$\Rightarrow \quad I_{\text{dc}} = \frac{V_r\cdot C}{T/2} = \frac{1\times 500\times 10^{-6}}{20\times 10^{-3}/2}$$

$$= 50 \text{ mA}$$

We have $V_{\text{dc}} = I_{\text{dc}}\cdot R_L$

$$R_L = \frac{V_{\text{dc}}}{I_{\text{dc}}} = \frac{9.5}{50\times 10^{-3}} = 190\ \Omega$$

4.8 Find the output voltage, current and ripple factor for the circuit shown in the figure.

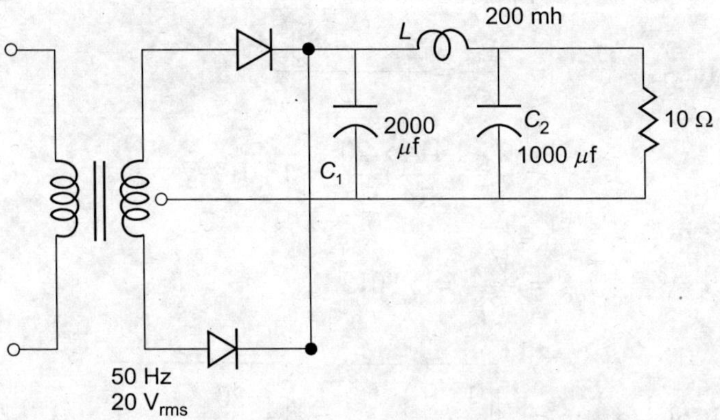

Solution

$$X_{C1} = \frac{1}{2\pi f C_1} = \frac{1}{2\times 3.14\times 50\times 2000\times 10^{-6}}$$

$$= 1.592\ \Omega$$

$$X_{C2} = \frac{1}{2\pi f C_2} = \frac{1}{2\times 3.14\times 50\times 1000\times 10^{-6}} = 3.184\ \Omega$$

$$X_L = \omega L = 2\pi f L = 2\times 3.14\times 50\times 200\times 10^{-3} = 62.8\ \Omega$$

$$V_{\text{dc}} = V_m = \sqrt{2}\,V_{\text{rms}} = 1.414\times 20 = 28.28 \text{ V}$$

$$I_{dc} = \frac{V_m}{R_L} = \frac{28.28}{10} = 2.828 \text{ A}$$

$$\text{For } \pi\text{-section filter ripple factor} = \frac{\pi}{2\sqrt{3}} \times \frac{X_{C1} \cdot X_{C2}}{X_L \cdot R_L}$$

$$= \frac{3.14 \times 1.592 \times 3.184}{2 \times \sqrt{3} \times 62.8 \times 10}$$

$$= 7.316 \times 10^{-3} = 0.007316$$

4.9 A full-wave rectifier with a centre-tapped transformer is present. The transformer is 20-0-20 V_{rms}, 50 Hz one. $R_L = 10\ \Omega$.

(a) Find V_{dc}, I_{dc} and ripple factor.

(b) Repeat part (a) if 10000 μF is considered and is shunted across the load.

(c) Repeat part (a) if 100 mh inductor is present in series with load.

(d) What is the minimum value of L in Part (c)?

Solution Ripple factor $= \dfrac{V_{rms,\,ac}}{V_{dc}} = \dfrac{V_{rms} - V_{dc}}{V_{dc}} = \dfrac{\dfrac{V_m}{\sqrt{2}} - V_{dc}}{V_{dc}}$

(a) We have $V_{dc} = \dfrac{2V_m}{\pi} = \dfrac{2 \times \sqrt{2}\, V_{rms}}{\pi}$

$$= \frac{2 \times 1.414 \times 20}{3.14} = 18.012 \text{ V}$$

$$I_{dc} = \frac{V_{dc}}{R_L} = 1.8012 \text{ A}$$

$$\text{Ripple factor} = \sqrt{\left(\frac{V_{rms}}{V_{dc}}\right)^2 - 1} = \sqrt{\left(\frac{20}{18.012}\right)^2 - 1} = 0.4825$$

(b) For capacitor filter $V_{dc} \approx V_m = \sqrt{2}\, V_{rms}$

$$V_{dc} = 20 \times \sqrt{2} = 28.28 \text{ V}$$

$$I_{dc} = \frac{V_{dc}}{R_L} = 2.828 \text{ A}$$

$$\text{Ripple factor} = \frac{1}{4\sqrt{3} f C R_L} = \frac{1}{4 \times \sqrt{3} \times 50 \times 10{,}000 \times 10^{-6} \times 10}$$

$$= 0.0288$$

(c) $X_L = \omega L = 2\pi f L = 2 \times 3.14 \times 50 \times 100 \times 10^{-3}$

$= 31.4\ \Omega$

$$V_{dc} = \frac{2V_m}{\pi} = \frac{2 \times 28.28}{3.14} = 18.012 \text{ V}$$

$$I_{dc} = \frac{V_{dc}}{R_L} = 1.8012 \text{ A}$$

$$\text{Ripple factor} = \frac{\sqrt{2}}{3} \cdot \frac{R_L}{X_L} = \frac{\sqrt{2}}{3} \times \frac{10}{31.4} = 1.501$$

(d) $$L_{min} = \frac{2R_L}{3W} = \frac{2 \times 10}{3 \times 2 \times 3.14 \times 10} = 106.15 \text{ mh}$$

EXERCISE PROBLEMS

4.1 The output of a Full Wave Rectifier is fed from 40-0-40 V transformer. The load current is 0.1 A. Two 40 μF capacitors are available. $R_L = 50\ \Omega$. (a) calculate the value of inductance for a two-stage L-sector filter ripple factor is 0.0001. (b) find the output voltage.

4.2 The voltage at the input capacitor of a π-section filter is given to the close approximation by V (t) = 525-40 sin 745 t. The output capacitor of the filter is 10 μF. If the filter d.c. Output voltage is 500 V for a 100 μA load with a ripple factor of 0.001. Determine the inductance and dc resistance of the filter choke.

4.4 A sinusoidal voltage of amplifier 40 V and frequency 50 Hz is applied to a Half Wave Rectifier. $R_L = 200\ \Omega$, $V_r = 0$, $R_f = 20\ \Omega$, $R_r = \infty$. Find V_{dc}, I_{dc}, I_m, I_{rms}, P_{dc}, P_{in}, η, ripple factor, % regulation.

4.5 In a centre-tapped Full Wave Rectifier, two-diodes with $R_f = 20\ \Omega$, $R_r = \infty$, $V_r = 0$ are used. $R_L = 2$ kΩ, find (a) I_m, I_{dc}, I_{rms}. (b) dc output voltage and dc power, (c) ac Input power (d) efficiency if a 48 V peak sinusoidal signal is applied.

4.6 A bridge rectifier uses $R_L = 1$ kΩ, each diode is Idealized to have $R_f = 10\ \Omega$ and $R_r = \infty$, $V_r = 0$ A sinusoidal voltage having an amplitude of 60 V is applied. Calculate (a) peak, dc and rms values of load current (b) V_{dc} and V_{rms} (c) P_{dc}, P_{in} (d) efficiency and percentage regulation.

4.7 A full-wave rectifier delivers 50 W to a load of 200 Ω. If the ripple factor is 1%, calculate the ac ripple voltage across the load.

4.8 design a CLC or π-section filter for $V_{dc} = 10$ V, $I_L = 200$ mA and ripple factor = 0.02.

4.9 Design a series voltage regulator with the following specifications:
$V_o = 10$ V, V_{in} = (22-30) V, $I_{L\,max} = 50$ mA.

4.10 design a filter for the full wave circuit with LC filter to provide an output voltage of 10 V with a load current of 20 mA and the ripple factor of 0.02.

REVIEW QUESTIONS

1. Explain the role of a filter in diode rectifiers. Name the different filters.
2. What is critical inductance? What is *L* section filter? Discuss.
3. Draw the circuit diagram of a full-wave rectifier employing bridge connection. Explain its working. Sketch the input and output wave forms.
4. What is a rectifier? Why are filters used? Name the filters used.
5. Draw the circuit diagram of a single-phase bridge rectifier with capacitor filter. Explain its operation with wave forms. Derive the expression for its output dc voltage and ripple factor.
6. Draw the circuits of half-wave and full-wave rectifiers. Compare their characteristics.
7. What is meant by TUF of a rectifier? Derive its value for that of a half-wave and a bridge rectifier.
8. Derive an expression for the ripple factor of a half-wave rectifier.
9. Compare a bridge rectifier with a centre-tapped full-wave rectifier.
10. Define regulation. Explain the operation of a voltage series regulator bringing out the advantages of this circuit over a Zener diode regulator.
11. Draw the circuit diagram of a full-wave rectifier and explain its operation with wave forms. Derive the expressions for its rectification efficiency and ripple factor with a capacitor filter.
12. Write a short note on the monolithic IC regulators.
13. What is a ripple? Derive the ripple factors of half-wave rectifier and full-wave rectifier circuits.
14. Draw the circuit diagram of a bridge rectifier and explain its working with wave forms.
15. Write a short note on the *L* and PI section filters.
16. Sketch the circuit for a full-wave rectifier. Derive the expression for the dc current, dc load voltage and the rms current.
17. Determine the efficiency of rectification for the half-wave rectifier circuit.
18. For a half-wave rectifier with a resistive load, determine the following: (i) average value of the current, (ii) rms voltage, (iii) PIV, (iv) dc load voltage and (v) ripple factor.
19. Draw the circuit of a bridge rectifier with a π section filter and derive the expression for ripple factor and output dc voltage.
20. What is the PIV of a bridge rectifier?

21. Explain the operation of a full-wave rectifier with capacitor filter and derive the expression for ripple factor.
22. What is a regulated power supply? Draw its block diagram and explain the function of each block in detail.
23. Give the parameters that give the performance of voltage regulators. What are IC regulators? Explain.
24. Define regulation and give its equation. Give various types of voltage regulators and discuss their performance.
25. What is PIV? What is its significance in rectifier circuit? Give its value for all the three rectifier circuits.

CHAPTER 5

Bipolar Junction Transistors

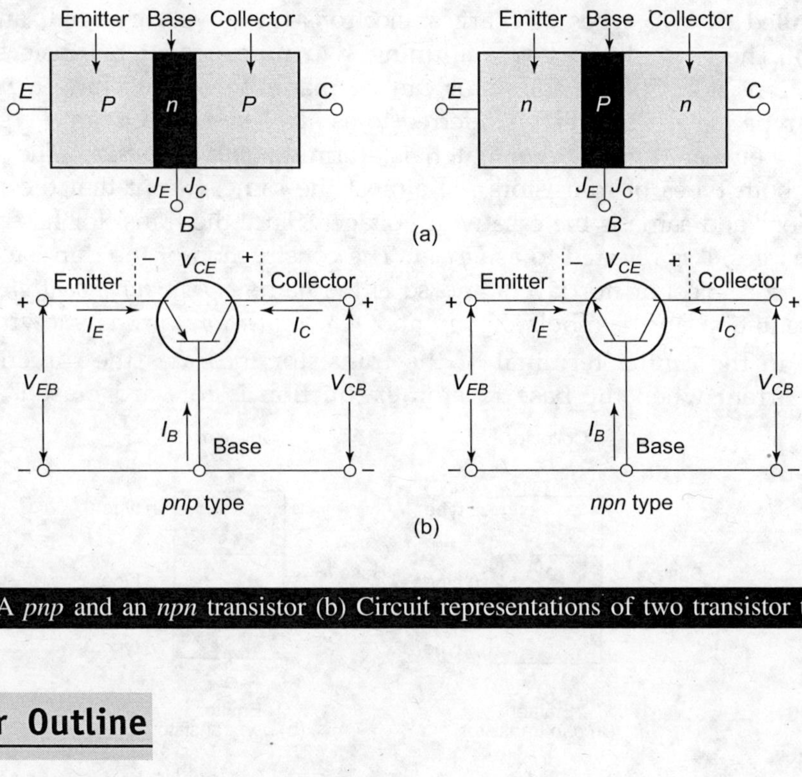

(a) A *pnp* and an *npn* transistor (b) Circuit representations of two transistor types

Chapter Outline

The concepts introduced in this chapter are:

- *Transistor, its types and applications*
- *Configurations and characteristics of the transistor*
- *Regions of operation*
- *Voltage ratings, current gains in the transistor*

5.1 INTRODUCTION

Bipolar junction transistor is an active device, which finds large number of applications in electronic circuits. Basically it is an amplifier that amplifies alternating sinusoidal signals. It also finds large applications in computer electronics. For example, it is the component that is responsible for memory, switching and so on. In a nutshell, without a bipolar junction transistor (BJT), also called as *transistor,* the electronics world has no meaning.

5.2 JUNCTION TRANSISTOR

The principle of operation of a bipolar junction transistor can be understood easily when the transistor is visualised as two *p-n* junction diodes connected back to back. The junction transistor has three terminals designated as *emitter, base* and *collector.* The current in the transistor is constituted by two types of charges, electrons (negatively charged) and holes (positively charged). There are three layers forming two junctions of extrinsic semiconductor materials in a transistor. So, the transistor can be made by sandwiching a *p*-type material between two *n*-type materials, which is referred to as *npn transistor.* Or, an *n*-type material can be sandwiched between two *p*-type materials, forming *pnp transistor.* The functions and applications of both types of transistors are almost the same, except that the currents are in opposite directions and same is the case with voltages. Since the transistor has currents due to two types of charges, it is referred to as *bipolar.* The construction of the transistor involves two junctions between semiconductor layers and so is the name *bipolar junction transistor.*

The representation and symbols of *pnp* and *npn transistors* are as shown in Fig. 5.1.

The arrow in the *emitter* terminal of the transistor indicates the direction of flow of conventional current when the base to emitter junction is forward biased. So, the arrow

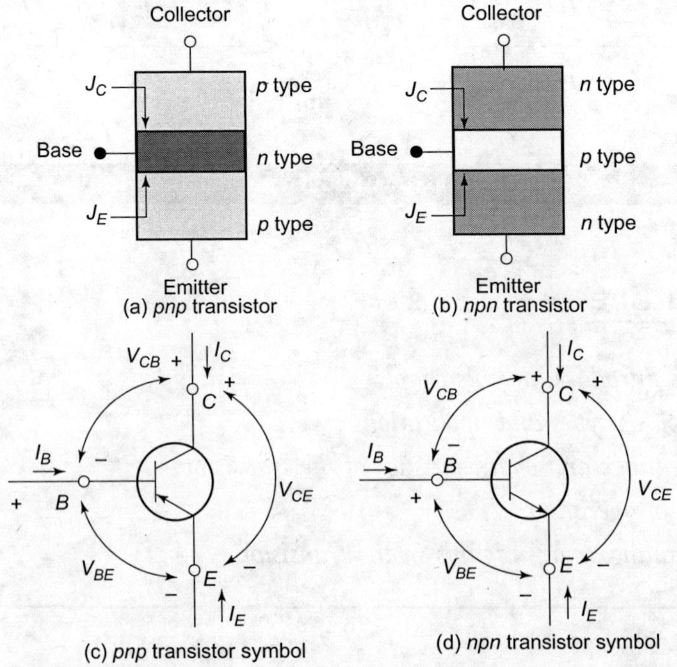

Fig. 5.1 *pnp and npn transistor*

will be pointing away from base in an *npn transistor*, whereas the arrow is pointing towards the base in a *pnp transistor*. The slanted line without an arrow indicates the *collector* of the device. The line that is perpendicular to the vertical line is the *base* of the device. Since base is common between both emitter and collector, the junction between base and emitter is referred to as emitter junction and is designated as J_E. The junction between collector and base is collector junction J_C. Conventionally, all the currents entering into the device are assumed to be positive. So, there are three currents entering into the three terminals of the device, known as *collector current* I_C, *base current* I_B and *emitter current* I_E. The voltages between the terminals are designated as V_{BE}, *base to emitter voltage*, V_{EC}, *emitter to collector voltage* and so on.

In principle, the emitter emits the current, the base modulates the current and the collector collects the remaining current. The current components in the transistor are similar to the current components in a *p-n junction diode*. So, the current components are made up of majority and minority charge carriers comprising both drift and diffusion phenomena.

In practice, transistors are available in different packages. When the transistor is encapsulated in a cylindrical metallic structure, this metallic structure is the collector of the transistor. Some of the transistors are available in a plastic package, which is, in general, flat type. The pin configuration of transistors, which is available commercially and widely used, is indicated in Fig. 5.2.

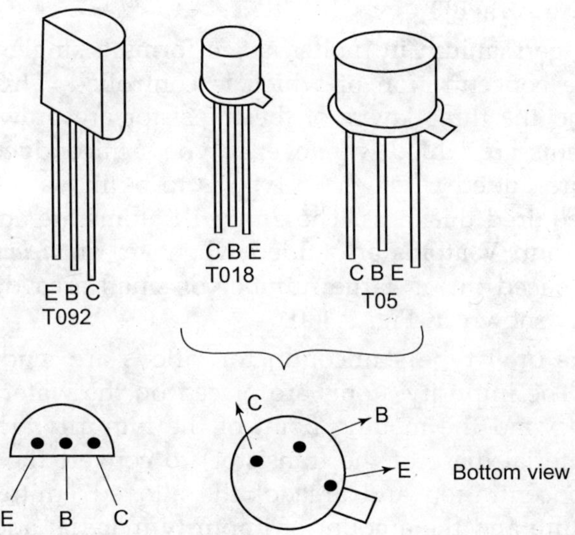

Fig. 5.2 *Pin configuration of commercial transistors*

Transistor coding: The European system for classifying transistors involves an alphanumeric code that employs either two letters and three figures (general purpose transistors) or three letters and two figures (special purpose transistors). The first two letters have the following significance:

(1) First Letter—Semiconductor Material

A germanium

B silicon

(2) Second Letter—Application

C low power, low frequency

D high power, low frequency

F low power, high frequency

L high power, high frequency

In the case of transistors for specialised applications, the third letter does not generally have any particular significance. A low-frequency transistor is designed specially for AF applications (below 100 kHz). High-frequency transistors are designed specially for RF applications (100 kHz and above). A power transistor operates at a significant power level. Switching transistors are designed for switching applications. High-voltage transistors are specially designed to handle high voltages. Low-noise transistors would have low-noise characteristics and are intended primarily for the amplification of low-noise applications.

5.3 TRANSISTOR FABRICATION

The transistor can be constructed using semiconductor materials. In general, silicon or germanium is taken as the base material in the construction of a bipolar junction transistor. There are various methods of sandwiching one type of extrinsic semiconductor material between two layers of the other type to form a transistor. The following is the brief constructional procedure of a BJT.

Grown Type: Silicon or germanium in molten stage forms a single crystal. In this process, impurity is added, the concentration of which is controlled. This converts the intrinsic material to extrinsic and the three layers of the transistor are drawn from the basic single crystal. Trivalent elements are added whenever *p* type semiconductor is to be made while pentavalent elements are added when the *n* type is to be made. The crystal is then made into the transistor of required thickness. The group III atoms are added whenever a *p layer is* to be formed and group V atoms are added whenever an *n layer* is to be drawn. The grown crystal is then sliced into a large number of small area devices and contacts are made to each region as shown in Fig. 5.3(a).

Alloy Type: In a *p* type or *n* type semiconductor, alloys are made on either side of the wafer of the material. The impurity atoms are placed on the wafer and the temperature is enhanced to a value beyond the melting point of the impurity. Thus the impurity atoms penetrate the basic material (base of the transistor) to convert the semiconductor into the opposite type. The concentration and area of the alloyed emitter and collector can be controlled by temperature and the amount of impurity material added. The alloying of the impurity with the original wafer produces the *p-n* junction to which leads are attached as shown in Fig. 5.3(b). The collector is made larger than the emitter.

Diffusion Type: This technique is closely related to the fabrication process of IC. A substrate of a given type of semiconductor is considered and the required layer conversions are made by diffusion of impurities in gaseous form. The depth of layer formation and concentrations are easily controlled with time and temperature variations. Masks of insulating materials are grown on the substrate and are etched out at portions where the layer transformations are required. Generally silicon dioxide is the insulating layer used in this process. After all the junctions are realized, a final insulating layer is

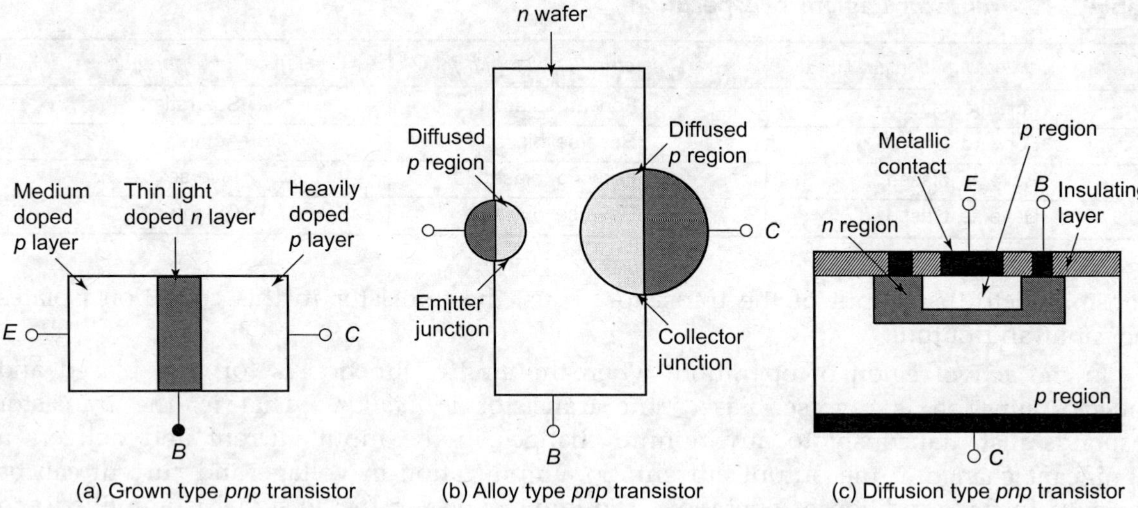

Fig. 5.3 *Transistor fabrication*

grown and the portions where electrode contacts are required is etched out. Aluminum contacts are made in the etched out portions to realize the three electrodes of the transistor.

Epitaxial Type: A heavily doped substrate of either Ge or Si semiconductor is taken and layers of the required type of semiconductor material is grown over the substrate. Highly pure thin layers are grown to realize the various layers of the transistor. The emitter and collector regions are diffused on this thin layer of semiconductor material. This structure has the advantage of low resistance and is suitable for operation at very high frequencies. This reduces the electrical series resistance in the collector, has large power-dissipation capability, can operate at very high frequencies and the structure is inherently rugged.

Various combinations of the above mentioned methods are also adopted in the construction of transistors.

5.4 REGIONS OF OPERATION

As there are two junctions in a transistor, there are four possible combinations of biasing the transistor. Each junction can be biased either in forward or reverse direction. When both the junctions are forward biased, the transistor is said to be in *saturation region.* When both the junctions are reverse biased, the transistor is in *cut off region.* When the emitter junction is forward biased and the collector junction is reverse biased, the transistor operates in *active region.* The condition of the collector junction being forward biased and the emitter junction reverse biased is sometimes referred to as *reverse active* or *inverse mode* and is rarely used. The transistor acts as an amplifier when it is biased in the active region. When the transistor is to be operated as a switch, it is biased in saturation and cut off regions. The regions of operation of a transistor are summarised in Table 5.1.

In the cut off region of operation both the junctions are reverse biased, and so minimum or ideally no current flows in the transistor, making it equivalent to an open-circuit switch. So, the response of the external circuitry in the input of the transistor is not

Table 5.1 ***Transistor Regions of Operation***

Emitter Junction (J_E)	*Collector Junction (J_C)*	*Region of Operation*
Forward biased	Forward biased	Saturation
Forward biased	Reverse biased	Active
Reverse biased	Forward biased	Inverse
Reverse biased	Reverse biased	Cut off

transmitted to the output of the transistor. Thus, the transistor in this condition isolates the input and output.

In the active region of operation, when the emitter junction is forward biased and collector junction is reverse biased, the transistor is *"active."* That is, the transistor responds instantaneously to any minute changes in the input current and delivers a significant change in the output current. So, amplification in voltage and current can be achieved in this condition of operation. The input power in the ac signal is amplified and delivered into the output of the circuit.

In the saturation region of operation both the junctions are forward biased and so maximum current flows in the device. In this region the transistor is saturated and once again it does not respond to any variations in the input current. Since maximum current is flowing in the device in this condition, the transistor is equivalent to a closed-circuit switch ideally.

5.5 TRANSISTOR CONFIGURATIONS

The transistor is a three-terminal device. If it were to be used as an amplifier or a switch, it is supposed to be a two-port device. For a two-port device, four terminals are required. One port is referred to as input port where the signals are fed into the network and the other port is output port, where the response of the network is available. So, when a transistor is supposed to be used as a two-port network, one of the terminals of the transistor is made common between input and the output.

When the emitter terminal of the transistor is made common between input and output of the transistor, the transistor is said to be operated in *common emitter configuration*. The signal is fed at the input between base and emitter of a transistor and the output is available between collector and emitter of the device. This is the most used configuration of the transistor, in practice, when power amplification is required.

The transistor is said to be operated in *common collector configuration* when the collector terminal is made common between input and output circuits. In this configuration, the input is fed between base and collector and the output is available between emitter and collector. This configuration finds application as *voltage buffer amplifier*.

In the *common base configuration,* the base is common between input and output circuits of a transistor. The input is fed between emitter and base of the circuit and the output is delivered between collector and base of the circuit. This configuration is best suited as *constant current source or current buffer amplifier.*

5.6 CURRENT COMPONENTS IN A TRANSISTOR

To understand various current components in a transistor, common base configuration in active region is considered. In active region the emitter junction is forward biased and the collector region is reverse biased. A PNP transistor is considered and so a positive voltage is applied to the emitter (*p* type) of the device with respect to the base (*n* type). The battery supply to this junction is designated, as V_{EE} and the voltage applied at J_E is V_{EB}, which is positive. Similarly, to reverse bias, the collector junction separate battery is used which is designated as V_{CC} and the voltage between collector and base, V_{CB}, is negative. The negative terminal of V_{EE} and positive terminal of V_{CC} are made common at base terminal of the transistor. The currents entering into the device at emitter and collector terminals are I_E and I_C, respectively, and are assumed to be positive. The schematic representation of the current components is shown in Fig. 5.4.

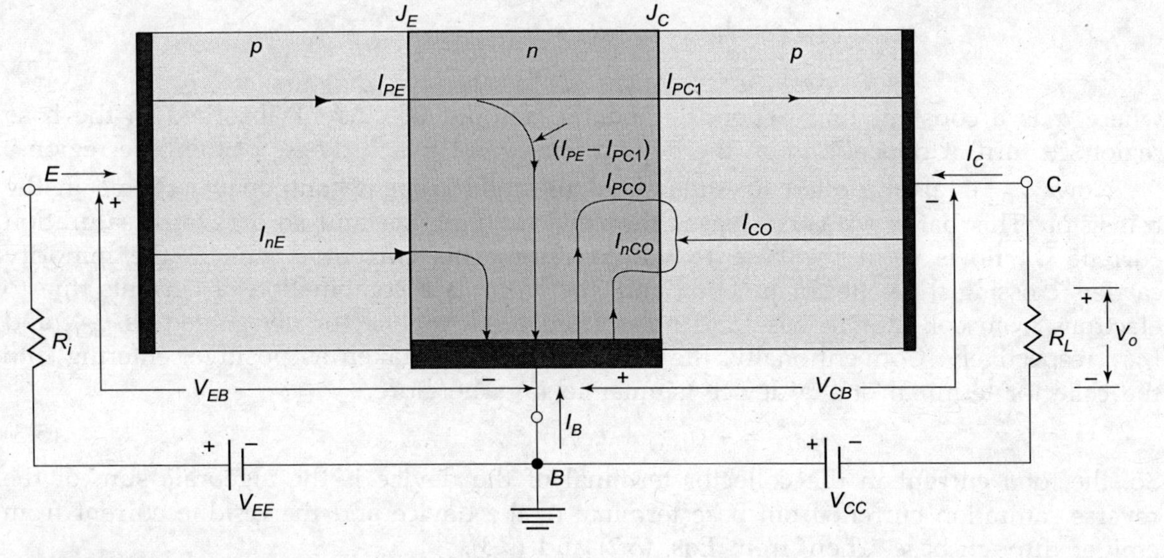

Fig. 5.4 *Current components in a transistor*

To simplify the analysis, it is assumed that only V_{EE} is applied and V_{CC} is not. So, the emitter junction is forward biased *p-n* junction diode. The current components in this junction are similar to that of a *p-n* junction diode in forward biased condition. The current I_E entering into the emitter will be split into two components in the emitter region of the transistor. The two components are due to majority carriers (holes) and minority carriers (electrons). These components are designated as I_{pE} and I_{nE}, respectively. Hence,

$$I_E = I_{nE} + I_{pE} \qquad ...(5.1)$$

I_{pE} is the current due to the holes crossing the junction between emitter and base and I_{nE} is the current due to the electrons crossing the junction from base to emitter. The magnitude of I_{nE} is very small compared to I_{pE} since, in general, the doping level of impurities in the emitter region is very large compared to that of the base and also that the emitter region is very large in volume compared with base. So, the holes, which are the majority carriers in the emitter, are very large in number compared to the electrons in the base region. The total current due to the electrons in the emitter region is I_{nE}. But the hole current in emitter, I_{pE}, originates in emitter, crosses the junction and reaches the base region. All the holes, which are penetrating the junction, are not absorbed into the base because the base region is very thin and lightly doped. The holes, upon entering into the base, become minority carriers in the base region and diffuse. The diffusion is not complete to make the current zero by the time the minority carriers reach the collector junction. So, a residue of the diffusion current due to holes reaches the junction between base and collector. The holes, which are not absorbed by the base region, penetrate the collector junction and reach the collector region to become majority carriers in this region. The residue current that reaches the collector can be designated as I_{pC1} and is due to the drift of the majority carriers in the collector region. Therefore, I_{pC1} is proportional to I_{pE}, or in other words, I_E(since $I_{pE} >> I_{nE}$, $I_E \approx I_{pE}$).

$$I_{pC1} \propto I_E$$

$$I_{pC1} = \alpha I_E \qquad \text{...(5.2)}$$

where α is a constant that depends upon the amount of current absorbed in the base region. In turn, it depends upon the doping levels and the thickness of the base region.

Now, V_{CC} is also applied to understand the total current component scheme in the transistor. This battery reverse biases the collector junction and so a reverse saturation current I_{CO} flows from the base to the collector. This current is due to the minority carriers crossing the collector junction and, therefore, is a combination of currents due to electrons from collector to base and holes from base to collector designated as I_{nCO} and I_{pCO}, respectively. Conventionally, the direction of I_{CO} is taken as positive, entering into the collector terminal of the device (similar to I_C). Therefore,

$$I_{CO} = -(I_{nCO} + I_{pCO}) \qquad \text{...(5.3)}$$

So, the total current in the collector terminal of the device is the algebraic sum of the reverse saturation current from base terminal of the device and the residue current from emitter through base. Then, from Eqs. (5.2) and (5.3),

$$I_C = I_{CO} - I_{pC1}$$

$$= I_{CO} - \alpha I_E \qquad \text{...(5.4)}$$

From the above equation, the constant α can be given as

$$\alpha = -(I_C - I_{CO})/(I_E - 0) \qquad \text{...(5.5)}$$

α is known as *large signal current gain in common base configuration.*

Now, α can be defined as negative of change in the collector current from cut off to the change in the emitter current from cut off (by definition, cut off condition is when both junctions are reverse biased, I_B is equal to I_{CBO} and I_E is equal to zero).

For a *pnp* transistor, which is active in common base configuration, the flow of current in the emitter is into the device and the current leaves out of the base and collector terminals. For an *npn* transistor all the voltages and currents are opposite in polarity to those of a *pnp* transistor. So, *large signal current gain* is the ratio of change of collector current to change of emitter current and is a positive quantity.

The value of a depends on the base region and its maximum is 1 (which is never possible). The values of a should be as large as possible and is typically around 0.90–0.995. In all the discussion the transistor is considered to be in active region. To derive a general equation for the collector current of common base configuration, all three regions of operation of the transistor are to be considered.

When the transistor is in cut off the emitter current is zero and so the collector current is only due to the reverse saturation current of collector junction. Therefore, from Eq. (5.4),

$$I_C = I_{CO} = I_{CBO} \qquad \text{...(5.6)}$$

When the transistor is in saturation the collector junction is also forward biased and I_{CO} in Eq. (5.4) is to be replaced by forward current of a *p-n* junction diode. Therefore,

$$I_C = -\alpha I_E + (-I_{CF}), \qquad \text{...(5.7)}$$

where I_{CF} is the forward current in the collector junction and can be given as

$$I_{CF} = I_{CO}\left[e^{V_{CB}/V_T} - 1\right] \qquad \text{...(5.8)}$$

Therefore the collector current can be given as

$$I_C = -\alpha I_E + I_{CO}\left[1 - e^{V_{CB}/V_T}\right] \qquad \text{...(5.9)}$$

where V_{CB} is the collector to base bias voltage.

Eq. (5.9) can be taken as a generalised equation of collector current in common base configuration. In active region the second term in brackets is very small compared to 1 since V_{CB} is negative and can be neglected. In cut-off region, in addition to the above condition, I_E is equal to zero. In saturation, Eq. (5.9) is a valid equation. Table. 5.2 gives the polarities of various voltages and currents for *pnp* and *npn* transistors.

Table 5.2 ***Polarities of Voltages and Currents in a Transistor***

Transistor	I_E	I_B	I_C	V_{CE}	V_{BE}	V_{BC}
pnp	positive	negative	negative	negative	negative	negative
npn	negative	positive	positive	positive	positive	positive

5.7 TRANSISTOR AS AN AMPLIFIER

The transistor in common base configuration has a current gain that is less than unity, but the voltage gain in this configuration is more than 1, and so the transistor in common base configuration is an amplifier for voltage.

Basically, the transistor handles current and is a current-controlled device. The input voltage to the amplifier is converted to the corresponding current at the input of the transistor and the transistor processes this current. The output current of the transistor is delivered into the load resistance, which develops the proportional output voltage across it.

The operation of the transistor as an amplifier can be understood when the transistor is visualised as a device that converts dc power into ac signals of the frequency same as that of input. This ac power is added in phase to the input ac signal and so the output of the transistor is an amplified version of input ac signal. So, the transistor basically is not an amplifying device but is a device that converts dc power into ac power. All the dc power from the battery is not converted into ac power, but some of it is dissipated in the form of heat. The conversion of dc to ac powers of a transistor depends upon its efficiency.

To understand the concept of amplification by a transistor theoretically, the input ac voltage is assumed to be a small increment in the dc biasing supply voltage. When the transistor is in common base configuration biased in active region, the J_E is forward biased by the virtue of proper selection of battery supply V_{EE}. A current I_E is drawn by this junction. Since the *V-I* characteristic of this forward biased junction is nonlinear, exponentially increasing, any small increase in voltage between the two regions will lead to a large increase in the current through the junction. This large current variation is transmitted to collector through the base and hence the collector current increases to a large extent. This increase of current in the collector is passed through output load resistance, which develops a large potential difference across it. Hence, a large increase in voltage in the output of the circuit is observed corresponding to a small increment of voltage in the input of the circuit. Analytically, the above discussion can be justified as follows. A load resistance R_L is in series with the collector and V_{CC}, as shown in Fig. 5.5.

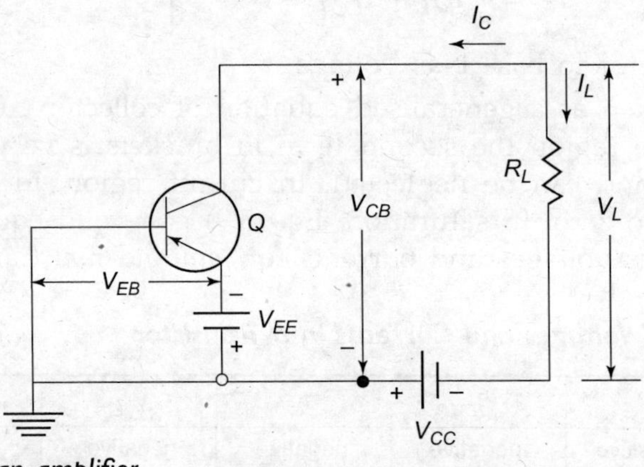

Fig. 5.5 *Transistor as an amplifier*

A small increment of voltage ΔV_i is given at the input of the circuit. This leads to a large increase in current ΔI_E. The fraction of the current reaching the collector through base is equal to $\alpha' \Delta I_E$. Therefore,

$$\Delta I_C = \alpha' \Delta I_E \qquad ...(5.10)$$

So, the increment in the output voltage can be given as

$$\Delta V_L = -R_L \Delta I_C = -R_L \alpha' \Delta I_E \qquad ...(5.11)$$

Therefore, the amplification in voltage can be given as

$$\begin{aligned} A &= \Delta V_L / \Delta V_i \\ &= (-R_L \alpha' \Delta I_E)/\Delta V_i \\ &= (-R_L \alpha' \Delta I_E)/r_e \Delta I_E \\ &= (-R_L \alpha')/r_e \end{aligned} \qquad ...(5.12)$$

where α' is known as *small signal current gain in common base configuration* and is defined as the ratio of small changes in output collector current to that of small changes in input emitter current, $\Delta V_i = r_e \Delta I_E$; r_e is the dynamic resistance of forward biased emitter junction and is given by $\eta V_T / I_E = 26 \text{ mV}/I_E$ at room temperature, 300 K. Therefore, the parameter α' can be given as

$$\alpha' = \left.\frac{\Delta I_C}{\Delta I_E}\right|_{V_{CB}} \qquad ...(5.13)$$

The name *transistor* is derived from the words transfer resistor.

5.8 *V-I* CHARACTERISTICS OF A TRANSISTOR

The *V-I* characteristics of a transistor are plotted for the output and input circuits separately. These characteristics are plotted for dc parameters and are referred to as *static characteristics*. In general, the input voltage and the output current of any transistor configuration are assumed to be dependent parameters and the input current and output voltage are taken as independent parameters. So, if port 1 is assumed to be input and port 2 as output, the output current is a function of output voltage and input current and the input voltage is a function of output voltage and input current. Therefore, the equations can be presented in an implicit form as follows:

$$I_o = \phi_2(V_o, I_i)$$

and,

$$V_i = \phi_1(V_o, I_i) \qquad ...(5.14)$$

where ϕ_1 and ϕ_2 are some functions of V_o and I_i

(The above equations can be read as "the output current is some function ϕ_2 of output voltage and input current.")

5.9 COMMON BASE CONFIGURATION

The transistor, when connected in common base configuration, is as shown in Fig. 5.6.

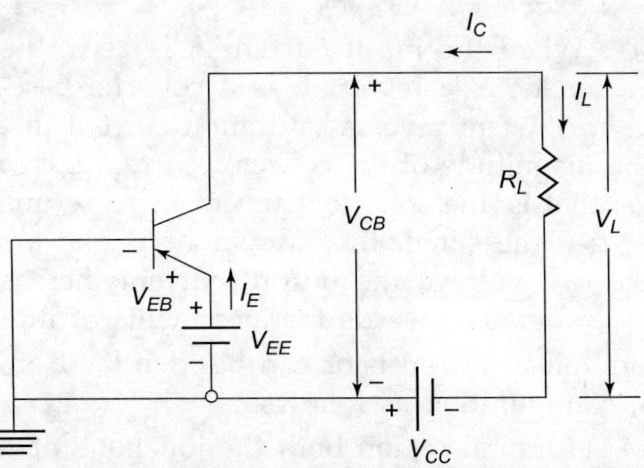

Fig. 5.6 *Transistor in common base configuration*

To understand the operation of the transistor in this configuration, the *V-I* characteristics are to be plotted. The output and input characteristics are plotted separately. The *output characteristic* is the variation of output collector current I_C as a function of output voltage V_{CB} with the input current I_E as a parameter. The *input characteristic* is plotted as the input current I_E as a function of input voltage V_{EB} with output voltage V_{CB} as a parameter. The typical characteristics of a *pnp* transistor are as shown in Fig. 5.7(a).

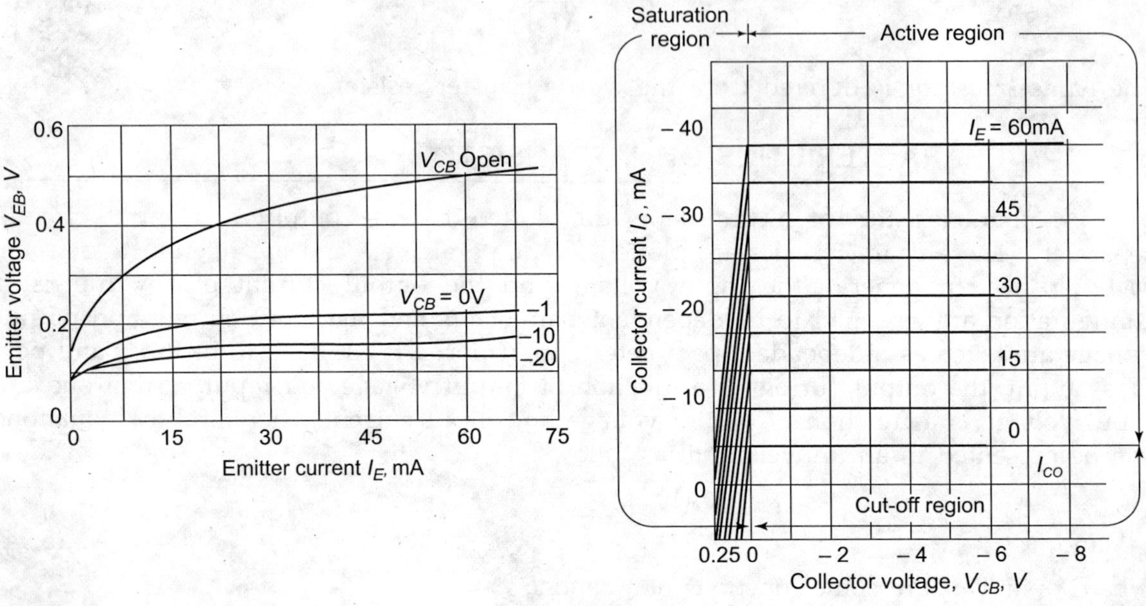

Fig. 5.7 *(a) Input characteristics of a transistor in common base configuration (b) Output characteristics of a transistor in common base configuration*

From Fig. 5.7(a) and Eq. (5.14) the governing equations can be written as

$$I_C = \phi_2(V_{CB}, I_E)$$

and,
$$V_{EB} = \phi_1(V_{CB}, I_E) \quad ...(5.15)$$

The output characteristics: When the input current I_E is zero, the output characteristic is nothing but a *V-I* characteristic of a reverse-biased collector-base junction. So, the current flowing in the device is a constant reverse saturation current in the collector junction I_{CO}. When I_E is increased, the magnitude of the collector current increases, but the shape of the curve remains the same; that is, the collector current increase initially with output voltage and remains constant with the same. Breakdown occurs at very high output voltages and beyond this breakdown voltage the output current increases very rapidly. So, the characteristics seem to be shifted upwards for increments of input current I_E.

The regions of operation of a transistor can be identified on output characteristics of the common base (CB) configuration as follows.

Saturation region: In the saturation region both the junctions of the transistor are forward biased and so a maximum current flows in this region of operation. For forward biasing the collector junction, both V_{CB} and I_E should be positive. The region to the left of the ordinate, $V_{CB} = 0$ and above $I_E = 0$ characteristics is saturation region. In this region, for a small change in collector voltage a large change in collector current is observed and V_{CB} is slightly positive. The above discussion is applied to a *pnp* transistor whose typical characteristics are as shown in the Fig. 5.7. For an *npn* transistor the regions of operation remain the same, except that the polarities of voltages and currents are reversed.

Cut off region: In this region of operation both the junctions are reverse biased and so a minimum current is expected to flow in the device. That is, both V_{CB} and I_E should be negative for a *pnp* transistor. So, the region below $I_E = 0$ and to the right of $V_{CB} = 0$ is the cut off region as shown in Fig. 5.7. In this region of operation the current flowing in the collector junction is I_{CO} and, by definition, I_E is zero.

Active Region: The transistor is said to operate in *active region* when the emitter junction is biased in forward direction and the collector junction in the reverse direction. The transistor is used as amplifier when it is biased in active region. For a *pnp* transistor to reverse bias, the collector junction V_{CB} is negative and is shown as a region to the right of $V_{CB} = 0$ in Fig. 5.7. The other condition for the active region is that the emitter junction should be forward biased. When the emitter junction is forward biased, the current in the junction flows from emitter through base to collector and is positive when entering the emitter. So, this region is identified as I_E is positive and greater than zero. The characteristics above $I_E = 0$, and right to $V_{CB} = 0$, in the output can be identified as active region, as shown in the figure.

This active region of operation in a transistor is the most sensitive region of operation in a transistor. The variation of output collector current depends on the input emitter current variations. Even though magnitude of output current is small compared to that of the input current, the variation is very sensitive.

Early effect or *base width modulation*: In order to increase the value of α, the base region is lightly doped and also the thickness is very small compared to either emitter region or collector region. Since the value of α is approximately equal to 1, the magnitude of the collector current and emitter current is almost the same. But since the emitter junction is forward biased, the potential drop across the junction is small (cutin voltage) while the collector junction is reverse biased and so the drop across the collector junction is very large. Due to this reason, the power dissipated in the emitter junction is small compared to the power dissipated in the collector junction. Hence, the collector should be designed to dissipate the unused energy efficiently. So, the area or, in other words, the volume of the collector is the maximum in a transistor. That is the reason why the metal can encapsulation of a general commercially available transistor is connected to the collector electrically. So, in general, the collector region has the maximum thickness and high doping levels and the emitter is slightly less doped and has a smaller area compared to collector. The base is the least doped region in the transistor and also the thickness is at the minimum.

When the collector-base junction is reverse biased, the depletion layer penetration into the base is very large compared to that into the collector region. As the value of V_{CB} increases, the penetration of the depletion layer into the base region increases. The penetration of depletion layer into the collector region is very small and can be neglected since the concentration is high. Due to this penetration of depletion layer into the base region, the region in the base where the recombination takes place reduces as the collector junction voltage increases. Therefore, as the collector junction voltage increases, the recombinations in base decrease and more current is delivered to the load, which ultimately increases the value of α.

If the physical base width of the transistor is W_B, then the effective electrical base width is $W_B' = W_B - W$, where W is the width of depletion layer in the base region. Thus, the effective electrical width of the base region changes with applied voltage. With increase in reverse voltage the base width decreases. This modulation of the base is referred to as *early effect* or *base width modulation*.

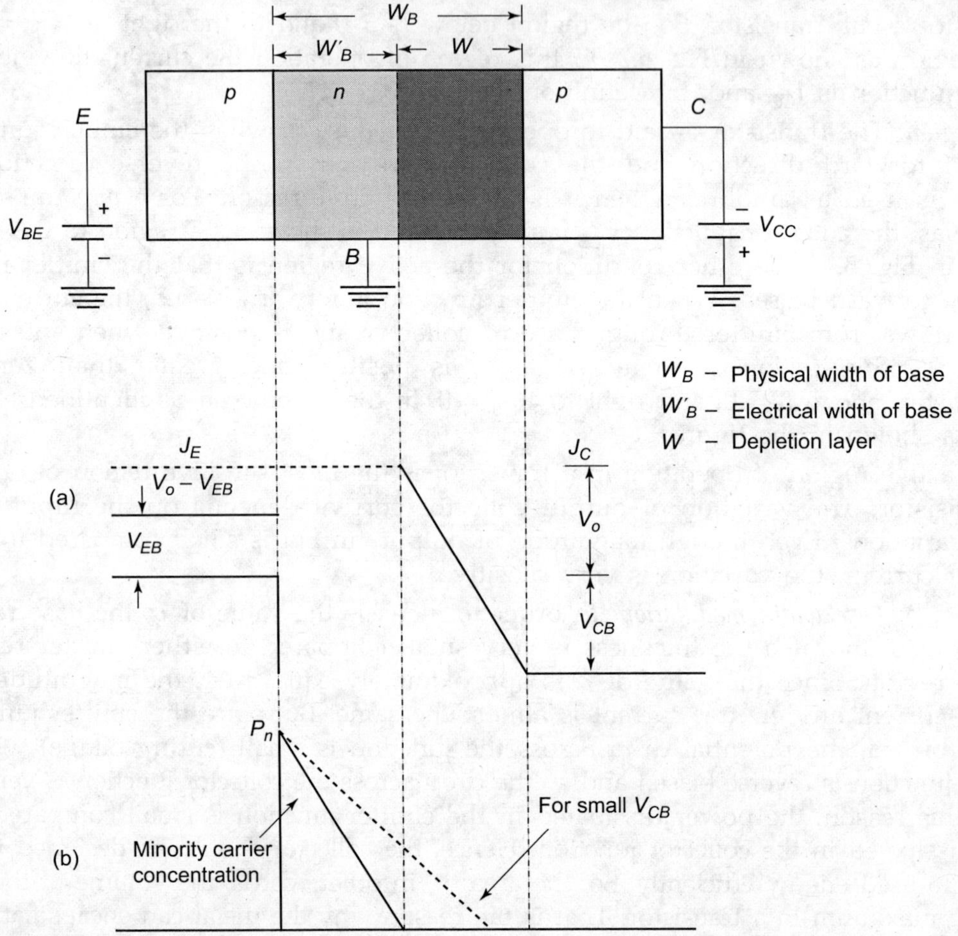

Fig. 5.8 *(a) The potential variation through a pnp transistor (b) The injected minority carrier charge density within the base*

The consequences of the base width modulation are the following:

- With increase in reverse collector voltage, the large signal current gain increases.
- In the base region, the concentration gradient of minority carriers increases.
- As the value of collector junction voltage increases, a voltage is reached at which the whole of the physical width of the base is occupied by the depletion layer. This condition is known as *punch through* or *reach through*.

In the condition of punch through, whole of the base region is depleted and there is no activity in the base, leading to a short circuit between collector and emitter regions. So, tremendously large amounts of current are prone to be flowing in the device, which dissipates large amounts of energy in the form of heat above the tolerable value of the device. So, the device burns out if the applied voltage at the collector is more than the rated voltage to cause punch through.

The input characteristics: In the active region the emitter junction is forward biased and so the characteristics are similar to that of the forward biased *p-n* junction diode. Conventionally, the abscissa is taken as the input emitter current I_E and the ordinate as the input emitter voltage $V_{EB.}$ In the saturation region also the junction is forward biased and the characteristics remain the same. But, in cut off region the emitter current is assumed to be zero and so the input characteristics do not have any significance.

The main concern of the input characteristics is the *cutin* or *offset* or *threshold voltage* V_γ below which the emitter current is very small and can be neglected. The value of V_γ is 0.1 V for a germanium transistor and 0.5 V for a silicon transistor.

When V_{EB} is constant, by the early effect, the emitter current increases with the increase in the magnitude of the collector voltage. The curves shift downwards as collector junction voltage increases. The curve with collector open represents the characteristics of the forward biased emitter diode.

5.10 COMMON EMITTER CONFIGURATION

The transistor connected in CE configuration is as shown in Fig. 5.9.

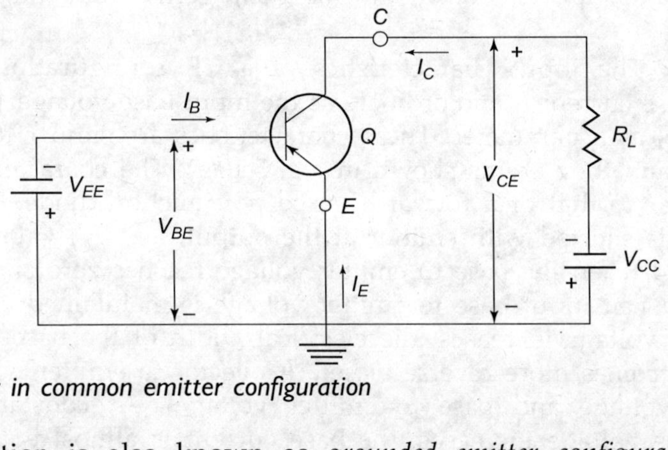

Fig. 5.9 *Transistor in common emitter configuration*

This configuration is also known as *grounded emitter configuration.* The amplifier connected in this configuration has ac voltage gain as well as current gain large compared to 1, and so there is a large amplification of power in this transistor circuit.

In this configuration, the input current I_B and the output voltage V_{CE} are taken as independent variables, whereas the input voltage V_{BE} and output current I_C are the dependent variables. So, the implicit relationship between the parameters can be written as

$$V_{BE} = f_1(V_{CE}, I_B)$$
$$I_C = f_2(V_{CE}, I_B) \quad ...(5.16)$$

The first equation describes the input characteristic curves of the CE configuration and the second equation describes the output characteristic curves.

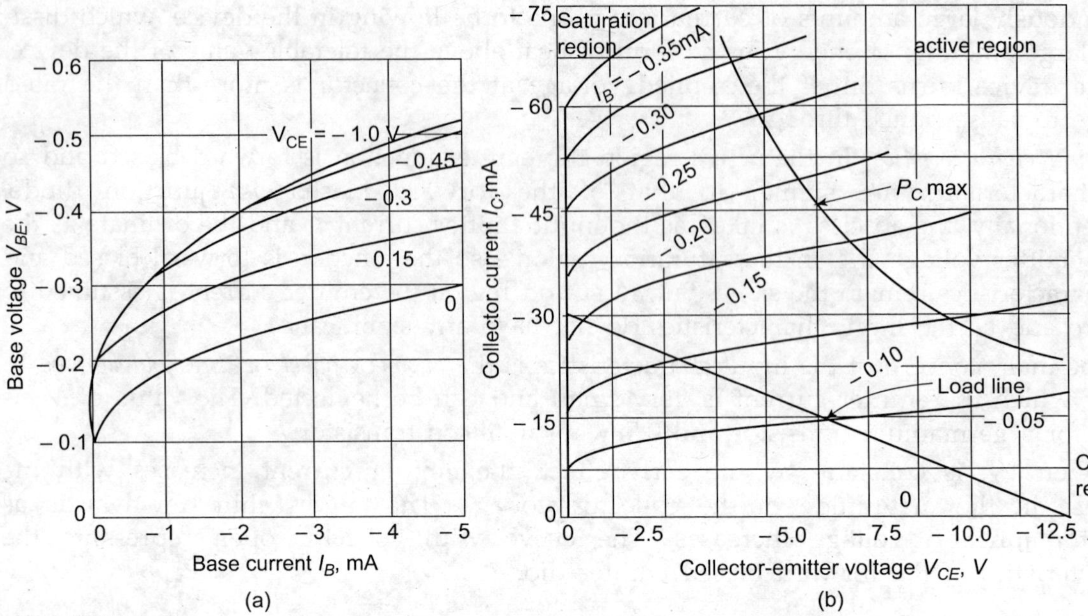

Fig. 5.10 *Input and output characteristics of common emitter configuration*

Typical output and input characteristics for a *pnp* germanium transistor are shown in Fig. 5.10.

Input characteristics: The input characteristics of a CE configuration are plotted with abscissa as input base current I_B and ordinate as the input base voltage V_{BE} with the output collector voltage V_{CE} as a parameter. These characteristics are similar to those of common base configuration and they are as shown in Fig. 5.10(a). The characteristics of input of a transistor are similar to that of a forward biased *p-n* junction diode. This is exactly true when the collector is shorted with emitter at the output, i.e., $V_{CE} = 0$. The current in the base reduces to zero when the base to emitter voltage reaches zero since both the biasing voltages are zero. When input base to emitter voltage is maintained constant and output collector to emitter voltage increases, the electrical width of the base decreases since the collector junction becomes more reverse biased. (Collector to emitter voltage is the sum of collector to base voltage and base to emitter voltage). At constant output collector voltage, as the base voltage increases, the base current is almost equal to zero initially until the cutin voltage is reached and above this the current increases exponentially as in the case of a *p-n* junction diode. When I_B is constant, increasing V_{BE} increases V_{CE}, as seen from the input characteristics. This is due to the reason that when V_{BE} increases, since base junction is forward biased, base current increases. In order to maintain a constant base current, V_{CE} is enhanced, which in turn increases V_{CB} and thus the electrical base width decreases, decreasing the current in the base to compensate for the initial increase in the base current.

The characteristics of a silicon transistor are similar to those of a germanium one, except that the cutin voltage is of the order of 0.5–0.6 V instead of 0.1–0.2 V for a germanium transistor.

Output characteristics: The output characteristics of the transistor in CE configuration are a family of curves that have a shape similar to that of CB configuration. But the output collector current is not constant with respect to output collector voltage when the input base current is held constant. The characteristics drift with a slight slope; this is due to the early effect. When base current is held constant and collector voltage is increased, the electrical base width decreases, leading to the increase in collector current.

Active Region: In this case, the collector junction is reverse biased and emitter junction is forward biased. The active region on the output characteristics is the region above $I_B = 0$ and to the right of $V_{CE} = 0$ and is as indicated in Fig. 5.10(b). The output current is very sensitive to the variations of input in the active region. For a small variation of input current, a large variation of current is obtained in this common emitter configuration transistor in the active region. Thus, amplification of current variation is observed and hence, the device can be used as an amplifying device.

The output collector current can be expressed in terms of base current with the help of Kirchchoffs current law (KCL) and the collector current expression [Eq. (5.4)] from CB configuration.

Applying KCL at the transistor,

$$I_C + I_B + I_E = 0$$

Therefore, the collector current in the CE configuration can be expressed as

$$I_C = \frac{I_{CO}}{1-\alpha} + \frac{\alpha I_B}{1-\alpha} \qquad ...(5.17)$$

$$= (1 + \beta)I_{CO} + \beta I_B \qquad ...(5.18)$$

where, $$\beta = \alpha/(1 - \alpha) \qquad ...(5.19)$$

β is called the *large signal current gain* in common emitter configuration and can be expressed in terms of output and input currents as

$$\beta = \frac{I_C - I_{CO}}{I_B - (-I_{CO})} \qquad ...(5.20)$$

The base current in cut off condition is $-I_{CBO}$. Now, the β, *the large signal current gain*, can be defined as the change in the output collector current from cut off to change in the input base current from cut off.

From Eq. (5.19) it can be seen that as the value of α tends towards 1, the value of β tends towards infinity. In other words, when typical value of $\alpha = 0.95$, $\beta = 19$, which is a large value compared to 1, and so a transistor in CE configuration has a large current gain, unlike that of a transistor in CB configuration.

For a slight change in the value of α, the change in the value of β is very large. For example, when α increases from 0.95 to 0.99, β increases from 19 to 99. In other words, for a 4.2 % increase in α, the percentage of increase in β is 421%. The value of β is very large because it is a ratio of collector current to base current. Since the magnitudes of collector current and emitter current are approximately equal, the magnitude of base current is very small. So, for a very small change in base current, a large change in collector current is anticipated.

Cut off region: In this region of operation both junctions are reverse biased and so the saturation current flows through the junctions. This region of operation can be identified on the output characteristics as shown in Fig. 5.10(b). Assuming collector junction is reverse biased, a reverse saturation current I_{CO} flows through this junction. If the base current is assumed to be zero, this current enters the emitter region and it can be designated as I_{CEO}. Therefore, the collector current equation when $I_B = 0$ is

$$I_C = -I_E = I_{CO}/(1-\alpha) = I_{CEO} \quad ...(5.21)$$

This current flows through the emitter junction and is very large when α tends towards 1. This means that a positive current is flowing from *p*-type base to *n*-type emitter, which makes the junction forward biased, and the transistor lands into active region. So, the definition of cut off region with $I_B = 0$ is not possible.

So, the cut off condition is redefined as with I_E=0. When I_E=0, the collector current enters the base and can be designated as I_{CBO}. Now, the base current I_B is equal to negative of I_{CBO}. In addition to $I_E = 0$, $I_B = -I_{CBO}$ and $I_C = I_{CBO}$, it is to be ensured that the emitter junction is reverse biased for a transistor to be in cut off region. This requires a minimum of zero volts for a silicon transistor and 0.1 V for a germanium to reverse bias the emitter junction.

The magnitude of I_{CBO} is large compared to that of I_{CO} due to the following reasons:

- There exists a leakage current across the surfaces of encapsulation due to the high voltage applied across the junction.
- The minority carriers, when crossing the junction, may collide with covalent bonds to create a new electron-hole pair, which initiates avalanche multiplication and eventual breakdown. Even though the breakdown is not complete, the magnitude of the current may be multiplied many times.

The temperature dependence of I_{CBO} is the same as that of the reverse saturation current of a *p-n* junction diode. The magnitude of the current doubles approximately for every 10 °C increase in temperature for both Ge and Si. However, because of the low absolute value of I_{CBO} in silicon these transistors may be used up to 200 °C, whereas germanium transistors are limited to about 100 °C.

The cut off condition of a transistor requires $I_E = 0$, $I_B = -I_{CBO}$ and the cut off voltage at base to emitter junction is $V_{BE} \approx -0.1$ V. This requires that

$$V_{BE} = -V_{BB} + R_B I_{CBO} \leq -0.1 \text{ V} \quad ...(5.22)$$

where V_{BB} is the battery supply at the base terminal, R_B is the baising resistant in the base circuit.

The manufacturer specifies the maximum voltage that can be applied between the emitter-base junctions when collector current is zero as BV_{EBO}, *breakdown voltage*. The breakdown voltage BV_{EBO} may be as high as some tens of volts or as low as 0.5 V. The value of the supply V_{BB} is so chosen that the emitter junction voltage never crosses BV_{EBO}.

The saturation region: Both the junctions of emitter and collector are forward biased by at least cutin voltage in the saturation region of operation of a transistor. The saturation region for CE configuration can be identified as shown in Fig. 5.10(b). Since the voltage V_{BE} is of the order of few tens of a volt, $V_{CE} = V_{BE} - V_{BC}$ is also only few tenths of a volt at saturation. So, the saturation region is very close to that zero voltage axis, where all the curves merge and fall rapidly toward the origin. In this region of operation, the collector current is approximately independent of the base current, for given values of V_{CC} and R_L.

For a transistor operating in the saturation region, the parameter of interest is *common emitter saturation resistance* R_{CS} or R_{CES} or $R_{CE\text{sat}}$. This is defined as the ratio of $V_{CE\,\text{sat}}/I_C$. The value of R_{CS} is of the order of few ohms.

5.11 COMMON COLLECTOR CONFIGURATION

The transistor connected in a common collector configuration finds application as a *buffer amplifier*. Since the voltage gain offered by this amplifier is approximately equal to 1, the output voltage is approximately the same as that of the input and there is no phase shift between input and output; the output voltage follows the input voltage. Or, in other words, the emitter voltage follows the input base voltage and so the circuit is referred to as *emitter follower*.

The circuit diagram of a transistor common collector configuration is as shown in Fig. 5.11.

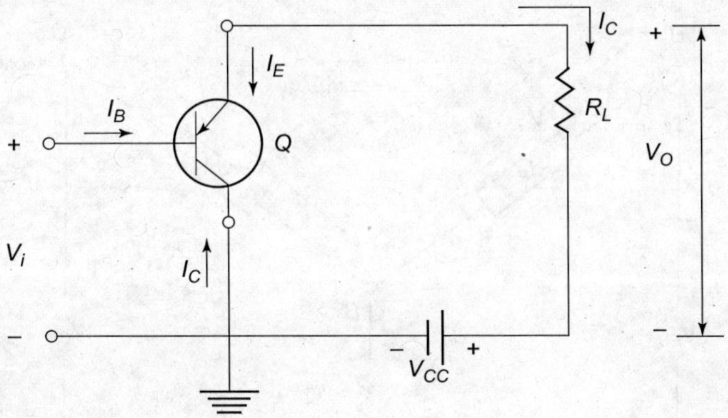

Fig. 5.11 *Common collector configuration*

The implicit functions for this configuration can be written as that of the CE configuration, with I_B and V_{EC} as independent parameters and I_E and V_{BC} as dependent parameters. So, the functions can be derived to be

$$V_{BC} = f_1(V_{EC}, I_B)$$
$$I_E = f_2(V_{EC}, I_B) \qquad \text{...(5.23)}$$

From the above equations, the output and input characteristics can be plotted and the various regions of operation can be identified similar to that of CE configuration.

5.12 EBERS-MOLL MODEL OF A TRANSISTOR

The general collector current depends on the junction voltages and can be written as

$$I_C = -\alpha_N I_E - I_{CO}\left[e^{(V_C/V_T)} - 1\right] \qquad \text{...(5.24)}$$

The subscript N in the above equation indicates that the transistor is operating in *normal manner*. However, in the transistor the roles of emitter junction and collector junction may be inverted and this interchanging of roles might not be as effective as normal mode of operation. In the inverted mode of operation, Eq. (5.24) may be recast as

$$I_E = -\alpha_I I_C - I_{EO}\left[e^{(V_E/V_T)} - 1\right] \qquad \text{...(5.25)}$$

where α_I is the *inverted* common base current gain. I_{EO} is the reverse saturation current and V_E is the emitter junction voltage and is positive for forward biased emitter.

From Eqs. (5.24) and (5.25) the parameters α_N, α_I, I_{CO} and I_{EO} are not independent and are related by the condition

$$\alpha_I I_{CO} = \alpha_N I_{EO}. \qquad \text{...(5.26)}$$

Base spreading resistance: In general, in a transistor the emitter is highly doped and the base is lowly doped. The current that is emitted from the emitter junction enters into the base where the charge carriers are very less in number and so this current encounters an inherent resistance when entering into the base. This resistance between the emitter and base can be represented as *base-spreading resistance* designated as $r_{bb'}$. The value of this resistance depends upon the concentrations of two regions and is of the order of 100 Ω.

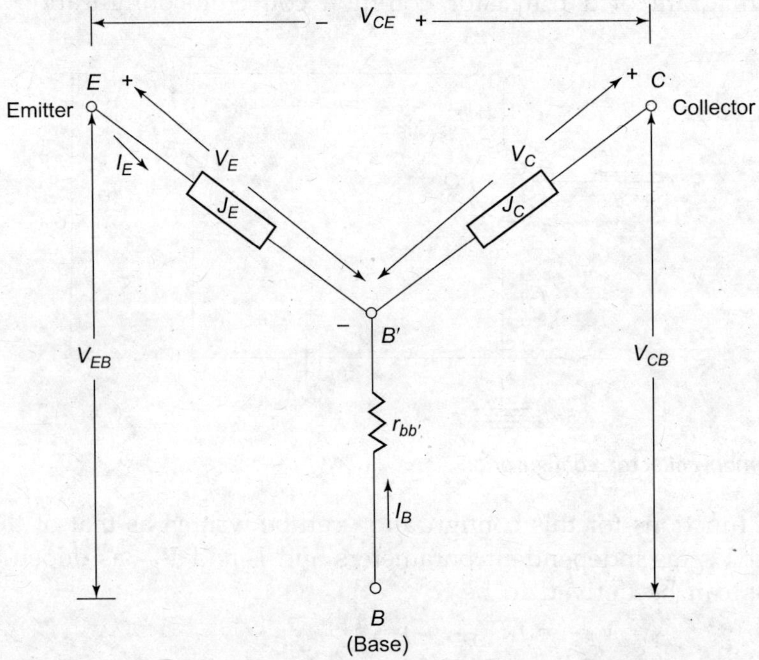

Fig. 5.12 *Diagram for defining the voltages and currents in Ebers-Moll model*

From Fig. 5.12, the voltage drop across collector-base junction, V_{CB}, can now be written as

$$V_{CB} = V_C - I_B r_{bb'} \qquad \text{...(5.27)}$$

The circuit drawn based on Eqs. (5.24) and (5.25), shown in Fig. 5.13, is known as *Ebers-Moll model.* This involves two ideal diodes placed back to back with reverse saturation currents I_{EO} and I_{CO} and two dependent current-controlled current sources shunting these diodes. For a *pnp* transistor the two currents are negative. The current sources account for the minority carrier transport across the base. Therefore,

$$I_C = -\alpha_N I_E + I_O\left[e^{(V_C/V_T)} - 1\right] \qquad \text{...(5.28)}$$

where $I_O = -I_{CO}$

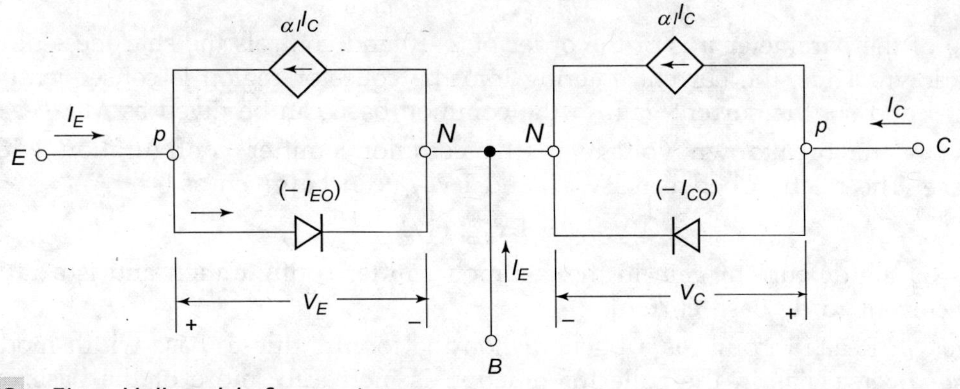

Fig. 5.13 *Ebers-Moll model of a transistor*

From Fig. 5.13, the dependent current sources can be eliminated when $\alpha_N = \alpha_I = 0$. If the base width is made very large compared to the diffusion length, the current from the emitter junction is totally absorbed by the base region and the current delivered to the collector is zero. Or, in other words, the large signal current gain $\alpha = 0$. A transistor is manufactured in such a way that the base region is very lightly doped and is very thin, so that the recombination in this region is minimal and major amount of current from emitter junction is delivered to collector, tending α towards 1. But, when one connects two *p-n* junction diodes back to back, this feature is absent and the base region is large compared to diffusion length and so all the current is absorbed and no charge carriers are left over to be delivered to the collector. This is the reason why one cannot build a transistor by connecting two diodes back to back.

5.13 VOLTAGE RATINGS OF A TRANSISTOR

The transistor has a capability of dissipating a given amount of power. This rated power depends upon the power dissipation capacity at the collector junction of the transistor. In active region, since the emitter junction is forward biased, the dissipation across this junction is very small and so can be neglected. While using a transistor one has to be sure that the applied collector voltage never crosses the maximum rated voltage for a given transistor. If the power dissipation in the collector junction is more than the rated value, the transistor burns out and is of no further use .

The maximum voltage rating of a transistor is fixed by the manufacturer in his catalogue based on two principles, which lead to maximum current in the device. The two principles are *avalanche multiplication* and *reach through.*

Avalanche multiplication: The collector junction is reverse biased and so saturation current is anticipated to be flowing through the junction. But, this current is not constant and increases drastically due to avalanche multiplication. The voltage at which the junction breaks down and the current multiplies manyfold is designated as *breakdown voltage,* BV_{CBO}. As a result of multiplication the current in the junction becomes MI_{CO}, where M is a factor by which the saturation current is multiplied by the process of avalanche. When the collector voltage reaches a value of BV_{CBO}, the value of M nominally tends to infinity. So, the current rises abruptly and for a small change in the applied voltage the current changes by a very large value. The value of M can be given as

$$M = 1/[1 - \{V_{CB}/BV_{CBO}\}^n] \quad ...(5.29)$$

The value of the parameter n is of the order of 2–10 and controls the sharpness of onset of the breakdown. Thus, the current entering into the collector region is consequently multiplied by M and so, the current gain in the common base can be given as $M\alpha$.

BV_{CEO} is the breakdown voltage in the common emitter configuration with open circuit base. The relation between BV_{CEO} and BV_{CBO} can be given as

$$BV_{CEO} = BV_{CBO}(1/h_{FE})^{1/n} \quad ...(5.30)$$

where h_{FE} is the dc current gain in the common emitter configuration and is a ratio of dc collector current to dc base current.

Reach through: Reach through is a phenomenon that occurs due to base width modulation. When the reverse bias in the collector junction is increased, more of the base region is depleted and a voltage is reached at which total base is depleted. At this voltage the transistor loses its action and draws maximum current, damaging the device permanently. So, caution is to be exercised so that this voltage is never reached.

Punch through or reach through occurs at a voltage when total base region is depleted. Thus, for a given transistor, it occurs at a particular voltage. This voltage has nothing to do with the circuit in which the transistor is connected. The maximum voltage that can be applied to the transistor is either of avalanche breakdown voltage or punch through voltage whichever is less.

5.14 TYPICAL JUNCTION VOLTAGES

The typical voltages for the junctions of a transistor are summarised in Table 5.3.

Table 5.3 ***Typical Junction Voltages for an npn Transistor***

	$V_{CE,\ sat}$	$V_{BE,\ Sat}$	$V_{BE,\ active}$	$V_{BE,\ cutin}$	$V_{BE,\ cutoff}$
Si	0.2	0.8	0.7	0.5	0.0
Ge	0.1	0.3	0.2	0.1	– 0.1

These values are for an *npn* transistor. For a *pnp* transistor all the polarities must be reversed.

The voltages referred to in Table 5.3 are typically for an *npn* transistor and for a *pnp* transistor the polarities of the voltages are to be reversed. Or, in other words, all the entries of the table will be negative except for $V_{BE\ \text{cut off}}$ of a Ge transistor. The voltage $V_{CE\ \text{active}}$ is a variable quantity that depends upon the biasing of the transistor in the active region.

5.15 CURRENT GAINS

α and β are defined as large signal current gain in common base and common emitter configuration, respectively, and are given as follows:

$$\alpha = -(I_C - I_{CO})/(I_E - 0) \quad ...(5.31)$$

$$\beta = \frac{I_C - I_{CBO}}{I_B - (-I_{CBO})} \qquad ...(5.32)$$

where I_{CBO} is the saturation current flowing from collector to base when emitter is open in cut off region. The relation between α and β is as given in the following:

$$\beta = \alpha/(1 - \alpha) \qquad ...(5.33)$$

Small signal current gain in common base and common emitter configurations are as given as follows:

$$\alpha' = \left.\frac{\Delta I_C}{\Delta I_E}\right|_{V_{CB}} = [\partial I_C/\partial I_E]_{V_{CB}} \qquad ...(5.34)$$

for common base configuration, and for common emitter configuration it is also designated as h_{fe} and is defined as

$$\beta' = h_{fe} = [\partial I_C/\partial I_B]_{V_{CE}} \qquad ...(5.35)$$

Another current gain of interest in a transistor is dc current gain and is given as follows:

$$\alpha_{dc} = I_C/I_E \text{ for common base configuration}$$

and

$$\beta_{dc} = I_C/I_B = h_{FE} \text{ for common emitter configuration.}$$

The transistor in common collector configuration is generally used as a buffer amplifier with voltage gain approximately equal to 1. The amplification of the power in this configuration is not a concern in practice. So, the concept of current gain in common collector configuration finds no place in the analysis of the circuit. On similar lines to those of common base and common emitter configurations, the current gain in common collector configuration may be defined in terms of γ.

SUMMARY

- A bipolar junction transistor can be visualised as two *p-n* junction diodes connected back to back. It consists of two junctions and three regions.
- The three terminals are known as emitter, base and collector. The two junctions are emitter-base junction and collector-base junction.
- Two types of transistors are *pnp* and *npn* transistors. The function is the same for both types, except the voltages and currents are exactly opposite.
- To realise the transistor as a two-port device, three configurations are possible—common base, common emitter and common collector.

- There are three regions of operation based on the biasing of two junctions. If both the junctions are forward biased, it is in saturation; if both are reverse biased, it is in cut off region and if emiter junction is forward biased and collector junction reverse biased, it is in active region.
- The transistor acts as an amplifier in active region. Transistor can be used as a switch when it is switched between saturation and cut off.
- In saturation, maximum current flows in the device. In cut off minimum or no current is allowed in the device and in active, the transistor responds to the changes in the output and amplifies the changes in input.
- Transistor is basically a current device. It amplifies the input current. Thus, current gain is the important parameter.
- Large signal current gain is the ratio of large change in output current to large changes in the input current. Small signal current gain is the ratio of small changes in output current to small changes in the input current. dc current gain is the ratio of dc output current to dc input current.
- α and β are the current gains in CB and CE configurations, respectively. They are related as $\beta = \alpha/(1 - \alpha)$. α is of the order of 0.9 or above and, thus, β is very large. Thus, current gain in CE is very large compared to CB configuration.
- The *V-I* characteristics of a transistor are plotted as output and input characteristics.
- Variation of output current versus output voltage with input current as parameter is the output characteristic and variation of input voltage versus input current with output voltage as parameter is the input characteristic. The characteristics can be drawn for all the three configurations.
- Early effect or base width modulation is the regulation of output current with respect to reverse voltage of collector current. As this voltage increases, the output current increases.
- Reach through or punch through is the voltage at which the total base is depleted and the transistor ceases its operation and is damaged totally. So, this is the maximum voltage that can be applied to the transistor.
- Cut off condition is the condition when the base current is the reverse saturation current of collector junction, the emitter current is zero and the emitter is slightly reverse biased.
- In saturation, the resistance offered by the device is least and the junction voltages are very small.
- Another limitation of the voltage at the reverse biased collector junction is the avalanche multiplication in the junction.
- Ebers-Moll model of the transistor interprets the current equation in a transistor. This involves two ideal diodes placed back to back with reverse saturation currents and two dependent current controlled current sources shunting the ideal diodes.

SOLVED PROBLEMS

5.1 Determine whether or not the transistor in the figure is in saturation. Assume $V_{CE}(\text{sat}) = 0.2$ V, and $\beta_{dc} = 50$ and $V_{BE} = 0.7$ V.

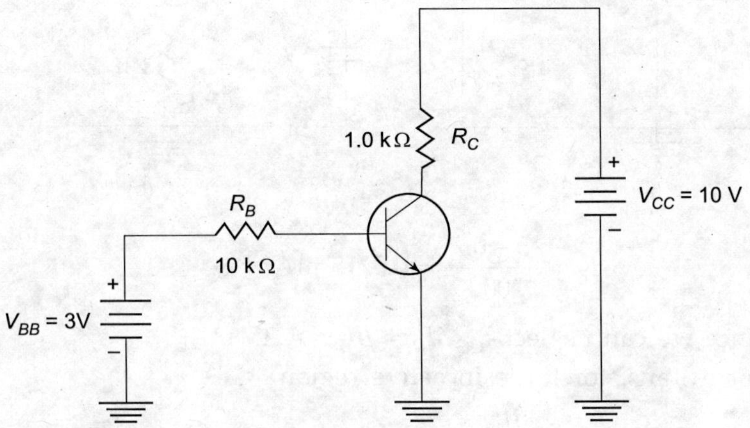

Solution First, determine $I_{C\text{sat}}$.

$$I_{C\text{sat}} = \frac{V_{CC} - V_{CE\,\text{sat}}}{R_C}$$

$$= \frac{10 - 0.2}{1.0\ \text{k}\Omega} = 9.8\ \text{mA}$$

Now, see if I_B is large enough to produce $I_{C\text{ sat}}$

$$I_B = \frac{V_{BB} - V_{BE}}{R_B} = \frac{3\,V - 0.7\,V}{10\ \text{k}\Omega} = 0.23\ \text{mA}$$

$$I_C = \beta_{dc} I_B = (50)\ (0.23\ \text{mA}) = 11.5\ \text{mA}$$

Since $I_C > I_{C\text{ sat}}$, we can say that the transistor is in saturation.

5.2 Determine whether the transistor shown in figure (a) is in active region or not. A silicon transistor with $\beta = 100$ and $I_{CO} = 20$ nA is under consideration. (b) Repeat part (a) if a 2 K emitter resistor is added to the circuit as in figure (b).

Solution (a) Since the base is forward biased, the transistor is not cut off. Hence, it must be either in its active region or in saturation. Assume that the transistor operates in the active region. Applying KVL to the base in figure (a),

$$-5 + 200\ KI_B + V_{BE} = 0$$

Since V_{BE} in active region = 0.7 V

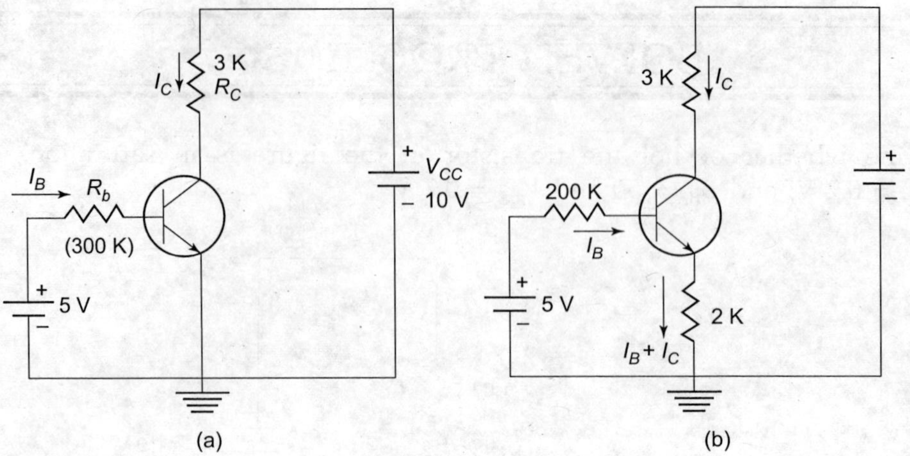

$$\therefore \qquad I_B = \frac{5-0.7}{200 \text{ K}} = 0.0215 \text{ mA}$$

Since we can neglect I_{CO}, $I_C \approx \beta I_B = 2.15$ mA.

For a transistor to be in active region

(1) $I_C \approx \beta I_B$.

(2) V_{CB} is to be + ve for an *npn* transistor,

because then only J_C is reverse biased (for a *pnp* transistor V_{CB} is to be negative).

Now, let us find V_{CB} and confirm that transistor is in active region.

Applying KVL to the collector loop

$3KI_C + V_{CE} - 10 = 0$

$\Rightarrow -10 + V_{CB} + V_{BE} + 3000 I_C = 0$

$\Rightarrow -10 + V_{CB} + 0.7 + 3000 \times 2.15 \times 10^{-3} = 0$

$\Rightarrow V_{CB} = +2.85$ V

Since V_{CB} is positive, transistor is in active region.

(b) Applying KVL to the base circuit yields

$-5 + 200 I_B + V_{BE} + 2(I_B + I_C) = 0$

$\Rightarrow -5 + 200 I_B + V_{BE} + 2(101) I_B = 0$

$(\because I_C = 100 I_B)$

$$\Rightarrow I_B = \frac{5-0.7}{402} = 0.0107 \text{ mA}$$

$\therefore \; I_C \cong \beta I_B = 1.07$ mA

Now, let us calculate the value of V_{CB}.

Applying KVL in the collector circuit,

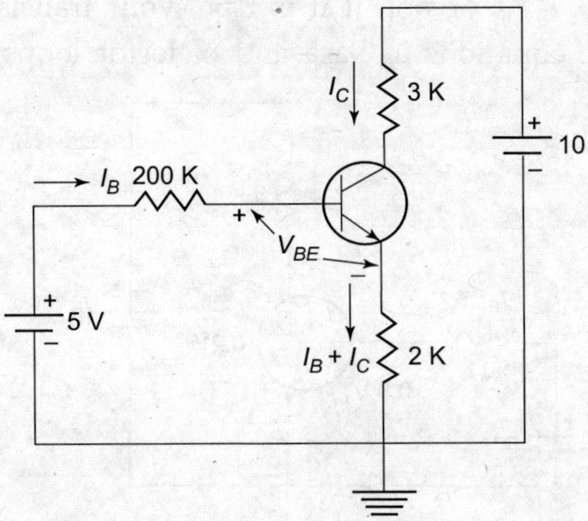

$$3I_B + V_{CB} + V_{BE} + 2(I_B + I_C) - 10 = 0$$

Substituting values of I_B and I_C, V_{BE}

$$V_{CB} = 10 - 3\ (1.07) - 2(101)\ (0.0107) - 0.7$$
$$= + 3.93 \text{ V}$$

Since V_{CB} is positive, *npn* transistor is in active region.

5.3 The transistors in figures (a) and (b) are having V_{BE} = 0.8 V, V_{CE} = 0.2 V and are of silicon type. If h_{FE} = 100, find whether they are in saturation or not and also find I_B and I_C.

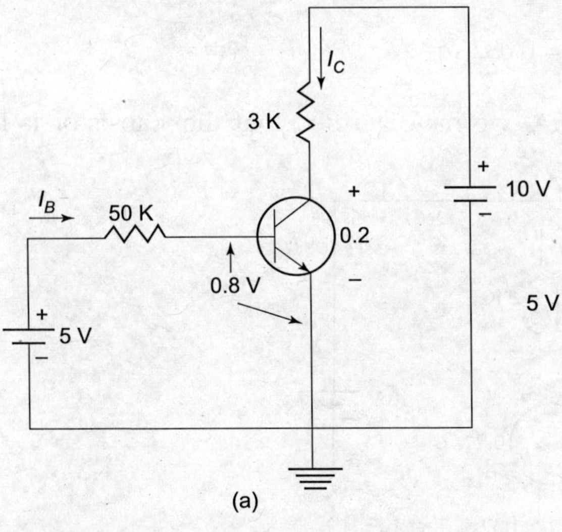

(a)

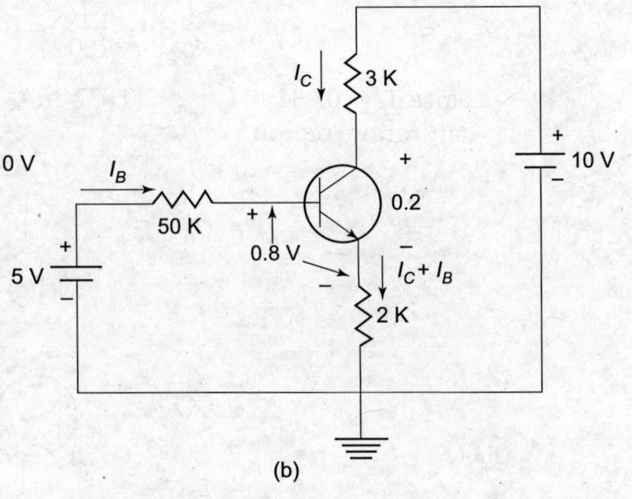

(b)

Solution To determine whether the transistor is in saturation or not

(1) Using KVL equations in base and collector loops, find I_B, I_C.

(2) Then find $I_{B\min} = \dfrac{I_C}{h_{FE}}$

(3) Check if $I_B > I_{B\min}$ or not, if it is true, your transistor is in saturation. Writing KVL equations in base and collector loops,

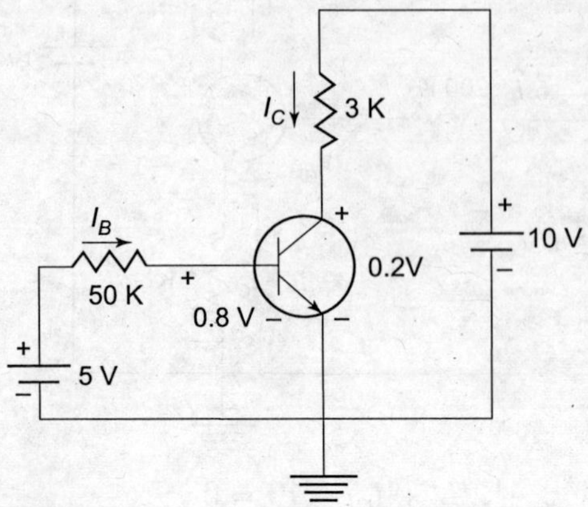

$$-5 + 50I_B + 0.8 = 0 \quad \text{...(1)}$$

$$\Rightarrow \quad I_B = \frac{4.2}{50} = 0.084 \text{ mA}$$

$$-10 + 3I_C + 0.2 = 0 \quad \text{...(2)}$$

$$\Rightarrow \quad I_C = \frac{9.8}{3} = 3.27 \text{ mA}$$

$$I_{B\min} = \frac{I_C}{h_{FE}} = \frac{3.27}{100} = 0.032 \text{ mA}$$

Since $I_B = 0.084 > I_{B\min} = 0.032$ mA, we have verified that the transistor is in saturation region.

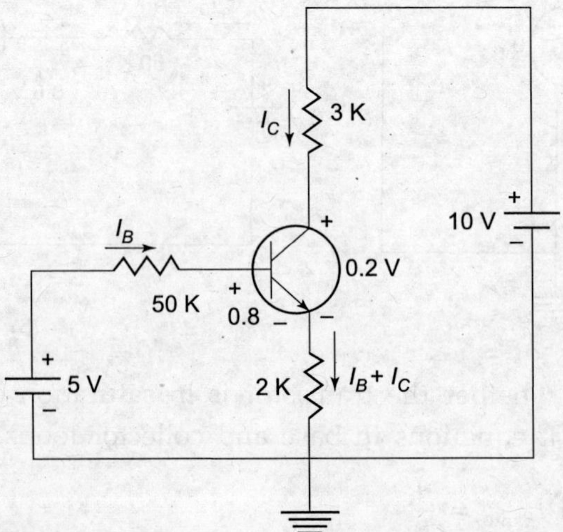

(c)

Applying KVL to the base and collector loops, we obtain

$$-5 + 50I_B + 0.8 + 2(I_B + I_C) = 0 \quad \text{...(1)}$$

$$-10 + 3I_C + 0.2 + 2(I_C + I_B) = 0 \quad \text{...(2)}$$

$$\Rightarrow \quad 52I_B + 2I_C = 4.2 \quad \text{...(1a)}$$

$$2I_B + 5I_C = 9.8 \quad \text{...(2a)}$$

To find I_B and I_C, let us use Cramer's rule.

$$\Delta = \begin{vmatrix} 52 & 2 \\ 2 & 5 \end{vmatrix} = 260 - 4 = 256$$

$$\Delta_1 = \begin{vmatrix} 4.2 & 2 \\ 9.8 & 5 \end{vmatrix} = 21 - 19.6 = 1.4$$

$$\Delta_2 = \begin{vmatrix} 52 & 4.2 \\ 2 & 9.8 \end{vmatrix} = 509.6 - 8.4 = 501.2$$

$$I_B = \frac{\Delta_1}{\Delta} = \frac{1.4}{256} = 5.468\ \mu\text{A}$$

$$I_C = \frac{\Delta_2}{\Delta} = \frac{501.2}{256} = 1.957\ \text{mA}$$

$$I_{B\min} = \frac{I_C}{h_{FE}} = \frac{1.957\ \text{mA}}{100} = 19.57\ \mu\text{A}$$

Since $I_B = 5.468\ \mu\text{A} < I_{B\min} = 19.57\ \mu\text{A}$

We can say that the transistor is not in saturation. So, it is in the active region.

5.4 (a) Find R_C and R_b in the figure, if $I_C = 10$ mA and $V_{CE} = 5$ V. A silicon transistor with $\beta = 100$, $V_{BE} = 0.7$ V and negligible reverse saturation current is used.

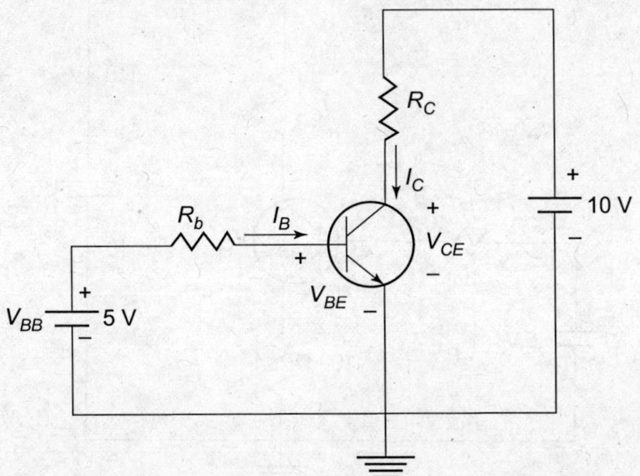

(b) Repeat part (a) if a 100 Ω resistor is added to the circuit.

Solution (a) First determine whether the transistor is in active, saturation or in cut off region. Since the base is connected to the positive of V_{BB} it is not in cut off. So, it should be either in active or in saturation.

$V_{CB} = V_{CE} - V_{BE} = 5 - 0.7 = 4.3$ V.

Since V_{CB} is +ve, it is in active region neglecting I_{CO}.

$$\beta = \frac{I_C}{I_B} \Rightarrow I_B = \frac{10}{100} = 0.1 \text{ mA}$$

Applying KVL to the base loop,

$$5 = I_B R_b + V_{BE}$$

$$\Rightarrow \quad R_b = \frac{5 - 0.7}{0.1 \text{ mA}} = 43 \text{ k}\Omega$$

From the collector circuit, applying KVL,

$$10 = R_C \times 10 \times 10^{-3} + V_{CE}$$

$$R_C = \frac{10 - 5}{10 \times 10^{-3}} = 500 \ \Omega$$

(b) Applying KVL to the base loop,

$$-5 + I_B \cdot R_b + V_{BE} + 100(I_B + I_C) = 0$$

$$R_b = \frac{5 - 0.7 - 100(10.1 \text{ mA})}{0.1 \text{ mA}}$$

$$= \frac{3.29}{0.1 \text{ mA}} = 32.9 \text{ k}\Omega$$

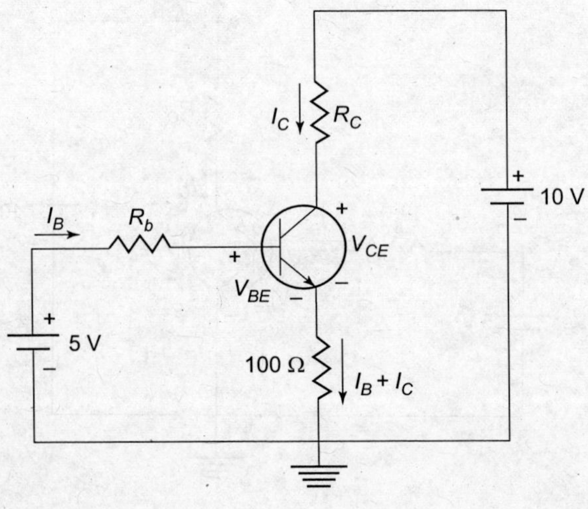

Similarly, applying KVL in the collector loop,

$$\Rightarrow I_C R_C + V_{CE} + (I_B + I_C)\, 100 - 10 = 0$$

$$\Rightarrow 10 \times 10^{-3} R_C + 5 + (0.1 + 10) \times 10^{-3} \times 100 - 10 = 0$$

$$\Rightarrow R_C = \frac{5 - 1.01}{10 \times 10^{-3}} = 399\ \Omega$$

5.5 In the circuit shown, V_{CC} = 24 V, R_C = 20 K, and R_e = 500 Ω, if a silicon transistor is used with β = 45 and if V_{CE} = 5 V, find R. Neglect the reverse saturation current.

Solution Since V_{CE} = 5 V, V_{BE} = 0.7 V

This is in active region.

Applying KVL to the collector to emitter circuit,

$$V_{CC} = I_E R_C + V_{CE} + I_E R_e$$

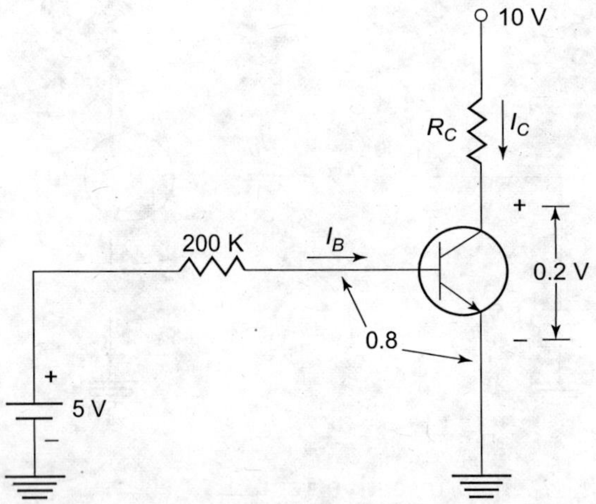

Applying KCL at node C, $I_E = I_B + I_C$

$$24 = (I_B + I_C)R_e + V_{CE} + (I_B + I_C)R_C$$

$$= 20.5\,(I_B + I_C) + 5$$

$$\therefore \quad I_B + I_C = \frac{24 - 5}{20.5} = 0.9268 \text{ mA} \qquad \text{...(1)}$$

Given that $\quad \beta = I_C/I_B = 45 \Rightarrow I_C = 45 I_B \qquad$ ...(2)

From (1) and (2)

$$46 I_B = 0.9268 \text{ mA} \Rightarrow I_B$$

$$I_B = 20.14\ \mu\text{A}.$$

$$I_C = \beta I_B = 906.3\ \mu\text{A}.$$

From the figure $V_{CB} = I_B \cdot R$ (voltage between base and collector).

$$R = \frac{V_{CE} - V_{BE}}{I_B} = \frac{5 - 0.7}{20.14} \times 10^6$$

$$= 213.505 \text{ k}\Omega$$

5.6 A silicon transistor with $V_{BE,\text{sat}} = 0.3$ V, $\beta = h_{FE} = 100$, $V_{CE,\text{sat}} = 0.2$ V is used in the circuit shown. Find the minimum value of R_C for which the transistor remains in saturation.

Solution We know that for a transistor to remain in saturation condition is

$$I_B > I_{B\min} \Rightarrow I_B > \frac{I_C}{h_{FE}}$$

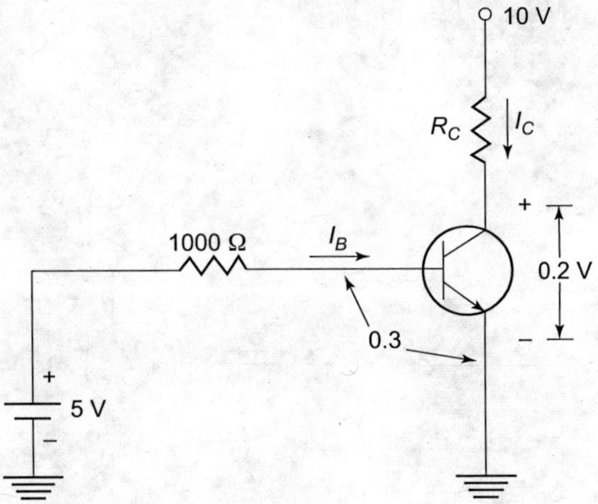

When $I_B = h_{FE} I_B$ it is in active region.
$I_C \approx I_E = 0$ it is in cut off condition.
From the figure, in the *B*–*E* loop.

$$I_B = \frac{5 - 0.3}{1000} = 4.7 \text{ mA.}$$

From the *C*–*E* loop,

$$I_C = \frac{10 - 0.2}{R_C} = \frac{9.8}{R_C}$$

$$\therefore \quad 4.7 \text{ mA} > \frac{9.8}{R_C \times 100}$$

$$\Rightarrow \quad R_C > \frac{9.8}{0.47} > 20.851 \ \Omega$$

5.7 For the circuit shown, assume $\beta = h_{FE} = 100$.

(a) Find if the silicon transistor is in cut off, saturation, or in the active region.

(b) Find V_o.

(c) Find the minimum value for the emitter resistor R_e for which the transistor operates in the active region.

Solution Since the n type base of a transistor is connected to 3 V negative terminal, J_E is forward baised. So, it can be in active or in saturation. It is not in cut off region. Assume transistor is in saturation region. From table $V_{BE} = 0.8$ V, $V_{CE} = 0.2$ V. Writing KVL in the base emitter loop circuit,

$$5 - 0.8 = 7I_B + 0.5\,(I_B + I_C)$$

$$4.2 = 7.5I_B + 0.5I_C \qquad \text{...(1)}$$

Similarly, from C–E circuit,

$$10 - 0.2 = 0.5(I_B + I_C) + 3I_C$$

$$9.8 = 0.5I_B + 3.5I_C \qquad \text{...(2)}$$

From (1) and (2)

$$I_B = 0.3769 \text{ mA}$$

$$I_C = 2.746 \text{ mA.}$$

$$I_{B\,\min} = \frac{I_C}{h_{FE}} = 27.46\ \mu\text{A}$$

Since $I_B > I_{B\,\min}$, transistor is in saturation.

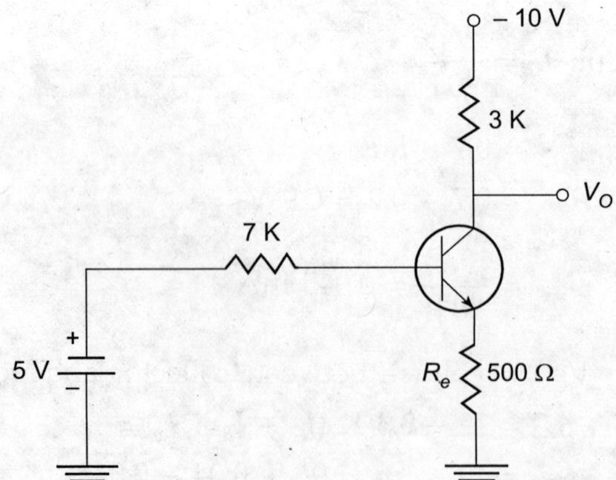

(b)
$$V_o = -10 + 3\text{ K} \times I_C$$
$$= -10 + 2.74 \times 10^{-3} \times 3 \times 10^{3}$$
$$= -1.762 \text{ V.}$$

5.8 If $\alpha = 0.98$, $V_{BE} = 0.7$ V, find R_1 in the circuit for $I_E = -2$ mA. Neglect reverse saturation current.

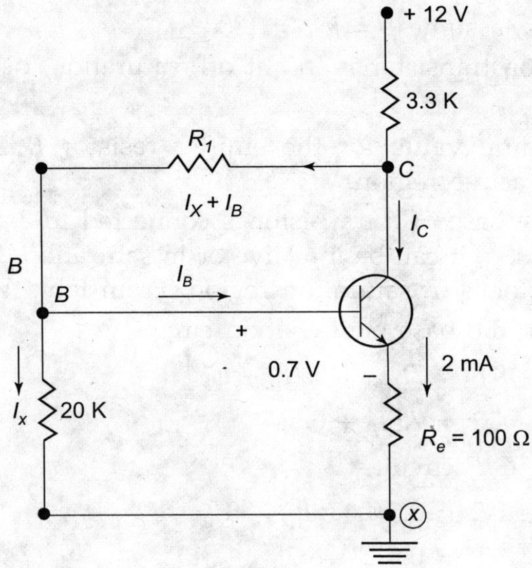

Solution

$$-\alpha = \frac{I_C}{I_E}$$

$$I_C = 0.098 \times 2 \text{ mA}$$

$$= 1.96 \text{ mA}$$

$$I_B = -I_E - I_C$$

$$= 2 - 1.96 = 0.04 \text{ mA}$$

From the circuit $I_x = \frac{V_{Bx}}{20\text{ K}}$

But, $V_{Bx} = 0.7 - I_E R_e$

$$= 0.7 + 0.2 = 0.9 \text{ V}$$

$$\text{I}_x = \frac{0.9}{20\text{ k}} = 0.045 \text{ mA}$$

V_{Cx} = voltage between collector and point x

$$= 12 - 3.3 \text{ K } (I_C + I_B + I_x)$$

$$= 12 - 3.8\ (1.96 + 0.04 + 0.045)$$

$$= 12 - 3.3\ (2.045)$$

$$= 12 - 6.7485 = 5.25 \text{ V}$$

$$R_1 = \frac{V_{Cx} - V_{Bx}}{I_x + I_B} = \frac{5.25 - 0.9}{0.04 + 0.045} = 51.19 \text{ k}\Omega$$

5.9 Show that the emitter volt–ampere characteristics of a transistor in the active region is given by $I_E = I_S e^{V_E/V_T}$ where $I_S = \dfrac{-I_{Eo}}{1-\alpha_N \cdot \alpha_I}$.

Solution In the active region J_E is forward biased and J_C is reverse bised. So, V_C is –ve, then $e^{V_C/V_T} << 1$. From Ebers-Moll equations,

$$I_C = -\alpha_N I_E - I_{Co}(e^{V_C/V_T} - 1) \quad ...(1)$$

$$I_E = -\alpha_I I_C - I_{Eo}(e^{V_E/V_T} - 1) \quad ...(2)$$

Put (1) in (2),

$$I_E = \alpha_I \alpha_N I_E + \alpha_I I_{Co}(e^{V_C/V_T} - 1) - I_{Eo}(e^{V_E/V_T} - 1)$$

$$\Rightarrow \quad (1 - \alpha_I \alpha_N) I_E = \alpha_I I_{Co}(-1) - I_{Eo}(e^{V_E/V_T} - 1)$$

$$= -\alpha_I I_{Co} - I_{Eo}(e^{V_E/V_T} - 1)$$

$$(\because e^{V_C/V_T} << 1).$$

But we have $\alpha_I I_{CO} = \alpha_N I_{Eo}$.

$$\therefore \quad (1 - \alpha_I \alpha_N) I_E = + I_{Eo}(-\alpha_N - e^{V_E/V_T} + 1)$$

$$= I_{Eo}((1 - \alpha_N) - e^{V_E/V_T})$$

$$\therefore \quad (1 - \alpha_I \alpha_N) I_E = I_{Eo}(-e^{V_E/V_T}) \; (\because 1 - \alpha_N \approx 0)$$

$$\approx -I_{Eo} e^{V_E/V_T}$$

$$\therefore \quad I_E = I_S \cdot e^{V_E/V_T}$$

where $$I_S = \frac{-I_{Eo}}{1-\alpha_I \alpha_N}$$

5.10 Show that $I_{CES} = \dfrac{I_{Co}}{1-\alpha_N \alpha_I}$ and $I_{CEO} = \dfrac{I_{Co}}{1-\alpha_N}$.

Solution $I_C = -\alpha_N I_E - I_{Co}(e^{V_C/V_T} - 1)$ ...(1)

When J_C is reverse biased, $e^{V_C/V_T} - 1 \approx -1$

$$\therefore \quad I_C = -\alpha_N I_E - I_{CO} \quad ...(2)$$

For base short circuited, $V_E = 0$,

$$\therefore \quad I_E = -\alpha_I I_C - I_{EO}(e^{V_E/V_T} - 1)$$

$$= -\alpha_I I_C \quad ...(3)$$

From (2) and (3)

$$I_C = \alpha_I \alpha_N I_C + I_{CO}$$

$$\therefore \qquad I_C = \frac{-I_C}{1-\alpha_N\alpha_I} = I_{CES}$$

For any transistor $I_B + I_C + I_E = 0$

For base open $I_B = 0 \Rightarrow I_C = -I_E$

From (2), $\quad I_C = -\alpha_N I_E + I_{CO}$

$$\Rightarrow \qquad I_C = \alpha_N I_C + I_{CO}$$

$$I_C = \frac{I_{CO}}{1-\alpha_N} = I_{CEO}$$

5.11 For the "floating point emitter" circuit shown, show that

$$V_{EBF} = V_T \ln(1 - \alpha_N)$$

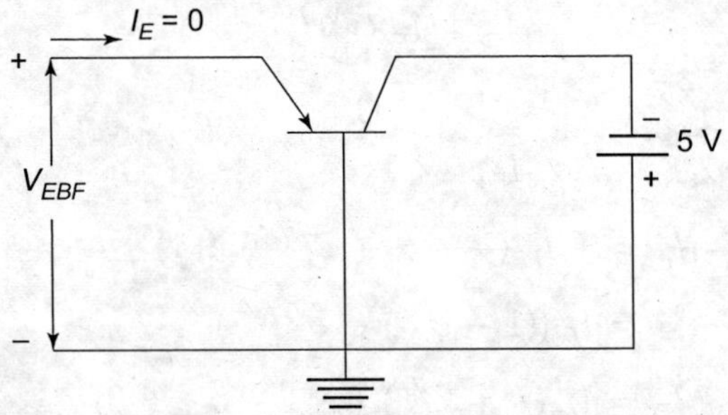

Solution Since J_C is reverse biased, $e^{V_C/V_T} = 0$, $I_C \approx I_{CO}$

Floating emitter means $I_E = 0$

$I_c = I_{CO}$. From Eber's-Moll equations

$$I_E = 0 = -\alpha_I I_C + I_{EO}(1 - e^{V_E/V_T})$$

$$\Rightarrow \qquad 1 - e^{V_E/V_T} = \frac{\alpha_I I_{CO}}{I_{EO}} = \alpha_N \; (\because \; \alpha_N I_{EO} = \alpha_I I_{CO})$$

$$1 - \alpha_N = e^{V_E/V_T}$$

$$\therefore \qquad V_E = V_T \log_e (1 - \alpha_N)$$

5.12 For the floating base, prove that $I_{CT} = \dfrac{2-\alpha_N}{(1-\alpha_N)^2} I_{CO}$

Assume that the transistors are identical.

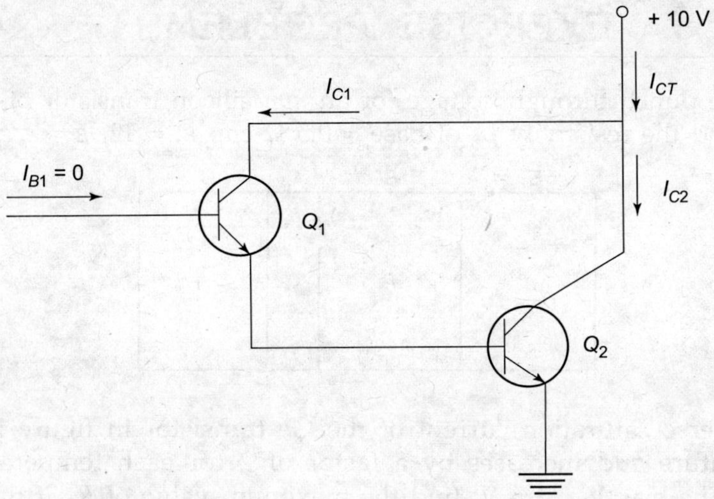

Solution Since

$$I_{B1} = 0$$
$$I_{C1} = -I_{E1}$$

Since collector junction of Q_1 is $R \cdot B$, $e^{V_{C1}/V_T} \approx 0$

$$I_{C1} = -\alpha_N(-I_{C1}) + I_{CO}$$

$$\therefore \quad \boxed{I_{C1} = \frac{I_{CO}}{1-\alpha_N}} \quad ...(1)$$

For transistor Q_2, $e^{V_{C2}/V_T} \approx 0$

$$I_{C2} = -\alpha_N I_{E2} + I_{CO}$$
$$= -\alpha_N(-I_{C2} - I_{C1}) + I_{CO}$$

$$\Rightarrow \quad I_{C2} = \frac{I_{CO} + \alpha_N I_{C1}}{1-\alpha_N}$$

$$= \frac{I_{CO} + \dfrac{\alpha_N \cdot I_{CO}}{1-\alpha_N}}{1-\alpha_N} = \frac{I_{CO}}{(1-\alpha_N)^2} \text{ (From (1))}$$

$$I_{CT} = I_{C1} + I_{C2} = \frac{I_{CO}}{1-\alpha_N} + \frac{I_{CO}}{(1-\alpha_N)^2}$$

$$= \frac{I_{CO}}{(1-\alpha_N)}\left(1 + \frac{1}{1-\alpha_N}\right) = \frac{I_{CO}(2-\alpha_N)}{(1-\alpha_N)^2}$$

EXERCISE PROBLEMS

5.1 Calculate the punch through voltage for an *npn* silicon transistor of the alloy type if $w = 1\ \mu m$ and the resistivity ρ_B of base = 0.5 Ω-cm, $\varepsilon_r = 12$, $\varepsilon = \varepsilon_o \varepsilon_r$.

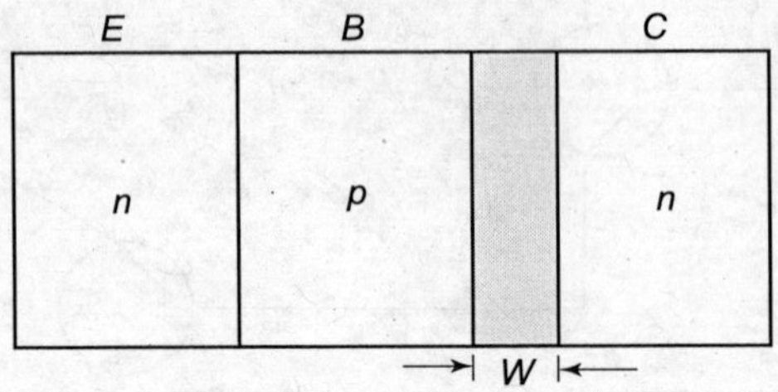

5.2 (a) The reverse saturation current of the Ge transistor in figure is 2 μA at room temperature and increases by a factor of 2 for each temperature increase of 10 °C. The bias $V_{BB} = 5$ V, find the maximum value of R_B if the transistor is to remain cut off at 75 °C.

(b) If $V_{BB} = 1$ V, $R_B = 50$ k, how much would be the temperature increase before the transistor comes out of cut off?

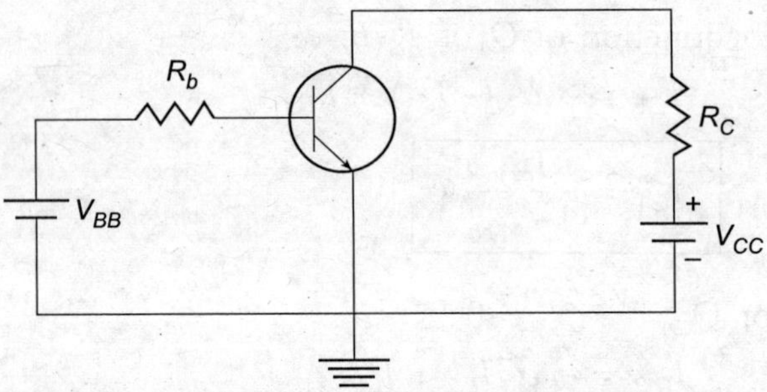

5.3 For the circuit shown, $\alpha_1 = 0.98$, $\alpha_2 = 0.96$, $V_{CC} = 24$ V, $R_c = 120\ \Omega$ and $I_E = -100$ mA.

Determine (a) I_{B1}, I_{C1}, I_{E1}, I_{B2}, I_{C2} and I_C. (b) V_{CE} (c) $\frac{I_C}{I_B}, \frac{I_C}{I_E}$

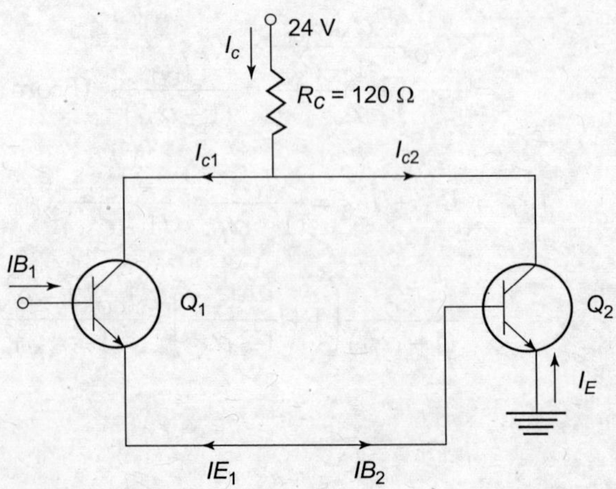

5.4 Using Ebers-Moll model equations find the expression for I_C and I_E in terms of V_C and V_E.

5.5 (a) Find the collector current for a transistor when both emitter and collector junctions are reverse biased. Assume $I_{Co} = 5\ \mu A$, $I_{Eo} = 3.57\ \mu A$ and $\alpha_N = 0.98$.

(b) Find I_E under the same conditions as in part (a).

5.6 ABJT has $I_B = 10\ \mu A$, $\beta = 99$ and $I_{Co} = 1\ \mu A$. Find I_C.

5.7 The Reverse Saturation Current in an *npn* transistor in *CB* circuit is 12.5 μA. For an $I_E = 2$ mA, $I_C = 1.97$ mA determine the circuit gain and base current.

5.8 A transistor is connected in *CE* configuration $V_{CC} = 10$ V and $R_L = 800\ \Omega$. Voltage across load resistor is 0.8 V and $\alpha = 0.96$. Determine V_{CE} and I_B.

5.9 The Reverse Saturation Current of an *npn* transistor in *CB* circuit is 12.5 μA, I_E = 4 mA. $I_C = 2.96$ mA. Find I_B and current gain.

REVIEW QUESTIONS

1. Sketch the family of *CE* input and output characteristics for a transistor and explain the shape of the characteristics qualitatively.
2. Define α and β parameters and state the relation between them.
3. Prove that the collector current is more sensitive to temperature in *CE* configuration than in *CB* configuration.
4. What are the configurations of a BJT? Compare their characteristics. What are the transistor specifications?
5. What are the applications of emitter follower? Why is this circuit referred to as emitter follower?
6. Which transistor configuration has highest power gain?
7. What is early effect? What are the effects produced by base width modulation?
8. Derive an expression for collector current and base current for a transistor in *CE* configuration in active region.
9. Explain why *CE* configuration is popular in amplifier circuits.
10. Write the Ebers-Moll equations of a transistor. Sketch the circuit that satisfies these equations.
11. Explain the terms avalanche multiplication and reach through in a transistor.
12. Describe various current components in a transistor.
13. Explain how a transistor can act as an amplifier.
14. Indicate the various operations of regions of a transistor in *CE* configuration on its output characteristics.

15. What are the characteristics of emitter follower? What are its applications?
16. Draw the Ebers-Moll model of a transistor and explain the model in detail.
17. What are dc current gain, small signal current gain and large signal current? Give their equations for a *CE* configuration transistor circuit.
18. What are the various regions that a transistor can operate? In which region should be a transistor operated if it were to be used as a switch?
19. What are the various fabrication techniques of BJT? Explain each in detail.
20. Draw a circuit diagram of a common base configuration transistor and explain the current components in active region.
21. Draw the circuit symbol for both *npn* and *pnp* transistors and indicate the voltage and currents along with polarities.
22. Describe the fabrication of an alloy type of transistor.
23. Define all the three regions of operation of a transistor in *CE* configuration. Discuss in detail the biasing conditions in each case.
24. Why is I_{CBO} greater than I_{CO}? Explain.
25. Discuss the two possible sources of breakdown in a transistor as the collector to emitter voltage is increased.

CHAPTER

6

Transistor Biasing

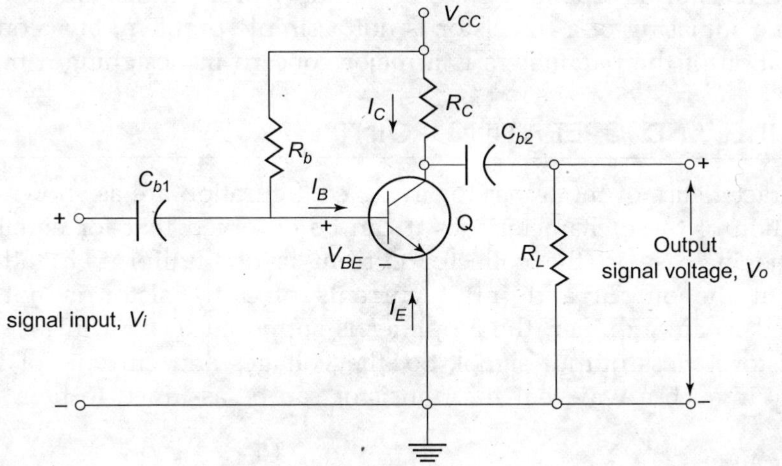

Fixed-bias circuit

Chapter Outline

The concepts introduced in this chapter are:

- *Load line and operating point*
- *Various biasing circuits*
- *Stability factors*
- *Stabilisation and compensation circuits*
- *Thermal runaway and condition to avoid it*

6.1 INTRODUCTION

In order to use a transistor in a circuit, the transistor needs to be given appropriate dc voltages and currents after selection of the region of operation. This is called the biasing of a transistor. The transistor has to be biased properly by providing required voltages at its two junctions. The problem with the transistor biasing is that the transistor operating dc voltages and currents are not constants but vary with ambient temperature. So, care is to be taken in selection of the voltages and currents such that the regions of operation of a transistor are not altered. This requires the stabilisation of the parameters of a transistor with respect to temperature. The variation of the parameters of the transistor is due to the fact that the basic qualities of a transistor rely on the temperature at which the transistor is operated. Or, in other words, transistor is a temperature sensitive device. So, biasing of a transistor is not a simple problem, but along with biasing thermal stabilisation of the parameters is a major concern in designing a transistor circuit.

6.2 LOAD LINE AND OPERATING POINT

The output characteristics of the transistor in *CE* configuration are as shown in Fig. 6.1. On careful observation of these characteristics it can be observed that for equal increments of base current the increase in the collector current is not uniform. In other words, the relation between the output and input currents of a transistor is not linear. If the transistor is used as an amplifier, the amplifier is supposed to be a linear device in order to achieve a distortionless output signal. So, the voltages and currents of a transistor are chosen carefully in such a way that the transistor can be assumed to be a linear device.

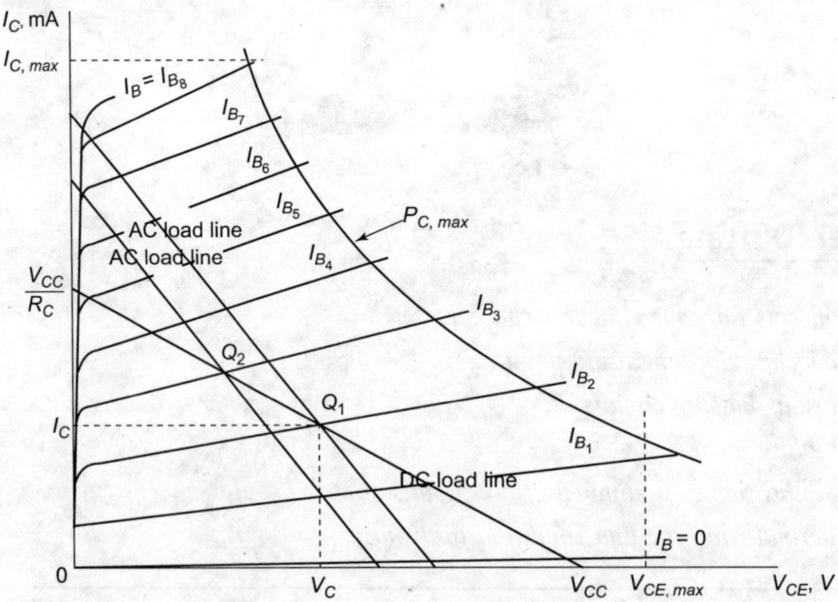

Fig. 6.1 *Typical output characteristics of a transistor in CE configuration*

In general, the common emitter configuration is most widely used as an amplifier due to a reason that it has both voltage and current gains large compared to unity. The other two configurations are not, in general, employed as amplifiers. So, the biasing of the transistor is discussed only for common emitter configuration. However, the same discussion can be extended even to the other two configurations of the transistor without any reservation.

Initially, the power supply to the collector and the biasing resistance of collector are selected for a given transistor to bias it in the active region of a *CE* configuration. The voltage of the power supply of collector is V_{CC} and the collector resistance is R_C. The locus of the point of constant V_{CC} and R_C is a straight line with a slope of $1/R_C$ on the output characteristics of a transistor. This locus is called *a load line* (*dc* or *static*). This load line intersects the x axis (V_{CE}) at a point (V_{CC}, 0) and the y axis at $(0, V_{CC}/R_C)$. The transistor will have to operate only on this load line once the collector voltage and resistor are selected. For selecting the operating region of a transistor, the base current is to be ascertained. When the base current is less than zero, the transistor is in cut off and when the base current is so chosen that the point is not in the constant current region and is located in the region where the current increases with small increments in voltage, the transistor is in saturation. If the base current is chosen between these two points, the transistor is said to be in active region. So, the operating region of a transistor can be selected by selecting appropriate values of the base current. The selection of the base current leads to a point on the load line of the characteristic. This point is called the *operating point* or *quiescent point*, and is designated as Q. The selection of Q is vital in any transistor circuit. The dimensions of Q on the output characteristics of a transistor are (V_{CE}, I_C, I_B). This quiescent point is plotted in the input characteristics of the transistor. The fourth dimension of this point, V_{BE}, is determined from these input characteristics. Having obtained the input and output coordinates of the operating point, the next step is to select the proper power supply and biasing resistors at the input of the circuit.

6.3 FIXED-BIAS CIRCUIT

To bias a transistor in active region, the operating point is selected as in the above discussion. The input parameters V_{BE} and I_B are to be established by a power supply at the base of a transistor, V_{BB}. The combination of V_{BB} and R_B is so selected that the input dimensions of the operating point are achieved. This is referred to as *fixed-bias circuit* and is as shown in Fig. 6.2.

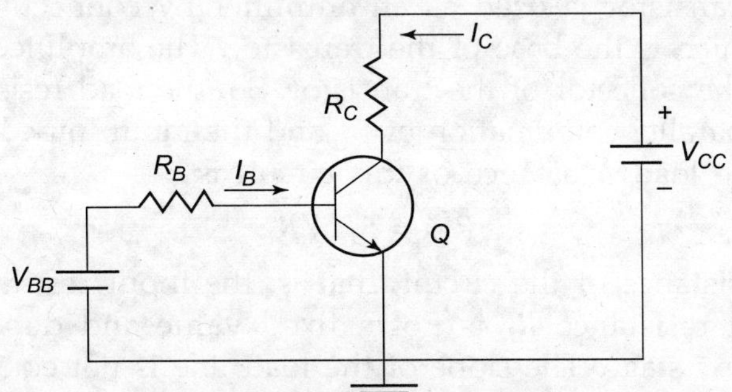

Fig. 6.2 *Transistor fixed-bias circuit using two power supplies*

The governing design equations of this circuit are as follows:

$$V_{CC} = I_C R_C + V_{CE} \quad \text{...(6.1)}$$

$$V_{BB} = I_B R_B + V_{BE} \quad \text{...(6.2)}$$

From the above equations it is observed that two power supplies are required to bias the transistor appropriately. This is costly and complex. The circuit can be simplified by using a single power supply and designing the input circuit from the power supply of the collector itself. The circuit which has only one power supply is as shown in Fig. 6.3.

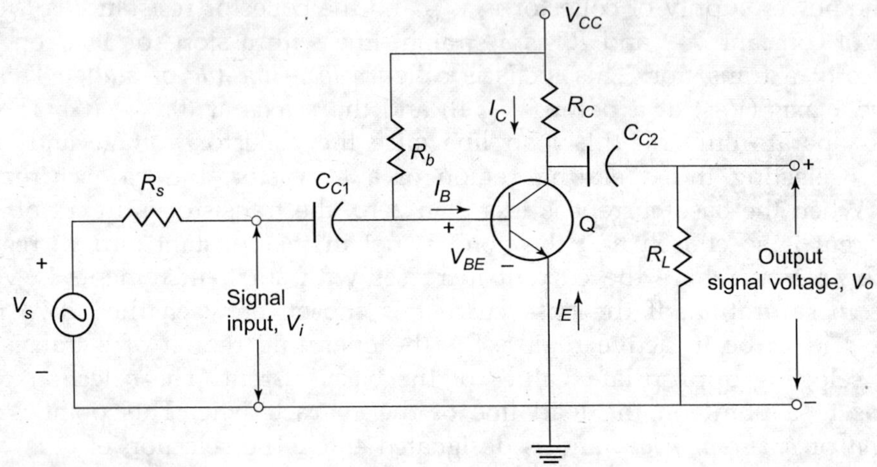

Fig. 6.3 *Fixed-bias circuit*

Now, the design equations are modified as follows:

$$V_{CC} = I_C R_C + V_{CE} \quad \text{...(6.4)}$$

$$V_{CC} = I_B R_B + V_{BE} \quad \text{...(6.5)}$$

This circuit biases the transistor in the active region at the required operating point. The operating point is, in general, selected midway of the load line in the active region. This is due to the reason that the characteristics are almost linear in the middle of the active region and the nonlinearity increases as one moves either towards the saturation or cut off on the load line from the mid-region. This ensures linear operation of the transistor.

The so-designed transistor is used as an amplifier by connecting an ac signal at the input of the circuit, which is the base of the transistor. The amplified version of this input signal is delivered to the collector of the transistor. So, the load resistance of the transistor is not R_C alone, but a parallel combination of R_C and the input impedance of the next stage. Or, in other words, the load resistance of this circuit is

$$R_L' = R_C // R_L \quad \text{...(6.6)}$$

where R_L is a load resistance of the circuit, that is, the input resistance of the next stage. The value of the load resistance R_L is not a fixed value and depends upon the circuit configuration of the next stage. The slope of the load line is not equal to $1/R_C$ and is now

$1/R_L'$. So, the slope of the load line is not a fixed value and is a variable depending upon the value of R_L. This load line is called *ac* or *dynamic load line.* Even though the slope of the load line is changed, the operating point Q remains the same as shown in Fig. 6.1. Care should be taken that the operating point should always be chosen in a safe region, that is, a region where the power dissipation in the collector junction should be less than the rated maximum value, P_{Cmax}. The locus of the P_{Cmax} is a hyperbola, as shown in Fig. 6.1. The manufacturer in the datasheet of the transistor gives the value of Pc_{max}. The x intercept of the load line is V_{CC} and this value of V_{CC} is to be chosen so that it is less than V_{Cmax}. The value of V_{Cmax} depends on avalanche multiplication and reach through and is specified by the manufacturer. Similarly, the y intercept of the load line, V_{CC}/R_L', should not be more than the rated $I_{C\,max}$. That is, the static load line, dynamic load line and the operating point should be carefully chosen so that at no time the transistor dissipates power more than the rated value.

6.4 STABILITY OF OPERATING POINT

The parameters of the transistor are not constant and vary with temperature. When the temperature increases, the parameters like h_{fe}, V_{BE} and I_{CBO} change which leads to the change in the output current I_C. In the circuit of Fig. 6.3, the coordinates of the operating point thus change and the operating point travels upward towards saturation on the dynamic load line, as shown in Fig. 6.4.

If the temperature rise is very large, the shift of the operating point will be large and there is a tendency that the transistor is driven into saturation for a given condition of operation. Even though the operating points do not tend towards saturation region, there is a possibility that the operating point will be relocated in the nonlinear region of

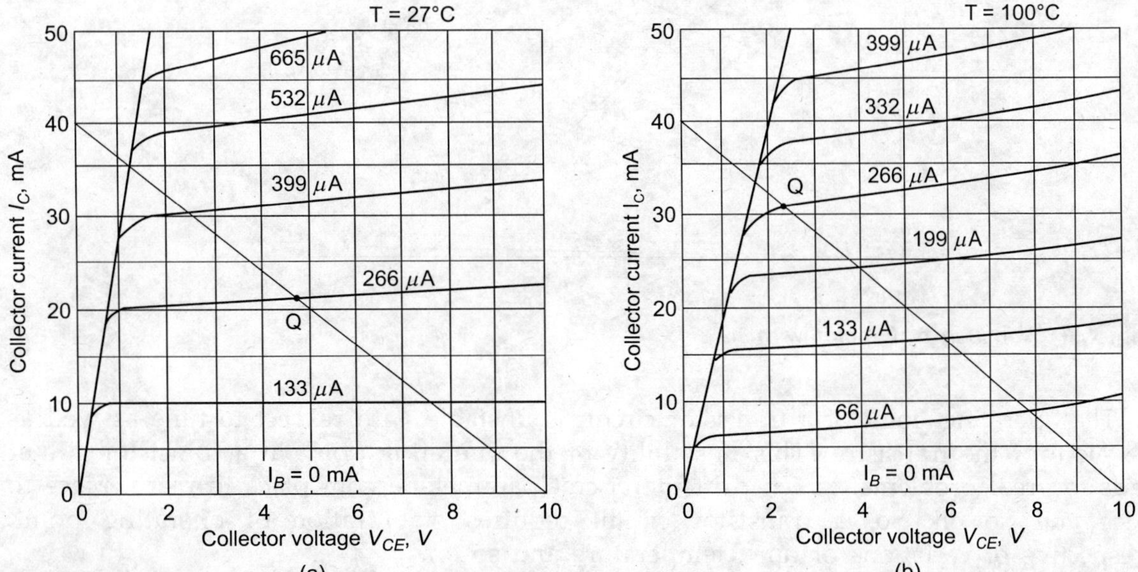

Fig. 6.4 *Movement of Q point with changes of temperature (a) at 27 °C (b) at 100 °C*

the transistor operation. When the temperature decreases, the situation is reversed and the operating point may travel towards cut off on the dynamic load line. Whatever may be the case, with temperature change in a fixed-bias circuit the nonlinearity sets in and the output signal is highly distorted.

The parameter h_{fe} is not a constant value for a given type of transistor and ranges between maximum and minimum values even for transistors manufactured in the same batch. The manufacturer of the transistor generally specifies these maximum and minimum values and also gives a typical value of the parameter. So, if a circuit is designed with a given h_{fe} value, in practice, this value may not be realisable due to this unpredictable value of the parameter. If, by chance, the transistor fails in a circuit, which is designed and realised for a given set of parameters, then even though the transistor is replaced with the same type of device, the circuit parameters need not be the same as in the previous case. To be more specific, the circuit parameters are not stable and vary from piece to piece of a transistor. This is due to the variation of the transistor parameter from piece to piece which leads to variation of operating point of a circuit (see Fig. 6.5).

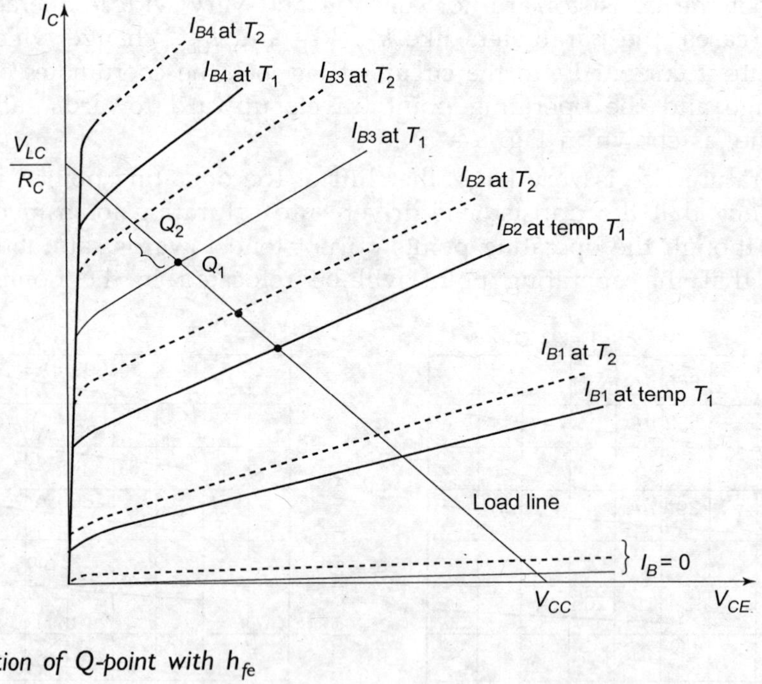

Fig. 6.5 *Variation of Q-point with h_{fe}*

The operating point of a transistor circuit is unstable with respect to time as well as the piece of the transistor. This unstability of the operating point in a transistor circuit leads to great problems, which drift the circuit parameters from place to place, piece to piece and so on. So, a transistor circuit requires stabilisation of operating point, irrespective of variations of time, temperature and so on.

The parameters of a transistor which change with respect to temperature are I_{CO}, V_{BE} and β. I_{CO} doubles for every ten degree rise in temperature, V_{BE} decreases at the rate of

2.5 mV for every degree rise in temperature and β increases with temperature. When these parameters depend on temperature, the operating point, in specific the output collector current I_C, is not stable and increases with temperature.

If in a given situation, temperature rises, I_{CO} increases leading to increase in I_C. This, in turn, increases the power dissipation in the collector junction of the transistor, which raises the temperature. Because of rise of this temperature at the collector-base junction, I_{CO} increases and consequently I_C, power dissipation and temperature rise and the cycle is completed. If the circuit is left unchecked, there is a tendency that temperature of the collector junction increases tremendously and it may lead to a situation of burnout. To avoid this condition, the biasing of the transistor is to be done in such a way that whatever may be the variation of temperature or any other parameter, the operating point drift should be ideally zero or, in practice, minimal. Various circuits are proposed for biasing the transistor and to achieve good stabilisation. The popular one is *self-bias circuit.*

6.5 SELF-BIAS CIRCUIT

The self-bias circuit, as shown in Fig. 6.6, stabilises the operating point to a very large extent. A resistor R_e is added in the emitter circuit and voltage divider resistors R_1 and R_2 in the base circuit of the transistor. These resistors act as feedback to control the variations of the operating point and stabilise it.

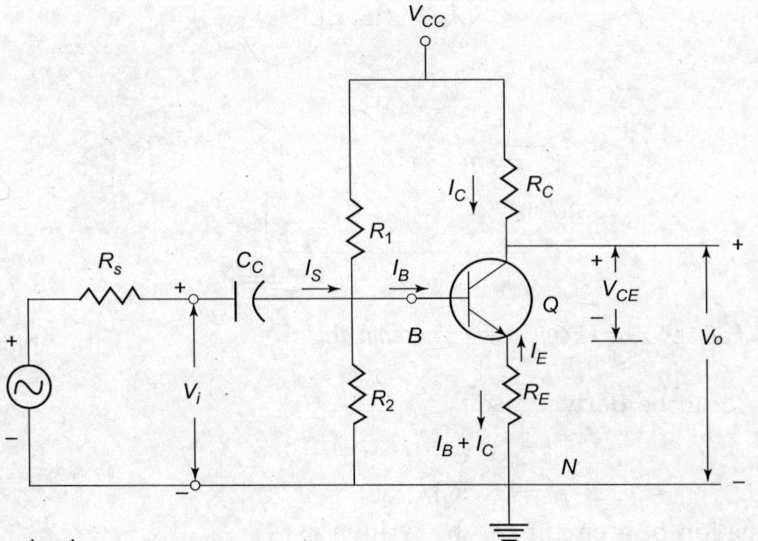

Fig. 6.6 *Self-bias circuit*

Whatever may be the reason for the rise in temperature, I_{CO} increases, which in turn increases I_C. The current through the resistor R_e is I_E, the sum of the base current I_B and the collector current I_C. Since I_C has increased due to rise in temperature, I_E increases and hence the voltage across the resistor R_e increases. With reference to the circuit shown in Fig. 6.6, the voltage across the resistor R_2 is constant, irrespective of variations of temperature. This voltage V_{R_2} is equal to the sum of voltages across the base-emitter junction V_{BE} of the transistor and the voltage across emitter resistance V_{R_e}. When I_E increase in R_e, increasing V_{Re}, since V_{R_1} is constant V_{BE} should come down by the same value as the increase in V_{Re}. Since the base-emitter junction is forward biased for the transistor to operate in the active region, the current in the base decreases exponentially, corresponding to the decrease in the V_{BE}. This decrease in I_B decreases the collector current I_C to compensate the increase due to temperature. Even though the compensation is not total, the increase in the current can be reduced to a very large extent and stabilise the operating point. When the temperature decreases, the above-mentioned procedure is reversed and so the increase in V_{BE} increases the base current to compensate the decrease in collector current.

Analysis: The base circuit of the self-bias circuit can be replaced with *Thevenin* equivalent circuit consisting of a series combination of a voltage source V_{TH} and a resistor R_{TH}, as shown in Fig. 6.7.

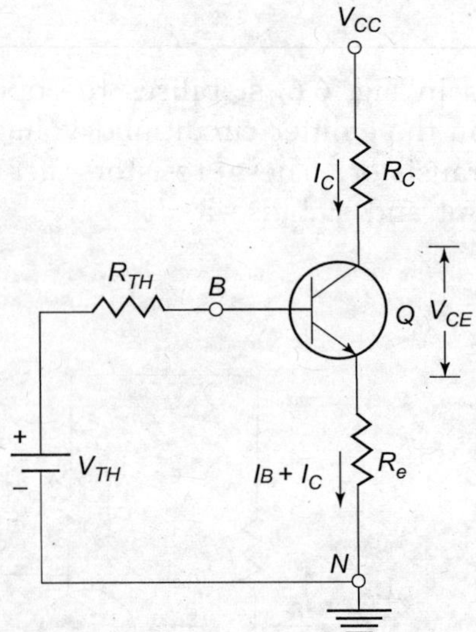

Fig. 6.7 *Simplified self-bias circuit using Thevenin theorem*

V_{TH} and R_{TH} can be derived as

$$V_{TH} = (V_{CC}R_2)/(R_1 + R_2) \quad \text{...(6.7)}$$

$$R_{TH} = (R_1R_2)/(R_1 + R_2) \quad \text{...(6.8)}$$

The output equation of a circuit can be written as

$$V_{CC} = I_CR_C + V_{CE} + R_e(I_B + I_C) \quad \text{...(6.9)}$$

The input equation is

$$V_{TH} = I_BR_{TH} + V_{BE} + R_e(I_B + I_C) \quad \text{...(6.10)}$$

The relation between the output and input currents is given as

$$I_C = (\beta + 1)I_{CO} + \beta I_B \quad \text{...(6.11)}$$

With the help of these four equations the self-bias circuit can be analysed or designed.

6.6 STABILITY FACTORS

To establish the stability of a given biasing circuit of a transistor amplifier, stability factors *S*, *S′* and *S″* are defined and determined for a given circuit. These factors are supposed to be as small as possible, tending towards zero for an ideal biasing circuit.

With respect to temperature the transistor parameters that vary are I_{CO}, V_{BE} and β. I_{CO} doubles for every ten degree rise in temperature, V_{BE} decreases by 2.5 mV for every one degree rise in temperature and β increases with temperature. These three are the parameters that cause the drift in the collector current and destabilise the circuit. When temperature varies, the above three parameters vary, causing the changes in the output collector biasing current. In order to estimate the exact variation of the output biasing current of a transistor circuit, the variations of three parameters, I_{CO}, V_{BE} and β are assumed to be independent and three stability factors known as *S* factors are defined as follows:

$$S = \left.\frac{\partial I_C}{\partial I_{CO}}\right|_{V_{BE},\,\beta\,\text{constant}} \qquad \text{...(6.12)}$$

$$S' = \left.\frac{\partial I_C}{\partial V_{BE}}\right|_{I_{CO},\,\beta\,\text{constant}} \qquad \text{...(6.13)}$$

$$S'' = \left.\frac{\partial I_C}{\partial \beta}\right|_{V_{BE},\,I_{CO}\,\text{constant}} \qquad \text{...(6.14)}$$

After determining the above three stability factors the change in the output collector current can be estimated as follows:

$$\Delta I_C = S\Delta I_{CO} + S'\Delta V_{BE} + S''\Delta\beta \qquad \text{...(6.15)}$$

6.7 STABILITY FACTOR *S*

We know that the collector current is related to the base current as

$$I_C = \beta I_B + (\beta + 1)I_{CBO} \qquad \text{...(6.16)}$$

Differentiating this equation with respect to I_C, we have

$$1 = \beta(\partial I_B/\partial I_C) + (1 + \beta)(\partial I_{CBO}/\partial I_C)$$

or,

$$S = \frac{\beta+1}{1-\beta(dI_B/dI_C)} \qquad \text{...(6.17)}$$

Self-bias circuit: For a self-bias circuit, differentiating Eq. (6.10) with respect to I_C, we have

$$0 = \frac{dI_B}{dI_C}R_{TH} + \left(1+\frac{dI_B}{dI_C}\right)R_e$$

So,

$$\frac{dI_B}{dI_C} = \frac{-R_e}{R_{TH} + R_e} \qquad ...(6.18)$$

Substituting Eq. (6.18) into Eq. (6.17),

$$S = \frac{\beta + 1}{1 - [\beta(-R_e/\{R_e + R_{TH}\})]}$$

$$= (1 + \beta)\frac{1 + R_{TH}/R_e}{1 + \beta + R_{TH}/R_e} \qquad ...(6.19)$$

The equation for S tends towards 1 as R_{TH}/R_e tends towards zero, and it tends to $1 + \beta$ as R_{TH}/R_e tends towards ∞ and S increases with R_{TH}/R_e. Ideally, the value of stability factor S should be as small as possible and the minimum value of S is 1. This can be achieved by choosing the value of R_{TH} very small or R_e very large.

Since the minimum value of S is 1, at this value, the variation of I_C is same as the variation of $I_{CBO.}$ In practice, the value of S can be only achieved when R_{TH} is equal to zero, which is not possible. In practice, the value of S lies between 5 and 10.

In order to improve the stability of the circuit there are two choices, either to decrease $R_{TH,}$ which is a parallel combination of R_1 and R_2 or to increase R_e. Decreasing the value of R_{TH} decreases the input resistance of the circuit which leads to drawing more current into the circuit, and thus the efficiency of the circuit deteriorates. In turn, if the value of R_e is increased, V_{CE} reduces and in order to maintain the quiescent point, V_{CC} has to be increased. Increasing V_{CC} increases the dissipation in the transistor, thus increasing the power loss in the circuit.

Fixed-bias circuit: For a fixed-bias circuit considering the base loop,

$$V_{CC} = I_B R_B + V_{BE} \qquad ...(6.20)$$

Differentiating the above equation with respect to I_C

$$0 = (dI_B/dI_C)R_B + 0$$

or, $$dI_B/dI_C = 0$$

So, from Eq. (6.17), $$S = 1 + \beta \qquad ...(6.21)$$

In this fixed-bias circuit the stability factor is constant, irrespective of circuit components and only depends on small signal current gain of the transistor. As β is very large in common emitter configuration, the value of S is also very large and so stability of fixed-bias circuit is very poor compared to self-bias circuit.

6.8 STABILITY FACTOR S'

For a self bias circuit, substituting Eq. (6.11) into Eq. (6.10) leads to

$$V_{TH} = R_{TH}[I_C - (1 + \beta)I_{CBO}]/\beta + V_{BE} + R_e\{[I_C - (1 + \beta)I_{CBO}]/\beta + I_C\}$$

Therefore,

$$V_{BE} = V_{TH} + (R_{TH} + R_e)\frac{1+\beta}{\beta} I_{CBO} - \frac{R_{TH} + R_e(1+\beta)}{\beta} I_C \quad \text{...(6.22)}$$

Differentiating this equation with respect to V_{BE} when I_{CBO} and β are constants,

$$1 = 0 - \frac{R_{TH} + R_e(1+\beta)}{\beta}\frac{\partial I_C}{\partial V_{BE}}$$

Therefore,

$$S' = (\partial I_C/\partial V_{BE}) = -\beta/[R_{TH} + R_e(1+\beta)]$$

$$= \frac{-\beta/R_e}{1+\beta+R_{TH}/R_e} \quad \text{...(6.23)}$$

From the above equation, when $\beta \gg 1$ and $\beta + 1 \gg R_{TH}/R_e$, it is observed that

$$S' \approx -(1/R_e) \quad \text{...(6.24)}$$

So, the stability factor S' decreases as R_e increases. For S' to tend to zero, R_e should tend to ∞, which is an impossible condition. The assumption that $\beta \gg 1$ and $\beta + 1 \gg R_{TH}/R_e$ is true in practice and S' will depend upon R_e.

6.9 STABILITY FACTOR S''

The variation of I_C with respect to β when V_{BE} and I_{CBO} are constant is referred to as stability factor S''. Therefore,

$$S'' = \partial I_C/\partial \beta \approx \Delta I_C/\Delta_\beta \quad \text{...(6.25)}$$

From Eq. (6.22) the current can be written as

$$I_C = \frac{\beta(V_{TH} + V' - V_{BE})}{R_{TH} + R_e(1+\beta)} \quad \text{...(6.26)}$$

where, $$V' = (R_{TH} + R_e)(\beta + 1)I_{CBO}/\beta \quad \text{...(6.27)}$$

Differentiating Eq. (6.26) with respect to β,

$$S'' = I_C S/[\beta(1 + \beta)] \quad \text{...(6.28)}$$

In Eq. (6.28), to determine the value of S″ one has to know the value of I_C, S and β. But there is an ambiguity in determining this value since all three parameters are variables here. So, the above equation cannot be directly used to determine the stability of the self-bias circuit.

To overcome the ambiguity of the above equation, the stability factor S ″ can be re-written as

$$S'' = \partial I_C/\partial \beta \approx \Delta I_C/\Delta \beta$$

$$= (I_{C2} - I_{C1})/(\beta_2 - \beta_1) \quad \text{...(6.29)}$$

where I_{C1} and β_1 are the initial values of the parameters, whereas I_{C2} and β_2 are the final values of collector current and small signal current gain, assuming the difference between the initial and final values is very small. Therefore, from Eq. (6.26),

$$\frac{I_{C_2}}{I_{C_1}} = \frac{\beta_2}{\beta_1}\frac{R_{TH} + R_e(1+\beta_1)}{R_{TH} + R_e(1+\beta_2)} \qquad \text{...(6.30)}$$

or,

$$\frac{I_{C_2}}{I_{C_1}} - 1 = \left(\frac{\beta_2}{\beta_1} - 1\right)\frac{R_{TH} + R_e}{R_{TH} + R_e(1+\beta_2)} \qquad \text{...(6.31)}$$

or,

$$S'' = \frac{I_{C_1} S_2}{\beta_1(1+\beta_2)} \qquad \text{...(6.32)}$$

where S_2 is the stability factor S at a constant value of β equal to β_2 and is given by Eq. (6.19).

It may be noted that Eq. (6.28) and Eq. (6.32) are the same. It can also be observed that as S is minimised, S'' also gets reduced. Thus, by reducing the ratio of R_{TH}/R_e, S and S'' both get reduced. However, in order to make S' small, one requires a large value of R_e. Therefore, on the whole, to optimise the values of S, S' and S'', R_e is to be selected as large as possible and a compromise value for R_{TH} is to be chosen.

The variation of the operating point with respect to temperature can thus be estimated by the determination of the three stability factors. The equations for the three stability factors for self-bias circuit are derived in the above discussion. The same discussion can be extended to any transistor circuit. One has to optimise the three stability factors such that the designed transistor circuit is stable with respect to temperature variations.

To simplify the process of estimation of operating point variation by transistor circuit, Eq. (6.15) can be expressed in terms of a parameter M as follows:

$$\frac{\Delta I_C}{\Delta I_{C_1}} = \left[1 + \frac{R_{TH}}{R_e}\right]\frac{M_1 \Delta I_{CBO}}{I_{C_1}} - \frac{M_1 \Delta V_{BE}}{I_{C_1} R_e} + \left[1 + \frac{R_{TH}}{R_e}\right]\frac{M_2 \Delta_\beta}{\beta_1 \beta_2} \qquad \text{...(6.33)}$$

where M is defined as

$$M = \frac{1}{1 + R_{TH}/[R_e(1+\beta)]}$$

$$\approx 1/[1 + R_{TH}/(\beta R_e)] \qquad \text{...(6.34)}$$

and M_1, M_2 correspond to β_1, β_2, respectively.

6.10 TECHNIQUES COMPENSATION

In the above discussion, the stability of the operating point for a given transistor circuit is estimated and it is understood that the design of the circuit should be such that the three stability factors are minimum. The bias stability can also be achieved by use of circuit components to compensate the increase or decrease of the collector current. This can be achieved simply by adding few resistive components as in the case of self-bias circuit, where the stability is improved to a large extent compared to fixed-bias circuit, or by adding temperature dependent components such as diodes, thermistors or sensistors. "The technique of improving stability by adding resistive components is referred to as *stabilisation technique,* whereas improving stability by the use of temperature-sensitive components is referred to as *compensation technique.*"

Compensation techniques involve compensating either the effect of one parameter at a time, or more, or the use of thermistor or sensistor compensates the variations of temperature as a whole.

Diode compensation circuits: The diode can be used as a component that compensates the variation of I_{CO} or V_{BE}. The diode to compensate the parameters is to be selected in such a way that the temperature dependence of the diode parameters should be same as that of the transistor that is used in the circuit. In other words, the material of the transistor and diode should be the same.

Compensation for variations in I_{CO}: The circuit using the diode that can compensate the variations in I_{CO} is as shown in Fig. 6.8. In the circuit the diode and the transistor are chosen to be of same type and material.

As temperature increases, I_{CBO} increases, doubles for every 10°C rise in temperature and also I_o reverse saturation current in the diode increases at the same rate since the material and type of transistor and the diode are the same.

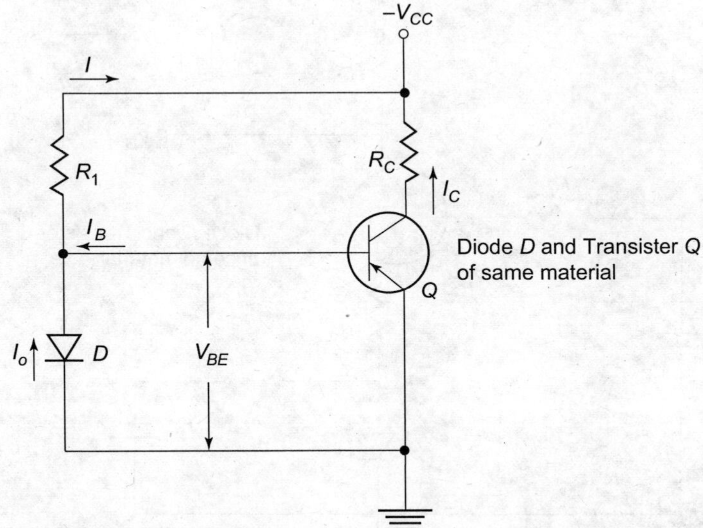

Fig. 6.8 *Diode compensation for I_{CO}*

From the circuit,

$$I = I_B + I_o = (V_{CC} - V_{BE})/R_1 \qquad ...(6.35)$$

The value of I is constant since V_{CC} and R_1 are constant, V_{BE} is assumed to be constant. The output collector current can now be written as

$$\begin{aligned} I_C &= \beta I_B + (\beta + 1) I_{CBO} \\ &= \beta(I - I_o) + (\beta + 1) I_{CBO} \\ &\approx \beta(I - I_o) + \beta I_{CBO} \text{ (since } \beta \gg 1) \\ &\approx \beta I + \beta(I_{CBO} - I_o) \end{aligned} \qquad ...(6.36)$$

Since the variation of I_{CBO} and I_o is the same with respect to temperature, the changes in I_{CBO} are tracked by I_o and so the effect of increase or decrease of I_{CBO} on the collector current is compensated.

Compensation for variations in V_{BE}: The base to emitter voltage V_{BE} is a voltage that biases the emitter junction in forward direction. This voltage is not constant with respect to temperature, but decreases by 2.5 mV for every 1 °C rise in temperature. In order to maintain the output collector current constant, the effect of this voltage is to be compensated.

The circuit shown in Fig. 6.9 employing a diode in the emitter circuit compensates the variations of V_{BE}. An ancillary source V_{DD} and a resistor R_d forward bias the diode. The diode is chosen to be of the same material as that of transistor. Since the emitter junction and the diode are both forward biased, both being of the same material, both have the same temperature dependence of parameters. Applying KVL to the base circuit,

$$I_C = \frac{\beta[V - (V_{BE} - V_0)] + (R_b + R_e)(\beta + 1) I_{CO}}{R_b + R_e(1 + \beta)} \qquad ...(6.37)$$

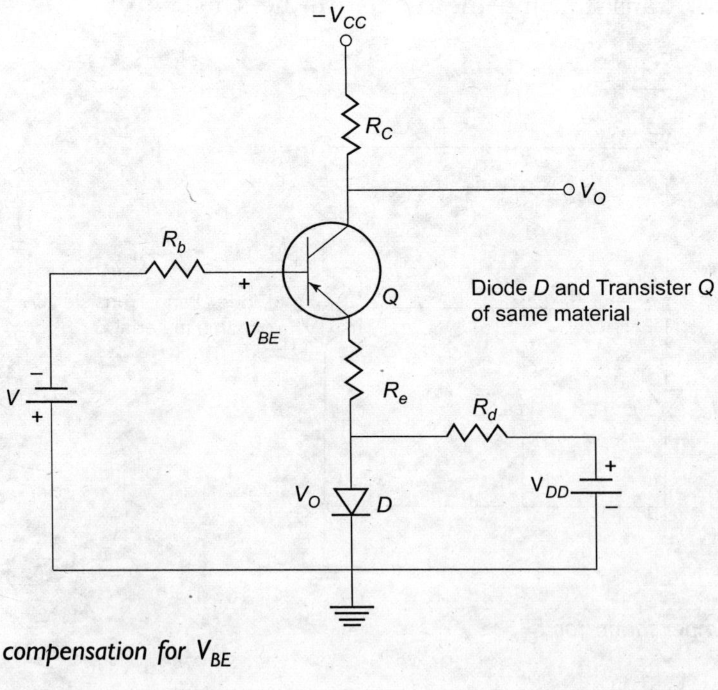

Fig. 6.9 *Diode compensation for* V_{BE}

Since the dependence of V_{BE} and V_o on temperature is the same, the changes in V_{BE} are compensated by the changes in V_o. Thus, the output collector current is independent of variations of V_{BE} with temperature.

Thermistor and sensistor compensation: Temperature sensitive devices such as thermistor and sensistor can replace the existing resistor in a transistor circuit or can be added at an appropriate position in the circuit to compensate the variation of the operating point with respect to temperature. A *thermistor* is a device whose resistance depends on temperature and decreases with increase in temperature. *A sensistor* is a device whose resistance increases with increase in temperature. *A thermistor* has negative temperature coefficient, because its resistance is decreasing exponentially with increasing temperature. *A sensistor* has positive temperature coefficient, because its resistance increases exponentially with increasing temperature. Both of them are together called as *varistors*.

The circuit shown in Fig. 6.10 employs a thermistor that is connected between the supply V_{CC}, and the emitter of the transistor. The current in the emitter resistance R_e is the sum of the current through R_T and the emitter current. As the temperature increases, the value of R_T decreases and so it draws more current from V_{CC}. This excess current flows in R_e to increase the voltage drop across it. This, in turn, decreases the voltage drop across the emitter junction since voltage across R_2 is constant. As the emitter junction is forward biased, the decrease in voltage across this junction decreases the base current, thus reducing the collector current. With this type of feedback, the increase in the collector current due to the increase in the temperature is compensated to a large extent.

The compensation may also be achieved by placing a thermistor in parallel to R_2. In this case, the thermistor between V_{CC} and emitter is not required. As the temperature increases, the effective resistance between base and ground decreases, which decreases

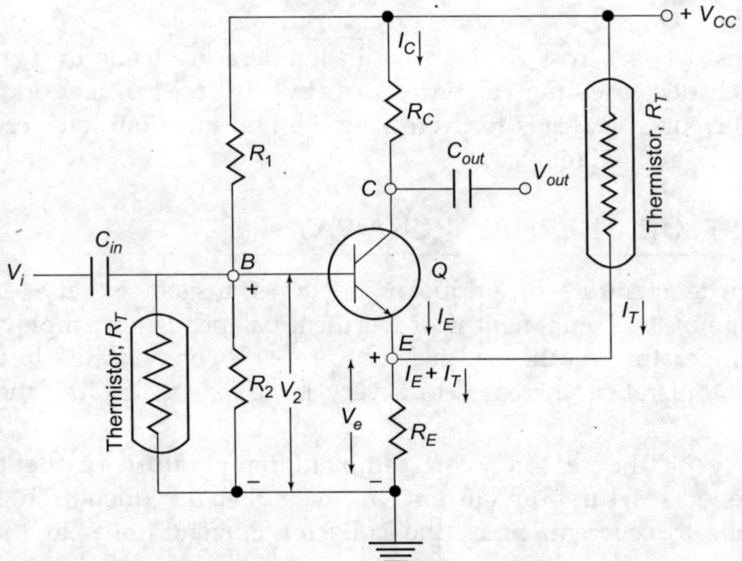

Fig. 6.10 *Thermistor compensation circuit*

the voltage across R_2. Thus, the voltage across the forward-biased emitter junction decreases, decreasing the collector current.

The compensation of the variations in the operating point can also be achieved by the use of a sensistor, as shown in Fig. 6.11. The sensistor may be either placed in shunt with R_1 or in shunt with R_e. When the sensistor is placed in shunt with R_e, the effective emitter resistance increases with increase in temperature, thus reducing the forward bias junction voltage. This reduces the collector current and compensation is achieved.

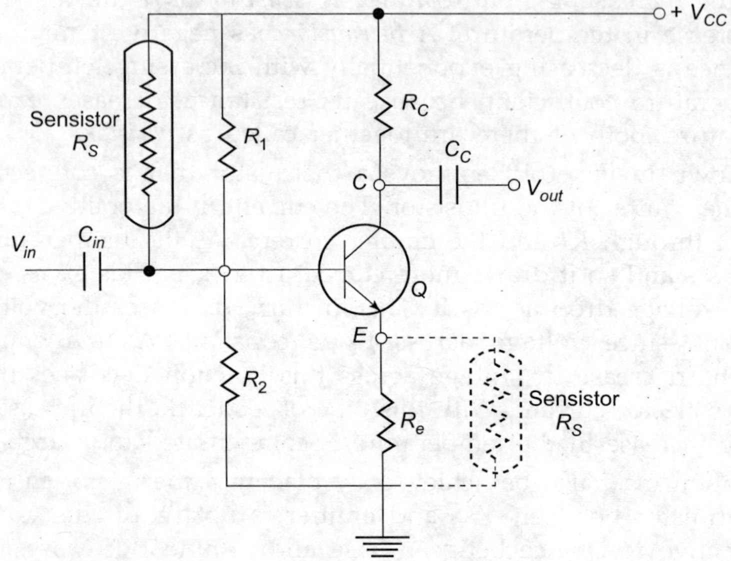

Fig. 6.11 *Sensistor compensation circuit*

When the sensistor is across R_1, increase in temperature leads to increase in effective value of R_1, which decreases the current drawn by the resistor network, R_1 and R_2. The decrease in the current in R_2 leads to decrease in emitter junction voltage, thus compensating the operating point variations.

6.11 CONCEPT OF THERMAL RUNAWAY

The effect of high temperature on a bipolar junction transistor is very serious, leading to permanent damage of the component if proper heat balance is not employed in the circuit. If the effect of temperature on the circuit is not properly compensated for, a closed loop is formed wherein temperature increases to a very large extent, tending the dissipated heat to infinity ideally.

Whatever may be the reason, when ambient temperature of the transistor circuit increases, the reverse saturation current in the collector junction increases and this increases the collector current. Since the collector current flows in the reverse-biased collector junction, whose voltage is a large value, the dissipation of power as heat in this junction is very large and tends to increase with rise in temperature. This, in turn, increases the temperature of the junction, which leads to increase of I_{CO} once again. This cycle of increase of temperature-I_{CO}-I_C-power dissipation-temperature forms a closed loop (as shown in Fig. 6.12) and thus the temperature at the collector junction of the transistor thus increases. This condition is impossible since the energy supplied by the battery V_{CC} is limited, the heat dissipation in the component crosses the maximum value for which it is designed. At this juncture, the transistor will not be able to tolerate the excess generated heat since it is not designed to dissipate the excess thermal energy effectively. This condition where the heat dissipation is slower compared to rate of heat generation leads to the burning of the transistor. This is known as *thermal runaway.*

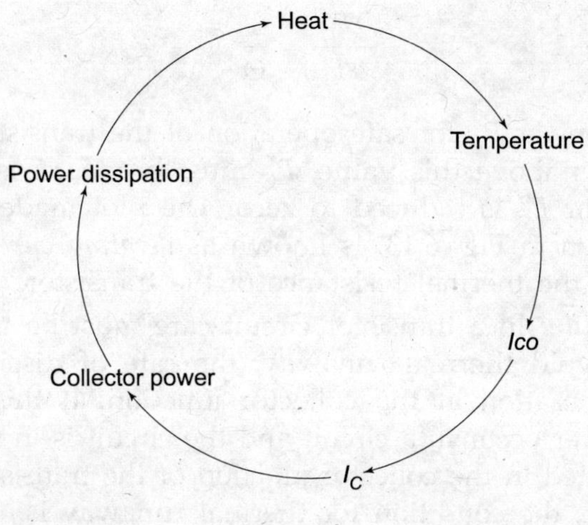

Fig. 6.12 *Thermal runaway*

In the design procedure of any transistor circuit, caution is to be exercised that *thermal runaway* never occurs. This may be achieved by making rate of power dissipation in the collector more compared to rate of power generation in the same junction.

The maximum average power $P_{D,\max}$ that a transistor can dissipate depends on the transistor construction and it may be in the range from few milliwatts to 200 W. The maximum power is limited by the temperature that the collector base junction can withstand. These values of maximum power dissipation depend upon the type of material and the type of the component for which it is designed. For a silicon transistor it is of the order of 150 °C to 225 °C and for a germanium transistor it is between 60 °C and 100 °C. *Thermal resistance*: From thermodynamics, heat flows from higher temperature to lower temperature and the rate of flow depends upon the type of media in which the heat is flowing. The constant of the media, which dictates the rate of flow of heat, can be defined as *thermal resistance*. This phenomena on of thermal resistance is governed by a law that can be given as

$$T_j - T_A = \Theta P_D \qquad ...(6.38)$$

where T_j is junction temperature, T_A is ambient (room) temperature, P_D is dissipated power and Θ is thermal resistance. This equation is similar to that of Ohm's law in electrical circuits. If $T_j - T_{A,}$ the temperature difference between two points, is similar to voltage drop between two points, electrical current is similar to power dissipation and so the electrical resistance is identical to thermal resistance. Thus, the electrical resistance and thermal resistance are the properties of the media. The value of the thermal resistance depends upon the size of the transistor, on convection or radiation to the surroundings on forced air cooling, and on thermal connection of the device to a metal chassis or a heat sink. Typical values of various transistor designs vary from 0.2 °C/W for high-power transistors with an efficient heat sink to 1000 °C/W for a low-power transistor in free air. The temperature rate at which power is dissipated under steady-state conditions is obtained by differentiating Eq. (6.38) with respect to T_j or

$$\frac{\partial P_D}{\partial T_j} = \frac{1}{\Theta} \quad \text{...(6.39)}$$

The maximum collector power P_C for safe operation of the transistor is specified at 25 °C. For ambient temperatures above this value, P_C must be decreased such that at extreme temperature, the transistor P_C is reduced to zero. The plot made between the case temperature and P_C, as shown in Fig. 6.13, is known as *derating curve.* The reciprocal of the slope of this curve gives the thermal resistance of the transistor.

Condition for thermal stability: In a transistor circuit care must be taken that thermal runaway never occurs. To avoid thermal runaway, the rate of dissipation should be large compared to rate of generation at the collector junction. If this condition is satisfied, thermal runaway will never occur in a circuit and the circuit is in safe region of operation. If P_C is the power generated in the collector junction of the transistor and P_D is the power dissipated by a transistor, the condition for thermal runaway is

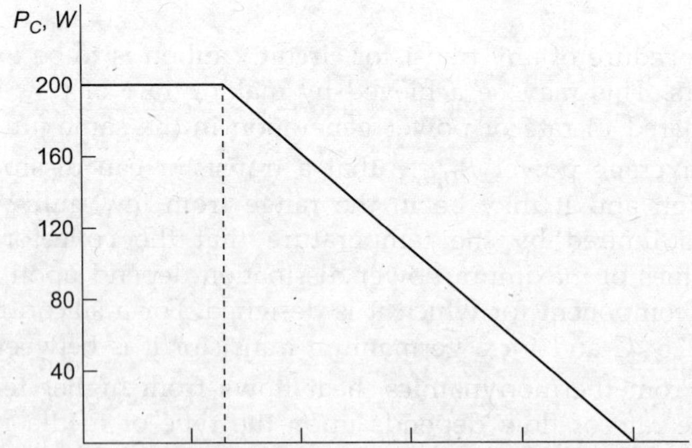

Fig. 6.13 *Derating curve of a transistor*

$$\frac{\partial P_C}{\partial T_j} < \frac{\partial P_D}{\partial T_j} \qquad ...(6.40)$$

or, from Eq. (6.39),

$$\frac{\partial P_C}{\partial T_j} < \frac{1}{\Theta} \qquad ...(6.41)$$

For a self-bias circuit, $V_{CB} = V_{CE} - V_{BE} \approx V_{CE}$ since V_{BE} is the forward bias voltage of the emitter junction. Therefore,

$$P_C = I_C V_{CB} \approx I_C V_{CE} \qquad ...(6.42)$$

Applying KVL to output circuit,

$$V_{CE} = V_{CC} - I_C(R_C + R_e) \qquad ...(6.43)$$

Substituting Eq. (6.43) into (6.42),

$$P_C = I_C[V_{CC} - I_C(R_C + R_e)] \qquad ...(6.44)$$

The Equation (6.41) can be rewritten as

$$\frac{\partial P_C}{\partial I_C}\frac{\partial I_C}{\partial T_j} < \frac{1}{\Theta} \qquad ...(6.45)$$

From Eq. (6.44),

$$\frac{\partial P_C}{\partial I_C} = V_{CC} - 2I_C(R_C + R_e) \qquad ...(6.46)$$

Since I_C increases with temperature, $\partial I_C/\partial T_j$ is positive and so, $\partial P_C/\partial I_C$ should be less than 1 since Θ is positive. Hence, to avoid thermal runaway it is necessary that

$$I_C > V_{CC}/[2(R_C + R_e)] \qquad ...(6.47)$$

From Eq. (6.43), the condition to avoid thermal runaway can be written as

$$V_{CE} < V_{CC}/2 \qquad ...(6.48)$$

If $V_{CE} > V_{CC}/2$, it can be seen that $\partial P_C/\partial I_C$ is positive and there is a possibility of thermal runaway. So, the designer of a transistor circuit should carefully design the circuit so that Eq. (6.48) is satisfied in a *self-bias* circuit always.

Operating point in thermal runaway: As discussed above, it is clearly understood that $V_{CE} < V_{CC}/2$ is the safe region of operation for a transistor circuit. The same fact can now be established with the help of selection of operating point on the load line. Figure 6.14 is considered where maximum dissipation curves for various powers are plotted along with the load line. The load line is tangential to 200 W curve and the operating point Q_2 bisects the load line. So, x ordinate of the operating point is $V_{CE} = V_{CC}/2$. If the operating point is so chosen such that V_{CE} is less than $V_{CC}/2$, say at Q_3, increase in I_C tends the operating point to move on load line towards lower power curve as shown in the figure, thus reducing the power dissipation in the circuit. But, if the operating point is chosen such that

$V_{CE} > V_{CC}/2$, at Q_1 as in the figure, increase in temperature leads to shift in the increase in I_C, which moves the operating point on the load line towards higher power curve as shown in the figure. Thus, increase in I_C increases the dissipation of power, which is an undesired condition. So, selecting the operating point such that $V_{CE} < V_{CC}/2$ ensures the safe region of operation for a transistor in self-bias circuit.

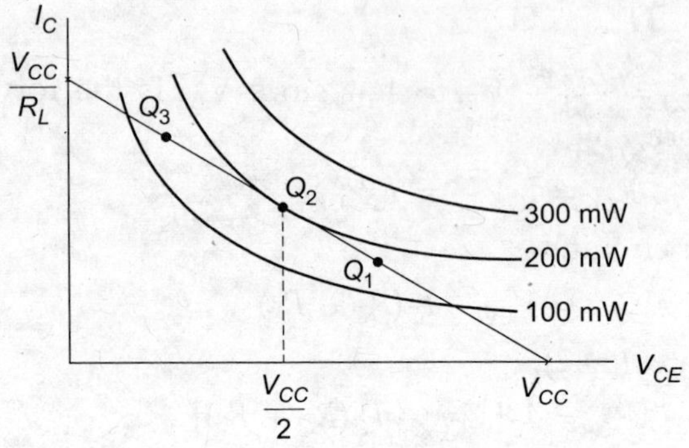

Fig. 6.14 *Selection of operating point in thermal runaway*

6.12 HEAT SINK

A heat sink is a metallic device that increases the effective dissipation area of the collector junction so as to increase the rate of dissipation far beyond the rate of generation. The design of a heat sink depends upon the type of transistor and dissipation capacity of the transistor. When the transistor is to dissipate large amounts of power, it is most convenient to connect the collector lead to the case of the transistor in metallic casing transistors. If the rate of dissipation is to be higher then heat sinks are used. The size, shape, orientation and type of material for the heat sink varies depending upon the application and type of transistor. To determine the qualities of the heat sink required, the following analysis is required.

Θ is the thermal resistance of the media. While calculating dissipation capacity, the difference in temperature between junction and ambience is directly related to dissipation power with the media constant as Θ. But the media is not the same between junction and ambience since the heat should travel from junction to casing and then to free space. So, this involves the heat flow in two media. The thermal resistance of the two media is not the same and so exact value of thermal resistance is to be determined properly. The manufacturers of the transistor generally specify the thermal resistance between junction and casing. In order to determine the size, shape and the capacity of the heat sink to enhance the dissipation capacity, the thermal resistance of casing to ambience is required. Therefore,

$$\Theta_{JA} = \Theta_{JC} + \Theta_{CA} \qquad ...(6.49)$$

where Θ_{JA} is the thermal resistance between junction and ambience, Θ_{JC} is the thermal resistance between junction and casing and Θ_{CA} is the thermal resistance between casing and ambience.

If heat sink is added to achieve the higher dissipation capacities, the above equation can be written as

$$\Theta_{JA} = \Theta_{JC} + \Theta_{CA} || \Theta_{HSA} \tag{6.49}$$

where Θ_{HSA} is thermal resistance between heat sink and ambience. The shapes of various types of heat sinks in practice are as shown in Fig. 6.15.

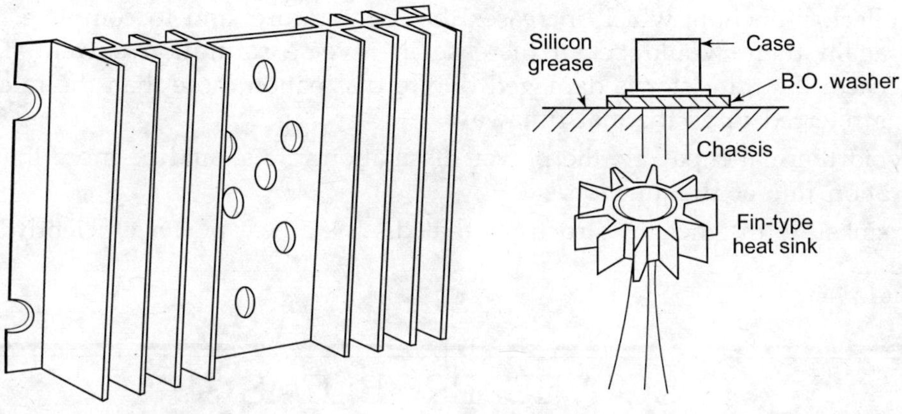

Fig. 6.15 *Different types of heat sinks*

SUMMARY

- If a transistor is to be used effectively, it is to be biased. Biasing the transistor is nothing but driving the transistor into the required region of operation.
- Load line is drawn on the output characteristic of the transistor, which has the slope proportional to load resistance. Thus, there are two types of load lines—static and dynamic.
- Operating point or quiescent point is selected on the load line such that the required region of operation is achieved. For a *CE* transistor, the selection of base current on the load line is the operating point.
- When two batteries or single battery biasing is used, the operating point is fixed and so, the stability of the circuit is very poor since the transistor parameters vary with temperature, time, piece and so on.
- A self-bias circuit stabilises the operating point with the help of emitter resistor.
- Stability of the transistor circuit is measured in terms of stability factors S, S' and S''.
- The variation of the collector current can be estimated in terms of these stability factors. The smaller these values, the better the stability of the circuit.

- Small signal current gain, reverse saturation current and base to emitter voltage are the parameters of a transistor that depend on temperature.
- Thermistors, sensistors or diodes can be used in transistor circuits to compensate the effect of temperature on operating point.
- Increase in temperature increases the reverse saturation current, which in turn increases the collector current, consecutively increasing the power dissipation in the collector junction, which increases the temperature, and to complete the cycle once again there would be an increase in reverse saturation current. If this is unchecked, the transistor is damaged due to dissipation more than the rated value. This is referred to as thermal runaway.
- To avoid thermal runaway, the power dissipation rate should be more than power generation rate at the junction.
- Heat sinks are the metallic structures that dissipate heat power efficiently into free space.

SOLVED PROBLEMS

6.1 Determine the quiescent currents and the collector to emitter voltage for a silicon transistor with β = 60 in the self-biasing arrangement of figure. The circuit component values are V_{CC} = 20 V, R_C = 2 K, R_1 = 100 K, R_e = 0.1 K, R_2 = 5 K.

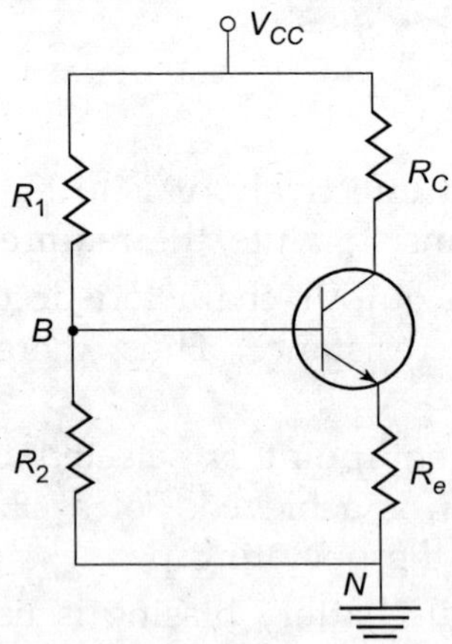

Solution Applying Thevenin's equivalent circuit between B and N,

$$V_{TH} = \frac{R_2 V_{CC}}{R_1 + R_2} = \frac{5 \times 20}{5 + 100}$$

$$= 0.952 \text{ V}$$

and, $R_{TH} = \dfrac{R_1 R_2}{R_1 + R_2} = \dfrac{100 \times 5}{100 + 5} = 4.76 \text{ k}\Omega$

Then, the circuit reduces as shown in the figure.

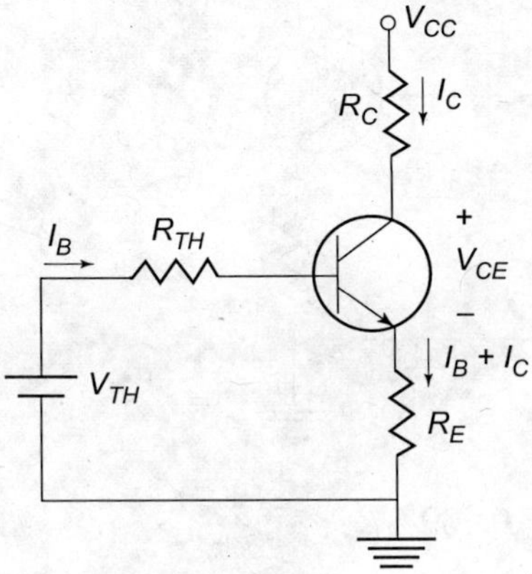

Applying KVL in base-emitter loop of this figure,

$$V_{TH} = R_{TH} I_B + V_{BE} + (I_B + I_C) R_e$$

$$\Rightarrow \quad V_{TH} = R_{TH} I_B + V_{BE} + (\beta + 1) I_B R_e$$

$$\Rightarrow \quad I_B = \frac{V_{TH} - V_{BE}}{R_{TH} + (\beta + 1) R_e} = \frac{0.95 - 0.7}{4.76 + (61) 0.1}$$

$$= 23.02 \ \mu\text{A}$$

$$\therefore \quad I_C = \beta I_B = 1.381 \text{ mA}$$

$$\therefore \quad I_E = -(I_B + I_C) = -1.4040 \text{ mA}$$

Applying KVL in *CE* loop

$$V_{CE} = V_{CC} - I_C R_C - (I_B + I_C) R_e$$

$$= 20 - 1.38 \times 10^{-3} \times 2 \times 10^3$$

$$- (1.404 \times 10^{-3}) \times 0.1 \times 10^3$$

$$= 20 - 2.76 - 0.1404$$

$$= 17.10 \text{ V}$$

6.2 A *pnp* Ge transistor is used in the self-bias arrangement. $V_{CC} = 4.5$ V, $R_e = 0.27$ K, $R_C = 1.5$ K, $R_2 = 2.7$ K, $R_1 = 27$ K, if $\beta = 44$,

(a) find the *Q* point.

(b) Recalculate all these values if the base-spreading resistance of 690 Ω is taken into account.

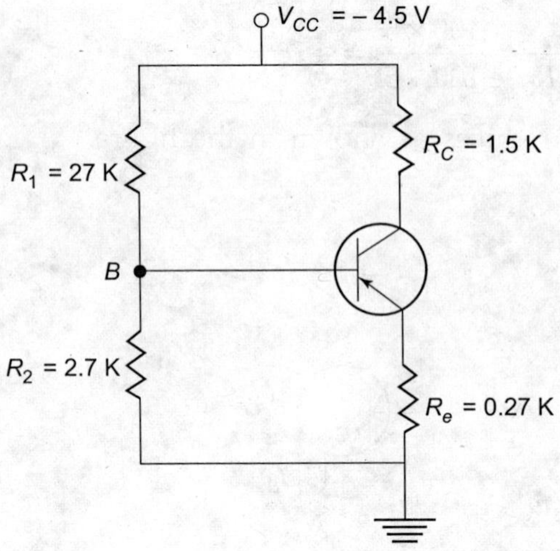

Solution (a) $V_{TH} = \frac{R_2}{R_1 + R_2} \cdot V_{CC} = 0.409$ V

$$R_{TH} = \frac{R_1 R_2}{R_1 + R_2} = 2.46 \text{ k}\Omega$$

For Ge transistor V_{BE} = 0.2 V
Applying KVL in the *B–E* loop,

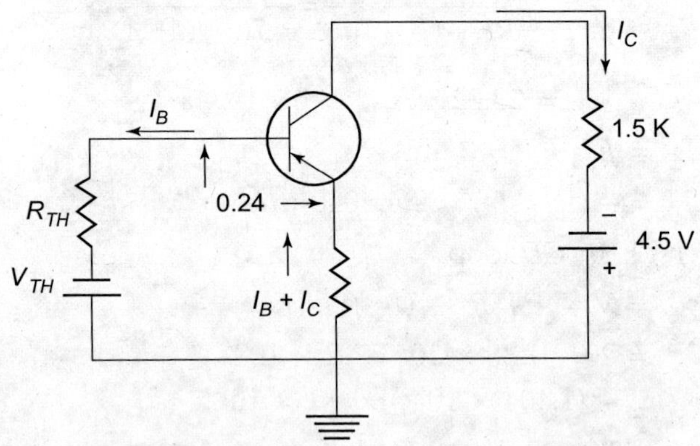

$$- V_{TH} = V_{BE} + I_B R_{TH} + (1 + \beta) I_B R_e$$

$$\Rightarrow \quad I_B = - 143 \ \mu\text{A}$$

$$I_C = \beta I_B = - 0.68 \text{ mA}$$

Applying KVL in *C–E* loop

$$V_{CE} = - V_{CC} - I_C R_C - (I_B + I_C) R_e$$

$$= - 3.38 \text{ V}$$

(b) If base spreading resistance of 690 Ω is taken into account, then R_{TH} is increased.

$$\therefore \qquad R_{TH} = 690 + 2.46 \text{ K} = 3.15 \text{ K}\Omega$$

$$\therefore \qquad I_B = \frac{-V + V_{BE}}{R_{TH} + (1+\beta)R_e} = -13.7\ \mu\text{A}$$

$$I_C = \beta I_B = -0.6 \text{ mA.}$$

Similarly, $V_{CE} = -3.4$ V.

6.3 Assume that a silicon transistor with $\beta = 50$, V_{BE}, active = 0.7 V, $V_{CC} = 22.5$ V and R_C = 5.6 K is used in the following figure. It is desired to establish a Q point at V_{CE} = 1.2 V, I_C = 1.5 mA and stability factor $S \le 6$. Find R_e, R_1 and R_2.

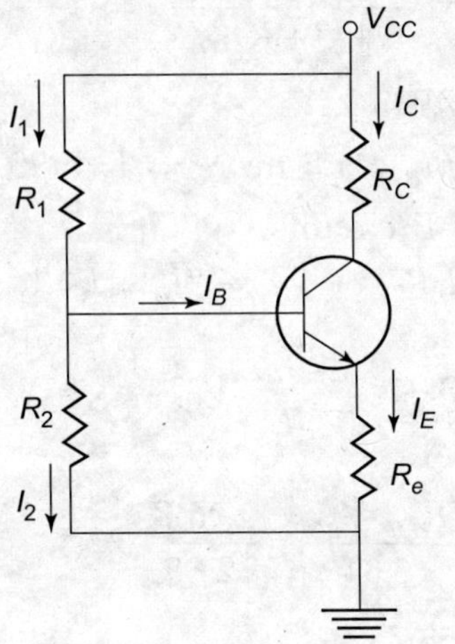

Solution Current in R_e is $I_E = I_B + I_C \approx I_C$

$= 1.5$ mA

Applying Thevenin's theorem, the circuit is redrawn as shown in the following figure:

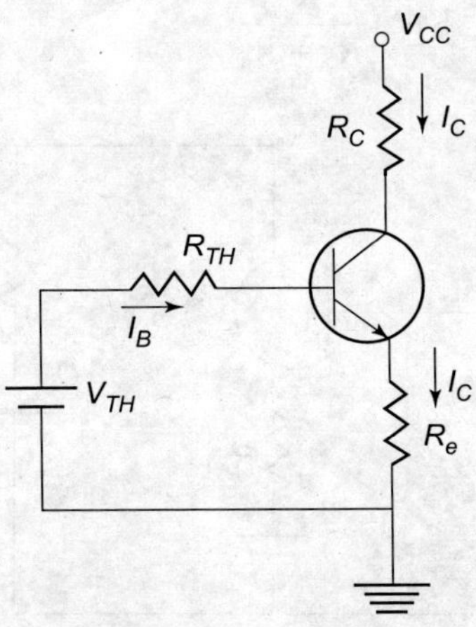

Applying KVL in *C–E* circuit,

$$R_C + R_e = \frac{V_{CC} - V_{CE}}{I_C} = \frac{22.5 - 12}{1.5} = 7.0 \text{ k}\Omega$$

or, $$R_e = 7.0 - 5.6 = 1.4 \text{ k}\Omega$$

We know that for a self-bias circuit

$$S = (\beta + 1)\ \frac{1 + R_{TH}/R_e}{1 + \beta + R_{TH}/R_e}$$

$$6 = 51 \cdot \frac{1 + R_{TH}/R_e}{1 + \beta + R_{TH}/R_e} \Rightarrow \frac{R_{TH}}{R_e} = \frac{51}{9}$$

$\therefore$ $$R_{TH} = 7.933 \text{ k}\Omega$$

And also $$I_C = \beta I_B = 1.5 \text{ mA} \quad \Rightarrow I_B = 30\ \mu\text{A}$$

Applying KVL to the *B–E* circuits,

$$V_{TH} = I_B R_{TH} + V_{BE} + (I_B + I_C)R_e \Rightarrow V_{TH} = 3.04 \text{ V}$$

$$V_{CC} \cdot \frac{R_2}{R_1 + R_2} = 3.04 = \frac{R_{TH}}{R_1} \cdot V_{CC}$$

$\therefore$ $$R_1 = \frac{R_{TH} \cdot V_{CC}}{3.04} = \frac{2.96 \text{ K}}{2.68} \cdot 22.5$$

$$= 58.75 \text{ k}\Omega$$

and, $$R_2 = 9.17 \text{ K}$$

6.4 For the two-battery transistor shown, prove that stabilisation factor S is given by

$$S = \frac{\beta + 1}{1 + \beta \dfrac{R_e}{R_e + R_b}}$$

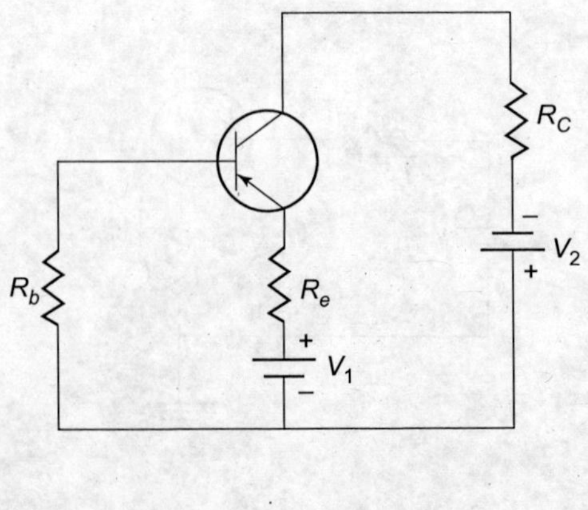

Solution Applying KVL to the *B–E* loop,

$$V_1 = I_E R_e - I_B R_b + V_{BE}$$

Neglecting V_{BE}, $V_1 = I_E R_e - I_B R_b$

But, $$I_E = -(I_B + I_C)$$

$\therefore$ $$V_1 = -(I_B + I_C)R_e - I_B R_b$$

$\Rightarrow$ $$I_B = -\left[\frac{I_C R_e + V_1}{R_e + R_b}\right]$$

We know that $$I_C = \beta I_B + (\beta + 1)I_{Co}$$

$$I_C = (I + \beta)I_{Co} - \frac{\beta}{R_e + R_b}(I_C R_e + V_1)$$

$$\Rightarrow I_C\left(1 + \frac{\beta}{R_e + R_b}R_e\right) = (1 + \beta)I_{Co} - \frac{\beta V_1}{R_e + R_b}$$

$\therefore$ $$S = \frac{\partial I_C}{\partial I_{Co}} = \frac{\beta + 1}{1 + \beta\dfrac{R_e}{R_e + R_b}}$$

6.5 In a *C–E* amplifier, a *pnp* Ge transistor is used. $\beta = 44$, $R_L = 1.5\ \text{K}\Omega$. $R_e = 0.27\ \text{K}\Omega$, $R_1 = 27\ \text{K}\Omega$, $R_2 = 2.7\ \text{K}\Omega$, $V_{CC} = -4.5$ V. Calculate *S* and *Q* point.

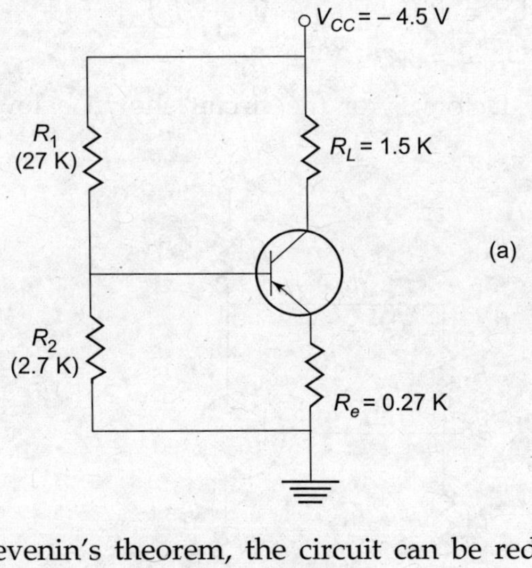

Solution Applying Thevenin's theorem, the circuit can be redrawn as shown in Fig. (b)

$$R_{TH} = R_1 || R_2 = 2.45\ \text{K}\Omega$$

$$V_{TH} = \frac{R_2}{R_1 + R_2}V_{CC} = -0.4\ \text{V}$$

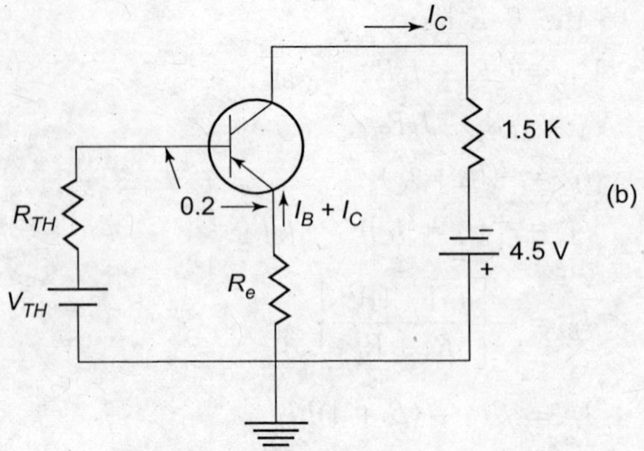

$$S = \frac{(\beta+1)(R_{TH}+R_e)}{(1+\beta)R_e + R_{TH}} = \frac{45\,(2.45+0.27)}{(46)\,(0.27)+2.45}$$

$$= 8.38$$

Applying KVL to the B–E loop in Fig. (b),

$$V_{TH} = -\,(I_B + I_C)R_e + V_{BE} - I_B R_{TH}$$

$$0.4 = \left(\frac{I_C}{44} + I_C\right)0.27 + 0.2 + \frac{I_C}{44}(2.45)$$

$$\Rightarrow \quad I_C = -\,0.6 \text{ mA}$$

Now, applying KVL to the C–E loop,

$$4.5 = R_e(I_C + I_B) + V_{CE} + I_C R_L$$

$$\therefore \quad V_{CE} = -\,3.435 \text{ V}$$

6.6 Determine the stability factors S for the circuit shown below:

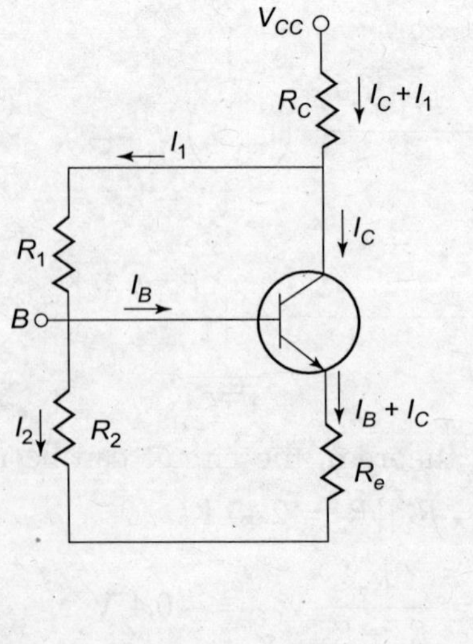

Solution $V_{CC}(I_C + I_1) = R_C + (I_C + I_B)R_e + V_{BE}$

Applying KVL in C–B–E loop, applying KCL at 'B' node,

$$I_1 = I_B + I_2.$$

$$I_2 = \frac{V_{BE} + (I_C + I_B)R_e}{R_2} \quad \text{...(1)}$$

$$I_1 = \frac{(I_B + I_C)R_e + V_{BE} + I_B R_2}{R_2} \quad \text{...(2)}$$

Put (2) in (1),

$$V_{CC} = I_C\left(R_C + R_e + R_e\frac{(R_1 + R_C)}{R_2}\right) + I_B\left(R_e + \frac{(R_e + R_2)(R_1 + R_C)}{R_2}\right)$$

$$+ \left(1 + \frac{R_1 + R_C}{R_2}\right)V_{BE} \quad \text{...(3)}$$

We know that $I_C = \beta I_B + (\beta + 1)I_{CO}$

$$I_B = \frac{I_C - (\beta + 1)I_{CO}}{\beta} \quad \text{...(4)}$$

Put (4) in (3),

$$V_{CC} = I_C\left[R_C + R_e + R_e\frac{(R_1 + R_2)}{R_2}\right] + \left[\frac{I_C - (\beta + 1)I_{CO}}{\beta}\right]$$

$$\left[R_e + \frac{(R_e + R_2)(R_1 + R_C)}{R_2}\right] + \left[1 + \frac{R_1 + R_C}{R_2}\right]V_{BE} \quad \text{...(5)}$$

But $$S = \left.\frac{\partial I_C}{\partial I_{CO}}\right|_{\beta, V_{BE}\text{ constants}}$$

Differentiating (5) w.r.t. I_{CO} $S = \dfrac{(1 + \beta)[R_2(R_C + R_1) + R_e(R_1 + R_2 + R_e)]}{R_1R_2 + (1 + \beta)[R_2R_C + R_e(R_1 + R_2 + R_e)]}$

6.7 (a) Calculate the thermal resistance for the 2N338 transistor for which the manufacturer specifies $P_{C\,max} = 125$ mW at 25 °C. Free air temperature and maximum junction temperature $T_J = 150$ °C.

(b) What is the junction temperature if the collector dissipation is 75 mW?

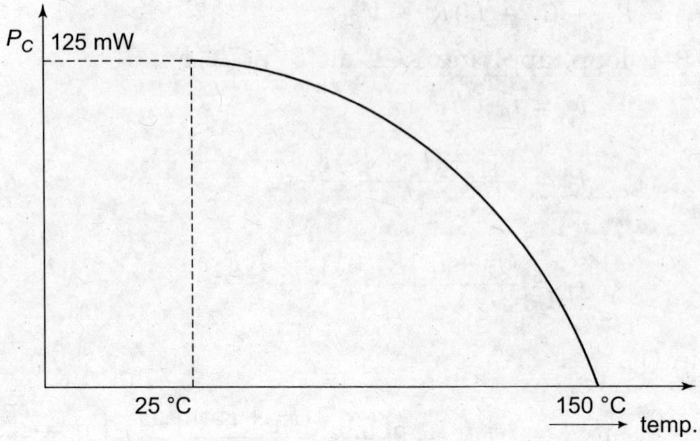

Solution (a) From the power temperature derating curve of given transistor,

$$-Q = \text{Thermal resistance} = \frac{150° - 25°}{(125 - 0)} = 1\ °\text{C/mW}$$

(b)
$$T_J - T_A = -QP_D$$
$$T_J = T_A - QP_D = 25° + (1)(75)$$
$$= 100\ °\text{C}$$

6.8 For the circuit shown below, find S.

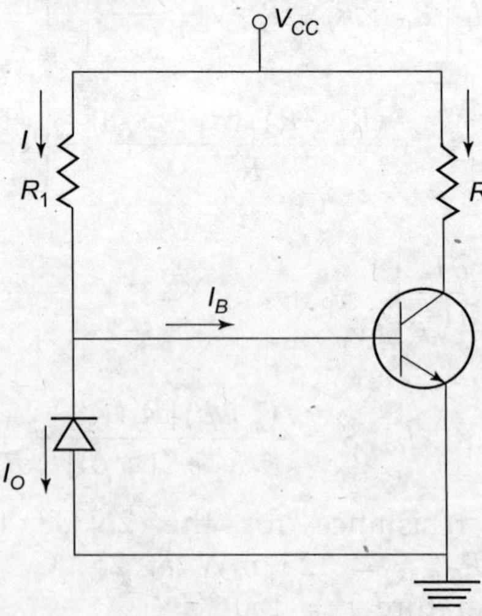

Solution
$$I = \frac{V_{CC} - V_{BE}}{R_1} \cong \frac{V_{CC}}{R_1}$$
$$= \text{Constant}$$
$$I_B = I - I_o$$

I_o is the Reverse Saturation Current in diode.

But $\quad I_C = \beta I_B + (\beta + 1)I_{CO}$,

$$= \beta(I - I_O) + (\beta + 1)I_{CO}$$

Since transistor and diode are of the same material $I_O = I_{CO}$

$\therefore \quad I_C = \beta I + I_{CO}$

$\therefore \quad S = \dfrac{\partial I_C}{\partial I_{CO}} = 1$

6.9 A silicon transistor with $\beta = 90$ is used in the fixed-bias configuration as shown in the following figure. Determine I_{BQ}, I_{CQ}, V_{CEQ}, V_C, V_B, V_E and $I_{C\ sat}$.

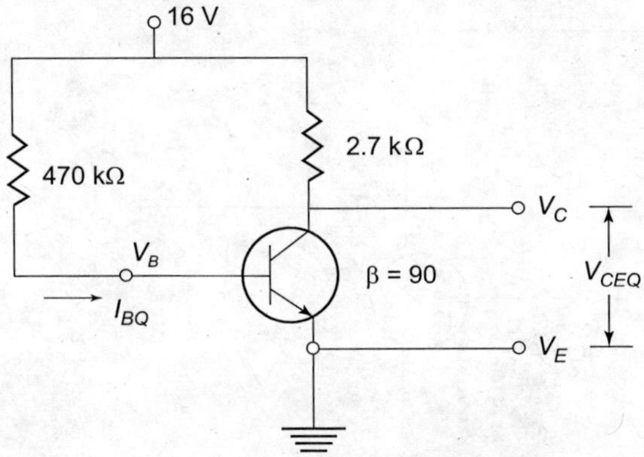

Solution Since emitter is grounded V_E =0. This circuit can also be rearranged as shown below:

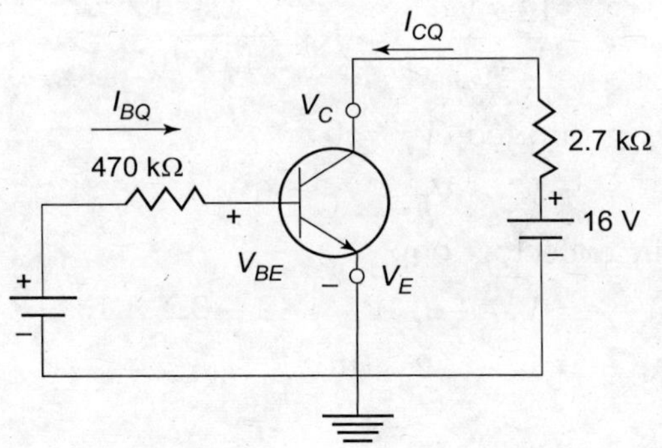

Writing KVL in the base loop,

$$I_{BQ} = \frac{16 - V_{BE}}{470\ \text{K}\Omega} = \frac{16 - 0.7}{470\ \text{K}} = 32.55\ \mu\text{A}$$

$$I_{CQ} = \beta \cdot I_{BQ} = 90 \times 32.55\ \mu\text{A} = 2.929\ \text{mA}$$

From the collector circuit, writing KVL,

$$V_{CEQ} = V_{CC} - I_C R_C = 16 - 2.929 \times 10^{-3} \times 2.7 \times 10^3$$
$$= 8.09 \text{ V}$$

Since $V_E = 0,\ V_{BE} = V_B - V_E = 0.7$ V

$$V_{CE} = V_C - V_E = V_C = 8.09 \text{ V}$$

$$I_{C\text{ sat}} = \frac{V_{CC} - V_{CE\text{ sat}}}{R_C} \cong \frac{V_{CC}}{R_C} = \frac{1.6}{2.7\text{ K}} = 5.925 \text{ mA.}$$

6.10 Find the values of I_C, R_C, R_B and V_{CE} from the following circuit.

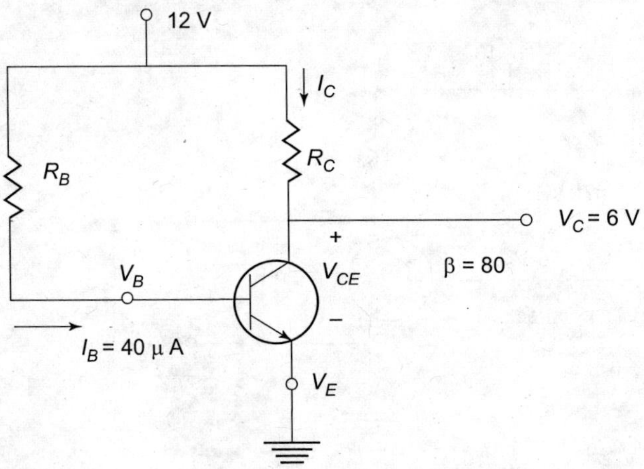

Solution Given that, $I_B = 40\ \mu\text{A},\ I_C = \beta I_B = 80 \times 40\ \mu\text{A} = 3.2$ mA

From figure, $I_B = \dfrac{12 - V_{BE}}{R_B} \Rightarrow R_B = \dfrac{12 - 0.7}{40 \times 10^{-6}} = 282.5\ \text{k}\Omega$

Since emitter is grounded, $V_E = 0$.

$$V_{CE} = V_C - V_E = V_C = 6 \text{ V} \qquad \text{...(1)}$$

Writing KVL in collector loop

$$V_{CE} = V_{CC} - I_C \cdot R_C = 12 - 3.2 \times 10^{-3} \cdot R_C$$
$$6 = 12 - 3.2 \times 10^{-3} R_C$$

$$\Rightarrow \quad \frac{12 - 6}{3.2 \times 10^{-3}} = R_C$$

$$R_C = 1.875\ \text{k}\Omega$$

6.11 Given the information appearing in the figure below, find I_C, V_{CC}, R_B and β.

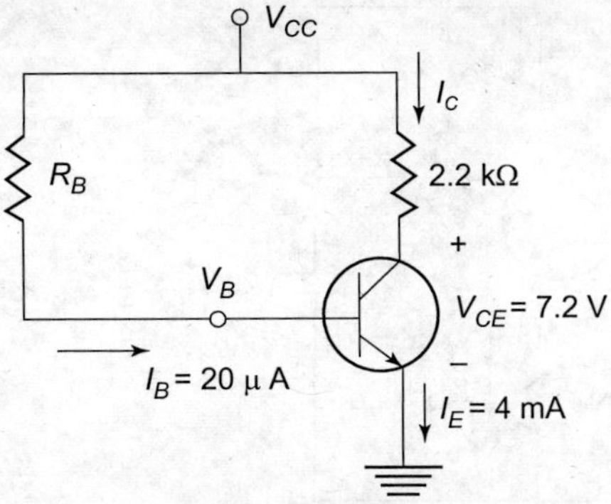

Solution Writing KCL by treating transistor as a node,

$$I_E = I_B + I_C \quad ...(1)$$

Given that I_B = 20 μA and I_E = 4 mA

$$\therefore \quad I_C = I_E - I_B = 4 \times 10^{-3} - 20 \times 10^{-6}$$

$$= 3.98 \text{ mA}$$

We have
$$\beta = \frac{I_C}{I_B} = \frac{3.98 \times 10^{-3}}{20 \times 10^{-6}} = 199$$

From the collector circuit, $V_{CE} = V_{CC} - I_C R_C$

$$7.2 = V_{CC} - 3.98 \times 10^{-3} \times 2.2 \times 10^3$$

$$V_{CC} = 7.2 + 8.756 = 15.956 \text{ V}$$

From the base circuit

$$I_B = \frac{V_{CC} - V_{BE}}{R_B}$$

$$R_B = \frac{15.956 - 0.7}{20 \times 10^{-6}} = 762.8 \text{ k}\Omega$$

6.12 For the following circuit find I_{BQ}, V_{CEQ}, I_{CQ}, V_C, V_B and V_E where subscript Q indicates Quiescent point condition. Given that β = 100.

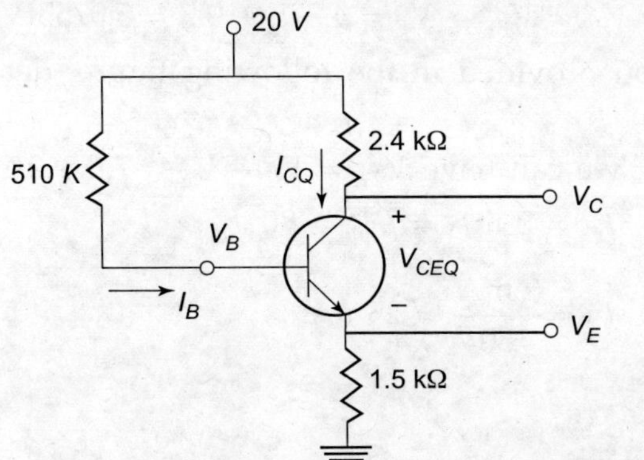

Solution The circuit can be drawn as below.

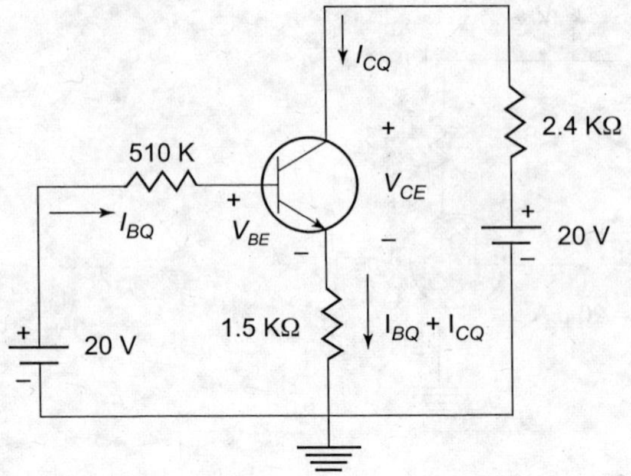

Writing KVL in the *B–E* loop,

$$20 = 510 I_B + V_{BE} + 1.5\,(I_B(\beta + 1))$$

$$20 - 0.7 = 510 I_B + 1.5 \times 101\, I_B$$

$$= 510\, I_B + 151.5\, I_B$$

$$I_B = 29.176\ \mu\text{A}$$

$$I_C \cong \beta I_B = 2.917\ \text{mA}$$

$\therefore$ $I_C >> I_B$. So, $I_E \approx I_C = 2.917$ mA

Writing KVL in *C–E* loop,

$$V_{CEQ} = V_{CC} - I_C(R_C + R_E)$$

$$= 20 - 2.917\ (1.5 + 2.4)$$

$$= 8.62\ \text{V}$$

$$V_C = V_{CC} - I_C \cdot R_C$$

$$= 20 - 2.917 \times 2.4$$

$$= 12.99\ \text{V}$$

$$V_E \approx I_E R_E \approx I_C \cdot R_C = 2.917 \times 1.5 = 4.375\ \text{V}$$

$$V_B = V_{BE} + V_E = 4.375 + 0.7 = 5.0755\ \text{V}$$

6.13 Given the information provided in the following figure, determine R_C, R_E, R_B, V_{CE} and V_B.

Solution From the circuit, we can have $V_{CE} = V_C - V_E = 7.6 - 2.4 = 5.2$ V.

$$I_C = 2\ \text{mA} = \beta I_B = 80 \cdot I_B$$

$$\therefore \qquad I_B = \frac{2\ \text{mA}}{80} = 25\ \mu\text{A}$$

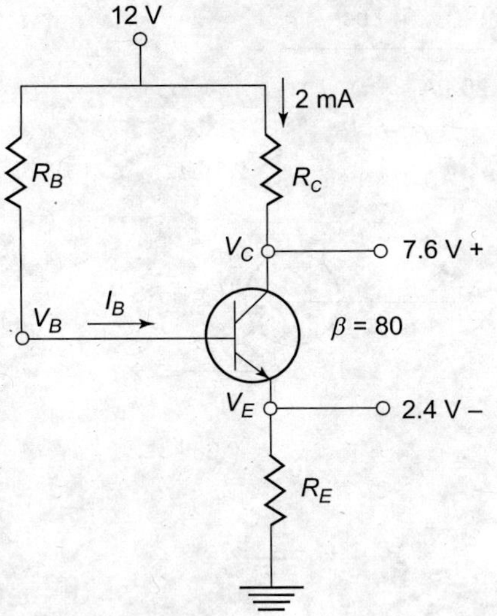

We can write $V_C = V_{CC} - I_C \cdot R_C$

$$7.6 = 12 - 2\text{ mA} \cdot R_C$$

$$\therefore \quad R_C = \frac{12 - 7.6}{2\text{ mA}} = 2.2\text{ k}\Omega$$

and also, $V_E = I_E \cdot R_E \approx I_C \cdot R_E$

$$\therefore \quad R_E = \frac{V_C}{I_C} = \frac{2.4}{2\text{ mA}} = 1.2\text{ k}\Omega$$

$$V_{BE} = V_B - V_E \text{ at the base of transistor}$$

$$V_B = V_{BE} + V_E = 0.7 + 2.4 = 3.1\text{ V}$$

$$\therefore \quad V_B = V_{CC} - I_B R_B$$

$$3.1 = 12 - 25 \times 10^{-6} \cdot R_B$$

$$\therefore \quad R_B = \frac{12 - 3.1}{25 \times 10^{-6}} = 356\text{ k}\Omega$$

6.14 Find β, V_{CC} and R_B for the following circuit with V_{BE} = 0.7 V.

Solution Given that $V_E = 2.1$ V

$$V_{CE} = 7.3\text{ V}$$

$$V_C = V_{CE} + V_E = 9.4\text{ V}$$

$$V_B = V_{BE} + V_E = 2.1 + 0.7 = 2.8\text{ V}$$

We have from collector circuit

$$V_{CE} = V_{CC} - I_C (R_C + R_E)$$

$$7.3 = V_{CC} - I_C (2.7 + 0.68)$$

$$= V_{CC} - I_C (3.38) \quad \text{...(1)}$$

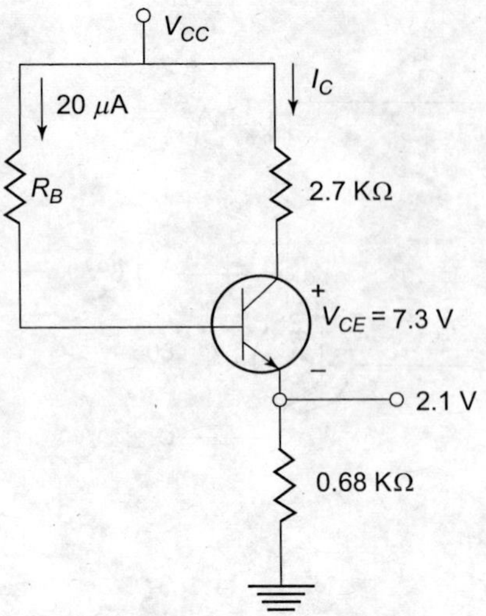

$$V_E = 2.1 = 0.68 \times I_C = 3.088 \text{ mA}$$

$$\therefore \qquad \beta = \frac{I_C}{I_B} = 154.41$$

Using the value of I_C in (1),

$$V_{CC} = 7.3 + 3.08 \times 10^{-3} \times 3.38 \times 10^3 = 17.71 \text{ V}$$

To find R_B, we have $V_B = V_{CC} - I_B \cdot R_B$ ...(2)

$$R_B = \frac{V_{CC} - V_B}{I_B} = \frac{17.71 - 2.8}{20 \times 10^{-6}} = 745.52 \text{ k}\Omega$$

6.15 For the voltage divider bias circuit, determine I_{BQ}, I_{CQ}, V_{CEQ}, V_C, V_E, V_B, and $I_{C\,sat}$.

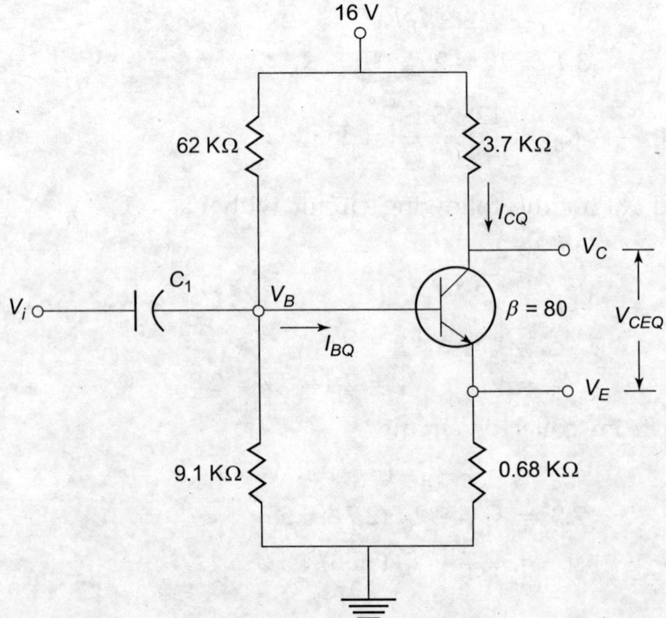

Solution Replacing capacitors as open circuit, for dc, the circuit can be redrawn as below:

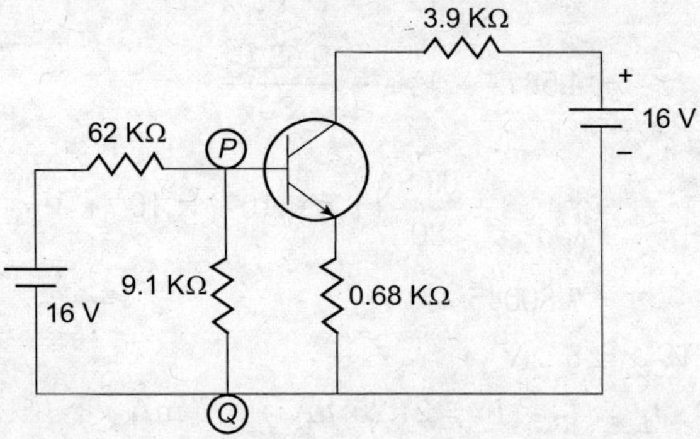

Using Thevenin's theorem, with respect to terminals P and Q,

$$V_{TH} = \frac{9.1}{9.1+6.2} \times 16 = 2.047 \text{ V}$$

$$R_{TH} = 9.1 \,||\, 6.2 = 7.935 \text{ k}\Omega$$

Then, the circuit becomes as below.

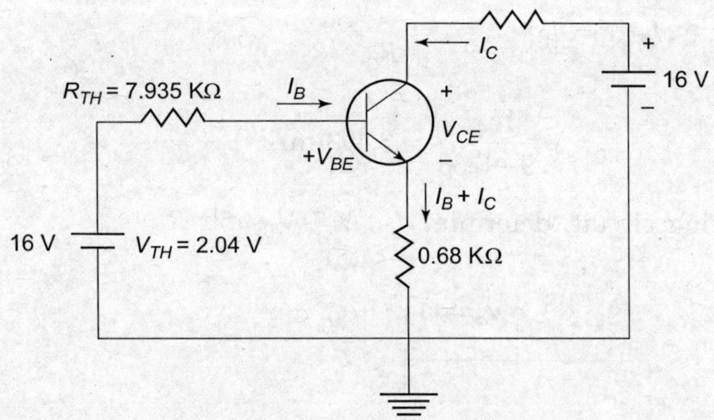

Writing KVL in B–E loop and using $I_C = \beta I_B$,

$$I_B = \frac{V_{TH} - V_{BE}}{R_{TH} + (\beta+1)R_E}$$

$$= \frac{2.04 - 0.7}{7.3935 + 81 \times 0.68}$$

$$= 21.26 \ \mu\text{A}$$

$$I_C = \beta I_B = 80 \times 21.26 = 1.7 \text{ mA}$$

Writing KVL in *C–E* loop,

$$16 = 3.9I_C + V_{CE} + 0.68\,(I_B + I_C)$$

$$= 4.58\,I_C + V_{CE} + \frac{0.68\,I_C}{80}$$

$$= \left(4.58 + \frac{0.68}{80}\right) 1.7 \times 10^{-3} \times 10^3 + V_{CE}$$

$$= 7.80045 + V_{CE}$$

$$V_{CE} = 8.2\text{ V}$$

$$I_E = I_B + I_C = 21.26\ \mu\text{A} + 1.7\text{ mA}$$

$$= 1.721\text{ mA}$$

$$V_E = I_E \cdot R_E = 1.17\text{ V}$$

$$V_{CE} = V_C - V_E \quad \Rightarrow V_C = V_{CE} + V_E$$

$$= 9.37\text{ V}$$

$$V_{BE} = V_B - V_E$$

$$V_B = V_{BE} + V_E = 0.7 + 1.17 = 1.87\text{ V}$$

To find $\qquad I_{C\text{sat}} = I_C\big|_{V_{CE} = 0\text{ V}} = \dfrac{V_{CC}}{R_C + R_E}$

$$= \frac{16}{3.9 + 0.68} = 3.493\text{ mA}$$

6.16 From the following circuit, determine I_C, V_E, V_B and R_1.

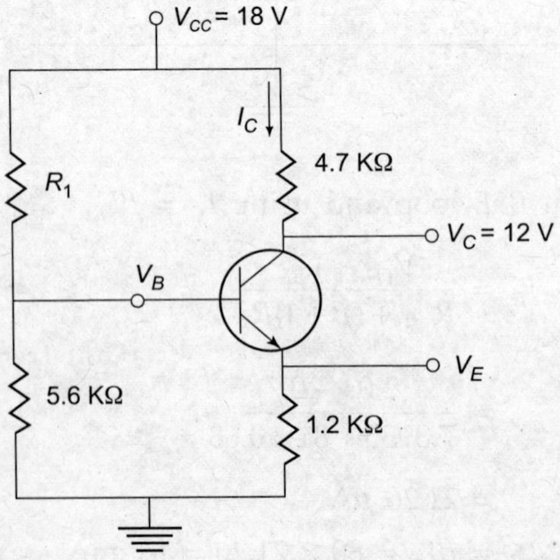

Solution We have

$$V_C = V_{CC} - I_C \cdot R_C$$

$$12 = 18 - I_C\ (4.7)$$

$$\Rightarrow \quad I_C = \frac{18-12}{4.7\ \text{k}\Omega} = 1.276\ \text{mA}$$

$$V_E = I_E \cdot R_E \approx I_C \cdot R_E$$

$$= 1.276 \times 1.2$$

$$= 1.532\ \text{V}$$

$$V_{BE} = V_B - V_E \quad \Rightarrow V_B = V_{BE} + V_E$$

$$= 0.7 + 1.532$$

$$= 2.23\ \text{V}$$

We have,

$$V_B = V_{CC}\frac{5.6}{5.6+R_1} = 2.23$$

$$\Rightarrow \quad 18 \times \frac{5.6}{5.6+R_1} = 2.23$$

$$5.6(18 - 2.23) = R_1 \times 2.23$$

$$R_1 = 39.6\ \text{k}\Omega$$

6.17 Find I_C, V_{CE}, I_B, V_E and V_B for the following circuit.

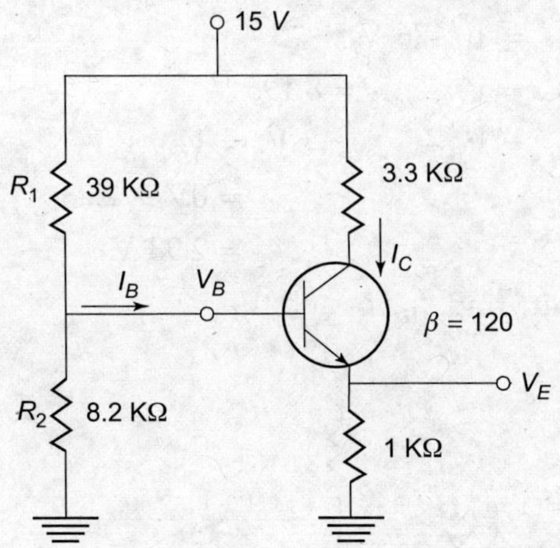

Solution Replacing R_1 and R_2 with their Thevenin's equivalent,

$$V_{TH} = \frac{8.2}{8.2+39} \times 18 = 3.127\ \text{V}$$

$$R_{TH} = 8.2 || 39 = 6.78\ \text{k}\Omega$$

Then, the circuit becomes

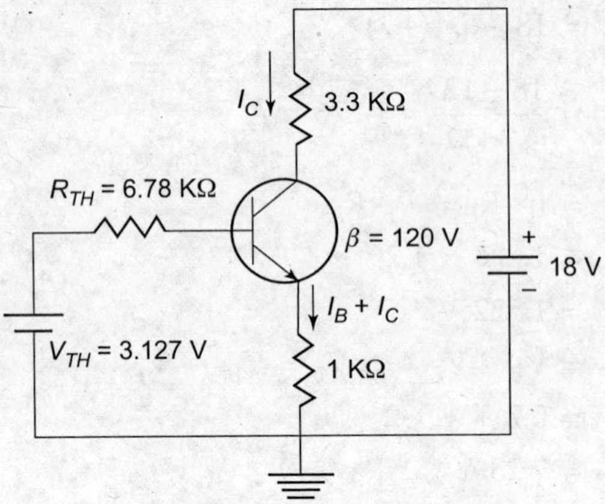

Writing KVL in *B–E* loop.

$$-3.127 + 6.78 \times I_B + V_{BE} + 1(120 + 1)I_B = 0$$

$$I_B = \frac{3.127 - 0.7}{6.78 + 121} = 18.99\ \mu\text{A}$$

$$I_C = \beta I_B = 2.28\ \text{mA}$$

$$V_E \cong I_C \cdot R_E = 2.28\ \text{V}$$

$$V_C = V_{CC} - I_C \cdot R_C = 18 - 2.28 \times 3.3$$

$$= 10.476\ \text{V}$$

$$V_{CE} = V_C - V_E = 8.196\ \text{V}$$

Since, $V_{BE} = V_B - V_E \Rightarrow V_B = V_{BE} + V_E$

$$= 0.7 + 2.28$$

$$= 2.98\ \text{V}$$

6.18 For the following circuit, find I_B, I_C and V_C.

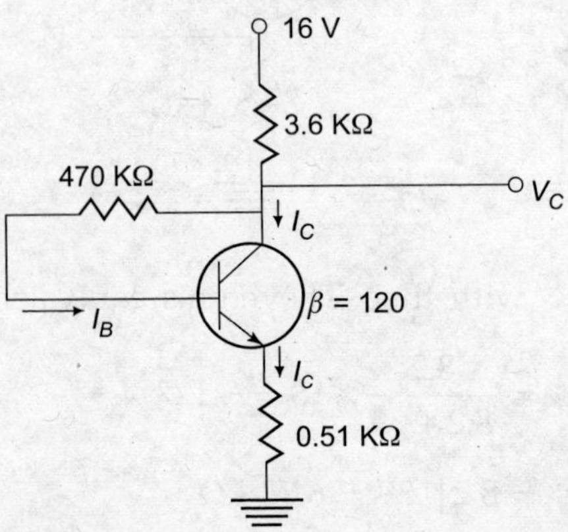

Solution Rearranging the circuit in the following manner,

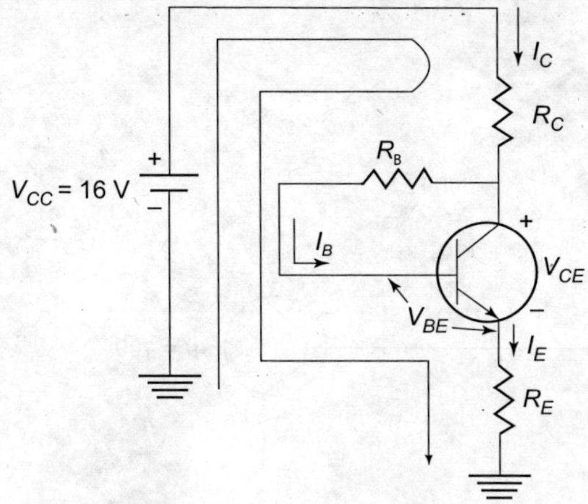

Writing KVL in the loop shown with a big arrow,

$V_{CC} - \beta I_B R_C - I_B R_B - V_{BE} - \beta I_B R_E = 0.$

$$\therefore \qquad I_B = \frac{V_{CC} - V_{BE}}{R_B + \beta(R_C + R_E)}$$

$$= \frac{16 - 0.7}{470 + 120(3.6 + 0.51)}$$

$$= 15.88\ \mu\text{A}$$

$$I_C = \beta I_B = 1.906\ \text{mA}$$

From the circuit $V_E \cong I_C \cdot R_E = 0.972$ V.

$$V_{CE} = V_{CC} - I_C(R_C + R_E)$$

$$= 8.1664 = V_C - V_E$$

$$V_C = V_E + V_{CE} = 9.138\ \text{V}$$

6.19 Find I_C, V_C, V_E and V_{CE} for the network shown below:

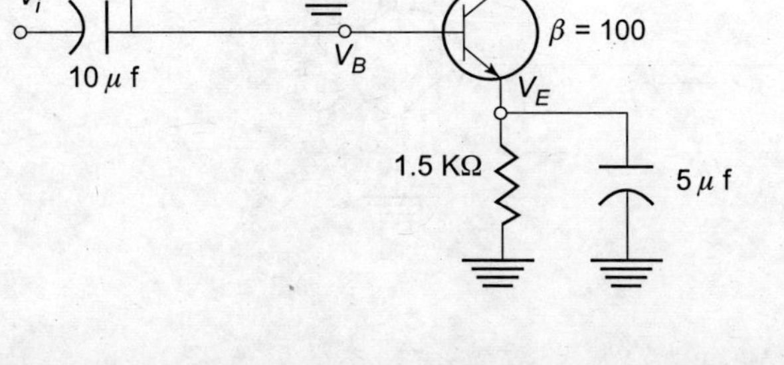

Solution While working for dc quantities, capacitors acts as open-circuit and the circuit becomes as below:

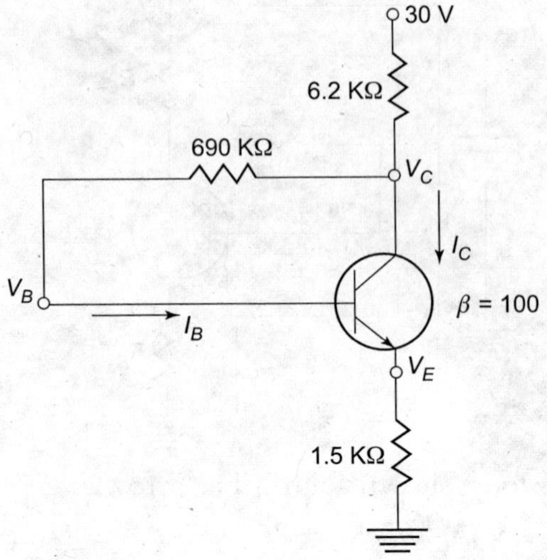

$$I_B = \frac{V_{CC} - V_{BE}}{R_B + \beta(R_C + R_E)}$$

$$= \frac{30 - 0.7}{690 + 100(6.2 + 1.5)}$$

$$= 20.06\ \mu\text{A}$$

$$I_C = \beta I_B = 2.006\ \mu\text{A}$$

$$V_E \cong I_C \cdot R_E = 3.01\ \text{V}$$

$$V_{CE} = V_{CC} - I_C(R_C + R_E) = 14.56\ \text{V}$$

$$V_C = V_{CE} + V_E = 17.57\ \text{V}$$

6.20 Given V_C = 8 V, determine I_B, I_C, β and V_{CE} for the network shown below:

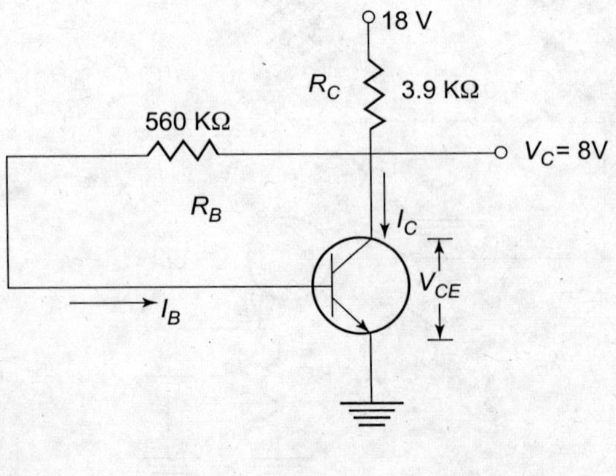

Solution Since emitter is grounded, $V_E = 0$

$\therefore \quad V_{CE} = V_C - V_E = V_C = 8\text{ V}$

and, $\quad V_{BE} = V_B - V_E = 0.7\text{ V} \Rightarrow V_B = 0.7\text{ V}$

We have $$I_B = \frac{V_{CC} - V_{BE}}{R_B + \beta R_C} \quad \text{...(1)}$$

$$V_{CE} = V_{CC} - I_C \cdot R_C \quad \text{...(2)}$$

From (2) $$I_C = \frac{V_{CC} - V_{CE}}{R_C} = \frac{18-8}{3.9\text{ K}} = 2.564\text{ mA}$$

From (1), we can write

$$I_B R_B + I_C \cdot R_C = V_{CC} - V_{BE}$$

$$I_B(560) + (2.56 \times 3.9) = 18 - 0.7$$

$$I_B \times 560 = 7.3$$

$$\therefore \quad I_B = \frac{7.3}{560} = 13.03\ \mu\text{A}$$

$$\beta = \frac{I_C}{I_B} = 196.7$$

6.21 Determine I_E, V_C and V_{CE} for the network below:

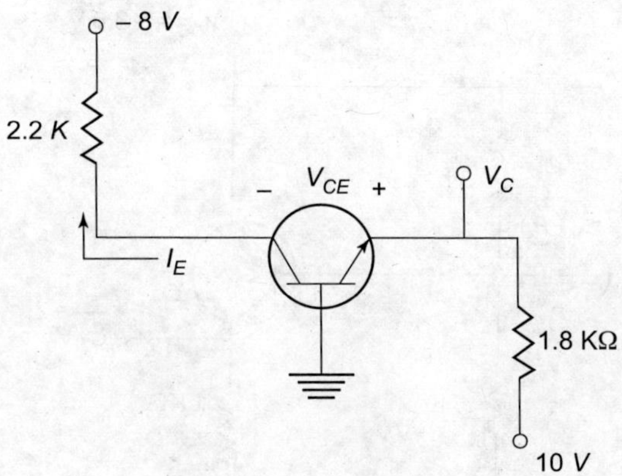

Solution The circuit can be drawn as below.

From circuit $$I_E = \frac{8 - V_{BE}}{2.2\text{ K}} = \frac{8-0.7}{2.2\text{ K}}$$

$$= 3.318\text{ mA}$$

$$\therefore \quad I_C \approx I_E$$

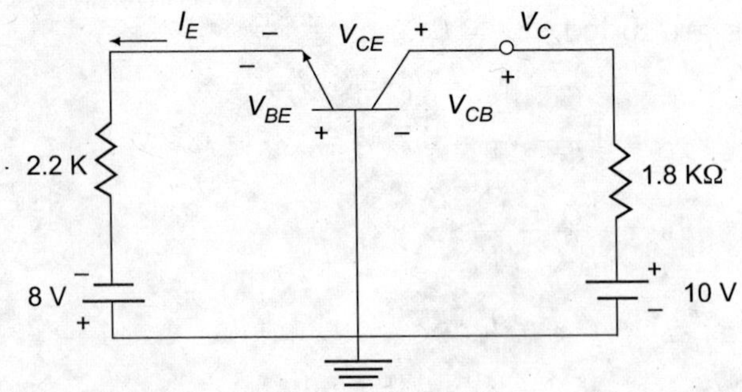

Since base is grounded, $V_B = 0$, $V_{CB} = V_C$

and, $V_{CB} = V_{CC} - I_C \cdot R_C$

$= 4.03$ V

From circuit, $V_{CE} = V_{BE} + V_{CB}$

$= 4.03 + 0.7$

$= 4.73$ V

6.22 A *pnp* silicon transistor is used in a common collector circuit. $V_{CC} = 3.0$ V, R_E = 1K, $R_1 = R_2 = 5$K and $\beta = 44$.

(a) Find *Q* point.

(b) Recalculate these values, taking the base-spreading resistance 690 Ω into account.

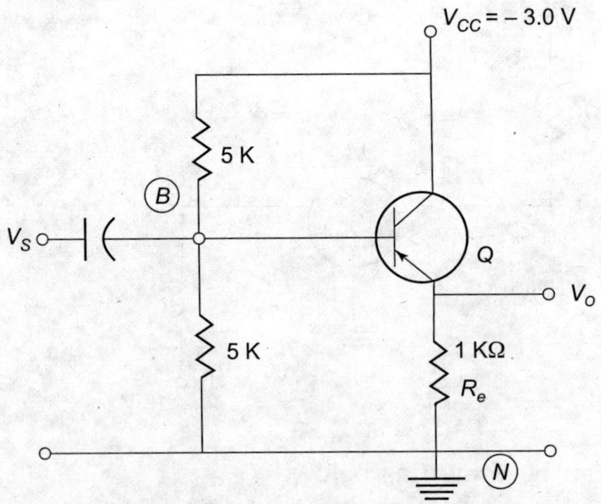

Solution (a) Using Thevenin's theorem, to the left of *B* and *N*,

$$V_{TH} = \frac{5}{5+5} \times 3 = 1.5 \text{ V}$$

$$R_{TH} = 5 \,||\, 5 = 2.5 \text{ K}\Omega$$

Applying KVL in the B–E circuit,

$$-V_{TH} = -V_{BE} + I_B R_{TH} + (I_B + I_C)R_e$$

$$\Rightarrow 1.5 - 0.7 + 2.5I_B + 45I_B = 0 \qquad ...(1)$$

$$\Rightarrow 0.8 + 47.5I_B = 0 \Rightarrow I_B = -\frac{0.8}{47.5} = -16.84\ \mu A$$

$$\therefore \quad I_C = \beta I_B = -0.741 \text{ mA}$$

Writing KVL in C–E loop,

$$V_{CC} = V_{CE} + (I_B + I_C)R_e \qquad ...(2)$$

$$-3 = V_{CE} + 45 \times 1 \times 10^3 \times (-16.84) \times 10^{-6}$$

$$\Rightarrow \quad V_{CE} = -2.242 \text{ V}$$

Hence, the Q-point is $I_C = -0.741$ mA

$$V_{CE} = -2.242 \text{ V}$$

(b) If base-spreading resistance 690 Ω is taken into account, then $R_{TH} = R_{TH} + 690 = 2.5 + 0.690 = 3.19$ K

∴ from (1),

$$1.5 - 0.7 + 3.19 \times 10^3 \times I_B + 45I_B \times 10^3 = 0$$

$$I_B = \frac{-0.8}{48.19} = -16.60\ \mu A$$

$$\therefore \quad I_C = \beta I_B = -0.7304 \text{ mA}$$

From (2),

$$V_{CC} = V_{CE} + (I_B + I_C)R_e$$

$$\Rightarrow \quad V_{CE} = -3 + 45 \times 16.60 \times 10^{-6} \times 10^3$$

$$= -2.253 \text{ V}$$

New Q point is $I_C = -0.7304$ mA

$$V_{CE} = -2.253 \text{ V}$$

6.23 For the circuit shown,

(a) Find I_B, I_C and V_{CE} if silicon transistor is used with $\beta = 50$.

(b) Find the value of R_b so that $V_{CE} = 7$ V.

(c) Show that $S = \dfrac{\beta + 1}{1 + \beta R_C/(R_C + R_b)}$

Solution (a) $V_{BE} = 0.7$ V (silicon)

Applying KVL to the loop of R_C, R_B and B–E junction, we get

$$V_{CC} = 2\,(I_B + I_C) + I_B R_b + V_{BE}$$

$$10 = 2\,(I_B + I_C) + 100I_B + 0.7 \qquad ...(1)$$

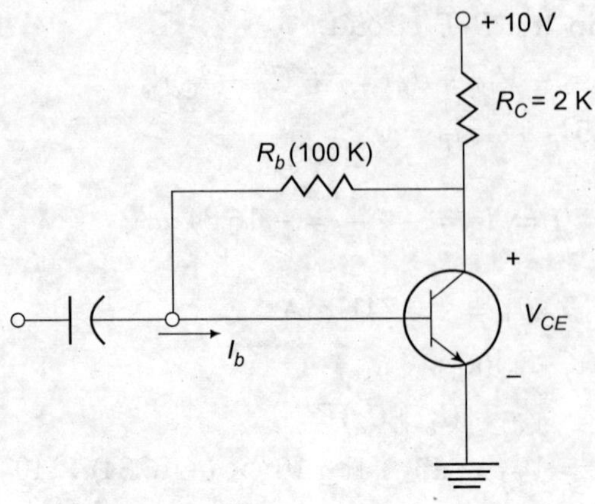

Since $\quad I_C = 50 I_B \qquad \text{...(2)}$

From (1) and (2),

$$10 - 0.7 = 2(51 I_B) + 100\, I_B$$
$$= 202\, I_B$$
$$I_B = \frac{9.3}{202} = 46.03\ \mu\text{A}$$
$$I_C = \beta I_B = 2.301\ \text{mA}$$
$$V_{CE} = V_{CC} - (I_B + I_C)R_C = V_{CC} - (\beta + 1) I_B R_C \qquad \text{...(3)}$$
$$= 10 - 2 \times 10^3 (51) \times 46.03 \times 10^{-6}$$
$$= 5.304\ \text{V}$$

(b) $\quad V_{CE} = 7\ \text{V}$

From (3), $R_C(I_B + I_C) = V_{CC} - V_{CE}$

$$I_B + I_C = \frac{V_{CC} - V_{CE}}{R_C} = \frac{10-7}{2\ \text{K}} = 1.5\ \text{mA}$$
$$51\, I_B = 1.5\ \text{mA}$$
$$I_B = 29.41\ \mu\text{A}$$

From the circuit, $R_B = \dfrac{V_{CE} - V_{BE}}{I_B}$

$$= \frac{7 - 0.7}{29.41} = 214.2\ \text{k}\Omega$$

(c) $\quad S = \partial I_C / \partial I_{CO}$

From the collector circuit, writing KVL

$$V_{CC} = V_{CE} + (I_B + I_C) R_C$$

$$\therefore \quad V_{CC} = V_{CB} + V_{BE} + (I_B + I_C)R_C$$

$$V_{CC} - V_{BE} = I_B R_b + (I_B + I_C)R_C \quad (\because V_{CB} = I_B R_b)$$

$$= (R_b + R_c)I_B + I_C \cdot R_C$$

$$= I_C \cdot R_C + (R_b + R_c)\left(\frac{I_C - (\beta+1)I_{CO}}{\beta}\right)$$

(Since $I_C = \beta I_B + (\beta + 1)I_{Co}$)

$$\therefore \quad V_{CC} - V_{BE} = I_C\left(R_C + \frac{R_b + R_C}{\beta}\right) - \frac{\beta+1}{\beta}\cdot(R_b + R_C)I_{CO}$$

$$\therefore \quad I_C = \left[V_{CC} - V_{BE} + \frac{\beta+1}{\beta}(R_b + R_C)I_{CO}\right]\left[\frac{\beta}{(\beta+1)R_C + R_b}\right]$$

$$\therefore \quad S = \frac{\partial I_C}{\partial I_{CO}}$$

$$= \frac{(\beta+1)(R_b + R_C)}{\beta}\cdot\frac{\beta}{(\beta+1)R_C + R_b}$$

$$= \frac{(\beta+1)(R_b + R_C)}{(\beta+1)R_C + R_b}$$

$$= \frac{\beta+1}{1+\dfrac{\beta R_C}{R_C + R_b}}$$

Hence proved.

6.24 If V_{CC} = 24 V, R_C = 10K and R_e = 270 Ω, for a silicon transistor with β = 45 and V_{CE} = 5 V. Find (a) R and (b) S.

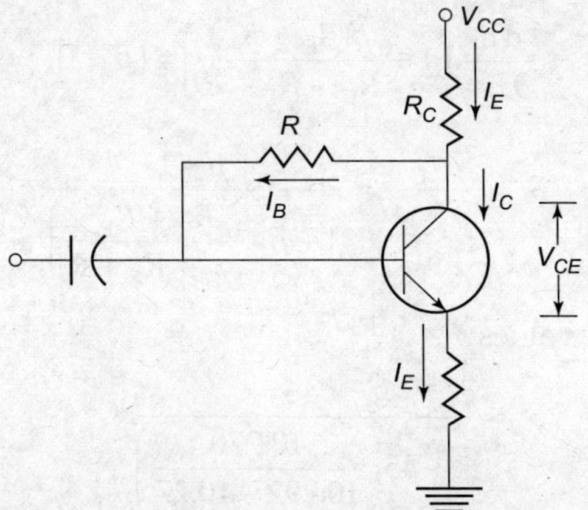

Solution Writing in the *C–E* loop,

$$V_{CC} = R_C \cdot I_E + V_{CE} + R_e \cdot I_E$$

$$\therefore \quad I_E = \frac{V_{CC} - V_{CE}}{R_C + R_e} = \frac{24 - 5}{10\,\text{K} + 270\,\Omega}$$

$$= 1.85 \text{ mA}$$

$$I_B = \frac{I_E}{\beta + 1} = 40.21\ \mu\text{A}$$

Applying KVL in *C–B–E* loop,

$$V_{CB} = V_{CE} - V_{BE} = 5 - 0.7 = 4.3 \text{ V}$$

$$R = \frac{V_{CB}}{I_B} = \frac{4.3}{40.21 \times 10^{-6}} = 106.92 \text{ K}\,\Omega$$

Applying KVL in *C–B* circuit,

$$V_{CC} = (I_B + I_C)(R_C + R_E) + I_B \cdot R + V_{BE}$$

$$I_B = \frac{V_{CC} - V_{BE} - I_C(R_C + R_E)}{R_C + R_E + R} \quad \text{...(1)}$$

We know that $I_C = \beta I_B + (\beta + 1) I_{CO}$ ...(2)

Put (1) in (2),

$$I_C = \beta\left[\frac{V_{CC} - V_{BE} - I_C(R_C + R_E)}{R_C + R_E + R}\right] + (\beta + 1) I_{CO}$$

$$\Rightarrow I_C\left[1 + \frac{\beta(R_C + R_E)}{R_C + R_E + R}\right] = \beta\left(\frac{V_{CC} - V_{BE}}{R_C + R_E + R}\right) + (\beta + 1) I_{CO}$$

$\therefore$ Differentiating with respect to I_{CO},

$$\frac{\partial I_C}{\partial I_{CO}}\left(1 + \frac{\beta(R_C + R_E)}{R_C + R_E + R}\right) = (\beta + 1)$$

$$\therefore \quad S = \frac{\partial I_C}{\partial I_{CO}} = \frac{\beta + 1}{1 + \beta\left(\dfrac{R_C + R_E}{R_C + R_E + R}\right)}$$

On substituting values

$$S = \frac{46}{1 + 45\left(\dfrac{10.270}{106.92 + 10.27}\right)}$$

$$= 9.3$$

6.25 Find R_e and S for the circuit, with $V_{BE} = 0.7$ V, $\beta = 50$ and $V_{CE} = 4$ V.

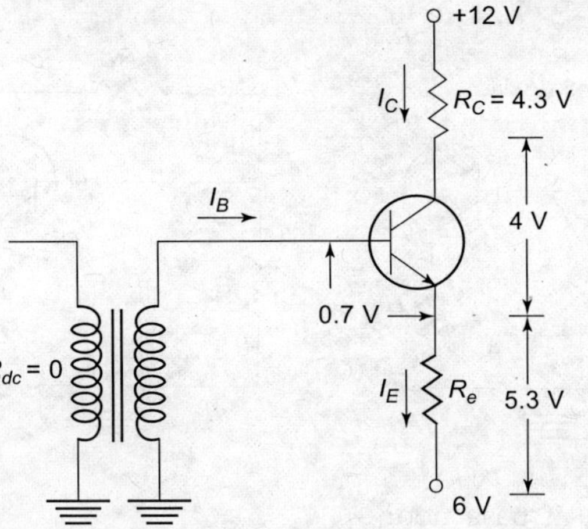

Solution Writing KVL in *B–E* loop,

$$0.7 + V_{Re} - 6 = 0$$

$$\Rightarrow \quad V_{Re} = 6 - 0.7 = 5.3 \text{ V}$$

Writing KVL in *C–E* loop,

$$4.3I_C + 4 + 5.3 - 6 - 12 = 0$$

$$\therefore \quad I_C = 2.02 \text{ mA}$$

$$I_B = \frac{I_C}{\beta} = 40.46\ \mu\text{A}$$

$$I_E = I_B + I_C = 2.06 \text{ mA}$$

We have $V_{Re} = I_E R_e = 5.3$ V.

$$\therefore \quad R_e = 2.572 \text{ k}\Omega$$

We have $S = \dfrac{(1+\beta)(Rth + R_e)}{R_{TH} + (\beta+1)R_e}$ for *CE* with R_e.

The given circuit is same as *CE* with R_e, but $R_{TH} = 0$

$$\therefore \quad S = \frac{(\beta+1)R_e}{(1+\beta)R_e} = 1$$

6.26 Determine the stability factor *s* for the circuit shown below.

Solution Let I_1 and I_2 be the current flowing through R_1 and R_2 and I_B, I_C be the currents through base and collector respectively.

Writing KVL in *C–B–E* loop,

$$V_{CC} = (I_C + I_1)R_C + V_{BE} + (I_B + I_C)R_e \qquad ...(1)$$

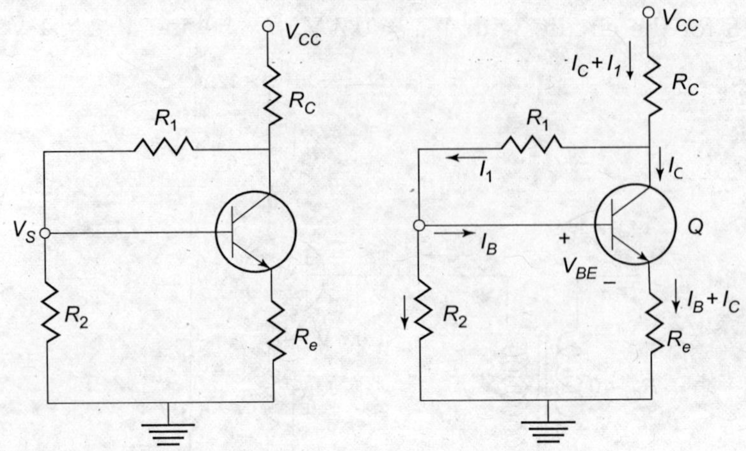

Writing KCL at base node

$$I_1 = I_2 + I_B \quad \text{...(2)}$$

and, $$I_2 = \frac{V_{BE} + (I_B + I_C)R_e}{R_2} \text{ by writing KVL in } B\text{–}E \text{ loop} \quad \text{...(3)}$$

From (2) and (3),

$$I_1 = I_B + \frac{V_{BE} + (I_B + I_C)R_e}{R_2}$$

$$= \frac{V_{BE} + I_B(R_2 + R_e) + I_C \cdot R_e}{R_2} \quad \text{...(4)}$$

Put (4) in (1),

$$V_{CC} = \left(I_C + \frac{V_{BE} + I_B(R_2 + R_e) + I_C \cdot R_e}{R_2}\right)R_C + V_{BE} + (I_B + I_C)R_e$$

$$= I_C\left[R_C + R_e + \frac{R_e \cdot (R_C + R_1)}{R_2}\right] + I_B\left[R_e + \frac{(R_e + R_2)(R_1 + R_C)}{R_2}\right]$$

$$+ \left[1 + \frac{R_1 + R_C}{R_2}\right]V_{BE} \quad \text{...(5)}$$

But we know that $I_C = \beta I_B + (\beta + 1)I_{CO}$

$$\Rightarrow \quad I_B = \frac{I_C - (\beta + 1)I_{CO}}{\beta} \quad \text{...(6)}$$

From (5) and (6),

$$V_{CC} = I_C\left(R_C + R_e + \frac{R_e(R_1 + R_C)}{R_2}\right) + \left[\frac{I_C - (\beta+1)I_{CO}}{\beta}\right]$$

$$\left[R_e + \frac{(R_e + R_2)(R_1 + R_C)}{R_2}\right] + \left[1 + \frac{R_1 + R_C}{R_2}\right] V_{BE}$$

Differentiating with respect to I_{CO} and $S = \dfrac{\partial I_C}{\partial I_{CO}}$

$$S = \frac{\partial I_C}{\partial I_{CO}} = \frac{(1+\beta)[R_2(R_C + R_1) + R_e(R_1 + R_2 + R_C)]}{R_1R_2 + (1+\beta)(R_2R_C + R_e(R_1 + R_2 + R_C))}$$

assuming β and V_{BE} as constants.

6.27 For the Darlington stage shown, V_{CC} = 24 V, β_1 = 24, β_2 = 39, V_{BE} = 0.7 V, R_C = 330 Ω, R_e =120 Ω; if at the Quiescent point V_{CE2} = 6V, determine (a) R (b) $S \equiv \partial I_C | \partial I_{CO}$.

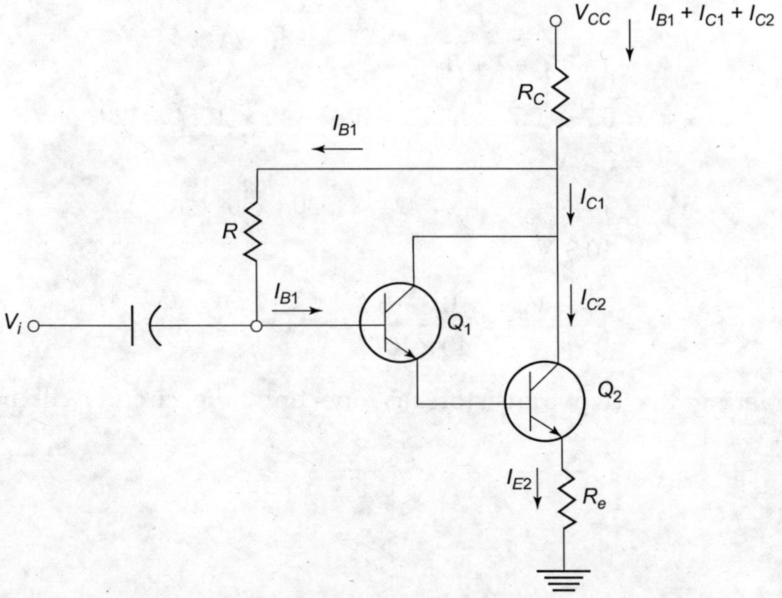

Solution (a) Applying KVL in the C–E_2 loop,

$$V_{CC} - V_{CE2} = (I_{B1} + I_{C1} + I_{C2})\,(R_C + R_e)$$

$$= (R_C + R_e)I_{E2}$$

$$\therefore \quad 24 - 6 = (450)I_{E2}$$

$$\Rightarrow \quad I_{E2} = \frac{24-6}{450} = 40 \text{ mA}$$

$$I_{B2} = \frac{I_{E2}}{1+\beta_2} = \frac{40\times10^{-3}}{40} = 1 \text{ mA}$$

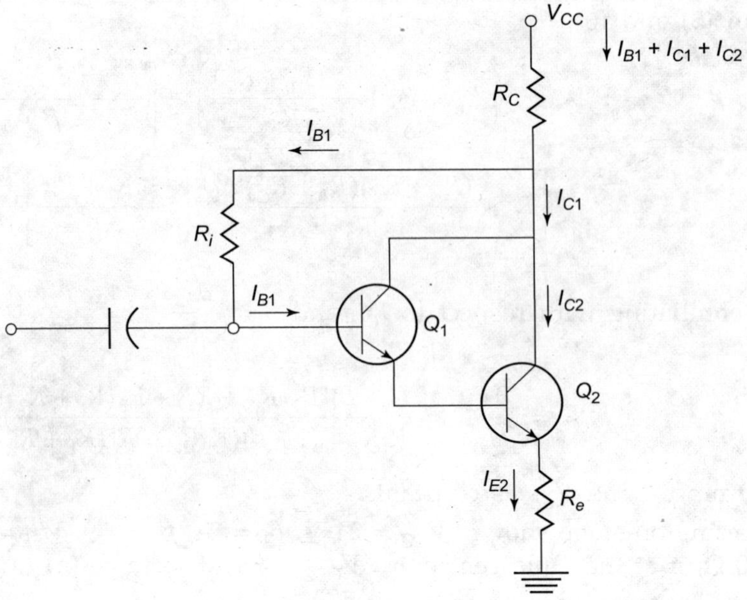

$$I_{B1} = \frac{I_{E1}}{1+\beta_1} = \frac{I_{B2}}{1+\beta_2} = \frac{1\,\text{mA}}{25} = 40\ \mu\text{A}$$

$$V_{B1} = 0.7 + 0.7 + I_{E2}R_e = 1.4 + 40 \times 10^{-3} \times 120$$

$$= 6.2\ \text{V}$$

$$V_C = V_{CC} - R_C \cdot I_{E2} = 24 - (330 \times 40 \times 10^{-3})$$

$$= 10.8\ \text{V}$$

$$\therefore \quad R = \frac{V_C - V_{B1}}{I_{B1}} = \frac{10.8 - 6.2}{40 \times 10^{-6}} = 115\ \text{K}\Omega$$

(b) Considering the two transistors as one unit, the circuit will become as below:

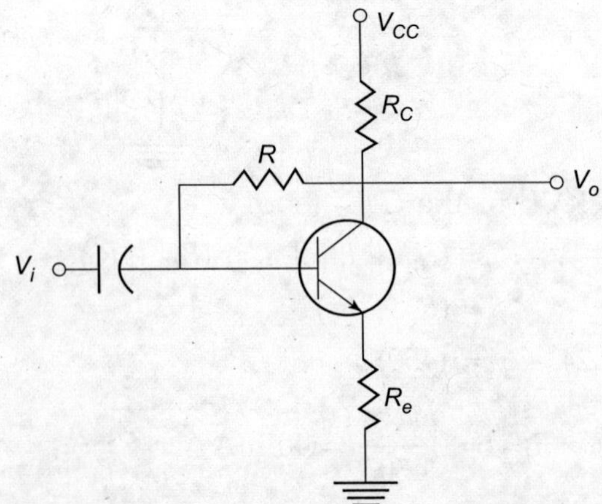

This is a collector to Base Bias Circuit, same as in 6.24.

$$\therefore \qquad S = \frac{1+\beta}{1+\beta\dfrac{R_C+R_e}{R_C+R_e+R}}$$

$$\beta + 1 = \frac{I_{E2}}{I_{B1}} = \frac{40\times10^{-3}}{40\times10^{-6}} = 1000$$

$$\therefore \qquad S = \frac{1000}{1+(999)\left(\dfrac{330+120}{120+330+115\times10^3}\right)}$$

$$= 204.5$$

EXERCISE PROBLEMS

6.1 A *pnp* silicon transistor is used in common-collector circuit. V_{CC} = 30 V, R_e = 1 kΩ, $R_1 = R_2$ = 5 k if β = 44. Find the Q-point.

6.2 Find (a) I_{BQ} and I_{cQ} (b) V_{CEQ} (c) V_B and V_C (d) V_{BC} for the following circuit:

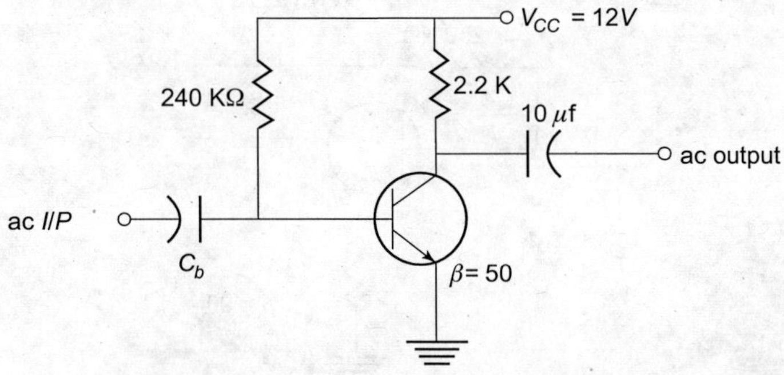

6.3 Find (a) I_B, I_C, V_{CE} (b) V_B, V_C, V_E (c) V_{BC} for the circuit shown below:

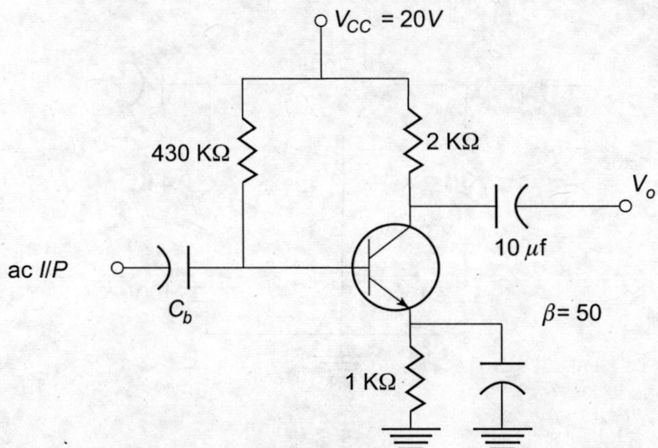

6.4 Find V_{CE} and I_C for the following circuit.

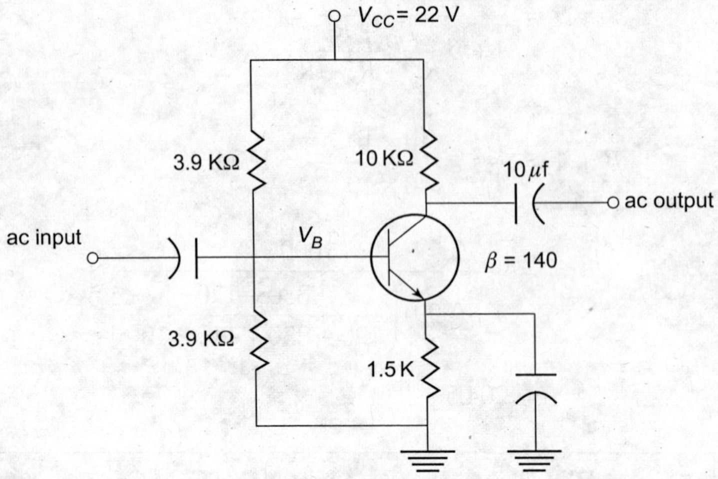

6.5 Find I_C and V_{CE} for the circuit shown below:

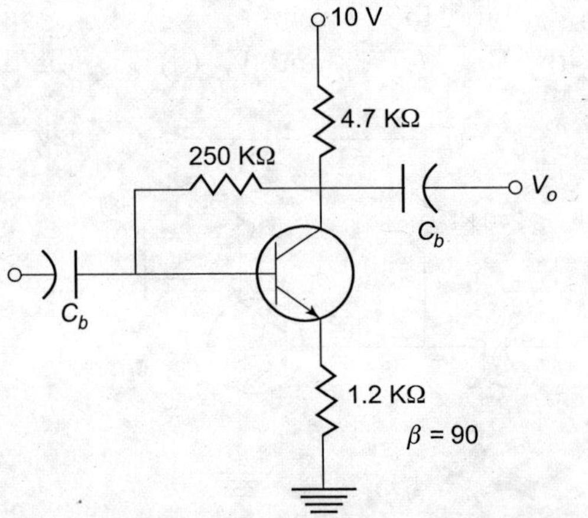

6.6 Find V_{CB} and I_B for the C_B configuration shown below:

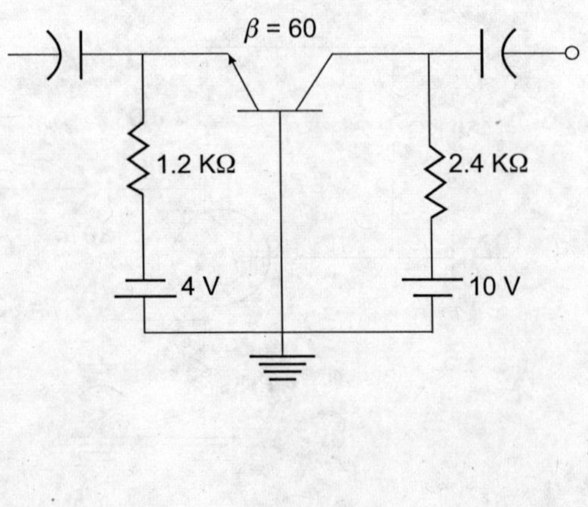

6.7 Find V_e and CB of the circuit shown below:

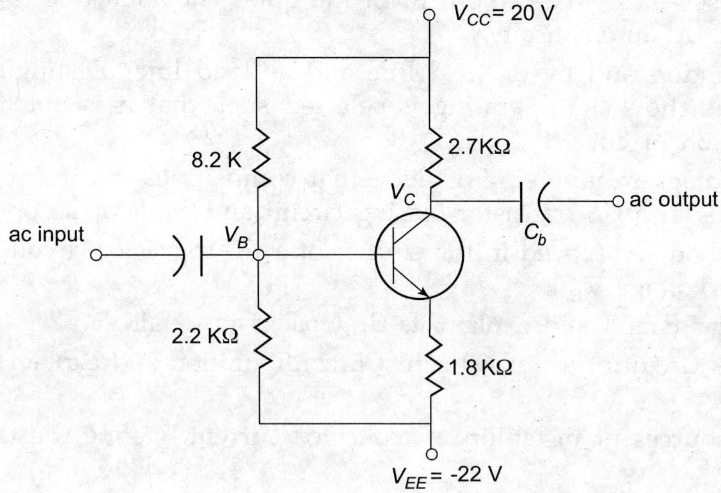

REVIEW QUESTIONS

1. Discuss the problem of selection of operating point for a transistor and also discuss how this operating point can be maintained stable.
2. Derive expressions for stability factors of a self-bias circuit.
3. Explain how stability is improved in a self-bias circuit.
4. What is the purpose of biasing of a transistor? Indicate the methods of biasing and compare them.
5. What do you understand by thermal runaway? Explain the term thermal resistance and derive the condition for thermal stability of BJT.
6. Define the three different stability factors and give an expression that determines the variation of collector current in terms of stability factors.
7. Explain the effect of temperature on various parameters of a transistor amplifier circuit and discuss the methods of stabilisation of the operating point.
8. Give a transistor circuit that employs a thermistor to compensate the variation of parameters with temperature and explain the circuit operation.
9. What is a sensistor? Explain how it can be employed in a transistor circuit to compensate the variation of operating point with temperature.
10. Derive the three stability factors for a fixed-bias transistor circuit.
11. Discuss the three factors that contribute thermal instability of the operating point in a BJT amplifier. Define the three stability factors.

12. What is thermal runaway? Discuss the method to prevent thermal runaway in a *CE* amplifier.
13. Show that in a self-bias circuit, the drift of the quiescent point due to temperature changes is made up automatically.
14. What do you understand by dc load line and ac load line? Distinguish between them and explain how the operating is selected such that it is not drifted either towards saturation or cut off.
15. Draw a circuit diagram involving a diode that compensates the reverse saturation current variations in a *CE* transistor biasing circuit and explain the operation.
16. Explain how a diode employed in the emitter of a *CE* transistor circuit compensates the variation of voltage V_{BE}.
17. Draw a fixed-bias circuit and explain its drawbacks in detail.
18. Draw a self-bias circuit of a transistor in *CE* configuration and explain its operation in detail.
19. List the three sources of instability of collector current. Define the three stability factors.
20. Distinguish between stabilisation and compensation techniques. Discuss about compensation techniques in detail.
21. Define thermal resistance. What is its significance in transistor circuits?
22. Derive a condition for avoiding thermal runaway in a transistor.
23. What is heat sink? Why is it employed? Draw diagrams of popularly used heat sinks.
24. Discuss about thermistor and sensistor compensation.
25. Define stability factor *M* and express variation of collector current in terms of *M*.

CHAPTER

7

Transistor Amplifier Circuits

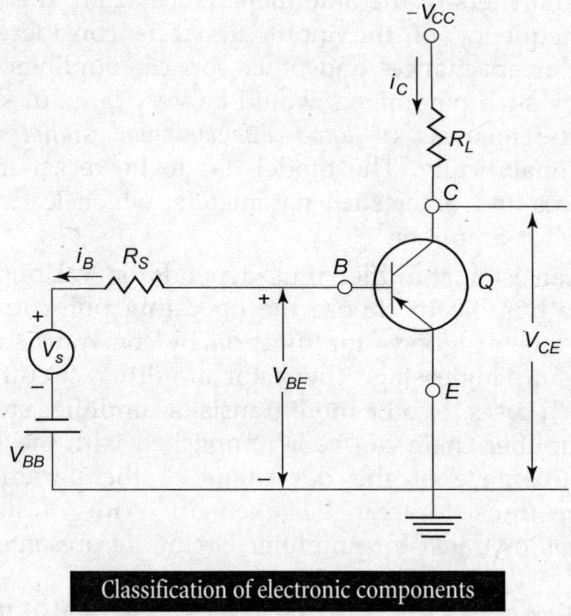

Classification of electronic components

Chapter Outline

The concepts introduced in this chapter are:

- *Graphical analysis of large signal amplifier*
- *Two-port hybrid model of a transistor*
- *Analysis of a transistor amplifier circuit*
- *Frequency response and bandwidth of an amplifier*
- *Effect of various capacitors on the analysis*
- *High-frequency model of a transistor*
- *Gain bandwidth product*

7.1 INTRODUCTION

The analysis of a transistor amplifier is considered in this chapter. Since the transistor is not a linear device, general analytical analysis is not possible. The analysis carried out is based on various practical assumptions. The *graphical analysis* is the way out when the input signal is large, that is, for large inputs, the nonlinearity of the device cannot be eliminated and so, the output of the amplifier can only be estimated by graphical method by the use of input and output characteristics of the transistor. Generalised analytical analysis is not possible in this case. However, the device can be assumed to be linear when the input signal is small. So, for the input signals, which are small, the device can be modelled as a two-port network and analytical equations can be developed for the estimation of various parameters of the amplifiers. Once again, the problem can be further simplified when the frequency of the input signal is considered. For low-frequency signals, the interelectrode capacitances and other sort of couplings can be neglected since the impedance offered by such parameters would be very large in shunt with regular two-port parameters. So, the analysis of *low-frequency small signal* model of a transistor amplifier can be done analytically. The model has to be recast taking into account the interelectrode capacitances and other such parameters, which leads to *high-frequency small signal* model of a transistor amplifier.

The analysis of a transistor amplifier thus depends on various constraints. Whatever may be the case, it is advisable to choose the operating point in the mid-region of the active region for a reasonably good amplification by the transistor. In many cases, the amplification offered by a single stage transistor amplifier circuit is not sufficient for a given application. In such cases, two or more transistor amplifier circuits are to be *cascaded*. The analysis of such amplifier chain can be accomplished with the above procedures or, in practice, when some tolerance of the deviations of the parameters can be allowed, simplified model of the transistors can be taken up. With all such practically possible assumptions, the analysis of transistor amplifier circuits is presented in this chapter.

7.2 LARGE SIGNAL ANALYSIS OF TRANSISTOR AMPLIFIER

When the input signal to a common emitter amplifier is large, analytical analysis of the amplifier cannot be performed. This is because the device is not linear or, in other words, the shift in the output characteristics of a *CE* amplifier is not equal for equal increments of the input base current. This introduces a sort of distortion in the output of the circuit, even though the input is assumed to be a perfect sinusoid. Due to this nonlinearity, modelling of a transistor using two-port network for large signals is not possible. Two-port network analysis is applicable only for linear, bilateral and finite networks. So, the analysis of such amplifiers with large signals at the input is to be performed graphically.

For the amplifier whose analysis is to be performed, the output and input characteristics of the transistor amplifier in *CE* configuration are plotted and based on the effective load resistance the load line is drawn on the output characteristics and the operating point is selected appropriately. The input ac signal is imposed at the base of the transistor, as shown in Fig. 7.1. So the input ac signal, which has a zero average value, is now applied to the base of the transistor, which is already biased at the dc base current, the operating point. Now, the input base current varies sinusoidally with an average value of biasing base current. The output collector current can then be estimated by superimposing the input sinusoid variations on the operating point that is located in the active region of the output characteristics of the transistors shown in Fig. 7.2(a). With the help of this analysis, the output current can be estimated and plotted to scale and the output voltage of the amplifier can also be plotted since the output current flows through a linear resistor.

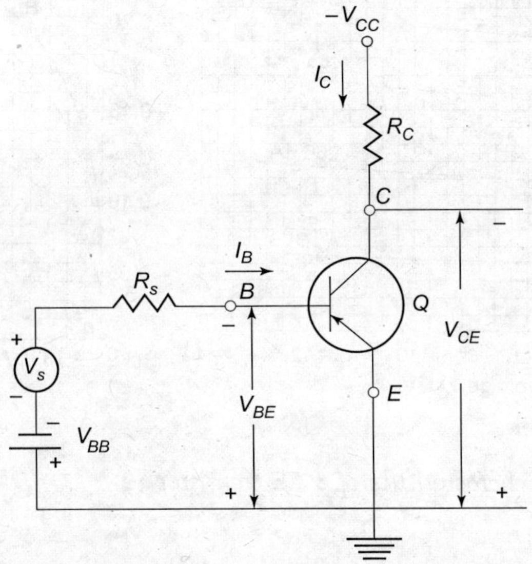

Fig. 7.1 *Basic amplifier circuit*

In the above analysis, the input current is assumed to be perfect sinusoid. But this condition is not satisfied since the input to the amplifier is voltage but not current and also that the relation between input current and voltage is not linear. If one assumes a perfect sinusoidal voltage to be applied at the input of the amplifier, the current variations can be estimated with the help of input characteristics as shown in Fig. (7.2b). These current variations are then transferred to the output characteristics to estimate the exact collector current variations.

In the above discussion, the output wave form has encountered two types of distortions, namely, *input nonlinear distortion* and *output nonlinear distortion*. The input nonlinear distortion is due to the nonlinearity of the voltage current relationship in the input of the circuit. The input voltage is applied between the forward biased base-emitter junctions. Since the *V-I* characteristics of the forward biased junction are not linear but exponential, even though one assumes the input voltage to be perfect sinusoid, the input current is not a linear replica of the input voltage. This causes the current at the input to be a distorted version of the sinusoidal signal and so this type of distortion is known as *input nonlinear distortion*. This distortion can be estimated and plotted as shown in Fig. 7.3(a).

The *output nonlinear distortion* in a *CE* amplifier is due to the fact that the shift in the output characteristics is not in proportion to that of input base current increments. The increase in the collector current is large for small variations in the base current when the base current is a small value, but the rate of increase in the collector current decreases as the base current increases. In other words, the amplification offered by the amplifier at low base currents is large when compared to the amplification when the base current is

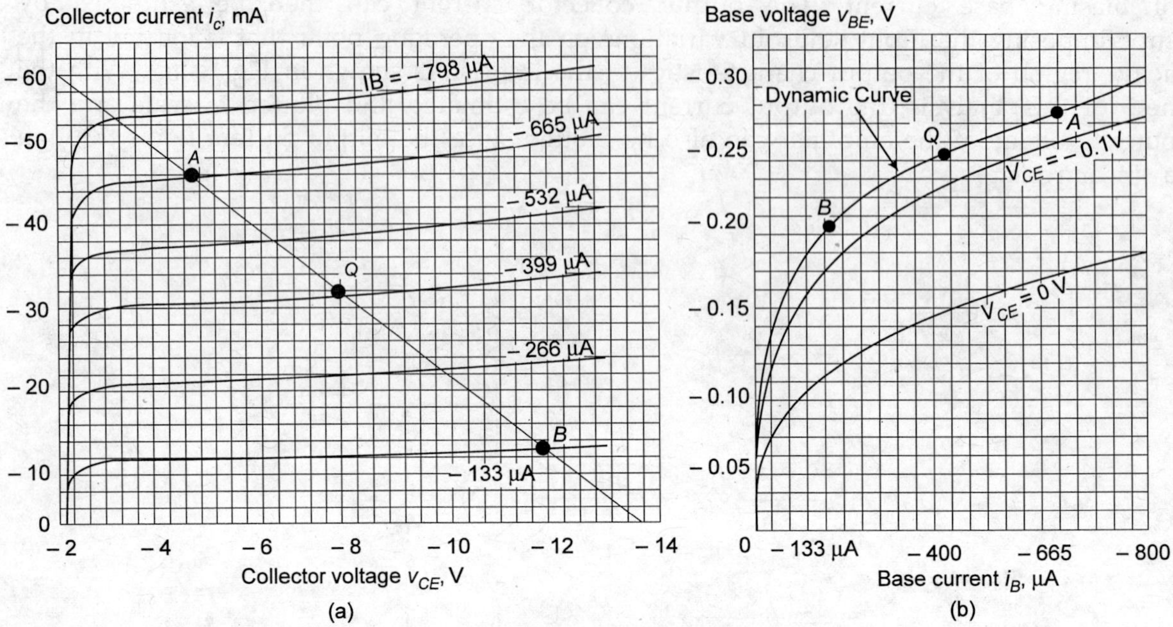

Fig. 7.2 *Output and input characteristics of a CE transistor*

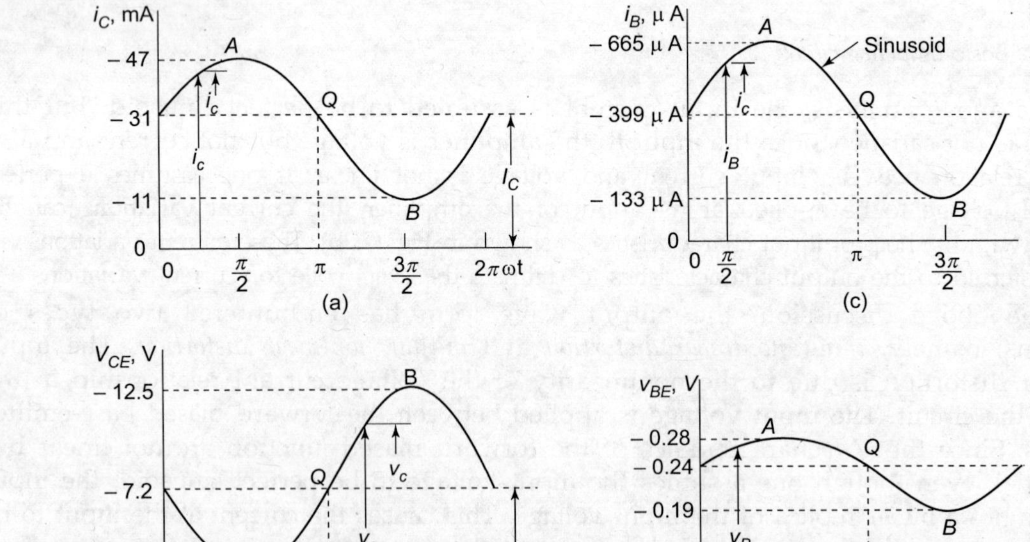

Fig. 7.3 *Input and output nonlinear distortion*

high. That is, the small signal current gain of a *CE* amplifier is not a constant value but depends upon the input base current, selection of the operating point. When the variation of the input base current is large enough around the operating point such that the collector current variation above the operating point is not the same as that below the operating point, even though the base current variation above and below the operating point is same, a distortion is introduced into the circuit at the output of the device as shown in Fig. 7.3(b). This distortion is termed as *output nonlinear distortion.*

In the above graphical analysis, various types of currents and voltages were encountered. So as to avoid confusion, a standard notation for identifying the instantaneous variation, quiescent values, instantaneous total value, effective value of the varying component and supply is to be followed and is tabulated in Table. 7.1

Table 7.1 ***Notation of Current and Voltages in Transistors***

Type of signal	Base voltage with respect to emitter	Collector voltage with respect to emitter	Base current towards electrode	Collector current towards electrode
Instantaneous total voltage	v_{BE}	v_{CE}	i_B	i_c
Quiescent value	V_{BE}	V_{CE}	I_B	I_C
Instantaneous value of varying component	v_{be}	v_{ce}	i_b	i_c
Supply voltage	V_{BB}	V_{CC}	–	–
Effective value of varying component	V_{be}	V_{ce}	I_b	I_c

7.3 TWO-PORT MODEL OF A TRANSISTOR

The transistor amplifier can be modelled as a linear bilateral network and two-port network analysis can be applied to this device when the transistor is a linear device. The transistor can be assumed to be operating in a linear portion when the input signal is very small such that the nonlinearity that is present in this small range of current variation can be neglected. One more assumption made in this type of analysis is that the frequency of the input signal is low. When a low-frequency signal is considered, the interelectrode capacitances and such coupling parameters tend towards a very large value and thus the impedance offered by them can be assumed to be infinity. So, these parameters can be neglected when low-frequency signals are assumed to be applied to the input of the transistor amplifier. Thus, the two-port network analysis can be applied to a transistor amplifier circuit when *low-frequency small signals* are assumed to be applied at the input of the transistor amplifier circuit.

Technical analysis of any circuit can be performed easily and analytically when the circuit is modelled. The reason for construction of models is that the physical systems are usually too complex and practical analysis is very complex. The complexity of analysis of transistor amplifier is that the transistor is a nonlinear device and establishing an analytical relation between output and input parameters is too difficult without modelling. Modelling of the transistor is nothing but a simple passive network representation of the basic component. When an equivalent circuit consisting of passive components and dependent or independent sources of either voltage or current can replace the basic component, the analysis of this circuit is nothing but simply applying the known network theorems to the equivalent circuits to derive a definite analytical relation between the output and input parameters.

The analysis of the two-port network can be performed by modelling the network into various bilateral linear components. There are various methods of this two-port network analysis where the input and the output currents and voltages can be related by analytical expressions. The parameters are expressed in terms of a 2–2 matrix whose elements are either impedances or admittances or *ABCD* or hybrid parameters and so on. Many such matrix elements can be adopted as per the convenience of the application of the concerned circuit. In this amplifier circuit analysis the parameters of importance are voltage gain, current gain and input and output impedances. The most suitable matrix of a two-port network to describe the transistor amplifier circuit is the *hybrid parameter model* or, in short, *h-parameter* model. This model is also referred to as *low-frequency small signal model* of a transistor amplifier circuit. So, in this analysis of transistor amplifier circuit, *h-parameter* model is adopted.

7.4 SMALL SIGNAL LOW-FREQUENCY MODEL OF A TRANSISTOR

As discussed in the above section, hybrid parameter model is the best model of two-port network, which can be adapted to a transistor amplifier circuit. The general equation relating the input and output parameters with reference to Fig. 7.4 can be written as

$$\begin{aligned} v_1 &= h_{11}i_1 + h_{12}v_2 \\ i_2 &= h_{21}i_1 + h_{22}v_2 \end{aligned} \qquad \text{...(7.1)}$$

The parameters h_{11}, h_{12}, h_{21} and h_{22} are called *hybrid* or *h-parameters*. The parameters can be defined as follows:

$$h_{11} = \frac{v_1}{i_1} = \text{input resistance (at port 1) when output is short circuited } (v_2 = 0).$$

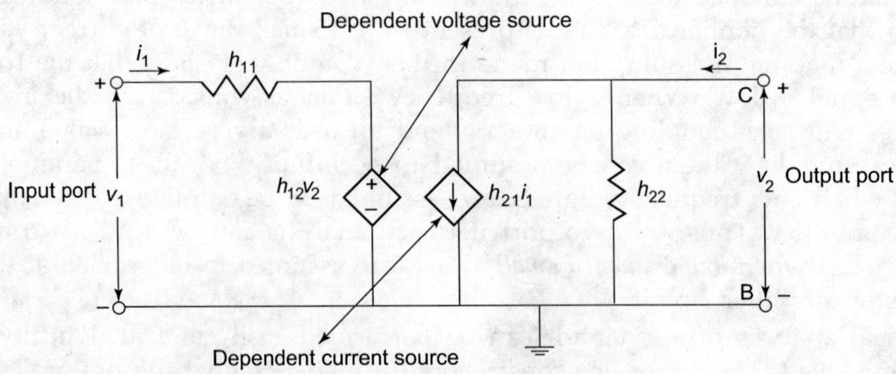

Fig. 7.4 *Hybrid two-port network model*

$$h_{12} = \left.\frac{v_1}{v_2}\right|_{i_1=0} = \text{reverse voltage gain with open input,}$$

$$h_{21} = \left.\frac{i_2}{i_1}\right|_{v_2=0} = \text{forward current gain with output short circuited,}$$

$$h_{22} = \frac{i_2}{v_2} = \text{output conductance (at port 2) when input open circuited } (i_1 = 0) \quad ...(7.2)$$

The above notation for the hybrid two-port network is the general notation that is adopted universally. When this model is adopted for transistor circuits, a third suffix is to be added so as to identify the parameter for a given configuration of the transistor. This leads to great confusion and the following standard notation for transistor circuits is accepted internationally. The first two suffixes are replaced with an alphabet *i*, *r*, *f* and *o* in place of 11, 12, 21 and 22, respectively. The second suffix is added as an alphabet once again (*e*, *b* and c identifying the configurations to be common emitter, common base and common collector configurations, respectively). This notation simplifies the analysis to a great extent. Thus,

$$\begin{aligned} h_i &= h_{11} = \text{input impedance,} \\ h_r &= h_{12} = \text{reverse voltage gain,} \\ h_f &= h_{21} = \text{forward current gain,} \\ h_o &= h_{22} = \text{output admittance.} \end{aligned}$$

For example, for common emitter configuration the following *h*-parameter model is accepted. For the common emitter configuration, the input voltage is v_{be}, input current is i_b, output voltage is v_{ce} and output current is i_c. Therefore, the governing equations can be written as

$$\begin{aligned} v_{be} &= h_{ie}i_b + h_{re}v_{ce} \\ i_c &= h_{fe}i_b + h_{oe}v_{ce} \end{aligned} \quad ...(7.3)$$

where,

$$h_{ie} = \left.\frac{v_{be}}{i_b}\right|_{v_{ce}=0} = \left.\frac{v_{be}}{i_b}\right|_{V_{CE}} = \text{input impedance,}$$

$$h_{re} = \left.\frac{v_{be}}{v_{ce}}\right|_{i_b=0} = \left.\frac{v_{be}}{v_{ce}}\right|_{I_B} = \text{reverse voltage gain,}$$

$$h_{fe} = \left.\frac{i_c}{i_b}\right|_{i_b=0} = \left.\frac{i_c}{i_b}\right|_{V_{CE}} = \text{forward current gain,}$$

$$h_{oe} = \left.\frac{i_c}{v_{ce}}\right|_{i_b=0} = \left.\frac{i_c}{v_{ce}}\right|_{I_B} = \text{output admittance.} \quad ...(7.4)$$

Similarly, the equations and the *h*-parameters for common base and common collector configurations can be drafted accordingly.

With the help of Eq. (7.3), the equivalent circuit of a transistor can be drawn. Considering the first equation of the two, this equation is a voltage equation, which satisfies KVL around a loop. So, the input of a transistor is a loop of a input resistor h_{ie} and a dependent voltage source, $h_{re}v_{ce}$ in series. Considering the second equation, which is a current equation, it will satisfy the KCL at the output node. So, the output circuit is a parallel combination of an admittance h_{oe} in shunt with a dependent current source $h_{fe}i_b$. So, the *h*-parameter model for a transistor in common emitter configuration is as shown in Fig. 7.5.

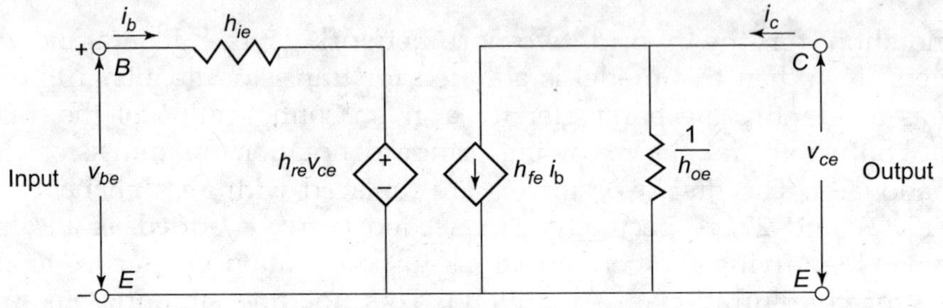

Fig. 7.5 *h-parameter model of a transistor in CE configuration*

Similarly, the governing equations and the equivalent *h*-parameter model can be drawn for common base and common collector. The input terminal for a common base configuration is between emitter and base terminals and the output port is between collector and base of the transistor as shown in Fig. 7.6. The four *h*-parameters for the common base transistor configuration are defined as

$$
\begin{aligned}
v_{eb} &= h_{ib}i_e + h_{rb}v_{cb} \\
i_c &= h_{fb}i_e + h_{ob}v_{cb}
\end{aligned}
\quad \text{...(7.5)}
$$

where

$$h_{ib} = \left.\frac{v_{eb}}{i_e}\right|_{v_{cb}=0} = \left.\frac{v_{eb}}{i_e}\right|_{V_{CB}} = \text{input impedance,}$$

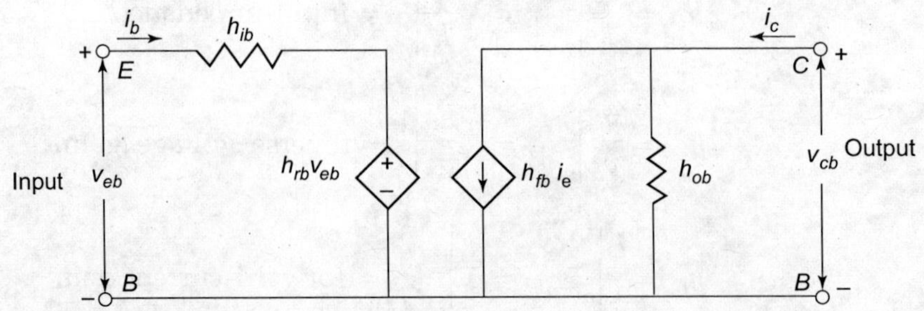

Fig. 7.6 *h-parameter model of a transistor in CB configuration*

$$h_{rb} = \left.\frac{v_{eb}}{v_{cb}}\right|_{i_e=0} = \left.\frac{v_{eb}}{v_{cb}}\right|_{I_E} = \text{reverse voltage gain,}$$

$$h_{fb} = \left.\frac{i_c}{i_e}\right|_{v_{cb}=0} = \left.\frac{i_c}{i_e}\right|_{V_{CB}} = \text{forward current gain,}$$

$$h_{ob} = \left.\frac{i_c}{v_{cb}}\right|_{i_c=0} = \left.\frac{i_c}{v_{cb}}\right|_{I_E} = \text{output admittance.} \qquad ...(7.6)$$

The input terminal for a common collector configuration is between base and collector terminals and the output port is between emitter and collector of the transistor, as shown in Fig. 7.7. The four h-parameters for the common collector transistor configuration are defined as

$$v_{bc} = h_{ic}i_b + h_{rc}v_{ec}$$
$$i_e = h_{fc}i_b + h_{oc}v_{ec} \qquad ...(7.7)$$

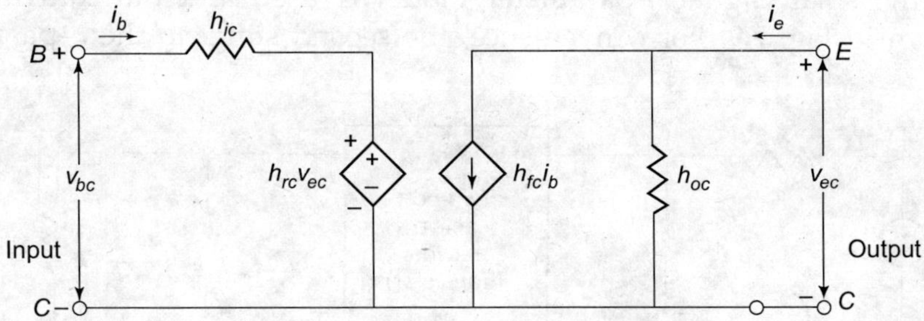

Fig. 7.7 *h-parameter model of a transistor in CC configuration*

where,

$$h_{ic} = \left.\frac{v_{bc}}{i_b}\right|_{v_{ec}=0} = \left.\frac{v_{bc}}{i_b}\right|_{V_{EC}} = \text{input impedance,}$$

$$h_{rc} = \left.\frac{v_{bc}}{v_{ec}}\right|_{i_b=0} = \left.\frac{v_{bc}}{v_{ec}}\right|_{I_B} = \text{reverse voltage gain,}$$

$$h_{fc} = \left.\frac{i_e}{i_b}\right|_{v_{ec}=0} = \left.\frac{i_e}{i_b}\right|_{V_{EC}} = \text{forward current gain,}$$

$$h_{oc} = \left.\frac{i_e}{v_{ec}}\right|_{i_b=0} = \left.\frac{i_e}{v_{ec}}\right|_{I_B} = \text{output admittance.} \qquad ...(7.8)$$

For each configuration, from KCL, one can note that

$$i_b + i_e + i_c = 0 \qquad \text{...(7.9)}$$

The circuit models and equations are valid for either an *npn* or a *pnp* transistor and are independent of the type of load or method of biasing.

7.5 ANALYSIS OF A TRANSISTOR AMPLIFIER CIRCUIT

In order to use the transistor as an amplifier, it is sufficient to provide signal source at the input of the transistor terminals and to drive the appropriate load at the output terminals of the transistor. The basic block diagram of such an amplifier is as shown in Fig. 7.8. Proper biasing of the transistor and selection of operating point are assumed and the transistor is available as a device that can amplify the input sinusoidal signals of low frequency and small signal amplitude. For all the ac analysis of transistor amplifiers, the dc biasing voltages and currents need not be considered and one can assume that the device is properly biased in the active region. So, any dc voltage source or current source can be assumed to be zero in such an analysis.

Considering Fig. 7.8, the R_s is the source resistance and Z_L is the load impedance. For analysing this amplifier, one has to replace the active device by its equivalent circuit and for transistor the small signal h-parameter model is used and the equivalent circuit is drawn as shown in Fig. 7.9. For convenience, the second suffix to the h-parameters is not

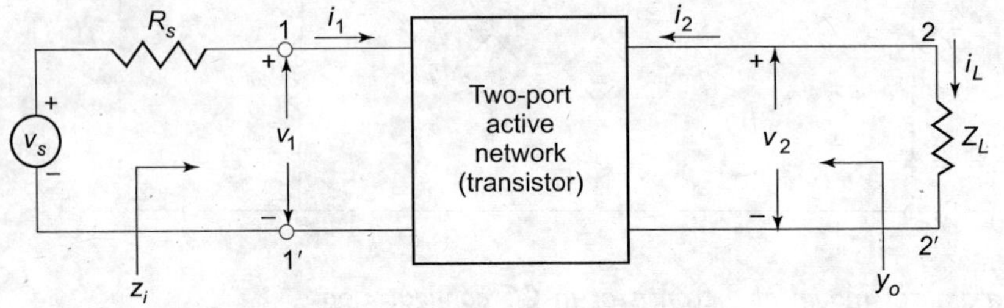

Fig. 7.8 *The basic amplifier circuit*

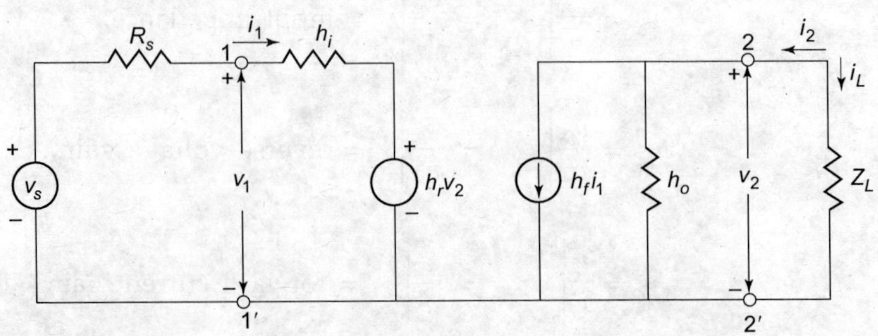

Fig. 7.9 *Equivalent amplifier circuit*

shown and so the derivation is for any general configuration of the transistor and suitable suffix can be added depending on the configuration of the transistor involved. It is assumed that the parameters are constant over the range of operation. For an amplifier, the parameters that are of interest are *voltage gain, current gain, input impedance* and *output impedance.* These parameters can be evaluated by simply applying known networks theorems to this equivalent circuit.

The current gain or current amplification: The *current amplification* A_I for a transistor amplifier stage can be defined as the ratio of output current to input current. Or,

$$A_I \equiv \frac{I_L}{I_1} = \frac{-I_2}{I_1} \qquad \text{...(7.10)}$$

From the circuit, applying KCL to the output node,

$$I_2 = h_f I_1 + h_o V_2 \qquad \text{...(7.11)}$$

and,
$$V_2 = -I_2 Z_L \qquad \text{...(7.12)}$$

Therefore,

$$A_I \equiv \frac{-I_2}{I_1} = \frac{-h_f}{1+h_o Z_L} \qquad \text{...(7.13)}$$

The negative sign in the above equation indicates that there is a phase shift of 180° between the input and output of the transistor amplifier.

The input impedance: The impedance seen between the two terminals 1 and 1′ is the amplifier *input impedance* Z_i and can be given as

$$Z_i \equiv \frac{V_1}{I_1} \qquad \text{...(7.14)}$$

Applying KVL to the input loop,

$$V_1 = h_i I_1 + h_r V_2 \qquad \text{...(7.15)}$$

and ,
$$V_2 = -I_2 Z_L = A_I I_1 Z_L \qquad \text{...(7.16)}$$

Dividing Eq. (7.15) by I_1 and substituting Eq. (7.16),

$$Z_i = h_i + h_r A_I Z_L \qquad \text{...(7.17)}$$

It can be observed that the input impedance is a function of load impedance.

The voltage gain or voltage amplification: The voltage gain or voltage amplification A_V is the ratio of output voltage to input voltage and can be given as

$$A_V \equiv \frac{V_2}{V_1} \qquad \text{...(7.18)}$$

From Eqs. (7.16) and (7.14),

$$A_V = \frac{A_I I_1 Z_L}{V_1} = \frac{A_I Z_L}{Z_i} \qquad \text{...(7.19)}$$

The voltage amplification taking source impedance into consideration: The voltage amplification taking source impedance into consideration, A_{Vs}, can be defined as the ratio of output voltage to the source voltage, that is,

$$A_{Vs} \equiv \frac{V_2}{V_s} = \frac{V_2}{V_1} \cdot \frac{V_1}{V_s} = A_V \frac{V_1}{V_s} \quad \text{...(7.20)}$$

From the equivalent circuit of the input of the amplifier, as shown in Fig. 7.10(a),

$$V_1 = \frac{V_s Z_i}{Z_i + R_s} \quad \text{...(7.21)}$$

Then,

$$A_{Vs} = \frac{A_V Z_i}{Z_i + R_s} = \frac{A_I Z_L}{Z_i + R_s} \quad \text{...(7.22)}$$

If $R_s = 0$, then $A_{Vs} = A_V$. Hence, A_V is the voltage gain for an ideal voltage source. In

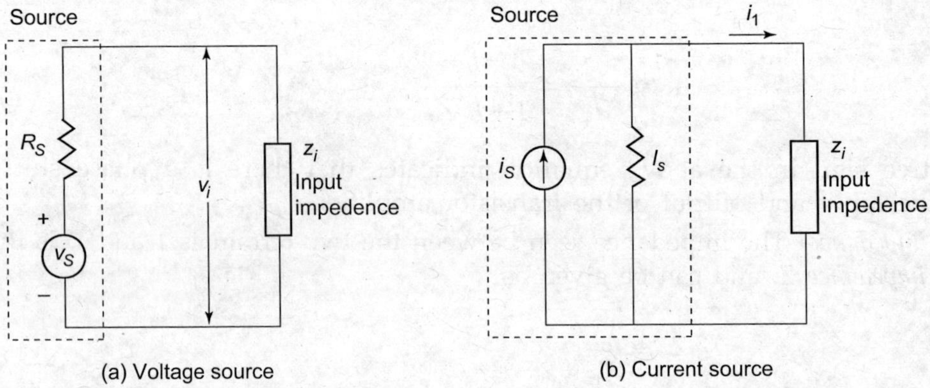

Fig. 7.10 *Voltage and current source circuits*

practice, the quantity A_{Vs} is more meaningful than A_v since the source resistance has an appreciable effect on the overall voltage amplification.

The current amplification taking source impedance into consideration: The current amplification taking source impedance into consideration, A_{is}, can be defined as the ratio of the output current to the source current and can be given as

$$A_{Is} \equiv \frac{-I_2}{I_s} = \frac{-I_2}{I_1} \cdot \frac{I_1}{I_s} = A_I \frac{I_1}{I_s} \quad \text{...(7.23)}$$

From Fig. 7.10(b),

$$I_1 = \frac{I_s R_s}{Z_i + R_s} \quad \text{...(7.24)}$$

Therefore,

$$A_{Is} = \frac{A_I R_s}{Z_i + R_s} \quad \text{...(7.25)}$$

If $R_s = \infty$, then $A_{Is} = A_I$. Hence A_I is the current gain for an ideal current source.
The output impedance: The *output impedance* Z_o can be defined as the ratio of voltage to current when a voltage source V_2 drives the output circuit with the load set to infinity ($Z_L = \infty$) and the source voltage set to zero ($V_s = 0$). If I_2 is the current drawn by the output circuit, then

$$Z_o \equiv \frac{V_2}{I_2} \text{ with } Z_L = \infty \text{ and } V_s = 0 \qquad ...(7.26)$$

Or, in our discussion, it is convenient to work with $Y_o = 1/Z_o$. From Eq. (7.11),

$$Y_o = h_f(I_1/V_2) + h_o \qquad ...(7.27)$$

with $V_s = 0$, from the input loop,

$$R_s I_1 + h_i I_1 + h_r V_2 = 0 \qquad ...(7.28)$$

or,

$$(I_1/V_2) = -h_r/(h_i + R_s) \qquad ...(7.29)$$

Thus,

$$Y_o = h_o - \frac{h_r h_f}{h_i + R_s} \qquad ...(7.30)$$

The output impedance is a function of source impedance.
Table 7.2 summarizes all the equations derived in the above analysis.

7.6 TRANSISTOR AMPLIFIER CONFIGURATIONS

It is sometimes necessary to convert the h-parameters of one configuration to any of the other two depending on application. For example, if the problem has a transistor in *CE*

Table 7.2 ***Small Signal Low-Frequency Analysis of Transistor Amplifier***

A_i	$-h_f/1 + h_o Z_L$	$-h_{fe}/1 + h_{oe} Z_L$
R_i	$h_i + h_r A_i Z_L$	$hi_e + hr_e A_I Z_L$
A_V	$A_I Z_L / Z_i$	$A_I Z_L / Z_i$
Y_o	$h_o - \dfrac{h_f h_r}{R_S + h_i}$	$h_{oe} - \dfrac{h_{fe} h_{re}}{R_s + h_{ie}}$
A_{Vs}	$\dfrac{A_V Z_i}{Z_i + R_S} = \dfrac{A_I Z_L}{Z_i + R_S}$	$\dfrac{A_V Z_i}{Z_i + R_S} = \dfrac{A_I Z_L}{Z_i + R_S}$
A_{IS}	$\dfrac{A_I R_S}{R_S + Z_i} = A_{Vs} \dfrac{R_S}{Z_L}$	$\dfrac{A_I R_S}{R_S + Z_i} = A_{Vs} \dfrac{R_S}{Z_L}$
A_P	$A_V \cdot A_i$	$A_V \cdot A_i$

configuration, but the h-parameters are specified in CC configuration, to analyze the amplifier circuit, one has to convert the available h-parameters into the other required configuration. In such a case very approximate conversion formulas as listed in Table 7.3 can be used and the analysis be carried out as usual.

The typical h-parameters for all the three transistor configurations for comparison of the three configurations are as listed in Table 7.4.

Table 7.3 ***Approximate Conversion for Hybrid Parameters***

$h_{ie} = h_{ic}$ $h_{fe} = -(1 + h_{fc})$	$h_{re} \approx 1$ $h_{oc} = h_{oe}$
$^*h_{ie} = \dfrac{h_{ib}}{1+h_{fb}}$	$^*h_{re} = \dfrac{h_{ib}\,h_{ob}}{1+h_{fb}} - h_{rb}$
$^*h_{fe} = \dfrac{-h_{fb}}{1+h_{fb}}$	$^*h_{oe} = \dfrac{h_{ob}}{1+h_{fb}}$

*The CB parameters in terms of CE parameters are obtained by interchanging the subscripts *e* and *b*.

Table 7.4 ***Typical h-Parameters for Three Configurations***

Parameter	*CE*	*CC*	*CB*
h_i, input impedance in ohms	1100 Ω	1000 Ω	25 Ω
h_r, reverse voltage gain	2.5×10^{-4}	– 0.99	3×10^{-4}
h_f, forward current gain	50	– 55	– 0.99
h_o, output admittance in mhos	24×10^{-3}	30×10^{-3}	0.5×10^{-3}
$1/h_o$, in ohms	42×10^{3}	33×10^{3}	2000×10^{3}

The h-parameters of a given configuration transistor can be determined from the output and input characteristics. The h-parameters do depend on the operating point selected. The variations of the parameters are to be taken over very small range over the operating point. Since the characteristics are nonlinear but they are assumed to be linear in this analysis, very small variations are to be considered. The forward current gain and output admittance parameters of the transistor can be obtained graphically from the output characteristics, while the input impedance and the reverse voltage gain can be obtained from the input characteristics.

From Fig. 7.11, the parameters h_{fe} and h_{oe} can be determined. By definition,

$$h_{fe} = \frac{\partial i_C}{\partial i_B} = \frac{\Delta i_C}{\Delta i_B} = \frac{i_{C2} - i_{C1}}{i_{B2} - i_{B1}}$$

$$h_{oe} = \frac{\partial i_C}{\partial v_{CE}} = \frac{\Delta i_C}{\Delta v_{CE}} = \frac{i'_{C2} - i'_{C1}}{v_{CE2} - v_{CE1}}$$

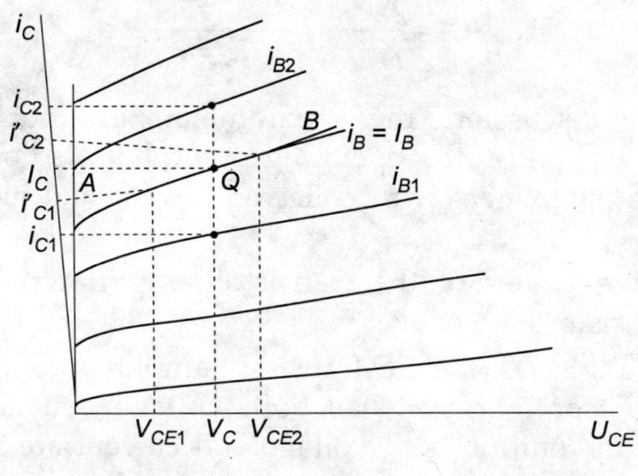

Fig. 7.11 *Determination of h-parameters from output characteristics*

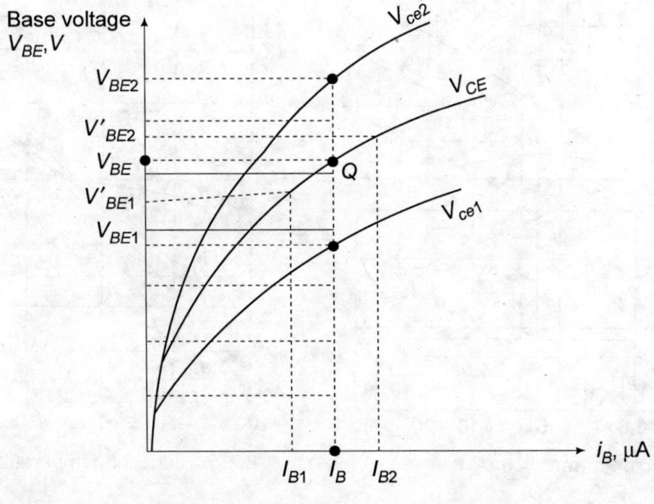

Fig. 7.12 *Determination of h-parameters from input characteristics*

As seen from the figure, these variations are taken over the operating point.

The rest of the two parameters can be similarly determined from the input characteristics as shown in Fig. 7.12.

The operating point is transferred to the input characteristics and then the variations are taken over this point. Thus, by definition,

$$h_{ie} = \frac{\partial v_{BE}}{\partial i_B} = \frac{\Delta v_{BE}}{\Delta i_B} = \frac{v'_{BE2} - v'_{BE1}}{i_{B2} - i_{B1}}$$

$$h_{re} = \frac{\partial v_{BE}}{\partial v_{CE}} = \frac{\Delta v_{BE}}{\Delta v_{CE}} = \frac{v_{BE2} - v_{BE1}}{v_{CE2} - v_{CE1}}$$

As seen from the above discussion, the h-parameters are not constant for a given transistor. They not only depend on the operating point, but also on the temperature. The typical variations of these parameters with collector current and junction temperature are as depicted in Fig. 7.13.

Table 7.5 gives relative values of the transistor amplifier parameters for the three configurations of the transistor.

As can be seen from Table 7.5, the *CE* transistor amplifier configuration is best suited as a power amplifier. *CE* configuration has both voltage and current gain very large compared to 1 and so, in the output, both voltage and current are very large compared to the input and thus, the power, which is the product of voltage and current, is very large compared to input power. In other words, the *CE* amplifier offers power gain, which is

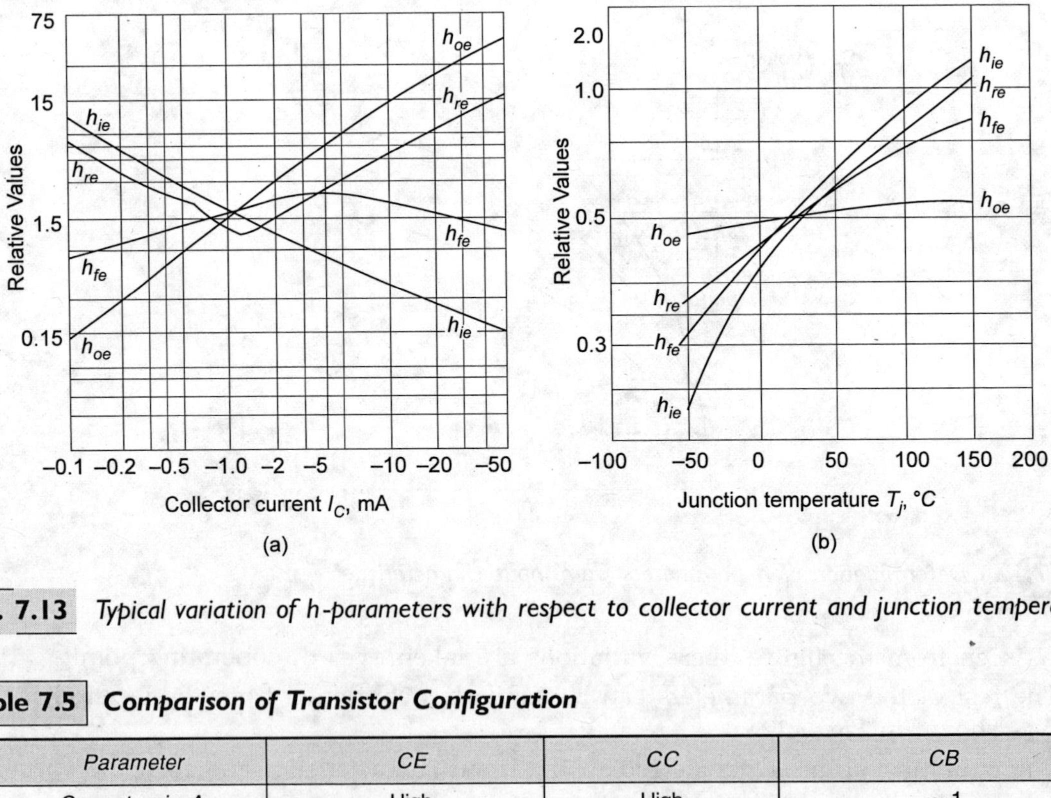

Fig. 7.13 *Typical variation of h-parameters with respect to collector current and junction temperature*

Table 7.5 ***Comparison of Transistor Configuration***

Parameter	*CE*	*CC*	*CB*
Current gain A_I	High	High	≈ 1
Voltage gain A_V	High	≈ 1	High
Input Resistance R_i	Medium	High	Low
Output Resistance R_o	Medium	Low	High
Application	Voltage or Current Amplifier	Voltage Buffer Amplifier	Constant current source (Current Buffer)

not possible either in *CB* or *CC* configuration since in *CB* configuration current gain is small compared to 1 and in *CC* configuration voltage gain is small compared to 1 and so, in both cases large power amplification is not possible in practice. The drawback of the *CE* transistor amplifier is that both the input and output impedances are of medium range, whereas one requires high input impedance and low output impedance for a good voltage amplifier.

CC configuration of transistor amplifier has a voltage gain of approximately 1, that is, the input and output voltages are approximately same. The advantage of this circuit is that the input impedance is very high and output impedance is very low, as required for that of a voltage amplifier. The *CC* configuration, also called as emitter follower circuit, finds application as buffer amplifier.

The transistor in *CB* configuration is not useful as a voltage amplifier, even though it has voltage amplification more than 1. This is due to the fact that input impedance is very low and output impedance is very high in contrast to what is required for a voltage amplifier. In this configuration, the current gain is approximately 1, that is, the output current would be approximately same as the input current. So, this transistor in *CB* configuration would be a good current buffer.

When very large amplification is required, a single transistor amplifier is not sufficient and a cascade of amplifiers is required. Cascading of amplifiers is nothing but connecting various single stage transistor amplifiers in series. The output of the first stage of the amplifier drives the input of the next stage and the output of this stage drives the consecutive stage. This cascading of different transistor amplifier stages is known as a multistage amplifier. The function of the multistage amplifier is not only large amplification but a good impedance match between the amplifier chain and the source and the last stage of amplifier with the load impedance. This matching of the impedances at the input and output of the amplifier chain can be achieved by appropriate selection of various stages in a multistage amplifier.

For example, when high source impedance is to drive the amplifier chain with low input impedance, the first stage of the chain can never be chosen as *CE* transistor amplifier, since the input impedance of this configuration is a medium value. The best choice in this case is the emitter follower, which has a high input impedance and low output impedance and so acts as a very good impedance transformation network. The voltage gain of this *CC* amplifier is unity and so one cannot expect the voltage to be amplified in the first stage of such an amplifier chain. The consecutive stages can be surely *CE* amplifiers since it is only amplifier configuration that gives both voltage and current gain. On similar argument, the final stage of the multistage amplifier can be chosen as emitter follower when a good matching of impedances of the output of the amplifier and the unknown load is required. So, in total, one can consolidate that in a poorly matched source and load impedance condition, the first and last stage of the amplifier chain need to be emitter follower and the intermediate stages can be *CE* configurations. One has to make a note that in such a chain, the first and last stages in the amplifier chain will not offer any voltage amplification.

The common base configuration of a transistor amplifier is rarely used in an amplifier chain where voltage gain is of prime concern. This configuration finds application wherever current is the prime parameter of interest. Common base configuration is a very good circuit, which produces a constant current in the output of the circuit, that is, the collector current of CB transistor amplifier is almost a constant value. This constant current is very useful for charging a capacitor so as to produce a sawtooth wave form.

7.7 LINEAR ANALYSIS OF A TRANSISTOR CIRCUIT

In practice, amplifier circuits are not simple as those discussed in the above section, consisting of simple *CE*, *CB* or *CC* configurations of transistor circuits. There may be resistors or any such components between the output and input networks or any other possible network combinations in the circuit configuration. In such cases the equations derived in the above sections cannot be applied directly to evaluate the amplifier parameters. The amplifier parameters can be evaluated by applying the known network tools to the equivalent circuit, as discussed in the following steps.

1. Draw the actual amplifier circuit diagram correctly.
2. Identify the transistor and its three terminals, emitter (*E*), base (*B*) and collector (*C*) on this circuit diagram.
3. Identify the transistor configuration by determining the input and output ports of the transistor.
4. Maintaining the relative positions of the three terminals of the transistor intact, replace the transistor with its small signal model.
5. Transfer all the remaining components from the basic circuit to this equivalent circuit, maintaining their positions in the circuit exactly.
6. Replace all ideal dc voltage sources by short circuit and all ideal dc current sources by open circuit. The source resistances, if any, are to be retained in the circuit.
7. Apply known network theorems like KVL, KCL, and Miller theorem to the resultant linear network. Solve for determining the amplifier parameters.

By following the above steps, the problem of analysis of transistor amplifier circuit can be brought down to simple linear network analysis. It is to be once again emphasized that this linear analysis is only applicable to low-frequency small signals.

7.8 MILLER'S THEOREM AND ITS APPLICATIONS

By applying the above-mentioned linear analysis steps, it is possible to analyse any transistor amplifier circuits. But, in some cases, when there is impedance connected between the output and input of the amplifier (for example, a resistor between base and collector of *CE* transistor amplifier) or when an impedance is connected in series with the common terminal of output and input of the amplifier (for example, emitter resistor in *CE* configuration), analysing the circuit with the help of KVL and KCL would be a bit complex and the problem can thus be simplified by the use of *Miller's theorem and its dual.* The Miller's theorem helps to find an equivalent circuit that does not have a resistor between the output and input of the amplifier or in the common terminal of the amplifier but its equivalent is either in series with or parallel to the output and input circuits of the amplifier.

Miller's theorem: Consider a two-port network with feedback impedance Z' between port-1 and port-2 of the network as shown in Fig. 7.14(a). The Miller's theorem helps to find two impedances, one each in shunt with each port as shown in Fig. 7.13(b), which is equivalent to the circuit in (a). The values of Z_1 and Z_2 can be given as

$$Z_1 = Z'/(1 - K) \quad \text{...(7.31)}$$

$$Z_1 = Z'K/(K - 1) \quad \text{...(7.32)}$$

where $K = V_2/V_1$, the voltage gain of the two-port network.

Dual of Miller's theorem: Consider a two-port network with feedback impedance Z' in series with terminal 3 of the network as shown in Fig. 7.15(a). The dual of Miller's theorem helps to find two impedances, one each in series with terminals 1 and 2, as shown in Fig. 7.15(b), which is equivalent to the circuit in (a). The values of Z_1 and Z_2 can be given as

$$Z_1 = Z'(1 - A_I) \quad \text{...(7.33)}$$

$$Z_2 = Z'(A_I - 1)/A_I \quad \text{...(7.34)}$$

where $A_I = - I_2/I_1$, the current gain of the two-port network.

Assuming the two-port network to be transistor amplifier circuit, the above two cases can be simplified for analysis to a large extent by applying the Miller's theorem and its dual. Now, after applying the Miller's theorem or its dual or both, if need be, the two-

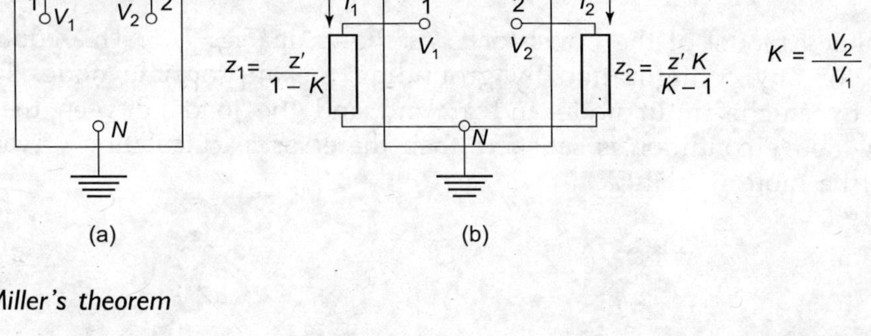

Fig. 7.14 *Miller's theorem*

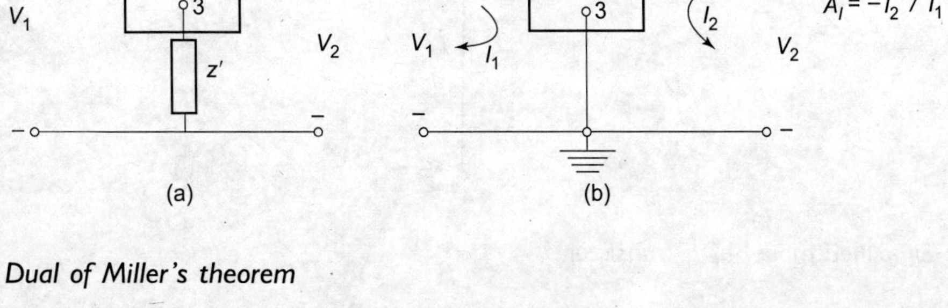

Fig. 7.15 *Dual of Miller's theorem*

port network is simple without any feedback impedance between the two ports or series impedance in the common terminal. The analysis discussed in Section 7.5 can be directly applied to the basic two-port network and the total analysis of the circuit can be performed by applying the KVL or KCL to the two ports of the network and determining the required amplifier parameters.

7.9 SIMPLIFIED MODEL OF A TRANSISTOR

In the analysis procedures discussed in the above sections for a transistor amplifier, the detailed calculations of all four parameters of the device are made. That is, exact values of voltage amplification, current amplification, input and output impedances can be determined. But, in practice, appropriate approximations can be allowed since analysing the circuit exactly gives no guarantee of the figures, which are achieved practically. This is because of the fact that the transistor parameters are not stable and depend on various factors like temperature, ageing and so on. So, even though exact analysis is performed and the parameters are determined accurately, in practice, one is not sure of achieving the same. So, approximations may be allowed within tolerable limits. In such a case detailed lengthy calculations need not be done and a simplified model of a transistor can be assumed and the mathematical analysis be simplified to a large extent.

Of the four h-parameters, only h_{ie} and h_{fe} are sufficient for the approximate analysis of low-frequency transistor circuits, provided that the load resistance is small enough to satisfy the condition

$$h_{oe} R_L < 0.1 \qquad ...(7.35)$$

The simplified model of the transistor is as shown in Fig. 7.16. This equivalent circuit can be used for any configuration by grounding the appropriate node. The source is connected between the input node and ground and the load between the output and ground. If the above condition is satisfied, then the error in calculation of various parameters will not be more than 10%.

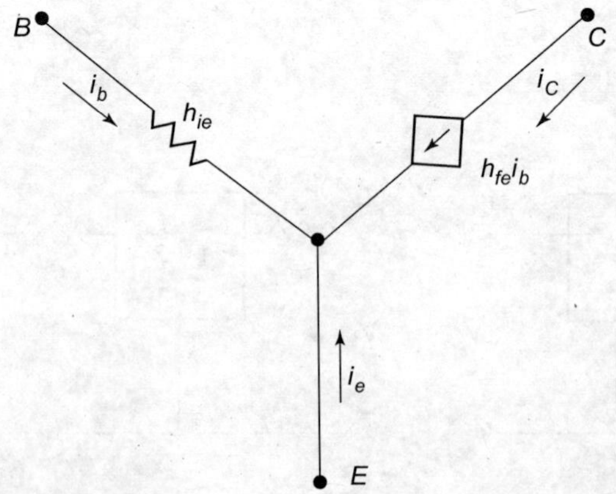

Fig. 7.16 *Simplified model of a transistor*

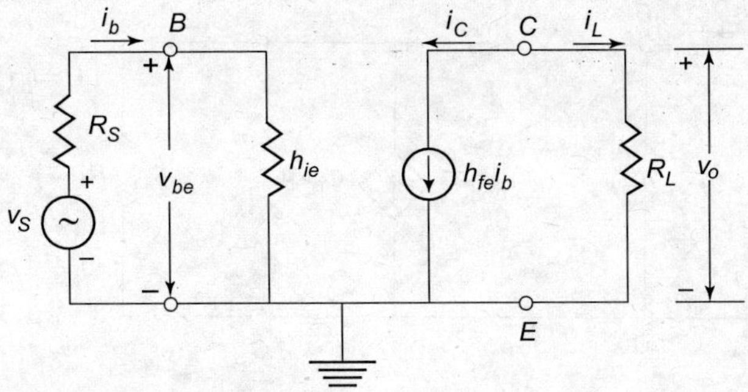

Fig. 7.17 *Simplified CE transistor circuit*

Simplified CE configuration: Figure 7.17 shows the equivalent simplified circuit of a *CE* transistor amplifier. The *current gain* of this circuit can be given as

$$A_I = -h_{fe} \quad ...(7.36)$$

Since $h_{oe}R_L < 0.1$, this term in the denominator of Eq. (7.13) can be neglected when compared to 1. This leads to not more than 10% error, which is tolerable.

For the *input impedance*, consider Eq. (7.17),

$$Z_i = h_i + h_r A_I Z_L = h_{ie}\left(1 - \frac{h_{re}\, h_{fe}|A_I|}{h_{ie}\, h_{oe}\, h_{fe}} h_{oe}\, R_L\right) \approx h_{ie} \quad ...(7.37)$$

Since $h_{re}h_{fe}/h_{ie}h_{oe} \approx 0.5$ for any transistor, $|A_I| \approx h_{fe}$ and $h_{oe}R_L < 0.1$. The second term will be less than 0.05 and so can be neglected compared to one in the brackets.

With the above two approximations, the *voltage gain* can be evaluated as

$$A_V = A_I(R_L/R_i) \approx -h_{fe}R_L/h_{ie} \quad ...(7.38)$$

All the above three expressions are well within the tolerance limit of 10% and so the approximations are valid.

To calculate the *output impedance*, if V_s is made zero and R_L is infinite, there is only one dependent ideal current source, $h_{fe}I_b$, in the output circuit and so, the output impedance would be infinity. The effective output impedance will be equal to the load resistance, that is, $R_o' \approx R_L$.

If the transistor amplifier is to drive a circuit, then the effective load resistance would be a shunt combination of various components of the load, R_L'. In such a case, the condition to be satisfied can be modified as

$$h_{oe}R_L' < 0.1 \quad ...(7.39)$$

Simplified CC configuration: The equivalent circuit for a transistor in CC configuration is as shown in Fig. 7.18.

The *current gain* can be derived to be

$$A_I = 1 + h_{fe} \quad ...(7.40)$$

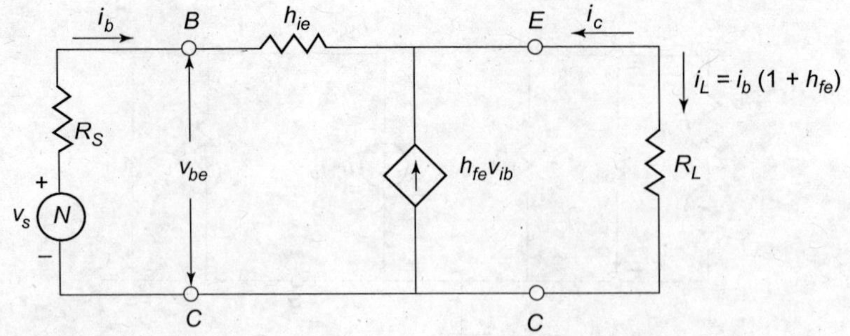

Fig. 7.18 *Simplified CC transistor circuit*

The *input impedance* be given as

$$Z_i = h_{ie} + (1 + h_{fe})R_L \quad \text{...(7.41)}$$

The *voltage gain* for an emitter follower is approximately unity and so,

$$A_V \approx 1 \quad \text{...(7.42)}$$

For *output impedance,* the ratio of open circuit voltage to the short circuit output current can be given as

$$I = (1 + h_{fe})I_b = \frac{(1+h_{fe})V_s}{h_{fe} + R_s} \quad \text{...(7.43)}$$

and,

$$Y_o = \frac{I}{V_s} = \frac{1+h_{fe}}{h_{ie} + R_s} \quad \text{...(7.44)}$$

and so,

$$Z_o = \frac{h_{ie} + R_s}{1+h_{fe}} \quad (7.45)$$

The tolerance limits of all these parameters are well within the specified value, 10%.

7.10 *CE* AMPLIFIER WITH EMITTER RESISTANCE

The selection of the operating point for stable operation of the amplifier in the active region of the transistor requires self-biasing of the transistor. This self-bias circuit requires an emitter resistor R_e in the emitter lead of the transistor circuit. This circuit achieves a very good stabilisation of the operating point, but since this emitter resistance is present in the common terminal of the *CE* transistor amplifier circuit, it acts as feedback network between the output and input. So, the amplification of the amplifier is reduced and one

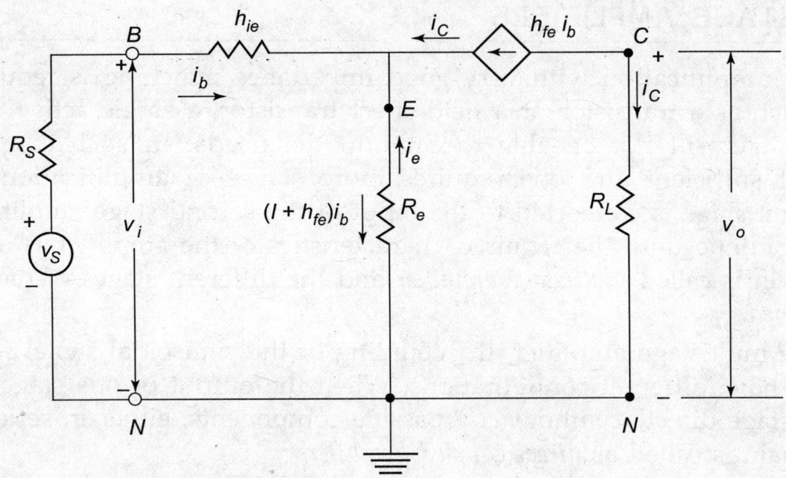

Fig. 7.19 *Simplified CE transistor amplifier with emitter resistance*

has to analyse the circuit afresh and determine the effect of this emitter resistance on various parameters of the *CE* transistor amplifier. An approximate solution assuming the condition $h_{oe}R_L < 0.1$ can be presented with the help of equivalent circuit as shown in Fig. 7.19. The *current gain* can be given as the ratio of $- I_c$ to I_b

$$A_I = - I_c/I_b = - h_{fe}I_b/I_b = - h_{fe} \qquad ...(7.46)$$

The *input impedance* is given as

$$R_i = V_i/I_b = h_{ie} + (1 + h_{fe})R_e \qquad ...(7.47)$$

Since h_{fe} is very large and so,

$$R_i \approx (1 + h_{fe})R_e \qquad ...(7.48)$$

And, the *voltage gain* is

$$A_V = A_I R_L/R_i = - h_{fe}R_L/[h_{ie} + (1 + h_{fe})R_L)] \qquad ...(7.49)$$

The *output impedance,* as in the previous case, can be taken as infinity and the effective output resistance taking load into consideration is the load resistance, R_L itself.

By applying the Miller's theorem dual for simplifying the analysis, a resistance R_2 comes in series with the load resistance and so effective load resistance would be the sum of R_2 and R_L. Thus,

$$R'_L = R_L + (A_I - 1)R_e/A_I \qquad ...(7.50)$$

And so, the condition for simplified analysis can be written as

$$h_{oe}R'_L \approx h_{oe}(R_L + R_e) \leq 0.1 \qquad ...(7.51)$$

7.11 MULTISTAGE AMPLIFIERS

Whenever large amplification with very good impedance matching is required using an active device such as a transistor or a field-effect transistor, a single active device and its associated circuitry will not be able to cater to the needs. In such case, single stage amplifier is not sufficient and one requires more stages of amplification. That is, the output of the first stage is connected to the input of the second stage amplifier circuit and the chain can continue until the required characteristics of the amplifier is achieved. Such an amplifier chain is called *multistage amplifier* and the different stages of the amplifier are said to be *cascaded.*

In a cascade multistage amplifier, the coupling of the sinusoidal signal from one stage to the next can have different configurations. When the output of one stage is coupled to the input next stage directly without any passive components, either in series or in shunt, the amplifier chain is called as *direct coupled amplifier.*

If coupling between the different stages of the amplifier chain is via an inductor or a transformer, the amplifier can be referred to as *transformer coupled.* In such amplifiers, analysis is bit complex and losses in the circuit would be large. The mutual inductance between the two windings is to be considered for exact analysis of the circuit. The advantage is that only ac is coupled but dc cannot be coupled via a transformer. So, dc isolation between the stages can be achieved in total.

If resistor and capacitor network is used for the interstage coupling, the chain is referred to as *RC coupled amplifier.* This *RC* coupled amplifier is most popular, in practice, since it is very good to adopt with minimum losses and also the biasing of the one stage does not affect the other stage since the capacitor, which is in series in the circuit acts as a dc *blocking* or ac *coupling* component. Such a capacitor is referred to as *coupling* or *blocking capacitor.* The capacitor, having an infinite reactance at zero frequency or dc current, does not allow the dc biasing current to flow from one stage to the other when they are cascaded. So, the consecutive stages of the amplifier are totally isolated in terms of dc biasing voltages and currents and so the biasing current of one stage will not affect the biasing of the other stage to disturb the operating point. But this being a capacitor, if the value of the capacitance is properly chosen such that the reactance of the capacitor at the operating frequency is very small or, if possible, negligible, this offers a short circuit to the ac sinusoid signal and so the coupling, of one stage to the other can be achieved without any problem. Thus, ac coupling and dc isolation between stages is achieved. Thus, such a coupling between different stages is very comfortable and so *RC* coupled amplifier finds a large application. Along with coupling, stability of the operating point is also of utmost importance. As we have already discussed and concluded, the self-bias circuit of *CE* configuration amplifier has a very good stability. So, such self-bias circuits are generally used in different stages of the amplifier chain. These circuits have an emitter resistance that acts as a feedback to bring down the amplification of the particular stage. One cannot afford such loss in gain of the stage but stability is also of prime concern. The way out is to have an *emitter capacitor* C_e in shunt with the emitter resistance R_e. The emitter capacitor acts as an open circuit, offering infinite impedance to dc currents and so, the dc currents are passed through the emitter resistor and not through the capacitor. Whereas selecting the capacitor value such that it offers negligible reactance at operating frequency allows the ac signal to take the minimum resistance path and, thus, ac signals do not pass through the emitter resistor but are bypassed by this emitter capacitor. Such a capacitor is referred to as *emitter bypass capacitor.* Since the ac signals are bypassed by this capacitor and are not allowed through the emitter resistor, the effect of the resistor on the amplification factor is not felt and so, the amplification of the stage is not much affected by the presence of the emitter resistor. Such a circuit of *RC* coupled amplifier is as shown in Fig. 7.20.

In the ac analysis of a *RC* coupled amplifier, all capacitors can be assumed to be short circuits and can be neglected while determining the parameters of the amplifier stages. Similarly, for determining the operating point and bias stabilisation, these capacitors can be assumed to have infinite impedance and can be treated to open circuits.

The analysis of a multistage amplifier involves various steps. The individual amplifier parameters are to be determined and the analysis for the whole of the amplifier chain is to be performed. The analysis is to be started with the last stage first and after determining those parameters, one has to take up the previous stage for analysis. This is due to the fact that the determination of the parameters such as current gain, input impedance and voltage gain requires the knowledge of the load resistance. The load impedance of a given stage is, in general, the collector resistance in shunt with the input impedance of the next stage and few other biasing resistors. Since the input impedance of the next stage is unknown, starting the exercise with the first stage is not possible. If one starts the exercise with the last stage, since the load impedance is known, the above three parameters can be determined without difficulty. Once the last stage parameters are determined, the load to the previous stage can be calculated and determination of the three parameters is a simple network problem. Thus, to determine current gain, input impedance and voltage gain of

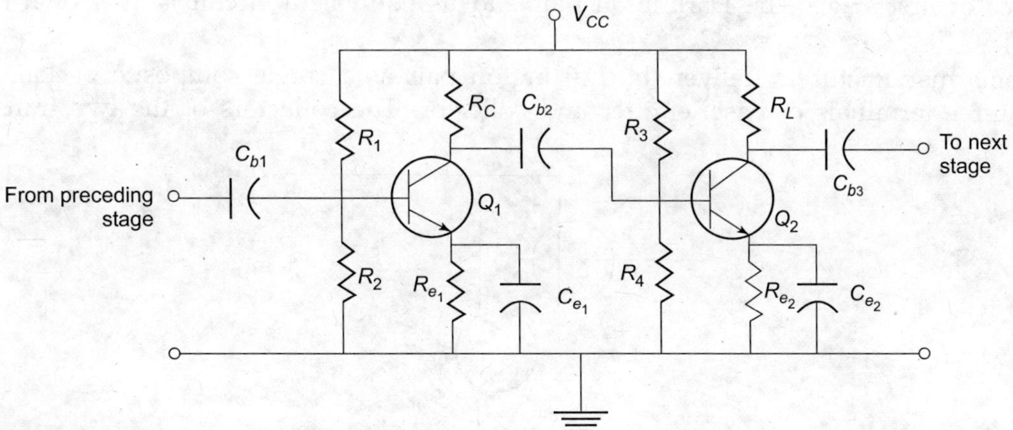

Fig. 7.20 *RC coupled amplifier*

various individual stages, one has to start the exercise from the last stage and proceed to first stage in the reverse order. In contrast, determination of output impedance requires the knowledge of source resistance. So, this parameter cannot be determined as the other three parameters, but the analysis of the first stage is to be taken first and one then proceeds to the consecutive stages in the forward direction.

When the amplifier stage parameters of all the individual stages are determined, the next step is to determine the overall performance of the amplifier chain. This can be achieved by simply following the known steps of the amplifier analysis as discussed in previous sections. The ratios of output current to input current, output voltage to input voltage, input voltage to input current and output voltage to output current can be determined by applying KVL and KCL appropriately.

While designing the multistage amplifiers, one has to be very careful with selection of operating point of each stage such that the required amplification is achieved with the selected transistor. Not only this, as the signal is proceeding in the various stages of the amplifier chain, its magnitude increases and so, the validity of the linear small signal model of the transistor for that particular stage is to be carefully checked. If proper checks are not done, one may lead to land in the nonlinear region of operation and so lot of distortion in the output of the signal may be introduced and also if the transistor is driven into saturation, clipping and other similar problems may creep in, spoiling the output signal totally. Also, one has to keep in mind the frequency of operation. All the analysis discussed so far is meant for low-frequency signals only. High-frequency analysis of a transistor amplifier is discussed in the following sections.

Darlington connection: The input impedance offered by the emitter follower circuit is not sufficient in some applications and one requires very high values. In such cases, *Darlington pair connection* gives the necessary very high input impedance values. Darlington connection is nothing but two emitter follower circuits in cascade with infinite load emitter resistance of first stage. The Darlington pair and its equivalent circuit is as shown in Fig. 7.21.

Some manufacturers deliver the Darlington pair as a single composite package with basic three terminals of base, emitter and collector. The collectors of the two transistors

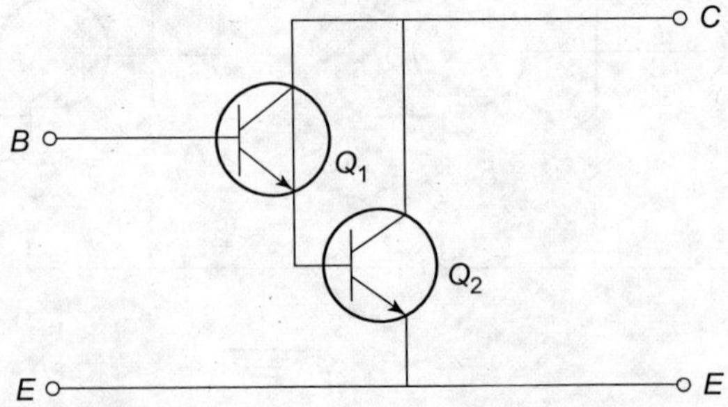

Fig. 7.21 *Darlington pair*

are tied up and brought out as collector terminal, the emitter of the first terminal is connected to the base of the second transistor internally, the emitter of the second transistor is brought out as the emitter output terminal and input terminal is the base of the first transistor.

Assuming the simplified analysis condition is satisfied, only two h-parameters, h_{ie} and h_{fe} are considered. For the second stage, the parameters can be derived to be

$$A_{I2} = 1 + h_{fe} \quad R_{i2} = (1 + h_{fe})R_e \qquad ...(7.52)$$

For the first stage, current gain is

$$A_{I1} = \frac{1+h_{fe}}{1+h_{oe}\,h_{fe}R_e} \qquad ...(7.53)$$

And now, the overall current gain can be given as

$$A_I = \frac{(1+h_{fe})^2}{1+h_{oe}\,h_{fe}\,R_s} \qquad (7.54)$$

and the input impedance of the first stage or the circuit can be given as

$$R_{i1} = h_{ie} + A_{I1}R_{i2} \approx \frac{(1+h_{fe})^2 R_e}{1+h_{oe}h_{fe}R_e} \qquad ...(7.55)$$

The above equation shows that the input impedance is very large to that of a single emitter follower circuit. For example, if R_e is 4 K, R_{i2} would be 205 K and R_{i1} will be 1.73 M from the typical values of h-parameters. Thus, for emitter follower if the input is of the order of hundreds of kiloohms, input of Darlington would be of the order of megaohms or higher. One can also note that even though the voltage gain of the pair is approximately unity, the current gain of the pair is very large compared to single emitter circuit.

The output impedance can be given as

$$R_o \approx \frac{R_s}{(1+h_{fe})^2} + \frac{2h_{ie2}}{1+h_{fe}} \qquad ...(7.56)$$

The output impedance of the Darlington pair may be greater or smaller than that of a single transistor emitter follower depending upon the value of R_e relative to h_{ie2}, where h_{ie2} is the h_{ie} of the second stage.

The drawback of the Darlington pair is that the leakage current of the transistor is amplified by the second. Hence, the overall leakage current may be high and a Darlington connection of three or more transistors is usually impractical.

The composite transistor pair may be used as a *CE* transistor amplifier. The advantage of this pair would be a very high overall h_{fe} nominally equal to the product of current gains of the individual transistors.

7.12 FREQUENCY RESPONSE OF AN AMPLIFIER

Frequency response of an amplifier is a plot of amplification versus frequency. This plot gives the ability of the transistor amplifier to amplify at a given frequency. It is to be noted that the amplifier will not be able to amplify signals of all possible frequencies. With the help of this frequency response of the amplifier, one can estimate the frequency range in which the designed amplification is achieved without distortion in the output of the amplifier. Fidelity of faithful reproduction of the signal at the output of the amplifier depends on the frequency of operation. The shape of the signal should remain unaltered, even though the amplitude and phase change. Then, the amplifier is said to have good fidelity.

Let the input signal to the amplifier be a sinusoidal waveform with angular frequency of ω, then $v = V_m \sin(\omega t + \phi)$. If the amplifier has a gain of A, then the output signal will be

$$AV_m \sin(\omega t + \phi + \theta) \qquad ...(7.57)$$

This output signal suffers a phase sift of θ and the total phase of he signal is $\theta + \phi$. To preserve the form of the signal, it is required that the amplification is independent of frequency and the phase shift θ is proportional to the frequency. Then the waveform shape is retained while there would be a shift in time by and amount θ/ω.

Lack of fidelity is the extent to which an amplifier's amplitude response is not uniform, and its time delay is not constant with frequency. It is not necessary to specify both amplitude and time delay response since the two are related and one having been specified, the other is uniquely determined.

The frequency response of an amplifier can generally be divided into three parts. First, the *midband,* where the amplification of the amplifier is constant or uniform with frequency. In this range of frequencies, the amplification is almost constant, equal to A_o and the time delay is also quite uniform. This is a useful range of frequencies where the amplifier has a good performance. Generally, the response is normalised and A_o (maximum amplification) will be made equal to 1. The amplification at other frequencies will be relative to this value. It is most practical to express the frequency response in decibels and so, the y axis is expressed in dB. 20 log (A) is the parameter of interest on such response graph.

At low frequency ranges, it is observed that the amplification is very low and the response rises as the frequency increases. This is the lower end of the response where the characteristic resembles the response characteristic of a high-pass filter. So, one can identify cut off frequency of this response where the filter allows signals above the cut off frequency and blocks the frequencies below this frequency. This cut off frequency would be 3 dB down the midband amplification factor A_o. This frequency f_L is the *lower cut off frequency* of the amplifier. The frequencies above this value would have good amplification with A_o. So, in the frequency range from zero to f_L, the amplifier does not offer optimum gain and the response starts with zero at dc frequency and increases towards A_o gradually.

As the frequency increase in the midband, the uniform characteristic of the response falls back towards zero. This is third part of the response curve, which is similar to that of low-pass filter. Once again, the cut off frequency of the low-pass characteristic is important. This determines the upper limit of the useful range of frequencies with the amplifier. Thus, the frequency f_H, called *higher cut off frequency*, is the frequency above which the response falls to zero gradually. The frequencies below this value are amplified with A_o in an amplifier. Above this frequency, the response is not optimum and gain falls down towards zero.

Figure 7.22 shows a complete frequency response curve of an amplifier. Thus, the range of frequencies that are amplified by the amplifier lie between the lower and higher cut off frequencies. In this useful range of frequencies the distortion of the signal would be minimum. This range is defined as the *bandwidth of the amplifier*. Thus,

$$\text{bandwidth} = BW = f_H - f_L \qquad ...(7.58)$$

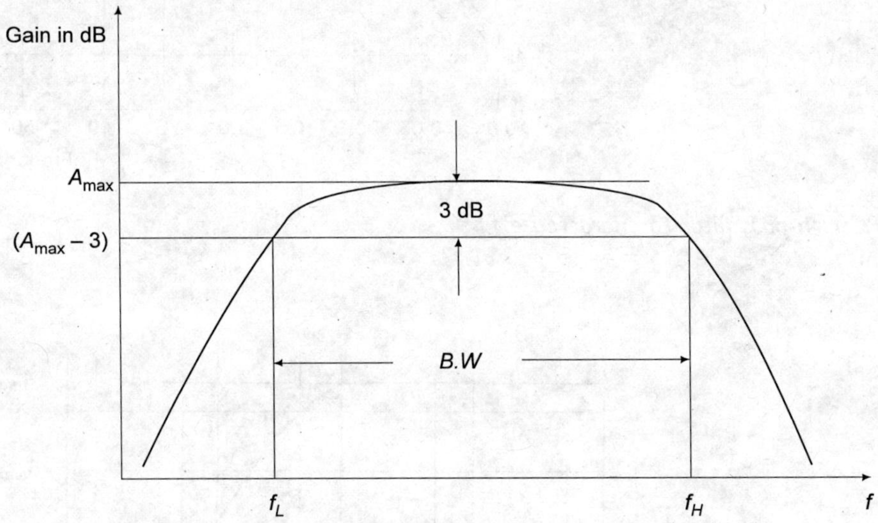

Fig. 7.22 *Frequency response of an amplifier*

The cut off frequencies of the high-pass and low-pass filters can be given and determined from the equivalent *RC* circuits. Consider a RC high-pass filter as shown in Fig. 7.23. The transfer ratio, the ratio of output response to input excitation, of such a circuit can be derived to be

$$A_L(f) = \frac{1}{1 - j(f_L/f)} \qquad ...(7.59)$$

where,

$$f_L = 1/(2\pi R_1 C_1) \qquad ...(7.60)$$

f_L is the frequency at which the response suffers by 3 dB from the maximum and is termed as *3 dB cut off frequency*. The absolute value of the response at this frequency would be 0.707 ($1/\sqrt{2}$) of the maximum (1). Since this is the lower edge of the frequency response of the amplifier, it is called as *lower cut off frequency*.

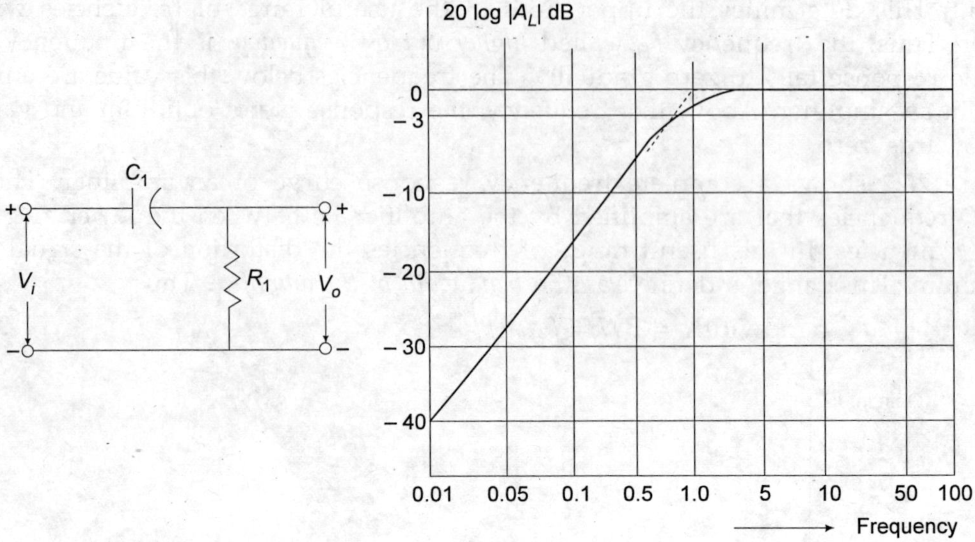

Fig. 7.23 *RC high-pass filter to characterise f_L*

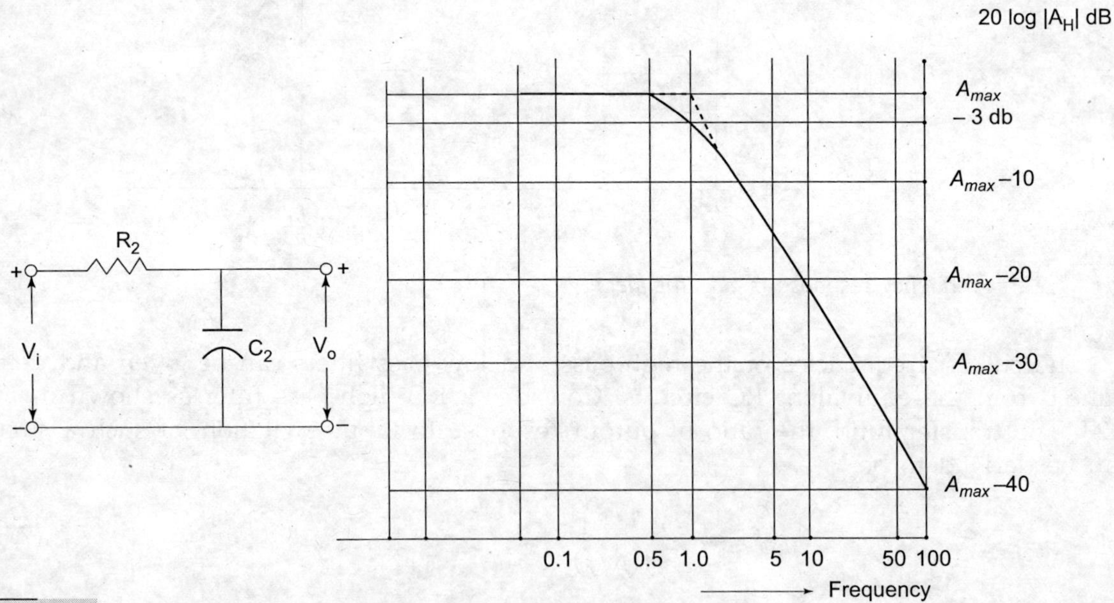

Fig. 7.24 *RC low-pass filter to characterise f_H*

Considering the *RC* low-pass filter as shown in Fig. 7.24, the higher cut off frequency can be characterised.

The transfer ratio of such a *RC* low-pass circuit can be derived to be

$$A_H(f) = 1 + J\,(f / f_H) \qquad ...(7.59)$$

where,

$$f_H = 1/(2\pi R_2 C_2) \qquad ...(7.60)$$

7.13 BANDWIDTH OF A CASCADED AMPLIFIER

When a multistage amplifier is cascaded, the amplifier parameters can be determined from the analysis steps, as discussed in the previous sections. The frequency response is also affected by the cascading of amplifier stages and the new cut off frequencies can be determined as follows.

The high 3 dB frequency for *n* cascaded stages is f_H^* and equals the frequency for which the overall voltage gain falls to $1/\sqrt{2}$ or by 3 dB of its midband value. To obtain the overall transfer ratio of the noninteracting stages, the transfer gains of individual stages are to be multiplied together. Thus,

$$\frac{1}{\sqrt{[1+(f_H^*/f_{H1})^2]}}\,\frac{1}{\sqrt{[1+(f_H^*/f_{H2})^2]}}\cdots\frac{1}{\sqrt{[1+)f_H^*/f_{Hn})^2]}}=\frac{1}{\sqrt{2}} \qquad ...(7.61)$$

For *n* stages with identical upper or higher 3 dB frequencies,

$$f_{H1} = f_{H2} = f_{H3} = f_{H4} = \cdots f_{Hn} = f_H \qquad ...(7.62)$$

Thus,

$$\left[\frac{1}{\sqrt{[1+(f_H^*/f_H)^2]}}\right]^n = \frac{1}{\sqrt{2}} \qquad ...(7.63)$$

and so,

$$f_H^* = f_H\sqrt{(2^{1/n}-1)} \qquad ...(7.64)$$

The lower 3 dB frequency for *n* cascaded stages is f_L^* and equals the frequency for which the overall voltage gain falls to $1/\sqrt{2}$ or by 3 dB of its midband value. To obtain the overall transfer ratio of the noninteracting stages, the transfer gains of individual stages are to be multiplied together. Thus,

$$\frac{1}{\sqrt{[1+(f_{L1}/f_L^*)^2]}}\,\frac{1}{\sqrt{[1+(f_{L2}/f_L^*)^2]}}\cdots\frac{1}{\sqrt{[1+(f_{Ln}/f_L^*)^2]}}=\frac{1}{\sqrt{2}} \qquad ...(7.61)$$

For *n* stages with identical lower 3 dB frequencies,

$$f_{L1} = f_{L2} = f_{L3} = f_{L4} = \cdots f_{Ln} = f_L \qquad ...(7.62)$$

Thus,

$$\left[\frac{1}{\sqrt{[1+(f_L/f_L^*)^2]}}\right]^n = \frac{1}{\sqrt{2}} \quad \text{...(7.63)}$$

and so,

$$f_L^* = f_L/\sqrt{(2^{1/n}-1)} \quad \text{...(7.64)}$$

From the above discussion, one can observe that the cascaded stages have a lower f_H and higher f_L and thus the effective bandwidth is lowered or reduced. That is, even though the amplification increases in cascaded stages, the bandwidth of the amplifier decreases.

7.14 EMITTER BYPASS CAPACITOR

An emitter bypass capacitor in shunt with the emitter resistor is generally used in the *CE* self-bias transistor amplifier circuit to bypass the effect of feedback on the amplification factor. In the general analysis, the reactance of the emitter bypass capacitor is assumed to be negligible, tending towards zero. This might be possible to some extent by choosing the appropriate capacitance value. But when exact analysis is to be considered, this reactance value cannot be neglected and so, the impedance, a resistor is shunt with a capacitance in the emitter lead of the amplifier is to be considered and the analysis steps are to be followed. Thus, the effect of the bypass capacitor on the low-frequency response of the amplifier can be estimated.

The circuit diagram of a *CE* amplifier with emitter resistor and capacitor in the common emitter terminal is as shown in Fig. 7.25. Also, the equivalent circuit of the same in a simplified transistor model is given in the same figure for simplified analysis of the circuit.

To simplify the analysis, the biasing resistors R_1 and R_2, which are in shunt, are assumed to be very large compared to the source resistance and so can be neglected in the

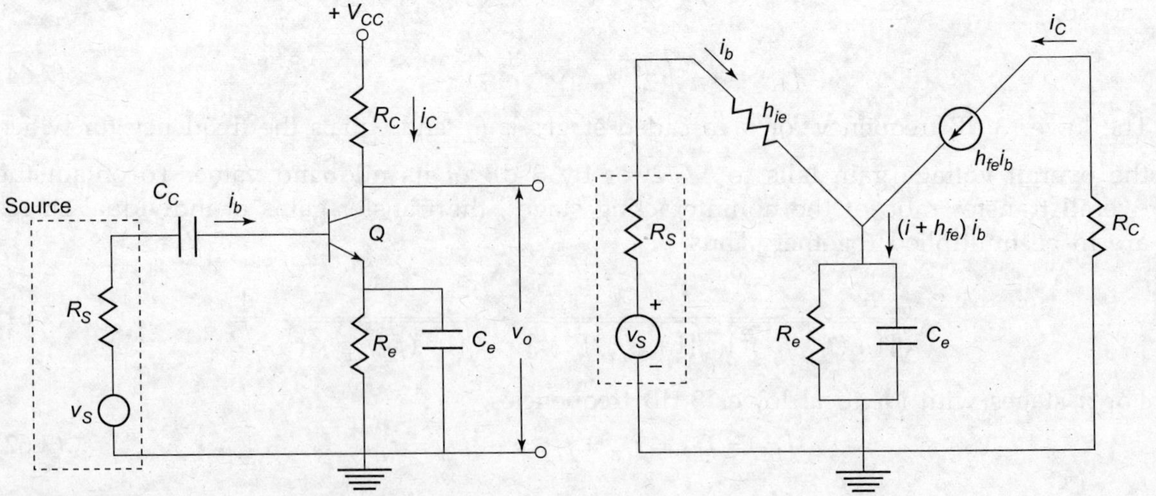

Fig. 7.25 *Effect of emitter bypass capacitor*

analysis. Also, the load resistance R_c is assumed to be of small value such that the simplified model of the transistor can be adopted. The blocking capacitor is also omitted and its effect will be separately considered in the following section.

With the above assumptions, from the simplified equivalent circuit of the amplifier, the output voltage V_o can be derived to be

$$V_o = - I_b h_{fe} R_c = \frac{- V_s h_{fe} R_c}{R_s + h_{ie} + Z_e'} \qquad ...(7.65)$$

where

$$Z_e' = (1 + h_{fe}) \frac{R_e}{1 + j\omega C_e R_e} \qquad ...(7.66)$$

Solving for voltage gain A_V,

$$A_V = - \frac{h_{fe} R_c}{R + R'} \frac{1 + j\omega C_e R_e}{1 + j\omega C_e [R_e R/(R + R')} \qquad ...(7.67)$$

where,

$$R \equiv R_s + h_{ie} \quad \text{and} \quad R' \equiv (1 + h_{fe}) R_e \qquad ...(7.68)$$

The midband gain A_o can be obtained with $\omega \to \infty$ or

$$A_o = - h_{fe} R_c / (R_s + h_{ie}) \qquad ...(7.69)$$

The effect of emitter bypass capacitor on the voltage gain can be estimated by the above equations. If more exact analysis is required, all the four h parameters are to be considered and after a complex network problem solution, the voltage gain of the amplifier stage can be determined.

7.15 COUPLING CAPACITOR

In a *RC* coupled amplifier a blocking capacitor is used in series with the two-cascaded stage to obtain dc isolation and ac coupling. In the previous analysis of the multistage amplifiers, the effect of this capacitor on the low frequency of the amplifier is neglected, assuming that the capacitor reactance tends to zero at operating frequency. But, in order to estimate the effect of this capacitor on the response, the following analysis is performed.

A single intermediate stage of any of the cascades is considered and is represented as shown in Fig. 7.26. The amplifier parameters may be expressed in h-parameters if the active device is transistor, or, in terms of mutual conductance, if it is a FET. In case of a FET, $I = g_m V_i$, $R_i = \infty$ and $R_o = r_d$ and when a BJT is considered, $I = h_{fe} I_b$, $R_i = h_{ie}$ and $R_o = 1/h_{oe}$. The equivalent circuit suggests a high-pass filter and so, the cutoff frequency of the *lower 3 dB frequency* would be affected by the presence of coupling capacitor. The 3 dB cut off lower frequency can now be estimated as

$$f_L = \frac{1}{2\pi (R_o' + R_i') C_b} \qquad ...(7.70)$$

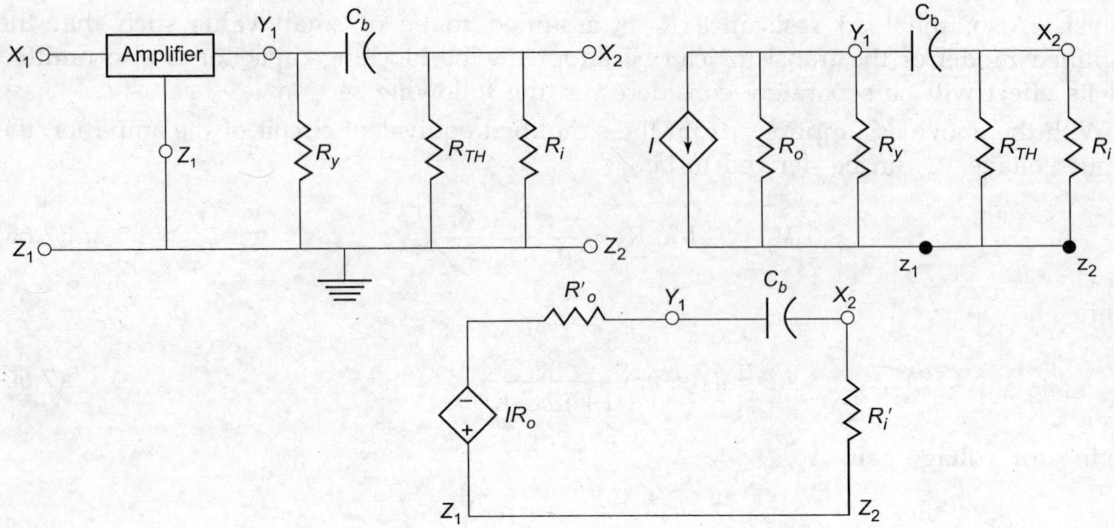

Fig. 7.26 *Effect of blocking capacitor*

where, $R_o' = R_o // R_y$ and $R_i' = R_i // R_b$ ...(7.71)

and R_y is biasing resistor, collector resistance R_c in case of BJT or drain resistance R_d in case of FET.

Thus, the lower 3 dB frequency of such an amplifier can be estimated with a time constant, which is equal to the blocking capacitor multiplied by the sum of the effective resistances R_o' to the left of the capacitor and R_i' to the right of the capacitor.

If in a given transistor amplifier both emitter bypass and blocking capacitors are present, first compute the lower 3 dB frequency assuming the emitter bypass capacitor to be infinity. Then, calculate the f_L assuming that only emitter bypass is present. If the two cut off frequencies are significantly different, the higher of the two is approximately the low 3 dB frequency of the stage.

7.16 HIGH-FREQUENCY MODEL OF A *CE* TRANSISTOR

In the analysis performed at low frequencies, it is assumed that the response of the active device is instantaneous. That is, the output responds immediately to any changes in the input of the circuit. The time delay for the charge carriers to travel from input to output is negligible in this case of low frequency since for low frequencies, the time period is very small and the response time of the device can be neglected. So, the storage action or time delay or diffusion time is not considered in the above analysis. But, as the frequency of operation is increased, the signal time period decreases and so, the charge carriers take significant time to respond compared to the time period of the signal. By the time the response for the change input reaches the output, the input might have changed and the response cannot be instantaneous. There is some time delay in the circuit leading to storage action and thus the capacitances involved in the terminals of the device cannot be neglected. These capacitances were totally neglected in the analysis of low-frequency signals. Also, diffusion mechanism cannot be neglected and consideration of all these parameters leads to very complex analysis of the transistor amplifier. A model based on the transmission line equations would be quite accurate, but the resulting equivalent circuit is too complicated to be of practical use. Hence, approximations are made and cruder the approximations, the simpler is the circuit. A compromise model, which is accurate as well as simple, is *hybrid pi* (Π) or *Giacoletto model.*

The simplicity of this model is that the resistive components can be derived from the low-frequency h-parameters of the transistor. The analysis of this hybrid Π model is in excellent agreement with that of the experimental results. All the parameters are assumed to be constant with frequency. The parameters may vary with the quiescent operating point but are assumed to be constant under the given condition.

The *Giacoletto* model of a *CE* transistor is as shown in Fig. 7.27. The model discussed is only valid for common emitter configuration. Since the other two find little application as amplifier at high frequencies, they are not discussed in this context. As one can observe, there is one dependent current source in the output circuit, which depends on the input voltage and there is no voltage source as in the low-frequency model. There are four more additional conductances and two capacitors. The understanding of the existence of the conductances and capacitances is as discussed in the following paragraphs.

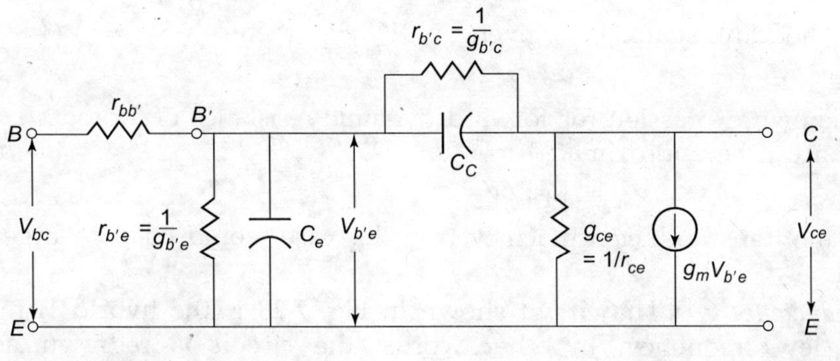

Fig. 7.27 *Hybrid Π model of a CE transistor*

There are two base terminals in the figure, B and B'. B is the physical base terminal and is accessible externally but the terminal B' is electrical and is not physical. All references to base terminal are made with respect to this terminal. Between these two terminals there is a resistor $r_{bb'}$, the base spreading resistor. This is the resistance that is felt by the charge carriers moving from the highly doped emitter terminal to thin, lightly doped base region. This is represented as a lumped element between the two terminals.

For small changes in the voltage $V_{b'e}$ across the emitter junction, the excess minority carrier concentration injected into the base is proportional to $V_{b'e}$ and, therefore, the resulting small signal collector current with the collector shorted to the emitter is proportional to $V_{b'e}$. This accounts for the current generator $g_m V_{b'e}$. g_m is the mutual conductance and can be defined as the ratio of output current to input voltage.

$g_{b'e}$ is the conductance between the base and emitter of the device which is due to the fact that increase in minority carriers in the base results in increase in recombination base current. The excess minority carrier storage in the base is represented by the diffusion capacitance C_e between the base and the emitter.

There is a feedback between the output collector and input base of the transistor. The cause for such a feedback is the early effect. The feedback is represented by a shunt combination of conductance $g_{b'c}$ and a collector junction barrier capacitance C_c. The conductance between the collector and emitter is g_{ce}.

The typical values of these parameters are as follows:

g_m = 50 mA/V, $r_{bb'}$ = 100 Ω, $r_{b'e}$ = 1 K, $r_{b'c}$ = 4 M, r_{ce} = 80 K, C_c =3 pF and C_e = 100 pF.

7.16.1 Hybrid Π Conductances

Transistor transconductance g_m: In the active region of operation for a transistor in CE configuration, the output collector current can be given as

$$I_C = I_{Co} - \alpha_o I_E \qquad \text{...(7.72)}$$

Since the short circuit current is $g_m V_{b'e}$, g_m can be defined as

$$g_m \equiv \left.\frac{\partial I_C}{\partial V_{BE}}\right|_{V_{CB}} = \alpha_o \frac{\partial I_E}{\partial V_{BE}} = \alpha_o \frac{\partial I_E}{\partial V_E} \qquad \text{...(7.73)}$$

If the emitter diode resistance r_e is given by $\partial V_E/\partial I_E$, then

$$g_m = \alpha_o / r_e \qquad \text{...(7.74)}$$

Or, with the known expression for forward dynamic resistance of a diode ($r = \eta V_T/I$), the above equation can be approximated as

$$g_m \approx |I_C| / V_T \qquad \text{...(7.75)}$$

The value of the transconductance achieved in this case would be very large compared to that of a FET.

The input conductance $g_{b'e}$: The circuit shown in Fig. 7.28 is the hybrid Π model of a *CE* transistor at low frequency. In other words, the circuit is redrawn neglecting the capacitances at low frequency. Comparing this circuit with that of the low-frequency *h*-parameter model, the values of the conductances can be derived in terms of *h*-parameters.

From the typical values of the hybrid Π conductances, $r_{b'c} \gg r_{b'e}$ and so, the input current I_b can be assumed to be flowing into $r_{b'e}$ and no current is coupled to $r_{b'c}$. Or,

$$V_{b'e} \approx I_b r_{b'e} \qquad \text{...(7.76)}$$

Therefore,

$$I_c = g_m V_{b'e} \approx g_m I_b r_{b'e} \qquad \text{...(7.78)}$$

By definition, the short circuit current gain h_{fe} is the ratio of collector current to base current at short-circuited collector to emitter. Thus,

$$h_{fe} = g_m r_{b'e} \qquad \text{...(7.79)}$$

Or,

$$r_{b'e} = h_{fe}/g_m = h_{fe} V_T / |I_C| \qquad \text{...(7.80)}$$

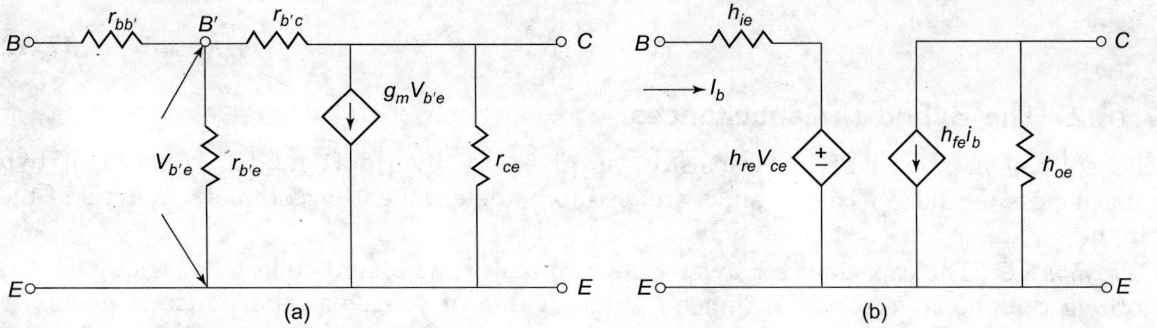

Fig. 7.28 *Hybrid Π model and h-parameter model at low frequencies*

Thus, the input resistance $r_{b'e}$ of the transistor is directly proportional to temperature and inversely proportional to the output current.

The feedback conductance $g_{b'c}$: With input open circuited, the ratio of input voltage to output voltage is defined as the reverse voltage gain h_{re},

$$h_{re} = \frac{V_{b'e}}{V_{ce}} = \frac{r_{b'e}}{r_{b'e} + r_{b'c}} \qquad \text{...(7.81)}$$

or,

$$r_{b'e}(1 - h_{re}) = h_{re}\, r_{b'c}$$

Since $h_{re} \ll 1$,

$$r_{b'e} = h_{re} r_{b'c}$$

$$g_{b'c} = h_{re} g_{b'e} \qquad \text{...(7.82)}$$

The base spreading resistance $r_{bb'}$: The ratio of input voltage to input current with output shorted is h_{ie} given by

$$h_{ie} = r_{bb'} + r_{b'e}$$

or,

$$r_{bb'} = h_{ie} - r_{b'e} \qquad \text{...(7.83)}$$

The output conductance g_{ce}: The output conductance with input open circuited is h_{oe} and, thus, when $I_b = 0$,

$$I_c = \frac{V_{ce}}{r_{ce}} + \frac{V_{ce}}{r_{b'e} + r_{b'c}} + g_m V_{b'e} \qquad \text{...(7.84)}$$

and

$$V_{b'e} = h_{re} V_{ce}$$

Thus,

$$h_{oe} = \frac{1}{r_{ce}} + \frac{1}{r_{b'c}} + g_m h_{re}$$

$$h_{oe} = \frac{1}{r_{ce}} + \frac{1}{r_{b'c}} + g_{b'e} h_{fe} \frac{g_{b'c}}{g_{b'e}} \qquad \text{...(7.85)}$$

or,

$$g_{ce} = h_{oe} - (1 + h_{fe})g_{b'c} \qquad ...(7.86)$$

7.16.2 The Hybrid Π Capacitances

The equivalent model of CE transistor, as given by hybrid Π model, consists of two capacitances C_c and C_e. These values are also to be determined for complete analysis of the circuit.

Capacitance C_c: The capacitance between the terminals base B' and collector C is $C_{b'c}$. This is nothing but the collector capacitance C_e. This value of collector capacitance is generally specified in the manufacturer catalogue of the transistor and is represented as C_{ob}. The variation of the collector capacitance with voltage depends on whether the junction is step or graded type. The variation would be V_{CE}^{-n} where n takes value of 1/2 or 1/3 for the step or graded junctions, respectively.

Capacitance C_e: The sum of the transition and diffusion capacitance at the emitter is the value of emitter capacitance C_e. In general, diffusion capacitance is very large compared to transition capacitance at the emitter junction and thus,

$$C_e = C_{De} + C_{Te} \gg C_{De} \qquad ...(7.87)$$

To calculate the emitter diffusion capacitance, the accumulated charge should be evaluated. From the Fig. 7.29, the excess minority injection into the region can be assumed to be linear since the thickness of the layer is very small. Thus, the area under the curve is p′(0)/2 and the volume is WA. The total charge accumulated in the base region is the electron charge q, i.e.,

$$Q_B = ½p'(0)\, AW_q \qquad ...(7.88)$$

and current

$$I = - AqD_Bp'(0)/W \qquad ...(7.89)$$

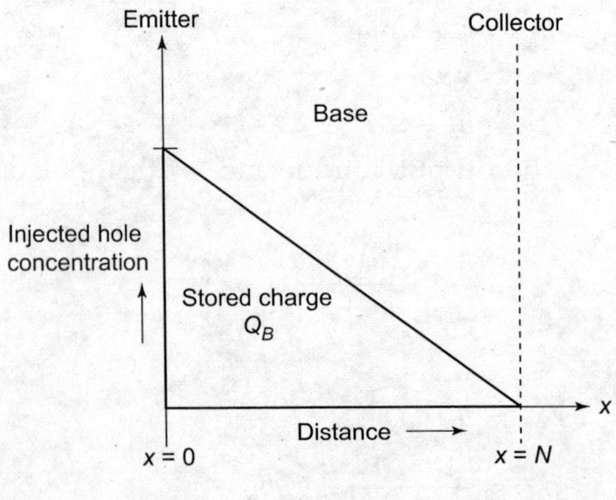

Fig. 7.29 *Minority carrier distribution in the base region*

D_B is the diffusion constant of minority carriers in the base. From the above equations,

$$Q_B = IW^2/(2D_B) \quad ...(7.90)$$

Then, C_{De} can be given as

$$C_{De} = \frac{dQ_B}{dV} = \frac{W^2}{2D_B} \cdot \frac{dI}{dV} = \frac{W^2}{2D_B}\frac{1}{r_e} \quad ...(7.91)$$

where $r_e = V_T/I_E$ is the emitter resistance. Then,

$$C_{De} = g_m W^2/2D_B \quad ...(7.92)$$

So, the diffusion current is proportional to the emitter bias current. It may be noted that the above equation gives the diffusion capacitance. But, determination of its value is difficult since no manufacturer specifies the value of the base width W and also the diffusion constant is difficult to get from the available data. The emitter capacitance can be determined experimentally as

$$C_e \approx g_m/(2\pi f_T) \quad ...(7.93)$$

where f_T is the frequency at which the *CE* short circuit current gain drops to unity. This value can be experimentally determined and used to establish the emitter capacitance value.

7.17 VALIDITY OF THE HIGH-FREQUENCY MODEL

The Giacoletto high-frequency model of a *CE* transistor assumes that the parameters defined are constant with frequency. This is possible under some specified conditions and so, the model is only valid under these conditions. From the discussion of mutual conductance and diffusion emitter capacitance, it is assumed that the minority charge carrier distribution in the base region is triangular. If this distribution remains triangular even when the voltage V_{BE} varies, the initial slope and final slope would be the same and so, the emitter and collector currents would be equal. Thus, the hybrid Π model is valid under dynamic conditions when the rate of change of V_{BE} is small enough so that the base incremental current I_b is small compared with the collector incremental current I_c. It can now be stated that the parameters are frequency independent, provided that

$$2\pi f \frac{W^2}{6D_B} \ll 1 \quad ...(7.94)$$

We know that

$$\frac{W^2}{6D_B} = \frac{C_e}{3g_m} = \frac{1}{6\pi f_T} \quad ...(7.95)$$

Thus,

$$f \ll \frac{6\pi f_T}{2\pi} = 3f_T \quad ...(7.96)$$

From the above, one can conclude that the high-frequency hybrid Π model of a *CE* transistor is only valid for frequencies up to approximately $f_T/3$.

7.18 HIGH-FREQUENCY CURRENT GAIN

The analysis of the *CE* transistor at high frequencies can be performed with the help of the above-discussed hybrid Π model. A single stage amplifier at high frequencies is considered with R_c, the collector resistance as the load impedance. For simplifying the problem under consideration, the load resistance $R_L = R_c$ is assumed to be zero such that the analysis is performed for maximum load current. Thus, $R_L = 0$, and from the typical values of the parameters, $r_{b'c} \gg r_{b'e}$, $r_{b'c}$ can be neglected since the input current flows entirely in the $r_{b'e}$ approximately. Since the collector is shorted to emitter, the collector is also at ground potential and thus the capacitances C_c and C_e will be in shunt between the base and emitter terminals. With all these simplifications and appropriate assumptions, the equivalent circuit only consists of $g_{b'e}$ in shunt with capacitance of value $C_c + C_e$ in the input between the base and emitter and a short-circuited dependent current source $g_m V_{b'e}$ in the output of the circuit as shown in Fig. 7.30.

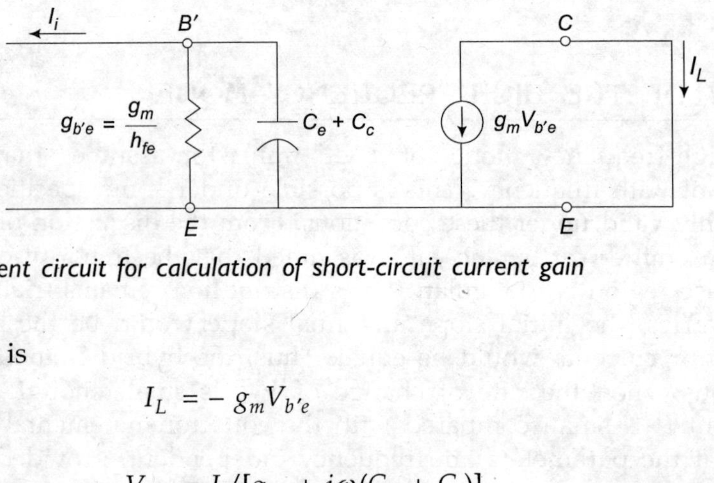

Fig. 7.30 *Equivalent circuit for calculation of short-circuit current gain*

The load current is

$$I_L = -g_m V_{b'e} \quad \text{...(7.97)}$$

and,

$$V_{b'e} = I_i/[g_{b'e} + j\omega(C_C + C_e)] \quad \text{...(7.98)}$$

Thus, the current amplification can be given as

$$A_i = \frac{I_L}{I_i} = \frac{-g_m}{g_{b'e} + j\omega(C_C + C_e)} \quad \text{...(7.99)}$$

or,

$$A_i = -h_{fe}/[1 + j(f/f_\beta)] \quad \text{...(7.100)}$$

where

$$f_\beta = \frac{g_{b'e}}{2\pi(C_C + C_e)} = \frac{1}{h_{fe}} \frac{g_m}{2\pi(C_C + C_e)} \quad \text{...(7.101)}$$

At $f = f_\beta$, $|A_i|$ is equal to $1/\sqrt{2}$ of its low-frequency value h_{fe}. The frequency range up to f_β is referred to as *bandwidth* of the circuit. At $\omega = 0$, the value of A_i is $-h_{fe}$, which is in agreement with the definition of the short-circuit low-frequency *CE* current gain.

7.19 GAIN BANDWIDTH PRODUCT

The parameter f_T as discussed in the preceding section is the *frequency at which the short-circuit current gain of a CE transistor equals unity*. From Eq. (7.100), the magnitude of the current gain can be given as

$$|A_i| = h_{fe}/[1 + (f/f_\beta)^2]^{1/2} \quad \text{...(7.102)}$$

Or, since $h_{fe} \gg 1$, at $f = f_T$, $|A_i| = 1$ and so,

$$f_T \approx h_{fe} f_\beta = \frac{g_m}{2\pi(C_C + C_e)} \approx \frac{g_m}{2\pi C_e} \quad \text{...(7.103)}$$

Thus,

$$A_i = -h_{fe}/[1 + j h_{fe}\,(f/f_T)] \quad \text{...(7.104)}$$

The parameter f_T is an important parameter of high-frequency characteristic of a transistor. The parameter, being the product of short-circuit current gain and the bandwidth (the lower 3 dB frequency is so small that it can be assumed to be zero), it represents the *gain bandwidth product* of a transistor. This product constant is a very important parameter for a given transistor since it gives the behaviour gain and bandwidth of the transistor. If the transistor is designed for optimum gain, the bandwidth would be a small value and if the amplifier is designed for a large bandwidth, the gain suffers. If two transistors have the same value of the product, the transistor with lower h_{fe} will have larger bandwidth and vice versa.

From Eq. (7.100),

when

$$f \ll f_\beta, \quad |A_i| \approx h_{fe}$$

and when

$$f \gg f_\beta, \quad |A_i| \approx h_{fe} f_\beta / f$$

or,

$$|A_i| f \approx h_{fe} f_\beta \approx f_T \quad \text{...(7.105)}$$

That is, the product of the current gain at operating frequency and the frequency of the amplifier is approximately equal to the *gain bandwidth product* of the transistor.

To determine the value of the emitter capacitance C_e, one needs to establish the value of f_T experimentally. But determining the value of f_T experimentally is very difficult in some cases since the value would be very large. In such cases, the value of f_T can be determined by the measurement of the current gain at a given frequency. The product of this gain with frequency gives the required value of f_T approximately.

SUMMARY

- Transistor finds a large application as amplifier.
- Analysis of a transistor circuit is complex since the device is nonlinear. Graphical analysis is carried out for large signals whereas the device is modelled as a two-port linear device when the signal is small.
- The nonlinearity of the transistor introduces input and output nonlinear distortions in the amplifier.
- Hybrid model of the two-port network can be conveniently adopted for a transistor circuit since the parameters voltage gain, current gain, input and output impedances are easy to be determined with this *h*-parameter model.
- The *h*-parameters of a transistor depend on the operating point.
- The behaviour and analysis of the transistor at low and high frequencies is different. At low frequency storage capacitive effects can be neglected but a special model is evolved at high frequency for ease of understanding.
- The analysis of a simple transistor amplifier circuit is nothing but determination of four amplifier parameters. This is done with the help of low-frequency *h*-parameter model of a transistor.
- Analysis of any transistor amplifier circuit involves identification of configuration, replacement of transistor by its hybrid model and evaluating the parameters with the help of network theorems.
- Miller's theorem and its dual help a lot in simplifying the analysis of complex transistor circuits.
- Amplifier stages are cascaded in order to improve the amplifier parameters to a large extent—this is a multistage amplifier.
- Analysis of a multistage amplifier should start with the final stage and proceed in the reverse direction to determine current gain, input impedance and voltage gain. The output impedance determination is to start with first stage and proceed to final stage.
- Simplified model of transistor is accepted when a condition is met. This helps in simplifying the analysis to a great extent within a specified tolerance.
- Capacitive coupling between two stages is usually the practice since the capacitor blocks the dc and couples the ac signals. This is referred to a *RC* coupled amplifier.
- The effect of coupling capacitor is on the lower cut off frequency of the frequency response. The effect of emitter resistor is on the voltage gain of the transistor stage.
- Very commonly the transistor in *CE* configuration is used as an amplifier since it has both voltage and current gain large. Thus, power gain is good.

- *CC* or emitter follower is widely used as a buffer amplifier wherever impedance matching between stages is required.
- *CB* configuration is very rarely used and it can be visualised as current buffer.
- Darlington pair transistor offers very large input impedance.
- The plot of amplification with frequency is the frequency response of an amplifier. It gives a lower and an upper cut off frequency, between which the response is smooth and is the usable frequency range. The difference between these two frequencies is the bandwidth of the amplifier.
- With cascading of stages, the bandwidth reduces.
- When the operating frequency is high, the storage capacitive effects are to be considered for analysis. Giacoletto model is a suitable simplified model of *CE* transistor.
- The hybrid Π model of the transistor is valid only when the frequency is less than one-third of unity crossover frequency.
- The gain bandwidth product of a given transistor amplifier is constant and is equal to unity cross over frequency. If gain is optimised, bandwidth suffers and vice versa.

SOLVED PROBLEMS

7.1 The following test results are obtained in a *CE* amplifier circuit while measuring *h*-parameters experimentally.

(a) With ac output short-circuited $i_b = 20\ \mu A$, $i_c = 1$ mA, $v_{be} = 22$ mV, and $v_{ce} = 0$

(b) With ac input open-circuited $v_{be} = 0.25$ mV, $i_c = 30\ \mu A$, $v_{ce} = 1$ V and $i_b = 0$.

Determine the *h*-parameters of the given transistor.

Solution From short-circuited data,

$$h_{ie} = \frac{v_{be}}{i_b} = \frac{22 \times 10^{-3}}{20 \times 10^{-6}} = 1.1\ \text{k}\Omega$$

$$h_{fe} = \frac{i_c}{i_b} = \frac{1 \times 10^{-3}}{20 \times 10^{-6}} = 50$$

From the open-circuit test data,

$$h_{re} = \frac{v_{be}}{v_{ce}} = \frac{0.25 \times 10^{-3}}{1\ \text{V}} = 2.5 \times 10^{-4}$$

$$h_{oe} = \frac{i_c}{v_{ce}} = \frac{30 \times 10^{-6}}{1} = 3 \times 10^{-5} = 30\ \mu\text{A/V}$$

7.2 Convert CE h-parameters to the CB h-parameters.

Solution To convert one set of parameter (X) to another set of parameters (Y), first draw the original diagram of Y. Then redraw Y as X. Apply KVL and KCL equations and find the required ratio.

In this example CE h-parameters are going to be converted into CB h parameters. So, we will draw CB model with emitter as a common one. This is as shown below as in Fig. (b).

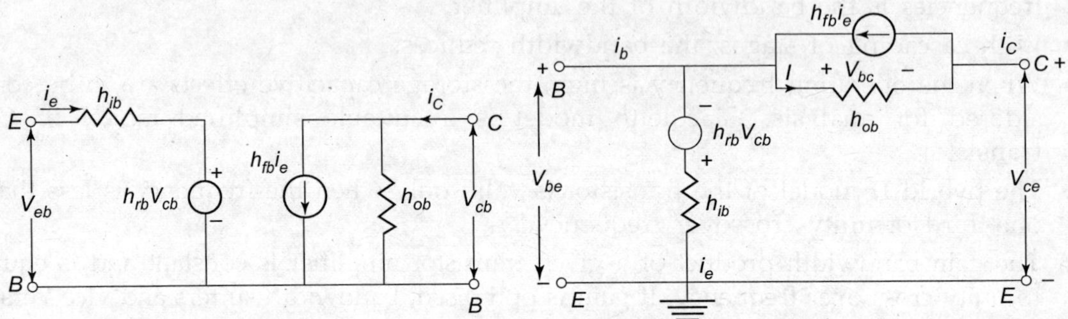

(a) CB Model (b) CB h-parameter model redrawn in CE configuration

Now, let us go for the definitions of CE h-parameters,

we have $h_{ie} = \left.\dfrac{V_{be}}{I_b}\right|_{V_{ce}=0}$ $\quad h_{re} = \left.\dfrac{V_{be}}{V_{ce}}\right|_{I_b=0}$

$$h_{fe} = \left.\frac{I_c}{I_b}\right|_{V_{ce}=0} \qquad h_{oe} = \left.\frac{I_c}{V_{ce}}\right|_{I_b=0}$$

$$h_{re} = \left.\frac{V_{be}}{V_{ce}}\right|_{I_b=0} = \left.\frac{V_{bc}+V_{ce}}{V_{ce}}\right|_{I_b=0}$$

$$= 1 + \left.\frac{V_{bc}}{V_{ce}}\right|_{I_b=0}$$

From circuit (b) we have

$$I_b + I_c + I_e = 0 \Rightarrow I_c = -\,I_e \text{ with } I_b = 0$$

The current I in h_{ob} is $= h_{ob} \cdot V_{bc} = (1 + h_{fb})I_e$ ...(1)

Applying KVL to the outer periphery of the circuit in (b),

$h_{ib}I_e + h_{rb}V_{cb} + V_{bc} + V_{ce} = 0$ ...(2)

Getting I_e from (1) and substituting in (2),

$$\frac{h_{ib} \cdot h_{ob}}{1+h_{fb}} V_{bc} + h_{rb} V_{cb} + V_{bc} + V_{ce} = 0$$

$$\therefore \quad \frac{V_{bc}}{V_{ce}} = \frac{-(1+h_{fb})}{h_{ib}h_{ob} + (1-h_{rb})(1+h_{fb})}$$

$$\therefore \quad h_{re} = 1 + \frac{V_{bc}}{V_{ce}} = \frac{h_{ib}h_{ob} - (1+h_{fb})h_{rb}}{h_{ib}h_{ob} + (1-h_{rb})(1+h_{fb})}$$

To find h_{oe}:

we have $\quad I_c = h_{fb} I_e + h_{ob} \cdot V_{bc}$

$$\frac{I_c}{V_{ce}} = h_{fb} \cdot \frac{I_e}{V_{ce}} + h_{ob} \cdot \frac{V_{bc}}{V_{ce}}$$

$$\therefore \quad \frac{I_c}{V_{ce}}(1 + h_{fb}) = \frac{h_{ob} \cdot (-(1+h_{fb}))}{h_{ib}h_{ob} + (1-h_{rb})(1+h_{fb})}$$

$$\therefore \quad \frac{I_c}{V_{ce}} = h_{oe} = \frac{-h_{ob}}{h_{ib}h_{ob} + (1-h_{rb})(1+h_{fb})}$$

To find h_{ie} and h_{fe}. Make $V_{ce} = 0$ for Fig. (b)
Then the circuit will become as shown below:

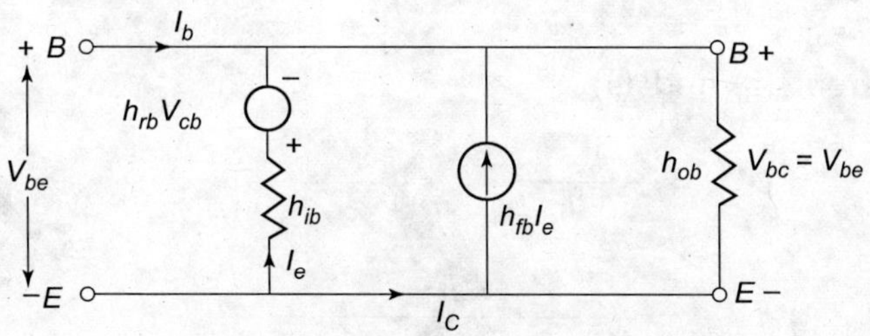

(c)

$$h_{ie} = \left.\frac{V_{be}}{I_b}\right|_{V_{ce}=0}$$

From circuit $\quad V_{cb} = -V_{bc} = -V_{be}$ ...(3)

Writing KVL in left-hand mesh,

$h_{ib}I_e + h_{rb}V_{cb} + V_{be} = 0$ ...(4)

$$I_e = -\frac{1-h_{rb}}{h_{ib}} V_{be} \qquad ...(5)$$

Writing KCL at node B

$I_e + I_b + h_{fb} I_e - h_{ob} \cdot V_{be} = 0$

$$\therefore \quad I_e = -(1 + h_{fb}) I_e + h_{ob} \cdot V_{be}$$

$$= h_{ob} \cdot V_{be} + \frac{(1+h_{fb})(1-h_{rb})}{h_{ib}} V_{be}$$

$$\therefore \quad h_{ie} = \frac{V_{be}}{I_b} = \frac{h_{ib}}{h_{ib} h_{ob} + (1+h_{fb})(1-h_{rb})} \qquad \ldots(6)$$

To find h_{fe}:

$$h_{fe} = \left.\frac{I_c}{I_b}\right|_{V_{ce}=0} \qquad \ldots(4)$$

From Fig. (c)

$$h_{fe} = \left.\frac{I_c}{I_b}\right|_{V_{ce}=0}$$

We have from Fig. (c)

$$I_c = h_{fb} \cdot I_e + h_{ob} \cdot V_{cb}$$

$$= h_{fb} \cdot I_e - h_{ob} \cdot V_{be}$$

$$\therefore \quad \frac{I_c}{I_b} = h_{fb} \cdot \frac{I_e}{I_b} - h_{ob} \cdot \frac{V_{be}}{I_b}$$

We have from (3) and (4)

$$Ie = -\frac{1-h_{rb}}{h_{ib}} \cdot V_{be}$$

$$V_{be} = \frac{h_{ib}}{h_{ib} h_{ob} + (1-h_{rb})(1+h_{fb})} I_b$$

$$\therefore \quad \frac{I_c}{I_b} = h_{fb} \cdot \frac{-(1-h_{rb})}{h_{ib}} \frac{h_{ib}}{h_{ib} h_{ob} + (1-h_{rb})(1+h_{fb})} - h_{ob} \cdot \frac{h_{ib}}{h_{ib} h_{ob} + (1-h_{rb})(1+h_{fb})}$$

$$= \frac{-\left\{(1-h_{rb}) h_{fb} + h_{ib} h_{ob}\right\}}{h_{ib} h_{ob} + (1-h_{rb})(1+h_{fb})}$$

$$\therefore \quad h_{fe} = \frac{-\left(h_{fb}(1-h_{rb}) + h_{ib} h_{ob}\right)}{h_{ib} h_{ob} + (1-h_{rb})(1+h_{fb})} = \frac{-h_{fb}}{1+h_{fb}}$$

(Using approximations that $h_{rb} << 1$, $h_{ib}h_{ob} << (1 + h_{fb})$).

$$\therefore \quad h_{re} = \frac{-h_{rb}}{1+h_{fb}}$$

$$h_{oe} = \frac{-h_{ob}}{1+h_{fb}}$$

$$h_{ie} = \frac{+h_{ib}}{1+h_{fb}}$$

$$h_{fe} = \frac{-h_{fb}}{1+h_{fb}}$$

In order to convert from *CE* to *CB* simply interchange subscripts *b* and *e*.

7.3 $h_{fe} = 50$, $h_{ie} = 0.83$ KΩ, find out the values of h_{fb} and h_{ib} for a transistor in *CB* configuration.

Solution

$$h_{fb} = \frac{-h_{fe}}{1+h_{fe}} = \frac{-50}{51} = -0.98$$

$$h_{ib} = \frac{h_{ie}}{1+h_{fe}} = \frac{0.83\times10^3}{51} = 16.27\ \Omega$$

7.4 Find *CE* parameters in terms of *CC h*-parameters.

Solution First draw the *CC h*-parameters diagram.

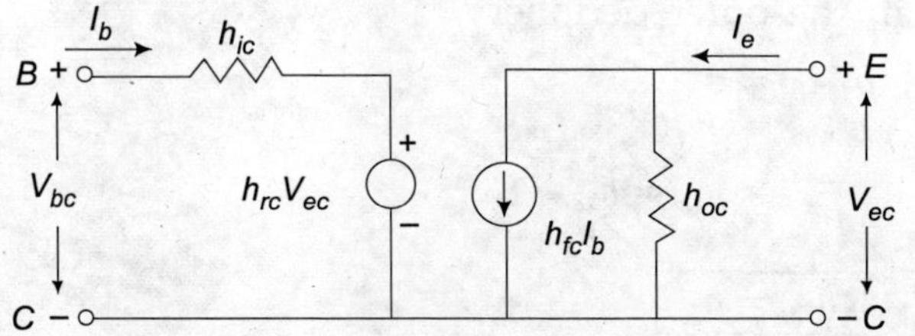

(a)

Now, redraw this diagram with emitter made as common between input and output,

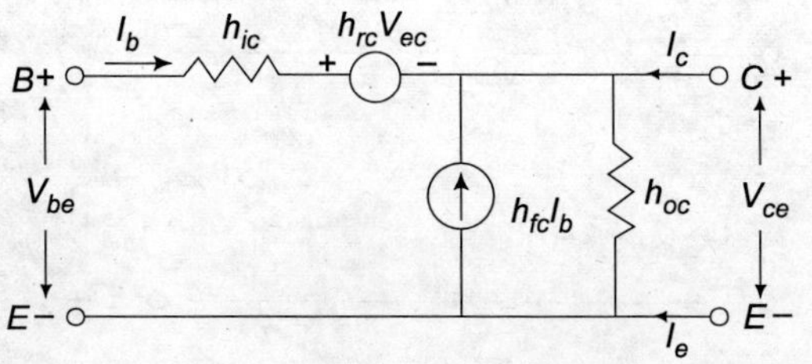

(b)

Now we have the definitions of *CE h* parameters as

$$h_{ie} = \left.\frac{V_{be}}{I_b}\right|_{V_{ce}=0} \qquad h_{re} = \left.\frac{V_{be}}{V_{ce}}\right|_{I_b=0}$$

$$h_{fe} = \left.\frac{I_c}{I_b}\right|_{V_{ce}=0} \qquad h_{oe} = \left.\frac{I_c}{V_{ce}}\right|_{I_b=0}$$

By making $V_{ce} = 0$ in Fig. (b)

$h_{re} V_{ec}$ source is replaced with a short circuit and h_{oe} in parallel with a short circuit also disappears the circuit becomes as below.

Writing KCL at node C

$$I_b + h_{fc} \cdot I_b + I_c = 0$$

$$\Rightarrow \qquad I_c = -(1 + h_{fc}) I_b$$

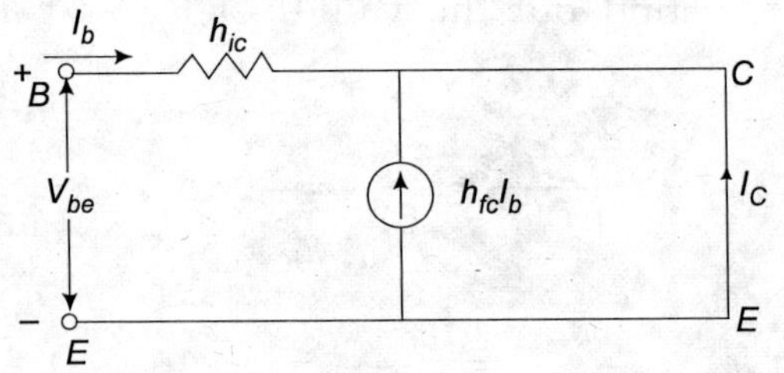

$$\therefore \qquad h_{fe} = \frac{I_c}{I_b} = -(1 + h_{fe})$$

$$\boxed{h_{fe} = -(1 + h_{fc})}$$

Writing KVL in the outer periphery

$$-V_{be} + h_{ic} I_b = 0$$

$$\therefore \qquad \frac{V_{be}}{I_b} = h_{ic} = h_{ie}$$

$$\boxed{h_{ie} = h_{ic}}$$

By making $I_b = 0$ in Fig. (b)

$h_{fe} \cdot I_b = 0$ so current source is open circuited and the circuit becomes,

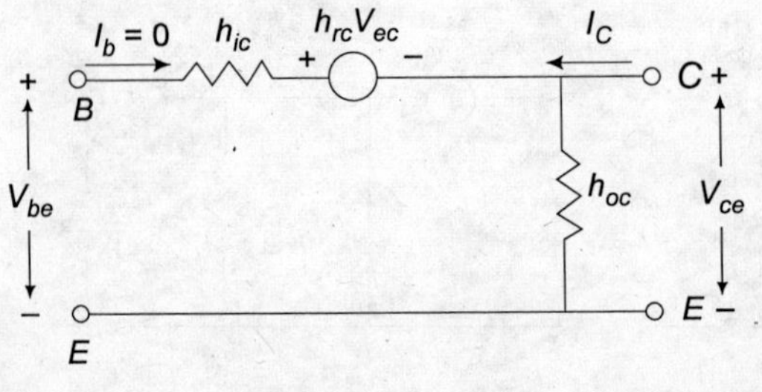

Writing KVL on the outer periphery,

$- V_{be} + h_{rc}V_{ec} + V_{ce} = 0$

$$V_{be} = (1 - h_{rc})V_{ce}$$

$$\frac{V_{be}}{V_{ce}} = 1 - h_{rc} = h_{re}$$

If approximation is made that $h_{rc} << 1$.

$\therefore$ $$\boxed{h_{re} = 1}$$

$$V_{ce} = \frac{I_c}{hoc} \Rightarrow \frac{I_c}{V_{ce}} = h_{oe} = h_{oc}$$

$\therefore$ $$\boxed{h_{oe} = h_{oc}}$$

If we want to convert from *CC* to *CE*, just interchange subscripts *c* and *e* in the above formulas.

7.5 The h parameters of a transistor used in single stage amplifier circuit are h_{ic} = 1100 r, h_{rc} = 1, h_{fc} = – 51 h_{oc} = 25 μA. Determine the amplifier parameters for CC configuration when $R_S = R_L$ = 10 K.

Solution

$$A_I = \frac{-h_{fc}}{1 + h_{oc} R_L}$$

$$= \frac{-(-51)}{1 + 25 \times 10^{-6} \times 10 \times 10^3} = 40.8$$

$$R_i = h_{ic} + h_{rc} \cdot A_I \cdot R_L = 1100 + 1 \times 40.8 \times 10 \times 10^{-3}$$

$$= 409.1 \text{ K}\Omega$$

$$A_V = \frac{A_I \cdot R_L}{R_i} = 0.9973$$

$$Y_o = h_{oc} - \frac{h_{fc} \cdot h_{rc}}{R_s + h_{ic}}$$

$$= 25 \times 10^{-6} + \frac{51 \times 1}{10 \times 10^3 + 1.1 \times 10^3}$$

$$= 4.62 \times 10^{-3} \text{ mho}$$

$$R_o = 1/Y_o = 216.45\ \Omega$$

$$A_{VS} = A_V \cdot \frac{R_i}{R_i + R_s} = 0.9973 \times \frac{409.1 \times 10^3}{409.1 \times 10^3 + 10 \times 10^3}$$

$$= 0.9735$$

7.6 For the circuit shown, verify that the modified h-parameters (indicated by primes) are

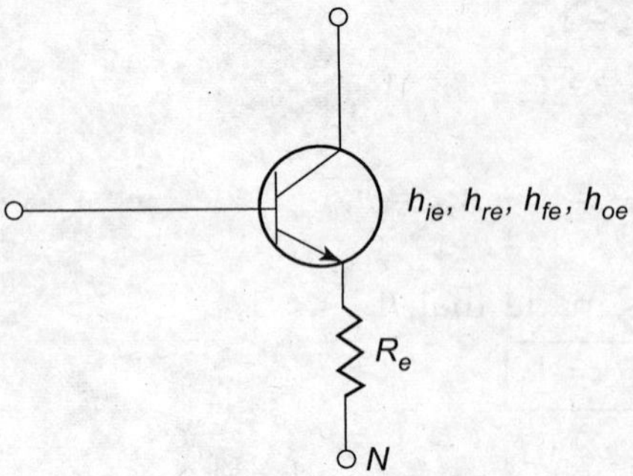

(a) $$h'_{ie} = h_{ie} + \frac{(1+h_{fe})R_e}{1+h_{oe}R_e}$$

(b) $$h'_{re} = \frac{h_{re}+h_{oe}R_e}{1+h_{oe}R_e}$$

(c) $$h'_{fe} = \frac{h_{fe}-h_{oe}R_e}{1+h_{oe}R_e}$$

(d) $$h'_{oe} = \frac{h_{oe}}{1+h_{oe}R_e}$$

(e) To what these expressions reduce if $h_{oe}R_e << 1$.

Solution Replacing the transistor with its h-parameter model,

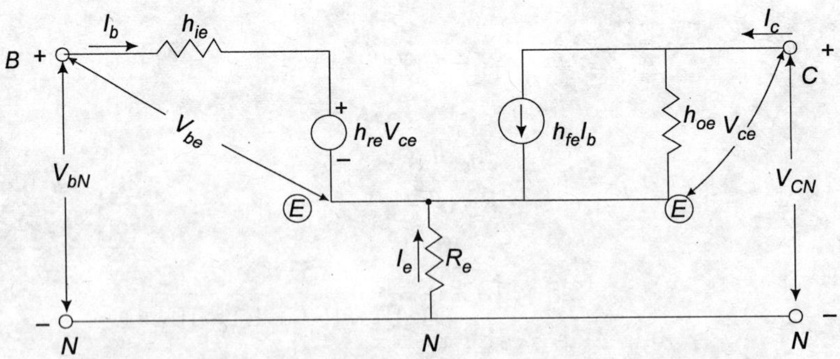

(a) $$h'_{ie} = \left.\frac{V_{bN}}{I_b}\right|_{V_{CN}=0}$$

when $\quad V_{CN} = 0,\ V_{ce} = -(I_b + I_c)R_e$ ...(1)

But $\quad I_c = h_{fe}I_b + h_{oe}V_{ce}$ ...(2)

$\therefore$ from (1) and (2) $V_{ce} = -I_bR_e - h_{fe}I_bR_e - h_{oe}V_{ce}R_e$

$$V_{ce}(1 + h_{oe}R_e) = -I_b(1 + h_{fe})R_e$$

$$\therefore \quad V_{ce} = \frac{-I_bR_e(1+h_{fe})}{1+h_{oe}R_e} = -V_{eN}$$

$$V_{bN} = V_{be} + V_{eN}$$

$$= h_{ie}I_b + h_{re}V_{ce} + \frac{I_bR_e(1+h_{fe})}{1+h_{oe}R_e}$$

$$= I_b\left(h_{ie} + \frac{R_e(1+h_{fe})}{1+h_{oe}R_e}\right) + hr_e\left(\frac{-I_bR_e(1+h_{fe})}{1+h_{oe}R_e}\right)$$

$$= I_b\left\{h_{ie} + \frac{(1-hr_e)(1+h_{fe})R_e}{1+h_{oe}R_e}\right\}$$

$$h'_{ie} = \frac{V_{bN}}{I_b} = h_{ie} + \frac{(1-h_{re})(1+h_{fe})R_e}{1+h_{oe}R_e}$$

Since $h_{re} << 1$.

$$\therefore \quad h_{ie} \cong h_{ie} + \frac{(1+h_{fe})R_e}{1+h_{oe}R_e}$$

(b) $$h'_{re} = \left.\frac{V_{bN}}{V_{cN}}\right|_{I_b=0}$$

$$V_{bN} = V_{be} + V_{eN} = h_{ie}I_b + h_{re}V_{ce} + V_{eN}$$

Since $I_b = 0,\ V_{bN} = h_{re}V_{ce} + V_{en}$

Again $I_c = h_{fe}I_b + h_{oe}V_{ce} = h_{oe}V_{ce} = \dfrac{V_{eN}}{R_e}$

$$\therefore \quad \boxed{V_{ce} = \frac{V_{eN}}{h_{oe}R_e}} \qquad ...(1)$$

From figure

$$V_{cN} = V_{ce} + V_{eN}$$

$$= \frac{V_{eN}}{h_{oe}R_e} + V_{eN} = V_{eN}\left(\frac{1+h_{oe}R_e}{h_{oe}R_e}\right)$$

$$V_{eN} = \frac{V_{cN}h_{oe}R_e}{1+h_{oe}R_e} \qquad ...(2)$$

Thus $V_{bN} = V_{be} + V_{eN}$

$$= h_{re}V_{ce} + V_{eN}$$

$$= \frac{h_{re} \cdot V_{eN}}{h_{oe} R_e} + \frac{h_{oe} R_e V_{cN}}{1 + h_{oe} R_e} \quad \text{(From (2))}$$

$$= \frac{h_{re} \cdot h_{oe} R_e \cdot V_{cN}}{(h_{oe} R_e)(1 + h_{oe} R_e)} + \frac{h_{oe} R_e V_{cN}}{1 + h_{oe} R_e}$$

$$\therefore \quad V_{bN} = V_{cN} \left\{ \frac{h_{re} + h_{oe} R_e}{1 + h_{oe} R_e} \right\}$$

$$\therefore \quad \boxed{h'_{re} = \frac{V_{bN}}{V_{cN}} = \frac{h_{re} + h_{oe} R_e}{1 + h_{oe} R_e}}$$

(c) $$h'_{fe} = \left. \frac{I_c}{I_b} \right|_{V_{cN} = 0}$$

We have $\quad I_c = h_{fe} I_b + h_{oe} V_{ce}$

With $\quad V_{cN} = 0, \; V_{ce} = \dfrac{-I_b R_e (1 + h_{fe})}{1 + h_{oe} R_e}$

$$\therefore \quad I_c = I_b \left[h_{fe} - \frac{h_{oe} R_e (1 + h_{fe})}{1 + h_{oe} R_e} \right]$$

$$= \left[\frac{h_{fe} - h_{oe} R_e}{1 + h_{oe} R_e} \right] I_b$$

$$\therefore \quad \boxed{h'_{fe} = \frac{h_{fe} - h_{oe} R_e}{1 + h_{oe} R_e}}$$

(d) $$h'_{oe} = \left. \frac{I_c}{V_{cN}} \right|_{I_b = 0}$$

From figure $\quad I_c = \dfrac{V_{cN}}{R_e}$

If $I_b = 0$, we have from eqn. (2)

$$V_{eN} = \frac{h_{oe} R_e V_{cN}}{1 + h_{oe} R_e}$$

$$\therefore \quad I_c = \frac{h_{oe} R_e V_{cN}}{(1 + h_{oe} R_e) R_e} = \frac{h_{oe} \cdot V_{cN}}{1 + h_{oe} R_e}$$

$$\therefore \quad \boxed{\frac{I_c}{V_{cN}} = h'_{oe} = \frac{h_{oe}}{1 + h_{oe} R_e}}$$

(e) If $hoe\,R_e << 1$, then these equations reduce to

$$h'_{ie} = h_{ie} + (1 + h_{fe}) R_e$$
$$h'_{re} = h_{re} + h_{oe} R_e$$
$$h'_{fe} = h_{fe} - h_{oe} R_e \approx h_{fe}$$
$$h'_{oe} = h_{oe}$$

7.7 For any single-stage amplifier express Input Resistance in terms of current gain and h-parameters only.

Solution We know that $R_i = h_i + h_r \cdot A_I \cdot R_L$

$$\text{and, } A_V = \frac{A_I R_L}{R_i}$$

$$\therefore \quad R_L = \frac{A_V \cdot R_i}{A_I}$$

Substituting R_L in the expression for R_i, we get

$$R_i = h_i + h_r \cdot A_I \cdot \frac{A_V \cdot R_i}{A_I}$$

$$\Rightarrow \quad R_i(1 - h_r \cdot A_I) = h_i$$

$$\therefore \quad R_i = \frac{h_i}{1 - h_r \cdot A_I}$$

7.8 Draw the equivalent circuit for the CE and CC configurations subject to the restriction that $R_L = 0$. Show that the input impedances of the two circuits are identical.

Solution CE hybrid model with $V_{CE} = 0$ is

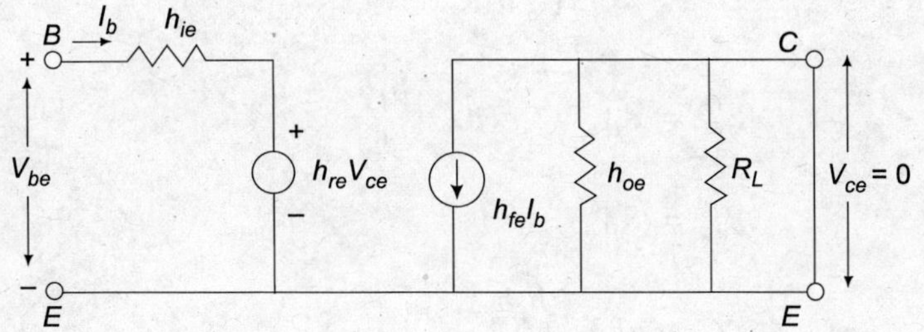

From the circuit input impedance

$$R_i = \left.\frac{V_{be}}{I_b}\right|_{V_{ce}=0} = h_{ie} \quad \text{...(1)}$$

CC h-parameter model with $R_L = 0$ is

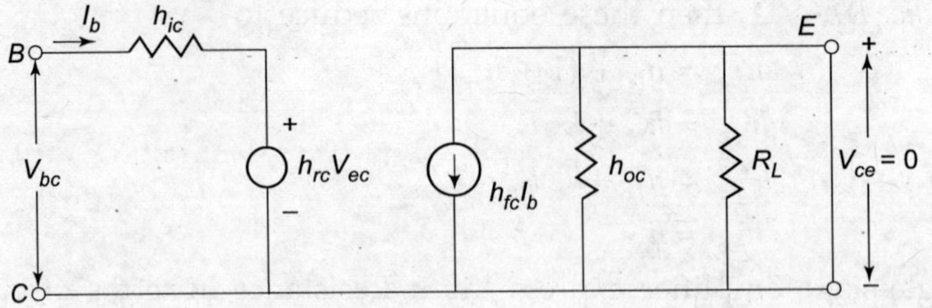

Writing KVL in the input loop

$$-V_{bc} + I_b h_{ic} = 0$$

$$\therefore \qquad R_i' = \frac{V_{bc}}{I_b} = h_{ic} \qquad \text{...(2)}$$

But we know that $h_{ic} = h_{ie}$

$\therefore$ From (1) and (2) $R_i = R'_i$

7.9 Prove that $Y_o = h_o\left(\dfrac{R_s + R_{i\infty}}{R_s + R_{io}}\right)$ where $R_{i\infty} \equiv R_i$ for $R_L \equiv \infty$ and $R_{io} \equiv R_i$ for $R_L = 0$.

Solution We have an expression for $Y_o = h_o - \dfrac{h_f h_r}{R_s + h_i}$

and,
$$R_i = h_i + h_r A_I R_L$$

$$\therefore \qquad R_i = h_i + h_r\left(\frac{-h_f}{1 + h_o R_L}\right)R_L$$

$$= h_i - \frac{h_f h_r}{1 + h_o R_L} \cdot R_L$$

$$R_{i\infty} = R_i\big|_{R_L = \infty} = h_i - \frac{h_f h_r}{ho} \qquad \text{...(1)}$$

$$R_{io} = R_i\big|_{R_L = 0} = h_i \qquad \text{...(2)}$$

$$\therefore \qquad Y_o = h_o - \frac{h_f h_r}{R_s + h_i}$$

$$= \frac{h_i h_o - h_f h_r + R_s h_o}{R_s + h_i}$$

From (1), $h_o h_i - h_f h_r = h_o R_{i\infty}$

$$\therefore \quad Y_o = \frac{h_o R_{i\infty} + R_s h_o}{R_s + R_{io}} \quad (\because\ R_{io} = h_i)$$

$$= h_o \left[\frac{R_s + R_{i\infty}}{R_s + R_{io}} \right]$$

7.10 For the emitter follower with R_s = 0.5 K, R_L = 5K, calculate A_I, R_i, A_V, A_{Vs} and R_o. Assume hf_e = 50, hi_e = 1 kΩ, $h_{oe} = h_{re} = 0$.

Solution

$$A_I = \frac{1 + h_{fe}}{1 + h_{oe} R_L} = \frac{1 + 50}{1} = 51$$

$$R_i = h_{ie} + A_I R_L = 1000 + 51 \times 5 \times 10^3$$

$$= 256 \text{ K}\Omega$$

$$A_V = 1 - \frac{h_{ie}}{R_i} = 1 - \frac{1 \times 10^3}{256 \times 10^3} = 0.996$$

$$A_{Vs} = A_V \cdot \frac{R_i}{R_i + R_s} = 0.9960 \times \frac{256}{256 + 0.5}$$

$$= 0.994$$

$$Y_o = h_{oe} + \frac{1 + h_{fe}}{R_s + h_{ie}}$$

$$= 0 + \frac{1 + 50}{(0.5 + 1) \times 10^3}$$

$$= 34 \text{ mA/V.}$$

$$R_o = \frac{1}{Y_o} = 29.411\ \Omega.$$

7.11 The transistor amplifier shown uses a transistor with typical h parameter values. Calculate $A_I = I_o/I_i$, A_V, A_{Vs}, R_o and R_i.

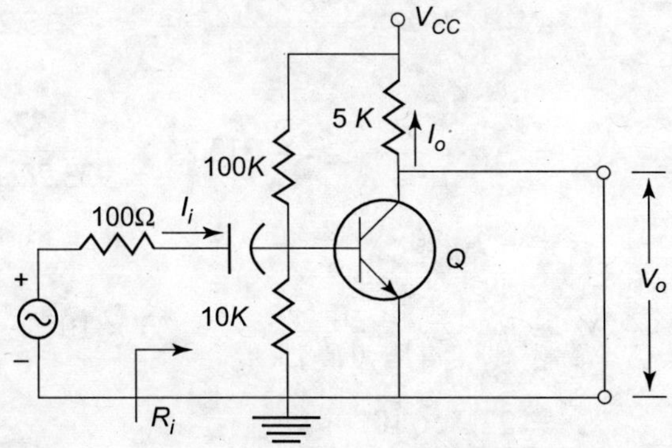

Solution By killing all dc sources, and short circuiting capacitances the circuit becomes.

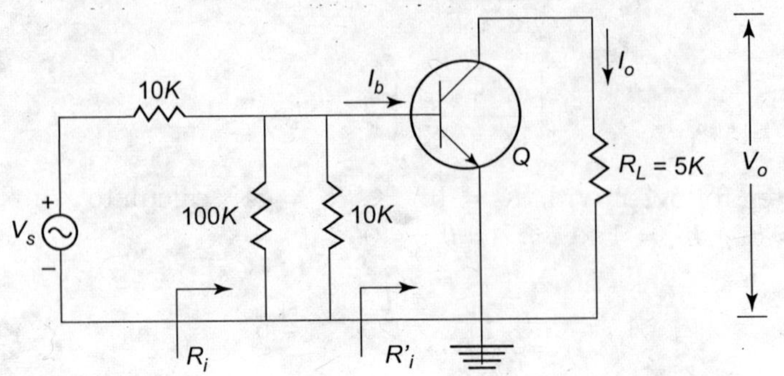

Let $$A_I' = \frac{I_o}{I_b} = \frac{-h_{fe}}{1 + h_{oe} R_L}$$

$$= \frac{-50}{1 + 25 \times 10^{-6} \times 5 \times 10^3} = -44.44$$

$$R'_i = h_{ie} + h_{re} A_I' R_L$$

$$= 1.14 + 2.5 \times 10^{-4} \times -44.44 \times 5 \times 10^3$$

$$= 1.04 \text{ k}\Omega$$

Parallel combination of 100 K and 10 K results 9.09 kΩ = R_{12}

$$A_I = \frac{I_o}{I_i} = \frac{I_o}{I_b} \cdot \frac{I_b}{I_i}$$

$$= A_I' \cdot \frac{R_{12}}{R_{12} + R_i'} = (-44.5) \frac{9.09}{9.09 + 1.04}$$

$$= -39.93$$

$$R_i = R_{12} || R'_i = 9.09 \text{ K} || 1.04 \text{ K}$$

$$= 933.22 \ \Omega$$

$$A_V = \frac{A_i R_L}{R_i} = \frac{-39.8 \times 5 \times 10^3}{933.22} = -213.24$$

$$A_{VS} = A_V \cdot \frac{R_i}{R_i + R_s} = -213.24 \times \frac{933.22}{933.22 + 10 \text{ K}}$$

$$= -18.20$$

$$Y_o = h_{oe} - \frac{h_{fe} h_{re}}{(R_{12} || R_s) + h_{ie}} = 25 \times 10^{-6} - \frac{50 \times 2.5 \times 10^{-4}}{1.1 \text{ K} + 4.76 \text{ K}}$$

$$= 22.9 \times 10^{-6} \ \mho$$

$$R_o = 1/Y_o = 43.7 \text{ k}\Omega$$

7.12 Find the input impedance for the circuit shown, in terms R_L and R_e, h-parameters. (using exact h-parameter model)

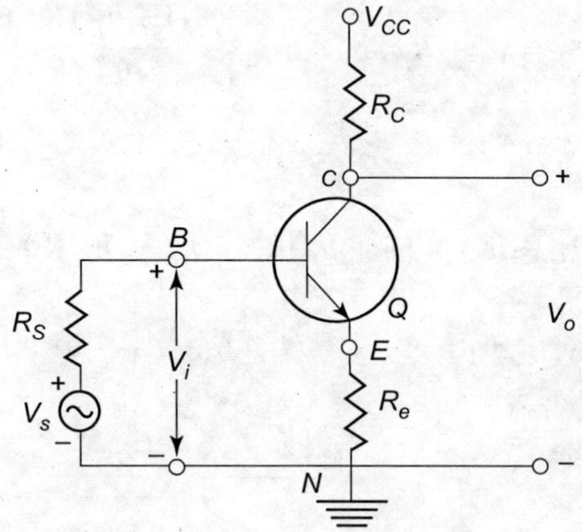

Solution Replacing transistor with its h-parameter model, the circuit will become as below:

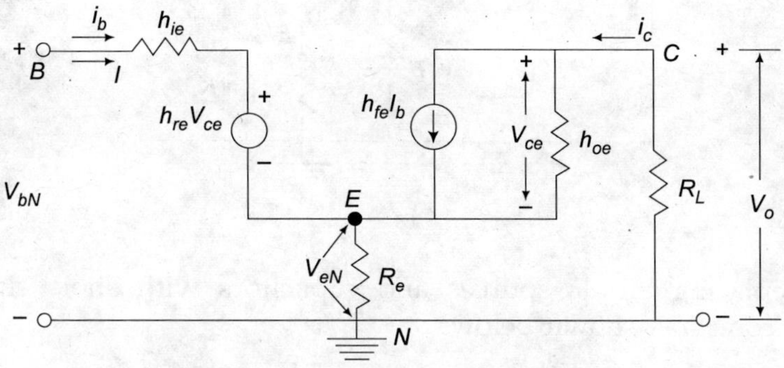

$$R_i = \frac{V_{bN}}{I} = \frac{h_{ie}\,I + h_{re}\,V_{ce} + V_{eN}}{I} = h_{ie} + h_{re} \cdot \frac{V_{ce}}{I} + \frac{V_{eN}}{I}$$

$$V_{eN} = (I + h_{fe}I + V_{ce}h_{oe})\,R_e$$

$$V_{ce} = -\,V_{eN} - R_L(h_{fe}I + V_{ce} \cdot h_{oe})$$

$$= -\,(I + h_{fe}I + V_{ce}h_{oe})\,R_e - R_L(h_{fe}I + V_{ce} \cdot h_{oe})$$

$$V_{ce}(1 + h_{oe}R_e + h_{oe}R_L) = -\,I((1 + h_{fe})R_e + h_{fe}R_L)$$

$$\therefore \quad V_{ce} = -\,I\,\frac{(1+h_{fe})R_e + h_{fe}R_L}{1+h_{oe}(R_e + R_L)} \qquad ...(1)$$

$$\frac{V_{ce}}{I} = \frac{-\,(1+h_{fe})\,R_e + h_{fe}\,R_L}{1+h_{oe}\,(R_e + R_L)} \qquad ...(2)$$

$$\frac{V_{eN}}{I} = (1 + h_{fe})R_e + h_{oe} \cdot \frac{V_{ce}}{I}$$

$$= (1 + h_{fe})R_e + h_{oe}\left\{\frac{-(1+h_{fe})R_e + h_{fe}R_L}{1+h_{oe}(R_e + R_L)}\right\} \quad ...(3)$$

From (2) and (3)

$$R_i = h_{ie} + (1 + h_{fe})R_e - (h_{re} + h_{oe}R_e)\,\frac{(1+h_{fe})R_e + h_{fe}R_L}{1+h_{oe}(R_e + R_L)}$$

7.13 For the amplifier shown, find $A_I = I_o/I_i$, A_V, A_{VS} and R_i use typical h-parameter values.

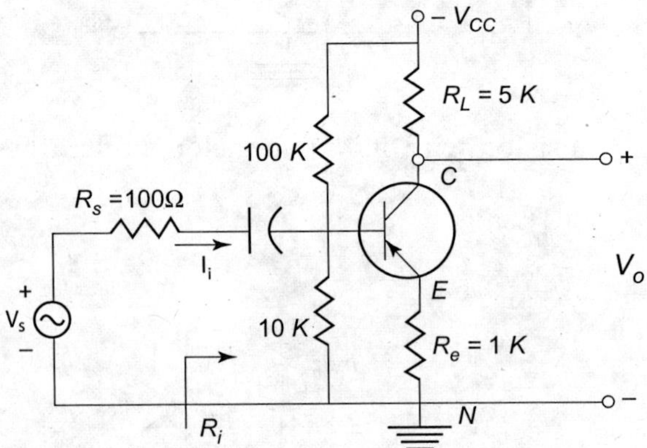

Solution 1. Replacing all dc sources and capacitors with short circuits, the circuit becomes as shown below:

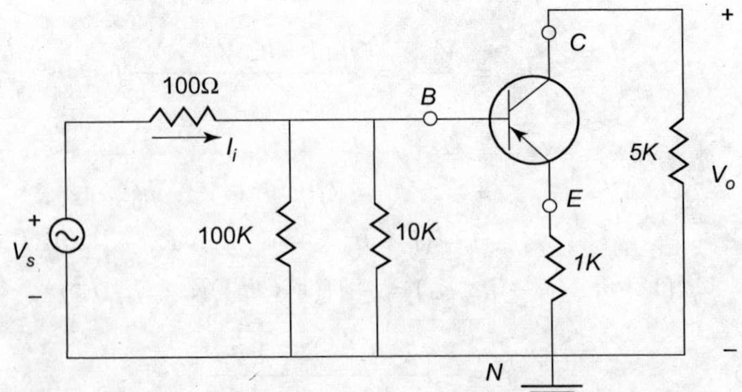

2. Replacing the transistor with its exact h-parameter model, and replacing 10 K || 100 K with an equivalent of 9.09 kΩ,

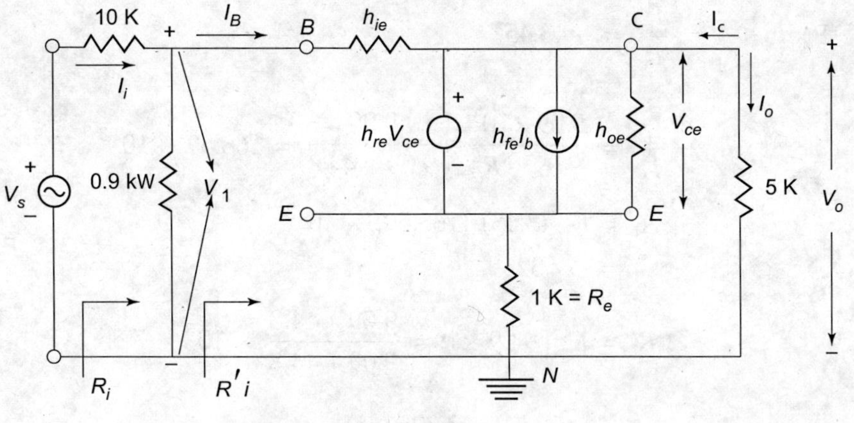

From the circuit $I_o = -(h_{fe}I_b + h_{oe}V_{ce})$

$$-I_o = 50\,I_b + 25 \times 10^{-6}V_{ce} \qquad ...(1)$$

$$V_{ce} = V_o - I_eR_e = V_o - (I_b - I_o)R_e$$

$$= V_o - 1000\,(I_b - Io) \qquad ...(2)$$

$$V_o = I_oR_L = 5000\,I_o \qquad ...(3)$$

Substituting equations (2) and (3) into (1).

$$-I_o = 50\,I_b + 25 \times 10^{-6}(5000\,I_0 - 1000\,I_b + 1000\,I_o)$$

$$= 50I_b + 25 \times 10^{-6}(6000\,I_0 - 1000\,I_b)$$

$$= 50\,I_b + 0.15\,I_o - 0.025\,I_b$$

$$-1.15\,I_o = 49.975\,I_b$$

$$I_o = -43.45\,I_b \qquad ...(4)$$

Again, $\quad V_1 = h_{ie}I_b + h_{re}V_{ce} + (I_b - I_c)R_e$

From (2), (3) and (4),

$$V_{ce} = 5000\,(I_o) - 1000\,(I_b - I_o)$$

$$= 6000\,I_o - 1000\,I_b$$

$$= -1888.18\,I_b - 1000\,I_b$$

$$V_{ce} = -2888.18\,I_b \qquad ...(5)$$

$$V_1 = 1100\,I_b + 2.5 \times 10^{-4}(-2.88 \times 10^3 I_b) + (44.45)\,I_b \times 10^3$$

$$= I_b(1.1 + 44.45 - 7.22 \times 10^{-4})\,10^3$$

$$= I_b(44.828) \times 10^3$$

$$R_i^1 = \frac{V_1}{I_b} = 44.8 \text{ k}\Omega$$

$$A_I = \frac{I_o}{I_i} = \frac{I_o}{I_b}\cdot\frac{I_b}{I_i}$$

From (4) $\dfrac{I_o}{I_b} = -\,43.45$

$$\frac{I_b}{I_i} = \frac{9.09 \times 10^3}{9.09 + R_i^1} = \frac{9.09}{9.09 + 44.8} = 168.59 \times 10^{-3}$$

$$\therefore \quad A_I = (-\,43.45)\ (168.59) \times 10^{-3}$$

$$= -\,7.325$$

$$R_i = R_i^1 \| 9.09\ \text{k}\Omega = \frac{9.09 \times 44.8}{9.09 + 44.8} \times 10^3 = 7.55\ \text{k}\Omega$$

$$A_V = \frac{A_I R_L}{R_i} = \frac{-7.325 \times 5}{7.55} = -\,4.84$$

$$A_{Vs} = A_V \cdot \frac{R_i}{R_i + R_s} = -\,4.84 \times \frac{7.55 \times 10^3}{(7.55 + 10) \times 10^3}$$

$$= -\,2.085$$

The resulting values are

$$A_I = -\,7.325$$

$$R_i = 7.55\ \text{k}\Omega$$

$$A_V = -\,4.84$$

$$A_{Vs} = -\,2.085$$

7.14 Derive the equation for voltage gain, V_o/V_s for the circuit shown in terms of R_s, R_e and R_L. Assume simplifial CE modle of transistor $[h_{oe}(R_e + R_L) \le 0.1.]$

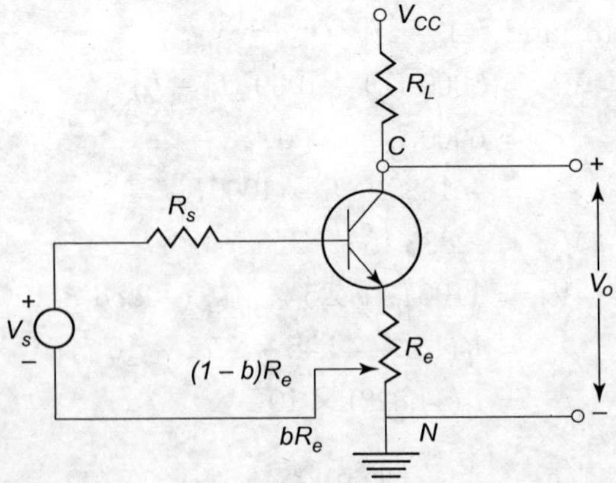

Solution Since $h_{oe}(R_e + R_L) \le 0.1$, approximate h-parameter model can be used.

$$V_o = -\,h_{fe} I_b R_L$$

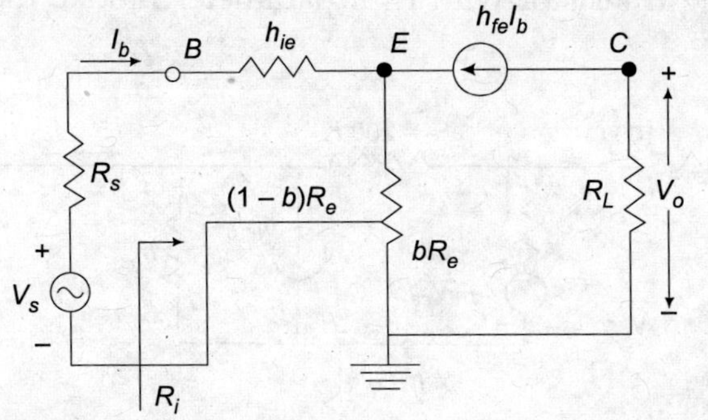

$$V_s = (R_s + h_{ie})I_b + h_{fe}I_b\,(1 - b)R_e + (1 - b)R_e I_b$$

$$A_V = \frac{V_o}{V_s} = \frac{-h_{fe}\,I_b\,R_L}{(R_s + h_{ie})I_b + h_{fe}\,I_b\,(1-b)R_e + (1-b)R_e\,I_b}$$

$$A_V = \frac{V_o}{V_s} = \frac{-h_{fe}\,R_L}{(R_s + h_{ie}) + (1 + h_{fe})(1-b)R_e}$$

Input impedance $R_i = \dfrac{V_s}{I_b}$

$$= \frac{I_b\left\{(R_s + h_{ie}) + (1 + h_{fe})(1-b)R_e\right\}}{I_b}$$

$$= (R_s + h_{ie}) + (1 + h_{fe})\,(1 - b)R_e$$

7.15 Calculate $A_I = I_o/I_i$, A_V, A_{Vs} and R_i for the circuit shown.

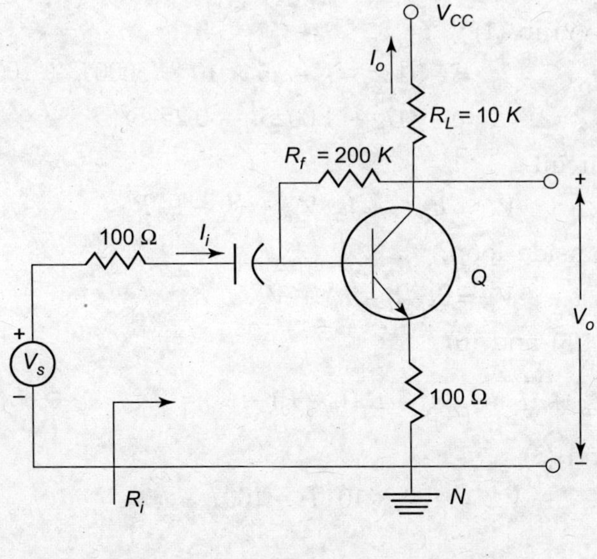

Solution After replacing transistor with its h parameter model, the circuit will become as shown below:

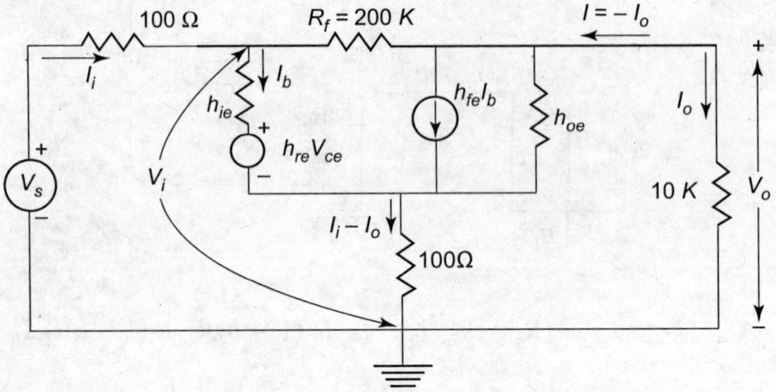

From the circuit

$$I_o = (I_i - I_b) - I$$

where $$I = h_{fe}\, I_b + h_{oe}\, V_{ce}$$

$$I_o = (I_i - I_b) - h_{fe}\, I_b - h_{oe}\, V_{ce}$$

$$= I_i - I_b - 50\,I_b - 25 \times 10^{-6} V_{ce}$$

$$= -\,51\,I_b + I_i - 25 \times 10^{-6}\; V_{ce} \qquad ...(1)$$

$$V_{ce} = V_o - (I_i - I_o)\,R_e$$

$$= V_o + (I_o - I_i)\,R_e$$

$$= V_o + 100\,I_o - 100\,I_i \qquad ...(2)$$

Also, from figure

$$V_o = I_o \times 10 \times 10^3 = 10^4\; I_o \qquad ...(3)$$

Put (3) and (2) in (1)

$$I_o = -\,51\,I_b + I_i - 25 \times 10^{-6}\;(5000\,I_o + 100I_o - 100I_i)$$

$$= -\;51I_b + 1.0025I_i - 0.2525I_o \qquad ...(4)$$

From the circuit,

$$V_i = h_{ie}I_b + h_{re}\,V_{ce} + (I_i - I_o)\,R_e \qquad ...(5)$$

From the outside loop,

$$V_i = R_f\,(I_i - I_b) + V_o \qquad ...(6)$$

Subtracting (5) and (6).

$$h_{ie}I_b + h_{re}\,V_{ce} + (I_i - I_o)\,R_e - R_f(I_i - I_o) - V_o = 0 \qquad ...(7)$$

From (2) and (3)

$$V_{ce} = 100 \times 101\,I_o - 100\,I_i \qquad ...(8)$$

Put (8) in (7)

$1100I_b + 2.5 \times 10^{-4}\,(101 \times 100\,I_o - 100\,I_i) + 100\,(I_i - I_o) - 200 \times 10^3(Ii - I_b) - 10^4 I_o =$ 0

$\Rightarrow I_b(1100 + 200 \times 10^3) + I_i\,(-\,2.5 \times 10^{-4} \times 100 + 100 - 200 \times 10^3)$

$+\,I_o\,(2.5 \times 10^{-4} \times 100 \times 101 - 100 - 10^4) = 0$

$\Rightarrow I_b(201.1 \times 10^3) + I_i(-\,0.025 + 100 - 200 \times 10^3)$

$+\,I_o(2.525 - 100 - 10^4) = 0$

$\Rightarrow I_b(201.1 \times 10^3) + I_i(-\,199900.025) + I_o(-\,10097.475) = 0$

$$\Rightarrow \quad \boxed{I_b = 0.99I_i + 0.0503I_o} \qquad \text{...(9)}$$

$\therefore$ From (9) and (4)

$$I_o = -\,51(0.99I_i + 0.0503I_o) + 1.0025\,I_i - 0.2525I_o$$
$$= -49.487I_i - 2.767\,I_o$$
$$I_o = -\,13.135\,I_i \qquad \text{...(10)}$$

From (9) and (10)

$$I_b = 0.99\,I_i + 0.0503\,(-\,13.135I_i)$$
$$= (0.99 - 0.6607)I_i = 0.3293I_i \qquad \text{...(11)}$$

So,

$$I_b = 0.33I_i$$
$$I_o = -\,13.135I_i$$

From (6) , (11) and (10)

$$V_i = 200 \times 10^3\,(1 - 0.33)\,I_i + 10^4\,(-\,13.135)\,I_i$$
$$= (200 \times 10^3 \times 0.67 - 131.35 \times 10^3)\,I_i$$
$$= 2.65 \times 10^3 I_i$$

$$R_i = \frac{V_i}{I_i} = 2.65\ \text{k}\Omega$$

$$A_V = \frac{A_I \cdot R_L}{R_i} = \frac{\frac{I_o}{I_i} \cdot R_L}{R_i} = \frac{-\,13.135 \times 10 \times 10^3}{2.65 \times 10^3}$$

$$= -\,49.56$$

$$A_{Vs} = \frac{A_V \cdot R_i}{R_i + R_s} = \frac{-\,49.56 \times 2.65}{2.65 + 10} = -\,10.383$$

7.16 Calculate $R_i = V_s/I_i$, A_V and $A_I = -\,I_o/I_i$ for the circuit shown here. $h_{ic} = 1.1$ kΩ, $h_{fc} = -\,51$, $h_{oc} = 0$ A/V and $h_{rc} = 1$

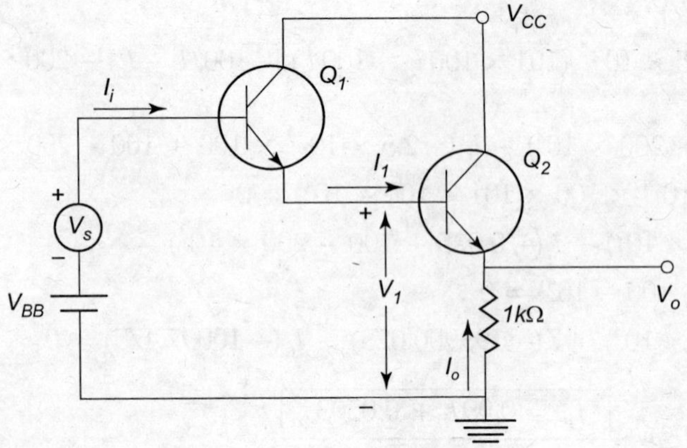

Solution First stage is an emitter follower, second is also an emitter stage. Let A_{I1}, A_{I2}, be the current gians of two stages. R_{i1}, R_{i2} be their input impedances.

$$A_{I2} = \frac{-h_{fc}}{1+h_{oc}\cdot R_L} = \frac{-(-51)}{1+0} = 51$$

$$R_{i2} = h_{ic} + h_{rc}A_{I2}R_L = 1.1 + 1 \times 51 \times 1$$
$$= 52.1 \text{ k}\Omega$$

$$A_{V2} = A_{I2}\cdot\frac{R_L}{R_{i2}} = (51)\left(\frac{1}{52.1}\right) = 0.9788$$

$$A_{I1} = \frac{-h_{fc}}{1+h_{oc}\cdot R_L}$$

Here, load resistance of first stage is the input resistance of second stage,

$$\therefore \quad A_{I1} = \frac{-(-51)}{1+25\times10^{-6}\times52.1\times10^3}$$
$$= 22.15$$

$$Ri_1 = h_{ic} + h_{rc}\cdot A_{I1}\cdot R_L$$
$$= 1.1 + 1 \times 22.15 \times 52.1$$
$$= 1155.10 \text{ k}\Omega$$

$$A_{V1} = A_{I1}\cdot\frac{R_{L1}}{R_{i1}} = 22.45 \times \frac{52.1}{1155.10} = 1.012$$

$$\therefore \quad A_V = A_{V1}\cdot A_{V2} = 0.9905$$
$$A_I = A_{I1}\cdot A_{I2} = 1129.65$$

7.17 When a transistor is biased at 20 mA, 20 V it has following h-parameters at room temperature $h_{ie} = 500\ \Omega$, $h_{fe} = 100$, $h_{re} = 10^{-4}$, $h_{oe} = 4 \times 10^{-5}\ \mho$. It has $f_T = 50$ MHz and $C_c = 3$ pF. Find all the values of hybrid π components.

Solution

$$g_m = \frac{|I_c|}{V_T} = \frac{20 \times 10^{-3}}{0.026} = 0.7692$$

$$r_{b'e} = \frac{h_{fe}}{g_m} = \frac{100}{0.3846} = 130\ \Omega$$

$$r_{bb'} = h_{ie} - r_{b'e} = 500 - 130 = 370\ \Omega$$

$$r_{b'c} = \frac{r_{b'e}}{h_{re}} = \frac{370}{10^{-4}} = 3.7\ \text{M}\Omega$$

$$g_{ce} = h_{oe} - (1 + h_{fe})\, g_{b'c}$$

$$= 4 \times 10^{-5} - (101) \times \frac{1}{3.7 \times 10^6}$$

$$= 12.70 \times 10^{-6}$$

$$\therefore \quad r_{ce} = 8.66 \times 10^5\ \Omega = 78.723\ \text{k}\Omega$$

$$C_e = \frac{g_m}{2\pi f_T} = \frac{0.7692}{2 \times 3.14 \times 50 \times 10^6} = 2452\ \text{pF}$$

7.18 A transistor biased at 5 mA, 10 V, $h_{ie} = 600\ \Omega$, $h_{fe} = 100$ $C_C = 3$ PF and current gain of 10 at a frequency of 20 MHz. Find β cut off frequency, gain band width product, C_e, $r_{b'e}$ and $r_{bb'}$.

Solution Magnitude of current gain as a function of frequency is given by

$$|A_i(f)| = \frac{h_{fe}}{\sqrt{1 + (f/f_\beta)^2}}$$

$$10 = \frac{100}{\left[1 + \left(\frac{20}{f_\beta}\right)^2\right]^{1/2}}$$

$$\therefore \quad \sqrt{1 + \left(\frac{20}{f_\beta}\right)^2} = 10$$

$$1 + \frac{400}{f_\beta^{\,2}} = 100$$

$$\therefore \quad \frac{20}{f_\beta} = \sqrt{99} = 9.949$$

$$f_\beta = 2.010 \text{ MHz}$$

$$f_T = h_{fe} \cdot f_\beta = 201.02 \text{ MHz}$$

$$g_m = \frac{I_c}{V_T} = \frac{5 \times 10^{-3}}{0.026} = 0.192 \text{ mho}$$

$$C_e = \frac{g_m}{2\pi f_T} = \frac{0.192}{2 \times 3.14 \times 100.5 \times 10^6}$$

$$= 304.21 \text{ pF}$$

$$r_{b'e} = \frac{h_{fe}}{g_m} = \frac{100}{0.1923} = 520.02\ \Omega$$

$$r_{bb'} = h_{ie} - r_{b'e} = 600 - 520 = 80\ \Omega$$

7.19 Find the base width of a silicon *pnp* trasistor whose figure of merit is 450 MHz. Diffusion length of holes is 15 cm^2/sec.

Solution Figure of merit $= f_T = 450$ MHz

$$f_T = \frac{g_m}{2\pi C_e}$$

We have $$C_e = \frac{g_m w^2}{2D_B}$$

$$\therefore \quad f_T = \frac{g_m \times 2D_B}{2\pi \cdot g_m w^2} = \frac{D_B}{\pi w^2}$$

$$\Rightarrow \quad w^2 = \frac{D_B}{\pi f_T} \Rightarrow w = \sqrt{\frac{D_B}{\pi f_T}}$$

$$w = \sqrt{\frac{15}{3.14 \times 450 \times 10^6}} = 1.030 \times 10^{-4} \text{ cm}$$

$$= 1.030 \text{ microns.}$$

7.20 When a Ge *pnp* transistor is biased at 2 mA, 15 V, it has a base width of 2 μm. Find C_e and f_T given that $D_B = 50$ cm^2/sec.

Solution $$g_m = I_c/V_T = \frac{2 \times 10^{-3}}{26 \times 10^{-3}} = 0.0769 \text{ mho}$$

$$C_e = \frac{g_m w^2}{2D_B} = \frac{0.0769 \times \left(2\times10^{-4}\right)^2}{2\times 50}$$

$$= 30.76 \text{ pF}$$

$$f_T = \frac{g_m}{2\pi C_e} = \frac{0.0769}{2\times 3.14\times 30.76\times 10^{-12}}$$

$$= 398.09 \text{ MHz}$$

7.21 The 3-db bandwidth of an amplifier extends from 20 Hz to 20 kHz. Find the frequency range over which the voltage gain differs by only 1 db from the mid band value.

Solution $-1 \text{ dB} = 10^{-1/20} = 0.8916$

We have $A_V = \dfrac{A_{vo}}{\sqrt{1+(f/fH)^2}}$

Taking logarithm on both sides,

$$20 \log A_V = 20\, A_{Vo} - 10 \log [1 + (f_1 + f_H)^2]$$

at $f = f_2$, gain falls by 1 db ...(1)

$\therefore$ $20 \log A_V(f_2) = 20 \log A_{Vo} - 1$ ...(2)

From (1) and (2)

$$-1 = -10 \log_{10} [1 + (f_2/f_H)^2]$$

$$1 + \left(\frac{f_2}{f_H}\right)^2 = 10^{1/10} = 1.258$$

$$f_2 = 0.5079\, f_H$$

$$= 0.5079 \times 20 \text{ K} = 10.16 \text{ kHz}$$

at $f = f_1$, gain falls by 1 db.

$\therefore$ $20 \log A_V(f_1) = 20 \log A_{Vo} - 1$ ...(3)

From (2) and (1)

$$-1 = -10 \log (1+ (f_L/f_1)^2$$

$$1 + \left(\frac{20}{f_1}\right)^2 = 10^{1/10} = 1.258$$

$$(20/f_1) = 0.5079$$

$$f_1 = \frac{20}{0\,5079} = 39.37 \text{ Hz}$$

$\therefore$ 1 db bandwidth is from 39.37 to 10.16 kHz

7.22 It is desired to build an audio amplifier with a pass band of 20 Hz to 20 kHz and a mid-band gain of 64000. Three stages are used in cascade for this purpose. What should be the mid band gain and band width of each stage?

Solution For three stages $n = 3$

$$\therefore \quad \sqrt{2^{1/3} - 1} = 0.5098$$

We have f_H = Individual high 3 dB frequency

$$= \frac{f_H^*}{\sqrt{2^{1/n} - 1}} = \frac{20 \text{ kHz}}{\sqrt{2^{1/3} - 1}}$$

$$= \frac{20}{0.5098} = 39.23 \text{ KHz}$$

$$f_L = f_L^* \sqrt{2^{1/n} - 1} = 10.196 \text{ Hz}$$

$$\text{gain} = (64000)^{1/3} = 40$$

7.23 It is desired that the voltage gain of an *RC*-coupled amplifier at 60 Hz should not decrease by more than 10% from its mid-band value. Show that the coupling capacitance C must be at least equal to $5.5/R'$, where $R' = R_o' + R_i'$ is expressed in kΩ, and C in μF.

Solution We have

$$A_L = \frac{A_o}{\sqrt{1 + (f_L/f)^2}}$$

$$\therefore \quad \frac{A_L}{A_o} = \frac{1}{\sqrt{1 + \left(\frac{f_L}{f}\right)^2}} = 0.9$$

$$\therefore \quad \frac{1}{1 + \left(\frac{f_L}{f}\right)^2} = 0.81 \Rightarrow 1.2345 = 1 + \left(\frac{f_L}{f}\right)^2$$

$$\Rightarrow \quad 0.2345 = \left(\frac{f_L}{f}\right)^2 \Rightarrow f_L = 0.4843\, f$$

$$\therefore \quad f_L = 0.4843 \times 60 = 29.05 \text{ Hz}$$

$$f_L = \frac{1}{2\pi C_b R'} = 29.05$$

$$\Rightarrow \qquad C_b = \frac{1}{2\times 3.14\times 29.05\times R'} = \frac{5.47\times 10^{-6}}{R'}\ \text{F}$$

$$= \frac{5.5}{R'}\ \mu\text{F if } R' \text{ is in k}\Omega$$

7.24 (a) For the transistor stage shown with $1/h_{oe} = \infty$ calculate the percentage tilt in the output if the input current I is a 200 Hz square wave.

(b) What is the lowest-frequency square wave which will suffer less than 2% tilt?

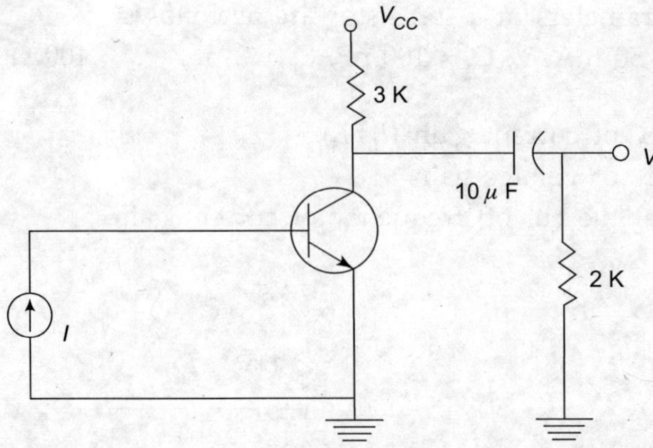

Solution (a) We have $f_L = \dfrac{1}{2\pi(R_o^1/R_i^1)C_b}$

$$R_o' = 3\ \text{K}\,||\,\infty = 3\ \text{k}\Omega$$

$$R_i' = 2\ \text{K}$$

$$C_b = 10\ \mu\text{F}$$

$$\therefore \qquad f_L = \frac{1}{2\pi\times(3+2)\times 10^3\times 10\times 10^{-6}}$$

$$= 3.1847$$

$$\%\ \text{tilt} = \frac{\pi f_L}{f}\times 100$$

$$= \frac{\pi\times 3.1847}{200}\times 100$$

$$= 4.99\% \approx 5\%$$

(b)

$$\% \text{ tilt} = \frac{\pi f_L}{f} \times 100$$

$$2 = \frac{100\pi f_L}{f}$$

$$\Rightarrow \quad f = 200\,\pi f_L$$

$$= 200 \times 3.14 \times 3.1847$$

$$\approx 2 \text{ kHz.}$$

7.25 The hybrid-π parameters of a transistor are as follows:
$r_{b'e} = 1$ kΩ, $g_m = 50$ mA/V, $C_e = 100$ pF, $C_c = 3$ pF, $r_{bb'} = 100$ Ω, g_{ce} and $g_{b'c}$ are very small. Find

(a) 3 dB frequency of current gain (I_L/I_s).

(b) 3 dB frequency of voltage gain.

(c) Voltage gain at the cut off frequency of current gain.

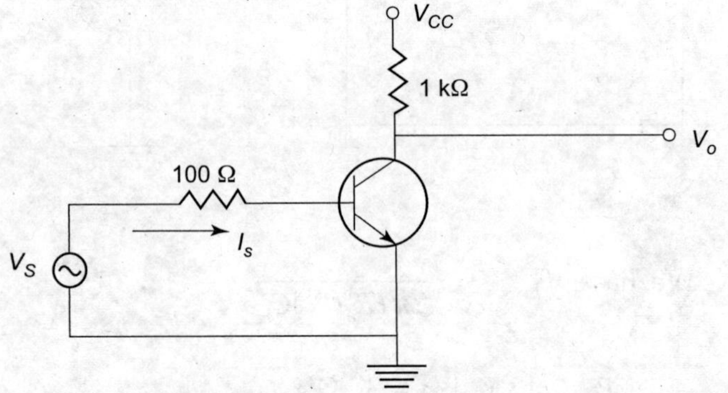

Solution Replacing transistor with its hybrid model, and using Miller's theorem,

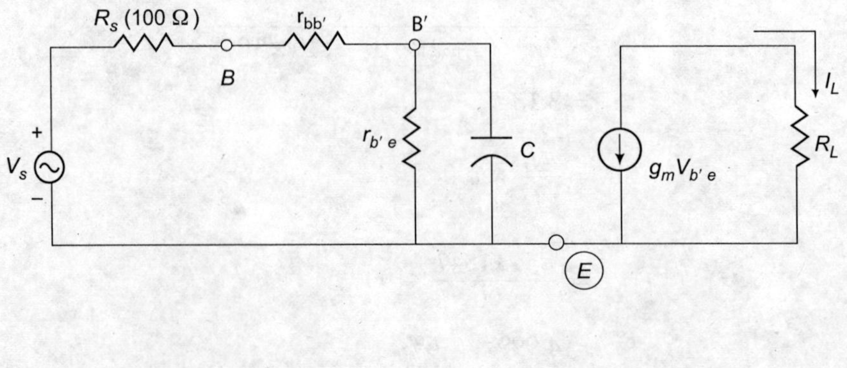

$$C = C_e + C_c(1 + g_m R_L)$$
$$= 100 + 3\ (1 + 50 \times 10^{-3} \times 10^3)$$
$$= 253 \text{ pF}$$

$$\text{Current gain} = \frac{I_L}{I_S} = \frac{-g_m V_{b'e}}{I_S}$$

$$V_{b'e} = \frac{r_{b'e} \cdot \dfrac{1}{SC}}{r_{b'e} + \dfrac{1}{Cs}} \cdot I_S$$

$$= \frac{r_{b'e}}{1 + r_{b'e}SC} I_S$$

$$\therefore \quad \text{Current gain } A_i = \frac{-g_m r_{b'e}}{1 + r_{b'e}SC}$$

Put $s = Jw$

$$A_i(Jw) = \frac{-g_m r_{b'e}}{1 + JwCr_{b'e}}$$

Let w_1 = cut off frequency

$$= \frac{1}{r_{b'e} \cdot C}$$

$$\therefore \quad A_i = \frac{-g_m r_{b'e}}{1 + \dfrac{Jw}{w_1}}$$

So,
$$f_1 = \frac{1}{2\pi r_{b'e} \cdot C} = \frac{1}{2 \times 3.14 \times 1000 \times 253 \times 10^{-12}}$$
$$= 0.629 \text{ MHz}$$

(b) Using Thevenin's theorem, with respect to $b'e$ terminals

$$R_{TH} = r_{b'e} \,||\, (R_s + r_{bb'})$$
$$= 1000 \,||\, (100 + 100) = 166.67\ \Omega$$

$$V_{TH} = V_s \cdot \frac{r_{b'e}}{r_{b'e} + R_s + r_{bb'}}$$

$$= V_s \cdot \frac{1\text{ K}}{2\text{ K}+200} = 0.833\ V_s$$

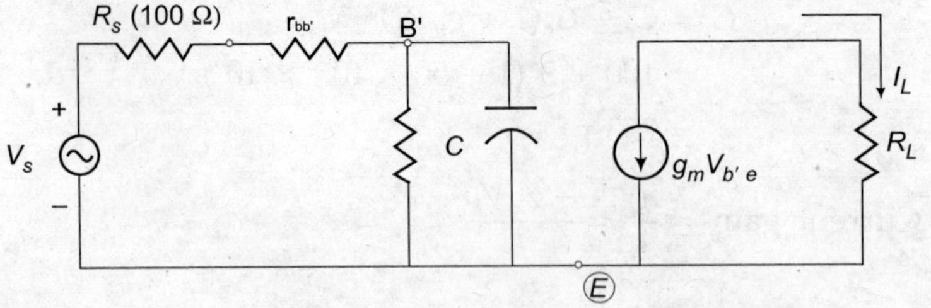

So, the circuit will become,

We can write f_H = 3 db freq. of voltage gain

$$= \frac{1}{2\pi R_{TH} \cdot C}$$

$$= \frac{1}{2 \times 3.14 \times 166.67 \times 253 \times 10^{-12}}$$

$$= 3.78 \text{ MHz}$$

(c) Voltage gain $\frac{V_o}{V_s} = \frac{-g_m V_{b'e} R_L}{V_s}$

We have $V_{b'e} = V_s \cdot \frac{r_{b'e}}{R_s + r_{b'e} + r_{bb'}}$

$$\therefore \quad \frac{V_o}{V_s} = -g_m R_L \cdot \frac{r_{b'e}}{R_s + r_{b'e} + r_{bb'}}$$

$$= -50 \times 10^{-3} \times 10^3 \times \frac{1000}{1000 + 200}$$

$$= -41.67$$

Writing this as a function of frequency $|A_v(f)| = \frac{25}{\sqrt{1 + (f/f_H)^2}}$

$$A_v(0.629) = \frac{41.67}{\sqrt{1 + \left(\frac{0.629}{1.259}\right)^2}} = 37.28$$

7.26 A transistor used in a *CE* amplifier has parameters h_{fe} = 100, g_m = 0.1 ℧, $r_{bb'}$ = 100 Ω, C_c = 1pf = f_T = 400 MHz.

(a) What should be the source resistance to get a 3 db bandwidth of 5 MHz with R_L = 500 Ω?

(b) Find midband voltage gain $\frac{V_o}{V_s}$.

Solution (a) We have $C = C_e + C_c(1 + g_m R_L)$ from the previous problem.

$$C_e = \frac{g_m}{2\pi f_T} = \frac{0.1}{2 \times 3.14 \times 400 \times 10^6}$$

$$= 0.0398 \text{ nF} = 39.80 \text{ pF}$$

$$\therefore \quad C = 39.8 + 1\ (1 + 0.1 \times 500)$$

$$= 90.8 \text{ pF}$$

$$\therefore \quad f_H = \frac{1}{2\pi RC} \text{ where } R = \frac{1}{2\pi f_H C}$$

$$R = \frac{1}{2 \times 3.14 \times 5 \times 10^6 \times 90.8 \times 10^{-12}}$$

$$= 350.\ 74\ \Omega$$

We have $R = (R_s + r_{bb'}) || r_{b'e}$

$$350.74 = \frac{(R_s + r_{bb'})\, r_{b'e}}{R_s + r_{bb'} + r_{b'e}}$$

$$350.74 = \frac{(R_s + 100)\,(r_{b'e})}{R_s + 100 + r_{b'e}} \qquad ...(1)$$

$$r_{b'e} = \frac{h_{fe}}{g_m} = \frac{100}{0.1} = 1000$$

Using the value of $r_{b'e}$ in (1).

$$350.74 = \frac{(R_s + 100)\, 1000}{R_s + 1100}$$

$$\Rightarrow R_s (1000 - 350 - 74) = -\ 100 \times 1000 + 1100 \times 350.74$$

$$\Rightarrow \quad R_s = 440.21\ \Omega$$

(b) Mid-band voltage gain $= \dfrac{V_o}{V_s} = \dfrac{V_o}{V_{b'e}} \cdot \dfrac{V_{b'e}}{V_s}$

$$= -\ g_m R_L \cdot \frac{r_{b'e}}{R_s + r_{b'e} + r_{bb'}}$$

$$= -\ 0.1 \times 500 \times \frac{1000}{1000 + 100 + 440.21}$$

$$= -\ 32.46$$

7.27 Consider a single stage *CE* amplifier with the load resistance R_L shunted by a capacitance C_L.

(a) Prove that the internal voltage gain $K = \dfrac{V_{ce}}{V_{b'e}}$ is

$$K \approx \frac{-g_m R_L}{1 + Jw(C_L + C_c)R_L}$$

(b) Prove that the 3-db frequency is given by

$$f_H = \frac{1}{2\pi\,(C_L + C_c)R_L}$$ provided that the condition is satisfied

$g_{b'e}R_L(C_c + C_L) >> C_e + C_c(1 + g_m R_L)$

Solution (a) The circuit will be as shown:

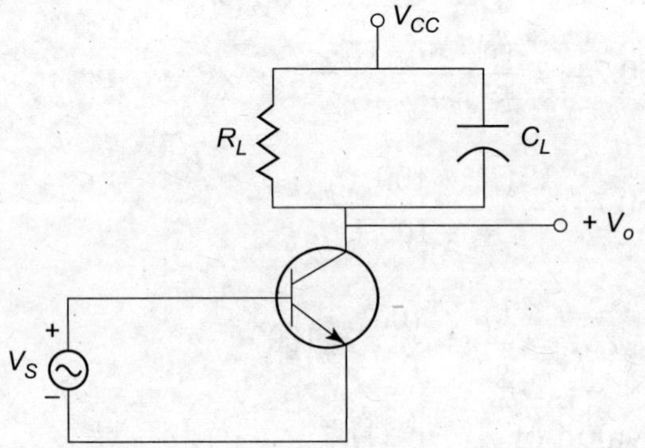

Replacing transistor with its hybrid-Π model and using Miller's theorem, we get

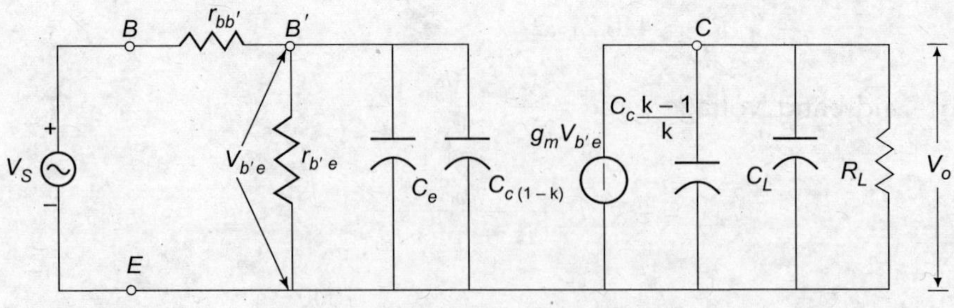

Since $K = \frac{V_{ce}}{V_{b'e}} >> 1, C_c \frac{K-1}{K} \approx C_c.$

$$\therefore \quad V_o = -g_m \cdot V_{b'e} \cdot \frac{R_L \cdot \frac{1}{Jw(C_c + C_L)}}{R_L + \frac{1}{Jw(C_c + C_L)}}$$

$$V_o = V_{ce} = \frac{-g_m R_L \cdot V_{b'e}}{1 + Jw(C_c + C_L)R_L}$$

$$\frac{V_{ce}}{V_{b'e}} = K = \frac{-g_m R_L}{1 + Jw(C_c + C_L)R_L} \quad \text{...(1)}$$

Hence proved.

(b) $$C_c(1 - K) = \frac{C_c\,(1 + g_m R_L + Jw(C_c + C_L)R_L)}{1 + Jw\,(C_c + C_L)R_L}$$

at the Input side = $C = C_e + C_c(1 - K)$

Input time constant = $r_{b'e} \cdot C$ (neglecting $r_{bb'}$).

$\therefore r_{b'e}(C_e + C_c(1 - K_o))$ ...(2)

where K_o is midband value of K.

Output time constant = $R_L \cdot (C_L + C_c)$ ...(3)

From equation (1), midband value of K is

$$K_o = K\big|_{w \to \infty} = -g_m R_L \quad \text{...(4)}$$

Using (4) in (2)

Input time constant = $r_{b'e}(C_e + C_c\,(1 + g_m R_L)$

Output time constant = $R_L(C_c + C_L)$ ...(5)

Given that $g_{b'e} R_L(C_c + C_L) >> C_e + C_c(1 + g_m R_L)$

$\Rightarrow R_L(Cc + C_L) >> r_{b'e}\,[C_e + C_c(1 + g_m R_L)]$ ...(6)

From equations (5) and (6), output time constant is more than input time constant. So, output time constant being larger will predominate and yet decide the 3 dB bandwidth.

$$\therefore \quad w_H = R_L(C_c + C_L)$$

$$\therefore \quad f_H \cong \frac{1}{2\pi\,(C_c + C_L)R_L}$$ is the equation for 3-db bandwidth.

7.28 For a single-stage *CE* amplifier with typical hybrid-π parameters, find the value of R_s which will give a 3-dB frequency which is (a) half the value for $R_s = 0$. (b) Twice the value for $R_s = \infty$. Use Miller's theorem and approximate analysis.

Solution (a) We have $$f_H = \frac{1}{2\pi RC}$$

where $$C = C_e + C_c(1 + g_m R_L)$$

and $$R = (R_s + r_{bb'}) || r_{b'e}$$

(b) with $$R_s = 0$$

$$R = r_{bb'} || r_{b'e}$$

$$= \frac{r_{bb'} \cdot r_{b'e}}{r_{bb'} + r_{b'e}} = \frac{r_{bb'} \cdot r_{b'e}}{h_{ie}}$$

$$C = C_e + C_c(1 + g_m R_L)$$

Let $$f_{H1} = \frac{1}{2\pi RC} = \text{new value of } f_H \text{ with } R_s = 0$$

$$R = \frac{100 \times 1000}{1100} = 90.90\ \Omega$$

$$C = C_e + C_c(1 + g_m R_L)$$

$$\therefore \quad f_{H1} = \frac{1}{2\pi \times 90.90 \times (C_e + C_c(1 + g_m R_L))} \quad ...(1)$$

$$f_H = \frac{1}{2\pi \times \dfrac{R_s + r_{bb'}}{R_s + h_{ie}} \cdot r_{b'e} \cdot \dfrac{(C_e + C_{c1} + g_m R_L)}{C_e + C_c[1 + g_m R_L]}} \quad ...(2)$$

$$\therefore \text{ given that } f_H = \frac{1}{2} f_{H1} \quad ...(3)$$

From (1), (2) and (3)

$$2 \times 90.90 = \frac{R_s + 100}{R_s + 1100}(1000)$$

$$2 \times (R_s + 1100)\, 90.90 = 1000\ (R_s + 100)$$

$$\Rightarrow (2000 - 180.80)R_s = -1000 \times 100 + 1100 \times 180.80$$

$$\Rightarrow \quad R_s = 122.19\ \Omega$$

(b) with $R_s = \infty$

$$R = \frac{(R_s + r_{bb'})r_{b'e}}{R_s + h_{ie}}$$

$$= \frac{\left(1 + \frac{r_{bb'}}{R_s}\right) r_{b'e}}{1 + \frac{h_{ie}}{R_s}}$$

if $R_s = \infty$

$$R = r_{b'e}$$

$$\therefore \quad f_{H1} = \frac{1}{2\pi\, r_{b'e}\left(C_e + C_c\left[1 + g_m R_L\right]\right)} \qquad \text{...(1)}$$

and

$$f_H = \frac{1}{2\pi \frac{R_s + r_{bb'}}{R_s + h_{ie}} r_{b'e}\,(C_e + C_c\,(1 + g_m R_L)} \qquad \text{...(2)}$$

given that $f_H = 2f_{H1}$ ...(3)

$\Rightarrow$ from (1), (2) and (3).

$$\frac{1}{\frac{R_s + r_{bb'}}{R_s + h_{ie}}} = 2$$

$$\Rightarrow \quad R_s + h_{ie} = 2\,(R_s + r_{bb'})$$

$$\Rightarrow \quad h_{ie} - 2r_{bb'} = R_s$$

$$R_s = 1100 - 200$$

$$= 900\ \Omega$$

7.29 Show that the low-frequency voltage gain of the FET stage with $r_d \gg R_s + R_L$ is given by

(a)

$$\frac{A_V}{A_o} = \frac{1}{1 + g_m R_s} \cdot \frac{1 + Jf/f_o}{1 + Jf/f_p}$$

where

$$A_o = -\,g_m R_L,\ f_o = \frac{1}{2\pi C_s R_s},\ f_p = \frac{1 + g_m R_s}{2\pi C_s R_s}$$

(b) If $g_m R_s >> 1$ and $g_m = 5$ mA/V, find C_s so that a 50 Hz square wave input will suffer no more than 10 % tilt.

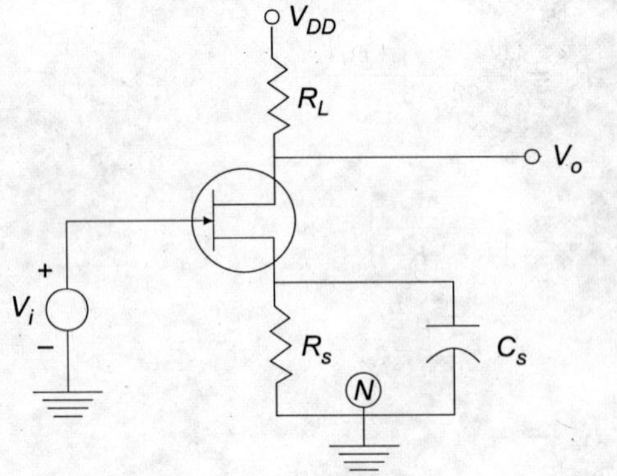

Solution (a) Replacing all dc sources with their internal resistances, and replacing FET with its small signal model, we have,

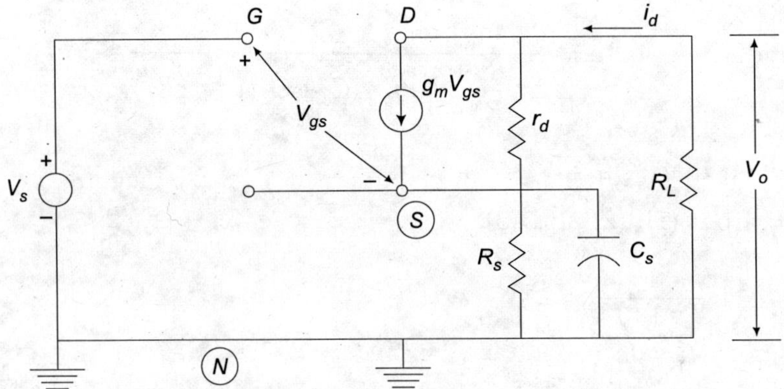

We are having R_s and C_s in parallel, call it as

$$Z_s = R_s \,||\, \frac{1}{JwC_s} = \frac{R_s \cdot \dfrac{1}{JwC_s}}{R_s + \dfrac{1}{JwC_s}}$$

$$= \frac{R_s}{1 + JwC_s R_s} \qquad ...(1)$$

Since $r_d \gg (R_L + R_s)$, it can be neglected.

$$V_o = -\, g_m V_{gs} \cdot R_L \qquad ...(2)$$

Writing KVL in the loop G–S–N–G,

$$-V_s + V_{gs} + g_m V_{gs} \cdot Z_s = 0$$

$$\therefore \quad V_{gs} = \frac{V_i}{1+g_m Z_s} \quad \text{...(3)}$$

$$V_{gs} = \frac{V_i}{1+\dfrac{g_m R_s}{1+JwC_s R_s}} = \frac{V_i(1+JwC_s R_s)}{(1+g_m R_s + JwC_s R_s)} \quad \text{...(4)}$$

From (2) and (4)

$$V_o = \frac{-g_m R_L \cdot (1+JwC_s R_s)}{(1+g_m R_s + JwC_s R_s)} V_i$$

$$\therefore \quad A_V = \frac{V_o}{V_i} = \frac{-g_m R_L \cdot (1+JwC_s R_s)}{(1+g_m R_s + JwC_s R_s)} \quad \text{...(5)}$$

To find A_o, midband gain take limit as w tends to infinity in A_V expression.

Then, $\quad A_o = -g_m R_L$

$$\therefore \quad \frac{A_V}{A_o} = \frac{1+JwC_s R_s}{1+g_m R_s + JwC_s R_s}$$

$$= \frac{1}{1+g_m R_s} \cdot \frac{1+JwC_s R_s}{1+\dfrac{JwC_s R_s}{1+g_m R_s}}$$

$$= \frac{1}{1+g_m R_s} \cdot \frac{1+J2\pi f C_s R_s}{1+\dfrac{J2\pi C_s R_f}{1+g_m R_s}}$$

$$= \frac{1}{1+g_m R_s} \cdot \frac{1+Jf/f_o}{1+Jf/f_p}$$

Since given that $f_o = \dfrac{1}{2\pi R_s C_s}$ and $f_p = \dfrac{1+g_m R_s}{2\pi C_s R_s}$

Hence proved.

(b) If $g_m R_s >> 1$, then $f_p >> 1$, so f_L is decided by f_p only.

$$\therefore \quad \% \text{ tilt} = \frac{\pi f_L}{f} \times 100 = \frac{\pi f_p}{f} \times 100$$

$$= \frac{\pi \times \dfrac{1+g_m R_s}{2\pi C_s R_s}}{f} \times 100$$

$$= \frac{g_m}{2C_s f} \times 100$$

$$\text{Per cent tilt } = 10 = \frac{g_m}{2C_s f} \times 100$$

$$C_s = \frac{5 \times 10^{-3}}{2 \times 10 \times 50} \times 100$$

$$= 0.5 \times 10^{-3} \text{ F}$$

$$= 500\ \mu\text{F}$$

7.30 For a two stage RC coupled amplifier, the parameters of the transistors are $h_{fe} = 50$, $h_{ie} = 1.1$ k, $h_{re} = h_{oe} = 0$. Find (a) midband gain (b) the value of C_b necessary to give a low 3-dB frequency of 20 Hz. (c) find the value of C_b necessary to ensure less than 10 % tilt for a 100 Hz square wave input.

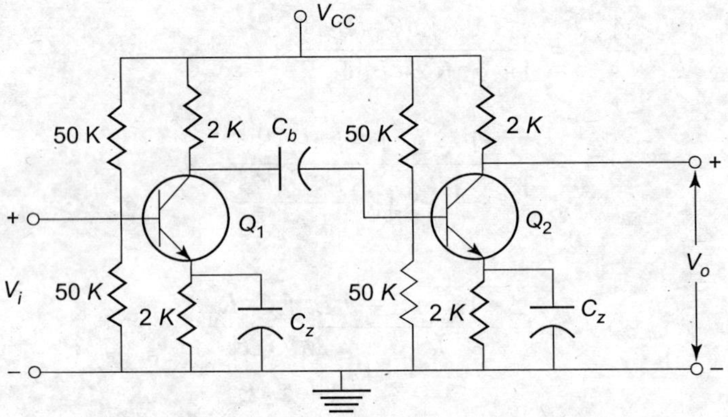

Solution (a) To find the value of gain, assume C_z and C_b as short-circuits.

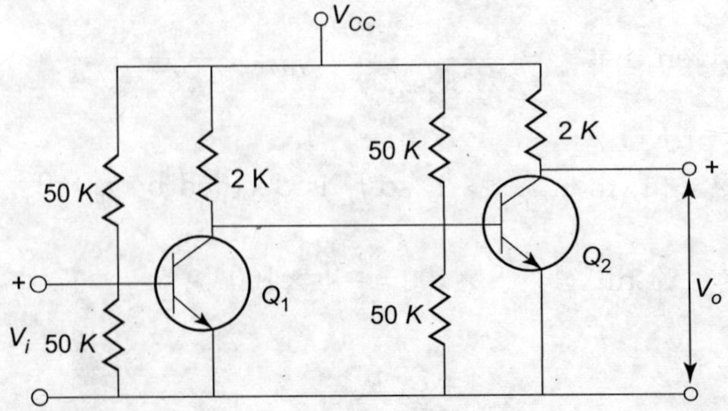

Now, Replacing transistor with their h-parameters,

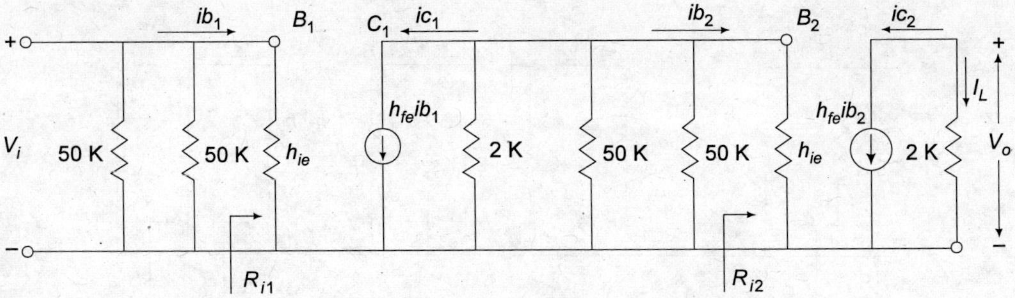

$$R_{L2} = \text{load resistance of 2nd stage} = 2 \text{ k}\Omega$$

$$R_{L1} = \text{load resistance of Ist stage}$$

$$= 2 \text{ K} \,||\, 50 \text{ K} \,||\, 50 \text{ K} \,||\, R_{i2}$$

$$= 0.69 \text{ k}\Omega$$

$$R_{i1} = \text{Input resistance of Ist stage} = h_{ie} = 1.1 \text{ k}\Omega$$

$$R_{i2} = h_{ie} = 1.1 \text{ k}\Omega$$

$\therefore$ *Second stage calculations*:

$$R_{L2} = 2 \text{ K}, R_{i2} = 1.1 \text{ k}\Omega$$

$$\therefore \quad A_{i2} = -h_{fe} = -50$$

$$A_{V2} = \frac{A_{i2}\, R_{L2}}{R_{i2}} = \frac{-50 \times 2}{1.1} = -90.90$$

First stage calculations:

$$R_{L1} = 0.69 \text{ k}\Omega, R_{i1} = 1.1 \text{ k}\Omega$$

$$A_{i1} = -h_{fe} = -50$$

$$\therefore \quad A_{V1} = \frac{A_{i1} \cdot R_{L1}}{R_{i1}} = \frac{-50 \times 0.69}{1.1}$$

$$= -31.37$$

$$\therefore \text{ Mid-band gain} = A_{V1} \cdot A_{V2}$$

$$= 2851.85$$

(b) *To find C_b*: We cannot make it as short circuit, only C_z is made short circuit. Then, the equivalent circuit will become,

We have $$f_L = \frac{1}{2\pi (R_o' + R_i')C_b}$$

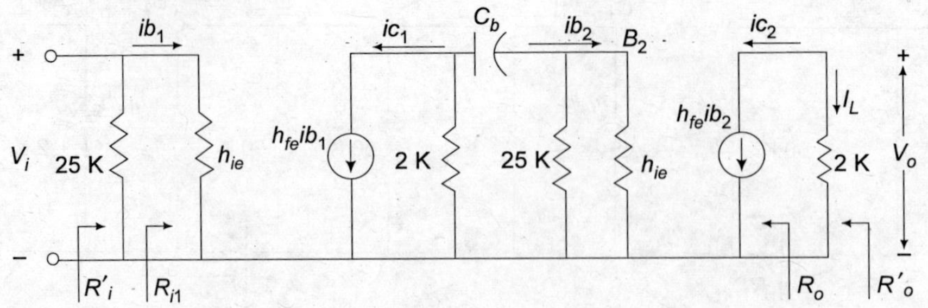

$$R_o' = 2\text{ K} \,||\, \infty = 2\text{ K}$$

$$R_i' = 25\text{ K} \,||\, h_{ie} = 1.053\text{ k}\Omega$$

Given that $f_L = 20$ Hz

$$\therefore \quad 20 = \frac{1}{2\pi(2+1.053)\times 10^3 \times C_b}$$

$$\therefore \quad C_b = \frac{1}{20\times 2\pi \times (3.053)\times 10^3}$$

$$= 2.607\ \mu\text{F}$$

So, we need a C_b of value 2.607 μF, to get a lower 3-db frequency of 20 Hz.

(c) We have % tilt $= \dfrac{\pi f_L}{f} \times 100$

$$10 = \frac{\pi \times \dfrac{1}{2\pi\times 3.053\times 10^3 \times C_b}}{100} \times 100$$

$$= \frac{1}{2\times 3.053\times 10^3 \times C_b}$$

$$\Rightarrow \quad C_b = 16.37\ \mu\text{F}$$

7.31 For the two-stage FET RC-coupled amplifier shown the following are the parameters $g_m = 10$ mA/V, $r_d = 5.5$ K $R_d = 10$ K, $R_g = 0.5$ M for each stage. Assume C_s to be arbitrarily large.

(a) What must be the value of C_b in order that the frequency characteristic of each stage be flat within 1 dB down to 10 Hz?

(b) Repeat part (a), if the overall gain of both stages is to be down 1 db at 10 Hz

(c) What is the overall midband voltage gain?

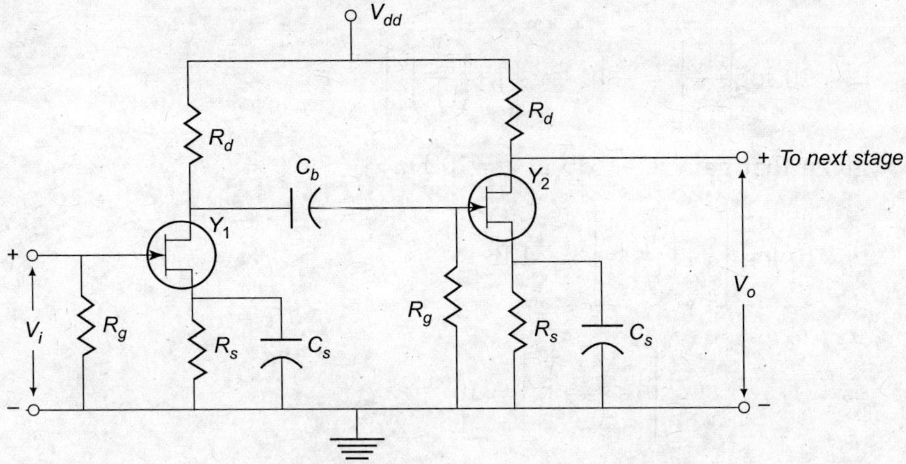

Solution (a) Since the value of C_s is assumed to be large, it acts as a short circuit. Replacing transistor (FET) with its small signal model, we have (killing all dc sources),

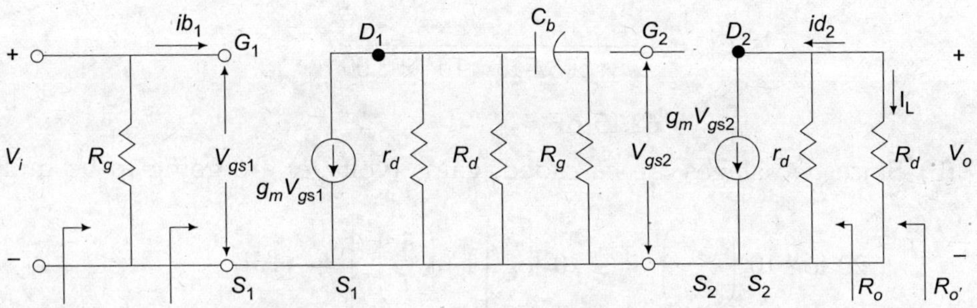

$$R_o = r_d = 5.5 \text{ k}\Omega$$

$$R_o' = r_d || R_d = 5.5 || 10 = 3.548 \text{ k}\Omega$$

$$R_i = \infty, \; R_i = \infty || R_g = 0.5 \text{ M}\Omega$$

$$\therefore \quad f_L = \frac{1}{2\pi (3.548 \times 10^3 + 0.5 \times 10^6) C_b}$$

$$= \frac{1}{2\pi \times 503.548 \times 10^3 \times C_b} \quad ...(1)$$

In the low frequency region, we can write

$$\left| \frac{A_V}{A_o} \right| = \frac{1}{\sqrt{1 + (f_L/f)^2}}$$ where A_o is mid-band value.

$$\Rightarrow \quad 20 \log\left|\frac{A_V}{A_o}\right| = -10 \log\left[1+\left(\frac{f_L}{f}\right)^2\right]$$

given that gain is –1 db at f = 10 Hz

$$\therefore -10 \log\left[1+\left(\frac{f_L}{f}\right)^2\right] = -1\text{dB}$$

$$\Rightarrow \quad \left[1+\left(\frac{f_L}{f}\right)^2\right] = 10^{0.1}$$

$$\Rightarrow \quad \frac{f_L}{f} = \sqrt{0.2589} = 0.509$$

$$\therefore \quad f_L = f \times 0.509 = 5.09 \text{ Hz} \qquad \text{...(2)}$$

Therefore, from (1) and (2)

$$C_b = \frac{1}{2\pi \times 503.548 \times 10^3 \times 5.09}$$

$$= 62.15 \text{ nF}$$

(b) Since two stages are cascaded, gains (voltage) are going to be multiplied.

$$\therefore 20 \log 10 \left(\frac{A_V}{A_o}\right)^2 = -20 \log\left[1+\left(\frac{f_L}{f}\right)^2\right] = -1\text{dB}$$

$$\left[1+\left(\frac{f_L}{f}\right)^2\right] = 10^{0.05}$$

$$f_L = 0.3493 \text{ f}$$

$$= 3.493 \text{ Hz}$$

$$\therefore \quad C_b = \frac{1}{2\pi \times 503.548 \times 10^3 \times 3 \times 3.493} = 90.53 \text{ nf}$$

(c) At the second stage

$$V_o = -g_m V_{gs2} \cdot \frac{R_d}{r_d + R_d} \cdot r_d$$

$$\therefore \quad \frac{V_o}{V_{gs2}} = -g_m (R_d || r_d) = A_{o2} = \text{midband gain of 2nd stage.}$$

$$\therefore \quad A_{o2} = -10 \times 10^{-3} \text{ (10 K || 5.5 K)}$$

$$= -35.48$$

Since the voltage gain of Ist stage is also same as second stage.

$$A_{o1} = \text{midband value}$$

$$= -35.48$$

$$\therefore \text{ overall midband value} = A_{o1} \cdot A_{o2}$$

$$= 1258.\ 83$$

7.32 A three stage RC-coupled amplifier uses FET'S with the following parameters g_m = 2.6 mA/V, r_d = 7.7 kΩ, R_d = 10 K, R_g = 0.1 MΩ, C_b = 0.005 μF and C_s = ∞. Evaluate (a) the overall mid-band voltage gain in decibels (b) f_L of each individual stage (c) the overall lower 3 dB frequency.

Solution (a) After replacing FET with its small signal model, the diagram will become as shown below:

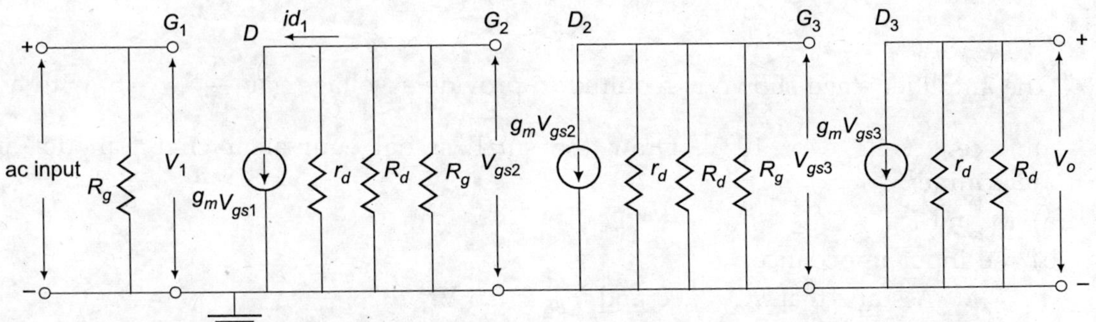

Since these three stages are cascaded, the overall gain is the product of individual gains. In decibels they are going to be added.

Since R_g of one stage shunts R_d, we can write

$$|A_o| = \text{mid-band value of gain one stage.} = g_m R.$$

where, $\quad R = r_d || R_d || R_g$

$$= 7.7 \text{ K} || 10 \text{ K} || 100 \text{ K}$$

$$= 4.1689 \text{ k}\Omega$$

$$\therefore \quad |A_o| = 2.6 \times 4.16789 = 10.84$$

$$A_{o\,\text{db}} = 20 \log |A_o| = 20.7 \text{ dB}$$

$$\therefore \text{ Overall mid-band gain of 3 stages} = 3 \times 20.7$$

$$= 62.10 \text{ dB}$$

(b) We have $$f_L = \frac{1}{2\pi(R_o' + R_i')C_b}$$

$$R_o' = r_d \,||\, R_d = 7.7\text{ K} \,||\, 10\text{ K} = 4.35\text{ k}\Omega$$

$$R_i' = R_g = 0.1\text{ M}\Omega = 100\text{ k}\Omega$$

$$C_b = 0.005\ \mu\text{F}$$

$$\therefore \quad f_L = \frac{1}{2\pi \times (104.35) \times 10^3 \times 0.005 \times 10^{-6}}$$

$$= 305.195\text{ Hz}$$

(c) $$f_L^* = \text{Overall 3-db frequency of 3 stages.}$$

$$= \frac{f_L}{\sqrt{2^{1/n} - 1}} = \frac{305.195}{\sqrt{2^{1/3} - 1}}$$

$$= \frac{305.195}{0.508} = 600.77\text{ Hz}$$

7.33 The amplifier stage shown is required to provide a voltage gain $\frac{V_o}{V_i} = -5$, with a dc quiescent voltage of 10 V. Draw the small signal equivalent circuit model and determine:

(a) R_1 and R_2.

(b) the input impedance.

Assume that $h_{fe} = 200$ and $V_{BE} = 0.6$ V.

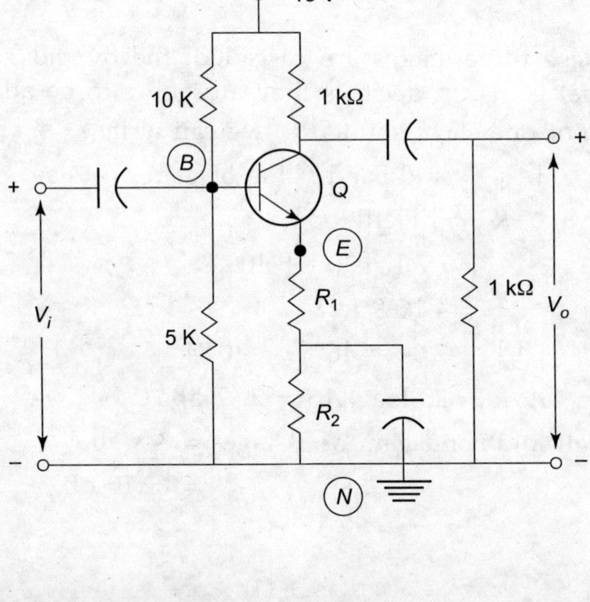

Solution Let the voltage at point B with respect to ground as V_B. Under dc conditions capacitors act as open circuit.

$$\therefore \quad V_B = \frac{5}{5+10} \times 15 = 5 \text{ V}$$

Then, V_E = voltage between E and N

$$= V_B - V_{BE} = 5 - 0.6 = 4.4 \text{ V}$$

given that $V_{CEQ} = 10$ V

$$I_C = \frac{V_{CC} - V_{CE}}{R_c} = \frac{15-10}{1\,\text{k}\Omega} = 5 \text{ mA}$$

$$\therefore \quad I_E \approx I_C = 5 \text{ mA}$$

$$V_E = I_E(R_1 + R_2)$$

$$\therefore \quad \frac{4.4}{5 \text{ mA}} = R_1 + R_2$$

$$R_1 + R_2 = 880\ \Omega \qquad ...(1)$$

Now, replacing transistor with its small-signal model.

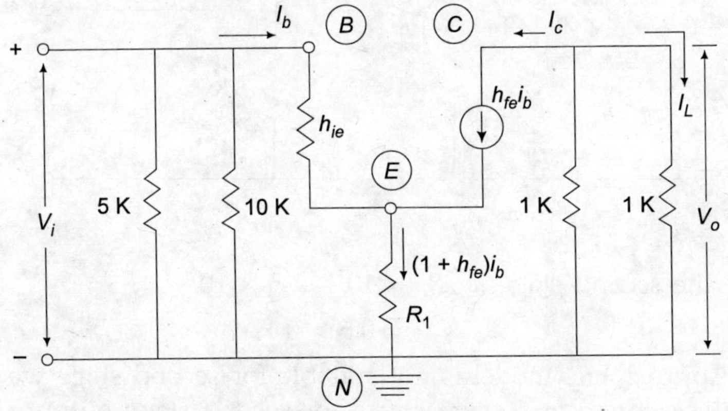

From the circuit $V_i = [h_{ie} + R_1 (1 + h_{fe})]I_b$ ...(2)

$$V_o = -500\ h_{fe} I_b \qquad ...(3)$$

From (2) and (3) $\dfrac{V_o}{V_i} = A_V = \dfrac{-h_{fe} \times 500}{h_{ie} + R_1(1+h_{fe})}$

given that $A_V = -5$

$$\Rightarrow \quad 5 = \frac{200 \times 500}{1100 + R_1(201)}$$

$$\Rightarrow \quad 200 \times 500 = 5500 + R_1(1005)$$

$$\Rightarrow \quad R_1 = 94.02\ \Omega$$

From equation (1) $R_2 = 880 - 94.02$

$$= 785.98\ \Omega$$

From equation (2) $\dfrac{V_i}{i_b}$ = Input resistance

$$= h_{ie} + (1 + h_{fe})R_1$$

$$= 19.99\ \text{k}\Omega$$

7.34 (a) Find the voltage gain A_{Vs} of the amplifier shown. Assume $hie = 1000\ \Omega$, $h_{re} = 10^{-4}$, $h_{fe} = 50$, $h_{oe} = 10^{-4}$ A/V.

(b) Find R'_o.

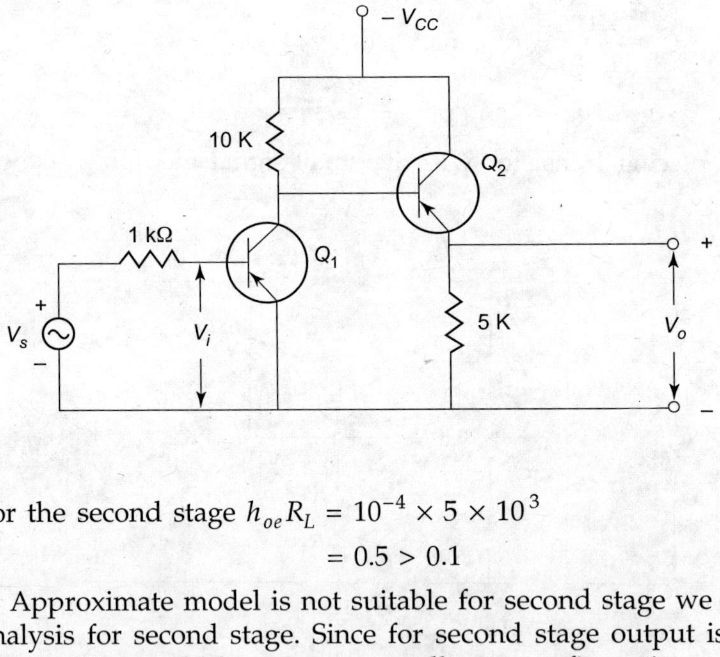

Solution (a) For the second stage $h_{oe}R_L = 10^{-4} \times 5 \times 10^3$

$$= 0.5 > 0.1$$

∴ Approximate model is not suitable for second stage we have to use exact analysis for second stage. Since for second stage output is taken at emitter terminal it should be a common-collector configuration

$$\therefore \quad A_{I2} = \frac{1 + h_{fe}}{1 + h_{oe} R_{L2}} = \frac{1 + 50}{1 + 10^{-4} \times 5 \times 10^3} = 34$$

$$R_{i2} = h_{ie} + A_{I2}R_{L2}$$

$$= 1000 + 34 \times 5 \times 10^3$$

$$= 171\ \text{k}\Omega$$

$$A_{V2} = 1 - \frac{h_{ie}}{R_{i2}} = 1 - \frac{1000}{171\ \text{k}\Omega} = 0.994$$

First stage is a common-emitter configuration.

$$R_{L1} = 10 \text{ K} \,||\, 171 \text{ K} = 9.44 \text{ k}\Omega$$

$$h_{oe} \cdot R_{L1} = 10^{-4} \times 9.44 \times 10^{3} = 0.944 > 0.1$$

Here also, we have to use exact analysis.

$$A_{I1} = \frac{-h_{fe}}{1 + h_{oe} R_{L1}} = \frac{-50}{1 + 0.944} = -25.71$$

$$R_{i1} = h_{ie} + h_{re} A_{I1} R_{L1}$$

$$= 1000 + 10^{-4} \times -25.71 \times 9.44 \times 10^{3} = 975.73 \ \Omega$$

$$A_{V1} = \frac{A_{I1} \cdot R_{L1}}{R_{i1}} = \frac{-25.71 \times 9.44 \times 10^{3}}{975.73}$$

$$= -248.739$$

$$\therefore \quad A_V = A_{V1} \cdot A_{V2} = -247.246$$

$$A_{Vs} = A_V \cdot \frac{R_{i1}}{R_{i1} + R_s} = -247.246 \times \frac{975.73}{975.73 + 1000}$$

$$= -122.105$$

(b) Since these two stages are cascaded, R_o is decided by the last stage.

$$R_o = \frac{1}{Y_{o2}} \text{ where } Yo_2 = h_{oe} + \frac{1 + h_{fe}}{h_{ie} + R_{s2}}$$

R_{s2} is the source resistance for 2nd stage.

$$R_{s2} = R_{o1} \,||\, R_{L1}$$

$$Y_{o1} = h_{oe} - \frac{h_{fe}\, h_{re}}{R_s + h_{ie}} = 10^{-4} - \frac{50 \times 10^{-4}}{1000 + 1000}$$

$$= 0.975 \times 10^{-4}$$

$$R_{o1} = \frac{1}{Y_{o1}} = 10.256 \text{ K}\Omega$$

$$R_{L1} = 9.44 \text{ K}\Omega$$

$$R_{s2} = 9.44 \,||\, 10.256$$

$$= 4.92 \text{ k}\Omega$$

$$\therefore \quad Y_{o2} = h_{oe} + \frac{1 + h_{fe}}{R_{s2} + h_{ie}}$$

$$= 10^{-4} + \frac{51}{(4.92+1)\times 10^3}$$

$$= 87.148 \times 10^{-4}\ \mho$$

$$\therefore \quad R_{o2} = 114.746\ \Omega$$

$$\therefore \quad R'_o = R_{o2} \,||\, R_{L2} = 114.746 \,||\, 5\ \text{K}\Omega$$

$$= 112.17\ \Omega$$

7.35 For the two-stage amplifier of the diagram, calculate the input and output impedances, and the individual and overall voltage gains. Assume $h_{fe} = 50$, $h_{ie} = 1.1$ kΩ, $h_{re} = h_{oe} = 0$.

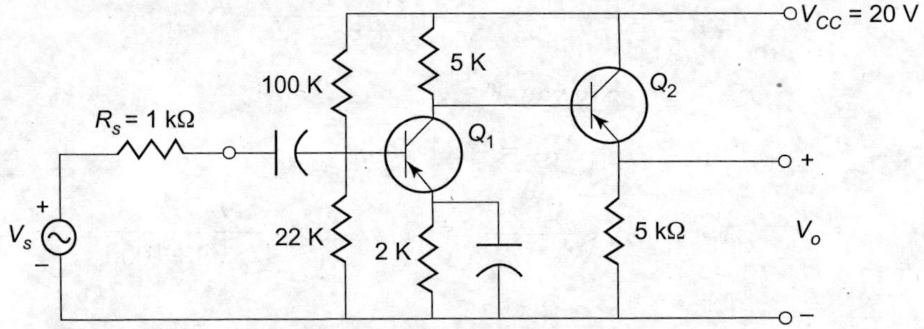

Solution Since $h_{oe} = 0$, $h_{oe}R_{L2} = 0$ so we can use approximate model for Q_2. As well as Q_1. Replacing transistor with their small signal model, by killing dc voltage sources,

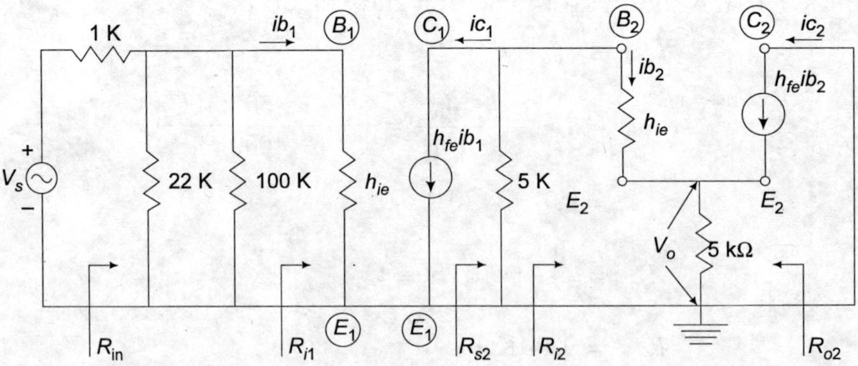

Second stage calculations

This is a CC stage,

$$A_{i2} = 1 + h_{fe} = 51$$

$$R_{L2} = 5\ \text{k}\Omega$$

$$R_{i2} = h_{ie} + (1 + h_{fe})R_{L2}$$

$$= 1.1 + (51) \times 5 = 256.1\ \text{k}\Omega$$

$$A_{V2} = \frac{A_{i2} \cdot R_{L2}}{R_{i2}} = 0.995$$

$$R_{o2} = \frac{R_{s2} + h_{ie}}{1 + h_{fe}}$$

$$R_{s2} = 5\text{ K} \,||\, R_{i2} = 5 \,||\, 256.1$$
$$= 4.90 \text{ k}\Omega$$

$$\therefore \quad R_{o2} = \frac{4.90 \times 10^3 + 1.1 \times 10^3}{51}$$
$$= 117.73\ \Omega$$

$$R'_{o2} = R_{o2} \,||\, R_{L2} = 117.73 \,||\, 5\text{ K}$$
$$= 115.02\ \Omega$$

First stage calculations

This is a *CE* amplifier stage

$$A_{i1} = -h_{fe} = -50$$
$$R_{i1} = h_{ie} = 1100\ \Omega$$

$$A_{V1} = \frac{A_{i1} \cdot R_{L1}}{R_{i1}} = \frac{-50 \times 5\text{ K} \,||\, R_{i2}}{1100}$$

$$= \frac{-50 \times 5\text{ K} \,||\, 256.1\text{ K}}{1100} = -222.93$$

$\therefore$ overall gain $A_V = A_{V1} \cdot A_{V2} = -221.80$

$$R_{\text{in}} = \text{overall input resistance} = R_{i1} \,||\, R_1 \,||\, R_2$$
$$= 1100 \,||\, 22\text{ K} \,||\, 100\text{ K} = 1036.59\ \Omega$$

EXERCISE PROBLEMS

7.1 Convert *CC* h-parameters into *CE* h-parameters.

7.2 In a single-stage Amplifier the h-parameter values are $h_{re} \approx 1$, $h_{fe} = -50$, $h_{oe} = 24$ μA/V. Find A_I, R_i, A_{Vs} and R_o for the *CC* configuration with $R_s = R_L = 5$ K.

7.3 For the emitter follower, with $R_s = 0.1$ kΩ, $R_L = 5$ k, $h_{fe} = 50$, $h_{ie} = 1000\ \Omega$, $h_{oe} = 25$ μA/V, calculate A_I, A_V, A_{Vs}, R_i and R_o.

7.4 For the transistor amplifier shown find A_I, A_V, A_{Vs} and R_o, R_i with $h_{oe} = h_{re} = 0$. $h_{fe} = 100$, $h_{ie} = 1.1$ kΩ.

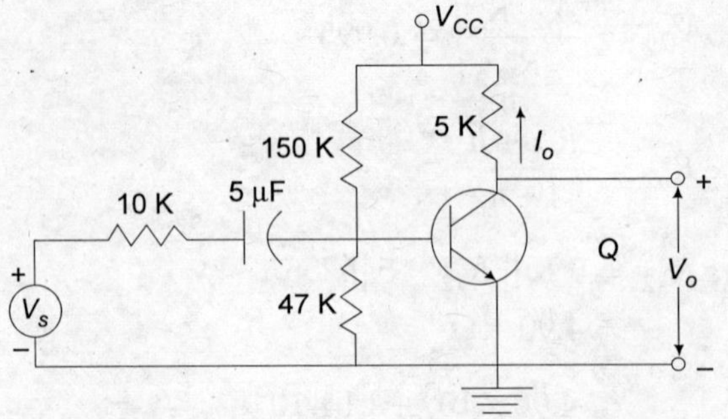

7.5 When a transistor is biased in 10 mA, 10 V it has following h-parameters at room temperature. h_{ie} = 600 Ω, h_{fe} = 50, $h_{re} = 2 \times 10^{-4}$, $h_{oe} = 4 \times 10^{-5}$. It has f_T = 100 Mhz and C_c = 3 pf. Find all the hybrid π-components.

7.6 Find the base width of a silicon *pnp* transistor whose figure of merit is 500 MHz. diffusion length of holes is 15 cm^2/sec.

7.7 When a Ge *pnp* transistor is biased at 10 mA, 10 V, it has a base width of 2 microns. Find Ce and f_T given that D_B = 50 cm^2/sec.

7.8 The 3-dB bandwidth of an amplifier extends from 50 Hz to 50 KHz. Find the frequency range over which the voltage gain differs by only 2 dB from its midband value.

7.9 (a) Show that the relative voltage of an amplifier with an emitter resistor Re bypassed by a capacitor C_z may be expressed in the form.

$$\frac{A_V}{A_o} = \frac{1 + JwC_z R_e}{B + JwC_z R_e} \text{ where } B = 1 + \frac{R'}{R},\ R = R_s + h_{ie} \text{ and } R' = (1 + h_{fe})R_e.$$

(b) Prove that lower 3-db frequency is $f_L = \dfrac{\sqrt{B^2 - 2}}{2\pi C_z R_e}$.

7.10 In a transistor circuit shown below, R_s = 500 Ω, $R_1 = R_2$ = 50 K, $R_c = R_e$ = 2 k, h_{ie} = 1.1 kΩ, h_{fe} = 50, $h_{re} = h_{oe}$ = 0, C_b = 5 μF.

(a) Neglecting the effect of C_z, find f_L.

(b) Neglecting the effect of C_b, find f_p and f_o due to C_z alone.

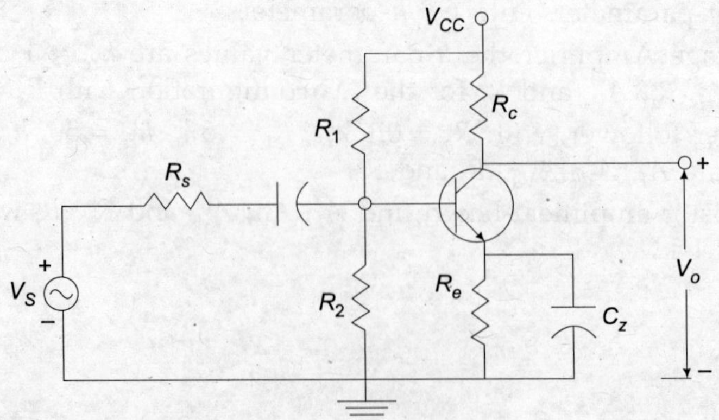

REVIEW QUESTIONS

1. Explain the different types of distortion in an amplifier. Sketch the frequency response of an *RC* coupled amplifier.
2. Draw the high-frequency model of one stage *RC* coupled amplifier and derive an expression for upper 3 dB frequency.
3. Discuss the effect of emitter bypass capacitor on the low-frequency response of a *CE* amplifier.
4. Draw the hybrid Π equivalent circuit of a transistor in *CE* configuration and explain the various parameters in it. In terms of *h*-parameters derive the equations for the various components of this circuit.
5. Draw a circuit diagram for a *CE* transistor amplifier and derive expressions for voltage and current gain and input and output impedance.
6. State Miller's theorem and its dual and give its applications.
7. Among the three configuration of *BJT*, *CE* configuration is widely used. Justify this statement and give its applications.
8. Classify amplifiers with respect to operating point selection and compare them in terms of efficiency, distortion and nonlinearity.
9. Compare *CB*, *CE* and *CC* configurations with respect to their characteristics and performance.
10. Draw the circuit diagram of Darlington emitter follower and derive the expression for input impedance.
11. Draw the equivalent circuit of a *CE* amplifier and derive the equation for all the four amplifier parameters.
12. Why is hybrid model of two-port network employed to represent a transistor? Explain each of the four *h*-parameters of the model and give their expressions.
13. State the linear analysis of a transistor amplifier circuit. State the Miller's theorem and its dual and give its applications.
14. When and why is the simplified model of a transistor employed? Give its circuit and derive the amplifier parameters of a *CE* transistor amplifier using this model.
15. Draw the simplified model of a *CC* transistor amplifier circuit and derive its parameters.
16. Discuss in detail about the selection of transistor configuration in each stage of a multistage amplifier.
17. Derive the bandwidth of a multistage amplifier, assuming that each stage has same upper and lower cut off frequencies.

18. Discuss the effect of emitter bypass and coupling capacitors on the frequency response of the *CE* amplifier.
19. Derive the equations for the parameters of high-frequency conductances in terms of *h*-parameters.
20. Derive the expression for the two capacitances associated with high-frequency model of transistor. Give the validity condition of Giacolletto model.
21. What is unity crossover frequency? Give its significance.
22. Derive an expression for short-circuit gain of a transistor in *CE* configuration at high frequencies.
23. What is the significance of gain bandwidth product of a transistor amplifier? Discuss in detail.
24. What are input and output nonlinear distortions? Explain and give the effect on the output signal of an amplifier.
25. Derive equations for current gain and voltage gain of a simple amplifier in terms of *h*-parameters at low frequency, taking into account the souce resistance.
26. What is a Darlington pair? Draw its circuit and explain the special features of the pair.
27. What is unity crossover frequency? What is its significance? Discuss about gain bandwidth product.
28. Sketch a typical frequency response curve of an amplifier and identify its bandwidth.
29. Discuss briefly about the choice of transistor configuration in a cascaded amplifier.
30. Under what conditions does an amplifier preserve the form of the input signal?

CHAPTER

8

Field Effect Transistors

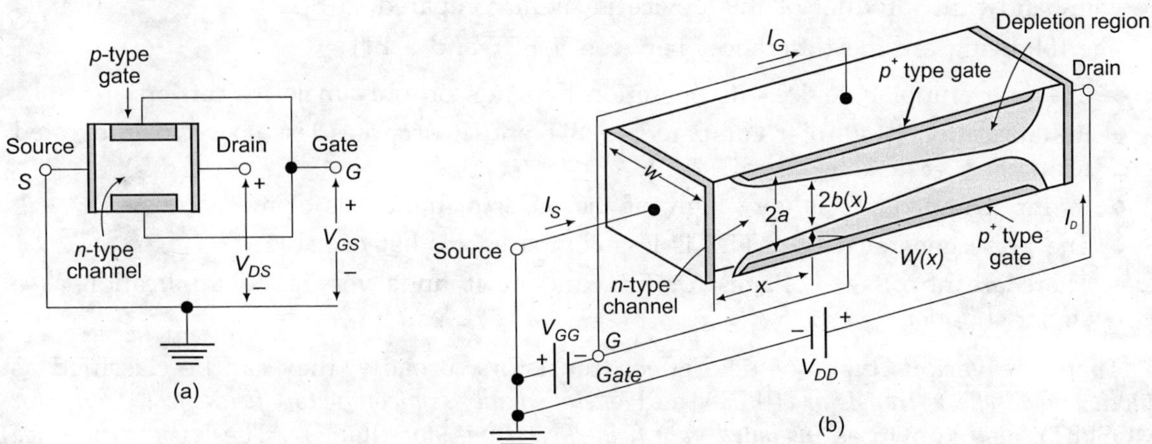

Structure of a JFET

Chapter Outline

The concepts introduced in this chapter are:

- *Comparison of a FET with a BJT*
- *Construction and characteristics of various types of FETs*
- *Small signal model of a FET*
- *Biasing circuits of a FET*
- *FET as a voltage variable resistor*

8.1 INTRODUCTION

Field-effect transistor (FET) is a popular active device of electronics which is widely used for amplification as well as switching similar to the bipolar junction transistor (BJT). The operation of this device depends on the control of flow of current by the virtue of electric field. This is a unipolar device, that is, the operation of the device is based on the flow of only one type of charge. Only majority carriers contribute to the flow of current and minority carriers have no role to play in the process of operation, unlike BJT. The greatest advantage with FET is that it has very high input impedance, which is the prime requirement for a voltage amplifier. The input impedance of the device can even tend to a very large value of the order of hundreds of megaohms. It comes with a drawback that the gain-bandwidth product of this device is small compared to BJT.

The following are the differences between a FET and a BJT:

- FET is a unipolar device—its operation depends on only majority carriers.
- Its fabrication is simpler compared to BJT and it occupies less space in integrated form.
- Its input impedance is very high, of the order of hundreds of megaohms.
- The noise generated by a FET is less compared to that of a BJT.
- There are no offset voltages for FET and so it finds very good application as a signal chopper.

There are various types of field-effect transistors. Broadly, they can be classified as *junction field-effect transistor* (JFET) and *metal oxide semiconductor field-effect transistor* (MOSFET), also known as *insulated gate field-effect transistor* (IGFET). The basic difference between the two is the construction and characteristics. The input impedance tending towards a value of infinity is possible in an IGFET.

8.2 WORKING PRINCIPLE OF JFET

The structure and construction of a junction field-effect transistor is as shown in Fig. 8.1. The channel may be n type or p type. Depending on the channel, the JFET is referred to as *n-channel or p-channel JFET*. The JFET consists of intrinsic semiconductor bar of a given type of material with opposite type of material fabricated around this bar by alloying, diffusion, or by any other techniques forming a p-n junction as shown in the figure. The concentration of this layer of semiconductor on the bar is very high. For example, p^+ region is alloyed to n type bar in an n-channel FET n-type. Three electrodes are brought out from this bar, two from each end of the bar and the other from outer layer on the surface of the bar. The three electrodes act as three terminals for the FET. The terminals are designated as *source, drain,* and *gate. Source* is the source of charge carriers, which is going to generate the majority charge carriers that flow in the channel of the FET. The source of the FET is identical to that of emitter in a BJT. *Drain* is the terminal that collects the remaining charge carriers after they have been regulated by the channel and gate. This terminal is identical to that of collector of BJT. Finally, *gate* of a FET is identical to that of the base of BJT, which is going to control the flow of charges from source to drain in the channel. The current is due to the majority carriers in the channel, that is, holes in a *p*-channel FET and electrons in an *n*-channel FET. The three currents entering the device are designated as I_D, I_S and I_G at drain, source and gate terminals of the FET, respectively. The voltages are designated by two-suffix terminology, as is the case with the transistor. For example, V_{DS} is the voltage between drain and source of the FET with source as the reference. Similarly, double suffix indicates that the supply voltage V_{DD} is the drain power supply.

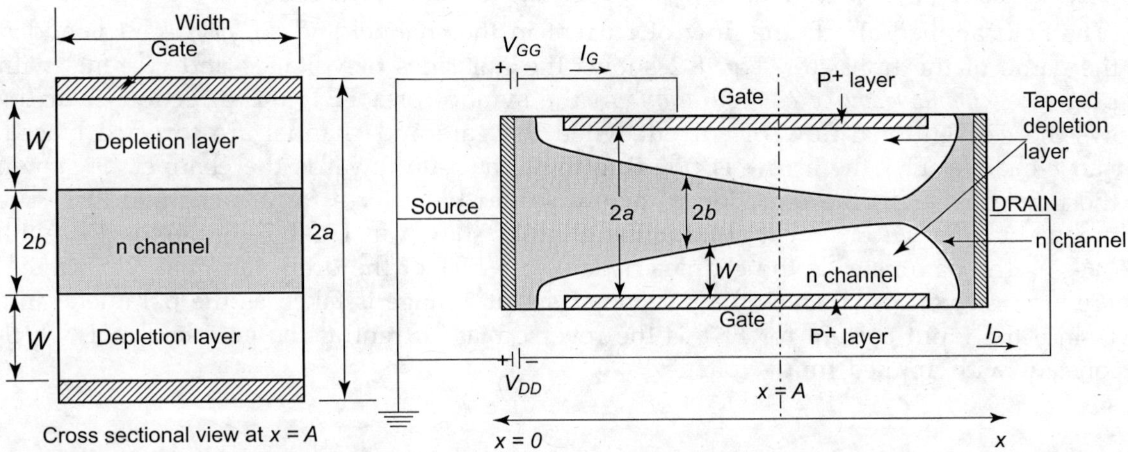

Fig. 8.1 *Structure of a JFET*

Operation of a JFET: When a power supply is connected between the drain and source of the FET, let us say, V_{DD}, assuming no bias is applied to the gate terminal, the channel of the FET acts as a simple resistor, and a constant current flows which is due to majority carriers in the channel. As the supply voltage increases, the current in the channel also increases almost linearly since the channel acts as a resistor offering resistance value depending on the conductivity and size of the channel under consideration. Considering an *n*-channel FET, as shown in Fig. 8.1, now if a battery V_{GG} is connected to gate of the device such that the junction between the gate and the channel is reverse biased, that is, negative voltage applied to the *p*-type gate, a depletion layer is formed at the junction. This layer penetrates very little into the gate whereas the penetration of the layer into the channel is large since the concentration of gate layer is very large compared to that of the channel. The penetration of the layer into the channel depends on the intensity of the bias applied at the gate terminal. The higher the amplitude of V_{DD}, the larger is the penetration into the channel. Since the depletion region is that region where there are no active charge carriers, as the penetration of this layer increases, the electrical dimensions of the channel decrease. In other words, as the reverse bias intensity is increasing, the channel is constricted and less number of majority carriers are allowed to pass from the source to drain through the channel, decreasing the drain current. Thus, maintaining the drain supply constant, if the gate reverse-bias supply is increased, the drain current decreases.

When a positive voltage is applied at the *p*-type gate of the FET, a forward bias is formed between the gate and channel, which enhances the majority carriers in the channel and so, the current in the channel, drain current increases. As the forward voltage increases, the drain current increases, but with a limitation that the forward voltage between the junction, V_{GS}, is restricted to cutin voltage of the junction. So, even though V_{GG} is increased beyond a value, V_{GS} is restricted to cutin voltage.

The field applied affects the flow of current in the channel and so, *field effect* is added in the name of the transistor. Fig. 8.2 shows the polarities of voltages and currents with respect to a *common source configuration.* In the symbol of a FET, the direction of arrow shows the conventional flow of current when the gate and channel are forward biased. For an *n*-channel FET the arrow is pointing from the gate towards the channel, as shown in the figure, whereas the direction of arrow would be reverse for a *p*-channel FET. The common source *drain* or *output characteristics* are as shown in Fig. 8.3. I_D versus V_{DS} with the V_{GS} as the parameter is the characteristic of the FET at the drain terminal. Unlike BJT, where input current is taken as the parameter, input voltage is taken as the parameter in a FET since the input current for FET is the reverse bias current at the gate terminal, which is constant with applied input voltage.

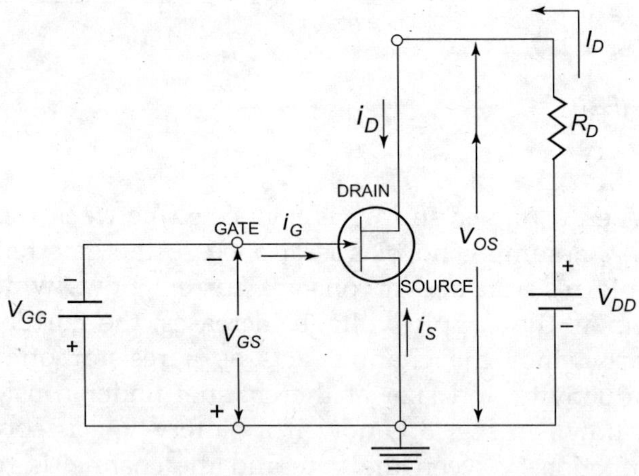

Fig. 8.2 *Common source FET*

8.3 DRAIN AND TRANSFER CHARACTERISTICS OF A JFET

To understand the drain characteristics of a JFET, consider the curve for $V_{GS} = 0$ for a common source configuration as shown in Fig. 8.3. When $I_D = 0$, the channel between the gate junction is entirely open. As V_{DS} is applied and increased, the *n*-type channel acts as a simple semiconductor resistor and I_D increases linearly with V_{DS}. As the current increases, the Ohmic voltage drop between the source and the channel region reverse

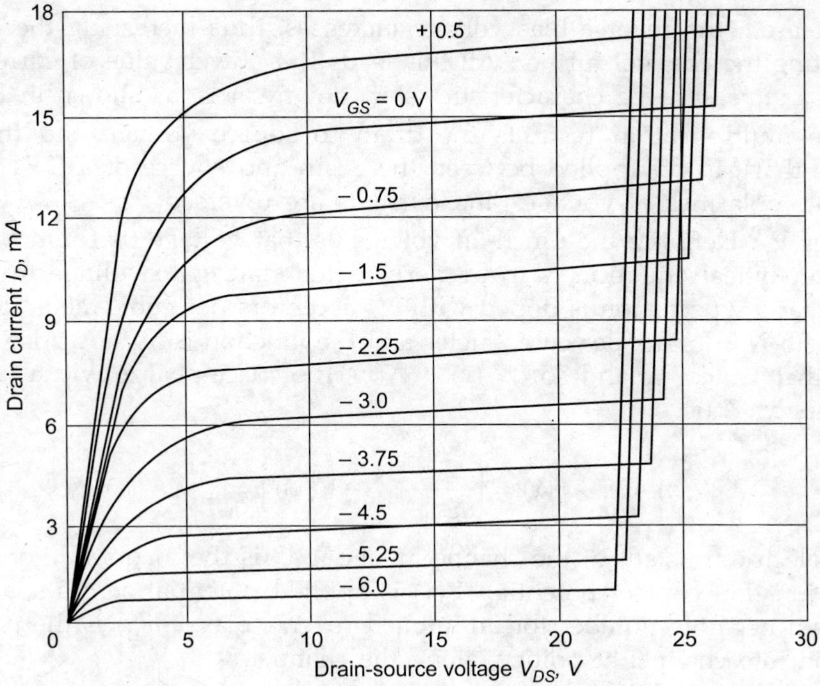

Fig. 8.3 *Drain characteristics of an n-channel JFET*

biases the junction and the conducting portion of the channel begins to constrict. The constriction of the channel is not uniform throughout the channel from source to drain due to the fact that the voltage drop is not uniform. When the reference is source, the applied voltage V_{DS} drops across the channel linearly from zero at the source to V_{DS} at the drain. So, voltage drop between the channel and the gate region is not uniform. When a constant voltage V_{GS} is applied on the gate region, the field between gate and channel is such that it reveres biases at the drain end compared to that at source end. In other words, the reverse bias voltage is more at the drain end of the channel compared to that at the source end. Since the width of the depletion layer depends on the extent of the reverse bias, the depletion layer will be very narrow at the source end and becomes wider as one moves towards the drain as shown in Fig. 8.1. As the voltage V_{DS} increases, the constriction of the channel increases and at a particular voltage, the *pinch-off* occurs. Beyond this voltage the current in the channel does not increase with V_{DS} but levels off and becomes constant. This voltage at which the current levels off is referred to as *pinch-off voltage* V_P. This voltage is not very well defined in the characteristic. In practice, total constriction of the channel is not possible but a constant current flows in the channel. As the voltage V_{DS} increases, the constriction remains constant, but expands towards the source, paving a way for a constant current to flow in the channel. So, the characteristic has two regions, one in which the channel is almost linear and acts as a resistor and the other where the current is almost constant. As the voltage V_{DS} increases to a very large value, as expected, avalanche breakdown occurs, causing the current in the channel to increase to a very large value at a given drain voltage.

As V_{GS} is increased, the reverse bias voltage increases, thus increasing the depletion layer and constricting the channel further. All this leads to a lower value of current in the channel. So, as V_{GS} increases, the characteristics shift downwards, as shown in Fig. 8.3.

The avalanche multiplication restricts the drain to source voltage and this is the maximum voltage that can be applied between the drain and source of a JFET. For a *p*-channel FET all the polarities of voltages and currents are reversed.

The pinch-off voltage V_P: Defining the pinch-off voltage as that voltage that removes all the free charges, an analytical equation is to be derived to relate gate voltage to pinch-off voltage. Assume that *p*-type region is doped with N_A acceptors per cubic meter, the *n*-type region is doped with N_D donors per cubic meter and the junction is abrupt. Also, $N_A >> N_D$ leads to the fact that $W_p \ll W_n$ and so, $W_n(x) \approx W(x)$, the space charge width which is a function of distance *x*. Thus,

$$W(x) = a - b(x) = \left(\frac{2\in}{qN_D}\{V_0 - V(x)\}\right)^{½} \qquad ...(8.1)$$

where $\in$ is the dielectric constant of the channel material. *Q* is the magnitude of electronic charge, V_o is the junction contact potential, $V(x)$ is applied potential across space charge region at *x* and is a negative number for an applied reverse bias, $a\text{-}b(x)$ is the penetration of depletion region into channel at point *x* along the channel.

If the drain current is zero, $b(x)$ and $V(x)$ are independent of *x* and $b(x) = b$. Substituting $b(x) = b = 0$ in Eq. (8.1) and solving for *V*, pinch-off voltage V_P is obtained. Hence,

$$|V_P| = [qN_D/2\in]a^2 \qquad ...(8.2)$$

Substituting for V_{GS},

$$V_{GS} = \left(1 - \frac{b}{a}\right)^2 VP \qquad ...(8.3)$$

The voltage V_{GS} represents the reverse bias across the gate junction and is independent of distance along the channel if $I_D = 0$.

The JFET volt-ampere characteristic: When a small voltage V_{DS} is applied between drain and source, the resulting small drain current I_D will have no appreciable effect on the channel profile. Under these conditions, the effective cross-section $A = 2bw$, $2b$ is the channel width corresponding to zero drain gate and *w* is the channel dimension perpendicular to the direction of *b*. Now, the drain current can be obtained as

$$I_D = AqN_D\mu_n E = 2bwqN_D\mu_n[V_{DS}/L] \qquad ...(8.4)$$

where *L* is the length of the channel. Substituting *b* from Eq. (8.3),

$$I_D = 2bwqN_D\mu_n = \left[1 - \left(\frac{V_{GS}}{V_P}\right)^{½}\right] V_{DS} \qquad ...(8.5)$$

The above equation gives the volt–ampere characteristics of JFET for very small V_{DS} and it behaves like an Ohmic resistance whose value is determined by V_{GS}. The ratio V_{DS}/I_D at the origin is called the ON *drain resistance* $r_{d\ ON}$. With $V_{GS} = 0$,

$$r_{dON} = L/[2awqN_D\mu_n] \quad ...(8.6)$$

In commercially available FETs, $r_{d\,ON}$ ranges from about 100 W to 100 kW. The bipolar junction transistor has an advantage over the FET that the R_{CS} is usually only few ohms and hence is much less than $r_{d\,ON}$. However, the FET has the advantage that it has no offset voltage as in the case of BJT and the characteristic passes through the origin, V_{DS} = 0 and I_D = 0.

The pinch-off region: Electric field E_x appears along the x axis and a substantial drain current I_D flows in the channel of the FET as already discussed in the previous sections. This leads to nonuniform depletion region. The current equation can be rewritten as follows:

$$I_D = 2b(x)wqN_D\mu_nE_x \quad ...(8.7)$$

As V_{DS} increases, E_x and I_D increase, where $b(x)$ decreases because the channel narrows and hence the current density $J = I_D/2b(x)w$ increases. It can be observed that now, complete pinch-off cannot occur since if it occurs, $b(x)$ would be zero and so J would tend towards infinity. This becomes a physically impossible condition and so is not possible. I_D and b become constant in this condition. It is found that the mobility is not constant with electric-field intensity and is inversely proportional to the electric field when the field is greater than 10^4 V/m. Or, in other words, the drift velocity v_x, which is the product of mobility and the electric-field intensity, is constant. Thus, as the electric field increases, the mobility decreases proportionately and so, the current remains constant. This is the reason for the constant current region in the drain characteristic of the JFET.

When V_{DS} is increased beyond this point when b is maintained to be very small constant, δ, this minimum width expands throughout the channel as shown in Fig. 8.4. The length of the minimum channel width L' increases as is indicated in the figure. With V_{DS}, L' increases when δ remains constant.

The transfer characteristic: The FET, when used as an amplifier, is biased beyond pinch-off, that is, in the *constant current region*. This region is also called *pentode or current saturation region*. Let the saturation drain current be I_{DS} and I_{DSS} be its saturation value when V_{GS} = 0. The relation between I_{DS} and V_{GS} is found to be approximated as parabola given as

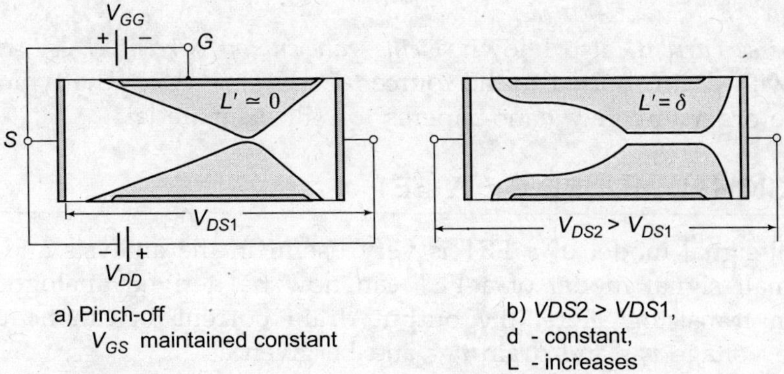

Fig. 8.4 *The pinch-off*

$$I_{DS} = I_{DSS}\left[1 - \frac{V_{GS}}{V_P}\right]^2 \qquad ...(8.8)$$

This simple parabolic fit is experimentally determined to be correct. The plot of this equation is referred to as *transfer characteristic* and is plotted as V_{GS} versus I_D with V_{DS} as a parameter and is as shown in Fig. 8.5.

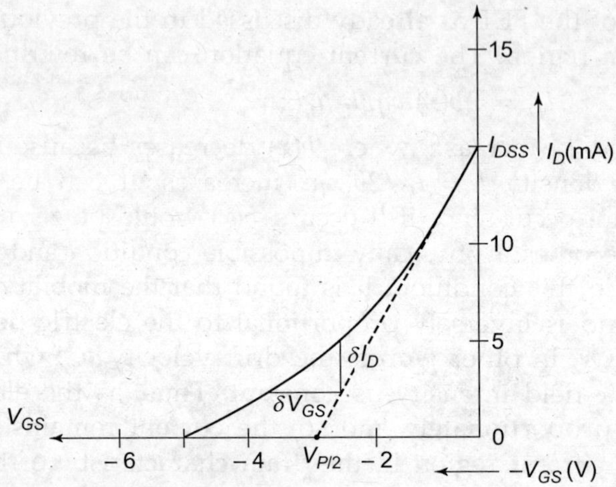

Fig. 8.5 *The transfer characteristics of a JFET*

Cut off: When V_{DS} is fixed in such a way that the FET is operating in the fixed current region, as V_{GS} is increased in the direction of reverse bias of the gate junction, the conducting channel will narrow. When $V_{GS} = V_P$, the channel width should reduce to zero theoretically, but in practice, a small leakage current, $I_{D\,\text{off}}$ flows even under cutoff condition $|V_{GS}| > |V_P|$. The manufacturer generally specifies the maximum value of $I_{D\,\text{off}}$ at a given values V_{DS} and V_{GS}. For a silicon FET, the value of $I_{D\text{off}}$ is of the order of nanoamperes.

The gate reverse current, also known as *the gate cut off current*. $I_{GSS,}$ gives the gate to source current with drain shorted to the source for $|V_{GS}| > |V_P|$. The typical value of this current is of the order of a few nanoamperes for silicon material.

8.4 SMALL SIGNAL MODEL OF A FET

The linear small signal model of a FET is very useful in the analysis and design of FET circuits. The small signal model of a FET can now be derived analogous to that of a bipolar junction transistor. Here, the output drain current i_D can be expressed as a function of gate voltage v_{GS} and drain v_{DS} and be given as

$$i_D = f(V_{GS,}\ V_{DS}) \qquad ...(8.9)$$

If both the gate and drain voltages are varied, the change in drain current is given approximated and can be given as

$$\Delta i_D = \left.\frac{\partial i_D}{\partial V_{GS}}\right|_{V_{DS}} \Delta V_{GS} + \left.\frac{\partial i_D}{\partial V_{DS}}\right|_{V_{GS}} \Delta V_{DS} \qquad \text{...(8.10)}$$

In the small signal notation, $\Delta i_D = i_d$, $\Delta V_{GS} = v_{gs}$ and $\Delta V_{DS} = V_{ds}$. And so,

$$i_d = g_m V_{gs} + (1/r_d)V_{ds} \qquad \text{...(8.11)}$$

where

$$g_m = \left.\frac{\partial i_D}{\partial V_{GS}}\right|_{V_{DS}} \text{ and } r_d = \left.\frac{\partial V_{DS}}{\partial i_D}\right|_{V_{GS}} \qquad \text{...(8.12)}$$

or

$$g_m = \left.\frac{i_D}{V_{GS}}\right|_{V_{DS}} \text{ and } r_d = \left.\frac{V_{DS}}{i_D}\right|_{V_{GS}} \qquad \text{...(8.13)}$$

where g_m is referred to as *mutual* or *transconductance.* It is also referred to as *common source forward transadmittance* g_{fs} or y_{fs}. r_d is the drain resistance.

For a FET, voltage amplification can be defined as the ratio of differential output voltage to differential change in the input when the output current is maintained constant. Thus, the *amplification factor* can be given as

$$\mu = \left.\frac{\partial V_{DS}}{\partial V_{GS}}\right|_{I_D} = -\left.\frac{V_{DS}}{V_{GS}}\right|_{id=0} \qquad \text{...(8.14)}$$

therefore,

$$\mu = g_m r_d \qquad \text{...(8.15)}$$

The transconductance of the FET can also be given another form as

$$g_m = g_m\left(1 - \frac{V_{GS}}{V_p}\right) = (2/V_p)(I_{DSS} I_{DS})^{1/2} \qquad \text{...(8.16)}$$

where g_{mo} is the transconductance at $V_{GS} = 0$. The short circuit transconductance can be derived as

$$g_{mo} = -2I_{DSS}/V_P \qquad \text{...(8.17)}$$

The above relationship among the three parameters is determined experimentally. It can be observed that the transconductance varies as square root of the drain current. G_{mo} takes a positive value as I_{DSS} and V_P are always of opposite polarity.

Curves of g_m and r_d versus temperature are as shown in Fig. 8.6. The drain current I_{DS} has the same temperature variation as does g_m. The principal reason for the negative temperature coefficient of I_{DS} is that the mobility decreases with increasing temperature. Since this majority carrier current decreases with temperature, the phenomenon of thermal runaway is not encountered with field-effect transistors.

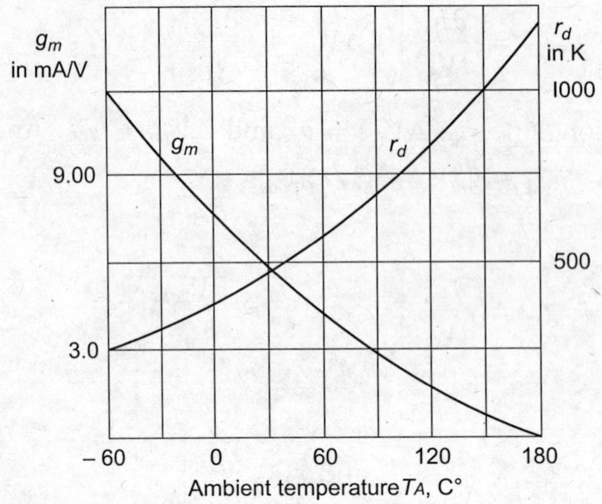

Fig. 8.6 *Temperature dependence of parameters*

Figure 8.7 shows a circuit that satisfies Eq. (8.11). This low-frequency small signal model has a Norton's output circuit with a dependent current generator whose current is proportional to the gate to source voltage. The proportionality factor is the transconductance g_m, which is consistent with the definition of g_m. The output resistance is r_d, and the input resistance between gate and source is infinite, since it is assumed that the reverse biased gate takes no current. For the same reason the resistance between gate and drain is assumed to be infinite.

The small signal model of a FET is as shown in Fig. 8.7. Since the output of this model is Norton's equivalent circuit, the output current is proportional to the input voltage, the proportionality constant being the transconductance. There is no feedback from output to input in a field-effect transistor at low frequencies like that in a BJT. In comparison with BJT, the FET is an ideal amplifier at low frequencies, but this is true only up to audio range of frequencies.

High-frequency model of FET: The high-frequency model of a FET is as shown in Fig. 8.8. This circuit is identical to that of the low-frequency model, as shown in Fig. 8.7, except that capacitances between the three pairs of terminals have been added. The intuitive capacitance between the gate and source is C_{gs} and similarly the capacitances C_{gd} and C_{ds}

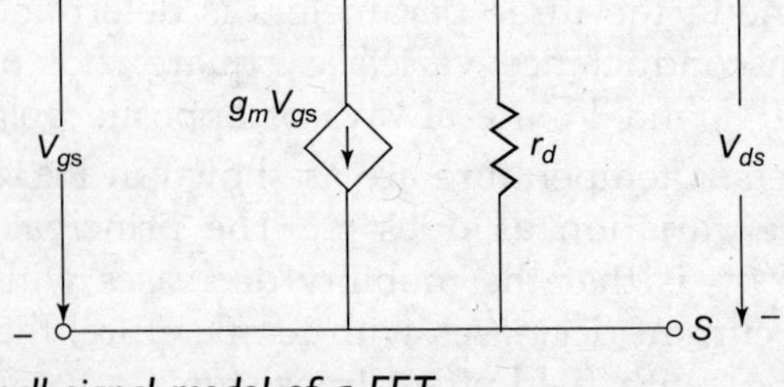

Fig. 8.7 *Low-frequency small signal model of a FET*

are intuitive capacitances between gate and drain and drain and source. These capacitances present a feedback between output and input internally and so the voltage amplification drops to a very small value as the frequency is increased. The typical values of these components for a JFET as well as MOSFET are presented in Table. 8.1.

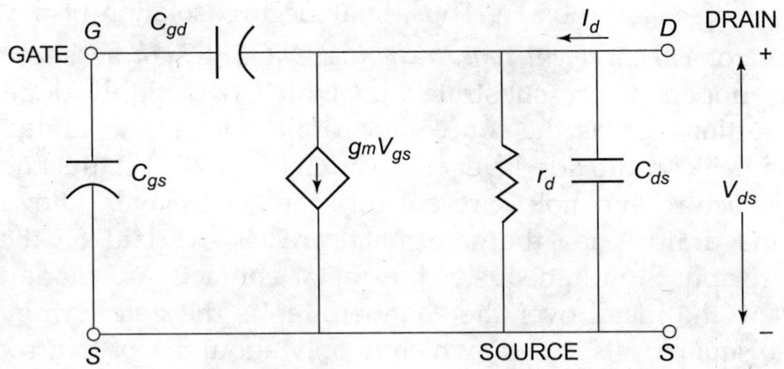

Fig. 8.8 *High-frequency model of a FET*

Table 8.1 ***Typical Values of FET small signal model parameters***

Parameter	*JFET*	*MOSFET*
g_m in mhos	7.5×10^{-3}	12×10^{-3}
r_d in ohms	450 K	60 K
C_{ds} in pF	0.45	0.6
C_{gs}	4.5	6.0
C_{gd}	6.5	8.0

Comparison of FET with BJT:

- The input impedance of a FET is very large compared to that of a BJT.
- FET is a voltage-controlled device whereas BJT is current controlled.
- There is no feedback from output to input in a FET; there is a feedback in a BJT through the parameter h_{re}.
- FET is an ideal amplifier up to audio frequencies.
- The switching speed of a FET is larger than that of a BJT.
- The gain bandwidth product of a FET is very small compared to that of a BJT.
- The high-frequency model of a FET is same as its low-frequency model except that three barrier capacitors between the three terminals are added. Whereas the high-frequency model of a BJT is very complicated and, in general, a simplified model only for *CE* configuration is presented.

8.5 MOSFET

The *metal oxide semiconductor field-effect transistor* (MOSFET), also known as *insulated gate field effect transistor* (IGFET), is another type of FET. The greatest advantage of this type of

FET is that it has very large input impedance compared to even that of a JFET. The order of this is extremely high, hundreds of megaohms, and even tending towards infinity. This feature is due to the fact that the gate of the device is *insulated* from the channel between the source and drain. The metal contact of the gate is separated from the semiconductor channel by an insulating layer of silicon dioxide. Thus, total electric isolation of the gate is achieved.

The construction of *enhancement-type p-channel MOSFET* is as shown in Fig. 8.9. It consists of a lightly doped *n*-type substrate into which two highly doped p^+ regions are diffused. These p^+ sections act as the source and drain and are separated by about 20–30 μm. A thin insulating silicon dioxide layer of about 1000 to 2000 Å thickness is grown over the surface of the structure and holes are cut into the oxide layer, allowing contact to be formed at source and drain. Then the gate metal area is overlaid on the oxide, covering the entire channel region. Simultaneously, the metal contacts are made to the drain and source. The contact to the metal over the channel area is the gate terminal. The chip area of a MOSFET is five square mils or less, which is only about 5% of that required by a polar junction transistor.

The enhancement MOSFET: Considering the MOSFET as shown in Fig. 8.9, if a voltage is applied between the source and drain, no current passes through the channel between the two terminals as the channel is *n* type between two *p*-type islands. Now, if a negative voltage is applied to the gate terminal, assuming that the substrate is at zero potential, the gate terminal is at negative potential and so attracts positive charge accumulation towards the gate edge in the insulating layer. Due to this an equal amount of negative charge is induced on the edge of the insulating layer near the semiconductor channel. This, in turn, attracts holes that are available in the substrate (*n* type) towards the insulating layer as shown in Fig. 8.9, forming an *inversion layer* between the two islands of drain and source. The holes accumulated at this end of the substrate depend on the intensity of the negative voltage at the gate terminal. The more is the voltage, the more is the accumulation of holes. This accumulation of holes between the two *p*-type islands forms a channel from the current to pass from the source to drain. The more the accumulation of holes, the more is the current in the channel, thus the name *enhancement* tagged with this type of MOSFET. Typical *drain and transfer characteristics* of enhancement MOSFET are as shown in Fig. 8.10.

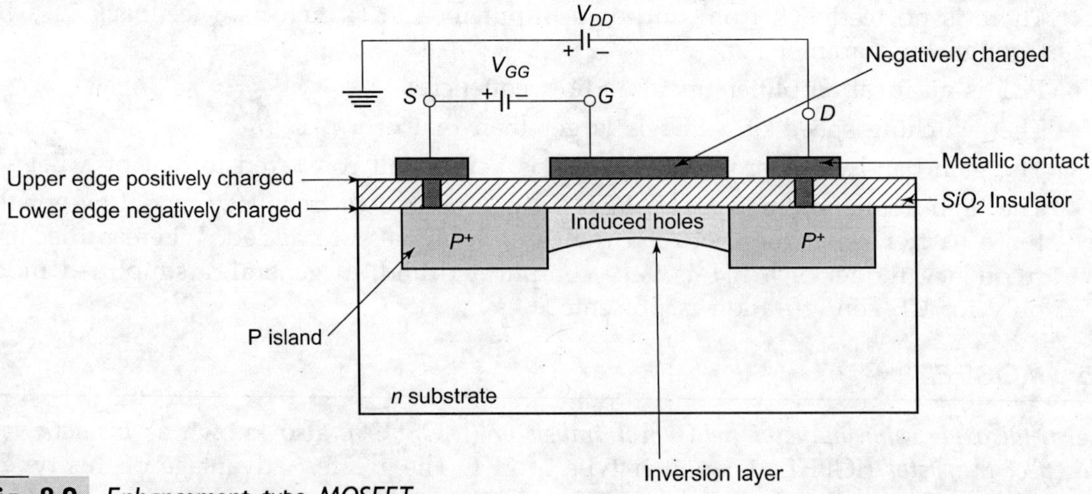

Fig. 8.9 *Enhancement type MOSFET*

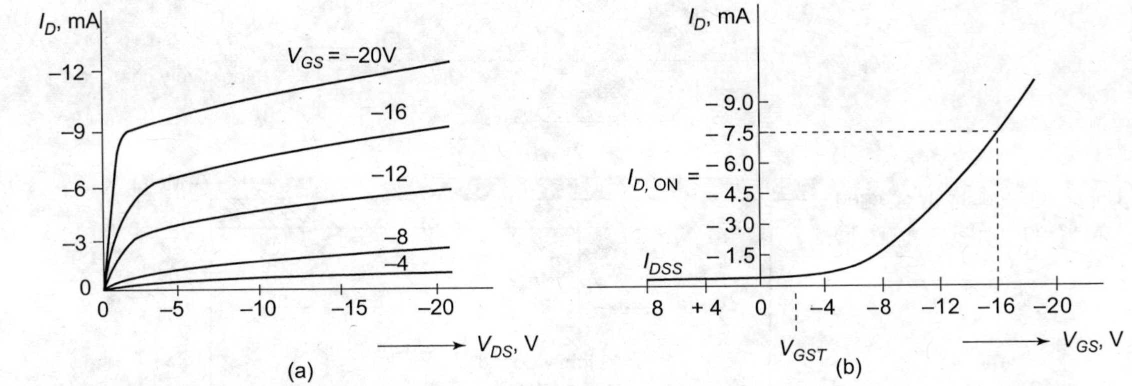

Fig. 8.10 *Drain and transfer characteristics of an enhancement MOSFET*

The transfer characteristic of the MOSFET shown above suggests that the increase in drain current is small for increments of gate to source voltage at small values, but at significant values of the gate voltage, the increase in the drain current is large. So, as observed from the curve, the drain current response is almost zero until the gate voltage reaches a *threshold voltage* V_{GST}. Typical value of V_{GST} for an enhancement MOSFET is –4 V. So, until the MOSFET is biased to be above –4 V or, in other words, above V_{GST}, the device is not useful. But in electronics, one requires the device to operate at very small voltages and the device is not very much useful. In order to use this device, the V_{GST} is to be reduced to a very small value. Smaller threshold voltages are required for the use of small power supply of drain, compatibility with bipolar devices, smaller switching time due to the smaller voltage swing during switching and adoptability of the device with the digital circuits. The threshold voltage can be reduced by choosing proper crystal orientation of the semiconductor, use of silicon nitride layer or use of polycrystalline silicon doped with boron as gate electrode instead of aluminum. With these techniques, the threshold voltage can be reduced to a large extent, up to 1.5 V.

The depletion MOSFET: The *depletion MOSFET* is another type of MOSFET, which has the drain characteristics similar to that of the JFET. The JFET cannot be operated beyond the cutin voltage of the forward biased gate-channel junction. This limitation is surpassed with that of depletion MOSFET and the current increases with increase in positive voltage for n-channel MOSFET and decreases with increase in negative voltage. Thus, the MOSFET is in enhancement mode when the gate voltage is positive and in depletion mode when the voltage is negative. One more advantage with this type of MOSFET is that the limitation with enhancement type MOSFET, the high value of threshold voltage, is also solved with this type of MOSFET. The depletion MOSFET can either be biased with negative or positive gate voltage without the limitation of threshold voltage.

The construction of *depletion MOSFET* is similar to that of *enhancement MOSFET* except that a lightly doped channel of same type as the source and drain islands is grown between the two islands as shown in Fig. 8.11. The figure shows the construction of n-channel depletion type MOSFET where two highly doped n^+ regions are grown in the p-type substrate. A lightly doped channel of n-type is also grown in the space between the two islands.

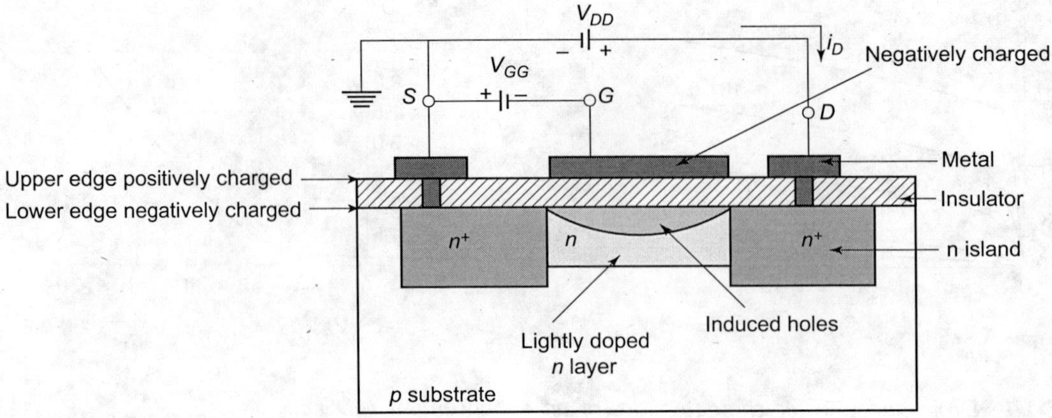

Fig. 8.11 *Depletion type MOSFET*

When the gate voltage is at zero potential and a supply is connected between the source and drain, the lightly doped layer acts as the channel between the source and drain and a current I_{DSS} flows between the drain and source terminals. When gate is made negative with respect to source, by the action of polarisation of the insulating layer, the holes in the channel are attracted towards the edge of the semiconductor towards the insulator as shown in Fig. 8.11. The accumulation of positive charges in the channel constricts the flow of charges between the drain and source and so the drain current decreases. One can observe that as the negative voltage to gate increases, the drain current decreases and the characteristics shift downwards as shown in Fig. 8.12. This action is similar to that of a JFET. On the other hand, as a positive voltage is applied to the gate of the device, the *n*-channel attracts more and more number of electrons from the substrate, adding more and more number of mobile charge carriers, electrons, into the channel. So, the drain current increases with increase of positive voltage at the gate and the characteristics shift upwards as shown in Fig. 8.12. This is similar to the characteristics of the MOSFET of enhancement type and so this type of MOSFET exhibits depletion mode when the gate voltage is negative and enhancement mode when the gate voltage is positive.

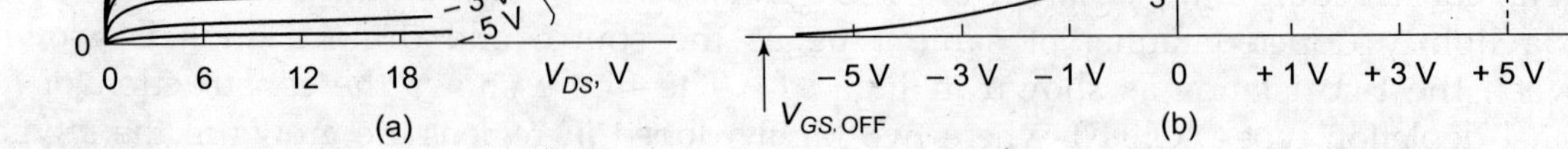

Fig. 8.12 *Drain and transfer characteristics of depletion MOSFET*

Figure 8.12 also shows the transfer curve of the depletion MOSFET. The depletion and enhancement regions are indicated in the figure. The gate source cutoff voltage is indicated in the catalogue of the FET supplier. At this voltage the drain current reduces to a pre-specified value at given drain source voltage. This gate source voltage is the pinch off voltage of the JFET. For a *p*-channel MOSFET the above discussion holds good, except the voltages and currents are to be reversed in polarities.

Since the insulating silicon-dioxide layer is extremely thin, it may easily be damaged by excessive voltage. An accumulation of charge on an open-circuited gate may result in a large-enough field to punch through the dielectric. To prevent this damage some MOS devices are fabricated with a Zener diode between gate and substrate. In normal operation, this diode is open and has no effect on the circuit. However, if the voltage at the gate becomes excessive, then the diode breaks down and the gate potential is limited to a maximum value equal to the Zener voltage.

Figure 8.13 shows various symbols for MOSFET. The symbols are for *n*-channel MOSFET. In some cases, the substrate terminal is accessible and in some cases it is not. Figure 8.13(c) shows the symbol for specifically enhancement MOSFET. For a *p*-channel MOSFET, the direction of the arrow is reversed.

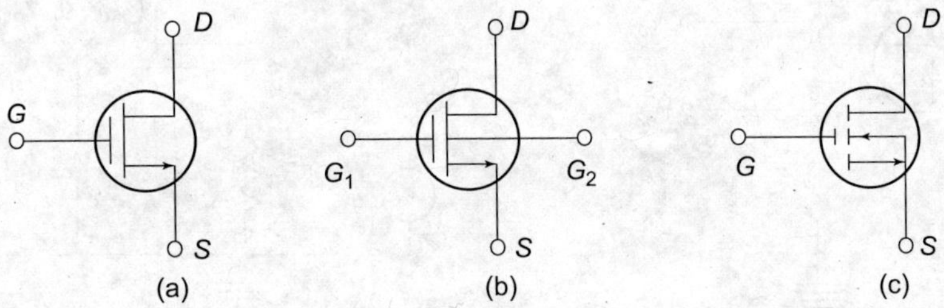

Fig. 8.13 *Three circuit symbols for an n-channel MOSFET*

The small signal model of a MOSFET is identical to that of a JFET except for the typical values of the parameters. The parameters for a MOSFET is listed in Table 8.1.

8.6 ANALYSIS OF FET AMPLIFIER

Amplifier circuits with FET can be either *common source* (*CS*) or *common drain* (*CD*) as common emitter or common collector in BJT, respectively. The circuits of the *CS* and *CD* configuration of FET amplifiers are as shown in Fig. 8.14.

For analysis of these two types of configurations, a generalised circuit, as shown in Fig. 8.15, is considered.

For *CS* configuration, the output is taken at v_{o1} at the drain with source resistance $R_s = 0$ and for *CD* configuration, the output is source v_{o2} with the drain resistance $R_d = 0$. The source resistance of the signal source at the input need not be considered and can be treated as negligible since the input impedance of the FET is assumed to be infinite. The device is assumed to be properly biased in the linear region of operation.

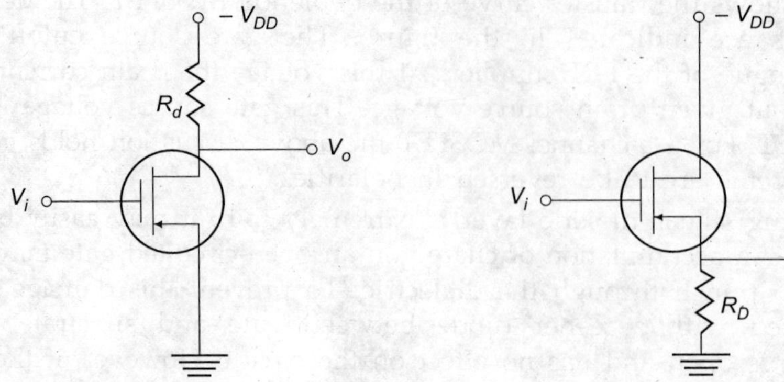

Fig. 8.14 *CS and CD configurations of a FET amplifier*

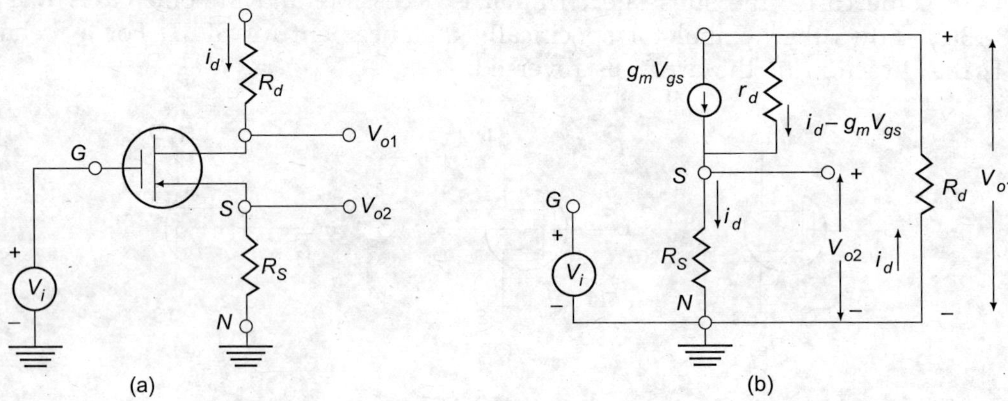

Fig. 8.15 *Generalised FET amplifier model for analysis*

Figure 8.15(b) shows the circuit when the FET is replaced with its small signal model. Applying KVL to the output,

$$i_dR_d + (i_d - g_mv_{gs})r_d + i_dR_s = 0 \qquad \text{...(8.19)}$$

And from input,

$$V_{gs} = v_i - i_dR_s \qquad \text{...(8.20)}$$

From the above two equations and Eq. (8.15),

$$i_d = \mu v_i \,/\, [\, \mathrm{r_d} + \mathrm{R_d} + (\mu + 1)\, \mathrm{R_s}] \qquad \text{...(8.21)}$$

CS amplifier with unbypassed source resistor: The output voltage for a CS amplifier is v_{o1}, which can be given as $-i_d\, R_d$. Then,

$$v_{o1} = -\,\mu v_iR_d / [r_d + R_d + (\mu + 1)R_s] \qquad \text{...(8.22)}$$

The voltage gain A_v can be given as $v_{o1}/\, v_i$. The minus sign in the equation indicates that the output is 180° out of phase with the input. If R_s is bypassed with a large capacitance, the above equation is valid with $R_s = 0$. Therefore,

$$A_v = v_{o1}/v_i = -\mu R_d/[r_d + R_d] = -g_m R_d' \quad ...(8.23)$$

where $\mu = g_m R_d$ and $R_d' = R_d || r_d$

The CD amplifier with a drain resistance: Since $v_{o2} = i_d R_s$, then from Eq. (8.20),

$$v_{o2} = \mu v_i R_s/[r_d + R_d + (\mu + 1)R_s]$$

$$= (\mu v_i/[\mu + 1])R_s/[(r_d + R_d)/(\mu + 1) + R_s] \quad ...(8.24)$$

The open-circuit voltage is $\mu v_i/[\mu + 1]$ and the output resistance is $(r_d + R_d)/(\mu + 1)$. The voltage gain is $A_v = v_{o2}/v_i$. Note that there is no phase shift between the input and output. If $R_d = 0$ and if $(\mu + 1)R_s \gg r_d$, then $A_v \approx \mu/(\mu + 1) \approx 1$ for $\mu \gg 1$. The voltage gain of unity means that the output at source follows the input at the gate. Hence, *CD* configuration is called *source follower*, which is analogous to emitter follower in a BJT.

8.7 FET BIASING CIRCUITS

The biasing of a field-effect transistor is not as critical as of biasing a bipolar junction transistor since the operating point selection in a FET is based only on three parameters, namely, I_D, V_{GS} and V_{DS}, whereas for a BJT the dimensions of operating point are four in number. If the FET were to be used as an amplifier, the operating point should be selected in the constant current region. The advantage of this type of amplifier is that it has very low output impedance and in the device is voltage controlled. There are various parameters that can affect selection of biasing the FET. The most important of them are total voltage change, distortion, change in current, voltage gain, power dissipation etc. It is not possible to bias the FET to satisfy all the parameter variations all at a time, but the FET is to be biased for given pre-specified application. There are various types of methods of biasing a FET, of which some of them are discussed as follows.

Fixed bias: The fixed-bias circuit of a field-effect transistor is as shown in Fig. 8.16. At the gate terminal of the circuit the current drawn by the FET is negligible and so the selection of the operating point depends only on the gate to source voltage V_{GS}. This is due to the

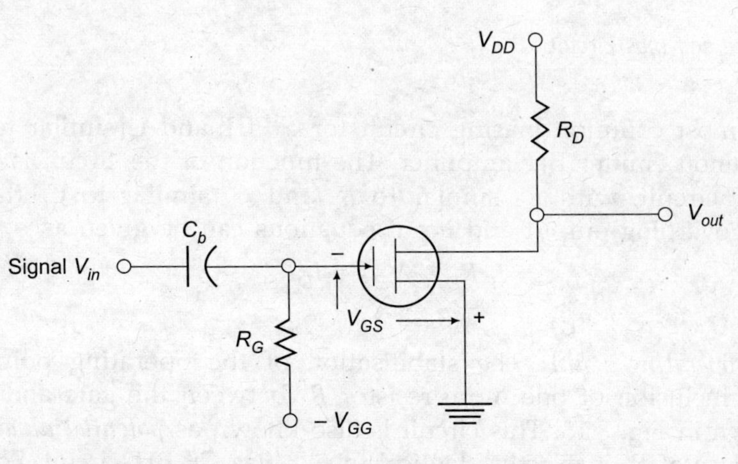

Fig. 8.16 *Fixed-bias circuit for FET*

fact that the FET is a high input impedance device. The *DC* bias at the gate is obtained using the supply V_{GG}. This ensures that the gate to channel junction is reverse biased and a resistor R_G is included in the circuit to allow the ac signal coupling through the capacitor *C*.

Thus, the gate to source voltage V_{GS} can be given as

$$V_{GS} = -V_{GG} \qquad ...(8.25)$$

The drain current I_D can be determined from the following equation:

$$V_{DD} = I_D R_D + V_{DS} \qquad ...(8.26)$$

Source self-bias: Figure 8.17 shows the circuit for biasing the JFET or depletion mode MOSFET. For a specified drain current I_D, the corresponding gate to source voltage V_{GS} can be obtained applying Eq. (8.8). Since the gate current is negligible, the source resistance R_S can be found as a ratio of V_{GS} to the desired I_D.

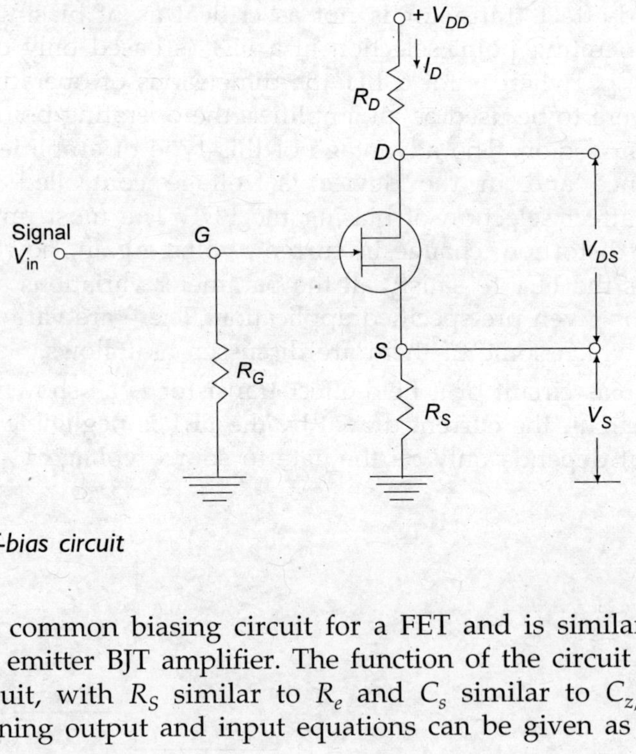

Fig. 8.17 *Source self-bias circuit*

This is the most common biasing circuit for a FET and is similar to that of self-bias circuit of a common emitter BJT amplifier. The function of the circuit is also analogous to that of the BJT circuit, with R_S similar to R_e and C_s similar to C_z, the emitter bypass capacitor. The governing output and input equations can be given as

$$V_{DD} = I_D R_D + V_{DS} + I_D R_S \qquad ...(8.27)$$

$$V_{GS} = -I_D R_S \qquad ...(8.28)$$

Stabilisation of operating point: The stabilisation of the operating point can further be improved with inclusion of one more resistor R_1 between the gate and the drain supply voltage as shown in Fig. 8.18. This circuit is also known as *potential divider biasing* and the resistor R_2 in place of R_g is exactly identical to self-bias *CE* BJT circuit. So, the input circuit can be visualised as a Thevenin equivalent and the Thevenin voltage and impedance can

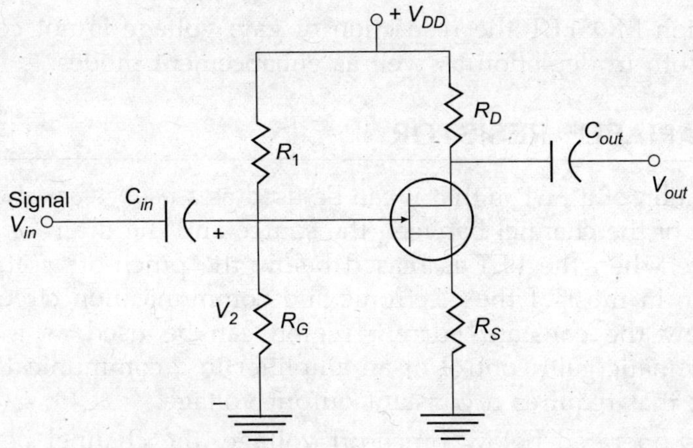

Fig. 8.18 *Potential divider biasing for FET*

be given as

$$V_{GG} = R_2 V_{DD}/(R_1 + R_2) \quad \text{...(8.29)}$$

$$R_g = R_1 R_2/(R_1 + R_2) \quad \text{...(8.30)}$$

Gate protection for enhancement MOSFET: The enhancement MOSFET cannot be biased for forward-biased gate channel junction since conduction between source and drain of the FET is not possible in this condition. When a very high forward bias voltage is applied at the gate of the device, since the thickness of insulating silicon dioxide layer is very small, the possibility of the breakdown of the layer is very high, which may damage the device permanently. So, one has to be very careful while working with enhancement MOSFET that a forward bias gate voltage is never applied to the device. This has to be taken care in the biasing circuit itself and the circuits as shown in Fig. 8.19 are the ones that do not allow this condition to take place. This type of circuits are only meant for enhancement

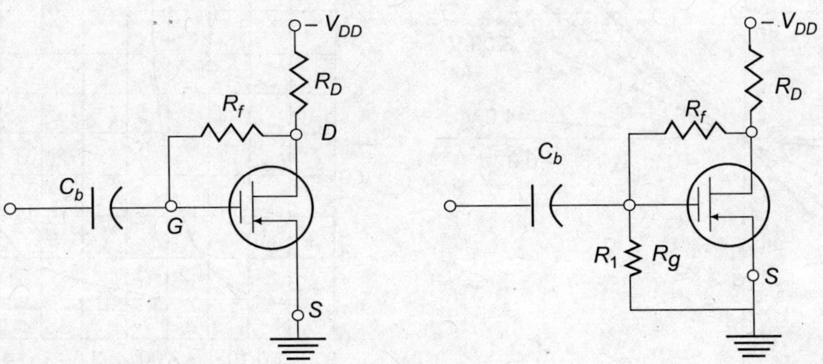

Fig. 8.19 *Biasing circuits for enhancement MOSFET*

MOSFET. For depletion MOSFET the restriction of gate voltage is not constant since this device can be used both in depletion as well as enhancement modes.

8.8 VOLTAGE VARIABLE RESISTOR

Another wide application of a FET is that it can be used as a *voltage variable resistor*. In other words, the resistance of the channel between the source and the drain of a FET is variable with the gate voltage when the FET is biased below the pinch-off voltage. This finds a very good application in most of the electronic and communication circuits. For example, the FET biased below the constant current region can be used as a voltage variable resistor (VVR) in automatic gain control of an amplifier in a communication receiver or in any electronic circuit that requires a constant output voltage.

When the FET is operated below pinch-off voltage, the channel between drain and source of a FET acts as a simple resistor. The resistance offered by this channel depends upon the thickness of the channel. The thickness of the channel, in turn, depends upon the biasing between the gate and channel. The larger the reverse bias to the gate, the smaller the channel width and larger is the resistance offered by the channel. As the reverse bias gate voltage varies, the slope of the characteristic changes in the output drain characteristic of the FET. When the FET is biased to be used in this region of operation, the resistance of the device can be controlled with the gate voltage and hence the name *voltage variable resistor* or *voltage dependent resistor (VDR).*

Fig. 8.20 shows the low-level bidirectional characteristic of a FET. The slope of these characteristics is r_d, which is a function of V_{GS}. In the figure the characteristics are extended into the third quadrant so as to get an understanding of the linearity of the device for small values of V_{DS}.

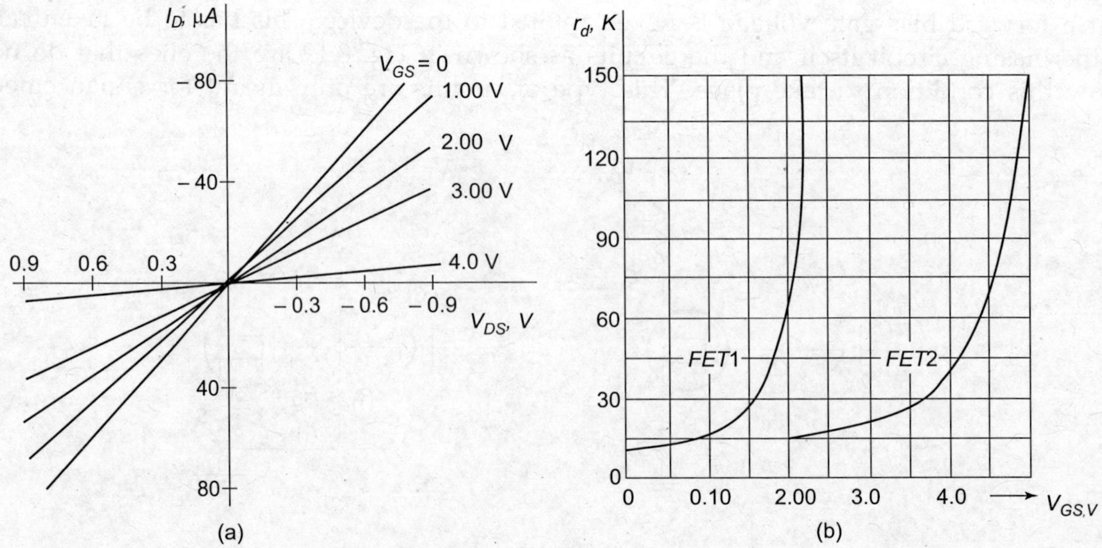

Fig. 8.20 *FET low-level drain characteristics*

The ratio of drain current to voltage is drain conductance, i.e., $g_d = I_D/V_{DS}$. This can be given, for small values as

$$g_d = g_{do}[1 - (V_{GS}/V_P)^{1/2}] \qquad ...(8.31)$$

where g_{do} is drain conductance at zero bias. The drain resistance varies with gate voltage and can be given as

$$r_d = r_o/[1 - KV_{GS}] \qquad ...(8.32)$$

where r_o is drain resistance at zero bias, K is a constant which depends on the type of FET and V_{GS} is the voltage between gate and source.

SUMMARY

- Field-effect transistor is a unipolar, easily fabricatable, low noise, no offset, high-input-impedance device. Drawback is low gain bandwidth product.
- The terminals drain, gate and source of a FET are identical to collector, base and emitter of a BJT.
- There are two types of junction FETs—*n* channel and *p* channel.
- The drain characteristics of FET are similar to that of output characteristics of a *CE* transistor, except that the parameter is not input current but input voltage.
- Since the input junction is reverse biased in a JFET, the current drawn is very small and so the input impedance is very high.
- The small signal model of a FET has two parameters—transconductance and drain resistance.
- Pinch-off voltage is the voltage beyond which the drain current levels off with drain voltage. Or, it is the voltage at which the constriction in the channel begins.
- MOSFET or IGFET has a very large input impedance, of the order of megaohms. Two types of MOSFET are depletion type and enhancement type.
- The threshold voltage is the voltage beyond which the FET is usable. This voltage poses restrictions on enhancement-type MOSFET, but is relaxed in depletion-type MOSFET.
- The FET can be used as amplifier in *CS* or *CD* configurations. *CS* is similar to *CE* and *CD* is similar to *CC* configurations in a BJT.
- Various biasing circuits are possible with a FET. Special care is required for biasing enhancement MOSFET.
- When the FET is operated below pinch-off voltage, it acts as a voltage variable resistor.

SOLVED PROBLEMS

8.1 Find the pinch-off voltage of a silicon *p*-channel FET having half channel height of 4 microns and channel resistivity of 10 Ω-cm. Dielectric constant of silicon is 12 and mobility of holes is 500 cm²/V-s.

Solution Expression for pinch-off voltage $V_p = \dfrac{a^2 qN_A}{2\varepsilon}$

$$qN_A = \frac{qN_A \cdot \mu_p}{\mu_p} = \frac{\sigma}{\mu_p} = \frac{1}{10 \times 500} = 2 \times 10^{-4}$$

$$\varepsilon = \varepsilon_o \varepsilon_r = 8.849 \times 10^{-12} \times 12$$

$$= 106.189 \times 10^{-12} \text{ F/m}$$

$$= 1.061 \times 10^{-12} \text{ F/cm}$$

$$a = 4 \times 10^{-6}\text{m} = 4 \times 10^{-4} \text{ cm}$$

$$\therefore \quad V_p = \frac{(2 \times 10^{-4}) \times 4.0 \times 10^{-14}}{2 \times 1.061 \times 10^{-12}}$$

$$= 7.54 \text{ V.}$$

8.2 Find the pinch-off voltage of a Ge *P*-channel FET with half channel height 1 μm and resistivity of 2 Ω-cm. The relative dielectric constant of Ge is 16 and mobility of holes is 1800 cm²/V-s.

Solution

$$qN_A = \frac{\sigma}{\mu_p} = \frac{1}{2 \times 1800} = 2.77 \times 10^{-4}$$

$$\varepsilon = \varepsilon_o \varepsilon_r = 16 \times 8.849 \times 10^{-12} = 1.416 \times 10^{-12} \text{ F/cm}$$

$$V_p = \frac{a^2 qN_A}{2\varepsilon} = \frac{(1 \times 10^{-4})^2 \times 2.77 \times 10^{-4}}{2 \times 1.416 \times 10^{-12}}$$

$$= 1.956 \text{ V}$$

8.3 Show that for small values of V_{GS} compared with V_P, the Drain current is given approximately as $I_D \approx I_{DSS} + g_{mo} V_{GS}$.

Solution We have

$$I_D = I_{DSS}\left(1 - \frac{V_{GS}}{V_P}\right)^2$$

$$= I_{DSS}\left(1 - 2 \cdot \frac{V_{GS}}{V_P} + \left(\frac{V_{GS}}{V_P}\right)^2\right)$$

Ignoring higher order terms,

$$I_D \approx I_{DSS} - 2\frac{I_{DSS}\cdot V_{GS}}{V_P}$$

We have $$g_{mo} = \frac{-2I_{DSS}}{V_P}$$

$\therefore$ $$I_D \approx I_{DSS} + g_{mo}\cdot V_{GS}$$

8.4 Show that the transconductance g_m of a JFET is related to the Drain Current I_{DS} as

$$g_m = \frac{2}{|V_P|}\sqrt{I_{DS}\cdot I_{DSS}}$$

Solution We have $$I_{DS} = I_{DSS}\left(1-\frac{V_{GS}}{V_P}\right)^2 \quad ...(1)$$

$$g_m = \frac{dI_{DS}}{dV_{GS}} = 2I_{DSS}\left(1-\frac{V_{GS}}{V_P}\right)\cdot\left(-\frac{1}{V_P}\right)$$

$$= \frac{-2I_{DSS}}{V_P}\left(1-\frac{V_{GS}}{V_P}\right)$$

$$= g_{mo}\left(1-\frac{V_{GS}}{V_P}\right) \quad ...(2)$$

where $g_{mo} = \frac{-2I_{DSS}}{V_P}$ is the transconductance at $V_{GS} = 0$

from (1) $$\left(1-\frac{V_{GS}}{V_P}\right) = \sqrt{\frac{I_{DS}}{I_{DSS}}} \quad ...(3)$$

Substituting (3) in (2)

$$g_m = g_{mo}\sqrt{\frac{I_{DS}}{I_{DSS}}} = \frac{-2I_{DSS}}{V_P}\sqrt{\frac{I_{DS}}{I_{DSS}}}$$

$$= \frac{-2}{V_P}\sqrt{I_{DS}\cdot I_{DSS}}$$

$$= \left|\frac{2}{V_P}\right|\sqrt{I_{DS}\cdot I_{DSS}}$$. Hence proved.

8.5 (a) For the FET draw the transfer characteristic curves with I_{DSS} = 10 mA and V_P = – 4 V.

(b) The magnitude of the slope of this curve at V_{GS} = 0 is g_{mo}. If the slope is extended as a tangent, show that it intersects the V_{GS} axis at the point $V_{GS} = V_P/2$.

Solution (a)

$$I_{DS} = I_{DSS}\left(1 - \frac{V_{GS}}{V_P}\right)^2$$

$$= 10\left(1 + \frac{V_{GS}}{4}\right)^2 \quad ...(1)$$

at $V_{GS} = 0$, I_{DS} = 10 mA

$V_{GS} = -1$, I_{DS} = 5.625 mA

$V_{GS} = -2$, I_{DS} = 2.5 mA

$V_{GS} = -3$, I_{DS} = 0.625 mA

$V_{GS} = -4$, I_{DS} = 0 mA

The transistor characteristics are plotted as shown below:

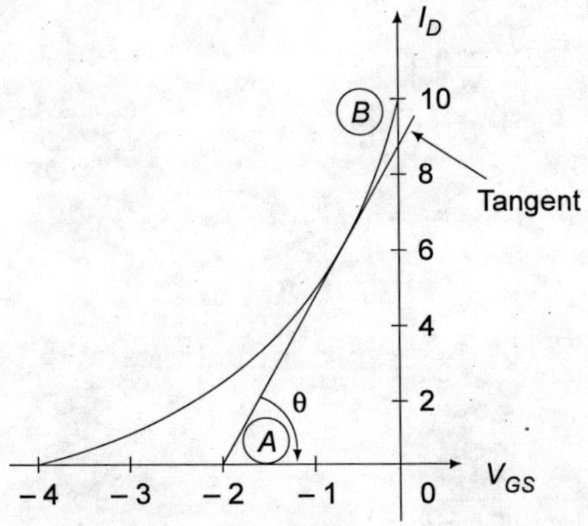

(b) Draw the tangent at point A, with V_{GS} = 0.

$$\text{Slope} = g_{mo} = \tan\theta = \frac{OB}{OA}.$$

$$\text{X intercept } OA = \frac{OB}{g_{mo}} = \frac{I_{DSS}}{-2I_{DSS}/V_p} = -\frac{V_p}{2}$$

8.6 (a) If two identical FET's are connected in parallel, show that g_m is doubled and r_d is halved. Since $\mu = r_d \cdot g_m$ then μ remains unchanged.

(b) If the two FETs are not identical, then show that $\frac{1}{r_d} = \frac{1}{r_{d1}} + \frac{1}{r_{d2}}$ and that

$$\mu = \frac{\mu_1 r_{d2} + \mu_2 r_{d1}}{r_{d1} + r_{d2}}.$$

Solution First consider that two FETs are not identical.

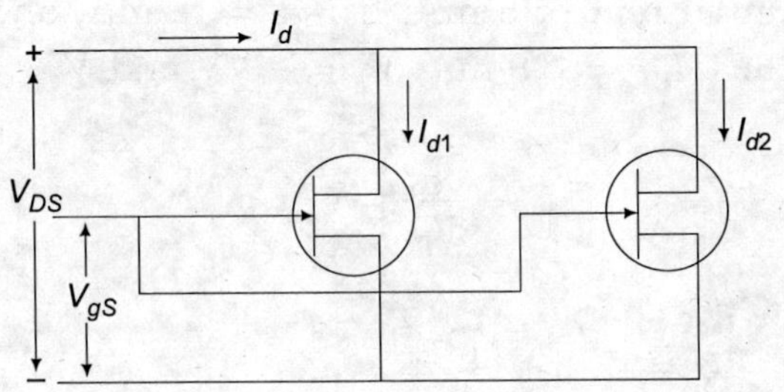

From circuit

$$\begin{aligned} I_d &= I_{d1} + I_{d2} \\ &= g_{m1}V_{gs} + g_{m2}V_{gs} \\ &= (g_{m1} + g_{m2})V_{gs} \qquad \text{...(1)} \end{aligned}$$

Considering $V_{ds} = 0$

$\therefore$ Total $g_m = g_{m1} + g_{m2} = 2g_m$ for identical stages

Consider $V_{GS} = 0$,

$$I_d = I_{d1} + I_{d2} = \frac{V_{DS}}{r_{d1}} + \frac{V_{DS}}{r_{d2}}$$

$$\therefore \quad \frac{1}{r_d} = \frac{I_d}{V_{DS}} = \frac{1}{r_{d1}} + \frac{1}{r_{d2}} \qquad \text{...(2)}$$

$$\therefore \quad \frac{1}{r_d} = \frac{1}{r_{d1}} + \frac{1}{r_{d1}} = \frac{2}{r_{d1}} \text{ for identical stages.}$$

$$r_d = r_{d1}/2 \text{ for identical stages.}$$

$\therefore$ g_m is doubled, r_d is halved, so μ does not change for identical stages.

(b) If FETs are not identical, then

$$g_m = g_{m1} + g_{m2}$$

$$\frac{1}{r_d} = \frac{1}{r_{d1}} + \frac{1}{r_{d2}}$$

$$\therefore \qquad \mu = g_m r_d = (g_{m1} + g_{m2})\frac{r_{d1} r_{d2}}{r_{d1} + r_{d2}}$$

$$= \frac{g_{m1} r_{d1} r_{d2} + g_{m2} r_{d1} r_{d2}}{r_{d1} + r_{d2}}$$

$$= \frac{\mu_1 r_{d2} + \mu_2 r_{d1}}{r_{d1} + r_{d2}}$$

8.7 (a) A JFET has a drain current of 6 mA. If I_{DSS} = 12 mA, and V_P = – 4 V, find V_{GS}.
(b) For an *n*-channel FET I_{DSS} = 5.8 mA, V_P = – 3 V and V_{GS} = – 2 V, find I_D, g_m, g_{mo}.

Solution (a)

$$I_{DS} = I_{DSS}\left(1 - \frac{V_{GS}}{V_P}\right)^2$$

$$6 = 12\left(1 + \frac{V_{GS}}{4}\right)^2$$

$$1 + \frac{V_{GS}}{4} = \left(\frac{1}{2}\right)^{1/2} = 0.707 \;\therefore\; V_{GS} = 4(0.707 - 1) = -1.17 \text{ V.}$$

(b)

$$I_{DS} = I_{DSS}\left(1 - \frac{V_{GS}}{V_P}\right)^2$$

$$= 5.8\left(1 - \frac{(-2)}{(-3)}\right)^2$$

$$= 0.65 \text{ mA.}$$

$$g_{mo} = \frac{-2I_{DSS}}{V_P} = \frac{-2 \times 5.8}{-3} = 3.87 \text{ mA/V}$$

$$g_m = g_{mo}\left(1 - \frac{V_{GS}}{V_P}\right) = 3.87\left(1 - \frac{2}{3}\right)$$

$$= 1.29 \text{ mA/V}$$

8.8 Determine the drain current I_D and V_{DS} for the fixed bias circuit with V_{GG} = – 2 V, R_g = 2 MΩ, V_{DD} = + 12 V, and R_d = 2 kΩ given that I_{DSS} = 8 mA and V_P = – 5 V.

Solution The fixed bias circuit is as shown below:

$$V_{GS} = V_{GG} = -2 \text{ V}$$

$$I_D = I_{DS} = I_{DSS}\left(1 - \frac{V_{GS}}{V_P}\right)^2$$

$$= 8\left(1-\frac{-2}{-5}\right)^2$$

$$= 2.88 \text{ mA}$$

$$V_D = V_{DD} - I_D \cdot R_d$$

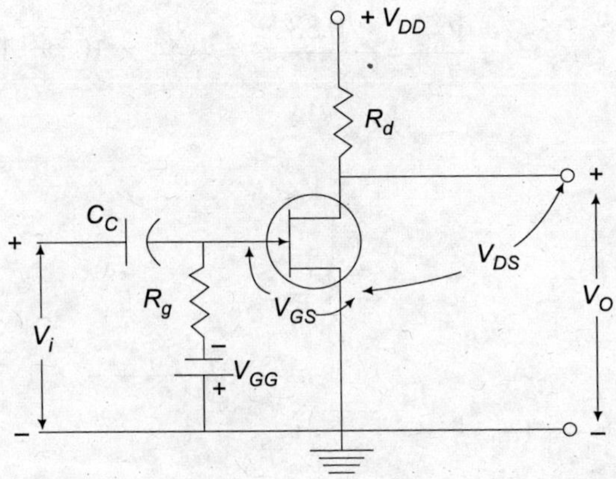

$$= 12 - 2.88 \times 10^{-3} \times 2 \times 10^3$$

$$= 6.24 \text{ V}$$

$$V_{DS} = V_D - V_S = 6.24 - 0 = 6.24 \text{ V}$$

8.9 For the circuit shown in the next page, $V_P = -3.0$ V, $I_{DSS} = 4$ mA. It is desired to bias the circuit at $I_D = 2$ mA using $V_{DD} = 24$ V. Assume $r_d \gg R_d$. Find (a) V_{GS} (b) g_m and R_s (c) R_d such that at least 32 dB of voltage gain to be present with R_s by passed by C_S.

Solution (a)

$$I_D = I_{DSS}\left(1-\frac{V_{GS}}{V_P}\right)^2$$

$$2 = 4\left(1-\frac{V_{GS}}{-3}\right)^2$$

$\Rightarrow$

$$V_{GS} = -0.88 \text{ V}$$

(b)

$$g_{mo} = -\frac{2I_{DSS}}{V_P} = \frac{-2\times 4}{-3}$$

$$= 2.67 \text{ mA/V}$$

$$g_m = g_{mo}\left(1 - \frac{V_{GS}}{V_P}\right)$$

$$= 2.67\left(1 - \frac{-0.88}{-3}\right)$$

$$= 1.87 \text{ mA/V}$$

$$R_s = -\frac{V_{GS}}{I_D} = \frac{0.88}{2 \times 10^{-3}} = 0.44 \times 10^3 = 440 \ \Omega$$

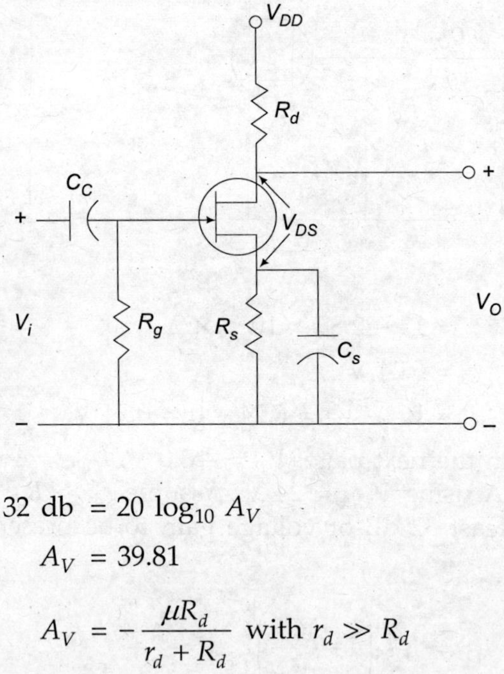

(c)

$$32 \text{ db} = 20 \log_{10} A_V$$

$$A_V = 39.81$$

$$A_V = -\frac{\mu R_d}{r_d + R_d} \quad \text{with } r_d \gg R_d$$

$$A_V = -\frac{\mu R_d}{r_d} = \frac{-g_m r_d \cdot R_d}{r_d} = -g_m R_d$$

Since 32 dB of voltage gain is to be present, $|A_V| \geq 39.81$

$\therefore \quad g_m R_d \geq 39.81$

$$R_d \geq \frac{39.81}{1.87 \times 10^{-3}} \geq 21.28 \text{ k}\Omega$$

8.10 (a) Calculate the voltage gain of the FET amplifier shown in the figure. Assuming blocking capacitor to be large. g_m = 4 mA/V, and r_d = 5 K.

(b) Repeat if C_b = 4 nF is taken into consideration at input of 2 kHz frequency.

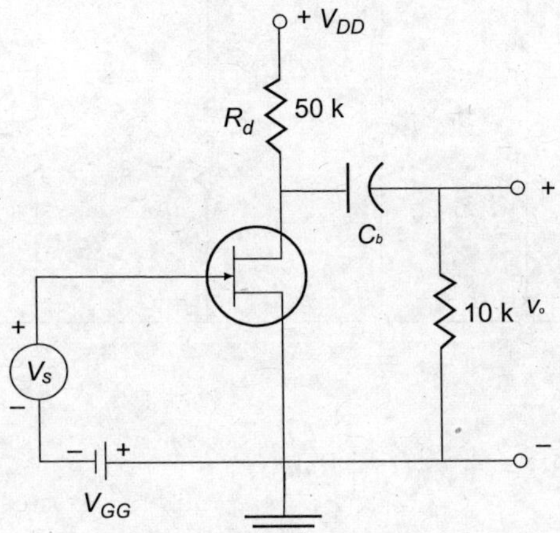

Solution (a) If C_b is large, it acts as short circuit. So, effective load = $r_d \| R_d \| R_L$.

$$= 10 \| 50 \| 5$$
$$= 4.545 \text{ k}\Omega$$

$$\text{Voltage gain} = -g_m \times \text{load resistance}$$
$$= -4 \times 10^{-3} \times 4.545 \times 10^3$$
$$= -18.18$$

Equivalent circuit used for the calculation of A_V.

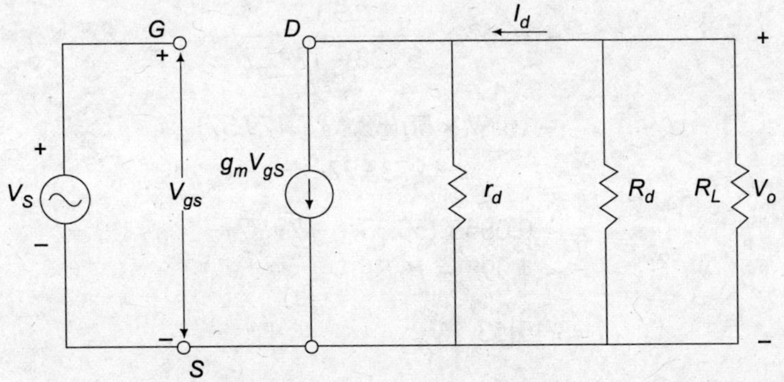

(b) For small values of C_b, the circuit becomes,

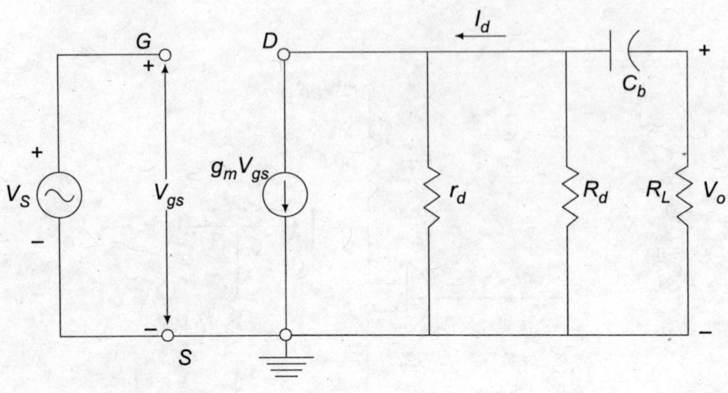

Ignoring R_L, $A_V = -g_m(r_d \| R_d)$.

$$= -4 \times 10^{-3} \times \frac{5 \times 50}{5+50} \times 10^3 = -18.18$$

$$R_o = \text{output impedance} = r_d \| R_d$$

$$X_c = \frac{1}{\omega C} = \frac{1}{2\pi \times 10^3 \times 2 \times 10^{-9}} = 79.57 \text{ k}\Omega$$

Considering the loading of capacitor and R_L,

$$A_V = -16.67 \times \frac{R_L}{R_L + R_o - JX_C}$$

$$= -16.67 \times \frac{50 \times 10^3}{(50 + 8.33 - J79.57) \times 10^3}$$

$$= -16.67 \times \frac{50}{58.33 - J79.57}$$

$$= \frac{-16.67 \times 50(58.33 + J79.57)}{9733.77}$$

$$= -0.0856\ (58.33 + J79.77)$$

$$= -4.993 - J6.811$$

$$= 8.45 \angle 53.76^\circ$$

8.11 (a) Find the voltage gain of the circuit shown in figure. FET parameters are $\mu = 24$, $r_d = 4$ K.

(b) Repeat if the output impedance of the signal source is 50 K.

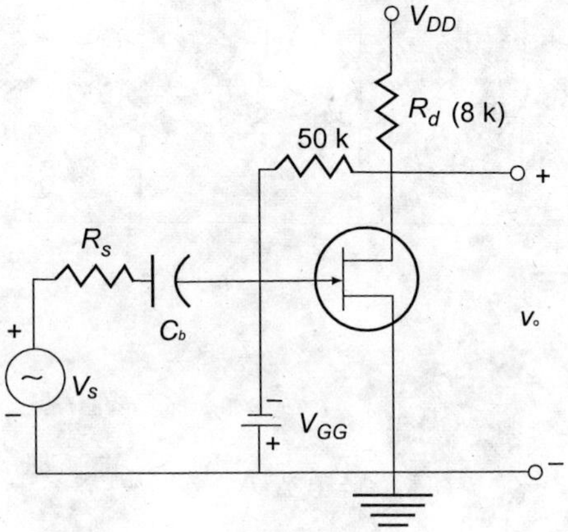

Solution (a) Using Miller's theorem, 50 K is split into two parts, R_1 and R_2.

$$R_1 = \frac{50}{1 - A_V}$$

$$R_2 = \frac{50 A_V}{A_V - 1} = \frac{50}{1 - \dfrac{1}{A_V}}$$

Then, the circuit becomes

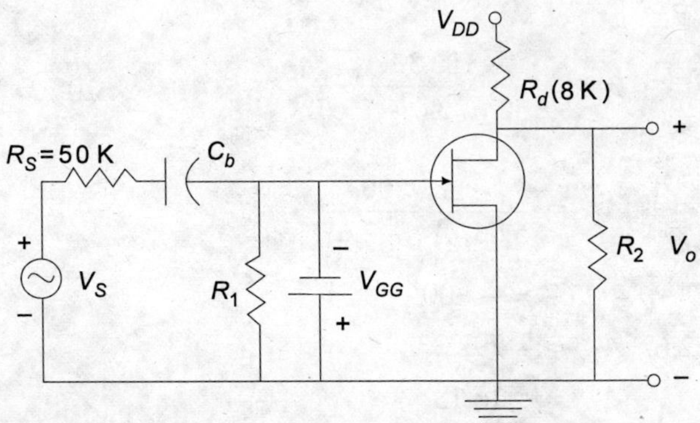

Replacing FET with its small-signal model by making dc sources and capacitors as short circuit,

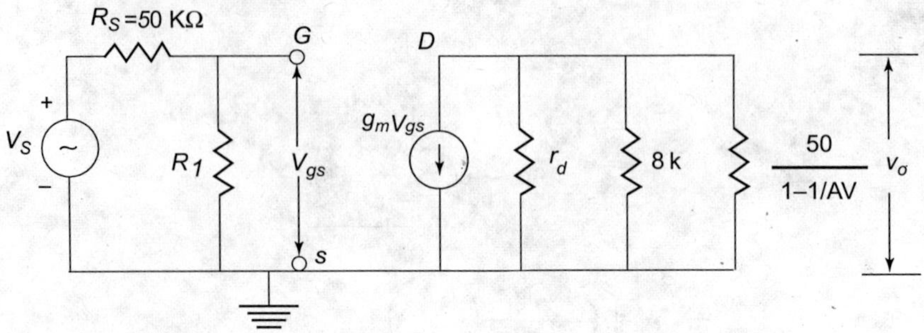

Given that $\mu = 24,\ r_d = 4$ K

$$g_m = \frac{\mu}{r_d} = \frac{24}{4\text{ K}} = 6 \text{ mA/V}$$

$$R_L = \text{load resistance} = r_d \,||\, R_d \,||\, R_2$$

$$= 4\text{K} \,\|\, 8\text{K} \,\|\, R_2$$

$$= \frac{2.67\, R_2}{R_2 + 2.67}\ \text{k}\Omega$$

$$= \frac{2.67 \times \dfrac{50\, A_V}{A_V - 1}}{2.67 + \dfrac{50\, A_V}{A_V - 1}} = \frac{133.33\, A_V}{52.67\, A_V - 2.67}$$

Voltage gain $= A_V = -\, g_m \cdot R_L$ (from circuit)

$$\Rightarrow \quad A_V = \frac{-\,133.33\, A_V}{52.67\, A_V - 2.67} \times 6 \times 10^{-3} \times 10^{3}$$

$$\Rightarrow \quad A_V = \frac{-\, A_V \cdot 800}{52.67\, A_V - 2.67}$$

$$\Rightarrow 52.67\ A_V - 2.67 = -\,800$$

$$\Rightarrow \quad 52.67\ A_V = -\,797.33$$

$$A_V = -\,15.13$$

$$\therefore \qquad R_1 = \frac{50}{1 - A_V} = 3.536 \text{ k}\Omega$$

$$R_2 = \frac{50 \times -15.13}{-16.13} = 46.90 \text{ k}\Omega$$

(b) From diagram (b) for (a) bit, signal is applied between gate and source. Now, signal is applied from signal generator itself.

Now, applying voltage decision

$$V_{gs} = \frac{R_1}{R_1 + R_s} \cdot V_s$$

$$= \frac{3.536}{3.536 + 50} \times V_s$$

$$= 0.066 \; V_s$$

$$\text{Now, gain } A_V = \frac{V_o}{V_{gs}} \cdot \frac{V_{gs}}{V_s}$$

$$= -\ 15.13 \times 0.066$$

$$= -\ 0.9993.$$

8.12 The circuit shown is called common-gate amplifier. For this circuit find (a) voltage gain, (b) input impedance (c) the output impedance. Power supplies are omitted for simplicity. Neglect capacitances.

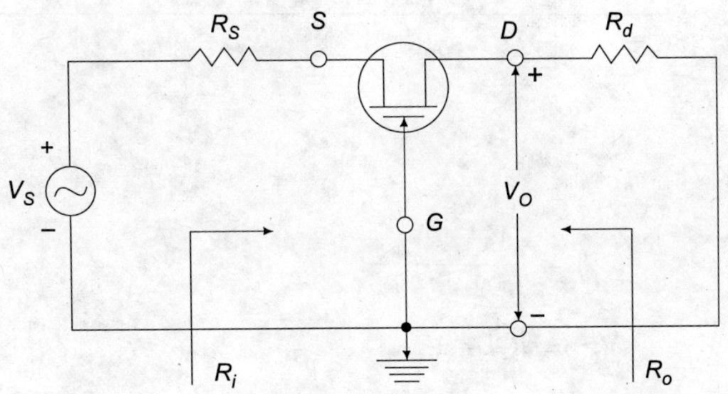

Solution Replacing FET with its small-signal model, the circuit becomes as shown below:

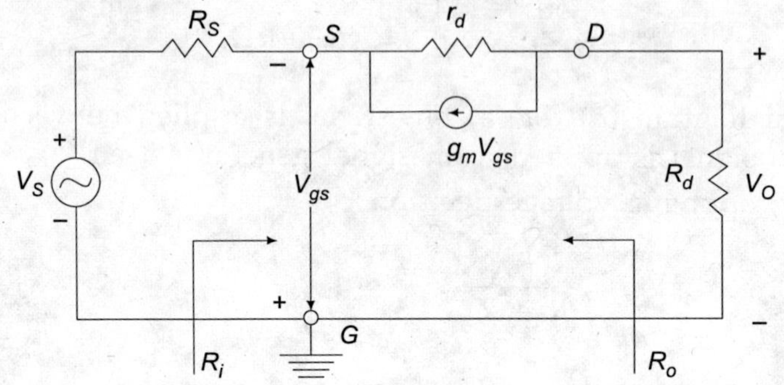

Converting Norton's model into Thevenin's model, the circuit becomes as

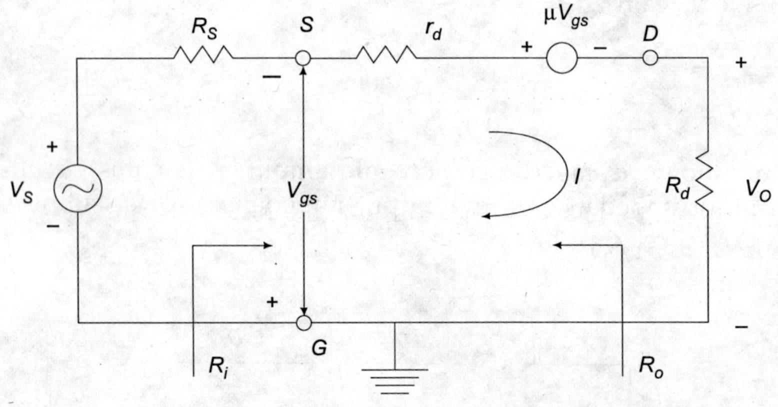

From circuit using KVL,

$$I = \frac{V_s - \mu V_{gs}}{R_s + r_d + R_d}$$

and, $$V_{gs} = R_s I - V_s$$

$\therefore$ $$I = \frac{V_s - \mu(R_s I - V_s)}{R_s + r_d + R_d}$$

$\therefore$ $$I\left(1 + \frac{\mu R_s}{R_s + r_d + R_d}\right) = \frac{V_s(1+\mu)}{R_s + r_d + R_d}$$

$$\therefore\ I\frac{((\mu+1)R_s + r_d + R_d)}{R_s + r_d + R_d} = \frac{V_s(\mu+1)}{R_s + r_d + R_d}$$

$$\therefore \quad \boxed{\frac{I}{V_s} = \frac{\mu+1}{(\mu+1)R_s + r_d + R_d}}$$

But $\quad V_o = R_d I$

(a) $$A_V = \frac{V_o}{V_s} = R_d \cdot \frac{I}{V_s} = \frac{(\mu+1)R_d}{(\mu+1)R_s + r_d + R_d}$$

(b) $$Ri = \frac{V_S}{I} = R_s + \frac{r_d + R_d}{\mu+1}$$

(c) To find R_o set $V_S = 0$. Disconnect R_d and apply a voltage V between D and ground, let the current through the circuit be I. Then, the circuit becomes as below and $R_o = \dfrac{V}{I}$.

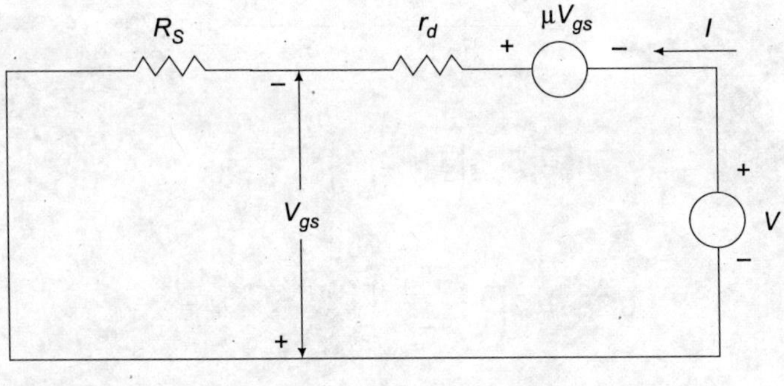

Writing KVL equation

$$-\mu V_{gs} + (R_s + r_d)I - V = 0$$

$$V + \mu V_{gs} = (R_s + r_d)I \qquad \text{...(1)}$$

But $\quad V_{gs} = -IR_s \qquad$ ...(2)

From (1) and (2)

$$V - \mu IR_s = (R_s + r_d)I$$

$$V = I[R_s(\mu + 1) + r_d]$$

$$\therefore \quad \frac{V}{I} = R_o = R_s(\mu + 1) + r_d$$

8.13 (a) The amplifier stage shown uses an n-channel FET having $I_{DSS} = 2$ mA, $V_P = -2$ V. If the quiscent drain to ground voltage is 10 V, find R_1.

(b) What is the effective input impedance?

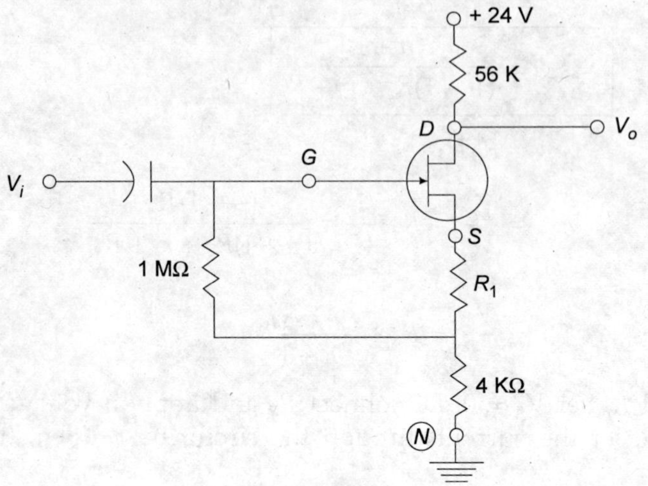

Solution (a)

$$I_D = \frac{V_{DD} - V_{DN}}{56\text{ K}} = \frac{24 - 10}{56\text{ K}} = 0.25\text{ mA}$$

But

$$I_D = I_{DSS}\left(1 - \frac{V_{GS}}{V_P}\right)^2$$

$$0.25 = 2\left(1 - \frac{V_{GS}}{-2}\right)^2$$

$$1 + \frac{V_{GS}}{2} = 0.125$$

$$V_{GS} = -1.75\text{ V.}$$

But

$$V_{GS} = -I_D R_1$$

$$\Rightarrow R_1 = -\frac{V_{GS}}{I_D} = \frac{1.75}{0.25 \times 10^{-3}} = 7\text{ k}\Omega$$

(b)

$$g_{mo} = -\frac{2I_{DSS}}{V_P} = \frac{-2 \times 2}{-2} = 2\text{ mA/V}$$

$$g_m = g_{mo}\left(1-\frac{V_{GS}}{V_P}\right) = 2\left(1-\frac{-1.75}{-2}\right)$$

$$= 7.03 \text{ mA/V}$$

The formula for the calculation of I/P impedance for this bootstrapped circuit will be,

$$R_i = \frac{R_G}{1-\dfrac{g_m R_s}{1+g_m R_s}\cdot\dfrac{R_s}{R_s+R_1}}$$

$$= \frac{1\,\text{M}\Omega}{1-\dfrac{1\times10^{-3}\times4\times10^{3}}{1+1\times10^{-3}\times4\times10^{3}}\cdot\dfrac{4\times10^{3}}{4\times10^{3}+2\times10^{3}}}$$

$$= \frac{1\,\text{M}}{1-\dfrac{4}{5}\times\dfrac{4}{6}} = \frac{30\,\text{M}}{14} = 2.14\ \text{M}\Omega$$

8.14 The amplifier of figure uses a FET with I_{DSS} = 3 mA, V_P = – 3 V, $r_d \gg R_d$. Find the quiescent drain current, quiescent drain to source voltage, and A_V.

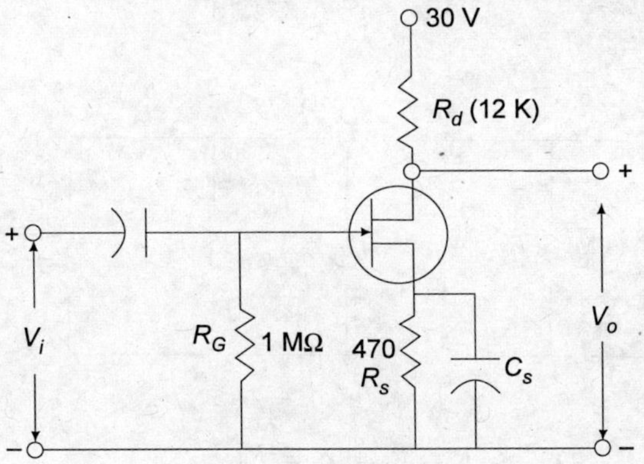

Solution We have $$I_D = I_{DSS}\left(1-\frac{V_{GS}}{V_P}\right)^2$$

To find Q-point capacitors are open circuited.

$$\therefore \quad V_{GS} = -\,470 \times I_D$$

$$\therefore \quad I_D = 3\left(1+\frac{0.47\,I_D}{-3}\right)^2$$

$$\Rightarrow \qquad \frac{1}{3} I_D = (1 - 0.157\, I_D)^2$$

$$= 1 + 0.0245 I_D^2 - 0.314 I_D$$

$$\Rightarrow \qquad 0.33\, I_D = 1 + 0.0245 I_D^2 - 0.314 I_D$$

$$\Rightarrow 0.0245 I_D{}^2 - 0.644 I_D + 1 = 0$$

$$\Rightarrow I_D{}^2 - 26.28 I_D + 40.82 = 0$$

$$\Rightarrow I_D = \frac{26.28 \pm \sqrt{(26.28)^2 - 4 \times 40.82}}{2}$$

$$I_D = \frac{26.28 \pm 22.96}{2} = 24.623,\ 1.66 \text{ mA}$$

$$\therefore \qquad I_D \approx 1.66 \text{ mA}$$

$$V_D = V_{DD} - I_D R_D = 30 - 1.66 \times 10^{-3} \times 12 \times 10^3$$

$$= 10.08 \text{ V}$$

$$V_S = I_D R_s = 470 \times 1.66 \times 10^{-3} = 0.7802 \text{ V}$$

$$V_{DS} = V_D - I_D R_S = 10.08 - 0.7802 = 9.3 \text{ V}$$

To find gain A_V

Replace FET with its small signal model.

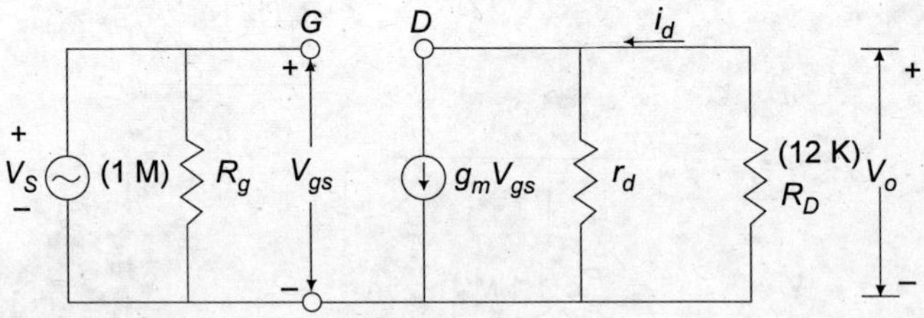

Since, $\qquad r_d \gg R_D,\ R_D || r_d = R_D$

$$A_V = -\, g_m R_D = -\, g_{mo} \left(1 - \frac{V_{GS}}{V_P}\right) R_D$$

$$= \frac{2 I_{DSS}}{V_P} \left(1 - \frac{V_{GS}}{V_P}\right) R_D$$

$$= \frac{2 \times 3 \times 10^{-3}}{-2.4} \left(1 - \frac{-0.7}{-3}\right) \times 12 \times 10^3$$

$$= -23.$$

8.15 Find the quiescent values of V_{DS}, I_D and V_{GS} for the amplifier shown. $I_{DSS} = -4$ mA, $V_P = +4$V.

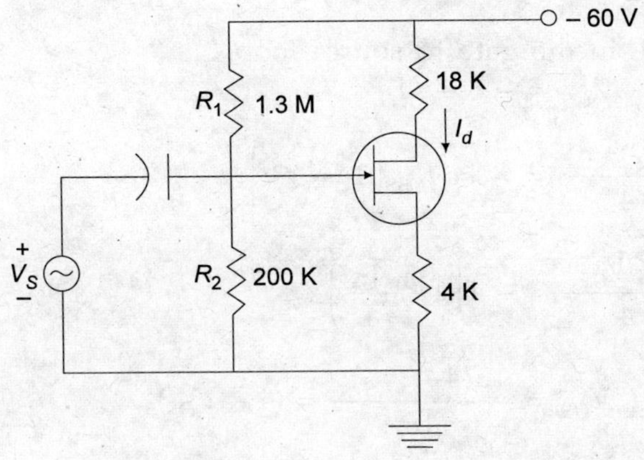

Solution Replacing R_1 and R_2 by their Thevenin's equivalent

$$V_{gg} = -60 \times \frac{200\text{ K}}{200\text{ K} + 1.3\text{ M}}$$

$$= -60 \times \frac{200}{1500}$$

$$= -8\text{ V}$$

$$R_G = 200\text{ K} \,||\, 1.3\text{ M} = \frac{200 \times 1300}{1500} \times 10^3 = 173.33\text{ k}\Omega$$

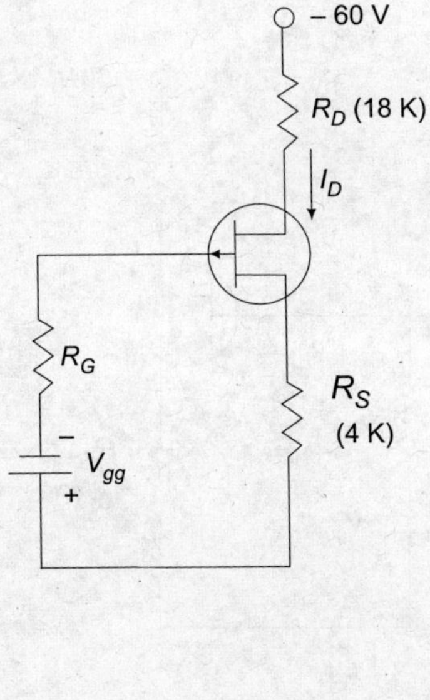

We have $$I_D = I_{DSS}\left(1 - \frac{V_{GS}}{V_P}\right)^2$$

Writing KVL in the gate to source loop,

$$V_{GS} = -8 - 4I_D$$

$$\Rightarrow \quad -\frac{(V_{GS}+8)}{4} = I_D = I_{DSS}\left(1 - \frac{V_{GS}}{V_P}\right)^2$$

$$-\frac{(8+V_{GS})}{4} = -4\left(1 - \frac{V_{GS}}{4}\right)^2$$

$$8 + V_{GS} = \frac{16(4 - V_{GS})^2}{16} = (4 - V_{GS})^2$$

$$16 - 8V_{GS} + V_{GS}^2 = 8 + V_{GS}$$

$$V_{GS}^2 - 9V_{GS} + 8 = 0$$

$$V_{GS} = \frac{9 \pm \sqrt{81 - 32}}{2} = \frac{9 \pm 7}{2} = 8, 1$$

$$\therefore \quad V_{GS} = 1 \text{ V}$$

$$\therefore \quad I_D = -\frac{(8+V_{GS})}{4} = -9/4 = -2.25 \text{ mA}$$

$$V_{DS} = V_{DD} - I_D(R_D + R_s)$$

$$= -60 + 2.25 \times 10^{-3}(18 + 4) \times 10^3$$

$$= -10.5 \text{ V}$$

8.16 For the fixed-bias circuit, determine I_D, V_{DS} and V_{GG}.

Solution Writing KVL in the gate-source loop,

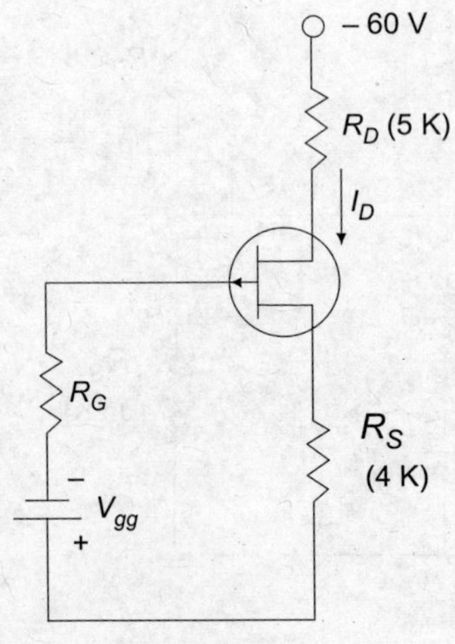

$$V_{GG} + V_{GS} = 0$$

$$\Rightarrow \quad V_{GG} = -V_{GS} \; (\because I_G = 0)$$

Writing KVL in drain-source loop

$$V_{DS} = V_{DD} - I_D R_D$$

$$\Rightarrow \quad V_D - V_S = V_{DD} - I_D R_D$$

$$9 - 0 = 14 - 1.6 \text{ K} I_D \; (\because V_s = 0, \text{ source is grounded})$$

$$\therefore \quad I_D = \frac{14-9}{5\text{K}} = 1 \text{ mA}$$

We have
$$I_D = I_{DSS}\left(1 - \frac{V_{GS}}{V_P}\right)^2$$

$$1 \text{ mA} = 8\left(1 - \frac{V_{GS}}{-4}\right)^2$$

$$\Rightarrow \quad 2 = (4 + V_{GS})^2$$

$$\Rightarrow \quad \pm 0.707 = 4 + V_{GS}$$

$$V_{GS} = -3.293, -4.707 \text{ V}$$

$$V_{GG} = -V_{GS} = 3.293 .$$

8.17 Find V_D for the following circuit shown below.

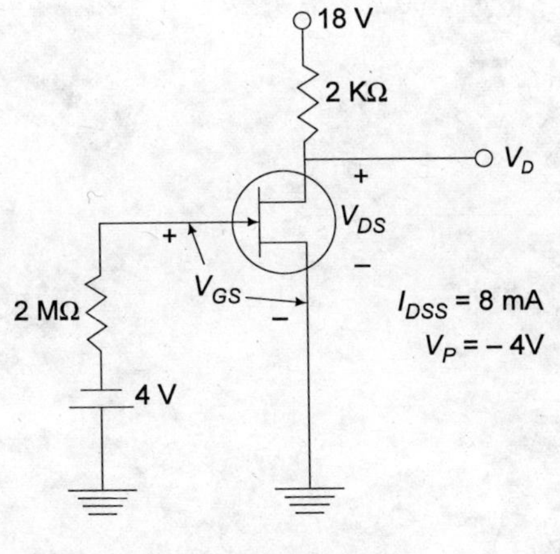

Solution From th KVL in G–S loop

$$V_{GS} = -4 \text{ V}$$

Since source is grounded, $V_S = 0$

$$\therefore \quad V_{DS} = V_D - V_S = V_D$$

$$V_{DS} = V_D = V_{DD} - I_D \cdot R_D$$

To find I_D

$$I_D = I_{DSS}\left(1 - \frac{V_{GS}}{V_P}\right)^2$$

$$= 8\left(1 - \frac{-4}{-4}\right)^2$$

$$= 0$$

$$\therefore \quad V_{DS} = V_{DD} - I_D R_D = 18 \text{ V}$$

$$V_D = V_{DS} = 18 \text{ V}$$

8.18 If $V_S = 1.7$ V, determine I_D, V_{GS}, I_{DSS}, V_D and V_{DS} for the circuit shown below.

Solution

$$V_{GS} = - I_D R_s$$

$$= - V_S = - 1.7 \text{ V}$$

$$\therefore \quad I_D = \frac{1.7}{0.51 \text{ K}}$$

$$= 3.33 \text{ mA}$$

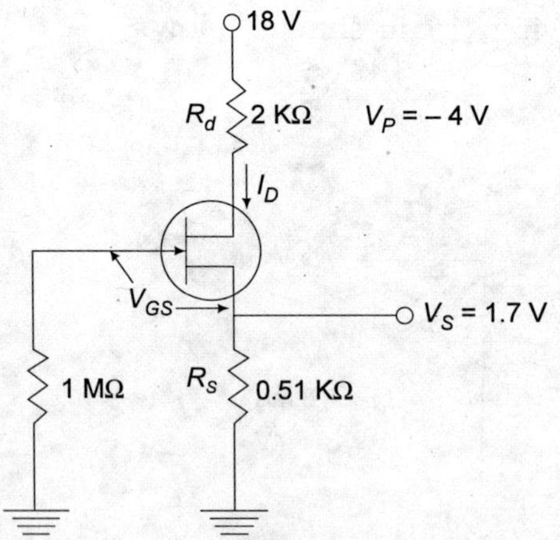

$$I_D = I_{DSS}\left(1 - \frac{V_{GS}}{V_P}\right)^2$$

$$3.33 = I_{DSS}\left(1 - \frac{-1.7}{-4}\right)^2$$

$$\Rightarrow \qquad I_{DSS} = 10.072 \text{ mA}$$

$$V_{DS} = V_{DD} - I_D(R_S + R_D)$$

$$= 18 - 3.33\,(0.5 \text{ K} + 2 \text{ K})$$

$$= 9.64 \text{ V}$$

$$V_{DS} = V_D - V_S$$

$$9.64 = V_D - 1.7 \Rightarrow V_D = 11.34 \text{ V}$$

8.19 Find I_D for the following circuit.

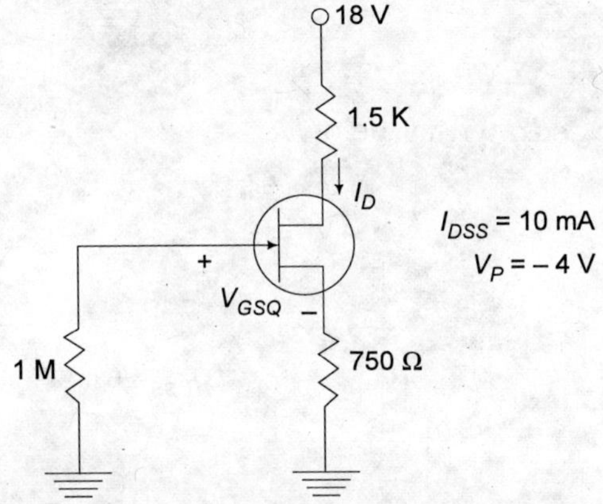

Solution Since $I_G = 0$, writing KVL in the gate to source loop,

$$V_{GSQ} = -\,750 I_D \qquad \text{...(1)}$$

We have
$$I_D = I_{DSS}\left(1 - \frac{V_{GS}}{V_P}\right)^2$$

$$= 10\left(1 + \frac{750 I_D}{-4}\right)^2$$

$$= 10\,(1 - 187.5 I_D)^2$$

$$= 10\,(1 + 35156.25 I_D^2 - 375 I_D)$$

$$\Rightarrow 351.5625 I_D^2 - 4.75 I_D + 10 \times 10^{-3} = 0$$

$$\therefore \qquad I_D = 10.9 \text{ mA}, 2.6 \text{ mA}$$

So, I_D should be 2.6 mA, since it should be less than I_{DSS}

$$V_{GSQ} = -\,750 \times I_D = -\,1.957 \text{ V}$$

8.20 Find V_S for the following network.

$$I_{DSS} = 6 \text{ mA}, V_p = -6 \text{ V}$$

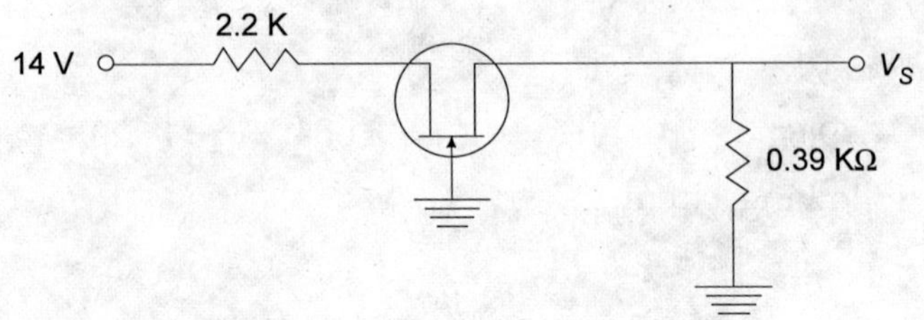

Solution The circuit can be redrawn as below.

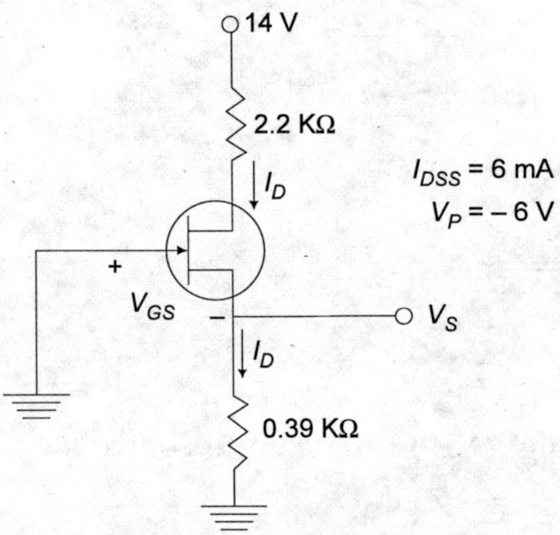

Writing KVL in gate source loop

$$V_{GS} + V_S = V_{GS} + 0.39 \text{ K } I_D = 0$$

$$V_{GS} = -0.39 \text{ K} I_D \qquad \text{...(1)}$$

$$\text{We have } I_D = I_{DSS}\left(1 - \frac{V_{GS}}{V_P}\right)^2$$

$$= 6 \times 10^{-3}\left(1 - \frac{-0.39 \times 10^3 I_D}{-6}\right)^2$$

$$= 6 \times 10^{-3}(1 - 65 I_D)^2$$

$$= 6 \times 10^{-3}(1 - 130 I_D + 4225 I_D^2)$$

$$\Rightarrow 25.35 I_D^2 - 1.78 I_D + 6 \times 10^{-3} = 0$$

Solving for I_D,

$$\Rightarrow \quad I_D = 66.67 \text{ mA}, 3.55 \text{ mA}$$

$$\therefore \quad I_D = 3.55 \text{ mA} \; (\because I_D \text{ should be less than } I_{DSS})$$

$$V_S = + I_D \times 0.39 \times 10^3 = 1.3845 \text{ V}$$

8.21 For the following network, determine I_D, V_{GS}, V_G, V_D, V_S and V_{DS}.

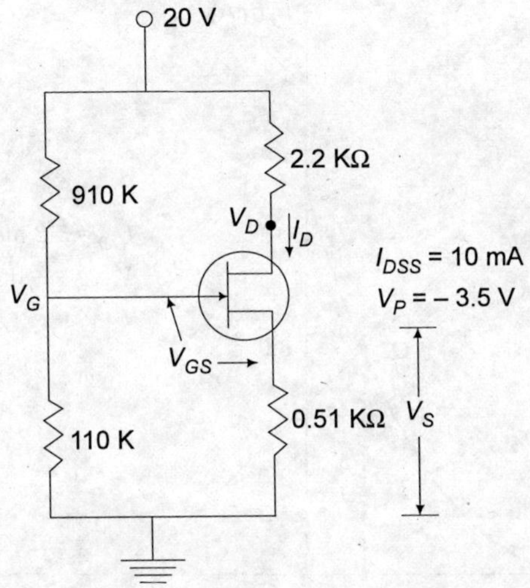

Solution From the circuit, using voltage division rule,

$$V_G = \frac{110}{110 + 910} \times 20 = 2.156 \text{ V}$$

Writing KVL in R_2–G–S loop,

$$V_{GS} = V_G - I_D R_S$$

$$= 2.156 - 510 \; I_D$$

When $\quad I_D = 0, V_{GS} = 2.156$ V

When $\quad V_{GS} = 0, I_D = \dfrac{2.156}{510} = 4.23$ mA

For transfer characteristic, if $I_D = \dfrac{I_{DSS}}{4} = \dfrac{10}{4}$

$$= 2.5 \text{ mA}$$

Then, $V_{GS} = V_P/2 = -1.75$ V. The resulting curve representing Shockey's equation appears as below. So, Q-point from graph is $I_D = 5.8$ mA

$$V_{GS} = -0.8 \text{ V}$$
$$V_S = I_D R_S = 5.8 \times 0.51 = 2.958 \text{ V}$$
$$V_D = V_{DD} - I_D R_D = 20 - 5.8 \times 2.2 = 7.2 \text{ V}$$
$$V_{DS} = V_D - V_S = 4.282 \text{ V}$$

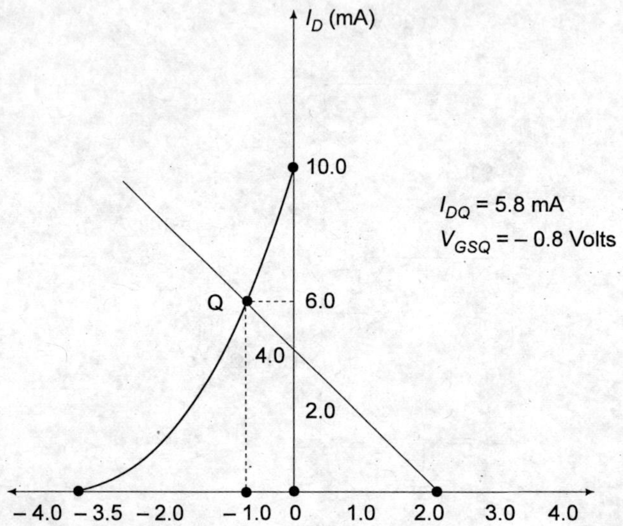

8.22 For the network if $V_D = 9$ V, determine I_D, V_S, V_{DS}, V_G, V_{GS} and V_P, given that $I_{DSS} = 8$ mA.

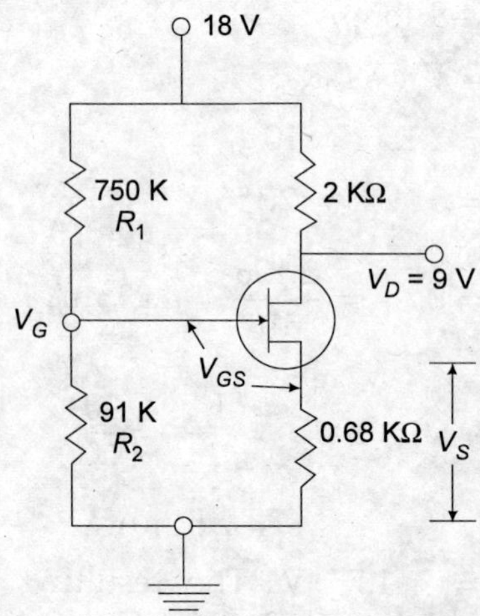

Solution

$$V_G = \frac{91}{91+750} \times 18 \text{ V}$$

$$= 1.94 \text{ V}$$

Writing KVL in R_2–G–S loop,

$$V_{GS} = V_G - I_D R_S = 1.94 - 680 I_D \qquad \text{...(1)}$$

$$V_D = V_{DD} - I_D \cdot R_D$$

$$9 = 18 - I_D(2 \text{ k}\Omega)$$

$$\Rightarrow \quad I_D = 4.5 \text{ mA}$$

$$V_{GS} = 1.94 - 680 \times 4.5 \times 10^{-3}$$

$$= -1.12 \text{ V}$$

$$V_S = I_D R_S = 4.5 \times 0.68 = 3.06 \text{ V}$$

$$V_{DS} = V_D - V_S = 9 - 3.06 = 5.94 \text{ V}$$

To find V_P

$$I_D = I_{DSS}\left(1 - \frac{V_{GS}}{V_P}\right)^2$$

$$\Rightarrow \quad 4.5 \times 10^{-3} = 8 \times 10^{-3}\left(1 - \frac{V_{GS}}{V_P}\right)^2$$

$$\Rightarrow \quad \pm\, 0.5625 = 1 + \frac{1.12}{V_P}$$

$$\Rightarrow \quad V_P = -0.7168,\ -2.56$$

So the value of V_P should be – 0.7168

8.23 For the network shown below, determine V_G, I_{DQ} and V_{GSQ}, V_D and V_S, V_{DSQ}, given that I_{DSS} = 10 mA, V_P = – 3.5 V.

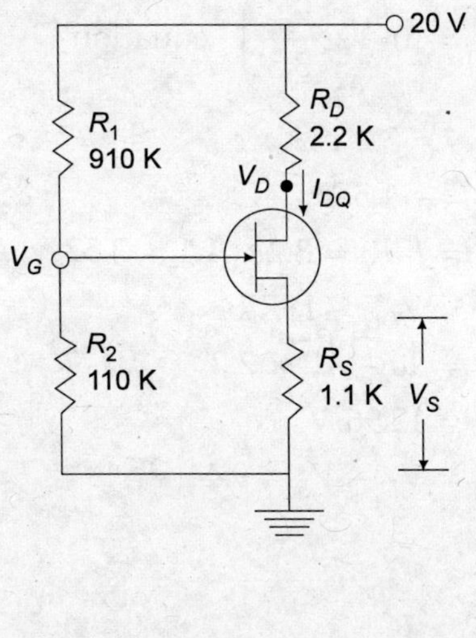

Solution

$$V_G = \frac{110}{110+910} \times 20 = 2.156 \text{ V}$$

Writing KVL in R_2–G–S loop,

$$V_{GS} = V_G - I_D \cdot R_S$$

$$= 2.156 - 1100 I_D \qquad \text{...(1)}$$

To draw the bias curve,

$$V_{GS} = 2.156 \text{ V if } I_D = 0$$

$$I_D = \frac{2.156}{1100} = 1.96 \text{ mA if } V_{GS} = 0$$

and the transfer characteristic follows,

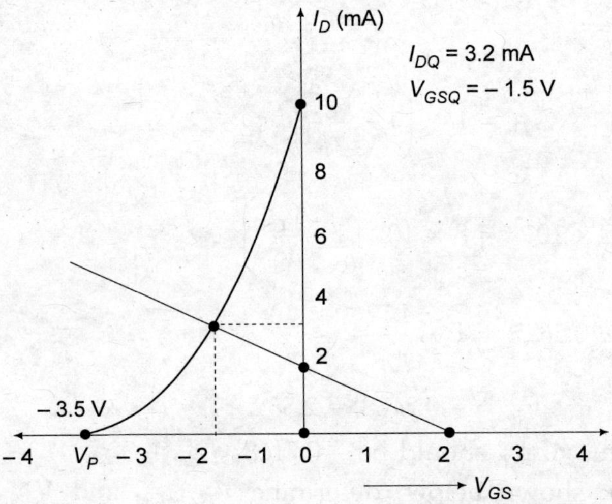

$$I_D = 10\left(1+\frac{V_{GS}}{3.5}\right)^2 \text{ equation}$$

From the graph $I_{DQ} = 3.2$ mA

$$V_{GSQ} = -1.5 \text{ V}$$

$$V_S = I_D R_s = 3.2 \times 1.1 = 3.52 \text{ V}$$

$$V_D = V_{DD} - I_D R_D$$

$$= 20 - 3.2 \times 2.2$$

$$= 12.96 \text{ V}$$

$$V_{DS} = V_D - V_S = 9.44 \text{ V}$$

EXERCISE PROBLEMS

8.1 Find the pich-off voltage of a silicon *p*-channel FET having half channel height of 2 microns and channel resistivity of 20 Ω-cm. Dielectric constant of silicon is 16 and μ_p = 1500 cm^2/V-s.

8.2 (a) A JFET has a drain current of 10 mA. If I_{DSS} =20 mA and V_P = – 4V. find V_{GS}.

(b) For an *n*-channel JFET I_{DSS} = 6.8 mA, V_P = – 6 V. and V_{GS} = – 4V. find I_D, g_m, g_{mo}.

8.3 Find I_D and V_{DS} for the circuit shown below. V_{GG} = – 4 V, Rg = 4 MΩ, V_{DD} = + 15 V, R_d = 3 kΩ, given that I_{DSS} = 10 mA and V_P = – 5 V.

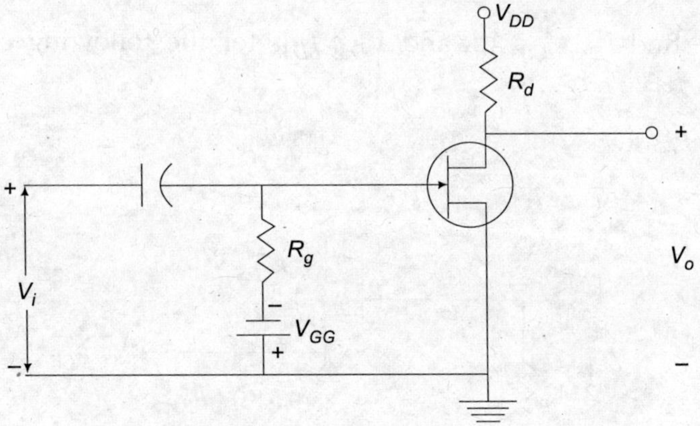

8.4 Find I_D, V_{DS}, V_{GG} for the following circuit:

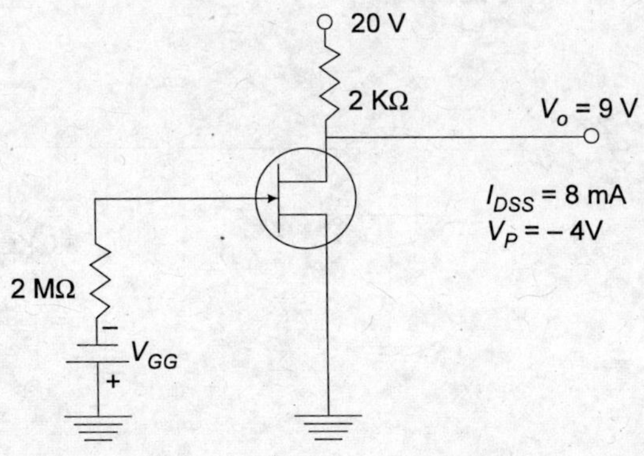

8.5 Find V_D for the following circuit:

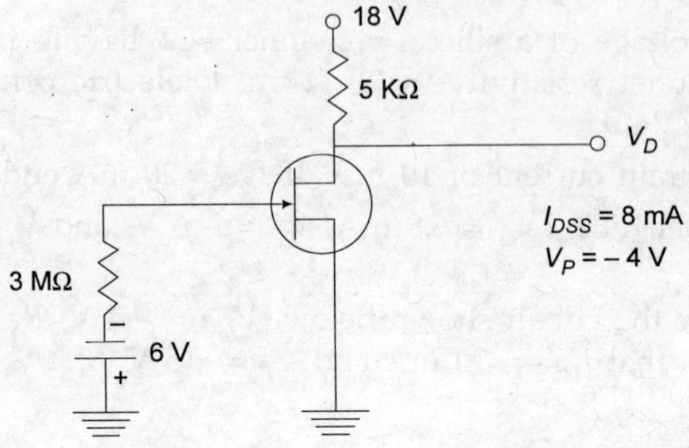

8.6 If V_S = 1.7 V, find I_D, V_{GS}, V_D and V_{DS}, I_{DSS} for the following circuit:

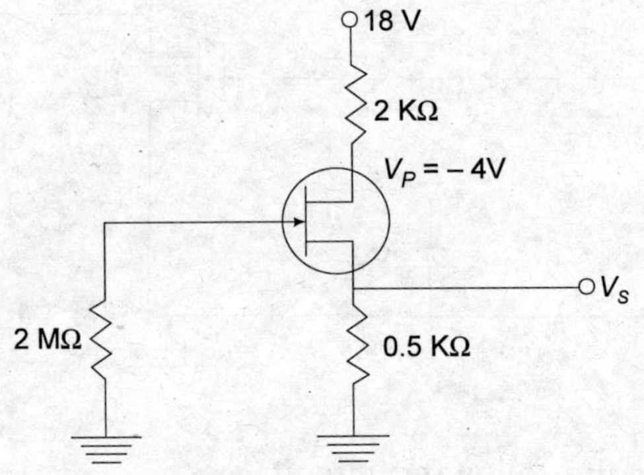

8.7 Find V_S for the following circuit:

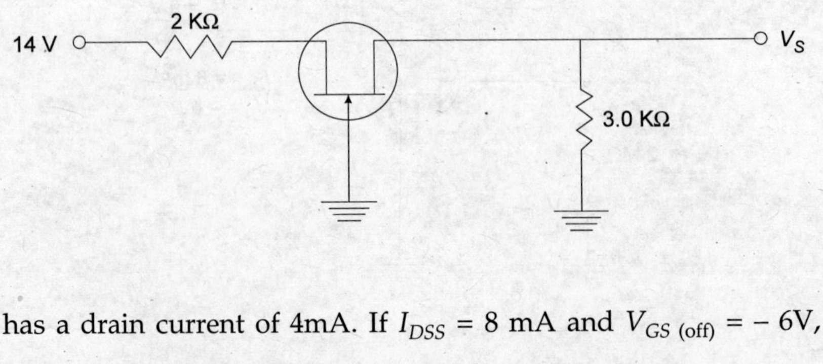

8.8 A FET has a drain current of 4mA. If I_{DSS} = 8 mA and $V_{GS\ (off)}$ = – 6V, find V_{GS} and V_P.

REVIEW QUESTIONS

1. How is a FET used as a voltage variable resistor?
2. With a neat construction diagram, explain the operation of enhancement and depletion MOSFET.
3. Explain why depletion region in a JFET is tapered.
4. Explain why FET is known as a unipolar device.
5. Draw the drain characteristics of a common source JFET and label different regions. Explain the reasons for its shape. Obtain all the parameters of JFET from drain characteristics.
6. Explain the significance of pinch-off voltage in effect.
7. Draw the small signal equivalent circuit for a JFET and derive the equation for voltage amplification.
8. Give the advantages of FET over BJT.
9. Mention the application of MOSFET.
10. Show how a FET can be used as voltage variable resistor.
11. Explain the modes of operation of a MOSFET.
12. Describe the construction and principle of operation of a JFET with the help of its output and transfer characteristics.
13. Distinguish between depletion and enhancement MOSFET.
14. Derive an equation for transconductance in terms of pinch-off voltage and saturation drain current.
15. Explain various methods of biasing JFET and MOSFET.
16. Draw the high-frequency model of a FET and explain each parameter of the circuit.
17. What is threshold voltage in a JFET? Give its limitations. Explain how these limitations are overcome in a MOSFET.
18. Define transconductance, drain resistance and amplification factor of a FET and derive the relation between them.
19. Derive an expression for pinch-off voltage in terms of gate to source voltage and saturation drain current.
20. Derive an expression for voltage gain of a common source FET amplifier with and without source resistance included in the circuit.
21. Determine the output voltage for a FET amplifier in common drain configuration and give the expression for voltage gain of the amplifier.
22. Sketch the cross section of a *p*-channel enhancement MOSFET and explain its operation. Give its circuit symbol.

23. Derive the equation for voltage gain of a *CS* FET amplifier.
24. Draw two biasing circuits for a JFET or depletion MOSFET. Explain under what circumstances each of these arrangements are used.
25. Sketch the small signal model of a *CD* FET amplifier and derive equation for the voltage gain.

CHAPTER

9

Thyristors

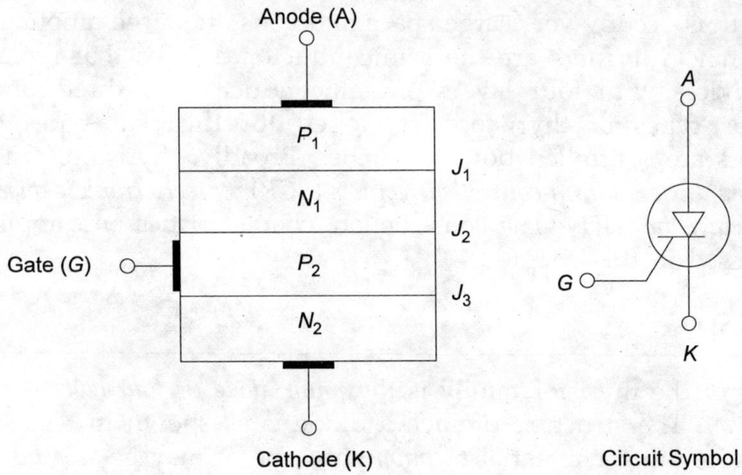

Structure and circuit symbol of a SCR

Chapter Outline

The concepts introduced in this chapter are:

- *Shockley diode*
- *Silicon-controlled rectifier*
- *TRIAC and DIAC*
- *Unijunction transistor*
- *Optoelectric devices and few examples of such devices—photodiodes, phototransistors, LEDs and LASCR*

9.1 INTRODUCTION

The electronic devices such as *p-n* junction semiconductor diode, bipolar junction transistor or field-effect transistor are the devices that can handle very low powers and are not suitable to be operated at high powers of the order of kilowatts or above. To control these high powers, special electronic devices are required. These devices control the input power and output power of an industrial appliance. Temperature controllers, illumination controllers, variable speed drives and voltage regulators are few examples where high power control is required. Power transistors and power diodes were initially developed to cater these needs of high power. But these devices were not very much useful and so consequently *thyratron*, a gas-filled triode, was developed. Thyratrons are tube devices and thus occupy very large space and consume large amounts of power with very small efficiency. *Thyristors* are the semiconductor devices that replaced thyratrons. Thyristors are devices with four layers of semiconductors or three junctions or more. There are various types of thyristors developed for different types of applications. Basically, thyristors are controlled power switches. Broadly, thyristors can be classified as *pnpn diode (Shockley diode), silicon-controlled rectifier (SCR), optoelectronic devices, TRIAC, DIAC* and *unijunction transistor (UJT)*. The construction, characteristics and applications of these devices are discussed in this chapter.

9.2 *pnpn* DIODE

The first member of the thyristor family is the *pnpn diode* or *four-layer diode* (also referred to as *Shockley diode*). The structure of such a diode is as shown in Fig. 9.1. It consists of four layers of semiconductor material of *pnpn* structure. It may be treated as equivalent to a combination of two transistors. This device works as a switch and is fabricated to have voltage ratings of several hundred volts and current rating of few amperes to thousands of amperes. It has three junctions identified as J_1, J_2 and J_3. The two terminals of the diode are identified as *anode* (*p* side) and *cathode* (*n* side).

When the device is biased such that the *p* side is more positive than the *n* side, the junctions J_1 and J_3 are forward biased. Then the applied voltage is directly available across the junction J_2. Thus, the current through the device is only the reverse saturation current. When the applied voltage is increased, the current increases slowly until the voltage called *breakover voltage* V_{BO} is reached. At this voltage, *firing* takes place and the current increases abruptly while the voltage drops across the device sharply. At this point, the diode switches over from OFF to ON state. Once the device is fired into conduction, a minimum amount of current known as *holding current* I_H is required to flow to keep the device in ON state. To turn the device OFF from ON state, the current has to be reduced below I_H by reducing the applied voltage close to zero, that is, below holding voltage. Thus, the device acts as a switch and the characteristics are as shown in Fig. 9.2.

The diode behaves as a simple *p-n* junction diode when reverse biased, the breakdown voltage being comparatively large.

The device is said to be operating in *saturation region*. The device in this condition is said to be *latched*. The current in the diode in this condition of latch is referred to as *latching* or *holding current*. Similarly, the voltage can be referred to as *holding* or *latching*

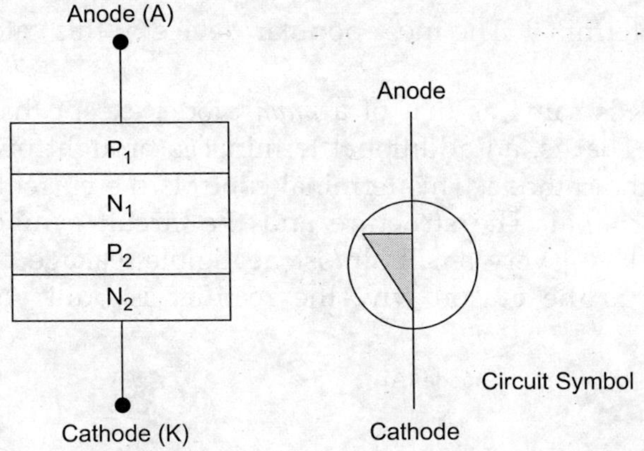

Fig. 9.1 *A pnpn diode structure and its circuit symbol*

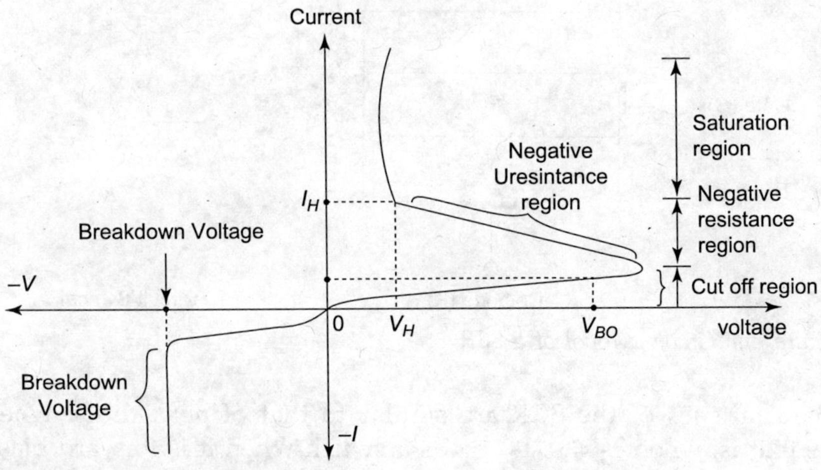

Fig. 9.2 *V-I characteristics of a Shockley diode*

voltage. To switch on the diode, the minimum current required is the latching current. The breakover voltage V_{BO} is the characteristic of a given *pnpn* diode and its value differs based on the type of diode and may range from several tens of volts to hundreds of volts. In the OFF condition the dynamic resistance of the diode may range from few megaohms to hundreds of megaohms. The holding current value varies from several tens of milliamperes to hundreds of milliamperes. The holding voltage varies from 0.5 V to 20 V. In the saturation region, the dynamic resistance is very small and very often crosses 10.

9.3 SILICON-CONTROLLED RECTIFIER

The four-layer Shockley diode, as discussed in the preceding section, is used to introduce the concept of four-layer devices. It is generally not used in practice since controlling terminal is absent in the device. There are various other devices that have control terminal like silicon controlled rectifier (SCR), TRIACS and so on in which the flow of current can be controlled by the third terminal. The most popular device of this category is SCR and is now considered at length.

The structure of a SCR is same as that of a *pnpn* diode, except that a gate terminal is attached to the P_2 region. That is, an additional terminal is brought out at the *p* region of the device, which is near the cathode. This terminal controls the current in the device. This acts as the controlling terminal. The structure and the circuit symbol of a SCR are as shown in Fig. 9.3. Silicon has a very small almost negligible leakage current compared to that of germanium. That is the reason why the rectifier is built with silicon and not germanium in practice.

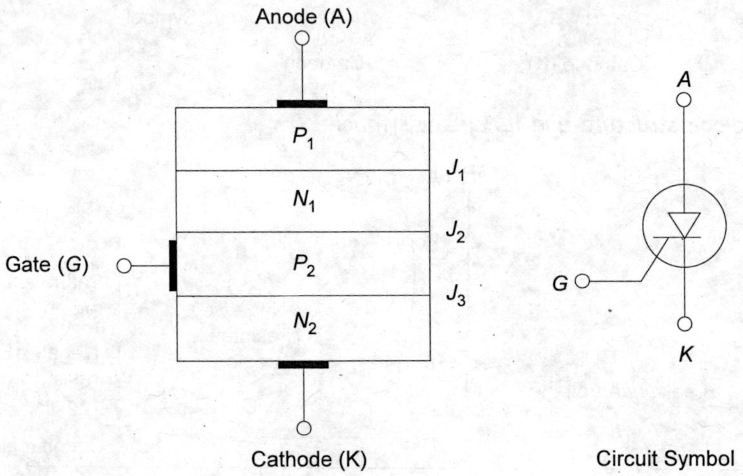

Fig. 9.3 *Structure and circuit symbol of a SCR*

The *V-I* characteristics of the SCR are similar to that of *pnpn* diode when no bias is applied to gate, that is, when $I_G = 0$. It acts as a switch when it is forward biased. When I_G is less than zero the amount of reverse bias applied to J_2 is increased, so the breakover voltage V_{BO} is increased. When I_G is greater than zero the amount of reverse bias applied to J_2 is decreased, thereby decreasing the breakover voltage. With very large positive gate current, breakover may occur at a very low voltage such that the characteristics of SCR are similar to those of the ordinary *p-n* junction diode. The voltage at which the SCR is switched ON can be controlled by varying the gate current. Thus, it is commonly known as *controlled switch*. Once SCR is turned ON, the gate loses its control, that is, the gate cannot be used to switch the device OFF. One way to turn OFF the device is to lower the anode current below the holding current by reducing the supply voltage below holding voltage, keeping the gate open. The *V-I* characteristics of such a SCR are as shown in Fig. 9.4.

When the SCR is biased in the forward region, and gate is given a positive voltage, a gate current of I_G flows between the gate and the cathode. This current introduces holes from the P_2 region into junction J_3 and N_2 region. This results in the reduction of overall resistance of P_2-N_2 region and hence the value of the breakover voltage decreases correspondingly. The curves show that the anode current can be controlled by controlling the gate current.

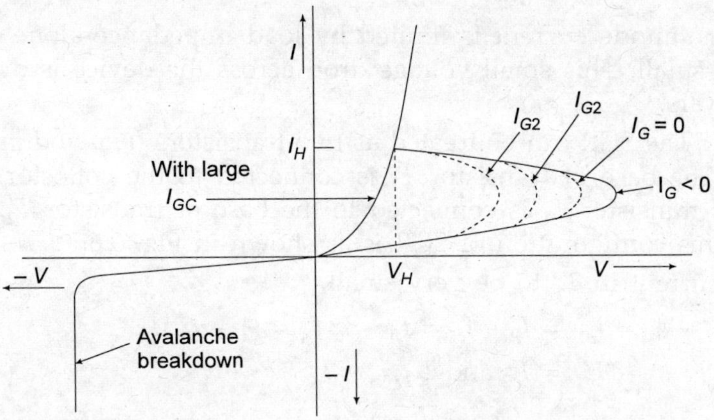

Fig. 9.4 *V-I characteristics of a SCR*

The *V-I* characteristics of the SCR reveal that the SCR can be operated in three modes. These modes can be categorised as *reverse blocking mode, forward blocking mode (OFF state)* and *forward conduction mode (ON mode).*

Reverse blocking mode is that mode when the SCR is reverse biased, that is, when anode is negative compared to cathode. The characteristics of this region are similar to those of an ordinary *p-n* junction diode. In this region, the junctions J_1 and J_3 are reverse biased whereas the junction J_2 is forward biased. The device behaves as if two diodes are connected in series with a reverse voltage applied across them. A small leakage current of the order of milliamperes or microamperes flows in the device. This reverse blocking mode is called the OFF state of the thyristor. When the reverse voltage of the SCR increases to a large extent breakdown occurs and the current in the device increases rapidly. Thus, when the SCR is biased in this region the power dissipated is very high. If the power dissipated is more than the rated value of the SCR, the SCR is permanently damaged. Thus, in the reverse bias condition the voltage should never cross the breakdown voltage.

Forward blocking mode is that mode when the SCR is forward biased. It is observed that the junctions J_1 and J_3 are forward biased and J_2 is reverse biased. In this mode a small current called the *forward leakage current* flows in the SCR. When the forward bias voltage is increased, the reverse bias junction J_2 will have avalanche breakdown at a voltage called *forward breakover voltage.* When the applied voltage is less than this voltage the SCR offers very high impedance and so it can be treated as an open switch in the forward blocking mode.

Forward conduction mode is that mode when the SCR conducts current from anode to cathode with a very small voltage drop across it. The SCR can be driven into this mode when the forward biased voltage is more than the breakover voltage. In this mode the thyristor offers the minimum resistance and behaves as a closed switch. Depending upon the rating of the SCR this voltage would be of the order of 1 V or 2 V. It may be observed that this voltage drop increases slightly with an increase in the anode current. In this conduction mode the anode current is limited by load impedance alone as voltage drop across SCR is quite small. This small voltage drop across the device is due to the Ohmic drop in the four layers.

Operation of the SCR: The SCR can be treated as two transistors (*pnp* and *npn*) connected as shown in Fig. 9.5. The base of transistor T_1 is connected to the collector of transistor T_2 and the collector of transistor T_1 is connected to the base of transistor T_2. The equivalent circuit diagram of this configuration of SCR is as shown in Fig. 9.6.

Assuming leakage current in T_1 to be very small,

$$I_{b1} = I_A - I_{c1} = I_A - \alpha_1 I_A = (1 - \alpha_1)I_A \quad ...(9.1)$$

But,

$$I_{b1} = I_{c2} \text{ and } I_{c2} = \alpha_2 I_K \quad ...(9.2)$$

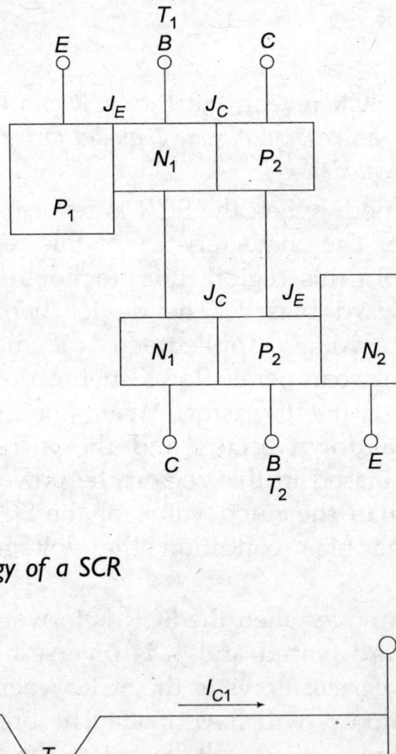

Fig. 9.5 *Two-transistor analogy of a SCR*

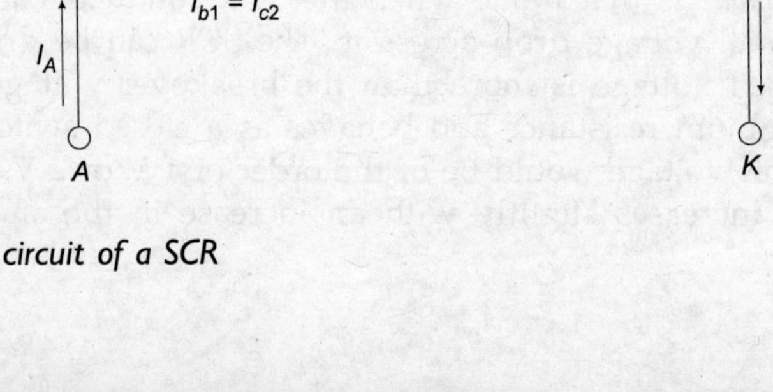

Fig. 9.6 *Equivalent circuit of a SCR*

Thus, $(1 - \alpha_1)I_A = \alpha_2 I_K$ ...(9.3)

From KCL, $I_K = I_A + I_g$ ...(9.4)

Thus, $(1 - \alpha_1)I_A = \alpha_2(I_A + I_g)$ ...(9.5)

$(1 - \alpha_1 - \alpha_2)I_A = \alpha_2 I_g$

or, $I_A = \alpha_2 I_g/[1 - (\alpha_1 + \alpha_2)]$ ...(9.6)

If $\alpha_1 + \alpha_2 = 1$, then $I_A = \infty$. That is, the anode current suddenly reaches a very high value approaching infinity. Therefore, the device suddenly triggers into ON state from the OFF state. This is known as *regenerative action.*

The value of $\alpha_1 + \alpha_2$ can be made almost equal to unity by proper selection of positive gate current for short duration. This signal applied at the gate will cause a flow of collector current I_{c2} by transferring T_2 to its ON state. At this stage, the transistor T_1 will also be switched ON. Now, the action is regenerative since each of the transistors would supply base current to the other. At this point even if the gate signal is removed, the device keeps on conducting until the current level is maintained to a minimum value of holding current.

A SCR can be turned on by any of the following methods.

1. *Application of high voltage*: If the anode to cathode voltage is increased beyond the breakover voltage, the device fires from OFF to ON state. This method is destructive and generally not followed.
2. *Gate current*: In a forward biased thyristor, the application of a gate current of proper magnitude will reduce the forward breakover voltage and trigger the device. The firing of a thyristor by application of a gate current pulse is the method generally employed for using the thyristor as a switch. The *turn on time* of a SCR is the sum of *delay time* and *rise time*. *Delay time* t_d is defined as the time interval that elapses between the application of 10% of gate current and 10% of the thyristor ON current. The *rise time* t_r is defined as the time interval between 10% of thyristor ON current and 90% of the full thyristor ON current. Since the gate loses control over the anode current once the device is fired, the gate pulse required for firing is either equal to or slightly more than turn-on time t_{ON}. If the gate pulse is maintained over long time, the gate junction power loss would be high. The gate pulse is not required in the reverse bias condition of the SCR.
3. *High temperature*: At high temperatures, the generation of electron-hole pairs would be large and thus increases the currents I_{e1} and I_{e2}. Thus, the values of α_1, α_2 increases and thus the value of $\alpha_1 + \alpha_2$ tends towards unity that the SCR may trigger. This may lead to thermal runaway and may damage the SCR permanently. So, it is not employed.
4. *Excitation by light*: If light is incident on the junctions of the SCR, many new electron-hole pairs are generated and thus, the value of $\alpha_1 + \alpha_2$ may be tending towards unity where the device is fired. This method is employed in devices known as *light-activated silicon-controlled rectifiers* (LASCR).
5. *Rate effect*: If the rate of rise of anode-cathode voltage is very high, in a forward blocking thyristor, a high current would flow through the junction capacitors. It may be sometimes destructive and so care should be taken while designing the circuit so that the rate specified by the manufacturer is never crossed.

To convert the SCR state from ON to OFF, the anode current should be made small compared to the holding current. This small value of current should be maintained for the time until all the excess carriers in the four layers are swept out or recombined. The *turn-off time* t_{OFF} is the minimum value of the time interval between the instant when the ON state current has decreased to zero and the instant when the thyristor is capable of withstanding forward voltage without turn on.

Thyristor Ratings

- *Forward breakover voltage* is the voltage at which the thyristor enters into the forward conduction region. This voltage is maximum when gate current is zero. As the gate increases, the breakover voltage decreases.
- *Average forward current* is the maximum anode current (dc) the SCR is capable of withstanding in ON state.
- *Latching current* is the minimum current required to latch or trigger the device from OFF to ON state.
- *Holding current* is the minimum value of current required to hold the device in ON state. For turning the device OFF, the anode current should be lowered below the holding current by increasing the external resistance.
- *Gate current* is the current applied to the gate of the device for control purposes. The minimum gate current is the minimum value of current required at the gate for triggering the device. The maximum current is the value that can be applied without damaging the gate.
- *Reverse breakdown voltage* is the reverse voltage at which the device breaks down into avalanche and conducts heavily.
- *Voltage safety factor* is a ratio of peak inverse voltage (PIV) to RMS value of the normal operating voltage. For safe operation, the normal working voltage of the device should be much below the PIV.

9.4 TRIAC

One of the major shortcomings of the SCR is that it cannot be used for control applications during the negative half-cycles of the applied ac input wave, unless special circuitry is used. TRIAC is a combination of two SCRs connected antiparallel to form a single device. The structure and the circuit symbol of the TRIAC are as shown in Fig. 9.7. TRIAC is also a four-layer silicon device with three terminals. However, the TRIAC is bidirectional, that is, it can block voltages of both polarities and conduct in either direction. A pulse of either polarity can switch a TRAIC from OFF to ON state. Gate terminal makes Ohmic nonrectifying contact with both p and n materials, thereby permitting either a positive or a negative trigger pulse to cause current flow.

Figure 9.8 shows the *V-I* characteristics of the TRIAC. It can be seen that it is an extension of *V-I* characteristics of the SCR. The characteristics are symmetrical about the first and third quadrants. The gate controls the current flow in either direction.

The TRIAC can be used in power frequency voltage controllers. Such a controller can be employed in a light dimmer or for controlling the power of a heating element. SCR and

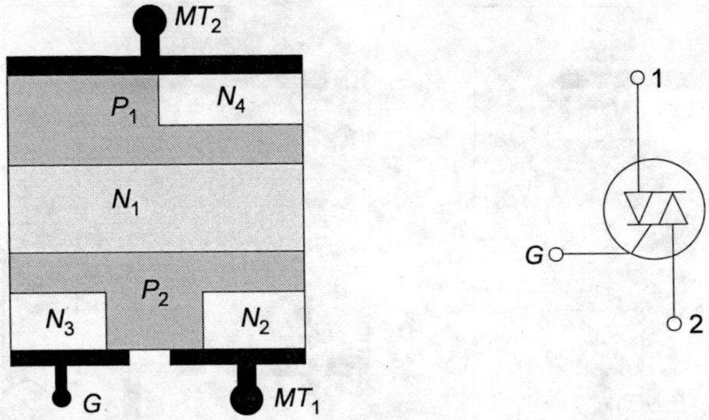

Fig. 9.7 *Structure and circuit symbol of a TRIAC*

TRIAC are used as controllers of speed of ac as well as dc motors. Fan regulators using TRIAC are small, compact and dissipate negligible amount of power. However, these regulators suffer from the generation of harmonics in the wave form. These harmonics give rise to noise in the operation of the fan. The TRIAC can be triggered in four different modes as shown in Fig. 9.9.

Mode 1: In this mode both MT_2 and gate current are positive. The operation in the first quadrant and device is very sensitive in this mode. This forward biases the P_2-N_2 junction. Thus, P_1-N_1-P_2-N_2 can fire the device ON.

Mode 2: In this mode MT_2 is positive and gate current is negative. The junction P_2-N_3 is forward biased. Thus, the electrons are injected from N_3 into P_2. A current thus flows in P_2 towards the N_3 gate and thus P_1-N_1-P_2-N_3 turns ON as the gain of $N_3P_2N_1$ transistor rises. This supplies current to N_2. Thus, P_1-N_1-P_2-N_2 conducts.

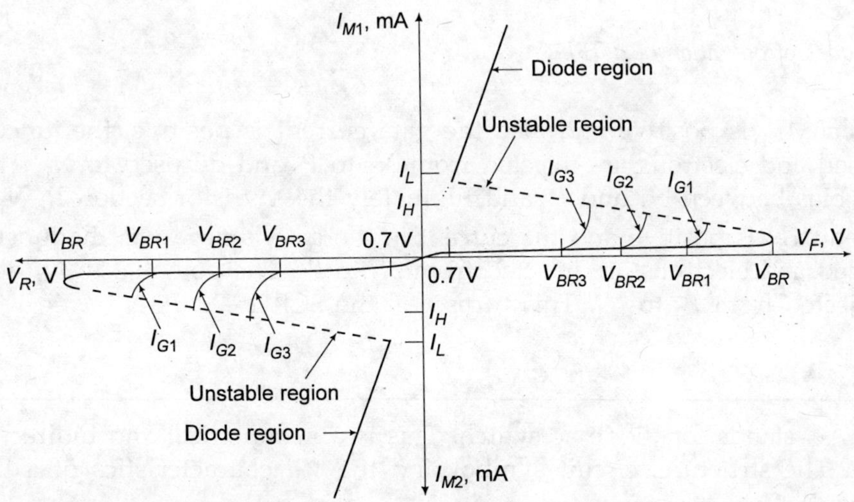

Fig. 9.8 *V-I characteristics of a TRIAC*

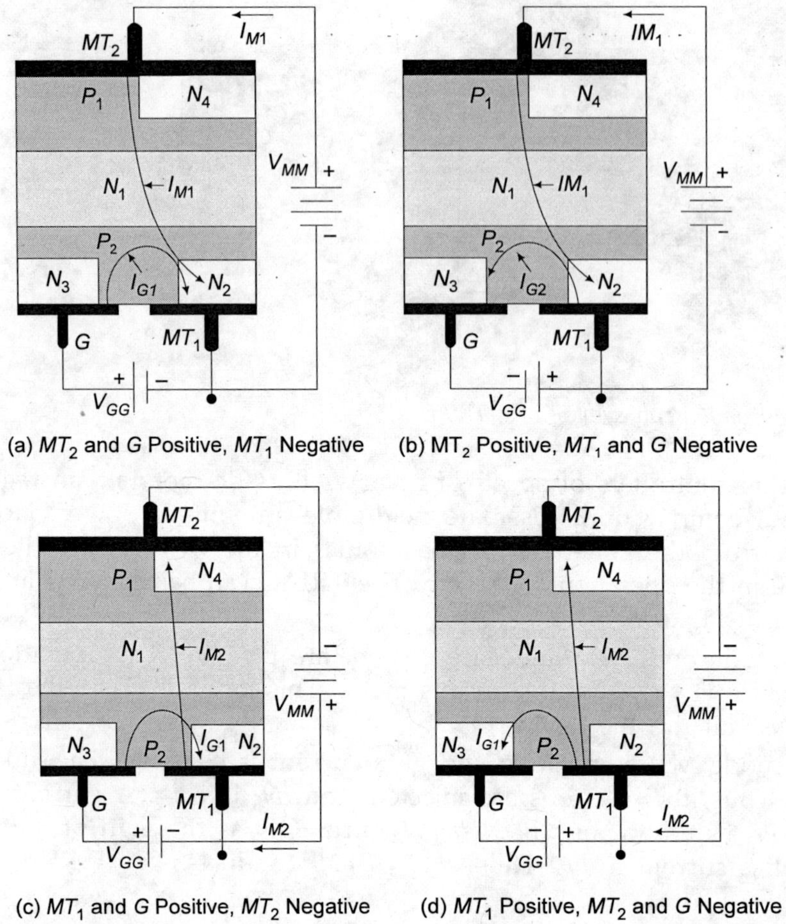

(a) MT_2 and G Positive, MT_1 Negative (b) MT_2 Positive, MT_1 and G Negative

(c) MT_1 and G Positive, MT_2 Negative (d) MT_1 Positive, MT_2 and G Negative

Fig. 9.9 *Modes of operation of a TRIAC*

Mode 3: When MT_1 is positive and also the gate current is positive, the junction N_2-P_2 is forward biased and electrons are injected from N_2 to P_2 and diffused to N_1. This increases the forward bias between N_1 and P_2 and ultimately the thyristor section P_2-N_1-P_1-N_4 fires.
Mode 4: When MT_1 is positive and the current in the gate is negative, the junction P_2-N_3 is forward biased and electrons are injected from N_3 to P_2. The potential in N_1 decreases and holes are injected from P_2 to N_1. This turns ON the SCR P_2-N_1-P_1-N_4.

9.5 DIAC

The term DIAC stands for diode ac switch. This is a symmetrical and bidirectional *pnp* or *pnpn* device. The structure, circuit symbol and the *V-I* characteristics of a DIAC are as shown in Fig. 9.10.

DIAC is a three-layer two-terminal silicon device and is referred to as *bidirectional diode*. It does not have any gate terminal. Switching from OFF to ON state may be done by

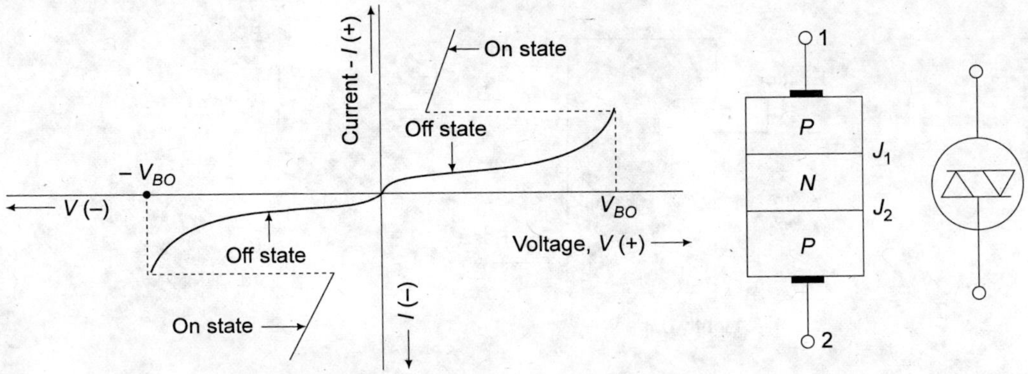

Fig. 9.10 *Structure, circuit symbol and V-I characteristics of a DIAC*

simply exceeding the avalanche breakdown voltage in either direction. The two *p* regions have similar doping characteristics resulting in symmetrical switching characteristics for both positive and negative voltages.

The applied voltage of any polarity results in a small saturation current across the reverse biased *p-n* junction. When the applied bias exceeds the avalanche breakdown voltage, the DIAC current rises sharply as shown in Fig. 9.10. In this ON state, the voltage across the DIAC decreases with increasing current and the device offers negative resistance.

DIAC is mainly used as a trigger device in TRIAC power control systems. It is used as triggering device in TRIAC phase control circuits used for light dimming, motor speed control and heat control. It is not a control device.

9.6 UNIJUNCTION TRANSISTOR

Unijunction transistor (UJT) is a three–terminal semiconductor-switching device. It has only one *p-n* junction and three terminals—emitter and two base terminals. The working of a UJT differs from that of a conventional BJT.

The structure and circuit symbol of a UJT is as shown in Fig. 9.11. The device consists of a base B in the form of a low-conductivity *n*-type silicon bar with two Ohmic contacts B_1 and B_2 attached to the bar at the two ends. These Ohmic contacts are brought out at right angles to the base and emitter terminal E of *p*-type silicon forming a small alloy-type *p-n* junction in the middle of the base bar length.

In the circuit symbol, the arrow head on the emitter points in the direction of conventional current and is inclined pointing towards B_1 as shown in Fig. 9.11. The UJT is usually operated with the emitter junction forward biased.

The resistance between B_1 and B_2 is known as *interbase resistance* and is denoted as R_{BB}. It is of the order of 5–10 kΩ with emitter open or with the junction reverse biased. The UJT may be represented as a potential divider consisting of two resistors R_{B1} and R_{B2} with *p-n* junction acting as a diode as shown in Fig. 9.12. Thus, $R_{BB} = R_{B1} + R_{B2}$. If a voltage V_{BB} is applied between B_2 and B_1, then the voltage between emitter and base B_1 can be given as

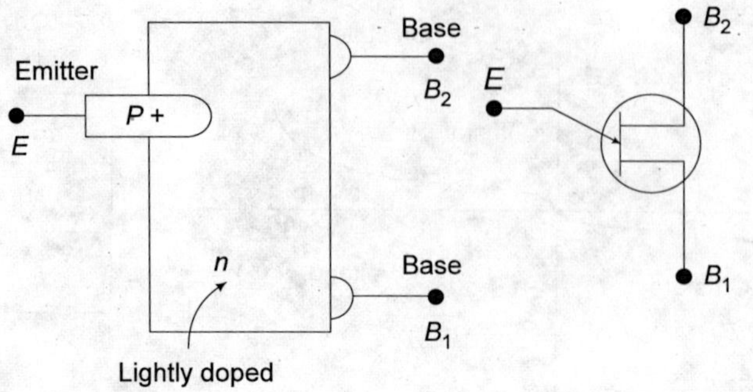

Fig. 9.11 *Structure and circuit symbol of a UJT*

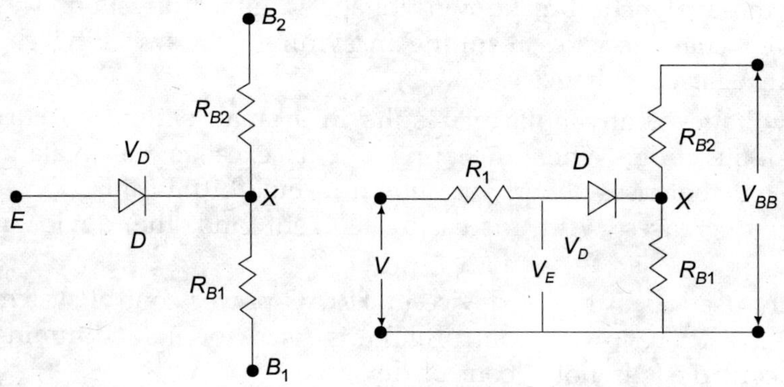

Fig. 9.12 *Equivalent circuit of a UJT*

$$V_E = \eta V_{BB} = V_{BB} R_{B1}/(R_{B1} + R_{B2}) \qquad ...(9.7)$$

where,
$$\eta = R_{B1}/(R_{B1} + R_{B2}) \qquad ...(9.8)$$

η is known as *intrinsic stand-off ratio.* This ratio has a value between 0.5 and 0.8.

If an external supply is connected to the emitter terminal, no current flows into the base as long as the applied voltage is less than $V_E = \eta V_{BB}$ since the *p-n* junction is reverse biased. A small leakage current flows from B_2 to emitter due to minority carriers. When once the applied voltage increases, when the drop across emitter base 1 equals cutin voltage and crosses it, the junction is forward biased and the current flows from the emitter towards base 1 and decreases the resistance R_{B1} greatly. This is referred to as *conductivity modulation.* The device is now in ON state. If a negative voltage is applied to the emitter, *p-n* junction remains reverse biased and the emitter current is cut off.

The device is now in OFF state. The *V-I* characteristics of a UJT are as shown in Fig. 9.13. Up to the peak point *P*, the diode is reverse biased and hence the region to the left of the peak is called *cut off region*. The diode starts conducting when the emitter base 2 voltage crosses the cutin voltage and so the resistance of base 2 region decreases, which

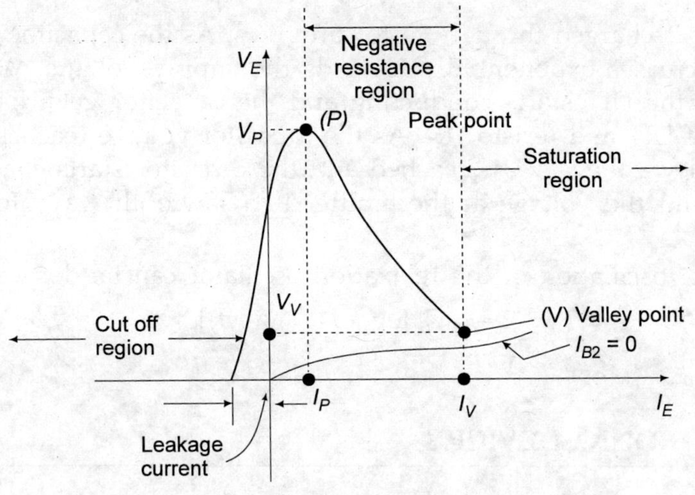

Fig. 9.13 *V-I characteristics of a UJT*

leads to negative resistance region as shown in Fig. 9.13. This negative resistance region is between the peak point and valley point. After the valley point, the device is driven into saturation and behaves like the conventional forward biased junction diode. This region is referred to as *saturation region*.

Relaxation oscillator: UJT finds a very wide application as a *relaxation oscillator*. It is a sawtooth generator and is very popularly used in electronics. The circuit of such an oscillator is as shown in Fig. 9.14. The figure also shows the associated wave forms.

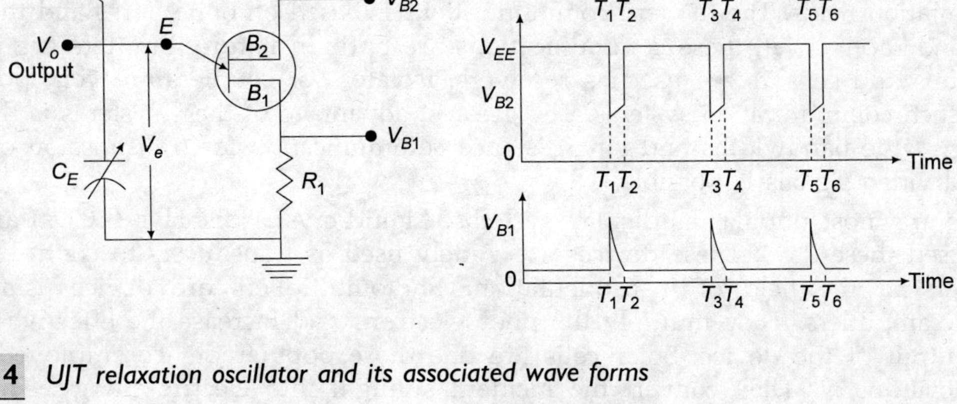

Fig. 9.14 *UJT relaxation oscillator and its associated wave forms*

The capacitor C_E is charged through R_E towards V_{BB}. As the capacitor gets charged, the voltage across it increases exponentially towards the supply voltage. When this voltage reaches a value V_P, the UJT starts conducting and the capacitor gets discharged through the emitter-base 1 of UJT and resistor R_1. As the capacitor voltage reaches zero, the device cuts off when a valley voltage V_V is reached and the capacitor starts charging again. This process continues and the voltage at the emitter is a sawtooth wave form as shown in Fig. 9.14.

The frequency of oscillation of the relaxation oscillator can be derived to be

$$T = 2.3 R_E C_E \log_{10} [1/(1 - \eta)] \quad \text{...(9.9)}$$

$$f = 1/T \quad \text{...(9.10)}$$

9.7 OPTOELECTRONIC DEVICES

Optoelectronic devices are electronic devices that work based on the principles of optics. Theses devices utilise energy in the visible and infrared regions of electromagnetic spectrum. The basic principle involved in these devices is that energy is emitted or absorbed in the form of light. If the device material is so chosen that the energy gap between two energy levels (equal to hf, h being the Planck's constant) is such that f is in the visible or infrared region, the emitted energy is in the form of light. Similarly, the light energy incident on a semiconductor breaks the bonds to create new charge carriers; the current thus produced is proportional to the incident light intensity. In other words, the optoelectronic devices are transducers which either convert light energy into electric energy or vice versa.

The Optoelectronic devices can be classified broadly as *emitters, sensors* and *couplers.* Emitters are LED or Laser or IR (infrared) devices. Sensors can be once again classified as photoemissive or photoconductive or photovoltaic. Photovoltaic devices generate a voltage, while photoconductive devices require an external source. Photoconductive devices are subdivided into bulk photoresistors and junction-type photoconductors. Photodiode, PIN diode, avalanche photodiode and the like fall under the category of junction-type photoconductive devices.

The optoelectronic devices find large applications in electronics, communications and instrumentation fields. The modern communications is based on optic fibres and the transmitters and receivers in these communications are optic in nature. Emitters are used as transmitters and sensors are used as receivers. Isolators or couplers are very popularly used in such communication systems. The greatest advantage of these systems is that they have very wide bandwidths and long-distance communication for transmission of voice, data, and video signals is possible.

LEDs are most popular as display systems. Liquid-crystal displays (LCDs) are most widely used these days. These devices are widely used as transducers to control various parameters in the field of instrumentation. Photomultipliers are devices similar to electronic amplifiers. They multiply the photoelectrons and increase the effective current in the output of the device. Solar cells are one more popular device employed in the electronic industry. They convert the incident sunlight into electric energy. They find application in the supply of electric power to the satellite-subsystems in the satellite communication system. Not only in satellite communications, but wherever solar energy is converted into electric energy—like solar heating systems, solar lighting systems and so on in consumer-electronics industry—these cells are very widely used.

Optoelectronic devices are used in computers because the optic fibres have very high data-handling capacity. They are used for mutual interfacing of central processor, linking them with other peripheral devices, data transmission and local area networks (LAN). In LAN, fibers are used to link the computers in various topologies. They find applications in office operations, private automatic branch exchanges and so on.

Optoelectronic devices are used in power plants, rail-road networks, metal industry for data acquisition control, signal processing and so on. In medical electronics they are very suitable since they are noise-free and least interference devices. They are ideal sensors in automobile industry since they are of less weight.

Optoelectronic devices *photodiodes* and *LEDs* are already discussed in Sec. 3.11 and 3.12, respectively. The *phototransistor* is another variant of a photodiode. This is a much more sensitive semiconductor photodevice than the *p-n* photodiode. The phototransistor is usually connected in a common emitter configuration with base open and radiation is concentrated on the region near the collector junction as shown in Fig. 9.15. The emitter junction is slightly forward biased and the collector junction is reverse biased. Thermally generated minority carriers cross the junction to constitute collector current when no radiation is incident on the junction,

$$I_C = (\beta + 1) I_{CO} \quad \text{...(9.11)}$$

When the light is turned on, additional charge carriers are photogenerated and these contribute to the reverse saturation current in exactly the same manner as do the thermally generated minority charges. Thus, the collector current is given as

$$I_C = (\beta + 1)\ (I_{CO} + I_L) \quad \text{...(9.12)}$$

The *V-I* characteristics of the photodiode are as shown in Fig. 9.15. They are similar to the output characteristics of a *CE* transistor configuration. The parameter in this phototransistor is light intensity instead of base current.

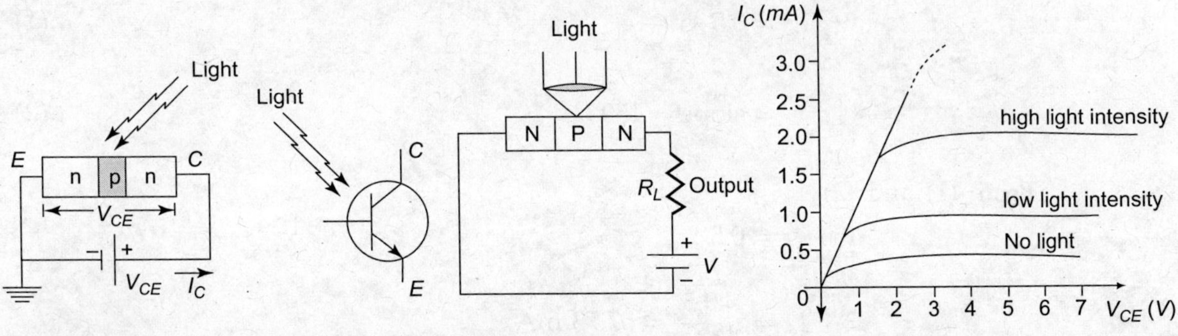

Fig. 9.15 *Structure, circuit symbol and characteristics of a phototransistor*

SUMMARY

- The four-layer *pnpn* structure with three junctions is treated as two transistors (*npn* and *pnp*) connected to each other.
- In a *pnpn* diode, *p* side is positive and *n* side is negative, the junctions J_1 and J_2 are forward biased and J_2 is reverse biased and so the current is small saturation current in the device.
- When the current in the Shockley diode is more than holding current, the device switches from OFF to ON.
- The gate terminal in the silicon-controlled rectifier controls the firing of the device.
- TRIAC is two SCRs connected back to back. It has controlling in both directions.
- DIAC is diode ac switch and is a bidirectional diode.
- A Unijunction transistor exhibits negative resistance and is widely employed as a sweep generator, also referred to as relaxation oscillator.
- Optoelectronic devices are those devices that are activated by or generate light. Photodiodes, phototransistors, LEDs, LASCR and laser diodes are a few examples of optoelectronic devices.

SOLVED PROBLEMS

9.1 For a SCR, the gate-cathode characteristic has a straight line slope of 150. For trigger source voltage of 20 V and allowable gate power dissipation of 0.5 W, compute the gate-source resistance.

Solution Given that

$$V_g \cdot I_g = 0.5 \text{ W} \quad \text{...(1)}$$

$$\frac{V_g}{I_g} = \text{slope} = 150 \quad \text{...(2)}$$

From (1) and (2)

$$150\, I_g^2 = 0.5$$

$$I_g = \sqrt{\frac{0.5}{150}} = 57.73 \text{ mA}$$

Writing KVL

$$E_s = I_g \cdot R_s + V_g$$

$$= I_g \cdot R_s + 150\, I_g$$

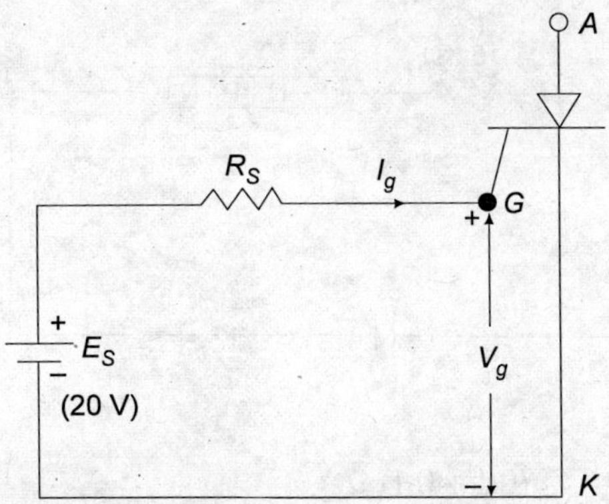

$$\therefore \quad 20 = R_s(57.73 \times 10^{-3}) + 150 \times 57.73 \times 10^{-3}$$

$$\Rightarrow R_s = \frac{20 - 8.66}{57.73}\ \text{K}\Omega$$

$$= 0.19642\ \text{K}\Omega$$

$$= 196.42\ \Omega.$$

9.2 A Schockley diode is biased in the forward blocking region with an anode to cathode voltage of 36 V. If $\alpha_1 = 0.25$ and $\alpha_2 = 0.35$ and the leakage currents are 160 nA. Determine the anode current and the forward resistance of the Schockley diode.

Solution Anode current $= \dfrac{I_{CO1} + I_{CO2}}{1-(\alpha_1 + \alpha_2)}$

$$= \frac{160}{1-(0.25+0.35)}$$

$$= \frac{160}{0.4}\ \text{nA} = 0.4\ \mu\text{A}$$

$$\text{Forward resistance} = \frac{\text{anode to cathode voltage}}{\text{anode current}}$$

$$= \frac{36}{0.4 \times 10^{-6}} = 90\ \text{M}\Omega$$

9.3 A Schockley diode is connected in series with a 15 KΩ resistor. Calculate (a) The anode current when the device is OFF if $V = 45$ V and $I_{CO1} = I_{CO2} = 0.1\ \mu$A, $\alpha_1 = \alpha_2 = 0.45$. (b) The anode current if $V_{BRO} = 48$ V, $V_{BE} = 0.7$ V and $V_{CE(\text{sat})} = 0.1$ V.

Solution (a) Anode current $= \dfrac{I_{CO1} + I_{CO2}}{1-(\alpha_1 + \alpha_2)}$

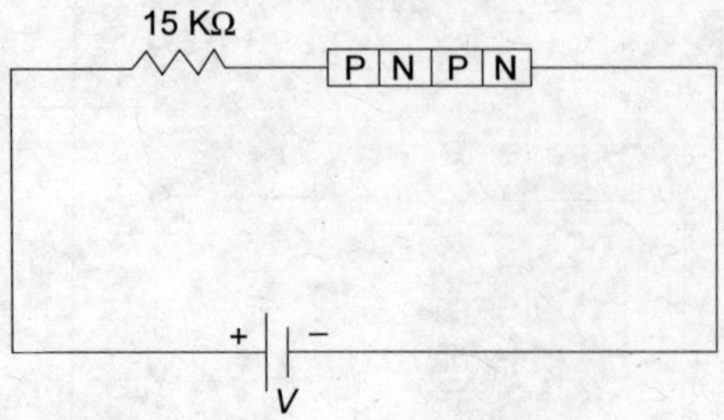

$$= \frac{(0.1 + 0.1)\ \mu A}{1 - (0.9)} = 2\ \mu A$$

(b) When the device is ON, the anode voltage

$$V_A = V_{BE} + V_{CE\,(sat)} = 0.7 + 0.1 = 0.8\ V$$

$$V_R = \text{Voltage across resistance} = 50 - 0.8$$

$$= 49.2\ V$$

$$\therefore \text{Anode current} = \frac{V_R}{R} = \frac{49.2}{15\ K} = 3.28\ mA$$

9.4 A SCR has $I_{CBO1} = I_{CBO2} = 0.075\ \mu A$, and a loop gain $G = 0.995$. Calculate the anode current if $I_G = 10$ mA and $\alpha_2 = 0.497$.

Solution Anode current $= \dfrac{\alpha_2 I_G + I_{CBO1} + I_{CBO2}}{1 - (\alpha_1 + \alpha_2)}$

$$= \frac{0.497 \times 10 \times 10^{-3} + (0.075 + 0.075) \times 10^{-6}}{1 - (0.995)}$$

($\because G = \alpha_1 + \alpha_2 =$ loop gain)

$$= 9940.3 \times 10^{-4}$$

$$= 0.994\ A$$

9.5 The V_g–I_g characteristic of a SCR is assumed to be a straight line passing through the origin with a gradient of 3×10^3. Calculate the gate source resistance, given that P_g = 0.012 W and E_S = 10 V.

Solution

$$P_g = 0.012$$

$$V_g \cdot I_g = 0.012 \qquad ...(1)$$

$$\frac{V_g}{I_g} = 3 \times 10^3$$

$$\Rightarrow \quad V_g = I_g \times 3 \times 10^3 \qquad ...(2)$$

From (1) and (2)

$$I_g \times 3 \times 10^3 \times I_g = 0.012$$

$$I_g^2 = \frac{0.012}{3\times10^3}$$

$$\therefore \quad I_g = 2 \text{ mA}$$

$$V_g = 3 \times 10^3 \times I_g = 6 \text{ V}$$

We have $E_S = V_g + I_g \cdot R_S$

$$\Rightarrow \quad R_S = \frac{E_S - V_g}{I_g} = \frac{10-6}{2 \text{ mA}}$$

$$= 2 \text{ K}\Omega$$

9.6 The trigger circuit of a thyristor has a source voltage of 15 V and the load line has a slope of 100 V per ampere. The minimum gate current to turn-ON the SCR is 25 mA. Compute:

(a) Source resistance required in the gate circuit

(b) The trigger voltage and trigger current for an average gate power dissipation of 0.4 W.

Solution (a) The slope of the load line gives the required gate source resistance.

$$\therefore \quad R_S = 100 \ \Omega$$

(b) $$V_g \cdot I_g = 0.4 \text{ W} \quad \text{...(1)}$$

and $$E_S = I_g \cdot R_s + V_g \quad \text{...(2)}$$

From (1) and (2) $$15 = 100 I_g + \frac{0.4}{I_g}$$

$$\Rightarrow 100\, I_g^{\,2} - 15 I_g + 0.4 = 0$$

$$\therefore \quad I_g = 115.31 \text{ mA or } 34.685 \text{ mA}$$

$$\therefore \quad V_g = \frac{0.4\times10^3}{115.31} \text{ or } \frac{0.4\times10^3}{34.685}$$

$$= 3.468 \text{ V or } 11.532 \text{ V}$$

$\therefore$ Choose $V_g = 3.468$ and $I_g = 115.31$ mA

9.7 Latching current of a thyristor circuit shown in figure is 50 mA. The duration of the firing pulse is 50 μs. Will the thyristor get fired?

Solution

$$\tau = \frac{L}{R} = \frac{0.5}{20}$$

$$= 0.025 \text{ sec.}$$

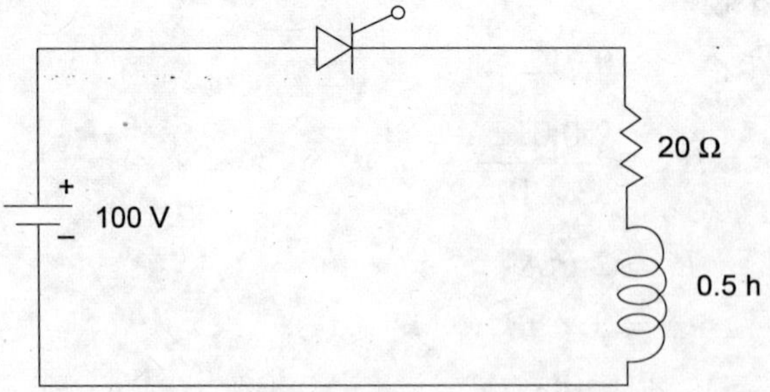

$$i(t) = \frac{V}{R}(1 - e^{-t/\tau})$$

$$= \frac{100}{20}\left(1 - e^{-t/0.025}\right)$$

$$= 5(1 - e^{-40t})$$

at $t = 50\ \mu s$, $i(50\ \mu s) = 9.99$ mA

Since the calculated latching current value is less than the given value, the SCR will not get fired.

9.8 Latching current of an SCR inserted in between a dc voltage source of 200 V and load is 100 mA. Compute the minimum width of gate pulse current required to turn-ON this SCR in case the load consists of (a) $L = 0.4$ h (b) $R = 20\ \Omega$ and an $L = 0.2$ h in series (c) $R = 20\ \Omega$ in series with L = 2.0 h.

Solution (a) When load is purely inductor,

$$V = L\frac{di}{dt}$$

$$\Rightarrow \quad di = \frac{V\,dt}{L} \Rightarrow i = \frac{V}{L}\cdot t$$

$$100 \times 10^{-3} = \frac{200}{0.4} \times t$$

$$\Rightarrow \quad t = \frac{100 \times 10^{-3} \times 0.4}{200} = 200\ \mu s$$

(b) When load is series connection of R and L,

$$V = Ri + L\frac{di}{dt}$$

$$\Rightarrow \quad i = \frac{V}{R}\left(1 - e^{-t/\tau}\right)$$

$$\tau = \frac{L}{R} = \frac{0.2}{20} = 0.01$$

$$\therefore \quad 100 \times 10^{-3} = \frac{200}{20}\left(1 - e^{-100t}\right)$$

$$10 \times 10^{-3} = 1 - e^{-100t}$$

$$\Rightarrow \quad e^{-100t} = 0.99$$

$$t = 100.503\ \mu s$$

(c) Here also $i = \frac{V}{R}(1 - e^{-t/\tau})$

$$\tau = \frac{L}{R} = \frac{2}{20} = 0.1$$

$$\therefore \quad 100 \times 10^{-3} = \frac{200}{20}\left(1 - e^{-10t}\right)$$

$$10 \times 10^{-3} = 1 - e^{-10t}$$

$$\Rightarrow \quad t = 1005.03\ \mu s$$

9.9 The data sheet of a certain UJT has $\eta = 0.7$, determine V_P if $V_{BB} = 20$ V.

Solution

$$V_P = \eta V_{BB} + V_D$$

$$= 0.7 \times 20 + 0.7 = 14.7 \text{ V}$$

9.10 A relaxation oscillator using UJT is to be designed for triggering an SCR. The UJT has the following data.
$\eta = 0.72$, $I_P = 0.6$ mA. $V_P = 18$ V, $V_V = 1.0$ V, $I_V = 2.5$ mA, $R_{BB} = 5$ KΩ. Normal leakage current with emitter open = 4.2 mA. The firing frequency is 2 KHz. For C = 0.04 μF, compute the values of R, R_1 and R_2.

Solution We have

$$T = RC \ln\left(\frac{1}{1-\eta}\right)$$

$$\Rightarrow \quad R = \frac{T}{C \ln\left(\frac{1}{1-\eta}\right)}$$

$$= \frac{1}{fC \ln\left(\frac{1}{1-\eta}\right)} = \frac{1}{2 \times 10^3 \times 0.04 \times 10^{-6} \ln\left(\frac{1}{1-0.72}\right)}$$

$$= 9.82 \text{ K}\Omega$$

Since V_D is not given, $V_P = \eta V_{BB}$.

$$V_{BB} = \frac{V_P}{\eta} = \frac{18}{0.72} = 25 \text{ V}$$

$$R_2 = \frac{10^4}{\eta V_{BB}} = \frac{10^4}{0.72 \times 25}$$

$$= 555.55 \ \Omega$$

With emitter open V_{BB} = leakage current $\times (R_1 + R_2 + R_{BB})$

$$\Rightarrow \quad 25 = 4.2 \times 10^{-3}(R_1 + 555.55 + 5000)$$

$$\therefore \quad R_1 = \frac{25}{4.2 \times 10^{-3}} - 5000 - 555.55$$

$$= 396.83 \ \Omega$$

9.11 Given the relaxation oscillator as shown in figure below.

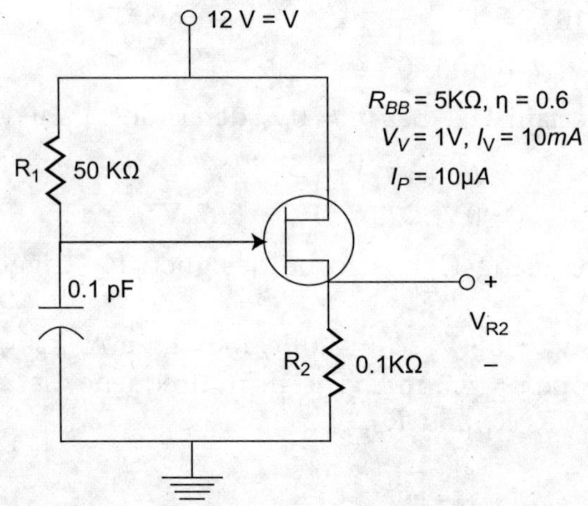

(a) determine R_{B1} and R_{B2} at $I_E = 0$ mA;
(b) calculate V_{P1} to turn on UJT;
(c) find whether R_1 is in the permissible range to ensure firing of UJT;
(d) find frequency of oscillator if $R_{B1} = 100 \ \Omega$ during discharge phase.

Solution (a)

$$\eta = \frac{R_{B1}}{R_{B1} + R_{B2}} = 0.6$$

$$\Rightarrow \quad R_{B1} = 0.6 R_{BB}$$

$$= 0.6 \ (5 \text{ k}\Omega) = 3 \text{ K}\Omega$$

$$\therefore \quad R_{B2} = R_{BB} - R_{B1} = 2 \text{ K}\Omega$$

(b) To find V_P, the following diagram results

$$V_P = 0.7 + \frac{(R_{B1} + R_2)12\text{ V}}{(R_{B1} + R_{B2} + R_2)}$$

$$= 0.7 + \frac{(3 + 0.1)12}{(5 + 0.1)}$$

$$= 0.7 + 7.294 = 8\text{ V}$$

(c) $$\frac{V - V_V}{I_V} < R_1 < \frac{V - V_P}{I_P}$$

$$\frac{12 - 1}{10\text{ mA}} < R_1 < \frac{12 - 8}{10\ \mu\text{A}}$$

$$1.1\text{ K}\Omega < R_1 < 400\text{ K}\Omega$$

The resistance $R_1 = 50\text{ K}\Omega$ falls within this range.

(d) $$t_1 = R_1 C \log_e \frac{V - V_V}{V - V_P}$$

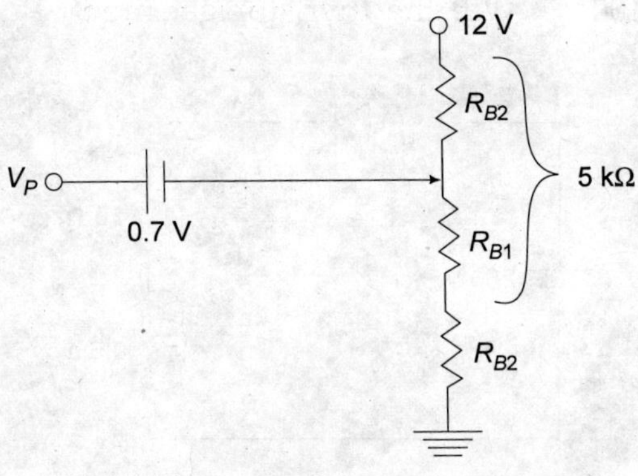

$$= (50\text{ K})\ (0.1\text{ pf})\log_e \frac{12 - 1}{12 - 8}$$

$$= 5.05\text{ ms}$$

$$t_2 = (R_{B1} + R_2)C \log_e \frac{V_P}{V_V}$$

$$= (0.1 + 0.1) \times 10^3 \times 0.1 \times 10^{-12} \log_e \frac{8}{1}$$

$$= 41.6\ \mu\text{s}$$

$$\therefore \qquad T = t_1 + t_2 = 5.092 \text{ ms}$$

$$\text{Frequency of oscillation} = \frac{1}{T} = 196 \text{ Hz}$$

EXERCISE PROBLEMS

9.1 For an SCR, the gate-cathode characteristic has a straight line slope of 130. For trigger source voltage of 15 V and allowable gate dissipation of 0.5 W, find the gate source resistance.

9.2 A Shockley diode is biased in the forward blocking region with an anode to cathode voltage of 30 V. If $\alpha_1 = \alpha_2 = 0.45$ and leakage currents are 120 nA. Find the anode current and the forward resistance of the Shockley diode.

9.3 An SCR has $I_{CBo1} = I_{CBo2} = 0.025\ \mu$A, and a loop gain $G = 0.9$. calculate the anode current if $I_G = 20$ mA and $\alpha_2 = 0.5$.

9.4 The V_g–I_g characteristic of an SCR is assumed to be a straight line passing through the origin, with a gradient of 5×10^3. Given that $P_g = 0.015$ W and $E_s = 15$ V, find the gate source resistance.

9.5 Latching current in a thyristor circuit shown in figure is 100 mA. The duration of the firing pulses is 100 μs. Will the thyristor get fired?

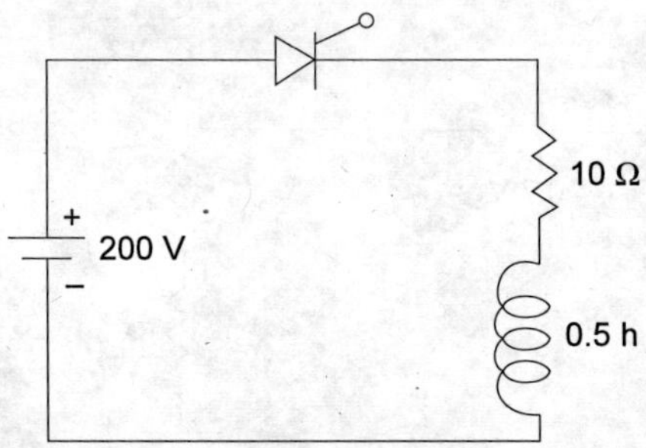

9.6 Desing a UJT relaxation oscillator circuit with $\eta = 0.51$, $I_p = 10\ \mu$A, $V_V = 3.5$ V, $I_V = 10$ mA. The frequency of oscillation is $f = 60$ Hz and the width of the triggering pulse is $t_g = 50\ \mu$s.

9.7 Derive $T = RC\log_e\left(\dfrac{1}{1-\eta}\right)$ for the UJT relaxation oscillator.

REVIEW QUESTIONS

1. What is meant by the term thyristors? Give the principal applications of thyristors.
2. Give the construction of a *pnpn* diode. Discuss the two-transistor version of a *pnpn* diode.
3. Derive the condition for breakover of a *pnpn* diode.
4. With reference to a *pnpn* diode, define the terms latching current, latching voltage and rate effect.
5. Give the construction and working of a bilateral *pnpn* diode switch.
6. Draw and discuss the volt–ampere characteristics of a *pnpn* diode.
7. Describe the basic structure and operation of a silicon-controlled rectifier and give two-transistor version of a SCR. Explain why always silicon, and not germanium, is used in the construction of a SCR.
8. Draw and discuss the *V-I* characteristics of a SCR.
9. Define the terms turn-on time and turn-off time of a SCR.
10. Draw a diagram showing the basic structure of a TRIAC. Draw its *V-I* characteristic.
11. Describe the basic structure of a DIAC and draw its *V-I* characteristic. Give the main uses of a DIAC.
12. Draw a diagram giving the basic structure of a UJT. Give the working of a UJT.
13. Define the term intrinsic stand-off ratio of a UJT.
14. Draw and discuss the static emitter characteristics of a UJT.
15. Draw a circuit diagram of a relaxation oscillator using a UJT and explain its operation.
16. What are the various modes of operation of a SCR? Explain in detail.
17. What is regenerative action of a SCR? Derive equation for anode current of a SCR with the help of two-transistor model.
18. Explain various methods of firing on the SCR. How do you turn off the SCR?
19. What are the various modes of operation of a TRIAC? Explain in detail.
20. What is an optoelectronic device? List the various components in this category and give their applications.
21. Draw the structure of a phototransistor and explain its operation in detail.
22. Draw the *V-I* characteristics of a phototransistor and explain their significance.
23. Compare the *V-I* characteristics and performance of a SCR and a TRIAC.
24. List thyristor ratings and explain their significance.
25. What is conductivity modulation in a UJT? Describe various regions of operation of a UJT.

CHAPTER

10

Feedback Amplifiers

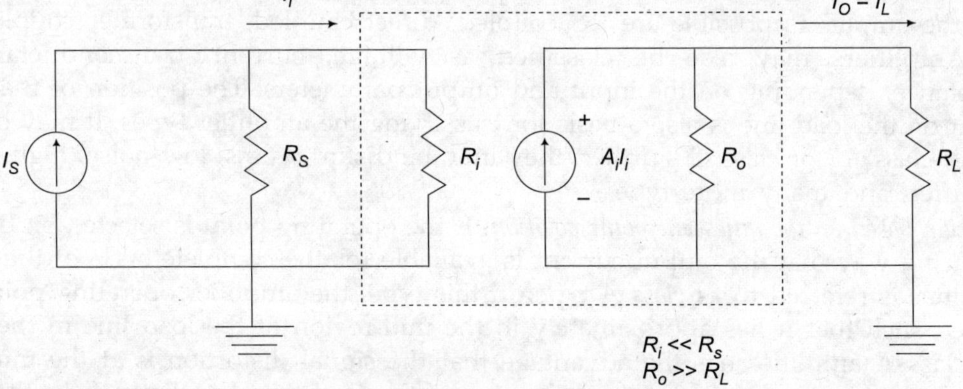

Equivalent circuit of a current amplifier

Chapter Outline

The concepts introduced in this chapter are:

- *Various types of amplifiers and distortion*
- *Need for feedback in an amplifier*
- *Types of feedback amplifiers*
- *Analysis of negative feedback amplifiers*

10.1 TYPES OF AMPLIFIERS

Transistor amplifiers can be of various types. The amplifiers are designed to suit a specific requirement. So, amplifiers can be classified under various heads. Classification of amplifiers can be done in many ways. For example, they may be classified based on frequency of operation, type of coupling between the stages, input and output parameters, selection of operating point, method of operation, types of applications, type of load, power delivered to the load, distortion, noise and so on.

Classification of amplifiers based on frequency of operation can be done as follows: dc (zero frequency), audio (20 Hz–20 kHz), video or wideband (few MHz of bandwidth), radio frequency (RF), ultrahigh frequency (UHF), microwave frequency and so on. Amplifiers may also be classified based on the coupling between the stages of the amplifier. The types of the amplifiers possible are RC coupled, direct coupled, transformer coupled and so on. Amplifiers may also be classified as voltage, current, transconductance or transresistance, depending on the input and output parameters. The position of the quiescent point on the load line is also a basis for classifying the amplifier types. It may be class *A*, class *B*, class *AB* or class *C*. Further, they may be distortionless, low noise, high power, linear, buffer, and many more types.

Classification based on the quiescent point position: If the operating point is selected on the load line in such a way that the output current is available for the complete cycle of the signal, the amplifier is referred to as *class A* type. In this type, the amplifier operating point is to be selected such that it lies approximately in the mid-region of the load line in the active region. Class A amplifier has the advantage that the signal distortion is at the minimum since the amplifier is operating in linear portion of the load line. In other words, the shape of the signal at the output is almost the same as that of the input with amplification. The limitation of the circuit is that its efficiency is very small, 50% maximum; large amount of input power is wasted in the form of heat. In this type, when the input signal is absent, the transistor continues to draw the same rated power and since the input signal power is not available, all the power is dissipated as heat. So, heat sink should be properly designed to dissipate total dc power drawn by the transistor rather than to dissipate only the excess power when the transistor is in operation. In other words, the transistor is cool when in operation rather than when it is idle.

When the operating point is selected towards one end of the active region on the load line of the transistor amplifier circuit, the signal is available only for 50% of the cycle and is grounded in the other half of the cycle. This type of amplifier is referred to as *class B*. The current in the output circuit flows only for 50% of the cycle and passing this pulsed current into a tank circuit develops the voltage signal in the output load. The distortion of the signal is large compared to the class *A* amplifier. The efficiency of the circuit is improved, and is approximately 78% maximum. Selection of the operating point midway between class *A* and class *B* has advantage of efficiency over class *A* and distortion over class *B*. So, selecting the operating point in such a way that the output current is available between 50% and 100% of the cycle is called *class AB*. This type of amplifier has the characteristics midway of class *A* and class *B*.

Efficiency of the amplifier circuit can be improved even up to 99.99% or more when the operating point is so selected that the current in the output of the amplifier circuit is far less than 50% of the cycle. In this *Class C* amplifier, the output current is of the form of spikes almost. When these spikes activate the resonant tank circuit, sinusoidal voltage is developed across it. But this signal at the output has lot of harmonics present, which distort the signal to a very large extent. Nonlinearity of the circuit operation is at its peak and so distorts the signal greatly. In this type of amplifier, heat dissipation is minimal since it has very good conversion efficiency. When the input signal is not present, the circuit never draws any current or, in other words, the power dissipated in the form of heat when the transistor in not in operation is zero.

The selection of the type of amplifier depends on the application. One has to compromise either efficiency or distortion. Various circuits are possible which can decrease distortion in the signal to a large extent in class *B* amplifiers also.

Classification based on the input and output parameter: The possible parameters either at input or output of the amplifier are *voltage* and *current*. Not all the amplifiers are of voltage type. They can be classified based on the parameter of interest at the input and output of the amplifier.

If one is interested in voltage both at the input as well as at the output of the amplifier circuit, the amplifier can be named as *voltage amplifier*. In this type of amplifier, the *transfer gain*, defined as the ratio of output parameter to that of the input parameter, is voltage gain. Hence, *voltage gain* A_V of an amplifier is the ratio of output voltage to input voltage. The equivalent circuit of the amplifier can thus be easily visualised with Thevenin's theorem. The equivalent circuit of such an amplifier is as shown in Fig. 10.1. As seen from the circuit diagram, if the amplifier is to have an amplification of A_V, the source resistance should be zero or the input impedance of the amplifier should be infinite so that the impressed voltage at the input of the amplifier, v_i, will be equal to v_s. Both conditions of source impedance being zero or input impedance being infinite are not practicable, the input resistance should be as high as possible as compared to the source resistance. This assures that the drop across the source resistance is very small and loss in the input signal will be at minimum. Looking at the output circuit of the amplifier, if the output impedance of the amplifier is large compared to the load impedance, lot of the signal is lost in this output impedance and so, one requires very low output impedance compared to the load resistance. Ideally, input impedance should be infinite and the output impedance should be zero in a voltage amplifier.

Fig. 10.2 shows the circuit of a *current amplifier* in which both the input and the output parameter of an amplifier is current. So, the transfer gain of such an amplifier is *current gain* A_I and can be defined as the ratio of output current to the input current. The input and output of the circuit are represented with the help of Norton's theorem. As seen from the circuit diagram, if the input impedance of the amplifier is very high compared to the source resistance, the current delivered to the amplifier would be very small and so rated current gain cannot be achieved. So, one has to design the current amplifier such that the input impedance should be as small as possible compared to the source resistance. With the same argument extended to output circuit, the output impedance should be as large as possible. Ideally, the current amplifier will have zero input impedance and infinite output impedance.

When the parameter of interest at the input of the amplifier is voltage and that of output is current, the transfer ratio is a ratio of output current to the input voltage, which can be designated as *transconductance* or *mutual conductance* G_M. The amplifier is referred to as *transconductance amplifier*. The circuit of such an amplifier, as shown in Fig. 10.3, suggests that both input and output impedance need to be very high compared to that of source and load impedance, respectively. Or, in other words, ideally, the transconductance amplifier has both the input as well as output impedance infinite.

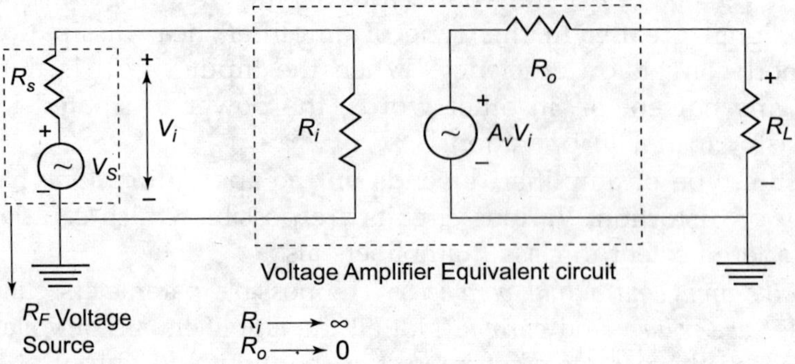

Fig. 10.1 *Equivalent circuit of a voltage amplifier*

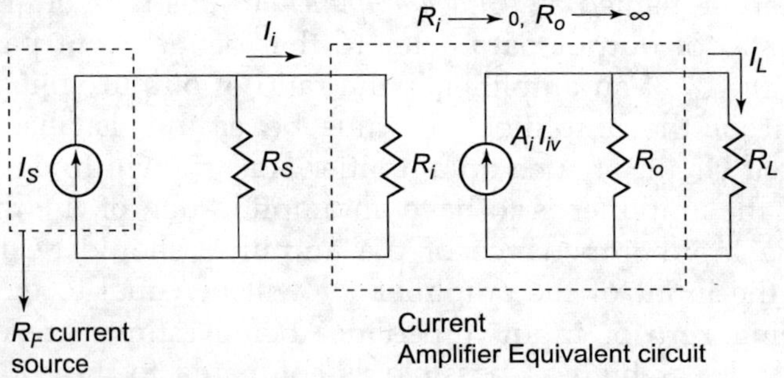

Fig. 10.2 *Equivalent circuit of a current amplifier*

Transresistance amplifier is that amplifier where the output parameter is voltage and that of input is current. The transfer gain is *transresistance* R_M, which can be defined as the ratio of output voltage to input current. Figure 10.4 shows an equivalent circuit of the transresistance amplifier where both the input and output impedances need to be very small compared to source and load impedances, respectively. Transresistance amplifier has both the impedances zero ideally.

Table 10.1 gives the ideal characteristics of various types of amplifiers along with their transfer ratios.

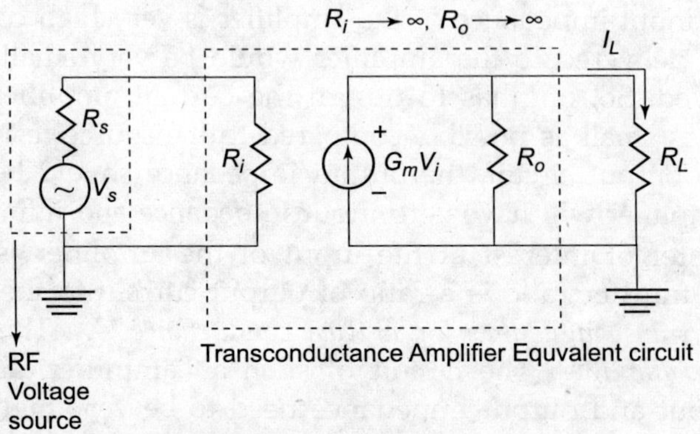

Fig. 10.3 *Equivalent circuit of transconductance amplifier*

10.2 TYPES OF DISTORTION IN AMPLIFIERS

The signal given at the input of an amplifier is available at the output of the amplifier after required amplification. This is not sufficient, the shape of the signal is also to be maintained intact. In addition, the signal should not have any other frequency components except those that are fed in the input of the amplifier. This ensures the perfect nature of the amplifier. But this type of amplifier is not practically possible and the shape of the signal is distorted and also additional components of frequency are present in the output of the circuit. The change in shape of the signal at the output of the amplifier is referred to as *distortion*. Distortion in amplifiers may be due to various sources and can be classified as *amplitude, frequency* and *phase distortion.*

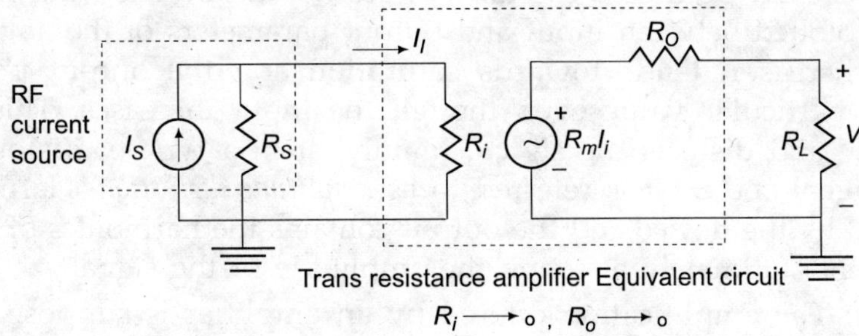

Fig. 10.4 *Equivalent circuit of a trans resistance amplifier*

Table 10.1 ***Ideal Characteristics of Various Types of Amplifiers***

Parameter	*Amplifier type*			
	Voltage	*Current*	*Transconductance*	*Transresistance*
R_i	∞	0	∞	0
R_o	0	∞	∞	0
Transfer characteristic	$V_o = A_V V_s$	$I_L = A_i I_s$	$I_L = G_m V_s$	$V_o = R_m I_s$
Input circuit	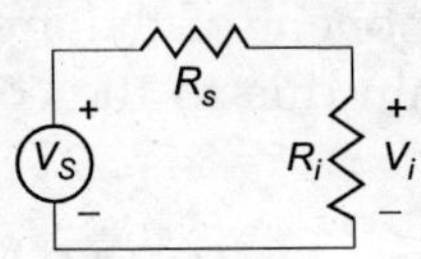 Thevinin	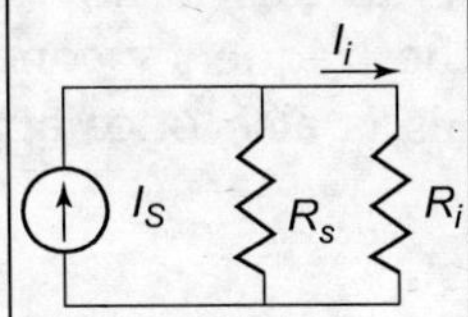 Norton	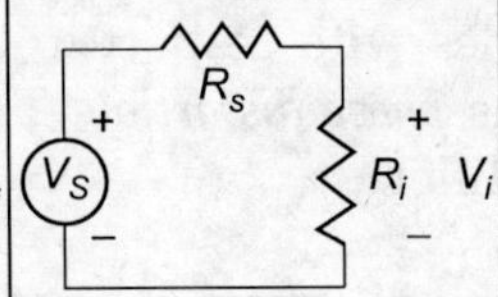 Thevinin	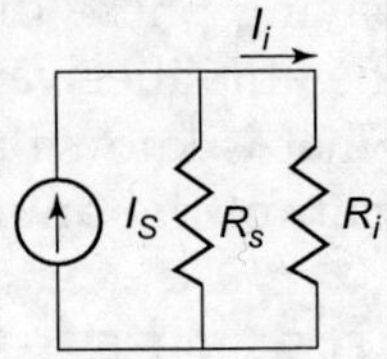 Norton
Output circuit	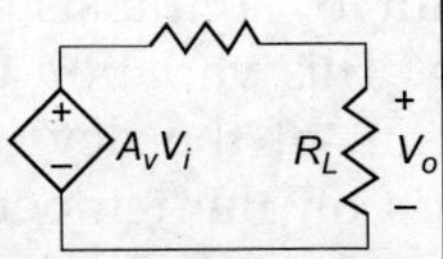 Norton	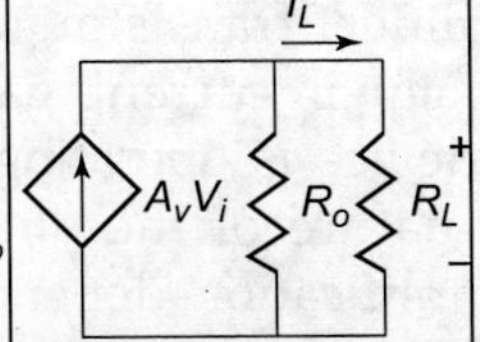 Norton	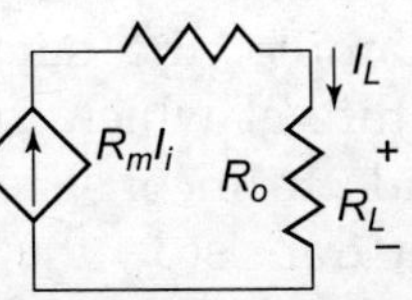 Thevinin	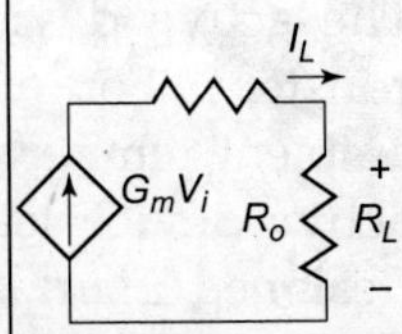 Norton

The additional frequency components available in the output of the amplifier can be termed as *noise*. The noise is the electrical random variations present all over the frequency band from zero to infinity cycles. Noise is present in all the electronic components, active or passive. The electrons, which are present in the device and not contributing to the signal processing, move randomly creating unwanted electrical variations in a device. The frequency of these unwanted electrons' movement depends randomly on the lifetime of the electron and so, the variations are omnipresent in the total frequency band.

Amplitude distortion: *Amplitude distortion*, also known as *nonlinear distortion*, is due to the nonlinearity of the device. The active device employed in the amplification of the signal is either transistor or a FET. Both of these devices are nonlinear. The transfer characteristic of the amplifier, if plotted between input and output parameters of the amplifier, is linear to some extent, whereas it tends towards saturation and the output power does not increase beyond a particular value even though the input increases. This is where the nonlinearity sets in, and distorts the signal. Mainly, such a type of distortion alters the amplitude of the signal and so it is referred to as *amplitude distortion*. Harmonics are generated in the output by the device and the output contains the harmonics of the basic input frequency components ultimately affecting the amplitude of the signal.

Frequency distortion: The amplification offered by an amplifier is suppose to be uniform throughout the frequency band of interest. If such an amplifier is designed, it does not have any frequency distortion. But, in practice, the amplification offered by an amplifier is not uniform throughout the frequency band of interest. That is, the transfer gain of the amplifier varies with frequency, leading to what is known as *frequency distortion*. Or, in other words, signal components of different frequencies are amplified differently. This is due to the internal capacitances of the device or due to the associated circuit being reactive. So, the gain of the amplifier is not a scalar quantity, but is complex, having both amplitude and phase. The frequency response of such an amplifier with frequency distortion is not a horizontal straight line.

Phase distortion: Additional phase is introduced into the input of the signal in an amplifier depending on the characteristic of the amplifier. If this additional phase introduced by the amplifier is same for all the frequency components of interest, then the amplifier is not having phase distortion. But, in general, this is not the case and the phase introduced by the amplifier varies with respect to the frequency component of the input signal leading to what is known as *phase distortion*. This is also once again due to the complex nature of the gain of the amplifier.

10.3 NEED FOR FEEDBACK

The active device used in an amplifier, transistor or FET, has parameters that are not constant. For example, the small signal current gain of a transistor is dependent on ambient temperature at which the device is operating. This variation of parameters of the active device leads to inconsistent nature of the output of the device. If an amplifier is designed using a transistor, the amplification depends on the temperature and increases with temperature and vice versa. This leads to a very unpleasant characteristic of the amplifier that the output is large when the amplifier is operating in a hot condition, but the output is very low when the amplifier is subject to very low temperatures. So, stability of the amplifier is required, that is, the output of the amplifier should be maintained almost constant irrespective of temperature variations. This requires a feedback system where the output of the amplifier is constantly monitored and the input is adjusted accordingly so that the output is constant. In other words, when temperature increases, increasing the gain of the device, the feedback system decreases the input signal level such that the output signal is brought back to approximately the previous value or vice versa. This type of feedback system involving simple passive circuit design is referred to as *negative feedback*. The negative feedback not only improves the stability of the device, but also improves various parameters of the amplifier. There are quite a large number of advantages of employing negative feedback in the amplifier circuits. The disadvantage of such a system is that the gain of the amplifier decreases.

10.4 NEGATIVE FEEDBACK IN AMPLIFIERS

The feedback employed in an amplifier has very good advantages. Negative feedback in an amplifier involves few more blocks to be added to the basic amplifier. The block diagram of the amplifier employing negative feedback is as shown in Fig. 10.5. The additional circuitry involves sampler at the output of the amplifier, feedback circuit consisting of passive components and a mixer at the input of the amplifier.

The *sampler* is at the output of the amplifier and it samples the signal of interest and presents it to the feedback circuit. The sampled signal may be voltage or current. If voltage is the signal that is sampled, it is referred to as *voltage* or *node sampling*. If current is sampled, it is known as *current or loop sampling*. The sampling is done in such a way that the output of the amplifier is not loaded and the output signal is tracked properly.

The *feedback network* is a circuit made up of passive components. Its input is the sample of the output of the amplifier and its output is fed to the input of the amplifier. The circuit processes the output signal from the amplifier and presents this feedback signal to the input out of phase to that of the source signal. The feedback signal may be either voltage

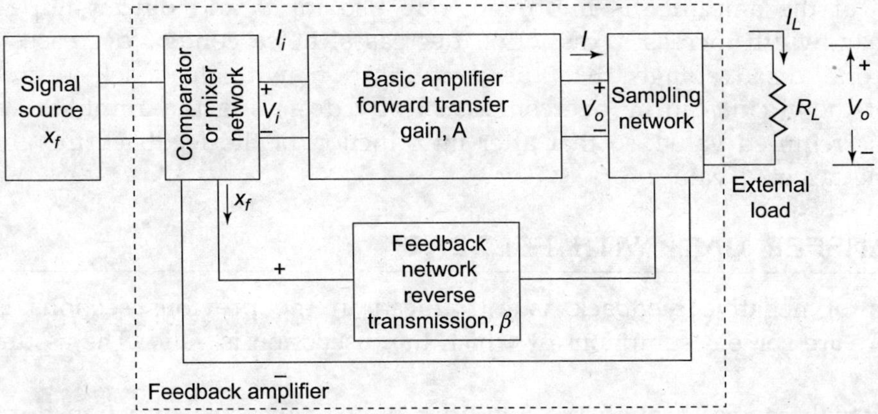

Fig. 10.5 *Negative feedback in an amplifier*

or current, depending on series mixing or shunt mixing employed at the input of the amplifier. Since the feedback signal and the source signal are out of phase by 180°, the difference of the signals is presented as the input signal to the amplifier. So, the amplifier amplifies the difference signal but not the source signal. This is the reason why such a type of feedback is referred to as *negative feedback*.

The *mixer*, also referred to as *comparator*, is a circuit that mixes the source signal and the feedback signal from the output of the feedback network out of phase. The mixing can be of two types. If the two are mixed in series, then it is referred to as *series* or *loop mixing*. If the mixing is in shunt, then it is *shunt* or *node mixing*. The difference of the two signals is presented to the amplifier as the input signal.

The *source signal* is derived from the source of the signal. This may be, once again, voltage or current. The amplifier is characterised by the type of parameter at the input and output of the amplifier of interest. So, the *transfer ratio* or *gain* of the amplifier may be *voltage gain, current gain, transconductance* or *transresistance*, depending on the signals of interest.

The first and foremost advantage of the negative feedback is *stability*. Whatever may the reason, if the output of the amplifier increases, the same increase is applied to the feedback circuit by sampling circuit. The feedback circuit processes the increase in signal and applies the proportionate ratio of the increase to the mixer. The mixer mixes this signal with source signal out of phase, or, in other words, the difference of the source and increased feedback signal is fed to input of the amplifier. Since the feedback signal is large compared to previous case, the input signal is decreased to the extent required. Thus, the output of the amplifier is decreased, compensating for the initial increase in the output signal of the amplifier. Thus, the output is stabilised, that is, maintained almost constant irrespective of variations of any type affecting the basic amplifier parameters. Even though the compensation may not be cent per cent, the output is constantly sampled and the output of the amplifier is maintained almost constant. The same argument can be extended when the output decreases. Decreasing output decreases the feedback signal to the mixer, which in turn increases the input to the amplifier. Thus, increasing the input to the amplifier compensates the decrease in signal at the output of the amplifier. Not only the stability of the amplifier is improved, but also input and output impedances are improved, bandwidth increases, distortion decreases, noise content decreases associated with only one disadvantage that the amplification with feedback decreases. This drawback of the circuit can be overcome easily by designing the amplifier with higher gain than the required value, so that after introduction of the feedback the amplification would be the targeted value.

10.5 TRANSFER GAIN WITH FEEDBACK

The concept of negative feedback as introduced in the previous sections is not true always. There are some assumptions at which the discussion is valid. The assumptions are as follows:

1. The signal travels through the amplifier only in forward direction but not in the feedback network. That is, the signal applied from source to input of the amplifier should only pass through the amplifier but should not enter the feedback network. Or, in other words, the feedback network should block the forward signal.

2. The feedback signal from the sampler should be only coupled to feedback circuit and the amplifier should not allow the signal to transmit through it in the reverse direction. That is, the amplifier should not be bilateral. It should allow forward signals but not the feedback signals.
3. The feedback circuit should not be loaded by either the source or load of the amplifier circuit. In other words, the transfer ratio of the feedback circuit should be independent of source and load impedances.

In a negative feedback amplifier configuration, since sampling can be either voltage or current, and mixing can be either in shunt or series, different types of feedback amplifiers can be evolved depending on the sampling and mixing. So, the feedback amplifiers can be broadly classified as voltage series, current series, current shunt and current series feedback amplifiers.

If voltage is sampled and the mixing is in series in an amplifier, then the type of feedback is referred to as *voltage series*. Since voltage is sampled, the output parameter of the amplifier monitored is voltage and since the mixing at the input is series type, the parameter that is affected at the input of the amplifier is also voltage. So, voltage is the parameter at the input that is affected and voltage is the parameter at the output that is sampled. So, the parameter that is stabilised in this configuration of feedback is *voltage gain* A_V. So, when voltage series feedback is employed in an amplifier, the parameter stabilised is voltage gain. The transfer ratio β, known as *reverse transmission factor*, of the feedback is defined as the ratio of output signal to the input signal of the feedback network, that is, ratio of feedback signal to the output signal of the amplifier, X_f/X_o. Then,

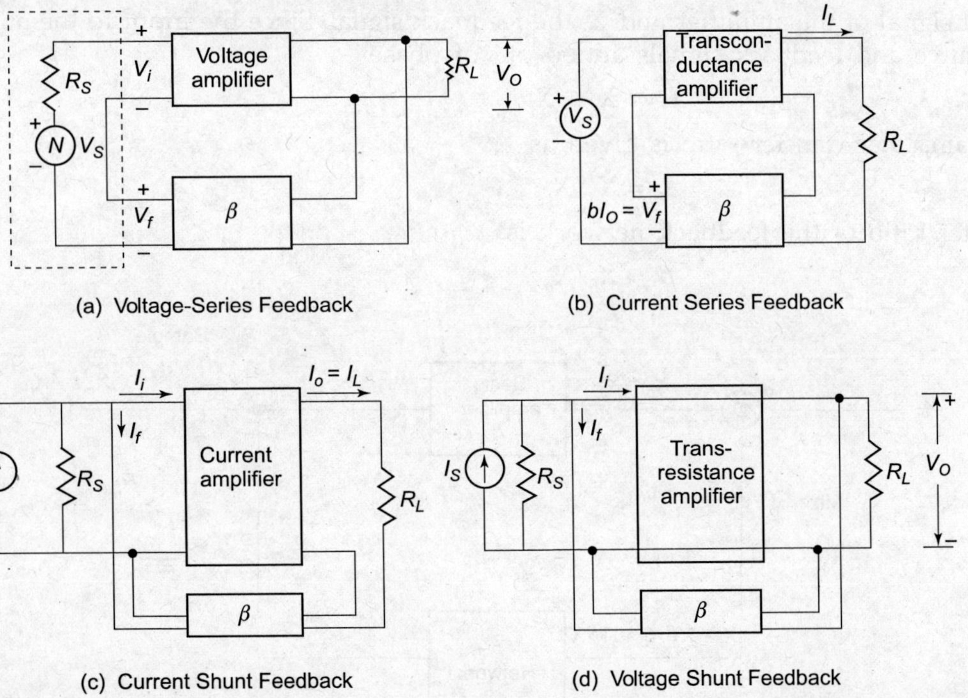

Fig. 10.6 *Types of feedback amplifiers*

the transfer ratio of the voltage series feedback amplifier will be the ratio of feedback voltage V_f and the output voltage of the amplifier V_o, or $\beta = V_f/V_o$. The schematic of such an amplifier is as shown in Fig. 10.6(a).

In a *current series feedback amplifier*, the parameter that is sampled at the output of the amplifier is current and the mixing at the input is in series. Since current is sampled, and the mixing is in series, the parameter at the output is current and that at the input is voltage. So, the parameter of the amplifier that is stabilised is *transconductance* G_M. The transfer ratio of the feedback network in this configuration has the ratio of voltage to current, that is, $\beta = V_f/I_o$. The block diagram of such a configuration is as shown in Fig. 10.6(b).

Figure 10.6(c) shows one more configuration of feedback amplifier, *current shunt*. In this configuration, the parameter that is sampled is current and mixing at the input of the amplifier is shunt. Since mixing is shunt, the parameter that is affected is current and so, the parameter that is stabilised is *current gain* A_I. The transfer ratio of feedback circuit can be given as the ratio of feedback current to output current, $\beta = I_f/I_o$.

The parameter of the amplifier which is stabilised in a *voltage shunt feedback amplifier* is *transresistance* R_M, since the sampled signal is voltage and mixing is shunt. As the schematic for such a configuration shown in Fig. 10.6(d) suggests, the transfer ratio β of the feedback is ratio of feedback current to output voltage, that is, $\beta = I_f/V_o$.

A general schematic of the negative feedback amplifier is as shown in Fig. 10.7. In this, X represents the signal; it can be either voltage or current depending on the configuration of the circuit. So, X_s represents the source signal, X_i the input signal to the amplifier, X_o the output signal of the amplifier and X_f the feedback signal. Since the input to the amplifier is the source and feedback signals added out of phase,

$$X_i = X_s - X_f \quad ...(10.1)$$

Basic amplifier transfer ratio is given as

$$A = X_o/X_i \quad ...(10.2)$$

Transfer ratio of the feedback network is

$$\beta = X_f/X_o \quad ...(10.3)$$

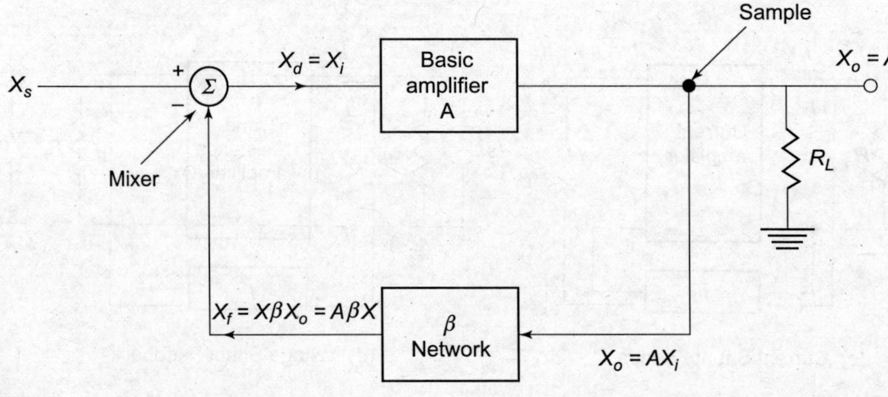

Fig. 10.7 *Feedback amplifier*

Therefore, *transfer ratio of the amplifier with feedback* can be expressed as

$$A_f = X_o/X_s \quad ...(10.4)$$

or, from Eq. (10.1),

$$A_f = X_o/(X_i + X_f) \quad ...(10.5)$$

Dividing the numerator and denominator by X_i

$$A_f = \frac{X_o/X_i}{(1+X_f/X_i)} \quad ...(10.6)$$

or

$$A_f = \frac{X_o/X_i}{\left(1+\frac{X_f X_o}{X_o X_i}\right)} \quad ...(10.6a)$$

$$A_f = \frac{A}{1+A\beta} \quad ...(10.7)$$

A in the above equation represents the transfer gain of the amplifier without feedback, but including the loading of β, R_L and R_s.

If $|A_f| < |A|$, the feedback is termed as *negative or degenerative* and if $|A_f| > |A|$, the feedback is referred to as *positive* or *regenerative.* For a negative feedback, it is observed that A_f is small compared to the basic amplifier transfer ratio or $|1 + A\beta|$ is greater than unity.

Starting at the input of the amplifier, the signal X_i is first multiplied by A and then by β and next by -1 to land at the same point in the closed loop in the feedback amplifier. Therefore the product $-A\beta$ is called *loop gain* or *return ratio*. The difference between 1 and loop gain is referred to as *return difference*, $D = 1 + A\beta$. The amount of feedback introduced into the amplifier is expressed in decibels as

$$N = \text{dB of feedback} = 20 \log |(A_f/A)|$$

$$= 20 \log |1/[1 + A\beta]| \quad ...(10.8)$$

N will be negative for degenerative feedback and positive for regenerative feedback circuits.

10.6 ADVANTAGES AND DISADVANTAGES OF NEGATIVE FEEDBACK

Use of negative feedback in an amplifier has lot of advantages and the only disadvantage is that the transfer gain of the amplifier is reduced. The following are the major advantages of employing negative feedback in an amplifier.

Desensitivity of transfer gain: As already discussed in the previous sections, the stability of the amplifier is improved to a very large extent by employing negative feedback in an amplifier. Or, in other words, the amplifier is made insensitive to any external variations. The output signal is stabilised and made constant irrespective of variations of any nature. To be more specific, the amplifier is desensitized and now the *desensitivity* of the amplifier is a measure of ability of the amplifier to maintain its output constant due to variations because of ageing, temperature, replacement and so on of the circuit components. The *sensitivity* of the amplifier can be expressed as the ratio of fractional change in amplification with feedback to the fractional change without feedback. The same can be expressed as

$$\left|\frac{dA_f}{A_f}\right| = \left|\frac{1}{1+A\beta}\right|\left|\frac{dA}{A}\right| \qquad ...(10.9)$$

or, *sensitivity* is $1/[1 + A\beta]$ and reciprocal of sensitivity is *desensitivity*. Hence,

$$D = 1 + A\beta \qquad ...(10.10)$$

Therefore, transfer with feedback can be given as

$$A_f = A/D \qquad ...(10.11)$$

and if $|A\beta| \gg 1$,

$$A_f \approx 1/\beta \qquad ...(10.12)$$

which shows that the gain of the amplifier entirely depends on the feedback network and if the network contains only passive components, the stability of the amplifier is very pronounced.

Since A represents transfer gain without feedback and A_f represents transfer gain with feedback, the parameter that is stabilised depends on the type of feedback employed in the circuit. If voltage series feedback is employed in the circuit, the parameter stabilised is voltage gain and so, Eq. (10.11) can be written as

$$A_{Vf} = A_V/[1 + A_V\beta] \qquad ...(10.13)$$

For an amplifier, all the four parameters can be measured and employing one particular feedback in the circuit stabilises the corresponding parameter of the amplifier. It is not that other parameters of the amplifier are not effected by the feedback, but they are altered and extent of change in those parameters should be only determined with the analysis of the amplifier circuit.

Similarly, employing current series feedback, current shunt feedback and voltage shunt feedback in the amplifier circuits, the parameters of the amplifier that are stabilised are transresistance, current gain and transconductance, respectively.

So, if one desires an amplifier with gain of A and desensitivity D, then he will have to design an amplifier with gain AD and desensitivity D to achieve the required gain as well as stability.

Frequency distortion: Considering Eq. (10.12), it can be easily seen that the amplifier response is independent of frequency. Under these circumstances a substantial reduction in frequency and phase distortion is obtained. It most practical cases it can be generalised that the distortion is reduced by a factor of desensitivity D with feedback.

If the feedback circuit is reactive, being frequency selective, β depends on frequency and so amplification may depend upon frequency. It is possible to obtain an amplifier with high Q band pass characteristic by using a feedback network that gives little feedback at the centre frequency and great deal of feedback at the edges of the frequency.

Amplitude or nonlinear distortion: The distortion due to nonlinearity of the amplifier affects amplitude at the output of the amplifier and generates harmonics. Harmonic distortion is defined as the ratio of amplitude of that particular harmonic amplitude to that of the fundamental. This harmonic distortion is decreased by a factor of desensitivity D with negative feedback in an amplifier. If B_2 is second-harmonic distortion without feedback and B_{2f} is the second-harmonic distortion with feedback, then

$$B_{2f} = B_2/D \qquad \text{...(10.14)}$$

where $D = 1 + A\beta$, A and β to be evaluated at the second-harmonic frequency since both are functions of frequency.

Noise: The amount of noise introduced by the amplifier is divided by desensitivity D when negative feedback is employed in the circuit. If D is a large value, a considerable reduction in noise is anticipated in the amplifier. However, in order to compensate the loss in the amplification, additional amplification of factor D is to be incorporated into the amplifier. This additional gain introduced is also applicable to the noise power amplification and also the feedback circuit adds some more noise, making the amplifier with feedback nosier than without feedback. The hum introduced into the circuit due to poor design of the dc power supply of the amplifier is reduced to a large extent.

Bandwidth: The gain bandwidth product of an amplifier is constant for a given amplifier. So, one can directly expect that the bandwidth of an amplifier with feedback will increase the bandwidth to a large extent since gain suffers with the feedback. This expectation is correct to a large extent and the influence of the feedback on the bandwidth is that the lower cutoff frequency is further lowered and upper cutoff frequency is further increased. The lower cutoff frequency with feedback is divided by desensitivity D whereas the upper cutoff frequency is multiplied by D, thus enhancing the operative bandwidth of the amplifier to a great extent. Thus,

$$f_{Lf} = f_L/D \qquad \text{...(10.15)}$$

and

$$f_{Hf} = f_H D \qquad \text{...(10.16)}$$

Thus,

$$BW_f = f_{Hf} - f_{Lf} \qquad \text{...(10.17)}$$

Consider the gain of an amplifier without feedback A to be given as

$$A = \frac{A_0}{1 + j(f/f_H)} \qquad \text{...(10.18)}$$

where A_0 is the midband gain without feedback and f_U is the upper 3 dB frequency. Now, the amplification with feedback

$$A_f = \frac{A_0/[1 + j(f/f_H)}{1 + \beta A_0/[1 + j(f/f_H)}$$

$$= \frac{A_0}{1 + \beta A_o + j(f/f_H)} \qquad \text{...(10.19)}$$

Therefore, the amplification with feedback can be expressed as

$$A_f = \frac{A_{0f}}{1 + j(f/f_{Hf})} \qquad ...(10.20)$$

where,

$$A_{of} = A_0/(1 + A_0\beta) \qquad ...(10.21)$$

and,

$$f_{Hf} = f_H(1 + A_0\beta) \qquad ...(10.22)$$

In the midband region the amplification with feedback, A_{0f} equals the midband amplification without feedback, A_0 divided by $1 + \beta A_0$. Therefore,

$$A_{0f}f_{Hf} = A_0 f_H \qquad ...(10.23)$$

Similarly, the effect of feedback on the lower cut off frequency can be estimated and determined as follows:

$$f_{Lf} = f_L/(1 + A_0\beta) \qquad ...(10.24)$$

The bandwidth of the amplifier is defined as the difference of the upper cut off frequency and the lower cutoff frequency and the effect of feedback on these two frequencies is as discussed above. In general, the value of lower cut off frequency is very small such that the bandwidth of the amplifier can be equated as the upper cut off frequency of the amplifier. So *one can conclude that with feedback the bandwidth is multiplied by a factor of desensitivity approximately.*

Input and output impedances: The effect of feedback on the input and output impedances of the amplifier is an added advantage. The impedances are increased wherever large values are required and decreased wherever small values are anticipated. The factor of increase or decrease is more or less the desensitivity *D*. In a voltage series feedback amplifier stabilised *h*-parameter is voltage gain and a voltage amplifier requires very large input impedance compared to that of source and a very small output impedance compared to that of load. So, employing this type of feedback in the circuit, the input impedance is multiplied by *D* and the output impedance is divided by *D* with feedback. Similar arguments hold good for other types of amplifiers and so, feedback in amplifier improves the impedances. The detailed discussion on this topic is presented in the subsequent section.

Drawback: The one and only drawback or limitation of the negative feedback, as already discussed, is that the amplification is divided by *D*. This can be overcome by designing the amplifier for a gain of *AD* such the loss is compensated. But, in practice, it may not be feasible to do so since *D* may be a very large value and the efficiency of the circuit may be a limitation. Higher powers would be dissipated requiring larger heat sinks and many more such design problems may creep in.

10.7 EFFECT OF FEEDBACK ON INPUT AND OUTPUT IMPEDANCES

Introducing negative feedback into an amplifier circuit affects the input and output impedances, but is a positive direction. The impedance increases by a factor of *D* wherever required and decreases by the same factor wherever small impedance is anticipated. The analysis of the impedances with feedback is presented in the discussion and derivation of these impedances for all types of feedback amplifiers is presented in Numerical Problems 10.18 and 10.19.

If the input mixing is in series, the input impedance increases regardless of the type of sampling and it decreases when the mixing is shunt. Similarly, the output impedance decreases when the sampling is voltage whatever may be the type of mixing of returned signal and output impedance increases for current sampling irrespective of the type of mixing. So, one can estimate the effect of feedback on the impedances for the type of feedback employed in a given circuit. The above discussion can be consolidated into a tabular form, which gives the effect of negative feedback on the amplifier characteristics. Table 10.2 gives these characteristics.

Table 10.2 ***Effect of Negative feedback on amplifier characteristics***

	Type of feedback			
	Voltage-series	*Current-series*	*Current-shunt*	*Voltage-shunt*
Reference	Fig. 10.6(a)	Fig. 10.6(b)	Fig. 10.6(c)	Fig. 10.6(d)
R_{of}	Decreases	Increases	Increases	Decreases
R_{if}	Increases	Increases	Decreases	Decreases
Improves charactersis of	Vlotage amplifier	Transconductance amplifier	Current amplifier	Transresistance amplifier
Desensitises	A_V	G_M	A_I	R_M
Bandwidth	Increases	Increases	Increases	Increases
Nonlinear distortion	Decreases	Decreases	Decreases	Decreases
Noise	Decreases	Decreases	Decreases	Decreases
Overall Transmission gain	Decreases	Decreases	Decreases	Decreases

10.8 ANALYSIS OF FEEDBACK AMPLIFIERS

Analysis of feedback amplifiers involves determination of the four parameters, namely, voltage gain, current gain, input and output impedances of the amplifier with feedback. In order to determine these parameters, the type of amplifier, the type of feedback involved in the circuit, the transfer ratio of the amplifier without feedback and transfer ratio of the feedback network is to be established. With the help of these parameters the analysis of the amplifier can be performed comfortably. The analysis of any type of feedback amplifier can be consolidated as the following steps.

1. Redraw the circuit neatly with a proper layout.
2. Identify the active devices and its terminals.
3. Identify the topology, that is, the type of feedback involved in the circuit using the following steps.
   a. The input connection is identified as *series* in the input loop when there is a circuit component W in series with the signal source V_s and the W is to be connected to the output of the amplifier. In other words, the mixing would be

series when the input voltage is affected by the feedback network element. Then the feedback signal would be voltage and $X_f = V_f$.

b. The input connection is identified as *shunt* in the input node when there is a connection between the input node and the output circuit. That is, the mixing is said to be in shunt when in the input current to the amplifier is affected by the feedback network. The feedback signal now is current and $X_f = I_f$.

c. By setting output voltage $V_0 = 0$ (load resistance $R_L = 0$) if X_f becomes zero, then the sampling involved in the circuit is *voltage sampling*.

d. If X_f becomes zero by setting $I_0 = 0$ ($R_L = \infty$), then the sampling will be *current sampling*.

4. Draw the circuit of the amplifier without feedback by using the following steps:
   a. To draw the input circuit, set $V_0 = 0$ for voltage sampling.
   b. To draw the input circuit, set $I_0 = 0$ for current sampling.
   c. To draw the output circuit, set $V_i = 0$ for shunt comparison.
   d. To draw the output circuit, set $I_i = 0$ for series comparison.
5. Use Thevenin's source if X_f is voltage and Norton's source if X_f is current.
6. Replace the active device by its appropriate equivalent model.
7. By applying KVL and KCL to the equivalent circuit, determine the transfer ratio A of the amplifier without feedback.
8. Indicate X_f and X_0 on the circuit and evaluate β.
9. From the values of A and β, find the desensitivity factor D, transfer ratio with feedback A_f, input and output impedances with feedback, R_{if} and R_{of}.

It may be noted that while analysing the feedback circuits, the three fundamental assumptions are not to be violated. The first two assumptions do not cause a great problem in the analysis but the third one, which states that the transfer ratio of the feedback network β should not be a function of either source or load impedances, does. By following the above steps, if one finds that β is a function of R_L or R_S, then the analysis is not valid and appropriate sampling and mixing are to be assumed at the output and the input of the amplifier, respectively.

The analysis of the feedback amplifier along with the type of signals at various stages of the amplifier and expressions for various parameters of the amplifier can be consolidated into a tabular form as shown in Table 10.3.

10.9 VOLTAGE SERIES FEEDBACK AMPLIFIER

The *voltage series feedback amplifier* is the amplifier where voltage at the output is sampled and the output of the feedback network is mixed in series. In this configuration, the parameter stabilised is *voltage gain* and the reverse transmission factor is the ratio of feedback voltage to output voltage. Thus, in this type of feedback, the input impedance is enhanced by desensitivity factor and the output impedance decreases by the same factor. The analysis of voltage feedback amplifier constitutes determining transfer gain of the amplifier without feedback A, reverse transmission factor β, and analysing for amplification with feedback, input impedance and output impedance as follows:

$$D = 1 + A_V\beta$$

$$A_{Vf} = A_V/D$$

Table 10.3 ***Feedback Amplifier Analysis***

Characteristic \ Topology	Voltage-series	Current-series	Current-shunt	Voltage-shunt
Feedback signal X_f	Voltage	Voltage	Current	Current
Sampled signal X_o	Voltage	Current	Current	Voltage
To find input circuit, set	$V_o = 0$	$I_o = 0$	$I_o = 0$	$V_o = 0$
To find output circuit, set	$I_i = 0$	$I_i = 0$	$V_i = 0$	$V_i = 0$
Signal source	Thevenin	Thevenin	Norton	Norton
$\beta = X_f/X_o$	V_f/V_o	V_f/I_o	I_f/I_o	I_f/V_o
$A = X_o/X_i$	$A_V = V_o/V_i$	$G_M = I_o/V_i$	$A_I = I_o/I_i$	$R_M = V_o/I_f$
$D = 1 + \beta A$	$1 + \beta A_V$	$1 + \beta G_M$	$1 + \beta A_I$	$1 + \beta R_M$
A_f	$A_V D$	G_M/D	A_I/D	R_M/D
R_{if}	$R_i D$	$R_i D$	R_i/D	R_i/D
R_{of}	$\frac{R_O}{1+\beta A_V}$	$R_o(1 + \beta G_m)$	$R_o(1 + \beta A_i)$	$\frac{R_O}{1+\beta R_m}$
$R'_{of} = R_{of} \| R_L$	$\frac{R'_O}{D}$	$R'_O \frac{1+\beta G_m}{D}$	$R'_O \frac{1+\beta A_i}{D}$	$\frac{R'_O}{D}$
Transfer gain	Decreases	Decreases	Decreases	Decreases
Distortion	$B_{2f} = B_2/D$	$B_{2f} = B_2/D$	$B_{2f} = B_2/D$	$B_{2f} = B_2/D$
Voisepower	$P_Nf = P_N/D$	$P_Nf = P_N/D$	$P_Nf = P_N/D$	$P_Nf = P_N/D$
Power cutoff frequency	$f_{Lf} = \frac{f_L}{D}$	$f_{Lf} = \frac{f_L}{D}$	$f_{Lf} = \frac{f_L}{D}$	$f_Lf = f_L/D$
Figher cutoff frequency	$f_uf = fu.D$	$f_uf = fu.D$	$f_uf = fu.D$	$f_uf = f_u/D$
Transfer gain	$A_V = \frac{A_v R_L}{R_D + R_L}$	$G_M = \frac{G_m R_o}{R_o + R_L}$	$A_I = \frac{A_i R_o}{R_o + R_L}$	$R_M = \frac{R_m R_L}{R_o R_L}$

$$
\begin{aligned}
R_{if} &= R_i D \\
R_{of} &= R_o/(1 + \beta A_v) \\
A_V &= A_v R_L/(R_o + R_L) \\
R_{of} &= R_o/D
\end{aligned}
\qquad \text{...(10.25)}
$$

Figure 10.8 shows a circuit of an *emitter follower* circuit employing a transistor. This circuit is an example of voltage series amplifier. The output voltage developed across the emitter resistance R_e is fedback to the input (voltage between the base and emitter of the transistor) in series. The input voltage between the base and emitter of the transistor is the source voltage without feedback. Now, since an emitter resistance is added into the circuit, the base to emitter voltage now is not simply the source voltage, but is the difference between the source and voltage developed across the emitter resistor. Thus, one can visualise that the voltage is affected in the input and thus series mixing is employed. Since the output voltage is same as the voltage across the emitter resistor, which is applied to input directly in opposition, the sampling is nothing but voltage. Thus, this circuit is an example of *voltage series feedback amplifier*.

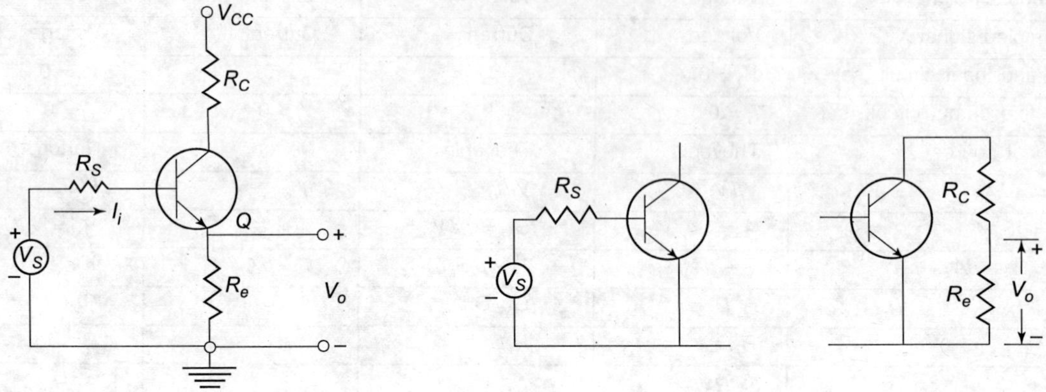

Fig. 10.8 *An emitter follower*

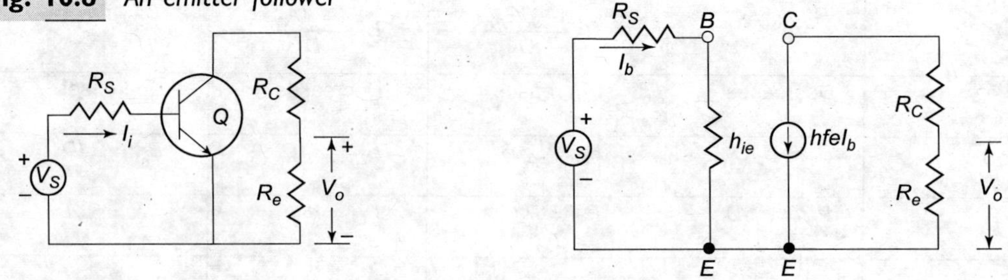

The analysis of this circuit can be done by following the steps outlined in Sec. 10.8. The amplifier without feedback is drawn and a simplified h-parameter model can be assumed wherever possible. In this case, the output and the fedback voltage is the same and thus

$$\beta = V_f/V_o = 1$$

and,

$$A_V = V_o/V_i = h_{fe}R_e/(R_s + h_{ie})$$

Thus,

$$D = 1 + A_V\beta = (R_s + h_{ie} + h_{fe}R_e)/(R_s + h_{ie})$$

Therefore,

$$A_{Vf} = A_V/D = h_{fe}R_e/(R_s + h_{ie} + h_{fe}R_e)$$

$$R_{if} = R_iD = R_s + h_{ie} + h_{fe}R_e$$

$$R_{of} = R_o/(1 + A_v\beta) = \infty$$

$$R_{of}' = R_o'/D = R_e(R_s + h_{ie})/(R_s + h_{ie} + h_{fe}R_e) \quad \text{...(10.26)}$$

Similarly, a source follower circuit using FET is also an example of voltage series feedback amplifier and the analysis can be performed in the same manner.

10.10 CURRENT SERIES FEEDBACK AMPLIFIER

The *current series feedback amplifier* is that amplifier where the current is sampled at the output of the amplifier and the output of β network is mixed with the input in series. In this configuration, the parameter stabilised is *transconductance* and the β of the feedback network is the ratio of the fedback voltage to the output voltage. The input impedance is enhanced by D, as in the voltage series amplifier, but the output impedance is also enhanced as required for a transconductance amplifier. The usual practice is performed for analysis of the circuit and the parameters are as follows:

$$D = 1 + G_M\beta$$
$$G_{Mf} = G_M/D$$
$$R_{if} = R_iD$$
$$R_{of} = R_o(1 + G_m\beta)$$
$$G_M = G_mR_o/(R_o + R_L)$$
$$R_{of}' = R_o'(1 + G_m\beta)/D \qquad ...(10.27)$$

Figure 10.9 shows a common emitter circuit with emitter resistor in the emitter circuit. This is a very good example of current series feedback amplifier. The current in the emitter resistor R_e develops a voltage, which is fed in series to the input source voltage. Thus, the output current (the base current being very small compared to collector current, emitter current can be assumed to be equal to the output collector current without loss in

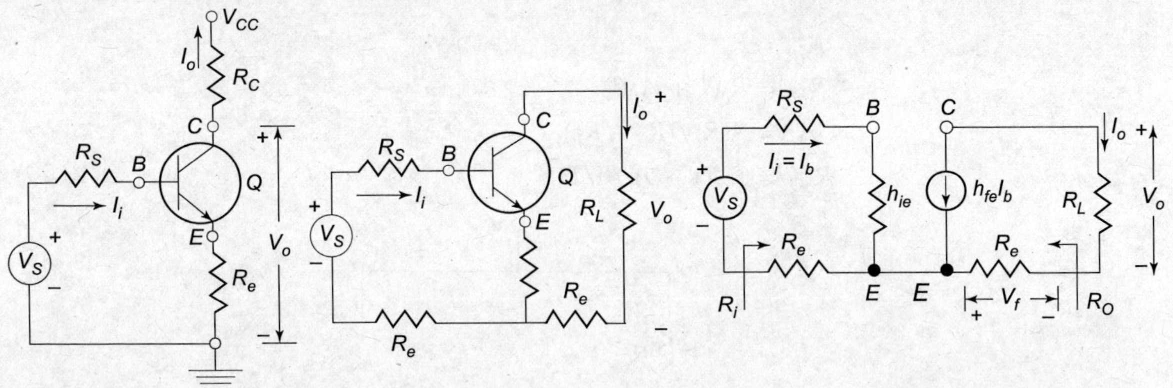

Fig. 10.9 *Common emitter with* R_e

the reality) is sampled and is let through the feedback network, the emitter resistor and the voltage developed across it is fed back out of phase to the input source voltage. Thus, the circuit can be justified as *current series feedback amplifier*.

One more argument can be made assuming the drop across the collector and emitter being small, the reverse transmission factor being equal to feedback voltage to output voltage assuming the configuration to be voltage series. But, if so, β would be the ratio of $V_f/V_o = -R_e/R_L$, which violates the assumptions made with negative feedback amplifiers that the β is a function of load. Thus, this argument is ruled out and the configuration is current series.

The analysis can be carried in the usual manner and the following can be derived:

$$\text{Feedback factor, } \beta = -R_e$$

$$\text{Trans conductance, } G_M = -h_{fe}/(R_s + h_{ie} + R_e)$$

$$\text{Desensitivity factor } D = \{R_s + h_{ie} + (1 + h_{fe})Re\}/(R_s + h_{ie} + R_e)$$

The parameter with feedback can this be determined to be

Trans conductance with feedback, $G_{Mf} = -h_{fe}/\{R_s + h_{ie} + (1 + h_{fe})R_e\}$

Input impedance with feedback $R_{if} = \{R_s + h_{ie} + (1 + h_{fe})R_e\}$

Output impedance with feedback $R_{of} = \infty$

Effective output impedance $R_{of}' = R_o'(1 + G_m\beta)/D$

and $R_o' = R_o || R_L = R_L$ ...(10.28)

10.11 CURRENT SHUNT FEEDBACK AMPLIFIER

When current is sampled and the fedback signal is mixed in shunt with the input, the feedback amplifier is referred to as *current shunt feedback amplifier*. In this configuration, the stabilised parameter is *current gain* and the ratio of fedback current to output current is the transfer ratio of feedback circuit. The input impedance decreases and the output impedance increases and the analysis of the amplifier is as follows:

$$D = 1 + A_I\beta$$

$$A_{If} = A_I/D$$

$$R_{if} = R_i/D$$

$$R_{of} = R_o(1 + A_i\beta)$$

$$A_I = A_iR_o/(R_o + R_L)$$

$$R_{of}' = R_o'(1 + A_i\beta)/D \qquad ...(10.29)$$

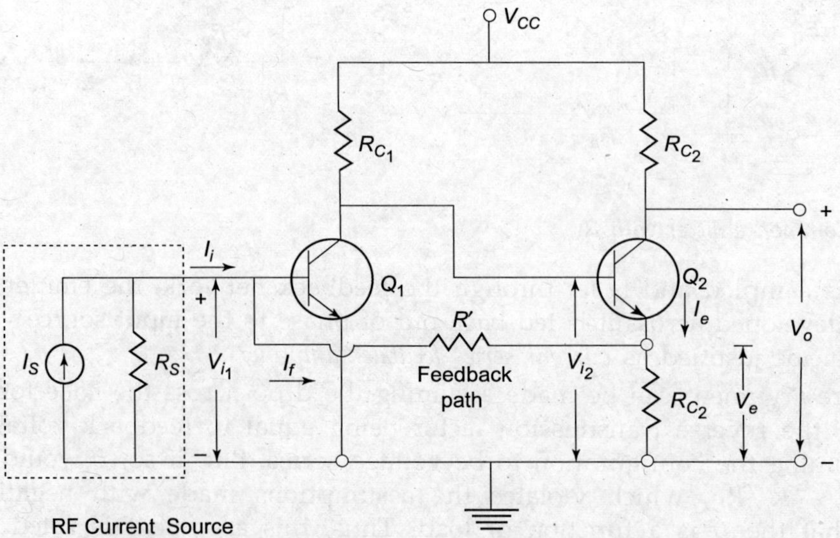

Fig. 10.10 *A current shunt feedback amplifier*

An example of current shunt amplifier is as shown in Fig. 10.10, employing a two-stage transistor amplifier. The output current of the second stage is sampled and passed through the feedback resistor and mixed in shunt with the input current of the first stage. Thus, the circuit is an example of *current shunt feedback amplifier*. The analysis can be performed as usual.

Feedback factor $\beta = R_e/(R' + R_e)$

Current gain with feedback $A_{If} = (R' + R_e)/R_e$

Desensitivity factor $D = 1 + A_I\beta$ ...(10.30)

10.12 VOLTAGE SHUNT FEEDBACK AMPLIFIER

In a *voltage shunt feedback amplifier*, the output voltage is sampled and the feedback network output is mixed in shunt at the input of the amplifier. The parameter stabilised is *transresistance* and the β is the ratio of feedback current to output voltage. Thus, analysis follows and can be consolidated as

$$D = 1 + R_M\beta$$
$$R_{Mf} = R_M/D$$
$$R_{if} = R_i/D$$
$$R_{of} = R_o/(1 + \beta R_m)$$
$$R_M = R_M R_L/(R_o + R_L)$$
$$R_{of}' = R_o'/D \quad \text{...(10.31)}$$

Figure 10.11 shows a circuit that employs voltage shunt feedback. The output voltage is applied across the feedback resistor R' (the output voltage is approximately equal to the

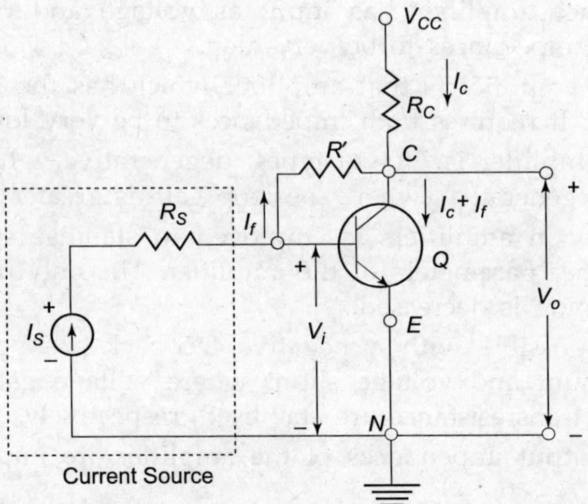

Fig. 10.11 *Voltage-shunt feedback amplifier*

collector to base voltage since the base to emitter voltage is very small and negligible compared to collector-base voltage) and the current resulting in this element is mixed in shunt with the input current. Thus, the feedback is voltage shunt and the analysis is as follows:

Feedback factor $\beta = -1/R'$

Mutual Resistance $R_M = -h_{fe}R_c'R/(R + h_{ie})$

where $R_c' = R_c || R'$

$R = R_s || R'$

$$R_i = Rh_{ie}/(R + h_{ie}) \quad \text{...(10.32)}$$

SUMMARY

- Amplifiers can be classified based on the frequency of operation, coupling between stages, operating point selection or input and output parameters.
- Distortion in amplifier can be classified as amplitude, frequency or phase.
- A voltage amplifier is that amplifier where the input and output parameters of interest are voltage. It requires large input impedance and small output impedance.
- A current amplifier input is current and output is also current. Here, the input impedance should be very small and output very large.
- A transconductance amplifier has input as voltage and output as current. It requires both the impedances to be very high.
- A transresistance amplifier is that amplifier which has the input as current and output as voltage. It requires both impedances to be very low.
- Feedback in an amplifier is of two types: degenerative, where loop gain is less than unity, and regenerative, where the loop gain is greater than unity.
- Negative feedback in amplifiers is employed to stabilise some parameters and improve some other parameters of the amplifier. The only drawback with this is that the amplification is decreased.
- Four topologies possible with a negative feedback are voltage series, current series, current shunt and voltage shunt where voltage gain, transconductance, current gain and transresistance are stabilised, respectively.
- The input and output impedances of the amplifier are improved with negative feedback.
- Distortion is decreased, noise is decreased, stability is improved and bandwidth is increased by employing negative feedback in an amplifier.

SOLVED PROBLEMS

10.1 An amplifier with open loop voltage gain of 1000 ± 20% is available. However, we desire to build an amplifier whose gain does not vary by more than 0.2%. Find the required feedback ratio β and the corresponding closed loop voltage gain.

Solution We have $$\frac{dA_f}{A_f} = \frac{dA/A}{1+A\beta}$$

$$0.2 = \frac{20}{1+A\beta}$$

$$\therefore \quad 1 + A\beta = 100$$

$$\beta = \frac{100-1}{1000} = 0.099$$

Closed loop gain $A_f = \dfrac{A}{1+\beta A} = \dfrac{1000}{100} = 10$

10.2 If an input of 0.035 V peak to peak is given to an open loop amplifier, it gives fundamental frequency output of 45 V_{pp} but it is associated with 9% distortion (a) if the distortion is to be reduced to 1% how much feedback has to be introduced and what will be the required input voltage? (b) If 1.5% of V_0 is feedback and the input maintained at the same level, what will be the output voltage?

Solution (a) Let K_{o2}, K'_{o2} be the harmonic distortion without and with feedback, respectively.

$$K'_{o2} = \frac{K_{o2}}{1+A\beta}$$

$$1 = \frac{9}{1+A\beta} \Rightarrow 1 + A\beta = 7 \qquad \text{...(1)}$$

We have $V_o = 45$ V, $V_i = 0.035$ V.

Open loop gain $A = \dfrac{45}{0.035} = 1285.7$

From (1) $\quad \beta = \dfrac{9-1}{1285.7} = 0.622\%$

Since the gain is reduced by (1 + Aβ) times, that is, by 9 times, new value of V_i should be

$$V_i = 0.035 \times 9 = 0.315 \text{ V}$$

(b)
$$A_f = \frac{A}{1+A\beta} = \frac{1285.7}{1+1285.7\times 0.015}$$
$$= 63.38$$

New $V_o = 0.035 \times 63.38 = 2.218$ V

10.3 An open loop amplifier has a midband gain of 500 and a pass band from 50 Hz to 50 kHz. Find voltage gain and cut off frequencies if 10% of output voltage is fed back.

Solution
$$A_f = \frac{A}{1+\beta A}$$

Here,
$$A = 500$$
$$\beta = 0.1$$
$$A_f = \frac{500}{1+500\times 0.1} = \frac{500}{51} = 9.8$$
$$f_{Lf} = \frac{f_L}{1+A\beta} = \frac{50}{51} = 0.98 \text{ Hz}$$
$$f_{Hf} = f_H(1 + A\beta) = 50 \times 51 = 2.55 \text{ MHz}$$

10.4 An amplifier has an open loop voltage gain of 1000 and delivers 10 W output with 10% second-harmonic distortion when the input is 10 mV. Find the distortion if 60 dB of negative feedback is applied.

Solution We have, $-60 \text{ dB} = -20 \log (1 + A\beta)$

$$\Rightarrow \quad 1 + A\beta = 10^3 = 1000$$
$$A_f = \frac{A}{1+A\beta} = \frac{1000}{1000} = 1$$

New value of second-harmonic distortion
$$K'_{o2} = \frac{K_{o2}}{1+A\beta} = \frac{10}{1000} = 0.01\%$$

10.5 An amplifier consists of three identical stages connected in series. The output voltage is sampled and returned to the input in series opposing. If it is specified that the relative change dA_f/A_f in the closed loop voltage gain A_f must not exceed ψ_f, show that the minimum value of the open loop gain A of the amplifier is given by $A = 3A_f \cdot \frac{|\psi_1|}{|\psi f|}$, where $\psi_1 = \frac{dA_1}{A_1}$ is the relative change in the voltage gain of each stage of the amplifier.

Solution We have $$\left|\frac{dA_f}{A_f}\right| = \frac{1}{1+A\beta}\cdot\left|\frac{dA}{A}\right| = |\psi_f| \qquad ...(1)$$

Let
$$A = \text{overall gain of three stages}$$
$$= A_1^3$$
$$dA = 3A_1^2 dA_1$$
$$\frac{dA}{A} = 3A_1^2\frac{dA_1}{A} = 3A_1^2\frac{dA_1}{A_1^3} = 3\cdot\frac{dA_1}{A_1} \qquad ...(2)$$

Substituting (2) in (1),

$$\left|\frac{dA_f}{A_f}\right| = |\psi_f| = \frac{1}{|1+\beta A|}3\cdot\frac{dA_1}{A_1} \qquad ...(1a)$$

We have $$A_f = \frac{A}{1+\beta A} \Rightarrow \frac{1}{1+\beta A} = \frac{A_f}{A} \qquad ...(3)$$

Substituting (3) in (1a)

$$|\psi_f| = \frac{A_f}{A}\cdot 3\cdot\frac{dA_1}{A_1}$$
$$= \frac{A_f}{A}\cdot 3\cdot|\psi_1|$$

$$\Rightarrow \qquad A|\psi_f| = 3A_f|\psi_1|$$

$$\therefore \qquad A = 3A_f\cdot\frac{|\psi_1|}{|\psi_f|}$$

Hence proved.

10.6 (a) An amplifier with voltage gain of 60 dB uses $\frac{1}{20}$ of its output in negative feedback. Calculate the gain with feedback in dB.

(b) Voltage gain of an amplifier without feedback is 60 dB. It decreases to 40 dB with feedback. Calculate the feedback factor.

(c) A single-stage amplifier has a voltage gain of 600 without feedback, and 50 with feedback. Calculate the percentage of output which is feedback to the input.

Solution (a) $A = 60 \text{ dB} = 20 \log_{10} A \Rightarrow A = 10^3 = 1000$

$$\beta = \frac{1}{20} = 0.05$$

$$A_f = \frac{A}{A + \beta A} = \frac{1000}{1 + 1000 \times 0.05} = 19.60 = 25.84 \text{ dB}$$

(b) Gain of the amplifier without feedback = 60 dB = A dB

$\Rightarrow$ $A = 1000$

Gain with feedback A_f = 40 dB or 100.

We have $A_f = \dfrac{A}{1 + \beta A}$

$$100 = \frac{1000}{1 + \beta \times 1000}$$

$$\frac{1000}{100} - 1 = \beta \times 1000 = \beta A = \text{feedback factor}$$

Feedback factor = $\beta A = 9$

(c) Gain without feedback = 600 = A

Gain with feedback = 50 = A_f

$$A_f = \frac{A}{1 + \beta A}$$

$$50 = \frac{600}{1 + \beta 600}$$

$$\beta = 0.01833$$

% of output voltage feedback to input = $\beta \times 100$
= 1.833%

10.7 A negative feedback of $\beta = 0.002$ is applied to an amplifier of gain 1000. Calculate the change in overall gain of the feedback amplifier if the internal amplifier is subjected to a gain reduction of 15%.

Solution Voltage gain without feedback $A = 1000$

$$\beta = 0.002$$

Gain with feedback $A_f = \dfrac{A}{1 + \beta A} = \dfrac{1000}{1 + 0.02 \times 1000}$

$$= 333.33$$

When open loop gain is reduced by 15%

$$A' = (1 - 0.15) \times 1000 = 850$$

$$\text{New voltage gain with feedback } A'_f = \frac{850}{1 + 0.02 \times 850}$$

$$= 314.814$$

$$\text{Percentage change in overall gain} = \frac{A_f - A'_f}{A_f} \times 100$$

$$= \frac{333.33 - 341.81}{333.33} \times 100$$

$$= 5.55\%$$

10.8 Prove that for voltage series feedback, with $R_s = 0$, Current gains with and without feed back are equal.

Solution For the voltage series feedback amplifier shown in the figure,

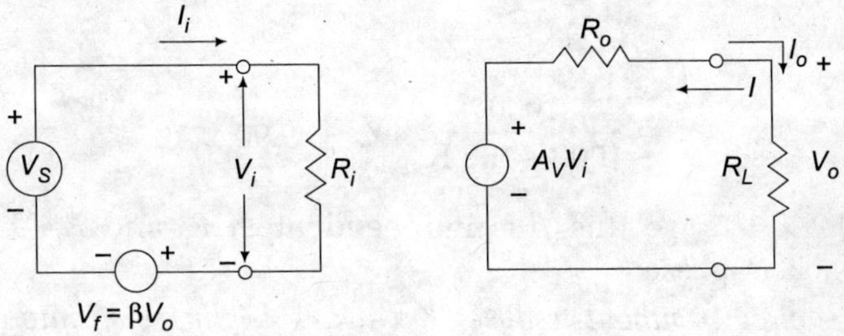

$$AI_f = \text{gain with feedback} = \frac{I_o}{I_i}$$

$$\text{From the second mesh } I_o = \frac{A_v V_i}{R_o + R_L}$$

$$V_i = I_i \cdot Ri$$

$$\therefore \quad I_o = \frac{A_v \cdot I_i R_i}{R_o + R_L}$$

$$\frac{I_o}{I_i} = \frac{A_v \cdot R_i}{R_o + R_L} = A_I$$

Since $\frac{I_o}{I_i}$ ratio is independent of β. So we can say that $A_{If} = A_I$.

10.9 The transistors in the feedback amplifier are identical, and their h-parameters are same. Calculate $R_{i_f} = \dfrac{V_s}{I_i}$, $A_{I_f} = -I/I_i$, $A'V_f = V_o/V_i$, $AV_f = V_o/V_s$ and $R_{o'f}$

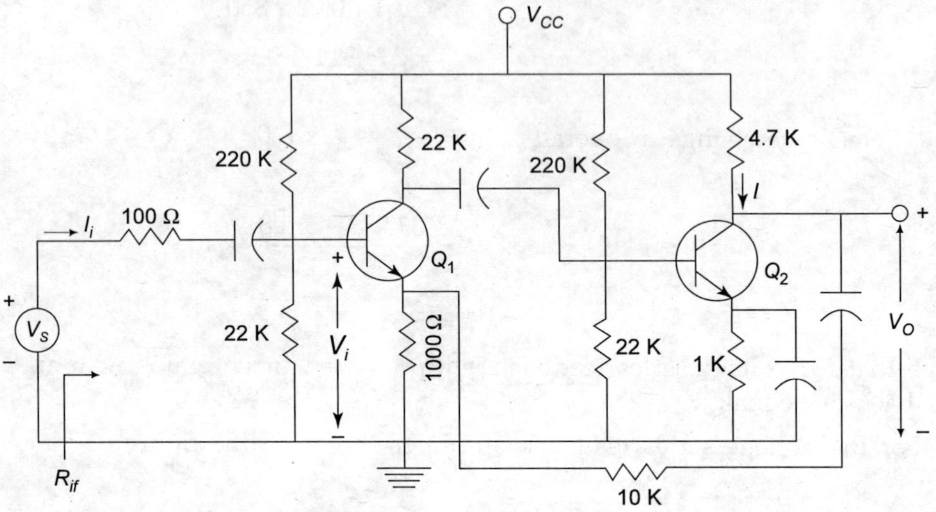

Solution

$$V_f = \frac{1000}{1000 + 100\text{ K}} \times V_o = 9.90 \times 10^{-3} V_o$$

$\beta = 9.9 \times 10^{-3}$ so this β is independent of R_s and R_L. Here, voltage-series feedback is present.

To draw the circuit without feedback but taking loading of β into account

(1) To find input circuit. Set $V_o = 0$. Making output voltage shorted, we get 1000 Ω and 10 K in parallel at the emitter of first transistor.

(2) To find output circuit. Set $I_i = 0$ for series camparison we will have 1000 Ω and 10 K in series, at the collector of second transistor, by making B–E of transistor Q_1 open.

Then, the circuit becomes as shown below:

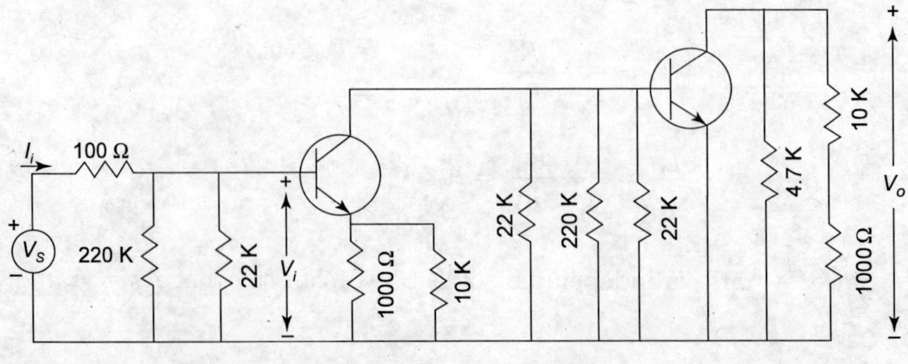

This can be redrawn as below:

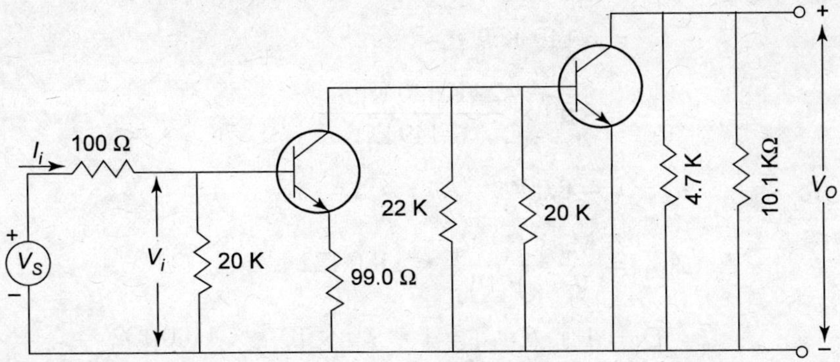

Upon replacing transistor with their approximate h-parameter models,

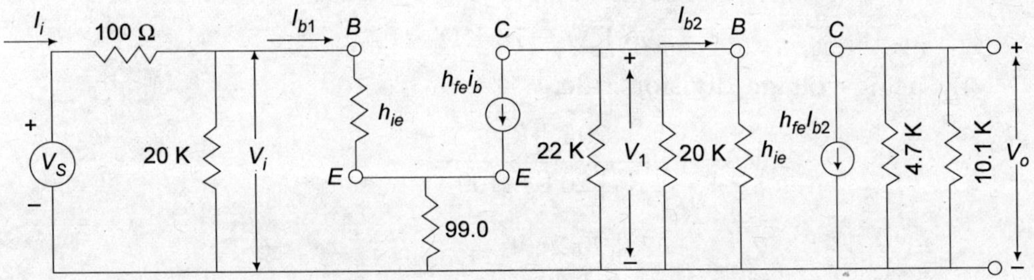

Gain of the amplifier without feedback $= \dfrac{V_o}{V_i}$

Let A_{V1} $\quad \dfrac{V_o}{V_i} = \dfrac{V_o}{V_1} \cdot \dfrac{V_1}{V_i}$

Here, second stage is *CE amp.*

$$\therefore \qquad \frac{V_o}{V_1} = \frac{-h_{fe} \cdot R_L}{h_{ie}} = -h_{fe} \cdot \frac{4.7 || 10.1}{h_{ie}}$$

$$= \frac{-50 \times 3.2 \times 10^3}{1.1 \times 10^3} = -145.79.$$

$\dfrac{V_1}{V_i}$ is obtained by using the formula for *CE* with R_e.

$$\frac{V_1}{V_i} = -h_{fe} \cdot \frac{R_{L1}}{R_{i1}}$$

$$R_{i1} = h_{ie} + (1 + h_{fe})R_e = 1100 + (1 + 50)99$$
$$= 6.149 \text{ K}\Omega$$

$$\therefore \quad \frac{V_1}{V_i} = -50 \times \frac{22\text{ K}||20\text{ K}||h_{ie}}{6.149\text{ k}\Omega}$$
$$= -8.094$$

$$A_{V1} = \frac{V_o}{V_i} = \frac{V_o}{V_1} \cdot \frac{V_1}{V_i} = 1180.132$$

$$D = 1 + \beta A_{V1} = 1 + 9.9 \times 10^{-3} \times 1180.132$$
$$=12.68$$

Since gain A_V is desensitised here, $A_{V1}f = A_{V1}/D$

$$\therefore \quad A'_{Vf} = AV_1/D = 93.04$$
$$R_{if} = R_i \times D = 6.149 \times 10^3 \times 12.68$$
$$= 77.96 \text{ K}\Omega = 78 \text{ K}\Omega$$

By using voltage division rule,

$$V_i = V_s \cdot \frac{20\text{ K}||R_{if}}{R_{if}||20\text{ K} + 100}$$

$$\frac{V_i}{V_s} = \frac{78||20.2}{0.1 + 78||20.2} = 0.993$$

$$A_{Vf} = \frac{V_o}{V_s} = \frac{V_o}{V_i} \times \frac{V_i}{V_s}$$

$$= A'_{Vf} \cdot \frac{V_i}{V_s} = 93.04 \times 0.993$$

$$= 92.43$$

$$R'_o = 4.7\text{ K}||\,10.1\text{ K} = 3.20\text{ k}\Omega$$
$$R'_{of} = R'_o\,|D = 252.96\ \Omega$$

10.10 A voltage shunt feedback amplifier as shown in figure has a transistor with parameter $h_{fe} = 100$, $h_{ie} = 1$ K. Assume approximate simplified model of the transistor. Assume $R_e = 0$ to determine.

(a) $R_{Mf} = \dfrac{V_o}{I_s}$ (b) $A_{Vf} = \dfrac{V_o}{V_s}$

(c) R_{if} (d) R'_{of}

(e) Repeat the four preceding calculations with $R_e = 1$ K.

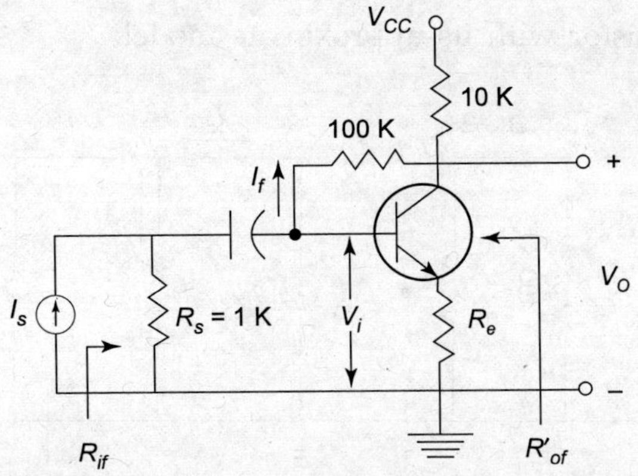

Solution For $R_e = 0$

Since the 100 K is common to both input node and output node, shunt connection exists.

$$I_f = \frac{V_i - V_o}{100\text{ K}} = \frac{-V_o}{100\text{ K}} \quad (\text{since } V_o \gg V_i)$$

$$\therefore \qquad \beta = -1/100\text{ K} = -10^{-5}$$

Since this β is independent of R_s and R_L, sampled signal should be voltage. So, the topology is voltage-shunt topology. We have to place a Norton's source at the input.

To find Input circuit: Set $V_o = 0$. This short circuit at the output places 100 K from base to ground.

To find output circuit: Set $V_i = 0$, so 100 K is placed from collector to ground.

So, the circuit without feedback is as shown below:

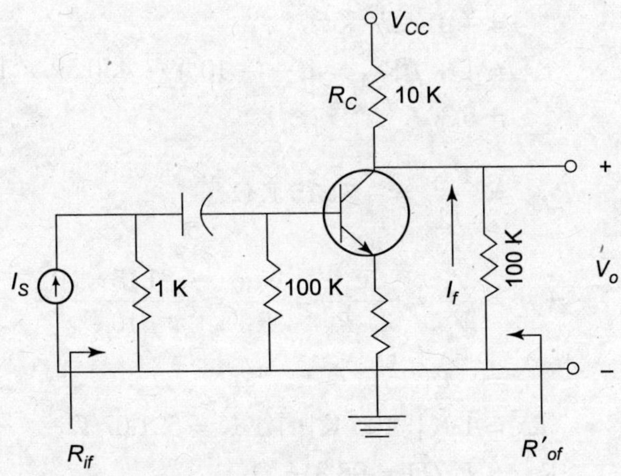

Replacing transistor with its approximate model:

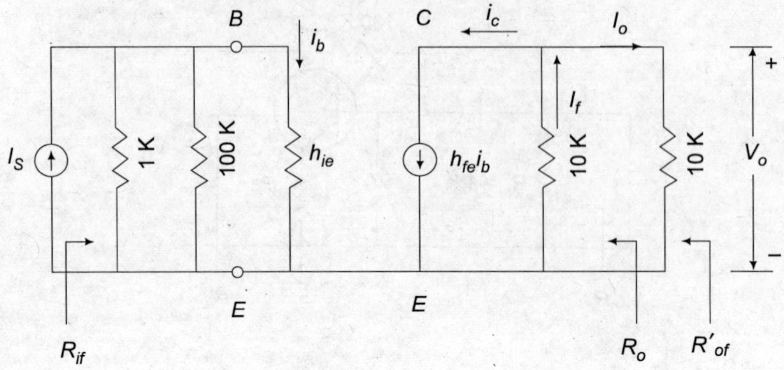

Since there is voltage-shunt feedback, it stabilises R_M.

$$R_M = \frac{V_o}{I_s} = \frac{R_L I_o}{I_s} = R_L \cdot \frac{I_o}{I_b} \cdot \frac{I_b}{I_s}$$

$$\frac{I_o}{I_b} = -h_{fe} \cdot \frac{100}{100+10} \text{ (using current division)}$$

$$= -90.90$$

$$\frac{I_b}{I_s} = \frac{1\,\text{K}||100\,\text{K}}{h_{ie} + 1\,\text{K}||100\,\text{K}} = \frac{990.09}{1.1\,\text{K} + 990.09}$$

$$= 0.4975$$

$$R_M = 10 \times 10^3 \times -90.90 \times 0.4737$$

$$= -452.27\ \text{K}\Omega$$

$$D = 1 + \beta R_M = 1 + (-10^5)(-430.59 \times 10^3)$$

$$= 5.30$$

$$R_{Mf} = \frac{R_M}{D} = -81.15\ \text{K}\Omega$$

$$A_{Vf} = \frac{V_o}{V_s} = \frac{V_o}{I_s \cdot R_s} = \frac{R_{Mf}}{R_s} = \frac{-81.15 \times 10^3}{10^3}$$

$$= -81.15$$

$$R_i = 1\,\text{K}||\,100\,\text{K}||\,1.1\,\text{K} = 521.07\ \Omega$$

$$R_{if} = R_i/D = 98.316\ \Omega$$

$$R_{of} = \frac{R_o}{1+\beta R_m} \text{ where } R_m = \underset{R_L \to \infty}{\text{Lt}} R_M$$

We have $R_o = 100\text{ K}, R_L = 10\text{ K}$

$$V_o = -h_{fe} \cdot I_b\, 100\text{ K}$$

$$R_m = \frac{V_o}{I_s} = -h_{fe} \cdot 100\text{ K} \cdot \frac{I_b}{I_s}$$

$$= -100 \times 100 \times 10^3 \times 0.4737$$

$$= -473.7 \times 10^4$$

$$R_{of} = \frac{100 \times 10^3}{1 + (-10^{-5})(-473.7 \times 10^4)}$$

$$= 2.067\text{ k}\Omega$$

$$R'_{of} = R_{of} \,||\, 10\text{ K} = 1.712\text{ K}\Omega$$

With $R_e = 1\text{ K}\Omega$. Here also voltage shunt connection exists. The diagram after replacing transistor with its equivalent is

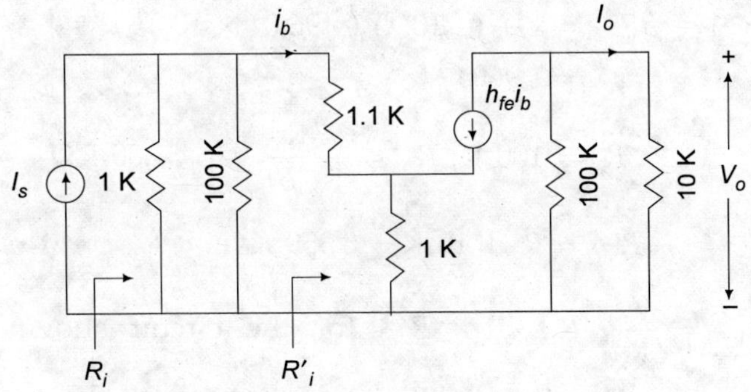

$$\frac{I_o}{I_b} = -h_{fe} \cdot \frac{100\text{ K}}{100\text{ K} + 10\text{ K}} = -90.90$$

$$\frac{I_b}{I_s} = \frac{R_s \,||\, 100\text{ K}}{R'_i + R_s \,||\, 100\text{ K}}$$

where $R'_i = h_{ie} + (1 + h_{fe})R_e = 102.1\text{ K}\Omega$

$$\frac{I_b}{I_s} = \frac{990.09}{102.1 \times 10^3 + 990.09} = 9.604 \times 10^{-3}$$

$$R_M = R_L \cdot \frac{I_o}{I_s} = 10 \times 10^3 \times -90.90 \times 9.604 \times 10^{-3}$$

$$= -8.73 \times 10^3\ \Omega$$

$$\beta = -10^{-5},\ D = 1 + \beta R_M = 1.0873$$

$$R_{Mf} = RM/D = -\ 8.029 \times 10^3\ \Omega$$

$$A_{Vf} = \frac{R_{Mf}}{R_s} = -\ 8.029$$

$$R_i = 1\text{ K} \,||\, 100\text{ K} \,||\, R'_i = 1\text{ K} \,||\, 100\text{ K} \,||\, 102.1\text{ K}\Omega$$

$$= 980.58\ \Omega$$

$$R_{if} = Ri/D = 901.856\ \Omega$$

$$R_{of} = R_o/(1 + \beta R_m) \text{ where } R_m = \underset{R_L \to \infty}{\text{Lt}} R_M$$

$$V_o = -\ h_{fe} \cdot I_b \cdot 100\text{ K}$$

$$\Rightarrow \frac{V_o}{I_s} = R_m = -\ h_{fe} \cdot \frac{I_b}{I_s} \cdot 100\text{ K}$$

$$= -\ 100 \times 9.60 \times 10^{-3} \times 100 \times 10^3$$

$$= -\ 96.04 \times 10^3\ \Omega$$

$$1 + \beta R_m = 1 + (-\ 10^{-5})\ (-\ 96.04 \times 10^3)$$

$$= 1.9604$$

$$R_o = 100\text{ K},\ R_{of} = \frac{R_o}{1 + \beta R_m} = 51.00\text{ k}\Omega$$

$$R'_{of} = R_{of} \,||\, 10\text{ K} = 8.359\text{ K}\Omega.$$

10.11 Calculate $A_{Vf} = \dfrac{V_o}{V_s}$, R_{if}, R_{of}, R'_{of} for the circuit shown below, use typical h-parameter values. $R_s = R_L = 10\text{ K}$, $R_e = 1\text{ K}\Omega$.

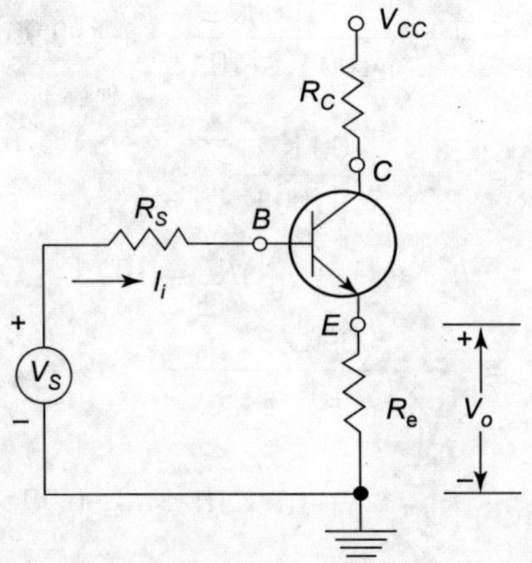

Solution 1. **Identify the topology.** Input loop is defined as loop containing B and E, output loop is the loop containing C and E. Since R_e is common between input and output loops we can say that there exists a series mixing voltage across R_e is the feedback signal V_f.

To find the signal sampled, assume first voltage is sampled. $\beta = \dfrac{X_f}{X_o} = \dfrac{V_f}{V_o} =$

1. Since β is independent of R_s and R_c we can say that there is voltage sampling. So, the topology is voltage-series topology.
2. *Draw the circuit without feedback.*
   (a) To find input circuit. Set $V_o = 0$ for voltage sampling, the circuit will be as shown below by looking into Base and Emitter terminals.

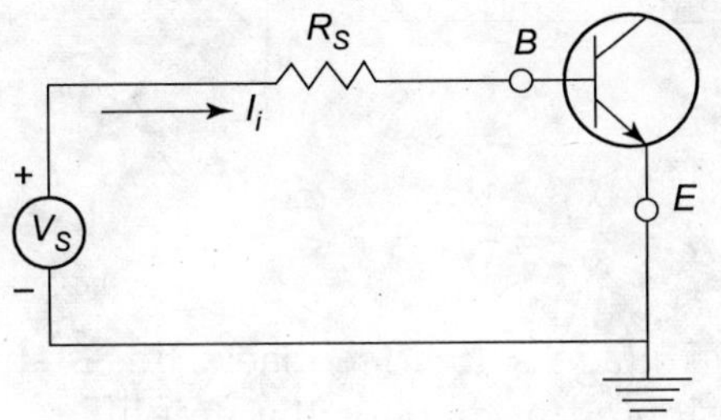

   (b) To find output circuit, set $I_i = 0$ for series mixing. Then, R_e and R_c will be series at the output terminals.

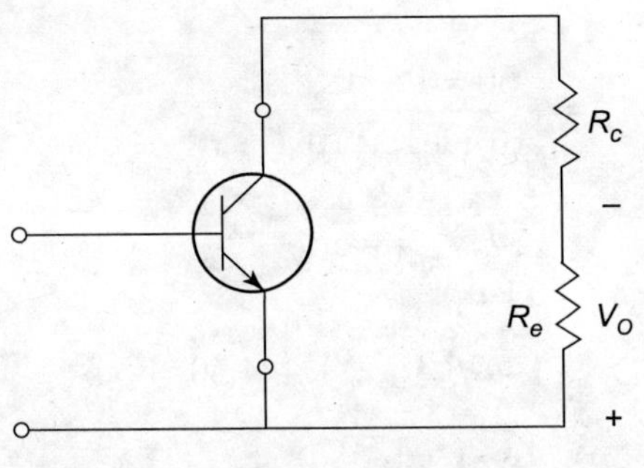

∴ the circuit without feedback will be as below:

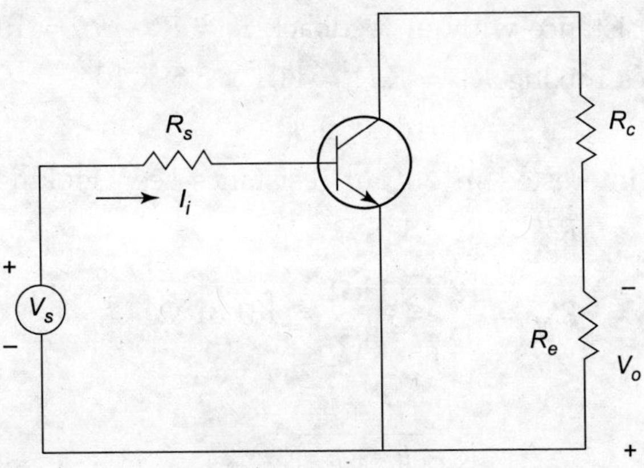

3. Since V_f is feedback signal, Thevenin's source is used.
4. Replacing transistor with its proper model:

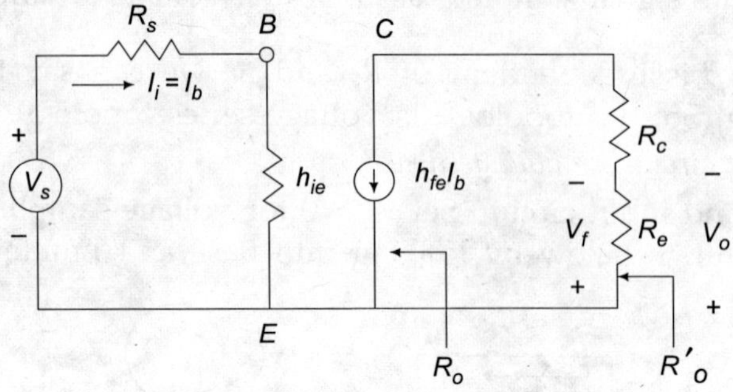

$$\beta = \frac{X_f}{X_o} = \frac{V_f}{V_o} = 1$$

Since there is voltage series feedback, this stabilises voltage gain. Here, R_s has to be considered as part of the amplifier

$$A_V = \frac{V_o}{V_i} = \frac{h_{fe}I_bR_e}{(R_s + h_{ie})I_b} = \frac{h_{fe}R_e}{R_s + h_{ie}}$$

$$= \frac{50 \times 1 \times 10^3}{(10 + 1.1) \times 10^3}$$

$$= \frac{50}{11.1} = 4.50$$

$$D = 1 + \beta A_V = 1 + 4.50 = 5.5$$

$$A_{Vf} = \frac{A_V}{D} = \frac{4.50}{5.5} = 0.819$$

Input resistance without feedback $R_i = R_s + h_{ie} = 10 + 1.1 = 11.1$ KΩ

For series mixing $R_{if} = R_iD = 11.1 \times 5.5 \times 10^3$

$= 61.05$ KΩ

We are interested in output resistance seen looking into the emitter

$$Ro' = R_e = 1\text{ K}\Omega$$

$$R'_{of} = \frac{R'_o}{D} = \frac{1\text{k}\Omega}{5.5} = 181.81\ \Omega$$

$$= \frac{R_e(R_s + h_{ie})}{R_s + h_{ie} + h_{fe}R_e}$$

$$R_{of} = \underset{R_e \to \infty}{\text{Lt}} R'_{of} = \frac{R_s + h_{ie}}{h_{fe}} = \frac{(10 + 1.1)}{50} \times 10^3$$

$$= 222\ \Omega$$

10.12 For the FET circuit, using feedback analysis techniques find A_{Vf}, R_{if}, R_{of}, R'_{of} if $R_s = 10\ \text{K}\Omega$, $g_m = 10$ mA/V, $r_d = 1\ \text{M}\Omega$.

Solution 1. Since R_s is common between G-S and D-S loops, we confirm that there is a series mixing. Assuming sampled signal as voltage $\beta = \dfrac{V_f}{V_o} = 1$. Therefore, there is a voltage series topology in the circuit.

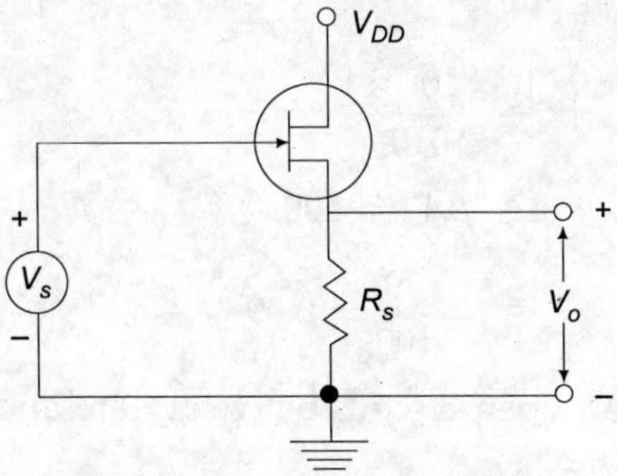

2. To find input circuit, set $V_o = 0$. To find output circuit, open circuit the input terminals. The circuit without feedback but taking loading of β network into account will be

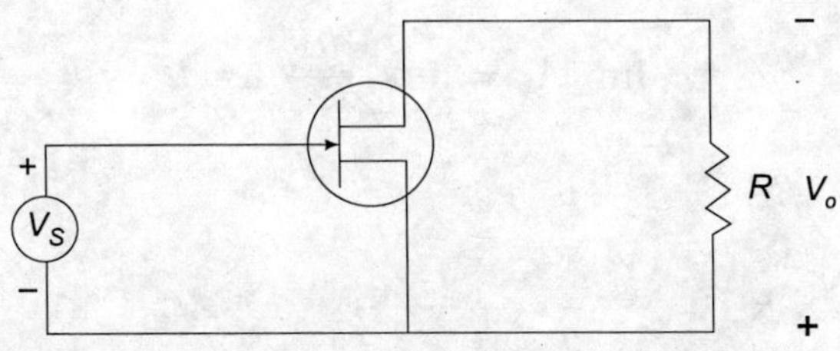

3. Replacing FET with its low-frequency model,

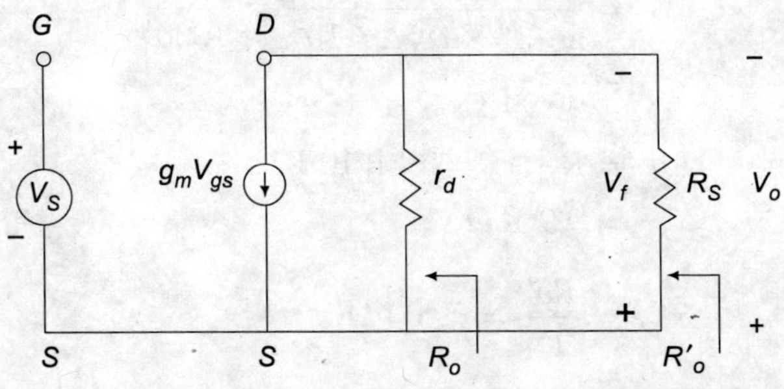

This topology stabilises voltage gain.

$$A_V = \frac{V_o}{V_i} = \frac{\frac{g_m V_{gs} r_d}{R_s + r_d} \cdot R_s}{V_{gs}} = \frac{g_m r_d R_s}{R_s + r_d}$$

$$= \frac{\mu R_s}{R_s + r_d} \quad \text{since } \mu = g_m r_d$$

$$= \frac{10\times10^{-3}\times1\times10^{6}\times10\times\cancel{10^3}}{10\times\cancel{10^3}+10^3\times\cancel{10^3}}$$

$$= \frac{10\times10^3\times10}{1010} = 99.00$$

$$D = 1 + \beta A_V = 100$$

$$\therefore \quad A_{Vf} = \frac{A_V}{D} = \frac{99}{100} = 0.99$$

Since there is an open circuit, between gate and source, $R_i = \infty$

$$\therefore \quad R_{if} = \infty$$

From the circuit $R_o = r_d$, $\beta = 1$

$$R_{of} = \frac{R_o}{1+\beta A_v}$$

$$A_v = \lim_{R_s\to\infty} A_V = \lim_{R_s\to\infty} \frac{\mu R_s}{R_s + r_d} = \mu$$

$$1 + \beta A_v = 1 + \mu$$

$$\therefore \quad R_{of} = \frac{r_d}{1+\mu} = \frac{r_d}{1+g_m r_d}$$

$$= \frac{10^6}{1+10\times10^{-3}\times10^6} = \frac{10^6}{1000^1}$$

$$= 99.99\ \Omega$$

$$Ro' = R_s \,||\, r_d = 10^6 \,||\, 10\text{ K}$$

$$= 9.9\ \text{K}\Omega$$

$$R'_{of} = \frac{Ro'}{D} = 99.00\ \Omega$$

10.13 For the two-stage amplifier circuit shown in figure, Assume $R_s = 0$, $h_{fe} = 50$, $h_{ie} = 1100\ \Omega$, $h_{re} = h_{oe} = 0$ and indentical transistors. Find A_{Vf}, R_{if}, R_{of}.

Solution 1. Input is applied at the base of first transistor and output is taken from the collector of second transistor. Since the feedback network consisting of the potential divider arrangement of R_1 and R_2, voltage across R_1 is fed back to the input. So, there exists a series mixing, and voltage across R_1 is the feedback signal.

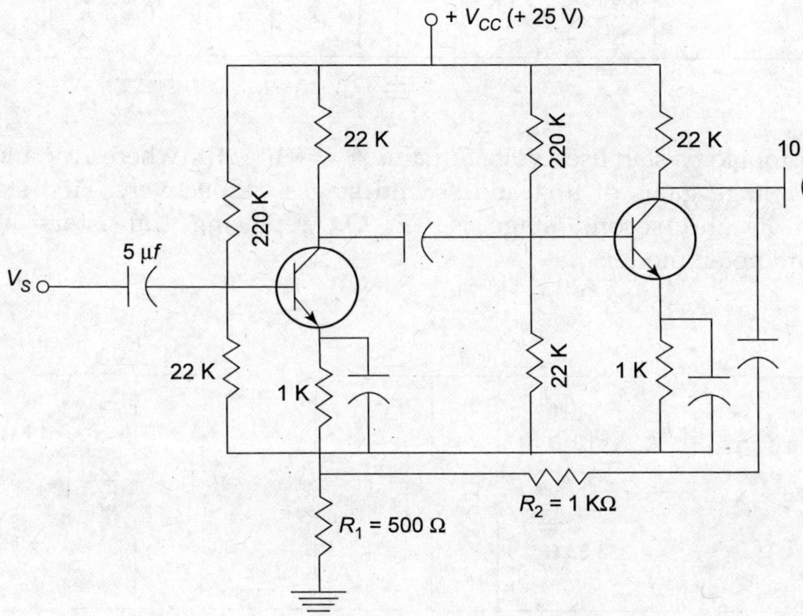

$$\therefore \qquad V_f = \frac{R_1}{R_1 + R_2} \cdot V_o \text{ using voltage division.}$$

$\beta = \dfrac{V_f}{V_o} = \dfrac{R_1}{R_1 + R_2}$ since this β is independent of R_s and R_L, we can confirm there is voltage sampling so, voltage series feedback is present.

2. To find input circuit, set $V_o = 0$. then R_1 and R_2 will be in parallel, at the emitter of transistor Q_1. To find output circuit set $I_i = 0$, then series connection of R_1 and R_2 will be at the collector of transistor Q_2. The circuit without feedback but taking loading of β network into account will be as shown.

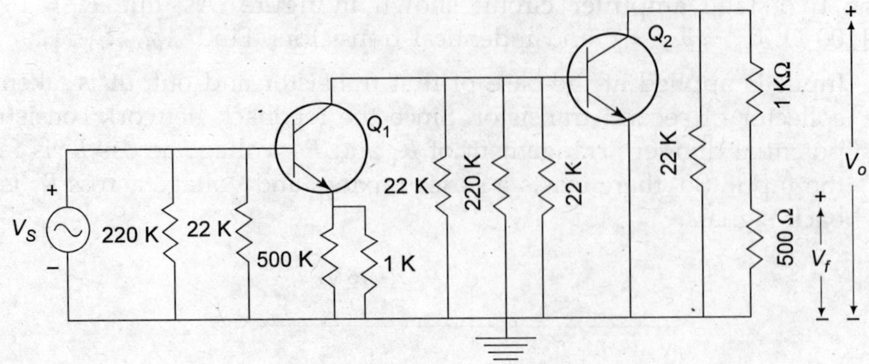

This topology stabilises voltage gain. $A_V = A_{V1} \cdot A_{V2}$ where A_{V1} and A_{V2} are the voltage gains of first and second stages, respectively. First stage is *CE* with R_e and second stage is *CE*. On replacing transistors with their *h*-parameter models.

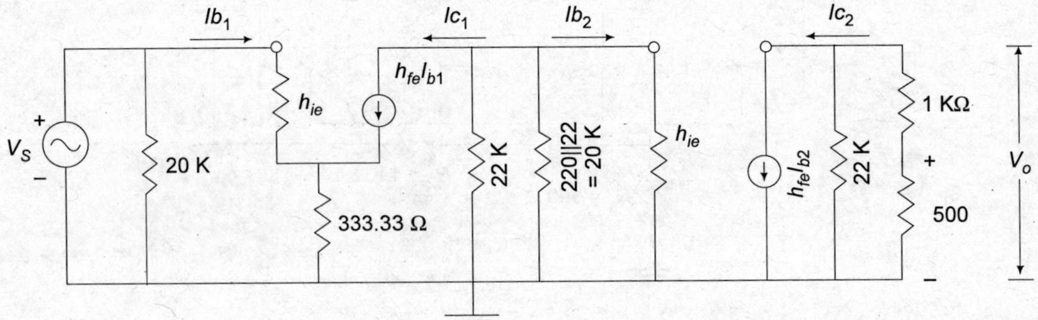

$$\text{Effective load resistance of } Q_1, R_{L1} = 22 \text{ K} \,||\, 220 \text{ K} \,||\, 22 \text{ K} \,||\, h_{ie}$$

$$= 22 \text{ K} \,||\, 20 \,||\, 1.1 \text{ K}$$

$$= 1.002 \text{ K}\Omega$$

$$\text{Effective load resistance of } Q_2 \text{ is } R_{L2} = 22 \text{ K} \,||\, 1500$$

$$= 1.404 \text{ K}\Omega$$

$$A_{V1} = \frac{-h_{fe} \cdot R_{L1}}{h_{ie} + (1 + h_{fe})R_e} = \frac{-50 \times 1.002 \times 10^3}{1.1 \times 10^3 + (51)(333.33)}$$

$$= -2.768$$

$$A_{V2} = \frac{-h_{fe} \cdot R_{L2}}{h_{ie}} = \frac{-50 \times 1.404 \times 10^3}{1.1 \times 10^3}$$

$$= -63.82$$

$$A_V = A_{V1} \cdot A_{V2} = 176.65$$

$$\beta = \frac{R_1}{R_1 + R_2} = \frac{500}{500+1000} = \frac{1}{3}$$

$$D = 1 + \beta A_V = 1 + \frac{1}{3}(176.65) = 59.88$$

$$A_{Vf} = \frac{A_V}{D} = \frac{176.65}{59.88} = 2.95$$

Total input resistance is decided by the input resistance of the first stage.

$$R_i = h_{ie} + (1 + h_{fe})R_e$$

$$= 1.1 \text{ K} + 51 \times 333.33 = 18.09 \text{ K}\Omega$$

$$R_{if} = R_i \cdot D = 1083.22 \text{ K}\Omega$$

Total output resistance is decided by the output resistance of second stage.

$$Ro = R_{L2} = 1.404 \text{ K}\Omega$$

$$R_{of} = \frac{R_o}{D} = \frac{1.404}{59.88} \times 10^3 = 23.44\ \Omega$$

10.14 In a *CE* amplifier with unbypassed R_e, $R_s = 0$, $R_L = 4$ K, $R_e = 1$ K, find A_{Vf}, R_{if}, use typical *h*-parameter values.

Solution 1. R_e is common between input loop and output loop. So, there is series mixing present, voltage across R_e is the feedback signal. Assume sampled signal is V_o,

$$\beta = \frac{V_f}{V_o} = \frac{-I_o R_e}{I_o R_L} = \frac{-R_e}{R_L}$$

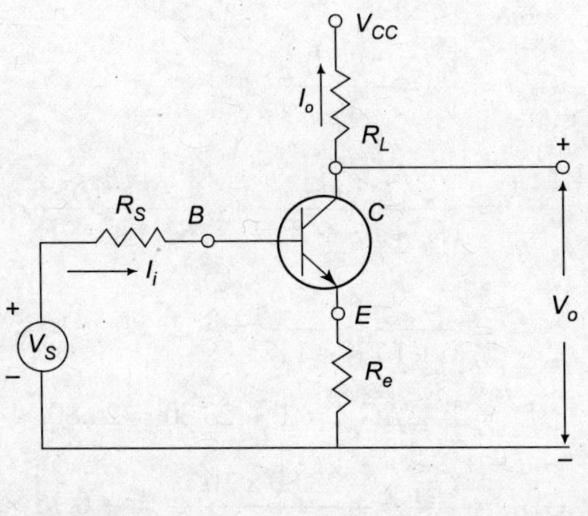

Here, β is dependent upon R_L, so voltage sampling is not possible. So current is the sampled signal. Here, the topology is current-series feedback which stabilises G_M.

2. To find input circuit, set $I_o = 0$. Then R_e is connected between emitter and signal source. To find output circuit, set $I_i = 0$. Then, R_c and R_e will be in series, at the collector voltage across R_e is the feedback signal V_f.

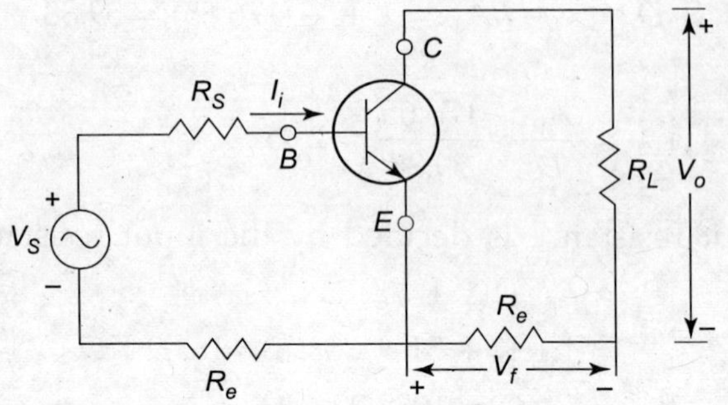

3. Replacing transistor with its approximate h-parameter model,

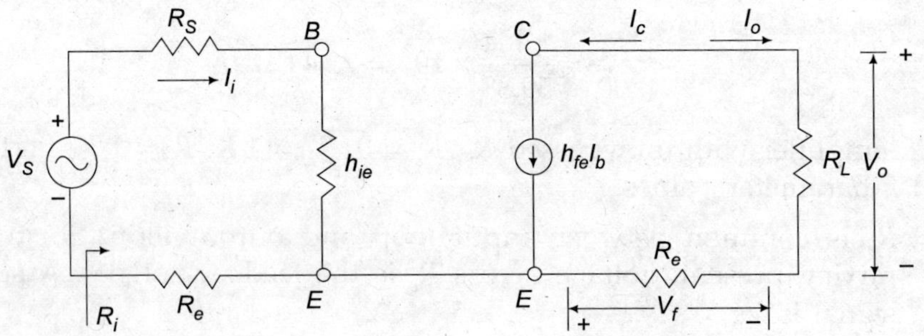

4. $$\beta = \frac{V_f}{I_o} = \frac{-I_o R_e}{I_o} = -R_e = -1000$$

$$G_M = \frac{I_o}{V_i} = \frac{-h_{fe} I_b}{V_s}$$

$$= \frac{-h_{fe} I_b}{(R_s + h_{ie} + R_e) I_b} = \frac{-h_{fe}}{R_s + h_{ie} + R_e}$$

$$= \frac{-50}{0 + 1.1\text{K} + 1\text{K}} = -23.80 \times 10^{-3}$$

$$D = 1 + \beta G_M = 1 + 23.80 = 24.80$$

$$G_{Mf} = \frac{G_M}{D} = \frac{-23.80 \times 10^{-3}}{24.80} = -0.96 \times 10^{-3}$$

$$A_{Vf} = \frac{I_o R_L}{V_s} = G_{Mf} \cdot R_L$$

$$= -0.96 \times 10^{-3} \times 4 \times 10^{3} = -3.84.$$

$$R_i = R_s + h_{ie} + R_e = 0 + 1.1\text{ K} + 1\text{ K} = 2.1\text{ K}\Omega$$

$$R_{if} = R_i \times D = 2.1 \times 24.80 \times 10^{3} = 52.08\text{ K}\Omega$$

10.15 The circuit has following parameters $R_{c1} = 2.5\text{ K}\Omega$, $R_{c2} = 600\ \Omega$, $R_e = 50\ \Omega$, $R_f = R_s = 1\text{ K}\Omega$, $h_{fe} = 50$, $h_{ie} = 1.1\text{ K}\Omega$, $h_{re} = h_{oe} = 0$. Find (a) A_{Vf} (b) R_{if} (c) The output resistance.

Solution 1. Since I_o current is following through emitter resistance R_e, and R_f is connected to the base node, there exists a shunt connection. To find the signal that is sampled, write an expression for feedback current.

$$I_f = \frac{V_{i1} - V_e}{R_f} \approx \frac{-V_e}{R_f} = \frac{(I_o - I_f)R_e}{R_f}$$

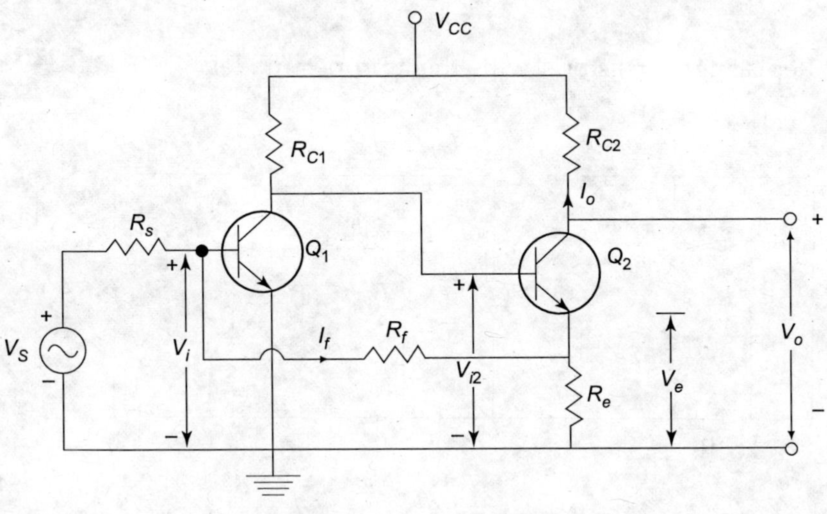

$$I_f\left(1 + \frac{R_e}{R_f}\right) = I_o\frac{R_e}{R_f}$$

$$I_f = \frac{R_e}{R_f + R_e} I_o = \beta I_o$$

$$\therefore\ \beta = \frac{R_e}{R_f + R_e}$$ is independent of R_s and R_L, so sampled signal is current.

There is current-shunt topology present.

2. To find input circuit, set $I_o = 0$ for current sampling. Then, R_f and R_e will be in series and appear at the base of transistor Q_1. To find output circuit, set $V_i = 0$, for shunt comparison. Then, R_f and R_e will be in parallel, at the emitter of Q_2. Since there is shunt mixing, we have to use Norton's source at the input.

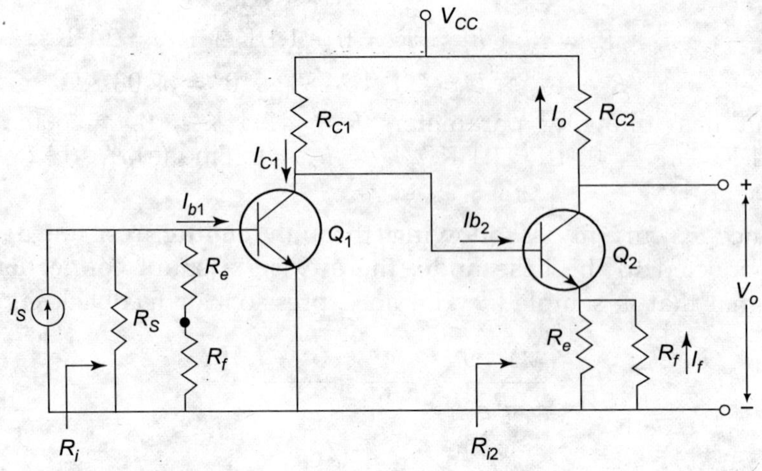

3. Replacing transistor's with their proper model,

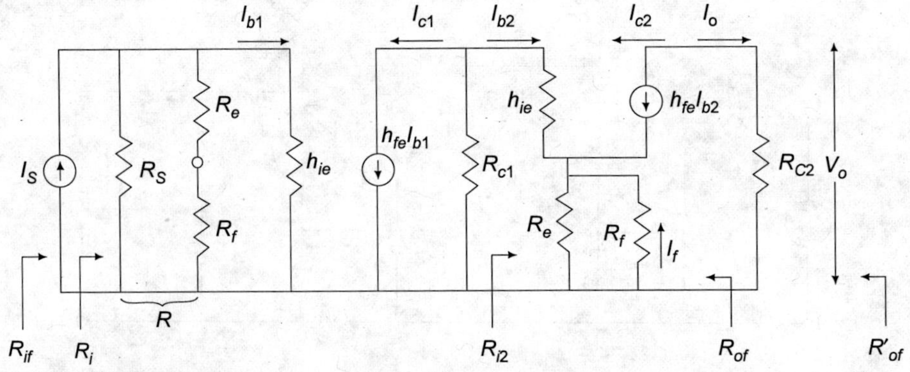

This topology stabilises current gain

$$A_I = \frac{I_o}{I_s} = \frac{-I_{c2}}{I_s} = \frac{-I_{c2}}{I_{b2}} \cdot \frac{I_{b2}}{I_{c1}} \cdot \frac{I_{c1}}{I_{b1}} \cdot \frac{I_{b1}}{I_s}$$

From the circuit, $\dfrac{-I_{c2}}{I_{b2}} = -h_{fe} = -50$

$$\frac{I_{c1}}{I_{b1}} = h_{fe} = +50$$

$$\frac{I_{b2}}{I_{c1}} = \frac{-R_{c1}}{R_{c1} + R_{i2}}$$

$R_{i2} = h_{ie} + (1 + h_{fe})(R_e \,||\, R_f)$ (Since second stage is *CE* with R_e)

$$= 1.1 + (51)(50 \,||\, 1000)$$

$$= 3528.5\ \Omega$$

$$\therefore \quad \frac{I_{b2}}{I_{c1}} = \frac{-2.5\times10^3}{(2.5+3.5)\times10^3} = -0.4167$$

$$\frac{I_{b1}}{I_S} = \frac{R}{R_e + h_{ie}} = \frac{500}{500+1100} = 0.3125$$

Since $R = R_s \| R_f = 1\text{ K} \| 1\text{ K} = 500$

$$\therefore \quad A_I = (-50)\,(-0.4167)\,(50)\,(0.3125)$$
$$= 325.54$$

$$\beta = \frac{R_e}{R_f + R_e} = \frac{50}{50+1000} = 0.0476$$

$$D = 1 + \beta A_I = 16.5$$

$$A_{If} = A_I/D = 19.72$$

$$A_{Vf} = \frac{V_o}{V_s} = \frac{-I_{c2}\cdot R_{c2}}{I_s\cdot R_s} = A_{If}\cdot\frac{R_{c2}}{R_s}$$

$$= \frac{(19.72)\,(600)}{1000} = 11.832$$

$$R_i = R \| h_{ie} = 500 \| 1100 = 343.75\ \Omega$$

$$R_{if} = \frac{R_i}{D} = 20.83\ \Omega$$

If R_{c2} is an external load, then $R_o = \infty$, $R_{of} = \infty$

$$R'_{of} = R'_o\left(\frac{1+\beta A_i}{1+\beta A_I}\right) = R'_o = R_{c2} = 600\ \Omega$$

Since $\quad A_i = \lim\limits_{R_L\to 0} A_I = A_I$

10.16 The circuit has following parameters R_c = 4 K, R' = 50 K, R_s = 6 KΩ, h_{ie} = 1.1 KΩ, h_{fe} = 60 and $h_{oe} = h_{re} = 0$. Find (a) A_{Vf} (b) R_{if} and (c) R_{of}.

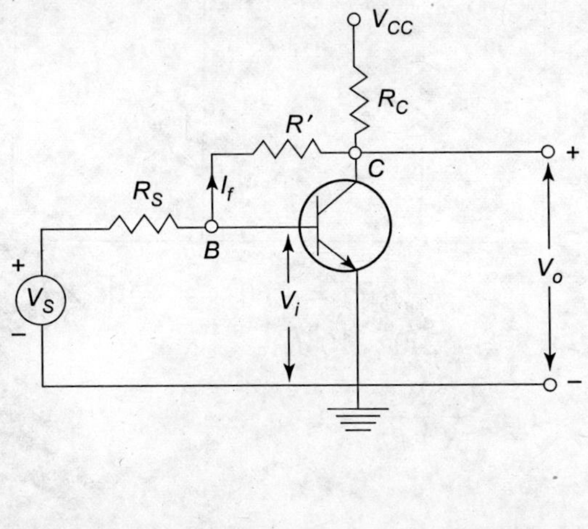

Solution 1. Since R' is common between input node (base) and output node (c), there exists a shunt connection. The feedback signal is current I_f.

To find the way of sampling, $I_f = \dfrac{V_i - V_o}{R'}$

Since $V_o \gg V_i$, $I_f = \dfrac{-V_o}{R'} = \beta V_o$. Since $\beta = -1/R'$ is independent of load R_c and R_s, there is voltage sampling. This is a voltage-shunt topology. This stabilises "Transresistance" (R_M).

2. To find input circuit, set $V_o = 0$. Then, R' appears between base and ground. To find the output circuit, set $V_i = 0$ for shunt comparison. Then R' appears between collector and ground.

The circuit will be,

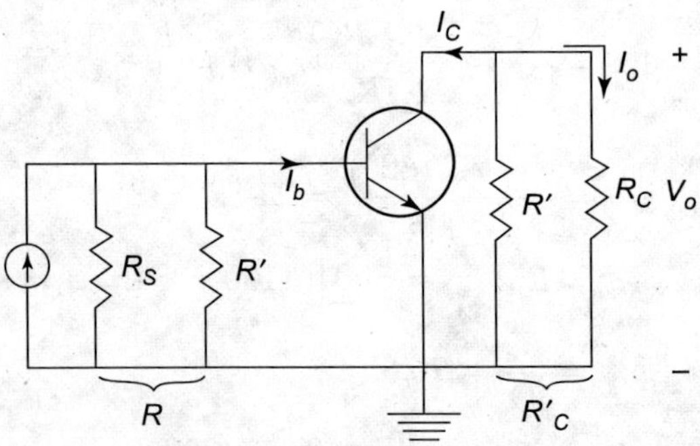

Here current source has been used since there is shunt mixing.

Let $R'_c = R_c \,||\, R' = 4\text{ K}\,||\,50\text{ K} = 3.70\text{ K}\Omega$

$R = R_s \,||\, R' = 6\text{ K}\,||\,50\text{ K} = 5.35\text{ K}\Omega$

Replacing transistor with its proper model,

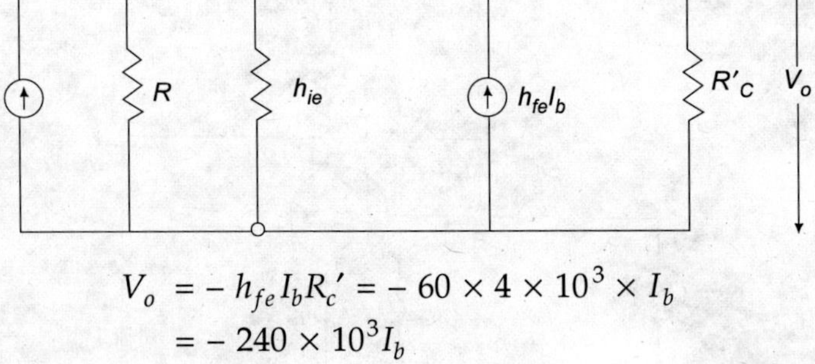

$$V_o = -h_{fe} I_b R_c' = -60 \times 4 \times 10^3 \times I_b$$

$$= -240 \times 10^3 I_b$$

$$I_b = \frac{R}{R + h_{ie}} \times I_s$$

$$\therefore \quad R_M = \frac{V_o}{I_s} = \frac{-240 \times 10^3 \times I_b \times R}{(R + h_{ie})\, I_b}$$

$$= \frac{-240 \times 10^3 \times 5.35 \times 10^3}{6.45 \times 10^3}$$

$$= -199.06 \times 10^3\ \Omega$$

$$R_{Mf} = \frac{R_M}{D}$$

$$D = 1 + \beta R_M = 1 + \left(-\frac{1}{50 \times 10^3}\right)(-199.06 \times 10^3)$$

$$= 4.98$$

$$R_{Mf} = -39.96\ \text{K}\Omega$$

$$A_{Vf} = \frac{V_o}{V_s} = \frac{V_o}{I_s R_s} = \frac{R_{Mf}}{R_s} = \frac{-39.96\ \text{K}\Omega}{6\ \text{K}\Omega}$$

$$= -6.66$$

$$R_i = R \,||\, h_{ie} \text{ from the circuit}$$

$$= \frac{R \times h_{ie}}{R + h_{ie}} = \frac{5.35 \times 1.1}{6.45} = 912.40\ \Omega$$

$$R_{if} = \frac{R_i}{D} = \frac{912.40}{4.98} = 183.21\ \Omega$$

If R_c is considered as load, $R_o = R' = 50\ \text{K}\Omega$

$$R_m = \lim_{R_L \to \infty} R_M = \frac{-h_{fe} \cdot RR'}{R + h_{ie}}$$

$$= \frac{-60 \times 5.35 \times 50 \times 10^3 \times 10^3}{(50 + 1.1) \times 10^3}$$

$$= -2.48\ \text{M}\Omega$$

$$\therefore \quad R_{of} = \frac{R_o}{1 + \beta R_m} = \frac{50 \times 10^3}{1 + \dfrac{314.09 \times 10^3}{50 \times 10^3}}$$

$$= 6.87\ \text{K}\Omega$$

10.17 Determine the input impedence of all types of feedback amplifiers.

Solution *Voltage series feedback:*

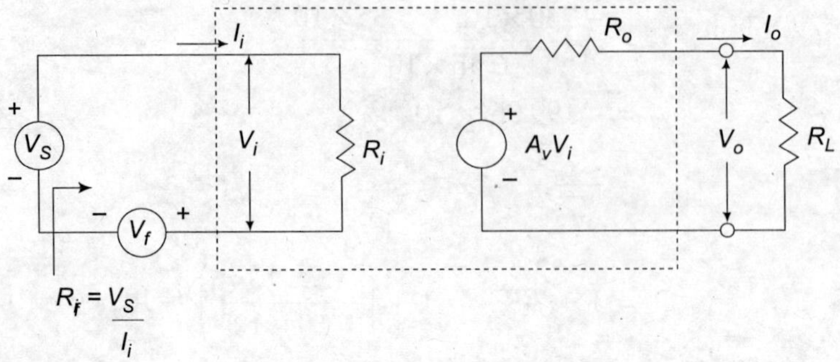

From the input circuit, we have

$$V_s = I_iR_i + V_f = I_iR_i + \beta V_o \quad ...(1)$$

Using voltage division in the output circuit

$$V_o = \frac{A_vV_iR_L}{R_o + R_L} = A_V \cdot I_iR_i \quad ...(2)$$

where $$A_V = \frac{V_o}{V_i} = \frac{A_vR_L}{R_o + R_L}$$

From (1) and (2)

$$V_s = I_iR_i + \beta A_V I_iR_i$$
$$= R_iI_i(1 + \beta A_V)$$

$$\therefore \quad R_{if} = \frac{V_s}{I_i} = R_i(1 + \beta A_V) \text{ so, with series mixing } R_{if} > R_i$$

A_v = open circuit voltage gain without feedback, and A_V = voltage gain without feedback taking R_L into account. Therefore,

$$A_v = \lim_{R_L \to \infty} A_V$$

Current series feedback:

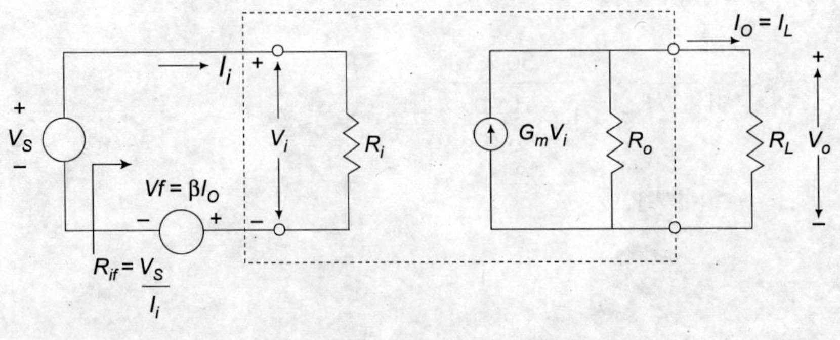

Writing KVL in the input circuit,

$$V_s = I_iR_i + V_f = I_iR_i + \beta I_0 \quad ...(1)$$

Using current division, I_o from output circuit is given by

$$I_0 = G_mV_i \frac{R_o}{R_o + R_L} = G_MV_i \quad ...(2)$$

where $G_M = \dfrac{G_mR_o}{R_o + R_L}$

From (1) and (2)

$$V_s = I_iR_i + \beta G_MV_i$$
$$= I_iR_i + \beta G_MI_iR_i$$
$$= I_iR_i(1 + \beta G_M)$$

$R_{if} = \dfrac{V_s}{I_i} = R_i(1 + \beta G_M)$. When there is series mixing input resistance increases with feedback G_m is short-circuit transconductance. G_M is the transconductance without feedback taking load into account.

$$\therefore \quad G_m = \lim_{R_L \to 0} G_M$$

Current-shunt feedback:

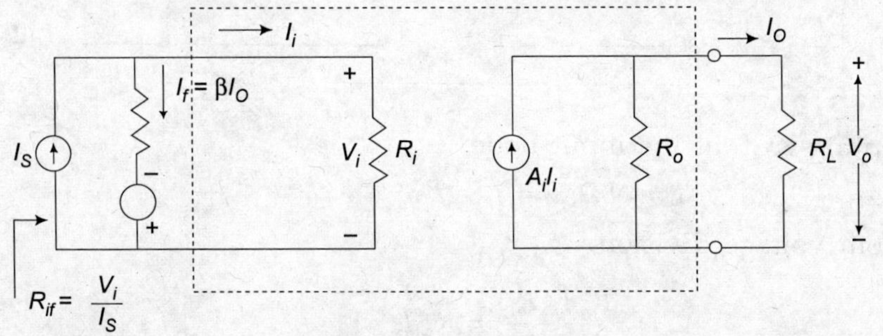

Writing KCL at the input node,

$$I_s = I_f + I_i = \beta I_o + I_i \quad ...(1)$$

Using current division, at the output circuit

$$I_o = \frac{A_iI_i \cdot R_o}{R_o + R_L} = A_II_i \quad ...(2)$$

where $\dfrac{A_iR_o}{R_o + R_L} = A_I$

$$I_s = I_i + \beta A_I \cdot I_i$$
$$= I_i(1 + \beta A_I)$$

We have $R_{if} = \dfrac{V_i}{I_s}$ and $R_i = \dfrac{V_i}{I_i}$

$$\frac{V_i}{R_{if}} = \frac{V_i}{R_i}(1 + \beta A_I)$$

$\therefore\ R_{if} = \dfrac{R_i}{1+\beta A_I}$ so we can say that with shunt mixing $R_{if} < R_i$

where A_i is the short circuit current gain taking R_s into account, A_I is current gain without feedback taking R_L into account.

$$\therefore \qquad A_i = \lim_{R_L \to \infty} A_I$$

Voltage-shunt feedback:

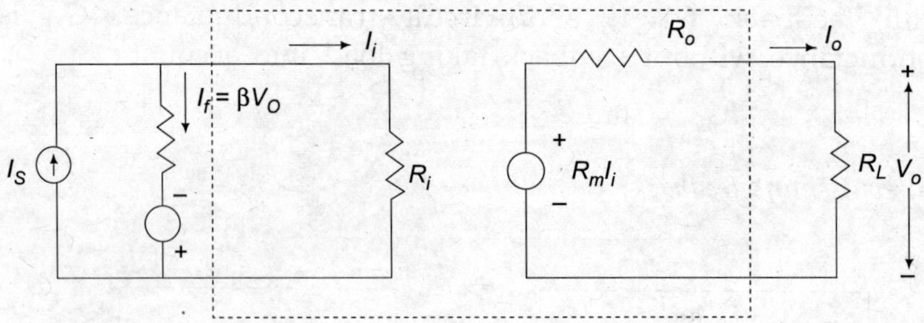

Writing KCL at the input node,

$$I_s = I_i + I_f = I_i + \beta_{Vo} \qquad \text{...(1)}$$

From voltage division,

$$V_o = \frac{R_m I_i \cdot R_L}{R_o + R_L} = R_M I_i$$

where $R_M = \dfrac{R_m R_L}{R_o + R_L}$

$$\therefore \qquad I_s = I_i(1 + \beta R_M)$$

We know that $R_{if} = \dfrac{V_i}{I_s}$ and $R_i = \dfrac{V_i}{I_i}$

$$\therefore \qquad \frac{V_i}{R_{if}} = \frac{V_i}{R_i}(1 + \beta R_M)$$

$$\Rightarrow R_{if} = \frac{R_i}{1+\beta R_M}$$

R_m is the transresistance under open-circuit. R_M is the transistance without feedback but taking the load into account.

$$R_m = \lim_{R_L \to \infty} R_M$$

Therefore, when there is shunt mixing $R_{if} < R_i$

10.18 Determine the output impedance of all types of feedback amplifiers.

Solution *Voltage series feedback*:

Put an arbitrary voltage source of value V and assume it is sending a current I, by making $R_L = \infty$, and $V_S = 0$ then the circuit becomes,

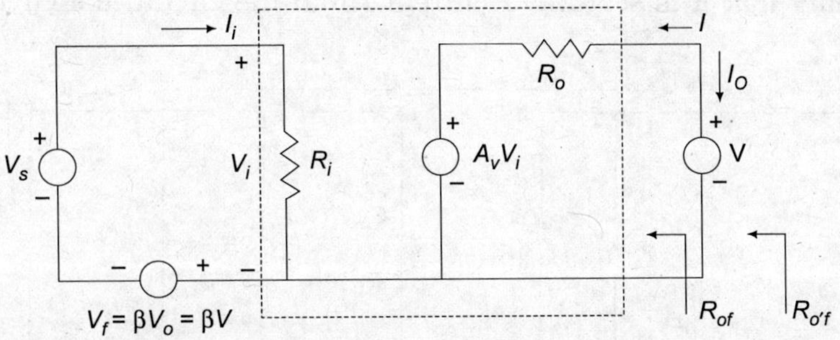

Then, $$R_{of} = \left.\frac{V}{I}\right|_{V_s=0,\, R_L=\infty}$$

From the output circuit

$$I = \frac{V - A_v V_i}{R_o} = \frac{V + \beta A_v \cdot V}{R_o}$$

($\because$ with $V_s = 0$, $V_f + V_i = 0$ $\therefore V_i = -V_f = -\beta V$)

$$\therefore \quad \frac{V}{I} = R_{of} = \frac{R_o}{1+\beta A_v}$$

$$R'_{of} = R_{of} \,||\, R_L$$

$$= \frac{R_{of} \cdot R_L}{R_{of} + R_L} = \frac{\dfrac{R_o}{1+\beta A_v} \cdot R_L}{R_L + \dfrac{R_o}{1+\beta A_v}}$$

$$= \frac{R_o R_L}{R_o + R_L(1+\beta A_v)}$$

$$= \frac{R_o R_L}{R_o + R_L} \cdot \frac{1}{1 + \frac{\beta A_v R_L}{R_o + R_L}}$$

Since $R_o' = R_o \| R_L$

and, $A_V = \frac{A_v \cdot R_L}{R_o + R_L}$

$R_o'{}_f = \frac{R_o'}{1 + \beta A_V}$ so we can say that with voltage sampling $R_{of} < R_o$

Current series feedback:

By making $V_s = 0$, $R_L = \infty$, and introducing a voltage source of value V, assuming that it is sending a current I into the circuit, the circuit becomes,

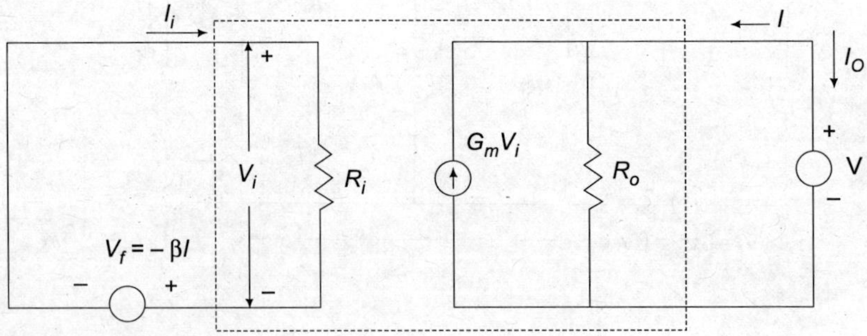

From the output circuit

$$I + G_m V_i = \frac{V}{R_o} \quad \text{...(1)}$$

Writing KVL in the input loop,

$$-\beta I + V_i = 0 \Rightarrow V_i = \beta I \quad \text{...(2)}$$

From (1) and (2),

$$I + \beta G_m I = \frac{V}{R_o}$$

$$\therefore \quad I(1 + \beta G_m) = \frac{V}{R_o}$$

$$\therefore \quad \frac{V}{I} = R_{of} = R_o(1 + \beta G_m)$$

Therefore, for current sampling $R_{of} > R_o$

R'_{of} = output resistance with feedback taking R_L into account

$$R'_{of} = R_{of} \| R_L$$

$$= \frac{R_o(1+\beta G_m)R_L}{R_o + R_L + R_o\beta G_m}$$

$$= \frac{R_o R_L}{R_o + R_L} \frac{1+\beta G_m}{1+\dfrac{\beta G_m \cdot R_o}{R_o + R_L}}$$

Since $G_M = \dfrac{G_m R_o}{R_o + R_L}$ and $R'_o = \dfrac{R_o R_L}{R_o + R_L}$

$$R'_{of} = R_o'\left(\frac{1+\beta G_m}{1+\beta G_M}\right)$$

Current-shunt feedback:

The circuit with $I_s = 0$, and $R_L = \infty$ will be

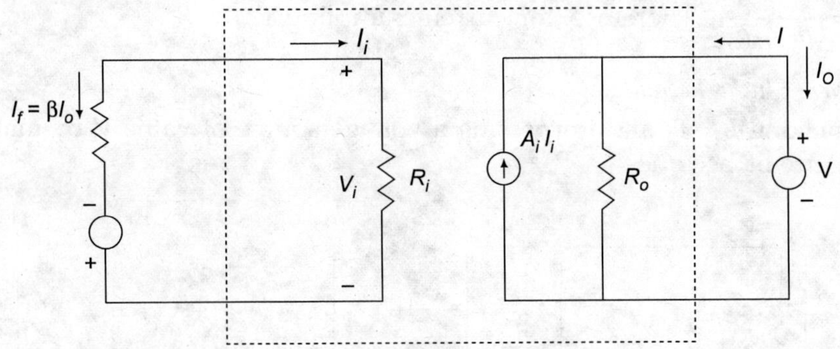

From the input circuit, writing KCL

$$I_i + I_f = 0$$

$$I_i + \beta I_o = 0 \Rightarrow I_i = \beta I \qquad ...(1)$$

From the output circuit,

$$I + A_i I_i = \frac{V}{R_o}$$

$$I = \frac{V}{R_o} - A_i \cdot \beta I$$

$$I(1 + \beta A_i) = \frac{V}{R_o}$$

$$\therefore \qquad R_{of} = \frac{V}{I} = R_o(1 + \beta A_i)$$

$$R'_{of} = R_{of} \,||\, R_L$$

$$= \frac{R_o(1+\beta A_i)R_L}{R_o(1+\beta A_i)+R_L}$$

$$= \frac{R_o R_L \cdot (1+\beta A_i)}{R_o + R_L + \beta A_i R_o}$$

$$= \frac{R_o R_L}{R_o + R_L} \cdot \frac{1+\beta A_i}{1+\beta A_i \dfrac{R_o}{R_o+R_L}}$$

$$= Ro'\left(\frac{1+\beta A_i}{1+\beta A_I}\right)$$

where $R'_o = Ro \,||\, R_L$

$A_I = \dfrac{A_i R_o}{R_o + R_L}$. Therefore, for current sampling $R_{of} > R_o$.

Voltage-Shunt Feedback:

By making $I_s = 0$, and impressing a voltage source of value V by making $R_L = \infty$. The circuit becomes

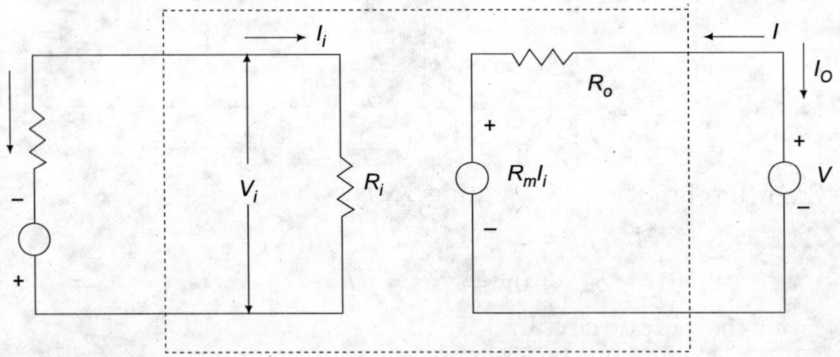

Writing KCL at input node

$$I_i + \beta V = 0$$

$$\therefore \qquad I_i = -\beta V \qquad \text{...(1)}$$

From the output circuit,

$$\frac{V - R_m I_i}{R_o} = I \qquad \text{...(2)}$$

From (1) and (2)

$$\therefore \qquad \frac{V + \beta R_m V}{R_o} = I$$

$$\therefore \quad \frac{V}{I} = R_{of} = \frac{R_o}{1+\beta R_m}$$

$$R'_{of} = R_{of} \,||\, R_L$$

$$= \frac{\dfrac{R_o}{1+\beta R_m} \cdot R_L}{\dfrac{R_o}{1+\beta R_m} + R_L}$$

$$= \frac{R_o R_L}{R_o + R_L(1+\beta R_m)}$$

$$= \frac{R_o R_L}{R_o + R_L + \beta R_m R_L}$$

$$= \frac{R_o R_L}{R_o + R_L} \cdot \frac{1}{1 + \dfrac{\beta R_m R_L}{R_o + R_L}}$$

Since $R_M = \dfrac{R_m R_L}{R_o + R_L}$ and $R'_o = R_o \,||\, R_L$

We have $R'_{of} = \dfrac{R_o'}{1+\beta R_M}$

$\therefore$ for voltage sampling $R_{of} < R_o$.

EXERCISE PROBLEMS

10.1 For the circuit shown, with R_c = 4 k = R_L, R_b = 20 k, R_s = 1 K and with typical h-parameter values. Find

(a) $A_I = I_L/I_s$ (b) $\dfrac{V_o}{V_s} = A_{Vs}$

(c) $G_M = \dfrac{I_L}{V_s}$ (d) $R_M = \dfrac{V_o}{I_s}$

(e) The input resistance seen by the source.

(f) The output resistance seen by the load.

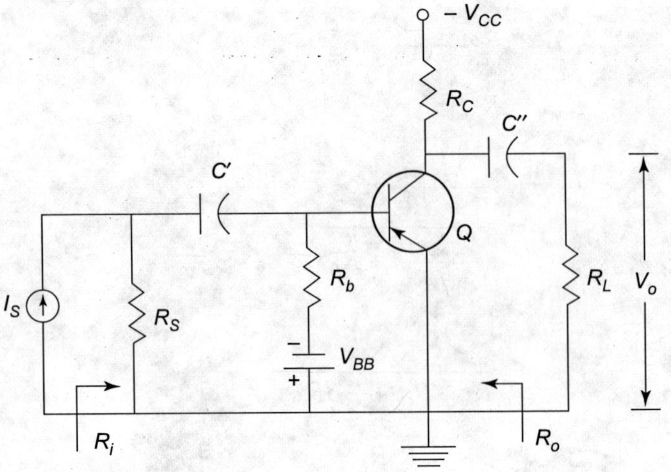

10.2 Repeat the above problem, with g_m = 5 mA/V, r_d = 100 K.

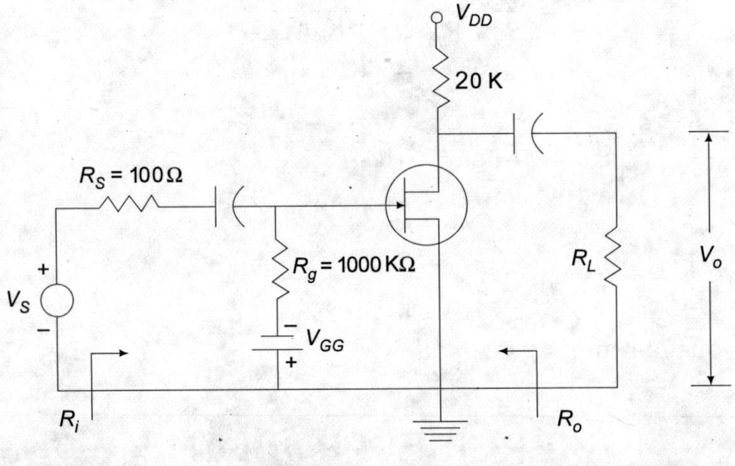

10.3 We have an amplifier of gain 100 dB. It has an output impedance Z_o = 100 KΩ. It is required to modify its output impedance to 500 Ω by applying negative feedback. Calculate the value of β. Calculate the percentage change in the overall gain.

10.4 For the transistor amplifier h_{fe} = 100, h_{ie} = 1 K and h_{re} = h_{oe} = 0. Find with R_e = 0.

(a) $R_{Mf} = \dfrac{V_o}{I_s}$ (b) $A_{vf} = \dfrac{V_o}{V_s}$

(c) R_{if} and (d) R_{of} (e) Repeat with R_e = 500 Ω

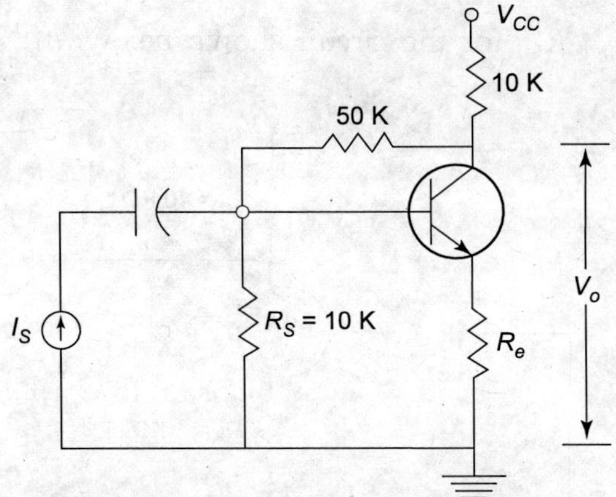

10.5 For the FET amplifier shown below has $g_m = 50$ mA/V, $r_d = 1$ MΩ. Find A_{vf}, R_{if}, R_{of}, with $R_s = 2$ KΩ.

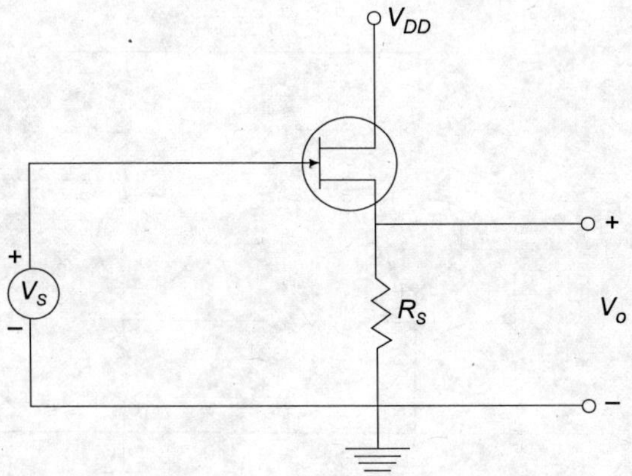

10.6 Find A_{vf}, R_{if}, R_{of} for the circuit shown below. If $R_s = 0$, $h_{fe} = 50$, $h_{ie} = 1100\ \Omega$, $h_{oe} = 0$. Assume Identical transistors.

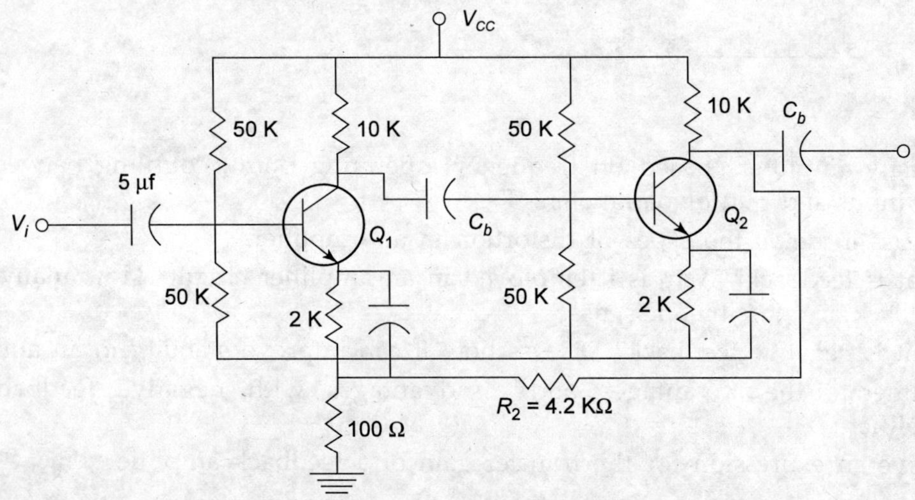

10.7 Find A_{vf}, G_{Mf}, R_{if}, R_{of}, R'_{of} for the circuit shown below. h_{fe} = 150, h_{ie} = 1000 Ω, h_{re} = h_{oe} = 0 and R_s = 1 KΩ.

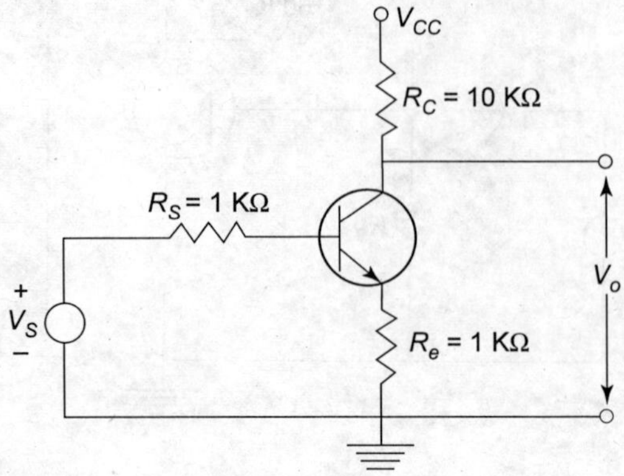

10.8 For the following circuit, R_{c1} = 3 K, R_c = 3 K, R_e = 50 Ω, R' = R_s = 1000 Ω, h_{fe} = 50, h_{ie} = 1.1 K and h_{re} = h_{oe} = 0. Find (a) A_{vf} (b) R_{if} (c) R_{of}.

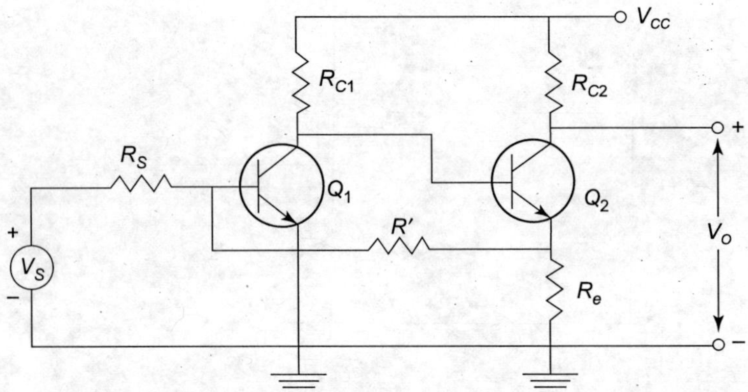

REVIEW QUESTIONS

1. Classify amplifiers based on frequency, operating point, coupling between stages and input and output parameters. Discuss.
2. Discuss in detail the types of distortion in an amplifier.
3. What is feedback? Why is it employed in an amplifier circuit? How many types of feedback are possible? Discuss.
4. What is negative feedback? Discuss how it can improve stability in an amplifier.
5. Enumerate the advantages and disadvantages with negative feedback in an amplifier.
6. Derive an expression for the transfer gain of a feedback amplifier.

7. What are the various types of amplifiers based on the output and input parameters? Give their equivalent circuits and ideal characteristics.
8. Discuss in detail about the types of negative feedback amplifiers giving the effect of each type of feedback on the parameters of the amplifier.
9. Derive an expression for the input impedance with feedback of a voltage shunt feedback amplifier.
10. Derive an expression for the input impedance and output impedance with feedback of a voltage series feedback amplifier.
11. Derive an expression for the output impedance with feedback of a current shunt feedback amplifier.
12. Derive an expression for the input and output impedance with feedback of a current series feedback amplifier.
13. Enumerate the procedure employed in the analysis of a feedback amplifier and discuss in detail the effect of feedback on the amplifier parameters.
14. Discuss how to separate basic amplifier circuit and the feedback circuit from a feedback amplifier circuit.
15. Derive an expression for the input impedance with feedback of a feedback amplifier, which stabilises current gain.
16. Define the following modes of operation of an amplifier: class *A*, class *B*, class *AB* and class *C*. Compare them based on the efficiency and distortion.
17. Draw a feedback amplifier in block diagram form and explain each block in detail giving its functions.
18. What are the four topologies of a feedback amplifier? Identify all the signals and transfer gains of each topology.
19. Distinguish between regenerative and degenerative feedback in amplifiers and give their applications.
20. Define desensitivity factor *D*. Give its significance in negative feedback amplifiers.
21. Draw a voltage series feedback amplifier employing FET and analyse it.
22. What are assumptions made in the negative feedback amplifiers?
23. What sort of feedback is employed in a *CE* amplifier with unbypassed emitter resistor? Discuss its analysis in detail.
24. Draw the circuit diagram of a voltage shunt feedback amplifier and explain the feedback and give the expression for *D*.
25. Classify amplifiers based on the coupling between two stages.

CHAPTER

11

Sinusoidal Oscillators

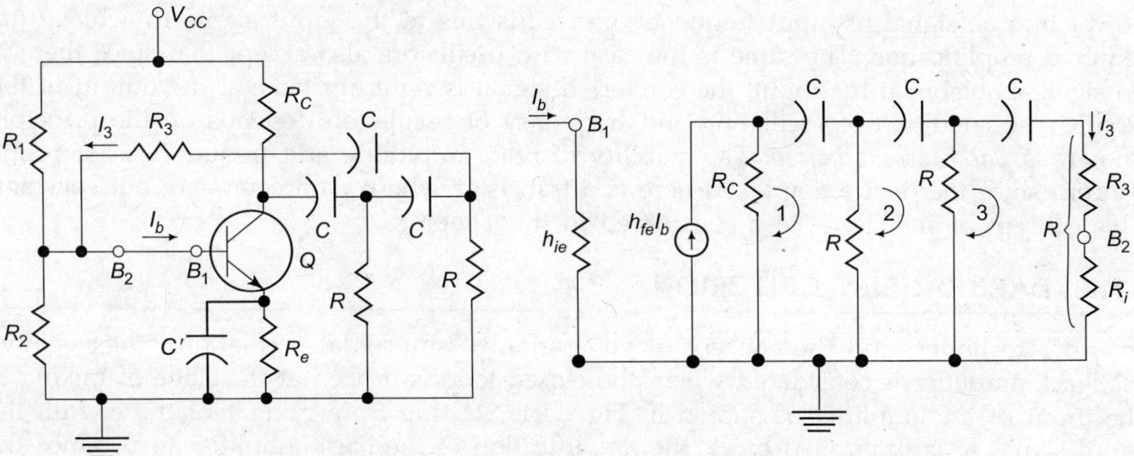

(a) BJT phase-shift oscillator (b) Equivalent circuit

Chapter Outline

The concepts introduced in this chapter are:

- *Positive feedback in amplifier circuits*
- *Condition and frequency of sinusoidal oscillations*
- *Various types of oscillator circuits*
- *Amplitude and frequency stability of sinusoidal oscillators*

11.1 INTRODUCTION

Generation of *sinusoidal oscillations* of a given frequency is highly essential in any electronic circuit. Using the concept of positive feedback generates such sinusoidal oscillations. Active devices such as transistor, FET or Op-Amp (operational amplifier) are employed in the generation of sinusoidal oscillations. The circuit configurations vary depending upon the frequency of operation, application and so on, but basically the principle of operation is the same for all such *sinusoidal oscillators*. Generating sinusoidal oscillations is nothing but producing the required signal without any given input to a transistor circuit. This is possible since a transistor is not an amplifying device but a device that converts dc energy into ac energy at required frequency. In a transistor amplifier, the device converts the dc power into ac signal of input frequency and adds this to the input signal to achieve the required amplification. The same is the case with oscillators also except that since there is no signal available at the input, the converted signal is right available at the output of the circuit. The condition of oscillation and frequency of oscillation depends on the principle known as *Barkhausen criterion*. The stability of both amplitude and frequency is of prime concern since the device employed here is a transistor whose parameters are not constant. This concept of stability is also considered in this chapter.

11.2 BARKHAUSEN CRITERION

In order to understand the concept of generation of sinusoidal oscillations, the negative feedback amplifier is considered where the closed loop is broken at the stage of mixing at the input of the amplifier as shown in Fig. 11.1. Starting from node 1 of the circuit, the input signal X_i is multiplied by A, the amplification of the basic amplifier to produce the output signal of the amplifier. This signal is further multiplied by β, the transfer ratio of the feedback circuit to generate the feedback signal. So, the input signal is multiplied by $A\beta$ to achieve feedback signal. In a negative feedback the input to the basic amplifier is the difference of the source signal and feedback signal. The same can be presented as the input signal is sum of the source signal and phase-shifted feedback signal (phase shift being 180°), that is, the new feedback signal, X_f', is equal to $-X_f$,

$$X_f' = -X_f \qquad ...(11.1)$$

In other words, the signal at node 2 of the circuit is achieved after multiplying the input signal at node 1 by $-A\beta$. That is, the loop gain of the network is

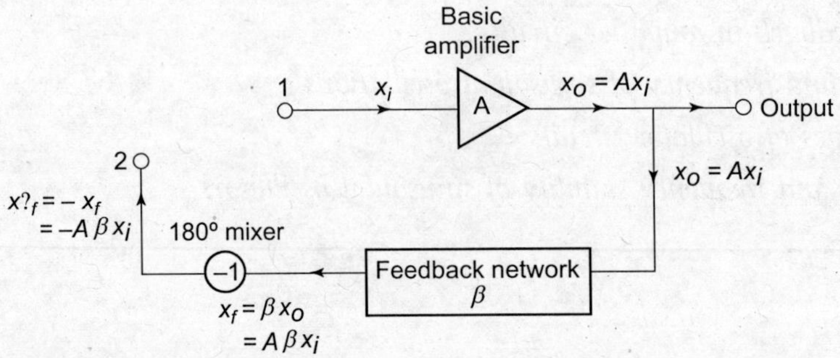

Fig. 11.1 *Barkhausen Criterion*

$$\text{Loop gain} = X_f'/X_i = -X_f/X_i = -A\beta \qquad ...(11.2)$$

If the loop and its gains are so adjusted such that X_f' is exactly identical to X_i, that is,

$$X_f' \equiv X_i \qquad ...(11.3)$$

implies that loop gain is equal to 1, that is,

$$-A\beta = 1 \qquad ...(11.4)$$

Then, node 2 can be connected to node 1 without disturbing the output and removing the input signal. As far as the basic amplifier is concerned, it is not concerned from where the source signal is derived. It is supposed to amplify the signal that is available at the input. So, if the above condition is satisfied for a given circuit, the output is available even though the input signal is removed. This is nothing but the concept of sinusoidal oscillations.

Considering Eq. (11.3), which states that the new feedback signal should be identical to the input signal, does not only imply that amplitudes are same, but along with *amplitude, frequency, phase* and also the *shape of the signal* should be *exactly* equal to that of the input signal. Then and only then, the above oscillations are possible for generation.

Basically the condition required for oscillations to sustain is that the *loop gain should be equal to unity.* This guarantees oscillations. If the loop gain is less than unity, then the feedback is said to be negative feedback, the one that is employed in amplifiers. In this type of feedback, if input is not connected, since the loop gain is less than 1, the signal slowly decreases and dies down to zero very soon. In a feedback amplifier, the input provides this loss in the signal, thus maintaining the output at a given level of the signal. Because of this reason, the gain of the feedback amplifier is small with feedback. Some of the power from the output is utilised to compensate for the loss in the signal due to low loop gain. When the loop gain is exactly unity, the oscillations are possible since no signal is lost and amplifier continues its action without any loss in the energy.

When the loop gain is large compared to 1, the signal is large compared to its initial value after the signal travels by one closed loop. This continues and as time extends to infinity, the output signal should build up in amplitude and will have to tend to infinity. In practice, infinity signal is not really possible since a source of infinite energy is not practicable. The energy drawn from the dc power supply of the circuit is limited and the amplitude of the signal has to be saturated at this energy level. So, when the loop gain is large compared to unity, the output increases to a level of saturation and is limited at this level.

The frequency of oscillation depends on the phase around the closed loop. The phase around the closed loop should be 0 or multiples of 2π radians. Since the feedback circuit employs reactive components, the phase introduced by this circuit depends on the frequency of operation. Or, putting it the other way round, the frequency at which the phase introduced by this circuit in association with the amplifier circuit is 0 or multiples of 2π radians, is the frequency of oscillation.

Based on the above discussion, the *Barkhausen criterion* can be stated as: To maintain oscillations in a transistor circuit, the loop gain should be more than unity. In other words, if the product of magnitude of transfer gain of the transistor amplifier and that of the feedback factor of the feedback network is more than unity. If the loop gain is less than unity, the oscillations will die and output will soon be zero.

The frequency of oscillation is determined by the phase shift condition. The phase shift around the closed loop starting from input of the amplifier to output of the amplifier to the input of the feedback circuit to the output of feedback circuit through input mixer back to input of the amplifier should be either zero or integral multiples of 360° (2? radians). In other words, the phase of the input signal and phase of the signal after traversing through amplifier, feedback circuit and mixer should be exactly the same, i.e., the two signals should be in phase.

To be more specific, the frequency at which the two signals will be in exactly same phase is the frequency of oscillation. The frequency component which satisfies this condition is picked up from the thermal noise available in the components of the circuit. The magnitude of the loop gain determines the condition of oscillation (loop gain unity) and the phase around the closed loop determines the frequency of operation.

When the loop gain is unity, the gain with feedback is

$$A_f = A/[1 + A\beta] = A/[1 - 1] \to \infty \qquad \text{...(11.5)}$$

Or, in other words, when A_f tends to infinity it means that the output is available even though the input signal is zero or absent, which is what oscillation means.

The *Barkhausen criterion* is true for whatever may be the signal shape, sine, triangular or square. But, the principle is only applicable for sine signals since even though the amplifier is able to handle any signal under consideration, the feedback network comprising the passive components cannot maintain the shape of the signal when the signal is transmitted through it. Only sinusoidal signals can maintain shape without any distortion, even though amplitude and phase suffer, when passed through a passive network. This is the reason why Barkhausen criterion is applicable to only sinusoidal signals.

Practical Limitation: If the minimum condition for oscillation, the loop gain equal to unity, is met sinusoidal oscillations are guaranteed theoretically. But, in practice, there is a limitation. The small signal gain of a transistor is not constant and depends on various factors and constraints such as temperature, ageing, voltage and so on. So, even though the circuit is designed to oscillate with loop gain unity, when circumstances change consequently, there is a chance that amplifier transfer gain A may droop down pulling the loop gain below unity. This causes the oscillations to damp down to zero very soon. This is not an impossible condition but will occur very frequently in practice. So, if loop gain is chosen to be exactly equal to unity, this limitation will make the oscillations die down.

To avoid the above limitation of sinusoidal oscillation, the loop gain is chosen slightly greater than unity. This selection will ensure that loop gain will never be less than 1 and oscillations are assured. Since the loop gain is more than 1, the output increases and is limited to saturation of the amplifier.

In practice, the magnitude of the loop gain around the closed loop consisting of amplifier, feedback network and mixer is chosen to be slightly more than unity. This leads to the fact that the magnitude of the output oscillations grow. But, this growth of the magnitude of oscillations is finite since the amplifier reaches saturation at a given voltage and the non linearity of the device restricts the growth of the signal at the output. Thus, the magnitude of the output signal is limited by the non linearity of the amplifier.

Types of oscillators: Broadly, the oscillators can be classified as *sinusoidal* and *relaxation—square or sawtooth*. They can be classified based on the mode of operation as *feedback type* and *negative resistance type*. The types of sinusoidal oscillators can be classified as:

- Tuned circuit or LC oscillators—Colpitt, Hartley, Clapp
- *RC* oscillators—phase shift, Wein bridge, tuned collector (resonant circuit)
- Crystal oscillators.

11.3 PHASE-SHIFT OSCILLATOR

The *phase-shift oscillator* is a very good example of the criterion discussed in the above sections. The basic amplifier, either BJT or FET, introduces a 180° phase shift between the input and output of the amplifier in *CE* or *CS* configuration. The remaining 180° phase shift, so as to satisfy the Barkhausen criterion, is to be introduced by the feedback network. The feedback network is chosen as resistor-capacitor network in case of a phase-shift oscillator. If the capacitor is ideal, the phase shift across it would be 90° and so, two capacitors in series might have been sufficient to introduce the additional phase. But, in practice, no capacitor is ideal and has some leakage conductance, the phase shift is always less than 90°. So, one requires at least three capacitors to achieve the requirement of 180° phase. Three capacitors in series with resistors in shunt is the feedback network chosen for this phase-shift oscillator. The oscillator oscillates at a frequency when the phase of this feedback network is exactly 180°. For any other frequency, the oscillations cannot be sustained since the phase condition is not satisfied.

Figure 11.2(a) shows the circuit diagram of a phase-shift oscillator using FET. In Fig. 11.2(b), FET is replaced by its small signal model for analysis. Applying the known network principles to the feedback network, its transfer ratio can be derived as the ratio of V_f' and V_o, which is nothing but β and be given as

$$-\beta = V_f'/V_o = 1/[1 - 5\alpha^2 - j(6\alpha - \alpha^3)] \qquad ...(11.6)$$

where $\alpha = 1/\omega RC$, $\omega = 2\pi f$. The phase shift of $-\beta$ would be 180° only when the above transfer function is real. So, equating the imaginary term in the denominator of the ratio to zero, the condition can be satisfied. Or, in other words, the frequency of oscillation is the condition at which the transfer ratio would be real. Imposing this condition on the above transfer ratio,

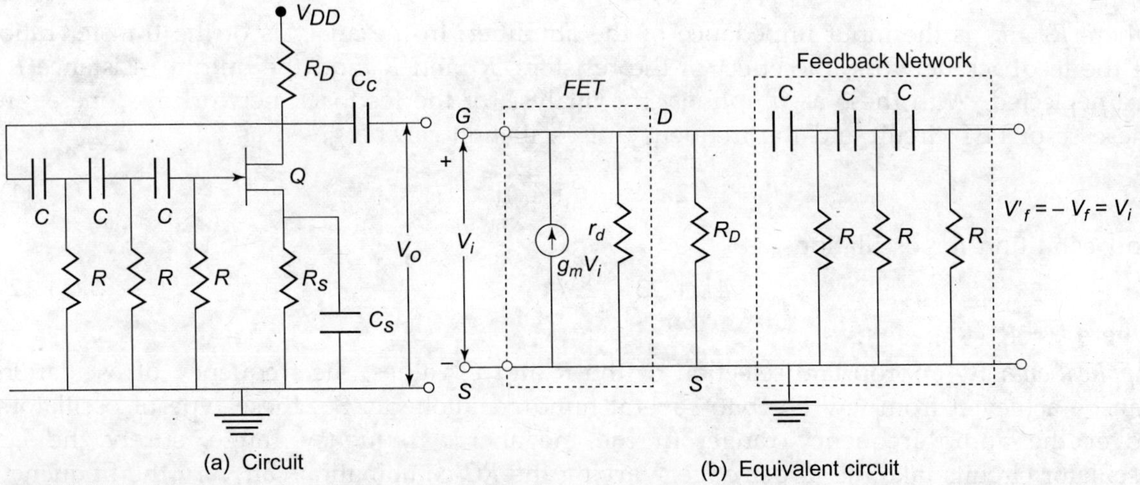

Fig. 11.2 *RC phase shift oscillator - FET*

$$\alpha^2 = 6$$

or,

$$f = 1/\left(2\pi RC\sqrt{6}\right) \qquad ...(11.7)$$

at this frequency of operation, the magnitude of the transfer ratio of the feedback network can be derived to be

$$\beta = 1/29 \qquad ...(11.8)$$

Therefore, applying the Barkhausen criterion to this circuit, which states that loop gain, $|A\beta|$ should be at least equal to unity, the gain of the amplifier without feedback should be 29 or slightly more. That is, the condition for oscillation of a FET phase-shift amplifier is

$$A \geq 29 \qquad ...(11.9)$$

In other words, a FET is to be chosen in this type of oscillator with an amplification factor $\mu > 29$.

In the above discussion, since a FET is employed as the active device, the input impedance of the device being very large, the loading of the feedback network is not considered. But when a phase-shift oscillator is to be realized with a BJT, whose input impedance is of medium range, the loading of the feedback network cannot be neglected and is to be taken care of. So, the circuit diagram of the oscillator with BJT is slightly modified such that the effective resistance offered by the third capacitor would be still R. So, the return resistor value is chosen to be equal to R_3 such that this resistor, in series with the input impedance of the basic amplifier circuit, would have a value equal to R. Such a configuration of the phase-shift oscillator along with its small signal equivalent circuit is as shown in Fig. 11.3.

As one can see from the circuit, voltage shunt feedback is employed in this circuit while voltage series feedback is employed in the FET circuit. The value of R_3 is chosen such that

$$R_3 = R - R_i \qquad ...(11.10)$$

where $R_i \approx h_{ie}$ is the input impedance of the amplifier. In the analysis of the transfer ratio of the feedback network, the effect of the resistors, R_1 and R_2 and the output resistance R_o are neglected. With these assumptions, the analysis of the feedback network is done as in the case of FET circuit and the frequency of oscillation can be given as

$$f = 1/[2\pi RC\sqrt{(6+4k)}] \qquad ...(11.11)$$

and condition for oscillations as

$$h_{fe} > 4k + 23 + 29/k \qquad ...(11.12)$$

where $k = R_c/R$.

Applications: By appropriate selection of the R and C values, the frequency of oscillation can be achieved from few hertz to several hundred kilohertz. So, these type of oscillators cover the audio frequency range. In the megahertz frequency range, surely the LC oscillator circuits take the upper edge. Varying the RC combination can vary the frequency of oscillation. The best way is to vary all three capacitor values simultaneously. So, ganged

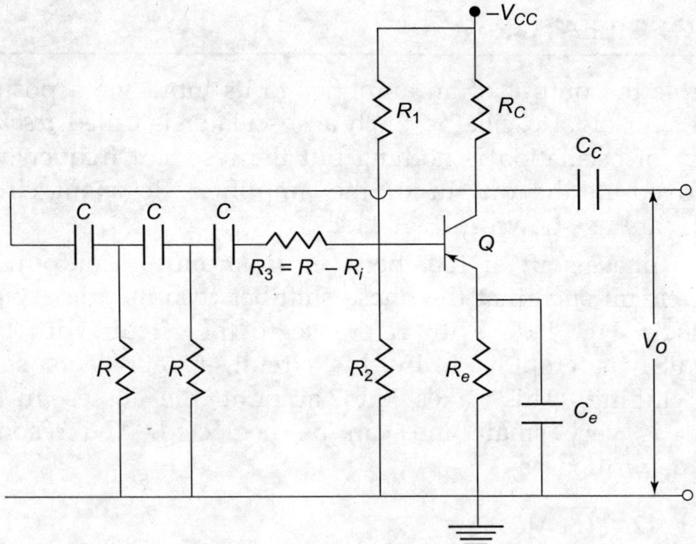

(a) Circuit

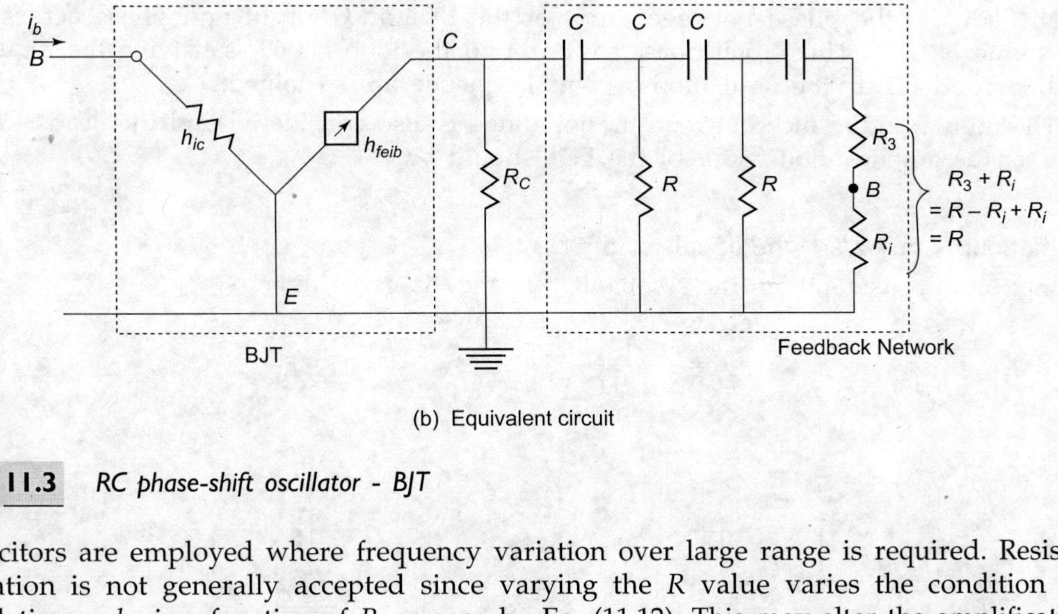

(b) Equivalent circuit

Fig. 11.3 *RC phase-shift oscillator - BJT*

capacitors are employed where frequency variation over large range is required. Resistor variation is not generally accepted since varying the R value varies the condition for oscillation as h_{fe} is a function of R, as seen by Eq. (11.12). This may alter the amplification and the condition for oscillations may not be satisfied for a given value of R. The transistor is operated in class A region in order to minimise distortion in the signal.

The RC phase-shift oscillator provides good stability. Circuit can be used to produce frequencies up to VLF. The circuit does not require inductances or transformers. The limitations of this circuit are that it cannot be used at high frequencies, as feedback is small it is not easy to start oscillations and the output is small comparatively.

11.4 RESONANT CIRCUIT OSCILLATOR

A RF transformer, which can couple the output of an amplifier to its input via a positive feedback, is another example of a sinusoidal oscillator. Such an oscillator is called *resonant circuit oscillator* since the frequency of oscillation is nothing but the resonant frequency of the tank circuit, which is employed in the output of the amplifier. Resonant circuit oscillators employing FET and BJT are as shown in Fig. 11.4.

Since the basic amplifier has a phase-shift of 180° between its input and output, the transformer is so connected in the circuit such that the phase shift between its primary and secondary is also 180°, as shown in Fig. 11.4. With reference to the circuit with FET, consider Fig. 11.4(a) in which FET is employed. In this circuit, r represents series resistance of the transformer winding. This takes into account the losses in the transformer. In general, value r is very small and can be neglected. The resonant frequency of oscillation can thus be written as

$$f = 1/[2\pi\sqrt{(LC)}] \qquad \text{...(11.13)}$$

At resonant frequency, the impedance of the resonant circuit is purely resistive and the reactance is zero. At this frequency of resonance the phase shift condition around the loop is satisfied, i.e., the phase between input of the FET and output and phase across the transformer is 180°. Thus, total phase shift around the loop is 360° satisfying the required condition and finally the oscillations at this frequency are sustained.

The mutual inductance between the windings is also considered and this leads to the fact that the amplification factor of the FET should be

$$\mu = L/M \qquad \text{...(11.14)}$$

for the loop gain $-A\beta$ to be equal to unity.

Taking r into consideration, the resonant frequency is given as

$$f = \frac{1}{2\pi\sqrt{LC}}\sqrt{1+\frac{r}{r_d}} \qquad \text{...(11.15)}$$

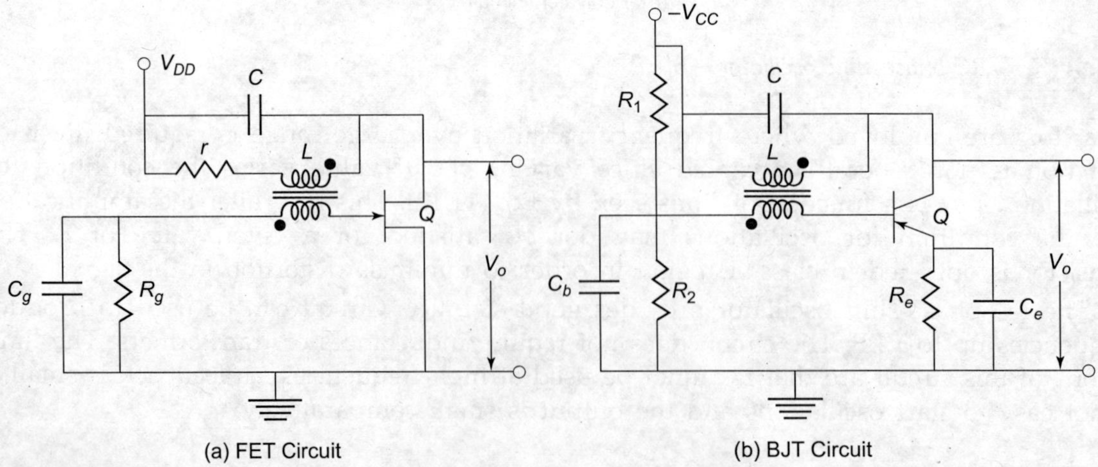

Fig. 11.4 *Resonant circuit oscillator*

and the condition for oscillation to sustain is

$$g_m = \mu r C/[\mu M - L] \qquad ...(11.16)$$

The frequency of oscillation and condition for oscillation are independent. The frequency of oscillation is the frequency at which the loop phase shift is zero or in multiples of 360°. Calling such an oscillator as *resonant circuit oscillator* instead of phase-shift oscillator is purely artificial. In fact, all types of oscillators discussed under this head are phase-shift oscillators.

The operation of the amplifier is class C. The bias for a resonant circuit oscillator is obtained from a parallel R_gC_g combination in series with the gate. The gate and source of the FET act as a rectifier and if the time constant R_gC_g is large compared with one period, the gate capacitor will charge up essentially to the peak gate swing. This voltage across C_g acts as the bias, and the gate is therefore driven positive only for a short interval at the peak of the swing. The voltage at the gate is a large sinusoid, and since its peak value is approximately at ground potential, the gate is clamped to ground.

The gate bias is zero initially when the circuit is switched on and the FET operates with large g_m. Then the loop gain is greater than unity and the amplitude of oscillation starts growing. The gate current is drawn and clamping takes place and the bias automatically adjusts itself so that its magnitude equals the peak of the gate voltage. As the bias becomes more negative, the value of g_m decreases and finally the amplitude stabilises itself at that value for which the loop gain for the fundamental frequency is reduced to unity. The value of g_m is considered to be the minimum value required at zero bias in order for oscillations to start. It also can be interpreted as the average value of transconductance, which determines the amplitude of oscillation.

Resonant circuit oscillator can also be built with a transistor and such a circuit is shown in fig. 11.4 (b). As in the case of FET circuit, the transistor operates in class C mode. The transistor is connected in self-bias configuration. R_2 C_b combination will control the base current.

11.5 HARTLEY AND COLPITT OSCILLATORS

Hartley, Colpitt or *Clapp oscillator* can be derived from a general configuration of the positive feedback circuit when Barkhausen criterion is satisfied. The general configuration of such an oscillator circuit is as shown in Fig. 11.5. The active device employed in this circuit may be a BJT or a FET or an operational amplifier. The active device is assumed to have infinite input impedance. The devices which have such high input impedance are FET or Operational Amplifiers. In Fig. 11.5(b), the device is assumed to have a gain of $-A_v$ and output resistance of R_o. The feedback employed in such a circuit is voltage series.

To determine loop gain, the transfer gain of the amplifier and feedback factor is to be calculated. $-A\beta$ is the loop gain where A is the amplifier gain and β is the feedback factor. The input terminals of the amplifier are 1 and 3 while the output terminals are 2 and 3. Thus, as seen from the figure, the load impedance of the circuit is Z_2 which is in parallel with series combination of Z_1 and Z_3.

Thus,

$$Z_L = Z_2(Z_1 + Z_3)/(Z_1 + Z_2 + Z_3) \qquad ...(11.17)$$

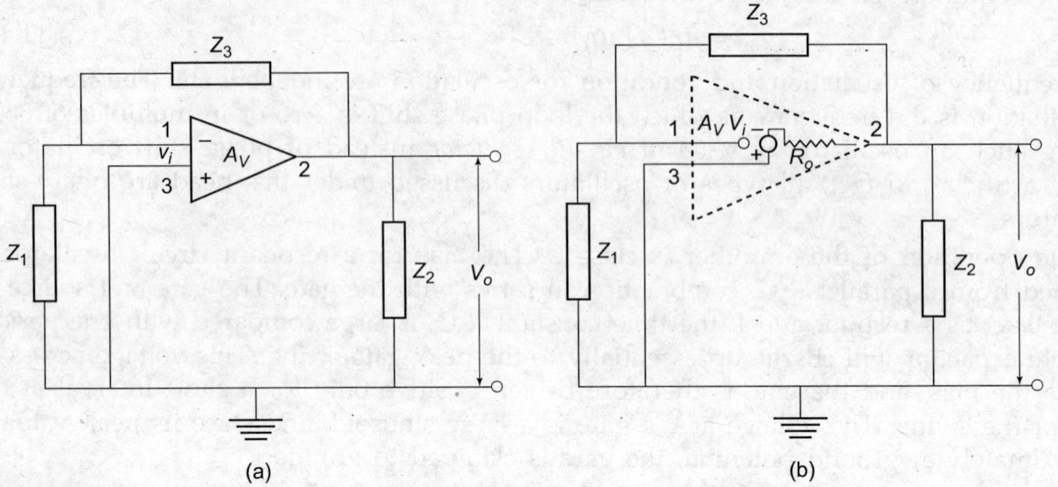

Fig. 11.5 *General form of sinusoidal oscillator*

The amplification without feedback can be given as

$$A = -A_v Z_L/(Z_L + R_o) \quad \text{...(11.18)}$$

and the feedback factor is

$$\beta = -Z_1/(Z_1 + Z_3) \quad \text{...(11.19)}$$

Thus, the loop gain can be derived to be

$$-A\beta = \frac{-A_v Z_1 Z_2}{R_o(Z_1 + Z_2 + Z_3) + Z_2(Z_1 + Z_3)} \quad \text{...(11.20)}$$

Assume that all the three impedances are pure reactive without resistive or conductive loss components, $Z = jX$, where X is the reactance. $X = \omega L$, if inductive or $X = -1/\omega C$, if capacitive. Then, Eq. (11.20) can be rewritten as

$$-A\beta = \frac{A_v X_1 X_2}{jR_o(X_1 + X_2 + X_3) - X_2(X_1 + X_3)} \quad \text{...(11.21)}$$

For loop phase shift to be zero, the loop gain should be real and this is possible when

$$X_1 + X_2 + X_3 = 0 \quad \text{...(11.22)}$$

which is the frequency at which the sustained oscillations are achieved.
Substituting this into Eq. (11.20),

$$-A\beta = \frac{A_v X_1 X_2}{-X_2(X_1 + X_3)} = \frac{A_v X_1}{-(X_1 + X_3)} \quad \text{...(11.23)}$$

or,

$$-A\beta = \frac{A_v X_1}{X_2} \quad \text{...(11.24)}$$

From the above equation it can be assessed that X_1 and X_2 should be of the same sign if the loop gain is to be unity or slightly more. That is, both reactances must be either inductive or capacitive. If one is inductive and the other is capacitive, the loop gain would be negative and the condition of positive feedback is not satisfied. Obviously, the other reactance X_3 should be of opposite sign. That is, when X_1 and X_2 are chosen to be inductive, X_3 should be capacitive or if X_1 and X_2 are chosen to be capacitive, X_3 should be inductive. Thus, the condition for oscillation can be derived to be

$$A_v > X_2/X_1 \quad ...(11.25)$$

If X_1 and X_2 are chosen to be inductive and X_3 capacitive, the oscillator is referred to as *Hartley oscillator* and if X_1 and X_2 are chosen to be capacitive and X_3 inductive, the oscillator is called *Colpitt oscillator.* While working with the Hartley oscillator, the mutual inductance between L_1 and L_2 is to be considered and cannot be neglected.

The essential features of LC oscillators are:

- resonant or tank circuit which determines the frequency of oscillations
- amplifier, which amplifies the oscillations produced by the tank
- network, which provides positive feedback to the tank.

Tuned *LC* oscillators are used at high frequencies. At low frequencies, the inductor and capacitors would be bulky and so, are not suitable. Thus, *RC* oscillators are employed at low frequencies. The limitation of both *RC* and *LC* oscillators is stability of oscillations.

Colpitt oscillator: Detailed analysis of the Colpitt oscillator is difficult since

- low input impedance shunts the Z_1 and expression is complicated
- oscillation frequency is beyond audio range and so simple low-frequency *h*-parameter model is not valid and a high-frequency hybrid Π model is to be employed.

For a Colpitt oscillator, the frequency of operation can be put as

$$-\frac{1}{\omega C_1} - \frac{1}{\omega C_2} + \omega L = 0 \quad ...(11.26)$$

or,

$$f = 1/[2\pi\sqrt{(LC)}] \quad ...(11.27)$$

where,

$$C = C_1C_2/(C_1 + C_2) \quad ...(11.28)$$

and the condition of oscillation is

$$A_v > C_1/C_2 \quad ...(11.29)$$

Colpitt oscillator finds large number of applications within a frequency range of 1 MHz to 500 MHz. It is generally used commercially as signal generator above 1 MHz frequency.

Hartley oscillator: Hartley oscillator is used a great deal in transistor radios and other receivers. For a Hartley oscillator, the frequency of operation can be put as

$$-\frac{1}{\omega C} + \omega L_1 + \omega L_2 = 0 \quad ...(11.30)$$

or,

$$f = 1/[2\pi\sqrt{(LC)}] \qquad ...(11.31)$$

where,

$$L = L_1 + L_2 + 2M \qquad ...(11.32)$$

$$M = \sqrt{L_1 L_2} \qquad ...(11.33)$$

and M is the mutual inductance between the two inductances.

And, the condition of oscillation is

$$A_v > L_2/L_1 \qquad ...(11.34)$$

or, considering the mutual inductance, it can be given as

$$A_v > (L_2 + M)/(L_1 + M) \qquad ...(11.35)$$

The transistor versions of both the circuits are as shown in Fig. 11.6. The procedure of analysis of such circuits is in similar lines as discussed above, but is very elaborate and complex as the input impedance of the BJT amplifier shunts with the Z_1. Also the range of frequency operation is above audio-frequency range. The situation is more complex when the high-frequency oscillators are to be analysed since hybrid Π model is supposed to be used and low-frequency h-parameter model is no more valid. If the performance of the two Hartley and Colpitt oscillators is same, Colpitt oscillator is preferred to Hartley since bulky inductances can be avoided and also the problem with mutual inductance can be neglected.

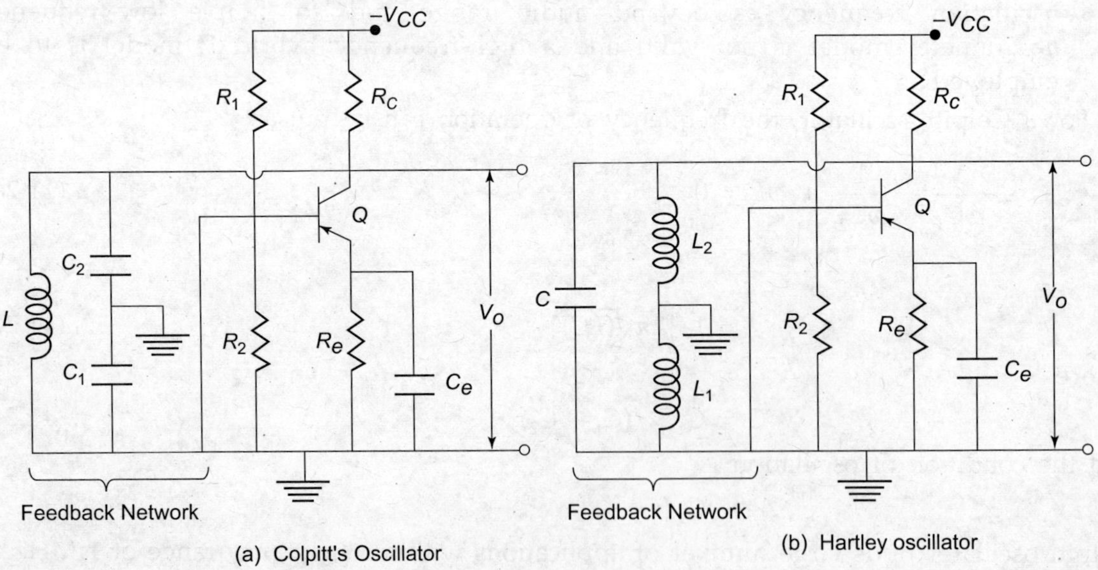

Fig. 11.6 *Transistor Colpitt and Hartley oscillators*

The advantages of Hartley oscillator are the following:

- Instead of two separate coils L_1 and L_2, a single coil of bare wire can be used and coil grounded at any desired point along it
- Frequency can be adjusted by variable capacitor C.

Clapp oscillator: Clapp oscillator also finds same applications as the Colpitt oscillator but its stability is very large compared to Colpitt oscillator. Thus, in practice, Clapp oscillator is used rather than Colpitt oscillator.

The drawback of Colpitt oscillator is that its frequency stability is very poor. This is because of the frequency of oscillation depends on the capacitances C_1 and C_2. But these two capacitances are shunted by the stray capacitances of the transistor. So, the output frequency is not the same as that of the designed value and also not stable since stray capacitance of a transistor is not stable. Thus, the frequency stability of Colpitt oscillator is very poor. This drawback is overcome by use of one more capacitor C_3 in series with the inductor and choosing C_3 to be very small compared to either C_1 or C_2. Thus, if $C_3 \ll C_1$, C_2, then the stability of the circuit is improved to a very large extent since the frequency of oscillation is only dependent on C_3 but not on other capacitors or stray capacitances of the transistor.

As shown in Fig. 11.7, the circuit of Clapp oscillator has an additional capacitor C_3 in series with the inductor. Then, assuming that all the three capacitors are in series, the frequency of oscillation can be given as

$$f = 1\Big/\left[2\pi\sqrt{(LC)}\right]$$

where,

$$\frac{1}{C_1} + \frac{1}{C_2} + \frac{1}{C_3} = \frac{1}{C} \qquad \text{...(11.36)}$$

when $C_3 \ll C_1, C_2, C \approx C_3$ and thus

$$f = 1/[2\pi\sqrt{(LC_3)}] \qquad \text{...(11.37)}$$

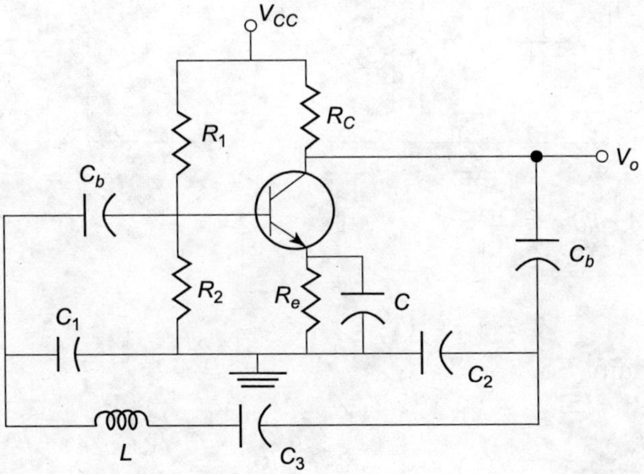

Fig. 11.7 *Clapp oscillator*

11.6 WIEN BRIDGE OSCILLATOR

An oscillator that uses a balanced bridge in the feedback path is another example of the sinusoidal oscillator. *Wien bridge* is ideally suited for this purpose and the frequency stability of the circuit is very much improved compared to the previous types of oscillators. Since the transfer gain of the amplifier with feedback tends to infinity for an oscillator, this condition can be achieved when the input to the oscillator tends to zero and the output is still a finite sinusoid. The two terminals of the bridge, which are equipotential in the balanced condition, are returned to the input of the active device, an Op-Amp. The bridge offers equipotential between the two terminals only when it is balanced and it is balanced when the condition on the arm impedances is satisfied. This condition on arm impedances is once again only satisfied at a particular frequency since capacitive reactances are involved. That is, the frequency of oscillation is the frequency at which the bridge is balanced.

The circuit of a *Wien bridge oscillator* is as shown in Fig. 11.8. The active device referred here is Op Amp which has infinite input impedance. Also Op Amp is assumed to have very large voltage gain. In addition, the output impedance of the device is negligible. Thus, voltage gain of the amplifier is constant since loading of the device is neglected. This is true over entire range of frequencies of interest. The loop gain of the circuit can be determined by breaking the loop at P and voltage V_o' is assumed to be applied between 3 and 4. The loop gain of the circuit can thus be calculated as

$$\text{Loop gain} = V_o/V_o' = A_v V_i/V_o' = -\beta A \quad ...(11.38)$$

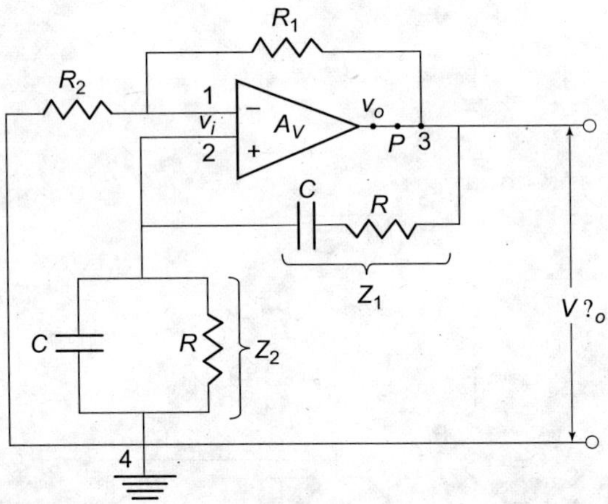

Fig. 11.8 *Wien bridge oscillator*

From Fig. 11.8,

$$V_i = V_2 - V_1 \quad ...(11.39)$$

and $A = A_v$, positive gain. So,

$$-\beta = V_i/V_o' = (V_2 - V_1)/V_o'$$

$$= \frac{Z_2}{(Z_1 + Z_2)} - \frac{R_2}{(R_1 + R_2)} \quad ...(11.40)$$

Z_1 is the series combination of R and C, and Z_2 is the parallel combination of R and C. Evaluating the above expression and equating the imaginary portion of the expression to zero since the phase of the expression is required to be zero for sustained oscillations, the frequency of oscillation can be derived to be

$$f = 1/[2\pi RC] \quad ...(11.41)$$

At this frequency,

$$Z_1 = (1 - j)R \quad \text{and} \quad Z_2 = (1 - j)R/2 \qquad \text{...(11.42)}$$

or,

$$V_2 = (1/3)V'_o$$

If a null is desired, $V_i = 0$ implies

$$R_2/(R_1 + R_2) = 1/3 \text{ or } R_1 = 2R_2 \qquad \text{...(11.43)}$$

But in this discussion, when the bridge is used as feedback network, the loop gain should be unity and the phase should be zero. Since A_v is positive, phase of $-\beta$ should be zero, but the magnitude should not be zero. If magnitude is zero, the loop gain magnitude would be zero and the oscillations are not at all possible. Since $Z_2/(Z_1 + Z_2) = 1/3$, the magnitude of $R_2/(R_1 + R_2)$ should be smaller than 1/3. Let us assume that δ is a positive number greater than 3 such that

$$V_1/V'_o = R_2/(R_1 + R_2) = (1/3) - (1/\delta) \qquad \text{...(11.44)}$$

Then,

$$-\beta = \frac{V_2 - V_1}{V'_o} = \frac{V_2}{V'_o} - \frac{1}{3} + \frac{1}{\delta} \qquad \text{...(11.45)}$$

or, at balance condition, $V_2 = (1/3)\ V'_o$ and so,

$$-\beta = 1/\delta \qquad \text{...(11.46)}$$

which implies from the fact that $-A\beta = 1$,

$$\delta = A_v \qquad \text{...(11.47)}$$

The frequency of oscillation is precisely the null frequency of the balanced bridge. At any other frequency, V_2 is not in phase with V'_o and, therefore, V_i is not in phase with V_o' and the condition of $-A\beta = 1$ is not satisfied. Varying simultaneously the two capacitors in gang can vary the frequency of oscillation. If the frequency range is to be changed, this can be accomplished by varying the two resistors R.

11.7 CRYSTAL OSCILLATOR

A very highly stable frequency signal can be generated with the help of a *crystal oscillator.* A crystal oscillator also utilises the principle of general oscillator circuit similar to that of Colpitt or Hartley oscillator except that instead of inductance, a *piezoelectric crystal* is employed. Such type of oscillators can generate very high frequencies, of the order of few megahertz. The drawback of the circuit is that the frequency of the sinusoidal signal generated is not variable. The oscillator can oscillate only at the natural resonant frequency of the crystal chosen and no oscillations are possible at any other frequency whatever may the changes of the other components of the circuit. So, the frequency stability of the circuit is high. If a change in frequency is required, one has to replace the crystal with the desired natural resonant frequency. Such oscillators find large applications in the communication equipment, for example, in the receiver, where very high stable frequency oscillations are required.

Piezoelectric effect is an elctromechanical phenomenon. Some natural crystals available exhibit such electromechanical effect. If mechanically the crystal is vibrated, it develops an ac electric signal variation of natural resonant frequency at the ends of the faces of the crystal. Or, if an ac signal is applied across the faces of the crystal, the crystal vibrates mechanically. Crystals such as Rochelle salt, quartz and tourmaline exhibit such piezoelectric effect. The natural resonant frequency of the crystal depends on the dimensions of the crystal as well as mechanical orientations of the crystal structure. The frequency is inversely proportional to the thickness of the crystal and so as the frequency of oscillation increases, the thickness of the crystal decreases. The mechanical stability of the crystal as well as the sensitivity of the crystal are vital in selection of crystal in sinusoidal oscillators. Mechanically, *Rochelle salts* are very weak even though their sensitivity is very high. Rochelle salts have greatest effect; for a given ac voltage they vibrate more than others, but they are mechanically the weakest and break easily. They are used in microphones, phonograph pick-ups, headsets, loudspeakers and so on. But, they are not adoptable in these oscillator circuits. On the other hand, *tourmaline* has high mechanical stability but very low sensitivity. Tourmaline is most expensive and is used in general at very high frequencies. A compromise is the *quartz crystal* whose mechanical stability and sensitivity are medium and reasonably good. Also, quartz is readily available. So, Quartz is very widely used as crystal in sinusoidal oscillators, filters and so on.

Natural shape of quartz is hexagonal with pyramids at ends. To get a usable crystal, it has to be recast into rectangular slab slices. There are different cuts employed in this process like *X* cut, *Y* cut, *AT* cut and so on. The frequency of the crystal is inversely proportional to the thickness of the slice. The resonant frequency and quality factor depend on the dimensions of the crystal, surface orientation with respect to axis, and orientation of the mount. Quartz crystals are available from few kHz to few MHz and quality factor from several 1000s to 100s of thousands. These extraordinary high values of the crystal *Q* make the crystal oscillator extremely stable with respect to time and temperature. The electric symbol, electric equivalent and the reactance characteristic of a quartz crystal are as shown in Fig. 11.9.

A look at the figure shows that the crystal has series resonance as well as parallel resonance. The three elements *R*, *L* and *C* are the characteristics of the natural crystal and C¢ is the capacitance of the two electrodes supporting the crystal. It can be observed easily

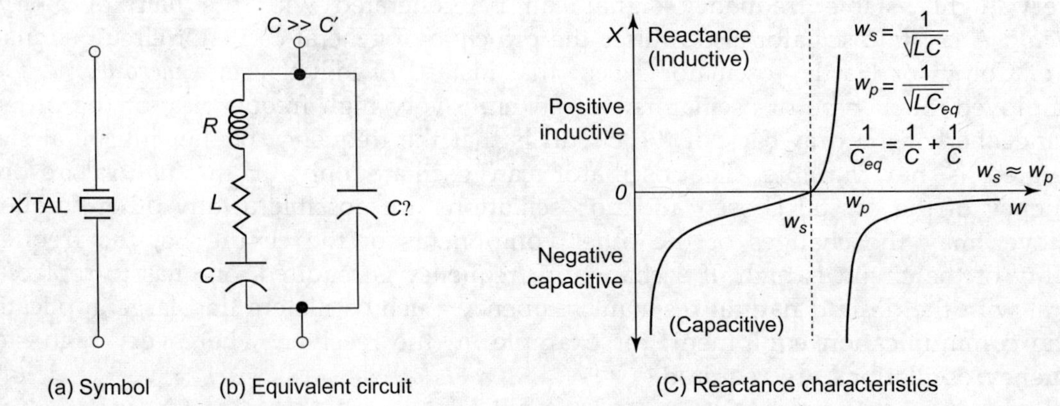

Fig. 11.9 *A piezoelectric crystal*

from the reactance characteristic that the crystal is capacitive up to series resonant frequency and also above a parallel resonant frequency. Between the two resonant frequencies, the crystal exhibits inductive reactance characteristics. The difference between the series and parallel resonant frequencies is extremely small and in the figure, it has been exaggerated for the purpose of understanding. In other words, the two resonant frequencies are too close such that one can say that almost the crystal exhibits inductive characteristic at a single frequency. This is the reason for a very high value of the quality factor Q. Since Q is extremely high, the bandwidth is very small and so, highly stable oscillations are possible. In practice, R is of the order of few kiloohms, L is of the order of few Henrys and value of C the is very large compared to C.

The impedance of the crystal can be derived as (neglecting the loss factor)

$$jX = -\frac{1}{\omega C'}\frac{\omega^2 - \omega_s^2}{\omega^2 - \omega_p^2} \qquad \text{...(11.48)}$$

where ω_S is the series resonant frequency taking R, L and C in series into consideration and ω_p is the resonant frequency taking into parallel electrode capacitance. The values of the two resonant frequencies can be given as

$$\omega_s^2 = 1/(LC) \text{ and } \omega_p^2 = (1/L)\,(1/C + 1/C') \qquad \text{...(11.49)}$$

since $C \gg C'$, $\omega_s \approx \omega_p$. Thus, when $\omega_s < \omega < \omega_p$, the reactance is inductive and outside it is capacitive.

Figure 11.10 shows a typical example of a crystal oscillator that employs a 1 MHz quartz crystal. In this circuit, which is basically a Hartley oscillator, Z_2 is the output tank circuit, which is a tank circuit tuned inductive, Z_3 is the stray drain to gate capacitance of the FET and Z_1 is the quartz crystal. So, the crystal only oscillates when Z_1 or crystal is inductive, that is, at resonant frequency of the crystal and nowhere else. This is possible since $\omega_s \ggg \omega_p$ and the frequency of oscillation is essentially determined only by this but not by any other parameter of the circuit.

The stability of a quartz crystal would be around 1 part in 10^6, that is, 0.0001% per day. Such crystals are employed in wristwatches. The stability of a crystal oscillator would be around 1 part in 10^{10} per day and so the variation is very low, there would be a variation of 1 in 300 years. A crystal oscillator is highly stable and is available up to 10 MHz of frequencies. However, the disadvantages are that the frequency cannot be varied and can be used only for low-power applications.

11.8 AMPLITUDE AND FREQUENCY STABILITY

Amplitude stabilization: Amplitude stabilisation of the sinusoidal oscillator is required to maintain the loop gain magnitude constant. With respect to temperature, ageing of the transistor and other components, even though the ratio, β is assumed to be constant, A is to be stabilized. The stabilization of this amplification parameter can be achieved by various methods. Use of temperature-sensitive resistors such as a sensistor or a thermistor is the best choice.

Considering the circuit of a Wien bridge oscillator, Fig. 11.9, a sensistor, a resistor with positive temperature coefficient, can replace R_2 in the circuit. The sensistor is a device

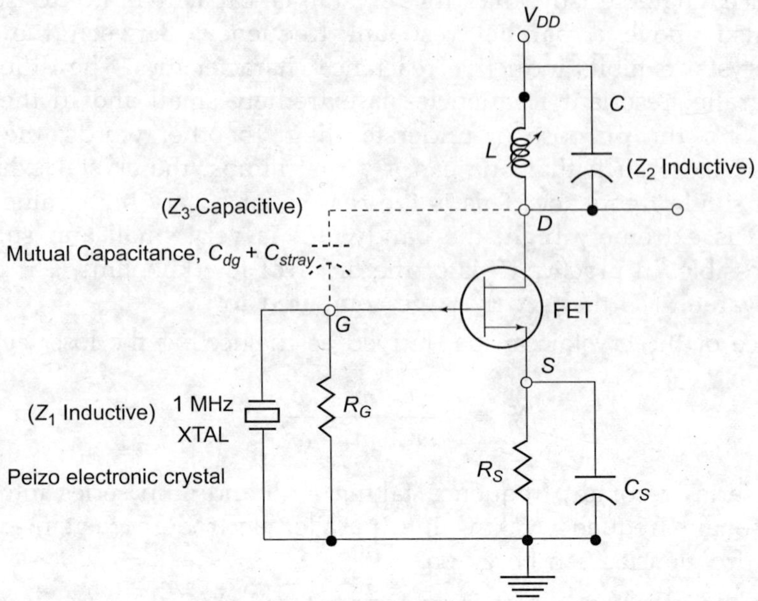

Fig. 11.10 *Crystal oscillator*

whose resistance increases with temperature. At room temperature, it operates as a simple resistor. When temperature increases, as one can anticipate, A increases, making the loop gain magnitude more than unity. This can be checked by the regulation mechanism of the sensistor. Since the resistance of the sensistor increases, β decreases and the loop gain is stabilised.

On the other hand, when temperature decreases, the amplification A decreases and because the resistance of sensistor decreases, β increases, compensating the decrease in the amplification and stabilising the loop gain $- A\beta$.

If a device that has negative temperature coefficient like a thermistor is to be employed, it should replace R_1 and same kind of stabilisation can be anticipated and the loop gain can be maintained almost constant at unity.

Frequency Stabilization: In any sinusoidal oscillator, the frequency of oscillation is determined by the condition that the phase shift around the loop should be zero or multiples of 2π radians. So, there is a possibility that the frequency of an oscillator may change erratically since the phase around the loop depends on the performance of various components in the circuit. The stability of oscillations in frequency is of prime concern in the design of any circuit and *frequency stability* of an oscillator can be defined as a measure of its ability to maintain as nearly fixed as possible over as long a time interval as possible. The values of the circuit parameters cannot be maintained constant with respect to time since, for example, with respect to a transistor, the parameters are time and temperature dependent. So, to achieve frequency stability by maintaining the component values constant is almost impossible. Also, the stray component values are a great problem at estimating the exact values of the components involved in the circuit. The solution is to make the frequency condition as far as possible to depend not on the temperature of time-dependent components, but on the passive components such as a resistor or a capacitor. For example, when a phase-shift oscillator is considered, the frequency of oscillation depends on the resistor and capacitor values, which are almost constant irrespective of time and temperature.

The frequency stability of the circuit depends on the ability of each component to maintain constant phase irrespective of frequency variation. Thus, variation of phase of the component θ with respect to frequency ω, that is, $d\theta/d\omega$ serves as a measure of the independence of the frequency of all other features of the circuit. The frequency stability improves as $d\theta/d\omega$ increases. In the limit $d\theta/d\omega$ tends to infinity. In other words, whatever may be the phase variation, the frequency variation is zero or negligible, the circuit becomes independent of these variations and the frequency of oscillation is maintained constant.

SUMMARY

- Sinusoidal oscillators are electronic circuits that produce oscillations of sinusoidal nature on their own.
- Sinusoidal oscillators work on the principle of Barkhausen criterion.
- Barkhausen criterion states that oscillations are possible only when the loop gain is unity or more.
- The frequency of oscillation of an oscillator is determined on the condition that phase around the closed loop should be zero or integral multiples of 2π radians.
- *RC* phase-shift oscillators can be built using either BJT or FET. These oscillators are useful at low frequencies. No inductors or chokes or transformers are utilised in these circuits.
- Resonant circuit oscillators operate at the resonant frequency of the tank circuit. The coupling between the output and input is via a *RF* transformer.
- At high frequencies, Hartley and Colpitt oscillators are used. Hartley oscillator has the drawback of mutual inductance and losses with bulky inductors. Colpitt oscillator's frequency stability is poor due to shunting of stray capacitance with feedback network.
- Clapp oscillator relaxes the limitation of Colpitt oscillator by the use of additional capacitor in series with the inductor.
- Wien bridge oscillator employs a balance bridge in the feedback circuit. Its amplitude stability is good.
- Piezoelectric effect is an electromechanical phenomenon. Natural crystal like quartz exhibits this effect. Mechanical vibrations are converted to electric oscillations or vice versa.
- Quartz crystals are employed in sinusoidal oscillators. These oscillators have extreme frequency stability and are available up to large frequencies of MHz range.

SOLVED PROBLEMS

11.1 Derive an expression for frequency of oscillation of a *RC* phase-shift oscillator using a transistor.

Solution In case of a transistor phase-shift oscillator voltage-shunt feedback is employed. The circuit from which loop gain is going to be calculated is as shown below:

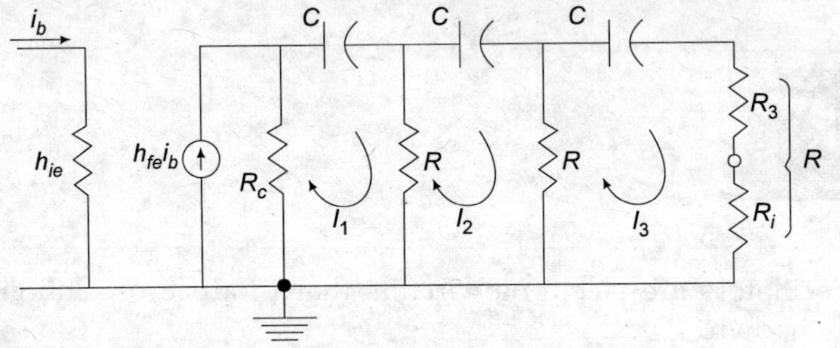

Assume $R_{TH} = R_1 || R_2 \gg R_i$ and $h_{oe} = h_{re} = 0$

For unity loop gain we need $i_b = I_3$.

Replacing the current source by its Thevenin's equivalent, we will get the circuit as below.

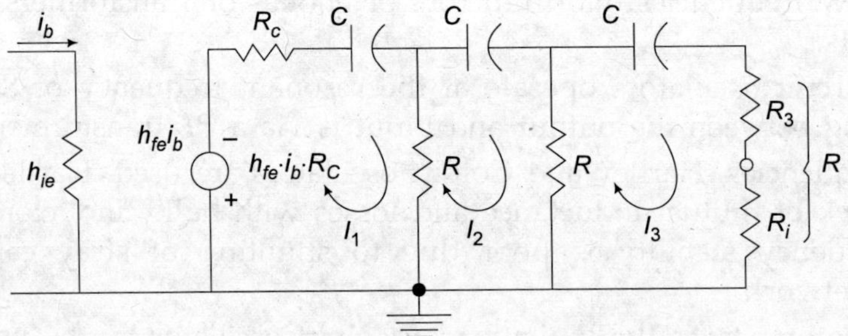

Writing three KVL equations,

$$-h_{fe} \cdot i_b \cdot R_C = (R + R_C - JX_C)I_1 - RI_2$$

$$O = - RI_1 + (2R - JX)I_2 - RI_3$$

$$O = - RI_2 + (2R - JX)I_3$$

These can be arranged in matrix form as

$$\begin{bmatrix} (R+R_C-JX) & -1 & 0 \\ -R & (2R-JX) & -R \\ 0 & -R & (2R-JX) \end{bmatrix} \begin{bmatrix} I_1 \\ I_2 \\ I_3 \end{bmatrix} = \begin{bmatrix} -h_{fe}i_bR_C \\ 0 \\ 0 \end{bmatrix}$$

where $X = \dfrac{1}{\omega C}$. Let $\alpha = \dfrac{1}{\omega RC}$ and $K = R_C/R$.

$$\Delta = \begin{vmatrix} R+R_C-JX & -1 & 0 \\ -R & 2R-JX & -R \\ 0 & -R & 2R-JX \end{vmatrix}$$

$$= R^3 \begin{vmatrix} 1+K-J\alpha & -1 & 0 \\ -1 & 2-J\alpha & -1 \\ 0 & -1 & 2-J\alpha \end{vmatrix}$$

$$= R^3(1 + K - J\alpha)[(2 - J\alpha)^2 - 1] + 1(-1(2 - J\alpha))$$

$$= R^3[(1 + K - J\alpha)[(4 - \alpha^2 - J4\alpha - 1) - 2 + J\alpha]$$

$$= R^3[(1 + K - J\alpha)[(3 - \alpha^2 - J4\alpha) - 2 + J\alpha]$$

$$= R^3[1 + 3K - (5 + K)\alpha^2] - J[(6 + 4k)\ \alpha - \alpha^3]$$

$$\Delta_3 = \begin{vmatrix} R+R_C-JX & -1 & -h_{fe}i_bR_C \\ -R & 2R-JX & 0 \\ 0 & -R & 0 \end{vmatrix}$$

$$= -h_{fe}i_bR_C\ (R^2)$$

$$= -h_{fe}\cdot i_b\cdot R^2\cdot R_C$$

$$\therefore \qquad I_3 = \frac{\Delta_3}{\Delta} = \frac{-h_{fe}i_bR^2R_C}{R^3[1+3K-(5+K)\alpha^2]-J[(6+4K)\alpha-\alpha^3]}$$

$$\therefore \qquad \frac{I_3}{i_b} = \frac{-h_{fe}\cdot R_C/R}{[1+3K-(5+K)\alpha^2]-J[(6+4K)\alpha-\alpha^3]} \qquad \text{...(1)}$$

To find frequency of oscillation, phase-shift should be zero.

For phase-shift to be zero, imaginary term is to be zero.

$$(6 + 4K)\alpha - \alpha^3 = 0$$

$$\alpha^2 = (6 + 4K)$$

$$\Rightarrow \qquad \alpha = \sqrt{6+4K}$$

$$\Rightarrow \qquad \frac{1}{\omega RC} = \sqrt{6+4K}$$

$$\omega = \frac{1}{RC\sqrt{6+4K}}$$

$$f = \frac{1}{2\pi RC\sqrt{6+4K}}$$

Substituting $\alpha = \sqrt{6+4K}$ in equation (1)

$$1 = \frac{-h_{fe}\cdot K}{1+3K-(5+K)\alpha^2}$$

$$= \frac{-h_{fe}\cdot K}{1+3K-(5+K)(6+4K^2)}$$

$$= \frac{-h_{fe}\cdot K}{1+3K-(30+26K+4K^2)}$$

$$\therefore \qquad -h_{fe}K = -29 - 23K - 4K^2$$

$$h_{fe} = 4K + 23 + 29/K$$

$\therefore h_{fe} > 4K + 23 + 29/K$ for sustained oscillations. To find minimum value of h_{fe}.

$$\frac{d}{dK}(h_{fe}) = 0$$

$$\Rightarrow \qquad 4 - \frac{29}{K^2} = 0 \Rightarrow K = \sqrt{29/4} = 2.693$$

$$\therefore \qquad h_{fe\,\min} = 4 \times 2.693 + 23 + \frac{29}{2.693}$$

$$= 44.54$$

So, in order to get sustained oscillations, a transistor with h_{fe} less than 45 should not be used.

11.2 Derive an expression for frequency of oscillation of a FET based *RC* phase-shift oscillator.

Solution A FET based *RC* phase-shift oscillator is as shown in the figure. It uses voltage-series feedback. The circuit used for the derivation of frequency of oscillation is as shown below:

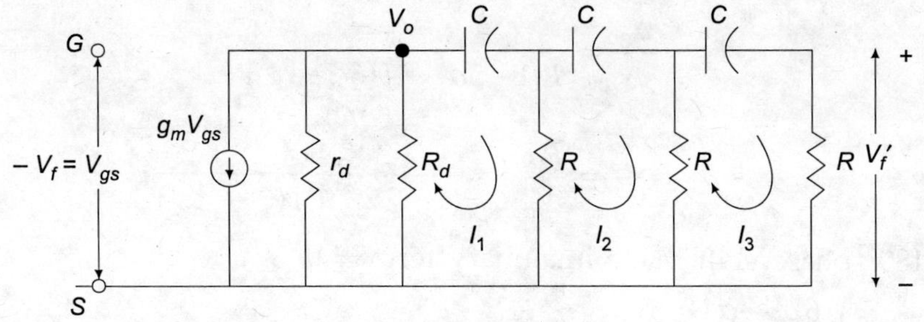

Writing KVL equations in meshes (1), (2), (3), letting $X = 1/\omega C$

$$V_o = (R - JX)I_1 - RI_2$$

$$O = -RI_1 + (2R - JX)I_2 - I_3R$$

$$O = -RI_2 + (2R - JX)I_3$$

These can be written in matrix form as below.

$$\begin{bmatrix} V_o \\ 0 \\ 0 \end{bmatrix} = \begin{bmatrix} R - JX & -R & 0 \\ -R & 2R - JX & -R \\ 0 & -R & 2R - JX \end{bmatrix} \begin{bmatrix} I_1 \\ I_2 \\ I_3 \end{bmatrix}$$

$$= R^3 \begin{bmatrix} 1 - J\alpha & -1 & 0 \\ -1 & 2 - J\alpha & -1 \\ 0 & -1 & 2 - J\alpha \end{bmatrix} \begin{bmatrix} I_1 \\ I_2 \\ I_3 \end{bmatrix}$$

where $\alpha = \dfrac{1}{\omega RC} = \dfrac{X}{R}$

$$\Delta = R^3 \begin{vmatrix} 1 - J\alpha & -1 & 0 \\ -1 & 2 - J\alpha & -1 \\ 0 & -1 & 2 - J\alpha \end{vmatrix}$$

$$= R^3\{[((1 - J\alpha)(2 - J\alpha)^2 - 1)] + 1\ (-2 + J\alpha)\}$$

$$= R^3[1 - 5\alpha^2 + J(\alpha^3 - 6\alpha)]$$

$$\Delta_3 = R^3 \begin{vmatrix} 1 - J\alpha & -1 & \dfrac{V_o}{R} \\ -1 & 2 - J\alpha & 0 \\ 0 & -1 & 0 \end{vmatrix}$$

$$= V_o \cdot R^2$$

$$I_3 = \frac{\Delta_3}{\Delta} = \frac{V_o}{R[1-5\alpha^2 + J(6\alpha - \alpha^3)}$$

$$-\beta = \frac{V_f}{V_o} = \frac{I_3 R}{V_o} = \frac{1}{(1-5\alpha^2) + J(6\alpha - \alpha^3)}$$

For 180° phase shift make imaginary term = 0.

$$6\alpha - \alpha^3 = 0$$

$$\alpha^2 = 6 \Rightarrow \alpha = \sqrt{6}$$

$$\frac{1}{\omega RC} = \sqrt{6} \Rightarrow f = \frac{1}{2\pi RC\sqrt{6}}$$

At this frequency of oscillation $\frac{V_f}{V_o} = \frac{1}{1-5(6)} = \frac{1}{29}$

$$\left|\frac{V_f}{V_o}\right| = \frac{1}{29} = \beta$$

We want $|-A\beta| \geq 1$ to get sustained oscillations

$$\left|A \cdot \frac{1}{29}\right| \geq 1$$

$$\Rightarrow \qquad |A| \geq 29$$

∴ The gain of the amplifier must be at least 29.

$A = g_m \cdot \frac{r_d \cdot R_d}{r_d + R_d}$. By using voltage division

$$\therefore \quad \frac{\mu \cdot R_d}{r_d + R_d} \geq 29.$$

11.3 For the tuned-drain oscillator, show that the frequency of oscillation is given by $\omega^2 = \frac{1}{LC}\left(1 + \frac{r}{r_d}\right)$ and that for the loop gain to be equal to unity, the FET must have mutual conductance given by $g_m = \frac{\mu rc}{\mu M - L}$ where μ is amplification factor, r is equivalent series resistance of the coil primary, M is the mutual inductance of the coil.

Solution

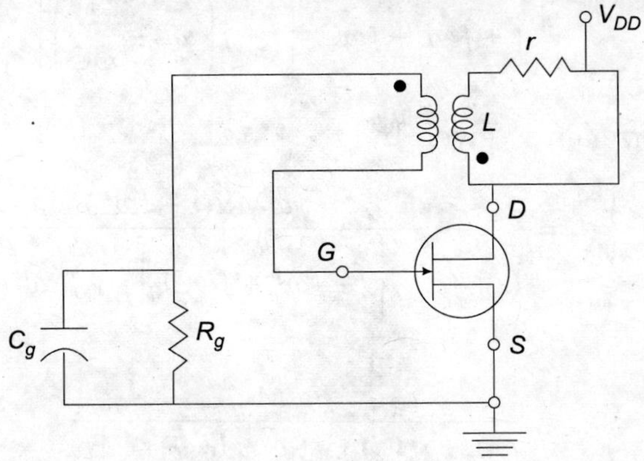

The equivalent diagram is as shown below, replacing FET with its equivalent circuit.

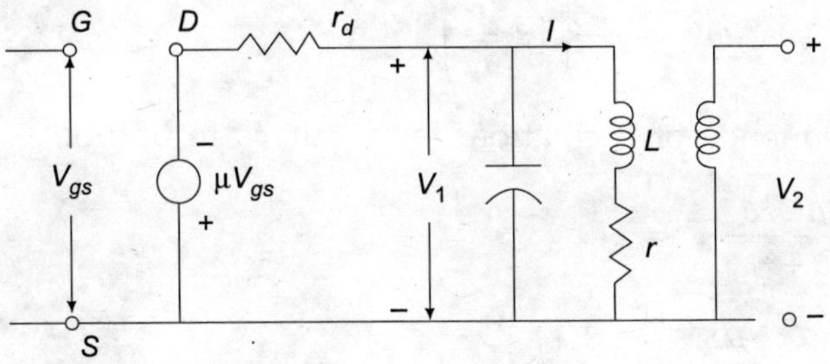

For unity loop gain $V_2 = V_{gs}$

$$V_2 = -J\omega M I = -\frac{J\omega M \cdot V_1}{r + J\omega L}$$

$$V_1 = \frac{-\mu V_{gs} \cdot z}{r_d + z}$$ where z is the impedance of the tuned circuit.

$$z = \frac{(r + J\omega L)\dfrac{1}{J\omega C}}{r + J\omega L + \dfrac{1}{J\omega C}}$$

$$= \frac{r + J\omega L}{1 + J\omega C(r + J\omega L)}$$

$$V_1 = \frac{-\mu V_{gs}(r + J\omega L)}{(r + J\omega L) + r_d[1 + J\omega C(r + J\omega L)]}$$

$$= \frac{-\mu V_g (r + J\omega L)}{r + J\omega L + J\omega C r_d \left[r + J\left(\omega L - \frac{1}{\omega C} \right) \right]}$$

Now, the loop gain is given by

$$\frac{V_2}{V_g} = \frac{J\omega M \mu (r + J\omega L)}{(r + J\omega L)\left\{ r + J\omega L + J\omega C r_d \left[r + J\left(\omega L - \frac{1}{\omega C} \right) \right] \right\}}$$

$$= \frac{\mu \omega M}{\omega(L + r r_d C) + J(\omega^2 r_d LC - r_d - r)}$$

The loop-phase shift is zero when, imaginary part = 0.

$$\omega^2 LC r_d = r_d + r$$

$$\omega^2 = \frac{1}{LC}\left(1 + \frac{r}{r_d}\right)$$

at this frequency of oscillation, loop gain is to be unity

$$\frac{\mu \omega M}{\omega(L + r r_d C)} = 1$$

$$\mu M = L + r r_d C$$

$$\mu M - L = r \cdot rd \cdot C = \frac{r \cdot \mu \cdot C}{g_m}$$

$$g_m = \frac{\mu r C}{\mu M - L}$$

Hence proved.

11.4 (a) A RC phase shift oscillator uses the feedback network shown. Prove that

$$\frac{V_f'}{V_o} = \frac{1}{3 + J\left(\omega RC - \frac{1}{\omega RC} \right)}$$

(b) Show that the frequency of oscillation is $f = \frac{1}{2\pi RC}$ and that the gain must exceed 3.

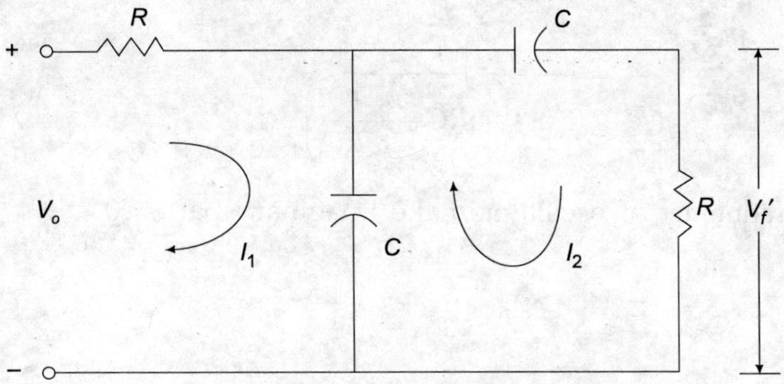

Solution Let $X = \dfrac{1}{\omega C}$

Writing KVL equations in two meshes.

$$V_o = (R - JX)I_1 + JXI_2$$

$$0 = +JXI_1 + (R - J2X)I_2$$

$$\Delta = \begin{vmatrix} R - JX & JX \\ JX & R - J2X \end{vmatrix}$$

$$= (R - JX)\ (R - J2X) + X^2$$

$$= R^2 - 2X^2 + X^2 + J(-3XR)$$

$$= R^2 - X^2 + J(-3XR)$$

$$\Delta_2 = \begin{vmatrix} R - JX & V_o \\ JX & 0 \end{vmatrix} = -JXV_o$$

$$I_2 = \frac{\Delta_2}{\Delta} = \frac{-JXV_o}{(R^2 - X^2) - J(3XR)}$$

$$V_f' = I_2 R = \frac{-JXV_oR}{(R^2 - X^2) - J(3XR)} \qquad \text{...(1)}$$

$$\frac{V_f'}{V_o} = \frac{-JXR}{(R^2 - X^2) - J3XR}$$

$$= \frac{1}{3 - \dfrac{R^2 - X^2}{JXR}} = \frac{1}{3 + J\left(\dfrac{R}{X} - \dfrac{X}{R}\right)}$$

$$= \frac{1}{3 + J\left(\omega RC - \frac{1}{\omega RC}\right)}$$

To get frequency of oscillator make imaginary part = 0.

$$\Rightarrow \omega RC - \frac{1}{\omega RC} = 0$$

$$\therefore \qquad f = \frac{1}{2\pi RC}$$

At this frequency of oscillation $\frac{V_f'}{V_o} = \frac{1}{3}$.

$$\therefore \qquad |A\beta| \geq 1$$

$$\left|A\frac{1}{3}\right| \geq 1$$

$$A \geq 3$$

$\therefore$ to get sustained oscillations A should be more than 3.

11.5 (a) Derive an expression for impedance of a crystal.

(b) Prove that the ratio of the parallel to series resonant frequencies is given by $1 + \frac{1}{2}\frac{C}{C'}$.

(c) If C = 0.02 pF and C′ = 4.0 pF, by what per cent is the parallel-resonant frequency greater than the series resonant frequency?

Solution (a) The electrical equivalent of a crystal is given by

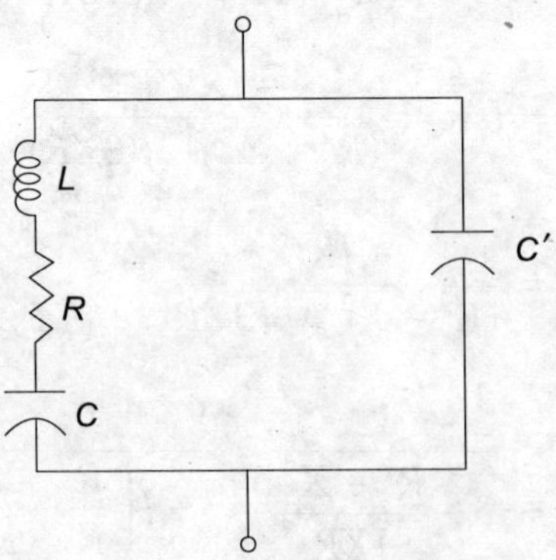

Neglecting R, the impedance is a reactive one,

$$z = JX = \frac{\left(J\omega L + \frac{1}{J\omega C}\right)\cdot\frac{1}{J\omega C'}}{J\omega L + \frac{1}{J\omega C} + \frac{1}{J\omega C'}}$$

$$= \frac{(1-\omega^2 LC)}{J\omega C' + J\omega C - J\omega^3 LCC'}$$

$$= \frac{-J\cdot 1 - \omega^2 LC}{\omega(C + C' - \omega^2 LCC')}$$

$$= \frac{-JLC\left(\frac{1}{LC} - \omega^2\right)}{\omega LCC'\left(\frac{C+C'}{LCC'} - \omega^2\right)}$$

$$= \frac{\frac{-J}{\omega C'}\cdot\left(w^2 - \frac{1}{LC}\right)}{\left(w^2 - \frac{1}{\frac{LCC'}{C+C'}}\right)}$$

Let $\omega s = \frac{1}{\sqrt{LC}},\ \omega_p = \frac{1}{\sqrt{L\frac{LCC'}{C+C'}}}$

$\therefore$ $JX = \frac{-J}{\omega C'}\cdot\frac{\omega^2 - \omega_s{}^2}{\omega^2 - \omega_p{}^2}$

(b) $$\frac{\omega_p}{\omega_s} = \frac{\frac{1}{\sqrt{L\frac{CC'}{C+C'}}}}{\frac{1}{\sqrt{LC}}} = \sqrt{\frac{\not{L}\not{C}(C+C')}{\not{L}\not{C}C'}} = \sqrt{1+\frac{C}{C'}}$$

$$= \left(1+\frac{C}{C'}\right)^{1/2} = 1+\frac{1}{2}\cdot\frac{C}{C'}+\cdots$$

Neglecting higher-order terms,

$$\frac{\omega_p}{\omega_s} \approx 1 + \frac{1}{2}\frac{C}{C'}$$

(c) If C = 0.04 pF and C′ = 2 pF, then

$$\frac{1}{2}\cdot\frac{C}{C'} \times 100\% = \text{per cent value.}$$

$$\Rightarrow \frac{1}{2}\times\frac{0.02}{4} \times 100\% = 0.25\%$$

11.6 A crystal has the following dimensions: L = 0.25 h, C = 2 pF, C′ = 15.38 pF and R = 4 KΩ.

(a) Find the series-resonant frequency.

(b) Find the Q of the crystal.

(c) By what per cent does the parallel-resonant frequency exceed series resonant frequency?

Solution (a)

$$\omega_s = \frac{1}{\sqrt{LC}} = 1.414 \text{ MHz}$$

$$f_s = \frac{W_S}{2\pi} = 0.225 \text{ MHz}$$

(b)

$$Q = \frac{\omega_s L}{R}$$

$$= \frac{1.414\times 10^6 \times 0.25}{4\times 10^3} = 88.375$$

(c)

$$\frac{f_P}{f_s} = \left(1+\frac{C}{C'}\right)^{1/2} = \left(1+\frac{2}{15.38}\right)^{1/2}$$

$$= 1.06303$$

$$\therefore \left(\frac{f_p}{f_s}-1\right)\times 100 = 6.303\%$$

f_p exceeds f_s by more than 6.303%.

EXERCISE PROBLEMS

11.1 In the Hartley oscillator, L_2 = 0.04 mH and c = 0.004 μF. If the frequency of the oscillator is 150 KHz, find L_1. Neglect mutual inductance.

11.2 In a Colpitts oscillator, C_1 = 0.2 μF and C_2 = 0.04 μF. If the frequency of oscillation is 10 KHz, find the value of Inductor. Also, find the required gain for oscillation.

11.3 In an *RC* phase-shift oscillator R = 200 KΩ and C = 200 pF. Find the frequency of the oscillator (a) BJT (b) for FET.

11.4 In wien-bridge oscillator R = 200 KΩ and f = 10 KHz. Find the value of C.

11.5 A crystal has the following parameters L = 0.1 h, C= 0.05 pF, C' = 2 pF and R = 5 K. Find series and parallel resonances frequency and *Q*-factor of the crystal.

11.6 A Colpitts oscillator is designed with C_1 = 100 pF and C_1 = 7500 pF. Find the range of inductance values if the frequency of oscillation vary between 950 and 2050 KHz.

11.7 A tuned collector oscillator in a radio receiver has a fixed inductance of 60 μh and has to be tunable over the frequency band of 500 to 1500 KHz. Find the range of variable capacitor to be used.

11.8 (a) Find $\dfrac{V_f'}{V_o}$ for the circuit shown.

(b) Find the expression for the frequency of oscillation.

(c) Find the minimum gain required for oscillation.

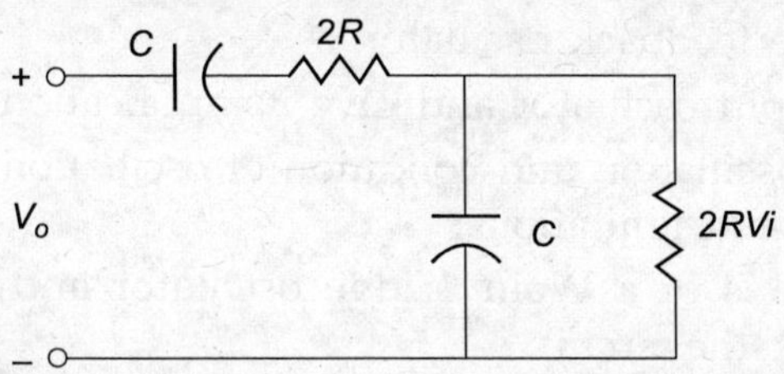

11.9 A MOSFET with μ = 60 and r_d = 4 K is used in a phase-shift oscillator.

(a) Find the minimum value of the R_d for which the circuit will oscillate.

(b) Find the product *RC*.

11.10 (a) For the Hartley oscillator shown, find the frequency of oscillation.

(b) Find the vlaue of R_s for which the value of loop gain will adjust equal unity.

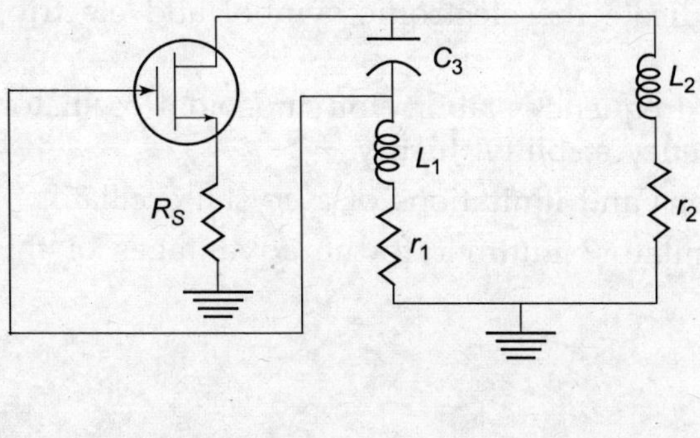

REVIEW QUESTIONS

1. State and explain the Barkhausen criterion for oscillations in a sinusoidal oscillator with feedback network.
2. Draw the circuit diagram of a *RC* phase-shift oscillator employing FET and derive the equation for the frequency of oscillation and condition of oscillation.
3. Draw the circuit diagram of a *RC* phase-shift oscillator employing BJT and derive the equation for the frequency of oscillation and condition of oscillation. Why is voltage shunt configuration employed in such a circuit?
4. What is resonant circuit oscillator? Draw its circuit employing FET and derive the condition and frequency of oscillation of the circuit.
5. Discuss the advantages and limitations of a resonant circuit sinusoidal oscillator.
6. Enumerate the advantages and applications of a *RC* phase-shift oscillator.
7. Discuss why Barkhausen criterion is only applicable to a sinusoidal oscillator. Also, give the practical limitation of applying Barkhausen criterion to the sinusoidal oscillator and how this is overcome.
8. Derive the equations for condition of oscillation and frequency of oscillation for a general format of positive feedback oscillator.
9. Draw the circuit of a Colpitt oscillator and give its applications.
10. Derive the frequency of oscillation and condition of oscillation of a Hartley oscillator and give its applications and limitations.
11. Enumerate the applications of a Wein bridge oscillator and derive the expression frequency of oscillation of the circuit.
12. Discuss about the amplitude stability of a sinusoidal oscillator and give a circuit that has maximum stability.
13. What oscillator is employed if very stable frequency oscillations are required? Why?
14. What is piezoelectric effect? What materials exhibit such effect? How is this effect useful in sinusoidal oscillators?
15. Draw the frequency response of a quartz crystal and give the expressions for resonant frequencies. Also, indicate the electronic symbol and electric equivalent circuit of such a crystal.
16. Discuss at length the frequency stability of sinusoidal oscillators. A crystal oscillator has the highest frequency stability. Justify.
17. Discuss the applications and limitations of a crystal oscillator.
18. What is a Clapp oscillator? Enumerate the advantages of this over the Colpitt or Hartley oscillators.

19. Draw the circuit of a Wein bridge oscillator using Op-Amp. Derive an equation for the condition of oscillation of a Wein bridge oscillator.
20. Draw the circuit diagram of a sinusoidal oscillator, which employs *RF* transformer for feedback using a BJT, and derive the equation for frequency of oscillation.
21. State the frequency stability criterion for sinusoidal oscillators.
22. Draw the electric model of a quartz crystal. Over which portion of the curve reactance curve the oscillations are possible?
23. Draw the circuit diagram of a crystal oscillator and explain its operation in detail.
24. What determines the frequency of oscillations in a Wien bridge oscillator?
25. Sketch the circuit of a tuned collector oscillator and explain its operation. Give the condition for oscillation and frequency of oscillation.

CHAPTER

12

Power Amplifiers

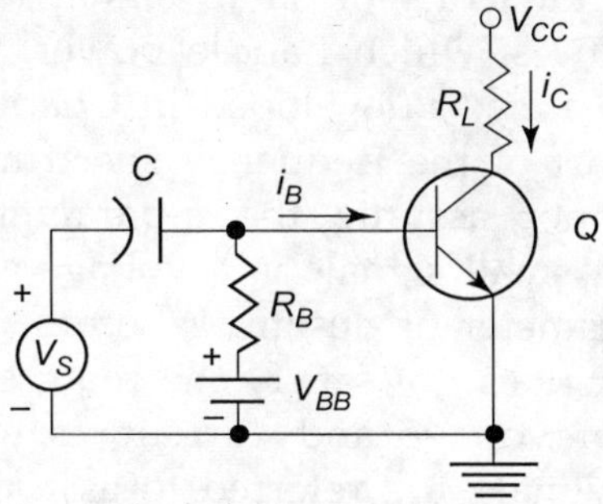

Series-fed class A amplifier

Chapter Outline

The concepts introduced in this chapter are:

- *Distortion due to harmonics*
- *Analysis of different types of class A amplifiers*
- *Class B amplifier analysis with minimum harmonic distortion*
- *Class AB amplifier configuration*
- *Different types of tuned amplifiers*
- *Instability in tuned amplifiers*

12.1 INTRODUCTION

An amplifying system consists of several stages in cascade. The input and intermediate stages amplify small signal excitations to a value large enough to drive the final device. The output stage feeds a transducer such as a cathode-ray oscilloscope, a loudspeaker, or a servomotor. Thus, the final amplifier stage must be capable of delivering a large voltage or current or appreciable amount of power. This requires an amplifier which is referred to as *power amplifier.* Power amplifiers are those amplifiers, that are the final stage of an amplifier chain. Power amplifiers find extensive use in audio frequencies. For example, in a public addressing system the amplification of the audio signal required is very large. The amplifier is to be designed to generate a very large signal at the output. This amplifier cannot be designed with the knowledge of small signal model of the transistor. So, a separate analysis of such amplifiers, which handle power, is required. The small signal linear model of the transistor as already developed in Chapter 8 is not valid. Not only at audio frequencies, but at any part of the frequency spectrum the design of the amplifier that handles large signals cannot assume the *h*-parameter model of the transistor. Basically, the transistor handling small signals is a voltage amplifier and the voltage gain of the amplifier is the basic parameter of design. Whereas when the signal is very large, the voltage gain of the amplifier does not satisfy the requirements of the designer and so power, which is the product of voltage and current, is to be considered. Thus, these amplifiers, which handle large signals, are referred to as *power amplifiers.*

A linear model of transistor cannot be developed for the power amplifiers. Thus, the equivalent linear circuit or set of linear two port network equations is not valid in the analysis of such power amplifiers. Since the input signal is large, the nonlinearity of the characteristics of the transistors cannot be neglected and so the analysis is based on the selection of operating point on the load line of the output characteristics of a transistor. Selection of operating point or quiescent point plays a vital role in the design of power amplifiers. If the operating point is so chosen that the signal lies in the centre of the load line, where the output characteristics are almost linear, the amplifier has a good amplification but its efficiency is very poor. When the operating point is chosen at one end of the load line, even though the efficiency is improved, harmonics are generated since the amplifier is biased into the nonlinear region. As the operating point moves away from the center, the efficiency increases with increase in the loss of harmonic distortion. The nonlinearity of the characteristics of the transistor generates harmonics at the output of the amplifier whose amplitude decays exponentially with frequency. In general, the existence of the first few harmonics affects the signal and it leads to distortion. The most predominant is the second harmonic since its amplitude is comparable to that of the fundamental. Thus, in the design of power amplifiers the harmonic suppression at the output of the circuit is the additional burden on an electronics engineer. Filters with very narrow band characteristics are required in the output circuitry of a power amplifier. Thus, the power amplifiers are, in general, not wide band amplifiers.

When an amplifier is designed as a power amplifier to have very good efficiency, as discussed above, the distortion due to harmonics is very high. Thus the harmonics in the output of a circuit can be suppressed by tank circuits, which are tuned to a particular frequency of operation. These tank circuits select the fundamental frequency while rejecting all unwanted signals at the output. Since the amplifier is tuned to amplify a particular frequency, these type of amplifiers are referred to as *tuned amplifiers.* As one can anticipate, the drawback of tuned amplifiers is very narrow bandwidth of the amplifier. Double-tuned and stagger-tuned amplifiers are variations of these amplifiers where the bandwidth can be improved to the required extent.

12.2 TYPES OF POWER AMPLIFIERS

The power amplifiers are classified based on the selection of operating point. When the operating point is so selected such that the signal is available for complete cycle of the signal at the output of the circuit, the amplifier is referred to as *class A amplifier.* In this type of amplifier, in general, the operating point is in the mid-range of the load line. The angle of conduction is 360° (for the full cycle of the signal). In this amplifier, the harmonic distortion is at minimum while the efficiency of the amplifier to convert the input dc power into the ac output power is very low. Theoretical maximum value of this conversion efficiency of a class A amplifier is 50%. In other words, more than half of the power drawn by the power amplifier is wasted as heat dissipation in this type of amplifier. Another drawback of this amplifier is that when the input signal is absent, the transistor continues to draw the same amount of power and this power drawn from the biasing source is dissipated as heat in the circuit. Thus, the class A amplifier is cooler when there is an input signal rather than in the absence of the input signal. The design of heat dissipation circuits, heat sinks, is also a part of class A amplifier design.

When the operating point is selected at one end of the load line, the output signal is only available for 50% of the complete cycle at the output of the amplifier and this type of amplifier is called *class B amplifier.* The conduction of the transistor is only for half of the cycle and in the remaining portion of the cycle the current at the output is either zero or maximum. In a class B amplifier, the output current is of the form of the current that of a half-wave rectifier. The transistor is in active region for 50% of the cycle, producing a sinusoid signal of conduction angle of 180° at the output of the circuit. The wave form of the current at the output of a class B amplifier is as shown in Fig. 12.1. Since the conduction period of the transistor has reduced to 50% compared to that of class A amplifier, the efficiency of the amplifier is increased and theoretical maximum of this efficiency is 78.5%. For the remaining 50% of the cycle, the transistor is either in cut off or saturation, thus leading to nonlinearity and hence the generation of harmonics. The harmonic content in the output of this circuit is high compared to that of class A. A configuration involving two transistors in *push-pull* mode reduces the content of harmonics by eliminating the even harmonics in the output of the circuit.

When the operating point is selected on the load line such that the conduction angle of the transistor lies between 50% and 100% of one complete cycle, the amplifier is called *class AB*. The characteristics of class *AB* amplifier lie between those of class A and class *B* amplifiers depending upon the selection of the operating point.

Class C amplifier is characterised by output current of small pulses of very short duration. The conduction angle of a transistor amplifier is very small (less than 50% of the cycle), tending towards zero. The greatest advantage of this type of amplifier is its

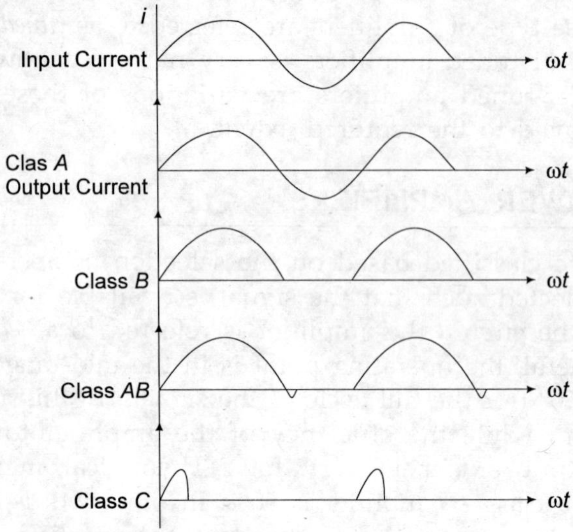

Fig. 12.1 *Output wave forms of power amplifiers*

efficiency, which is tending towards 99.99999%. The lesser the conduction angle, the greater the efficiency, and the higher is the harmonic distortion. Another advantage of this amplifier is that it draws no power from the dc source in the absence of the input signal. Since the efficiency is very large, the heat dissipation in the transistor is at minimum. Thus, natural air-cooling of the device is more than sufficient and no special design of heat sinks is required. However, the spectrum of the output signal consists of large number of harmonics, multiples of the fundamental frequency. Thus, the distortion due to the harmonics is maximum in this configuration. This distortion corrupts the shape of the wave form in the output, which is no more a simple sinusoid but is a superposition of sinusoids of multiple frequencies. To eliminate this distortion, one has to pickup the fundamental frequency and suppress all the harmonic frequencies at the output of the circuit. This requires a filter with very narrow band characteristics, which is a tank (parallel resonant) circuit. Thus, the bandwidth of the class C amplifier is very narrow and is tuned to given a particular fundamental frequency and thus these type of amplifiers are also referred to as *tuned amplifiers.* In order to enhance the bandwidth of such amplifiers, double-tuned or stagger-tuned configurations can be used wherever required.

12.3 HARMONIC DISTORTION

In an amplifier, when harmonics are present in the output along with the input fundamental frequency, the amplitude of the signal is affected to a large extent and this distortion of the signal is referred to as *amplitude distortion* or *harmonic distortion.* In power amplifiers, this distortion is the most dominant phenomenon since nonlinearity of the characteristics of the transistor generates harmonics in the output. Along with the amplification of the amplifier, the measure of harmonic distortion is also a design parameter of a power amplifier. Thus, the harmonic distortion of an amplifier is to be determined for analysing or designing a power amplifier. The measurement of the harmonic distortion is as follows:

For simplicity, only second harmonic is considered and the distortion due to this harmonic can be determined assuming that the output current is

$$i_c = k_1 i_b + k_2 i_b^2 \qquad \text{...(12.1)}$$

where k_1 and k_2 are constants. In this equation, only first two terms of the power series expansion of the collector current are considered and higher-order terms are neglected. If the input instantaneous base current is sinusoid of the form

$$i_b = I_{Bm} \cos \omega t \qquad \text{...(12.2)}$$

substituting Eq. (12.2) into Eq. (12.1), the output instantaneous collector current can be expressed as

$$\begin{aligned} i_c &= k_1 I_{Bm} \cos \omega t + k_2 (I_{Bm} \cos \omega t)^2 \\ &= A_0 + A_1 \cos \omega t + A_2 \cos 2\omega t \qquad \text{...(12.3)} \end{aligned}$$

where A_0, A_1 and A_2 are constants, that can be evaluated in terms of k's

The total instantaneous current at the output collector of the transistor amplifier can be given as

$$\begin{aligned} i_C &= I_C + i_c \\ &= I_C + A_0 + A_1 \cos \omega t + A_2 \cos 2\omega t \qquad \text{...(12.4)} \end{aligned}$$

Thus, the output current has a dc component $I_C + A_0$, a fundamental frequency component of magnitude A_1 and a second-harmonic frequency component of magnitude A_2. Thus, the output consists of three components of frequency. Since A_0 is the additional dc component added, the power amplifier not only amplifies but also converts a small portion of the input signal as dc and thus it has rectification action also. Since there is a 2ω term in the output, the amplifier generates second harmonic of the input frequency.

The absolute ratio of amplitude of second harmonic A_2 to the amplitude of the input frequency A_1 is defined as *second-harmonic distortion* and is given as

$$D_2 = \left| \frac{A_2}{A_1} \right| \qquad \text{...(12.5)}$$

The distortion can also be expressed in percentage when the above expression is multiplied by 100.

To understand the distortion due to higher harmonics, the complete power series expansion of the output collector current is considered,

$$i_c = k_1 i_b + k_2 i_b^2 + k_3 i_b^3 + k_4 i_b^4 + \cdots \qquad \text{...(12.6)}$$

Substituting the equation for input base current (12.2) into the above equation and rearranging the terms, the output collector current of an amplifier can be determined as

$$i_C = I_C + A_0 + A_1 \cos \omega t + A_2 \cos 2\omega t + A_3 \cos 3\omega t + A_4 \cos 4\omega t + \cdots \qquad \text{...(12.7)}$$

Thus, the distortion due to each harmonic present in the output of the circuit can be defined as

Third-harmonic distortion,

$$D_3 = |A_3/A_1| \quad ...(12.8)$$

Fourth-harmonic distortion,

$$D_4 = |A_4/A_1| \quad ...(12.9)$$

Fifth-harmonic distortion,

$$D_5 = |A_5/A_1| \quad ...(12.10)$$

And in general *nth-harmonic distortion,*

$$D_n = |A_n/A_1| \quad ...(12.11)$$

The *total distortion,* also known as *distortion factor,* is defined as

$$D = \sqrt{[D_2^2 + D_3^2 + D_4^2 + D_5^2 + \cdots]} \quad ...(12.12)$$

Since the amplitude of the harmonic frequencies is not negligible, the power contained in these frequencies is to be determined in order to estimate the fraction of power available at the output of the circuit at the input fundamental frequency. The power contained in each harmonic and fundamental frequency can thus be determined as

power at fundamental frequency,

$$P_1 = (A_1^2 R_L/2) \quad ...(12.13)$$

power at second harmonic,

$$P_2 = (A_2^2 R_L/2) \quad ...(12.14)$$

power at third harmonic,

$$P_3 = (A_3^2 R_L/2) \quad ...(12.15)$$

power at fourth harmonic,

$$P_4 = (A_4^2 R_L/2) \quad ...(12.16)$$

and so on. Thus, the total power at the output can be given as

$$P = (A_1^2 + A_2^2 + A_3^2 + A_4^2 + \cdots)(R_L/2) \quad ...(12.17)$$

or,

$$P = (1 + D_2^2 + D_3^2 + D_4^2 + \cdots)P_1 \quad ...(12.18)$$

$$= (1 + D^2)P_1 \quad ...(12.19)$$

It can be observed that the higher is the distortion, the lesser is the fraction of fundamental frequency component of the total power. Ideally, if distortion is zero, or if the signal is distortionless, then the total output power would be equal to the power in the fundamental frequency.

The magnitudes of the harmonic frequencies can be determined graphically as follows. Assuming input to be sinusoidal as given in Eq. (12.2), considering the general expression

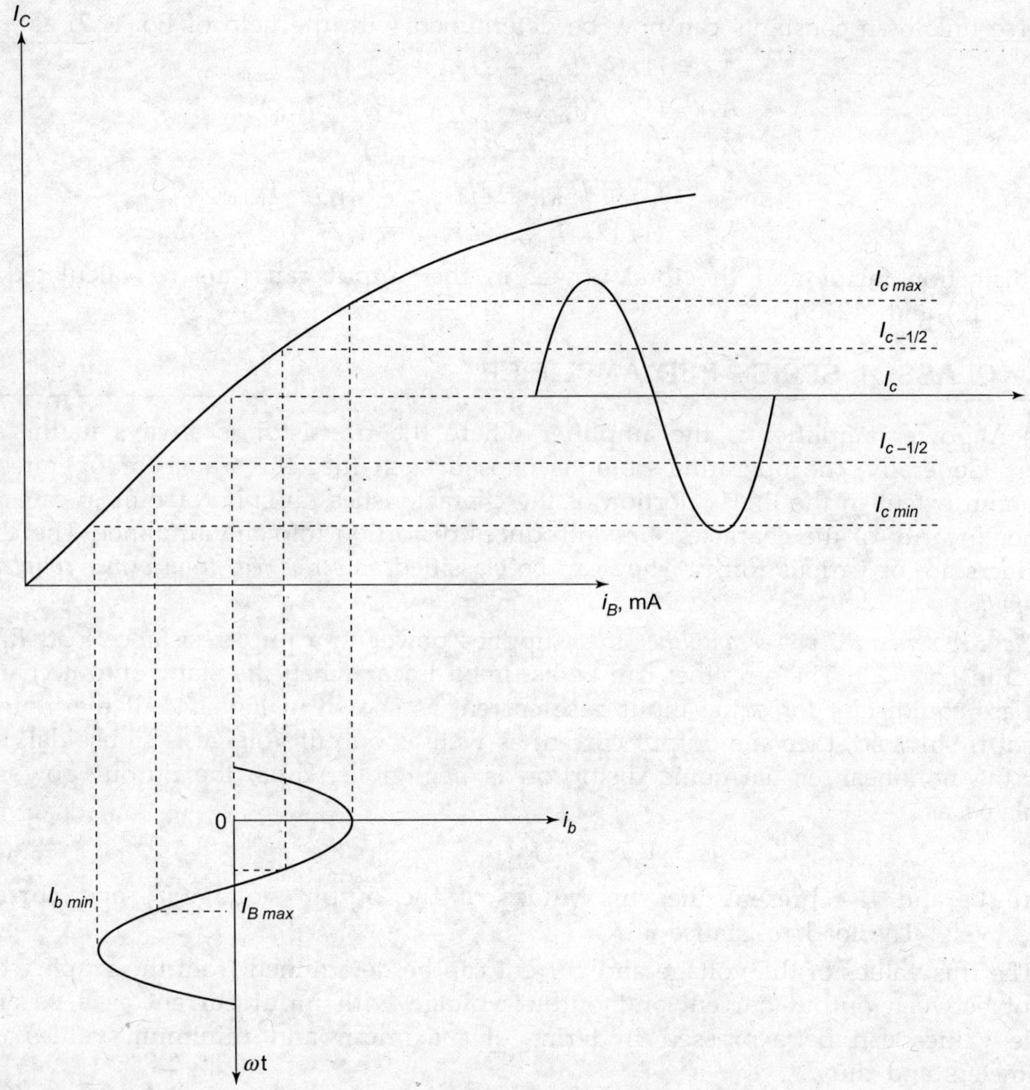

Fig. 12.2 *Transfer characteristics*

for the output collector current (12.6) from the transfer characteristic between output current and input current as shown in Fig. 12.2,

$$\begin{aligned}
&\text{when} && \omega t = 0;\ i_C = I_{\max} \\
&\text{when} && \omega t = \pi/3;\ i_C = I_{1/2} \\
&\text{when} && \omega t = \pi/2;\ i_C = I_C \qquad \text{...(12.20)} \\
&\text{when} && \omega t = 2\pi/3;\ i_C = I_{-1/2} \\
&\text{when} && \omega t = \pi;\ i_C = I_{\min}
\end{aligned}$$

The five unknown constants can now be determined with the help of Eq.(12.7) as

$$\begin{aligned} A_0 &= (1/6)\,(I_{max} + 2I_{1/2} + 2I_{-1/2} + I_{min}) \\ A_1 &= (1/3)\,(I_{max} + I_{1/2} - I_{-1/2} - I_{min}) \\ A_2 &= (1/4)\,(I_{max} - 2I_C + I_{min}) \\ A_3 &= (1/6)\,(I_{max} - 2I_{1/2} + 2I_{-1/2} - I_{min}) \\ A_4 &= (1/12)\,(I_{max} - 4I_{1/2} + 6I_C - 4I_{-1/2} + I_{min}) \end{aligned} \qquad ...(12.21)$$

The distortion factor and the total power in the output can thus be calculated with Eqs. (12.12) and (12.19).

12.4 CLASS *A* SERIES-FED AMPLIFIER

Class A power amplifier is the amplifier where the transistor is always in the active region. Generally, the operating point is chosen such that the transistor operates to a maximum extent in the linear portion of the characteristics. In spite of utmost care taken, the nonlinearity of the characteristics introduces distortion into the amplifier. The class *A* amplifiers are of various forms. They can be classified as *series fed, transformer coupled* and *push-pull.*

Class A series-fed power amplifier that supplies power to a pure resistance load R_L is as shown in Fig. 12.3. The amplifier can be assumed linear when the static output characteristics are equidistant for equal input base current as shown in Fig. 12.4. If the input base current is sinusoid, then the output current as well as output voltage is sinusoidal. In this case, the nonlinear or harmonic distortion is negligible. Thus, the output power may calculated as

$$P = V_c I_c = I_c^2 R_L \qquad ...(12.22)$$

where V_c and I_c represent the rms values of the output voltage v_c and current i_c, respectively. The load resistance is R_L.

The rms values of the voltage and current can be determined from the graph which is a plot between output current and output voltage with input current as a parameter. These values can be expressed in terms of maximum and minimum values of the parameters and thus,

$$I_c = (I_{max} - I_{min})/(2\sqrt{2}) \qquad ...(12.23)$$

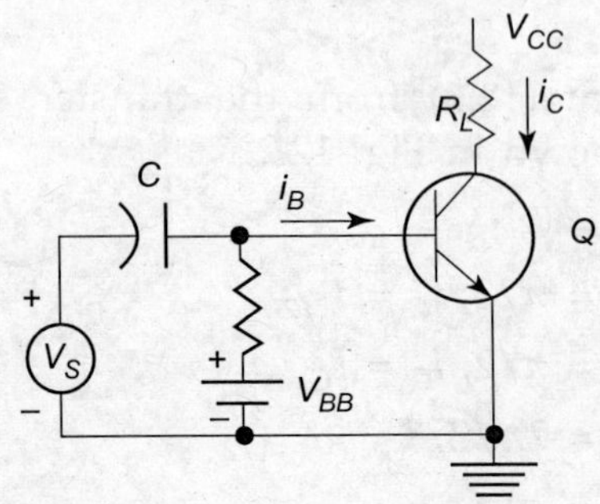

Fig. 12.3 *Series-fed class A amplifier*

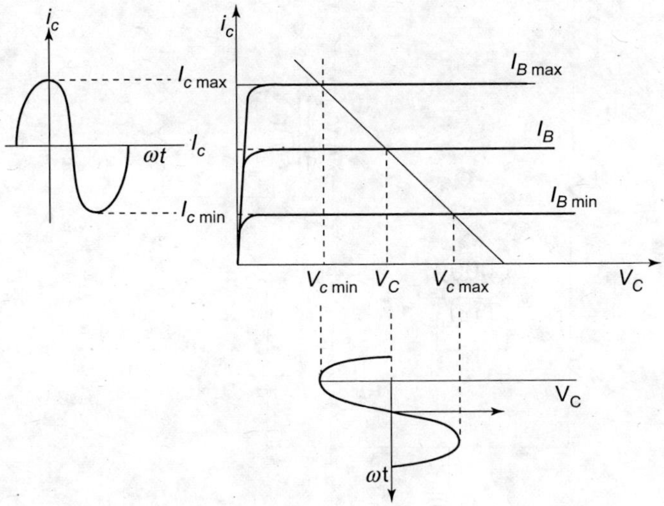

Fig. 12.4 *Output characteristics and wave forms*

$$V_c = (V_{max} - V_{min})/(2\sqrt{2}) \quad ...(12.24)$$

and the power,

$$P = (I_{max} - I_{min})\ (V_{max} - V_{min})/8 \quad ...(12.25)$$

$$= V_m^2/(2R_L) = I_m^2 R_L/2 \quad ...(12.26)$$

where V_m and I_m are peak sinusoidal voltage and current swings.

12.5 CLASS *A* TRANSFORMER-COUPLED AMPLIFIERS

In case of a series-fed amplifier, the output current is delivered to the load resistance. The output collector current not only consists of the ac component, but also along with this component, a dc quiescent current is delivered to the load resistance. This component is unwanted and it is inadvisable to pass this dc current into the output device, for example, voice coil of a load speaker. This component of dc current does not contribute to ac power. Thus, this current component is to be avoided in the load of the amplifier. A transformer is used to couple the output collector of the transistor to the load. This transformer couples only the ac power while dc is not allowed to pass into the load of the amplifier. The circuit configuration of such an arrangement is as shown in Fig. 12.5. This configuration of power amplifier is referred to as *transformer-coupled power amplifier*. This type of amplifier is used at audio frequencies and this is the final stage in the public address systems.

The input coupling need not be always transformer but RC coupling can also be used. This type of transformer coupling reduces the wastage of dc power since the dc resistance of transformer is less. It also provides impedance matching to transformer a significant amount of power to a load such as a loudspeaker, whose voice coil impedance is 5–15 Ω. The impedance matching of an ideal transformer can be given as

$$V_1 = (N_1/N_2)V_2 \text{ and } I_1 = (N_2/N_1)I_2 \quad ...(12.27)$$

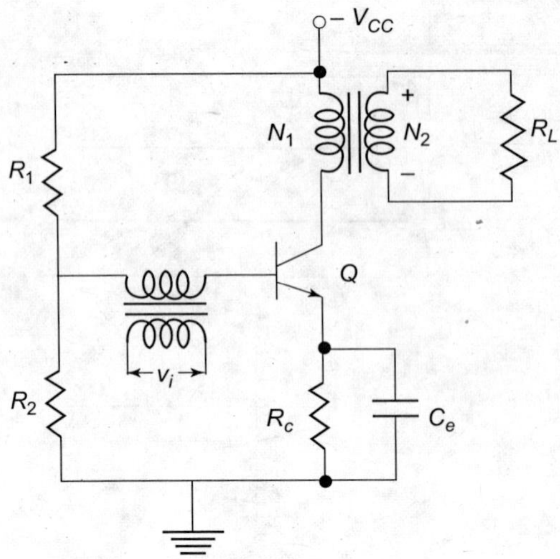

Fig. 12.5 *Transformer-coupled power amplifier*

where V_1 and V_2 are the voltages at primary and secondary of transformer, I_1 and I_2 are the current at primary and secondary of transformer and N_1 and N_2 are the windings of primary and secondary of transformer. If $n = N_2/N_1$ is the turns ratio,

$$\frac{V_1}{I_1} = \frac{1}{n^2}\frac{V_2}{I_2} \qquad ...(12.28)$$

Thus, the effective input resistance $R'_L = V_1/I_1$ can be given as

$$R'_L = (1/n^2)R_L \qquad ...(12.29)$$

The design of such an amplifier can be given as follows:

- The quiescent operating point is to be selected on the dc load line. Since the dc resistance of the transformer is very small, the dc load line is almost parallel to the *y*-axis as shown in Fig. 12.6.
- The ac or dynamic load line is to be drawn on the same characteristics through the operating point with a slope of R_L.
- The base current corresponding to the operating point is to be determined.
- With the help of the input characteristics, the swing of the base current is to be determined in terms of maximum and minimum values of the current. The quiescent base current is to be selected such that the relation between the base current and voltage is linear.
- The base current minimum and maximum are to be transferred to the output characteristics to determine the output collector current swing.
- The voltage swing in the output of the amplifier is then determined.
- The output ac signal power of the amplifier is to be calculated graphically.
- The harmonic distortion for each harmonic as well as total harmonic distortion can also be determined from this graph with the help of Eq. (12.21) and (12.11).

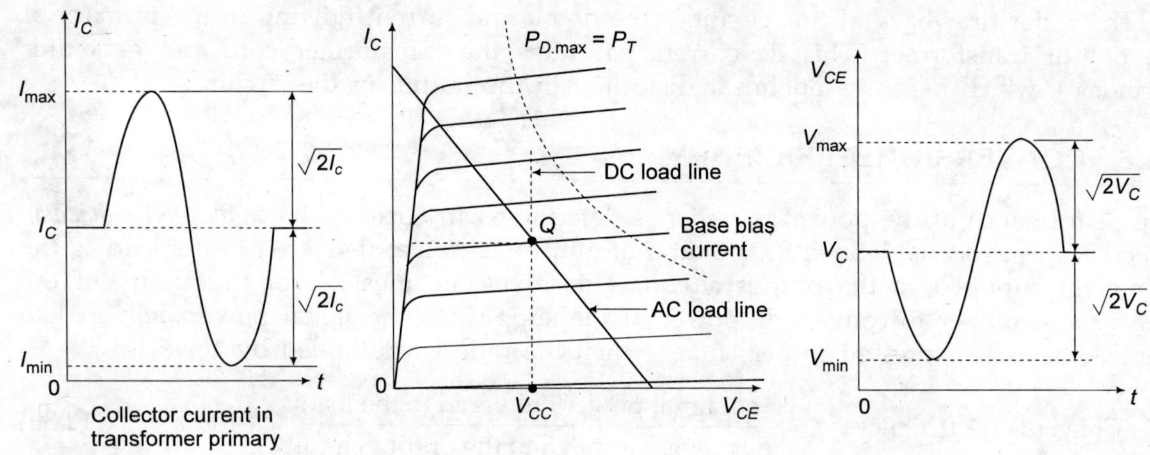

Fig. 12.6 *Operating point*

- The variation of the output power and harmonic distortion with load resistance R_L can be determined for various load resistance values and Fig. 12.7 gives the general variation of these two parameters.

From Fig. (12.7), it can be observed that the total harmonic distortion is almost equal to the second harmonic distortion and so the second harmonic is the major component, which contributes the distortion. For R_L very small, the voltage swing, and hence the power output, approach zero. For R_L very large, the current swing is small, and again P approaches zero. Therefore, for a given value of load resistance the output power has a maximum value. But this value is spread over a broad range of resistance. It is seen from the curve that the load that provides minimum total harmonic distortion does not yield maximum output power. So, the power and harmonic distortion depend upon the load resistance and a compromise between these two parameters is to be made.

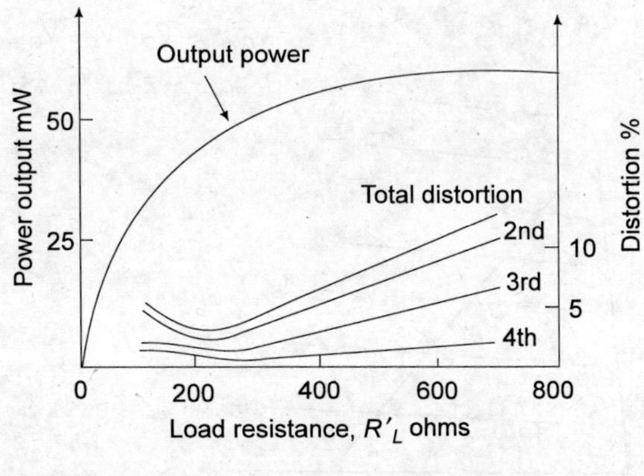

Fig. 12.7 *Variation of power and distortion with load resistance*

The major drawback of this circuit is the dc biasing current flowing in the primary of the output transformer. This dc current saturates the transformer core and generates harmonics, which increases nonlinear distortion in the output of the circuit.

12.6 CONVERSION EFFICIENCY

The performance of the power amplifier is determined in terms of its *conversion efficiency*. *Conversion efficiency* is defined as the ratio of output ac power delivered to the load to the dc power supplied to the output circuit. This gives the measure of the ability of the transistor amplifier to convert dc power of the supply into ac signal power delivered to the load. This is expressed in percentage when the ratio is multiplied by 100. Thus,

$$\text{Conversion efficiency, } \eta = \frac{\text{ac signal power delivered to the load}}{\text{dc power supplied to the output circuit}} \times 100\% \quad \text{...(12.30)}$$

$$= \frac{\frac{1}{2} B_1^2 R_L}{V_{CC}(I_C + B_0)} \times 100\% \quad \text{...(12.31)}$$

If distortion is negligible as assumed, then

$$\eta = \frac{\frac{1}{2} 2 V_m I_m}{V_{CC} I_C} \times 100\% = 50 \frac{V_m I_m}{V_{CC} I_C} \% \quad \text{...(12.32)}$$

This conversion efficiency is defined at the output collector. Thus, collector circuit efficiency differs from the overall efficiency since the total dc power supplied to the amplifier is partially taken by the base circuit of the amplifier and this is not considered in the above calculations.

The approximate expression for the conversion efficiency can be obtained with the assumption that the device is operating in the linear portion of the characteristic as shown in Fig. 12.8.

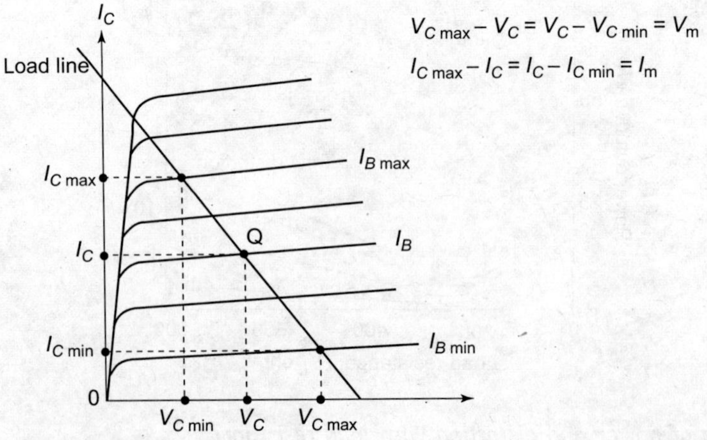

Fig. 12.8 *Conversion efficiency calculations*

In a *series-fed power amplifier*, $I_C = I_m$, $V_m = (V_{max} - V_{min})/2$, $V_{max} = V_{CC}$, and the conversion efficiency of such a circuit is

$$\eta = \frac{\frac{1}{2}V_m I_m}{V_{CC} I_C} \times 100\% = 50\,\frac{\frac{1}{2} I_C (V_{max} - V_{min})}{V_{max}\, I_C}\,\%$$

$$= [(V_{max} - V_{min})/V_{max}] \times 25\% \qquad \text{...(12.33)}$$

Hence, the maximum efficiency is 25% in ideal case when $V_{min} = 0$.

In case of a *transformer-coupled power amplifier*, $I_C = I_m$, $V_m = (V_{max} - V_{min})/2$, $V_{CC} = (V_{max} + V_{min})/2$ and the conversion efficiency in this case can be given as

$$\eta = \frac{\frac{1}{2}V_m I_m}{V_{CC} I_C} \times 100\% = 50\,\frac{(V_{max} - V_{min})}{(V_{max} + V_{min})}\,\% \qquad \text{...(12.34)}$$

The conversion efficiency of the transformer-coupled amplifier tends towards a maximum of 50% in ideal case when V_{min} occurs near saturation and $V_{min} \ll V_{max}$. Thus, it can be seen that the conversion efficiency of the transformer-coupled amplifier is twice that of the series-fed amplifier.

The performance of a power amplifier can also be evaluated in terms of the *figure of merit*, which is defined as the ratio of maximum transistor collector dissipation to maximum power dissipated in the load. The value of this parameter depends upon the transistor characteristics, which are specified by the manufacturer as P_{Cmax}.

12.7 CLASS *A* PUSH-PULL AMPLIFIER

The two types of power amplifiers considered in the previous sections were designed assuming that the transistor operates in linear portion of the load line, that is, the transfer characteristic between input and output currents is a straight line. But as the input current swing increases, it is not possible to assume that the characteristic is a straight line and one has to live with nonlinearity, which leads to generation of harmonics in the output. So, in addition to the design considerations as discussed previously, harmonic suppression is to be considered. Of all the harmonics present in the output, the most predominant one is the second harmonic, which causes largest distortion, and a scheme is now proposed which suppresses all even harmonics in the output. This scheme utilises two identical transistors in *push-pull* configuration. Second harmonics along with all even harmonics are totally suppressed and thus the content of distortion is reduced to a very large extent.

The circuit diagram of *push-pull class A amplifier* employing two identical transistors is shown in Fig. 12.9. The transformer at the input of the circuit produces voltage at the secondary windings according to the polarity as indicated in the figure. This provides phase-splitting function in which two voltages are out of phase at the input base of the two transistors. If the two windings of secondary are equal in number, the two voltages would be equal and at any given instant of time the voltages at the base of the transistors will be exactly out of phase by 180° that is, if one transistor base is positive, the other would be negative with equal amplitude. Thus, if $i_{b1} = I_{bm} \cos \omega t$, then

$$i_{b2} = -\, i_{b1} = I_{Bm} \cos (\omega t + \pi) \qquad \text{...(12.35)}$$

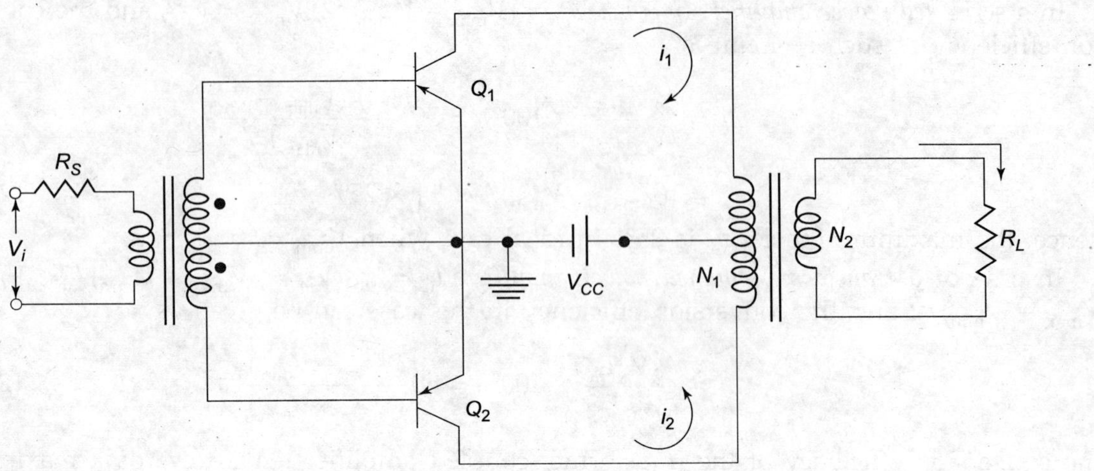

Fig. 12.9 *Class A push-pull amplifier*

Thus, the output current at the collector of transistor $Q1$, i_1 will be

$$i_1 = I_C + A_0 + A_1 \cos \omega t + A_2 \cos 2\omega t + A_3 \cos 3\omega t + A_4 \cos 4\omega t + \cdots \qquad ...(12.36)$$

and the output current at the collector of transistor $Q2$, i_2, can be given as

$$i_2 = I_C + A_0 + A_1 \cos(\omega t + \pi) + A_2 \cos 2(\omega t + \pi) + A_3 \cos 3(\omega t + \pi) + A_4 \cos 4(\omega t + \pi) + \cdots$$

$$= I_C + A_0 - A_1 \cos \omega t + A_2 \cos 2\omega t - A_3 \cos 3\omega t + A_4 \cos 4\omega t - \cdots \qquad ...(12.37)$$

As shown in the figure, the output currents i_1 and i_2 of the two transistors are in opposite directions through the output transformer primary windings. The total output current in the secondary of the output transformer is proportional to the difference between the currents in the two transistors. Therefore,

$$i = k(i_1 - i_2) = 2k(A_1 \cos \omega t + A_3 \cos 3\omega t + \cdots) \qquad ...(12.38)$$

It can be observed from the above equation that the output current in the secondary of the transformer consists of frequency components of fundamental and odd multiples of the fundamental. In other words, the second, fourth, sixth ... harmonics are totally suppressed. Since the even harmonics are suppressed, the power in the output signal is increased. Not only this, the dc component is also absent in the output. Since the most predominant distortion is due to second harmonic, which is totally suppressed, this configuration of amplifier reduces harmonic or nonlinear distortion to a very large extent. This was possible by the *push-pull* action of the transistors. Since the input voltages of the transistors were out of phase by 180°, in the process of amplification, while one transistor pushes the phase, the other transistor pulls it and since these two are added appropriately in the output transformer, even harmonics are eliminated totally. The core saturation of the transformer due to dc current present is avoided in this circuit. This is due to the fact that the currents i_1 and i_2 in the primary of the output transformer are in the opposite direction and of the same magnitude. So, the effective total dc current flowing in the primary of the transformer is zero. Thus, the problem of core saturation due to dc current is eliminated in this configuration.

The ripple present in the dc power supply of the circuit may cause instability in the circuit. This ripple may be due to improper filter in the power-supply circuit. This problem is avoided in the push-pull amplifier since this is balanced out and the operating point stability is achieved. It may be noted that the hum present in the power supply cannot be suppressed in this circuit. The presence of hum when the amplifier is operating at audio frequencies is an undesired audio effect, that reduces the quality of the audio signal. When the power amplifier is operated at higher frequencies the effect of ripple and hum on the signal can be totally neglected.

12.8 CLASS *B* AMPLIFIER

The performance of the class *A* amplifier is limited since its maximum theoretical efficiency is 50%. The push-pull configuration improves the performance of the amplifier in terms of distortion, but does not improve the efficiency. In order to improve the efficiency of the circuit, *class B amplifier* is proposed. In the class *B* amplifier, the operating point is so chosen that the output current is present only for half of the cycle of the input signal. In the remaining half of the cycle the current is zero. This can be achieved by choosing the operating point on the load line at the knee of cut off. The general configuration adopted in the class *B* amplifier is push-pull. The circuit of the class *B* push-pull amplifier along with the wave forms is as shown in Fig. 12.10.

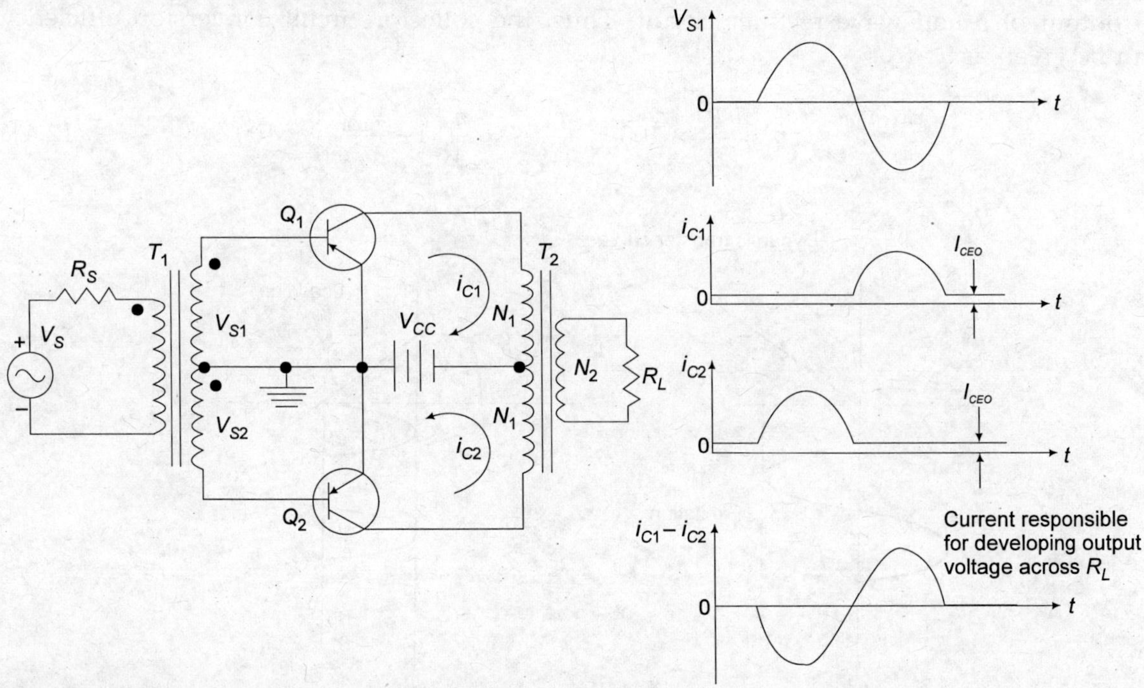

Fig. 12.10 *Class B push-pull amplifier and its wave forms*

The operation of the class *B* push-pull amplifier is similar to that of the class A push-pull amplifier. The input base currents of the two transistors are out of phase and the two output collector currents are in opposite direction in the primary of the output transformer. However, unlike the class *A* push-pull amplifier where the current is available throughout the complete cycle, the output current in the circuit is only for 50% of the cycle. Since the two currents in the transistor are out of phase when one transistor is in active region, the other is in cut off region. In other words, when one transistor conducts the other is OFF. Thus, the current delivered to the primary of the transformer at any given instant of time is only due to one transistor and the current in the other transistor is minimum (saturation current). Thus, the current in the secondary of the output transformer is a complete sinusoid as shown in the figure.

The conversion efficiency of the class *B* amplifier can be determined with the help of the dynamic transfer characteristic as shown in Fig. 12.11.

The output power can be given as the product of rms values of output voltage and current. Thus, the output power from the figure can be given as

$$P = V_m I_m/2 = I_m(V_{CC} - V_{min})/2 \quad \text{...(12.39)}$$

The input power can be calculated as the product of average dc collector current and the supply voltage and is given by

$$P_i = 2I_m V_{CC}/\pi \quad \text{...(12.40)}$$

The factor 2 is due to the presence of two transistors in the push-pull configuration. The dc average collector current is I_m/π for each transistor since the current is that of the form of output of a half wave rectifier circuit. Thus, the collector circuit conversion efficiency can be given as

$$\eta = \frac{P}{P_i} \times 100 = \frac{\pi V_m}{4V_{CC}} = \frac{\pi}{4}\left(1 - \frac{V_{min}}{V_{CC}}\right) \times 100\% \quad \text{...(12.41)}$$

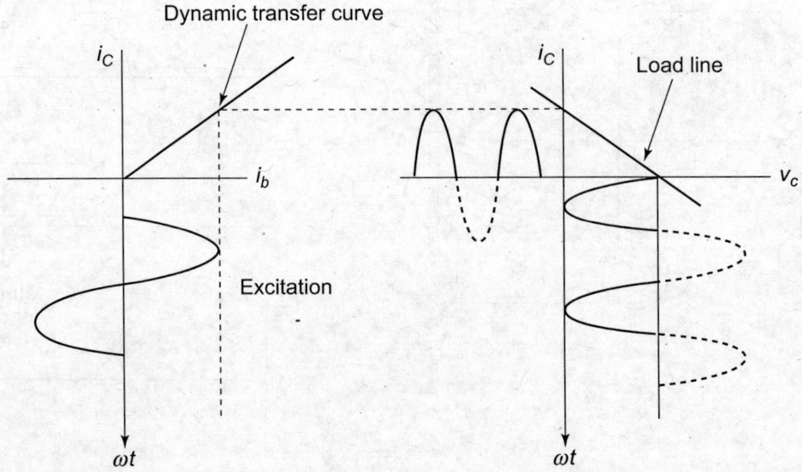

Fig. 12.11 *Class B amplifier efficiency calculations*

From the above equation it can be observed that the maximum value of conversion efficiency is $(\pi/4) \times 100\%$, which is 78.5%. Thus, the conversion efficiency of the class *B* amplifier is higher than that of the class *A* amplifier (50% max).

12.9 OTHER TYPES OF CLASS *B* AMPLIFIERS

The other possible class B push-pull amplifiers are *push-pull amplifier without output transformer and complementary symmetry push-pull power amplifier.*

The problems and losses with transformers can be minimised to the extent possible by avoiding these elements in the circuit of class B push-pull amplifiers. There are two possible configurations in which the transformers can be avoided while achieving the same characteristics if a class *B* push pull amplifier. The circuit that avoids the transformer in the output is as shown in Fig. 12.12.

It can be observed from the above circuit that the centre-tapped transformer at the input of the circuit is replaced with a transformer, which has two similar secondary windings. As in the previous case, the two currents in the input of the transistor are out of phase by 180°. The output transformer is avoided and one end of the load resistance is grounded as shown in the figure. The two output collector currents flow in the opposite directions in the common load resistance R_L. Hence, the output voltage across this load resistance is proportional to the difference of two collector currents.

Complementary symmetry push-pull power amplifier avoids transformers in both output and input of the circuit as shown in Fig. 12.13. In this circuit, the two transistors are not identical. One transistor is of *pnp* type while the other is *npn* type. These two transistors are identical and complementary symmetrical. In the positive half-cycle the emitter junction of the *npn* transistor is forward biased while that of the *pnp* is reverse

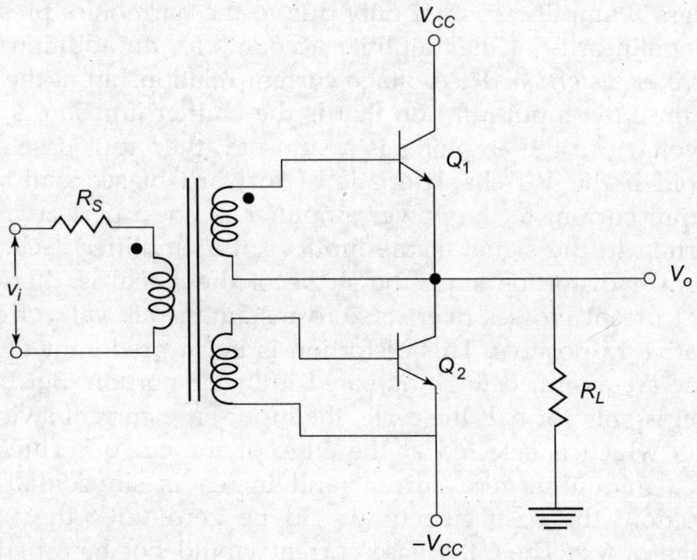

Fig. 12.12 *Class B amplifier without output transformer*

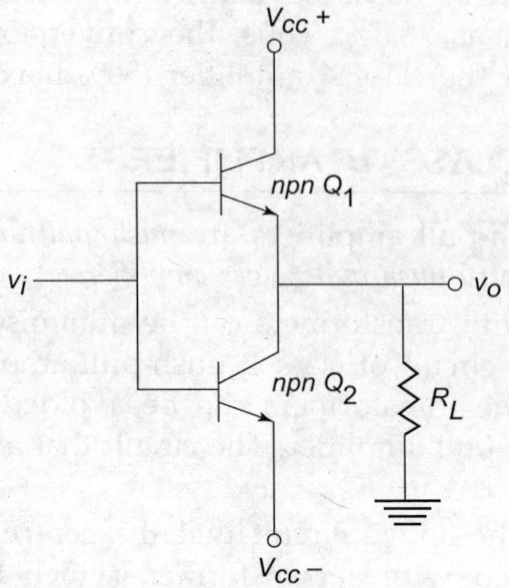

Fig. 12.13 *Complementary symmetry push-pull power amplifier*

biased. Thus, the *npn* transistor is in active region, while *pnp* is in cut off. In the negative half-cycle the roles of the two transistors are reversed. Thus, the output collector currents of the two transistors would be same as that of the class *B* push-pull amplifiers. The performance of the circuit is same as that of the class *B* push-pull amplifier except that obtaining the matched complementary pair of transistors is difficult.

12.10 CLASS *AB* AMPLIFIERS

The distortion in class *B* amplifiers is not only due to the harmonics present in the output, but also the input nonlinearity of the amplifier accounts for an additional distortion factor which is called *crossover distortion.* The voltage current relationship at the input of the transistor is not linear since the input junction that is the emitter junction is forward biased. If the applied input voltage to the amplifier is a sinusoid, the input base current would not be a perfect sinusoid if the *V-I* characteristic of forward biased emitter junction is not linear. Thus, the input current to the power amplifier is not a perfect sinusoid. This leads to additional distortion in the signal at the input of the amplifier itself. This distortion is referred to as crossover distortion since the shape of the signal is altered or the signal is distorted when the current crosses over its zero instantaneous value from either positive to negative or negative to positive. This distortion is not a predominant feature in class *A* amplifiers since they are assumed to be operated in linear portion. But in class *B* amplifier since the conduction is only for half the cycle, the input base current swing is applied on to the operating point, which is selected at the edge of the cut off. Thus, the base current swing starts from a minimum zero current and increases sinusoidally. In the forward biased emitter junction, the base current would be zero until the input base voltage reaches cutin voltage of $V\gamma$. Thus, the base current would not be a perfect sinusoid even though the input base voltage is a sinusoid, as shown in Fig. 12.14. This distortion is most predominant in class *B* amplifiers.

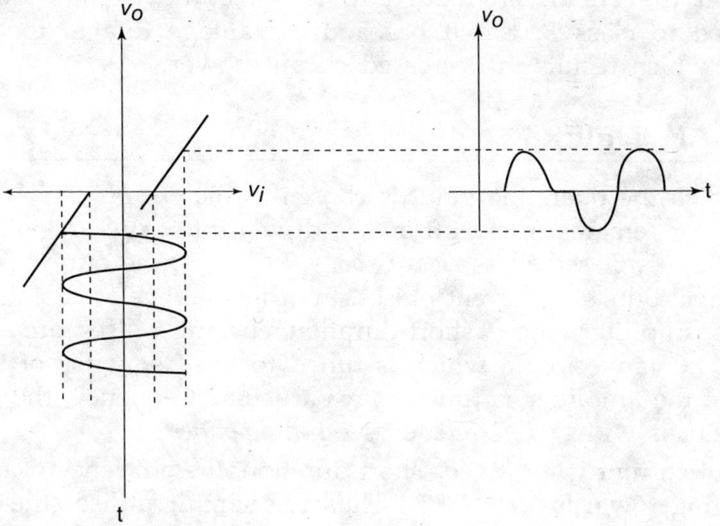

Fig. 12.14 *Crossover distortion*

Class *AB* amplifier eliminates this crossover distortion by selecting the operating point not at the edge of the cut off but slightly above this value of cut off so as to overcome the effect of crossover distortion. The circuit diagram of class *AB* amplifier that minimizes the crossover distortion is shown in Fig. 12.15.

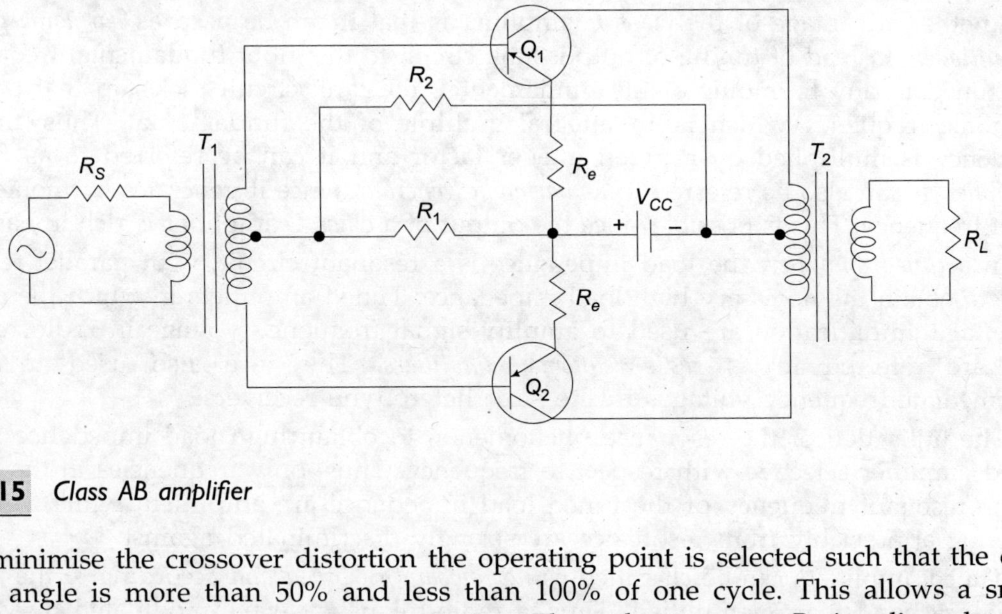

Fig. 12.15 *Class AB amplifier*

To minimise the crossover distortion the operating point is selected such that the conduction angle is more than 50% and less than 100% of one cycle. This allows a small standby current to flow under zero excitation. The voltage across R_2 is adjusted to be equal to V_γ. This biasing is such that the transistors are driven slightly towards class *A* and driven away from class *B*. Hence, the resulting circuit is referred to as class *AB*. This biasing of the transistors results in flow of current immediately after the signal is applied.

Emitter resistors in the circuit are added to stabilize the circuit. This leads to less distortion compared to class *B*, but it has a disadvantage due to the introduction of negative feedback, which results in decreased output power.

12.11 TUNED AMPLIFIERS

An amplifier in which the operating point is chosen in the cut off region and for most of the time the transistor remains in cut off and is driven into active region for only a very short duration of time is classified as *class C amplifier.* Thus, the output current of a class *C* amplifier is of small bursts of current pulses. In order to develop a sinusoidal voltage at the output of the amplifier, these short-duration current pulses are allowed to pass through a parallel *LC* tank circuit, which is tuned to the frequency of the input signal. Since the output of the amplifier is tuned to a particular frequency, the input frequency, this type of amplifier is widely referred to as *tuned amplifier.*

Since the output current is for a very short duration, the efficiency of the circuit is very large, ideally tending towards 99.99999%. Thus this circuit utilises the input dc power very effectively in converting into output a.c signal power. The harmonic distortion is highest in this circuit since pulsated current produces harmonic frequencies. In order to avoid the unwanted harmonic frequencies a narrowband filter is used which filters out harmonics and passes the fundamental unattenuated. A very good example of such a narrowband filter is parallel resonant circuit. The frequency selection of the tank circuit depends upon the quality factor *Q* of the circuit. As the quality factor *Q* increases, the bandwidth of the amplifier decreases.

Another advantage of the class *C* amplifier is that it can be used as *harmonic generator* or *multiplier.* Instead of tuning of output tank circuit to the input fundamental frequency, if it is tuned to any harmonic of the fundamental, the tank circuit picks up that particular harmonic frequency which is an integral multiple of the fundamental. Thus, the input frequency is multiplied by a given integer factor and it can be referred to as *frequency multiplier.* It can also be referred to as *harmonic generator* since it generates harmonics of the input frequency. This is possible since the output of a class *C* amplifier is rich in harmonics.

In a *tuned amplifier,* the load impedance is a resonant circuit, with parallel resonance used to obtain the necessary high load impedance. Tuned amplifiers in which the objective is voltage amplification are used to amplify signal frequency voltage in radio receivers, and are referred to as *radio-frequency amplifiers.* They are also used to amplify intermediate-frequency voltages in the superheterodyne receivers.

The utilisation of the resonance phenomenon to obtain high load impedance makes a tuned amplifier selective with respect to frequency. Thus, only frequencies in the viability of the resonant frequency of the tuned load impedance are amplified, while frequencies differing appreciably from resonance are strongly discriminated against.

Tuned amplifiers can be classified as *narrowband* or *broadband* according to the width of frequency band that is amplified. Thus, a radio-frequency wave modulated by voice frequency currents would be handled by a narrowband amplifier, because the frequency band occupied by such a wave, including the sideband frequencies, would be relatively small. In contrast, a *RF* wave modulated a TV, video signal or by a short pulse would involve sidebands extending over such a wide frequency band as to require a wideband amplifier. The important characteristics of tuned amplifiers are the amplification at the resonance, the variation of amplification with frequency in the immediate vicinity of resonance and discrimination against frequencies differing appreciably from resonance.

12.12 SINGLE-TUNED AMPLIFIER

A resonant circuit is used in place of a load resistance in *single-tuned amplifiers* as shown in Fig. 12.16. Tuned resonant frequency offers very high impedance at the resonant frequency. At other frequencies, the impedance is small. If, therefore, the signal frequency is equal to resonant frequency of the tank circuit, a large amplification will result because of the high impedance. If the signal consists of several frequencies the amplifier will select and amplifies strongly only those frequencies, that are close to its resonant frequency. All other frequencies are rejected. The output may be taken out by inductive or capacitive coupling. The voltage gain of such an amplifier is given as

$$A = v_o/v_i = (i_c Z_L)/(i_b Z_i) = (\beta L)/(Z_i RC) \quad ...(12.42)$$

For a tuned circuit, L/RC is very large, hence the voltage gain in a tuned amplifier is very high. At frequencies away from resonance the impedance decreases rapidly and hence voltage gain falls off. The bandwidth of the amplifier can be determined as

$$\text{Bandwidth} = f_2 - f_1 = f_p/Q_p \quad ...(12.43)$$

It is clear that larger the quality factor, smaller the bandwidth and sharper the resonance.

In an amplifier system consisting of more than one stage of amplification, the overall characteristic is the product of amplification curves of the individual stages. As a result, the 3 dB bandwidth of a multistage amplifier composed of identical single-tuned stages is

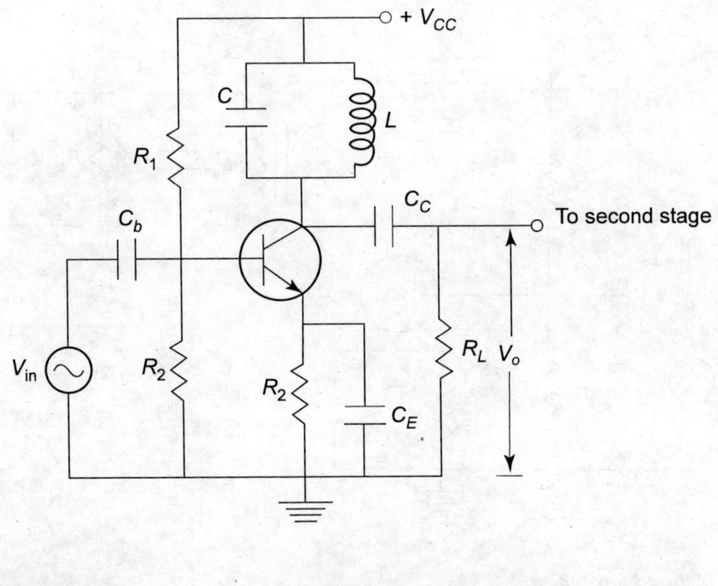

Fig. 12.16 *Single-tuned amplifier*

less than the bandwidth of a single stage. Thus, in order to maintain a given bandwidth, it is necessary that effective quality factor Q_{eff} of the individual stages be decreased as more stages are incorporated in the amplifier system. It is also to be noted that as the number of stages increases, the response curve becomes flatter on the top and steeper on the sides. The half-power bandwidth of an n-stage tuned amplifier is

$$B = \Delta\sqrt{[2^{1/n} - 1]} \qquad ...(12.44)$$

where Δ is the bandwidth of individual stage and n is the number of stages.

The advantage of the tuned amplifier is that the power loss in the load is very small. This is since the load resistance is very small and the load consists of inductance and capacitance. The conversion efficiency is very high and thus, the power to be supplied by the dc power supply is very small.

These tuned amplifier circuits have a limitation at audio frequencies. As the frequency of operation reduces, the inductance and capacitance become bulky. Also, as these are narrowband amplifiers, they cannot cover the entire audio-frequency range.

The value of the quality factor becomes extremely high as the selectivity of the circuit increases. At high values of Q, the reproducibility of the signal is poor. Thus, a *double-tuned amplifier* is suggested as discussed in the following section.

12.13 DOUBLE-TUNED AMPLIFIER

The *double-tuned amplifier* employs load impedance consisting of two circuits resonant at the same frequency and coupled together as shown in Fig. 12.17.

The frequency response of the double-tuned amplifier depends on the degree of coupling between the inductors in the two tank circuits. This degree of coupling is

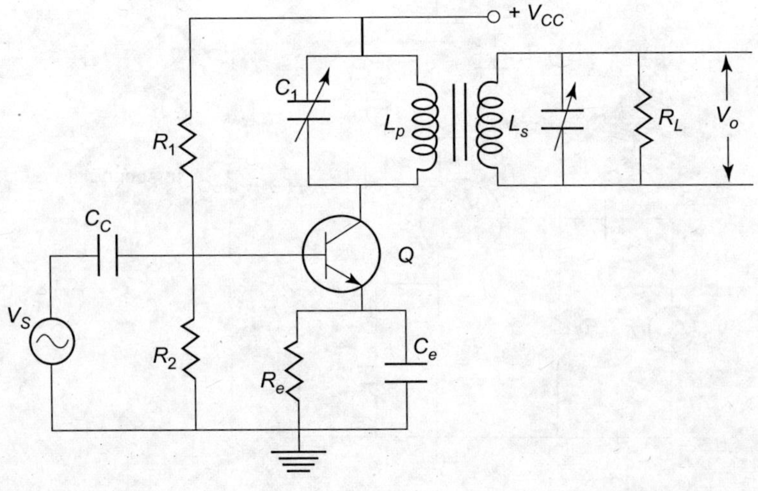

Fig. 12.17 *Double-tuned amplifier*

specified by the coefficient of coupling. If the inductances of the two inductors are L_p and L_s and M is the mutual inductance, then the coefficient of coupling is given by

$$k = M/\sqrt{[L_p L_s]} \quad ...(12.45)$$

The power transferred to the secondary circuit is maximum when the resistance that the secondary circuit reflects into the primary is equal to the resistance of the primary circuit. The value of the coefficient of coupling when this happens is called *critical coefficient of coupling.* This is equal to

$$k_c = 1/\sqrt{[Q_p Q_s]} \quad ...(12.46)$$

Therefore, $$(\omega M)^2 = R_p R_s \quad ...(12.47)$$

where R_p and R_s are resistances of primary and secondary circuits.

The exact shape of the frequency response curve depends on the ratio k/k_c. When this ratio is greater than 1, the system is *overcoupled,* when it is less than 1, it is *undercoupled* and when $k = k_c$, it is referred to as *critically coupled.* The three conditions of coupling are as shown in Fig. 12.18.

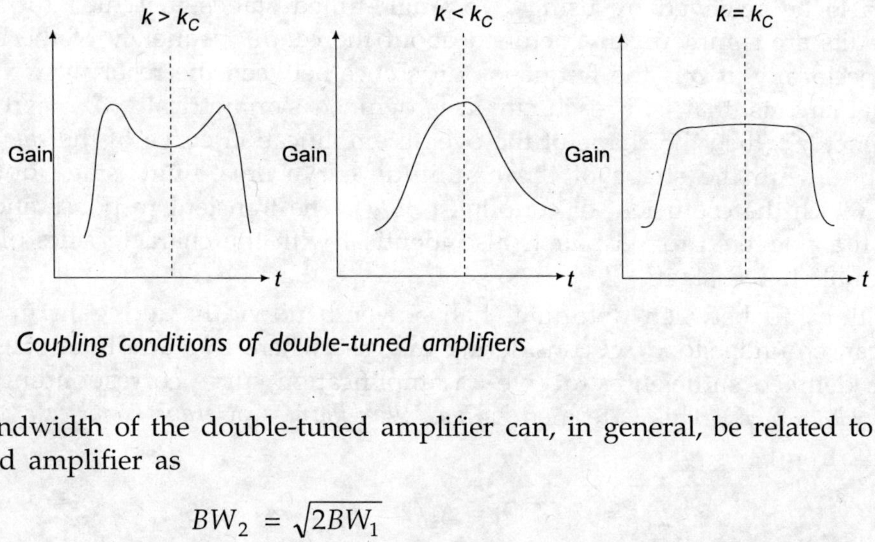

Fig. 12.18 *Coupling conditions of double-tuned amplifiers*

The bandwidth of the double-tuned amplifier can, in general, be related to that of the single-tuned amplifier as

$$BW_2 = \sqrt{2BW_1} \quad ...(12.48)$$

where BW_2 and BW_1 are bandwidths of double- and single-tuned amplifiers, respectively.

Compared to a single-tuned amplifier of the same bandwidth, the frequency response of a double-tuned amplifier is flatter near the resonant frequency and the roll off at the edges is steeper. These are the important characteristics of a double-tuned amplifier.

When a number of identical double-tuned stages of amplifiers are connected in cascade, the overall bandwidth of the system is thereby narrowed and steepness of the sides of the characteristic is increased. Thus,

$$B = \Delta_2 \sqrt[4]{[2^{1/n} - 1]} \quad ...(12.49)$$

where Δ_2 is the bandwidth of single stage of such a system and n number of stages are cascaded. It can be seen that the percentage narrowing of the bandwidth that occurs with increasing number of stages is much less in the double-tuned amplifier than when single-tuned stages are employed. This is the result of flatter top and steeper sides of the characteristic of individual double-tuned stages.

As the number of stages is increased, the superiority of the single-tuned system rapidly disappears because of the greater bandwidth narrowing in the single-tuned systems. Also, for a given number of stages of double-tuned system always gives more off-channel selectivity than does the corresponding single-tuned amplifier.

Double-tuned amplifiers find extensive use as intermediate frequency amplifiers in radio receivers. In such applications, it is customary to make the stage slightly undercoupled. This gives slightly better off-channel selectivity and by causing the response curve to have a slight peak makes it easier to adjust the individual circuits to resonance at a common frequency.

12.14 STAGGER-TUNED AMPLIFIER

A curve of amplification as a function of frequency such as that obtained with double-tuning can also be achieved by using two single-tuned stages in which the individual resonant circuits are symmetrically detuned about the centre frequency of the band to be amplified. Specifically, if δ is the frequency difference between the resonant frequencies of the individual circuits, that is, if each circuit is detuned symmetrically $\delta/2$ hertz from the centre frequency f_o, then the shape of the overall amplification curve of the *staggered tuned pair* is identical with the shape of the amplitude curve of a single-stage double-tuned amplifier in which the coefficient of coupling is δ/f_o. The transient response and envelope delay of such a staggered tuned pair is also identical with the characteristics of the corresponding double-tuned stage.

The relationship between detuning, half-power bandwidth, and gain in a stagger-tuned pair can be understood by considering the case where two amplifier stages initially identical are detuned sufficiently to give an amplification curve corresponding to critical coupling ($k = k_c = \delta/f_o$). This is termed as *maximum flatness* or *transitional case.*
Detuning $\delta/2$ from

$$f_o = f_o/(2Q) = \Delta_1/2 = \Delta_2/2\sqrt{2} \qquad \text{...(12.50)}$$

where Q is the value applicable to the individual resonant circuits of the amplifier, Δ_1 is half-power bandwidth of a single-tuned circuit and Δ_2 is half-power bandwidth of staggered tuned pair. The amplification characteristics of the individual detuned circuits are then related to each other as shown in Fig. 12.19. It can be observed that the centre frequency f_o is now at 70.7% of individual amplification since the difference $\delta/2$ between the resonant frequency and the centre frequency is $f_o/(2Q) = \Delta_1/2$.

The overall response of the two-stage tuned pair is compared with the corresponding two-stage nonstaggered pair having same resonant circuits in Fig. 12.18(c). It can be seen that staggering reduces the total amplification of the centre frequency to 50% of the value obtained before detuning, since at the frequency each detuned stage has an amplification

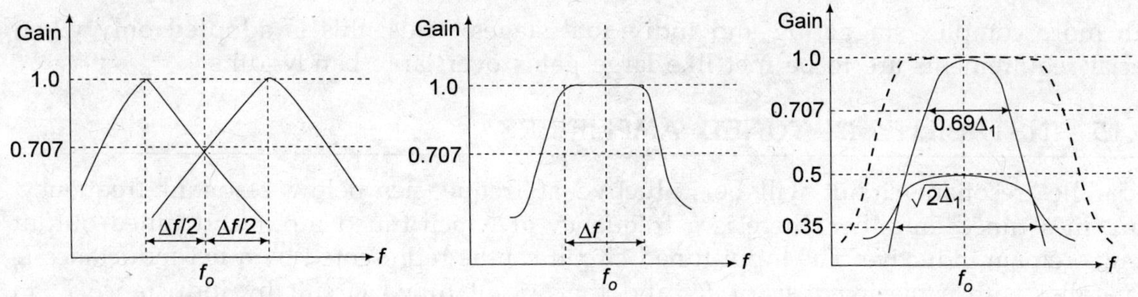

Fig. 12.19 *Frequency response of stagger-tuned amplifier*

that is 0.707 of the peak amplification of an individual stage. Thus, the equivalent voltage amplification per stage of the staggered pair is 0.707 times as great as when the same two stages are used without staggering. However, since the amplification curve of the stagger-tuned pair has a bandwidth corresponding to that of a critically coupled double-tuned stage, the half-power bandwidth Δ_2 of the staggered pair is $\sqrt{2}$ times as great as the half-power bandwidth $\Delta_1 = f_o/Q$ of an individual single tuned stage. Hence, the equivalent gain bandwidth product per stage of a staggered tuned pair is 0.707 $(\sqrt{2})$ = 1.0 times that of the individual single-tuned stages.

By readjusting circuit Q's and detuning to give the staggered pair, the same bandwidth Δ_1 previously possessed by an individual single-tuned stage, one can obtain a staggered pair that has the same equivalent voltage amplification per stage as does the single-tuned stage. Thus, two-stage staggered system will have the characteristics as shown in Fig. 12.19. In comparison, the two-stage identically tuned system with same gain per stage as this staggered pair will have a bandwidth only 0.644 times as great as does the staggered pair. This is because of the bandwidth narrowing that results when two identically tuned single-tuned stages are cascaded. In addition to being greatly superior to the two-stage identically tuned single-tuned system, the staggered pair has a performance almost as good as two double-tuned stages, and is both simpler and less expensive than using double-tuning.

The staggering of single-tuned stages introduces very little practical complications. To align the staggered system, a test voltage is applied separately to the two stages and these two stages are tuned to the required response. The gain and the bandwidth of the stages are adjusted for the required characteristics. Then, the two are cascaded and readjusted to achieve the required total response of the staggered pair amplifier.

The staggered tuning concept can be extended to more complicated systems when the gain achievable with a stagger pair amplifier gives very little gain compared to the required level. In place of each amplifier stage as in case of a stagger pair, a multistage arrangement consisting of two or more transistors is used. Staggering of frequency would be now done between of two similar multistage pairs. For example, double-tuned amplifier pair systems can be staggered with different Q's and coupling coefficients for different stages. When staggering is carried out properly, greater gains can be achieved with more complex staggering and individual stages. Thus, this is adopted only when special requirements are to be met like large gains over large bandwidths.

12.15 INSTABILITY IN TUNED AMPLIFIERS

A parallel resonant circuit will be inductive at frequencies below resonant frequency. Thus, in a tuned amplifier, there is a frequency at which tuned input and tuned output circuits remain inductive. The input tuned circuit can be represented by a net inductance L_i in parallel with a biasing resistor R_B and the output tuned circuit by inductance L_o in parallel with collector resistance R_C. The resistance values at input and output can be neglected when their values are large compared to inductances. Thus, the equivalent circuit of the amplifier consists of L_i, L_o and $C_{be,}$ forming a resonant circuit with resonant frequency

$$\omega_o = 1/\sqrt{[L_o + L_i)C_{b'e}]} \qquad ...(12.51)$$

At this resonant frequency, the loop of the system develops self-sustained oscillations. Noise, the random voltage variations, pushes the system into oscillatory mode: all the required conditions for oscillations are satisfied. Then, the amplifier is said to be *instable.* In order to avoid this instability, the resistance values at the input and output of the tuned circuit should be made small compared to the reactance of the inductances. This does not satisfy the condition for oscillation and so, the oscillations are not possible. Since the resistance values are decreased, the gain of the amplifier decreases.

In addition to the above, the stability of the amplifier is affected by the capacitance C_{be} which provides positive feedback. This can be avoided by applying proper coupling between the input and output circuits. This can be achieved with small values of C_{be}. Since the capacitance, $C_{b'e}$, the capacitance between the base and collector, is of significant value at high frequencies, the circuits operating at these high frequencies are unstable and oscillations are possible.

This effect can be reduced by any one of the following methods:

- neutralisation
- unilaterlisation
- mismatch technique

Neutralisation: *Neutralisation* is a technique, that reduces the oscillations and improves stability by reducing the feedback. In this method, the feedback is cancelled by introducing a current that is equal to but out of phase by 180° with the feedback signal at the input of the active device. Thus, these two signals are cancelled out and the effect of the feedback can be eliminated. This neutralises the energy feedback through the inter-electrode capacitance.

Adding a neutralising coil L_n and capacitor C_n from the output to the base, the effect of feedback can be neutralised in *hazeltine or neutrodyne method of neutralisation*. Such a circuit is as shown in Fig. 12.19 along with its equivalent circuit. By choosing proper vales of L_n and C_n, the bridge can be balanced to tune out the effect of feedback in the circuit.

When a push-pull configuration is used in a tuned amplifier, the neutralisation can be achieved by cross-feeding the energy from one electrode to the other as shown in

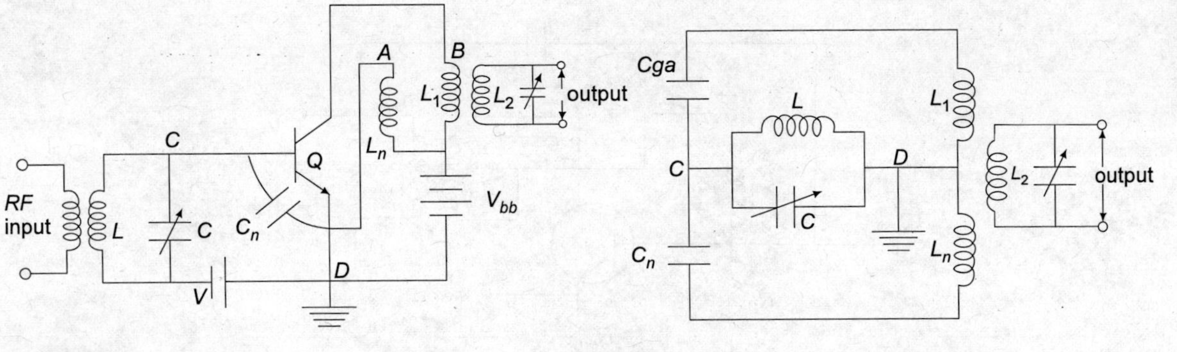

Fig. 12.20 *Hazeltine or neutrodyne method of neutralisation*

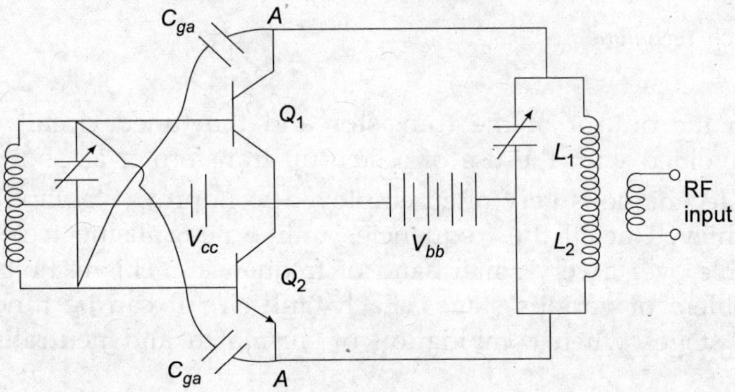

Fig. 12.21 *Cross-neutralisation method*

Fig. 12.21. This method is referred to as *cross-neutralisation*. The required phase-shift 180° results from the fact that the two transistors have energies at the collector, which are out of phase by 180°.

Unilaterlisation: In neutralisation, only the effect of reactive component of the feedback is neutralised. But, in *unilaterlisation*, both reactive as well as resistive effects are removed. Reflecting equivalent impedance across input and output ports can is the method adopted in unilaterlisation. The reflected capacitance across the input is known as *Miller capacitance*, $C_f(1 + |A_v|)$, where C_f is the feedback capacitance. Any impedance Z_f which is a feedback impedance will get reflected across the input as $Z_f/(1 + |A_v|)$. This effect is called *Miller effect*. This reduces the input voltage v_i. Thus, the output voltage, as well as the gain, is also considerably reduced. Thus, this method reduces the feedback effect and also the gain of the amplifier.

Mismatch Technique: The load impedance of a tuned amplifier is generally very small and hence a large current flows through this impedance. This results in a very little feedback to the input through the neutralising capacitance C_n. But, one requires high impedance to be presented to the tank circuit in order to achieve a high Q value. Thus, there is a

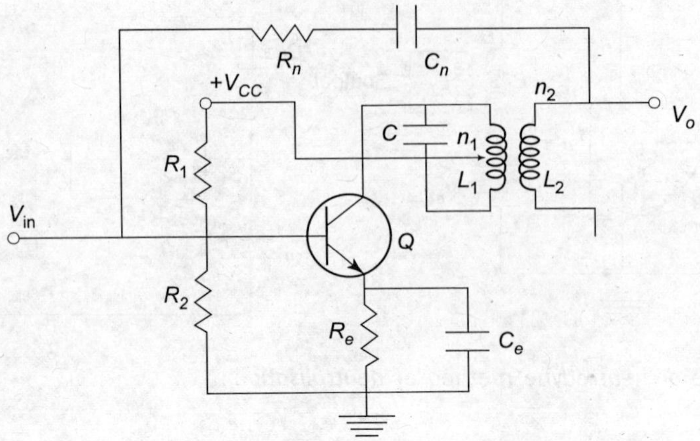

Fig. 12.22 *Mismatch technique*

mismatch between the output of the transistor and the tuned circuit. This problem of mismatch can be avoided with the use of a step-up transformer as shown in Fig. 12.22.

This mismatch technique is very often employed to improve stability. The advantage is that stability is achieved at all the frequencies unlike neutralising techniques where the stability is achievable over a very small band of frequencies. This technique also decreases the alignment problem of various stages. Each tank circuit can be tuned without much effect on adjacent stages when combination of mismatch and neutralisation techniques are employed.

SUMMARY

- Power amplifiers are those that involve large signal amplification.
- Linear approximation of the active devices is not possible and so graphical analysis is employed and the analysis depends on the selection of operating point.
- The conversion efficiency is the parameter that gives the amplifier the ability to convert the dc power to ac signal power.
- When the current flows for 100% time of the full cycle at the output of the amplifier, the amplifier is class *A* type. In this, the conversion efficiency of the amplifier is very small and distortion due to harmonics is minimum.
- To improve the efficiency of the class *A* amplifier, transformer coupling is used. To reduce the harmonic distortion, two transistors in push-pull configuration are employed. This eliminates the even harmonics totally.

- Class B amplifier has current only for 50% of the cycle and thus conversion efficiency is improved to a large extent. The nonlinearity in the input of the transistor causes crossover distortion.
- Crossover distortion can be eliminated by the use of class AB amplifier where a small current is allowed to be passed above 50% of the cycle, which eliminates the threshold voltage effect of the input of the transistor.
- Efficiency of the amplifier tends to a very large value when the output current of the amplifier is in form of short bursts. This is achieved in the class C amplifier where the harmonic distortion is at its peak.
- Tuned parallel resonant circuits are used to generate the required voltage signals in the class C amplifier. These circuits are referred to as tuned amplifiers.
- Single-tuned amplifiers have the drawback of small bandwidth and high Q of the tank circuit makes the reproducibility of the signal poor.
- Double-tuned amplifier utilizes two tuned circuits that are coupled properly.
- Staggered tuned amplifier increases the bandwidth of the tuned amplifier by use of two tuned circuits, which are slightly tuned away from resonant input frequency.
- Various techniques are employed in the tuned amplifiers to reduce instability, which is caused due to the possibility self-sustained oscillations in the amplifier.

SOLVED PROBLEMS

12.1 Input signal consisting of two frequency components, ω_1 and ω_2, $i_b = I_1 \cos \omega_1 t + I_2 \cos \omega_2 t$ is applied on a network which is characterized as $i_c = k_1 i_b + k_2 i_b^2$. Show that the output will contain a dc term and frequency components of ω_1, ω_2, $2\omega_1$, $2\omega_2$, $\omega_1 + \omega_2$ and $\omega_1 - \omega_2$.

Solution Given that

$$
\begin{aligned}
i_b &= I_1 \cos \omega_1 t + I_2 \cos \omega_2 t \\
i_c &= k_1 i_b + k_2 i_b^2 \\
&= k_1(I_1 \cos \omega_1 + I_2 \cos \omega_2 t) + k_2(I_1 \cos \omega_1 t + I_2 \cos \omega_2 t)^2 \\
&= k_1(I_1 \cos \omega_1 t + I_2 \cos \omega_2 t) + k_2(I_1^2 \cos^2 \omega_1 t + I_2^2 \cos^2 \omega_2 t \\
&\quad + 2I_1 I_2 \cos \omega_1 t \cos \omega_2 t) \\
&= k_1(I_1 \cos \omega_1 + I_2 \cos \omega_2 t) + k_2\left(\frac{I_1^2(1-\cos 2\omega_1 t)}{2}\right. \\
&\quad \left. + \frac{I_2^2(1-\cos 2\omega_2 t)}{2} + I_1 I_2(\cos(\omega_1 + \omega_2)t + \cos(\omega_1 - \omega_2)t)\right) \\
&= A_0 + A_1 \cos \omega_1 t + A_2 \cos \omega_2 t + A_3 \cos 2\omega_1 t + A_4 \cos 2\omega_2 t \\
&\quad + A_5 \cos(\omega_1 + \omega_2)t + A_6 \cos(\omega_1 - \omega_2)t
\end{aligned}
$$

where A_0, A_1, A_2, A_3, A_4, A_5, A_6 are constants.

So, we can say that the output will contain a dc term, and sinusoidal terms of frequency ω_1, ω_2, $2\omega_1$, $2\omega_2$, $\omega_1 + \omega_2$ and $\omega_1 - \omega_2$.

12.2 A transistor supplies 0.85 W to a 5 K load. The zero signal dc collector current is 30 mA, and the dc collector current with signal is 36 mA. Determine the per cent second-harmonic distortion.

Solution Power output = 0.85 W

$$\text{Load } R_L = 5\text{ k}\Omega$$

$$I_C = 30\text{ mA}$$

We know that $$P = \frac{A_1^2 R_L}{2}$$

$$\Rightarrow \quad 0.85 = \frac{A_1^2 \times 5 \times 10^3}{2}$$

$$A_1^2 = 0.34 \times 10^{-3}$$

$$\therefore \quad A_1 = 18.43 \times 10^{-3}$$

$$A_0 = A_2 = (36 - 30) \times 10^{-3}$$

$$= 6 \times 10^{-3}$$

$$\therefore \quad D_2 = \left|\frac{A_2}{A_1}\right| \times 100$$

$$= 32.56\%.$$

12.3 The input excitation of an amplifier is $i_b = I_{bm} \sin \omega t$. Prove that the output current can be representated by a Fourier series which contains only odd sine components and even cosine components.

Solution Assume that the dynamic curve can be represented by a second-order relationship given by

$$i_c = k_1 i_b + k_2 i_b^2$$

$$i_b = I_{Bm} \sin \omega t$$

$$\therefore \quad i_c = k_1 I_{Bm} \sin \omega t + k_2 I_{Bm}^2 \sin^2 \omega t$$

$$= k_1 I_{Bm} \sin \omega t + \frac{k_2 I_{Bm}^2}{2} (1 - \cos 2\omega t)$$

$$= A_0 + A_1 \sin \omega t + A_2 \cos 2\omega t$$

So, we can say that for a second-order relationship only, output will contain a dc term, an odd sine component, and even cosine component. Similarly, if the dynamic curve is represented by a higher-order terms the output will contain all odd sine components and even cosine components.

12.4 A sinusoidal signal $V_s = 1.95 \sin 400t$ is applied to a power amplifier. The resulting current is $i = 12 \sin 400t + 1.2 \cos 800t + 0.9 \sin 1200t + 0.4 \cos 1600t$. Find (a) the total harmonic distortion (b) percentage increase in power because of distortion.

Solution (a)

$$D_2 = \left|\frac{A_2}{A_1}\right| = \frac{1.2}{12} = 0.1$$

$$D_3 = \left|\frac{0.9}{12}\right| = 0.075$$

$$D_4 = \left|\frac{0.4}{12}\right| = 0.033$$

$$\therefore \quad D = \sqrt{D_2^2 + D_3^2 + D_4^2} = \sqrt{(0.1)^2 + (0.075)^2 + (0.033)^2}$$

$$= 0.1294 \text{ or } 12.94\ \%$$

$\therefore$ Total harmonic distortion = 12.94%.

(b)

$$P = (1 + D^2)P_1 = (1 + (0.1294)^2)P_1$$

$$= 1.01674\, P_1$$

$\therefore$ Percentage increase in power because of distortion $= \dfrac{P - P_1}{P_1} \times 100$

$$= 1.674\ \%$$

12.5 Calculate the input and output power, circuit efficiency for the circuit shown below, if the input base current is 5 mA peak.

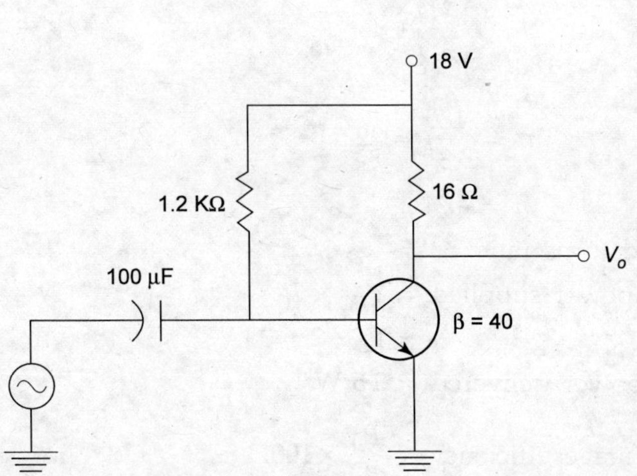

Solution First, we have to determine *Q*-point value.

$$I_B = \frac{V_{CC} - V_{BE}}{R_B} = \frac{18 - 0.7}{1.2\,\text{k}\Omega} = 14.416 \text{ mA}$$

$$I_C = \beta I_B = 40 \times 14.416 \times 10^{-3} = 0.576 \text{ A}$$

$$\therefore \quad V_{CE} = V_{CC} - I_C \cdot R_C = 8 - 0.576 \times 16 = 8.784 \text{ V}$$

ac collector current is $i_c = \beta i_b = 40 \times 5$ mA

$$= 200 \text{ mA}$$

$$P_{ac}, \text{ ac output power } = \frac{i_c^2 R_C}{2} = \frac{200 \times 200 \times 10^{-6} \times 16}{2}$$

$$= 32 \times 10^{-2} = 320 \text{ mW} = 0.32 \text{ W}$$

$$P_{dc}, \text{ dc input power } = V_{CC} \cdot I_C$$

$$= 18 \times 0.576$$

$$= 10.368 \text{ W}$$

$$\text{circuit efficiency } = \frac{P_{ac}}{P_{dc}} \times 100$$

$$= 3.08\%$$

12.6 For a particular amplifier, the optimum load impedance is 180 Ω. Calculate the turns ratio required to a match an 8 Ω load to this transistor. If the amplifier takes a mean collector current of 2 A from a 15 V supply, and delivers an ac load power of 2.5 W to the transformer coupled-load, calculate the efficiency and the collector dissipation (neglecting the losses).

Solution In a transformer-coupled amplifier

$$R_L' = n^2 \cdot R_L$$

$$180 = n^2 \cdot 8$$

$$\Rightarrow \quad n^2 = \frac{180}{8} = 22.5$$

$$\Rightarrow \quad n = 4.7$$

$\therefore$ required turns ratio = 4.7

$$P_{dc}, \text{ dc power supplied } = V_{CC} \times I_C$$

$$= 15 \times 2 = 30 \text{ W}$$

$$P_{ac}, \text{ ac power delivered } = 2.5 \text{ W}$$

$$\text{percentage efficiency } = \frac{P_{ac}}{P_{dc}} \times 100$$

$$= \frac{2.5}{30} \times 100 = 8.33\%$$

Collector dissipation is $= P_{dc} - P_{ac}$

$$= 30 - 2.5 = 27.5 \text{ W}.$$

12.7 A power transistor operating in Class A in figure below is to deliver a maximum of 5 W to a 4 Ω load. The supply voltage is $V_{CC} = 20$ V and $V_{min} = 0$.

(a) Find the transformer turns ratio n.
(b) What is the peak collector current I_m?
(c) What is the Q-point value?
(d) What is the collector circuit efficiency?

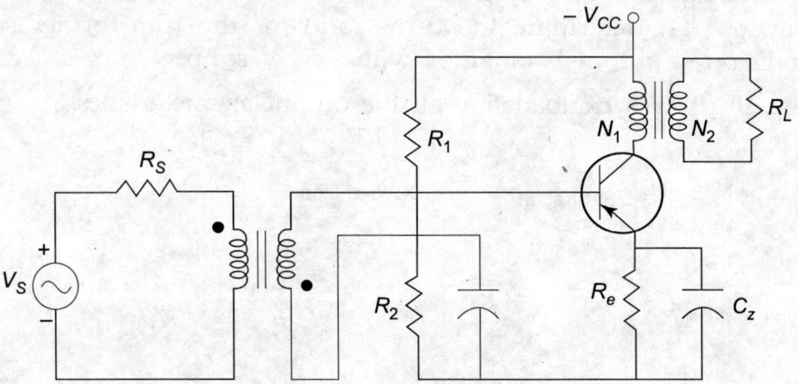

Solution (a)

$$\text{Power} = \frac{V_m^2}{2R_L'} = \frac{1}{2} \cdot \frac{V_{CC}^2}{R_L'} = 5 \text{ W}$$

$$\Rightarrow \quad R'_L = \frac{V_{CC}^2}{2 \times 5} = \frac{400}{100} = 40 = \frac{R_L}{n^2}$$

$$\Rightarrow \quad n^2 = \frac{R_L}{R_L'} = \frac{4}{40} = \frac{1}{10}$$

$$\therefore \quad n = \frac{1}{\sqrt{10}}$$

(b) ac power, $P_{ac} = \frac{1}{2} I_m^2 R_L'$

$$\Rightarrow \quad I_m^2 = \frac{2P_{ac}}{R_L'} = \frac{2 \times 5}{40} = \frac{1}{4}$$

$$\Rightarrow \quad I_m = 1/2 = 0.5 \text{ A}$$

(c) $$I_c = I_m = 0.5 \text{ A}$$

$$V_{CE} = V_{cc} = 20 \text{ V}$$

$\therefore$ *Q*-point is 20 V, 0.5 A

(d) $$\% \eta = \text{collector circuit efficiency} = \frac{P_{ac}}{P_{dc}} \times 100$$

$$= \frac{5}{20 \times 0.5} = 50\%$$

12.8 A single-ended class *A* amplifier is to drive a 15 Ω load via a transformer. The transistor has a maximum dissipation of 10 W, and the characteristics are ideal, apart from a saturation region defined by a line through the origin at a slope corresponding to 5 Ω, determine the turns ratio for the transformer so that the maximum undistorted output is obtained with a 50 V supply.

Solution First, we will draw a dc load line on the output characteristics.

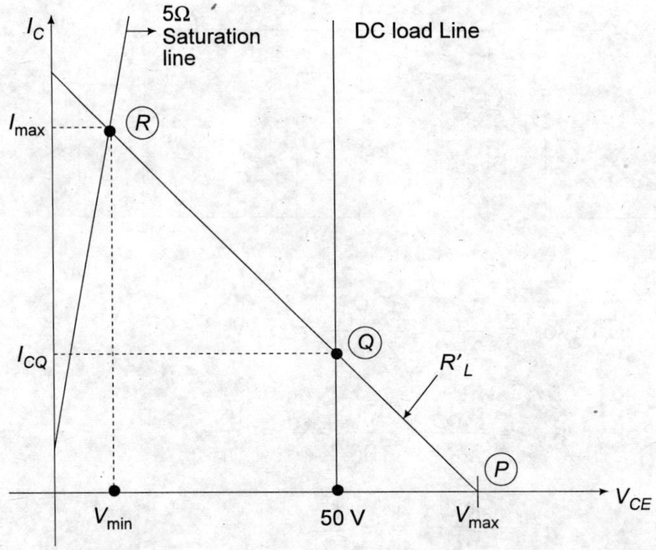

Power rating of the transistor is equal to

$$P_{dc} = V_{CC} \cdot I_{CQ}$$

$$10 \text{ W} = 50 \cdot I_{CQ}$$

$\Rightarrow$ $$I_{CQ} = 0.2 \text{ A}$$

From the graph, $V_{min} = 5 \times I_{max}$

$$= 5 \times 2I_{CQ}$$

(Assuming that the curves are equidistant)

$$\therefore \quad V_{min} = 5 \times 2 \times 0.2 \text{ A}$$
$$= 2 \text{ V}$$

$$R'_L = \frac{50 - V_{min}}{I_{CQ}} = \frac{50 - 2}{0.2}$$
$$= 240 \ \Omega$$

We have $\quad R'_L = n^2 \cdot R_L$

$$\Rightarrow \quad 240 = n^2(15)$$

$$\Rightarrow \quad n^2 = \frac{240}{15} = 16$$

$$\therefore \quad n = 4$$

For maximum power, $PQ = RQ$ (from graph)

$$\frac{V_{max} - 50}{I_{CQ}} = R'_L = \frac{50 - V_{min}}{I_{CQ}}$$

$$\therefore \quad V_{max} = 50 + R_L^1 \, I_{CQ}$$
$$= 50 + 240 \times 0.2$$
$$= 98 \text{ V}$$
$$V_{min} = 50 - I_{CQ} R_L = 50 - 240 \times 0.2$$
$$= 2 \text{ V.}$$

Maximum ac output power is given by

$$P_{ac,\ max} = \frac{(V_{max} - V_{min})^2}{8R'_L}$$

$$= \frac{(98-2)^2}{8 \times 240} = 4.8 \text{ W}$$

dc input power $P_{dc} = V_{CC} \cdot I_{CQ}$

$$= 50 \times 0.2 = 10 \text{ W}$$

$$\therefore \quad \% \, \eta = \frac{P_{ac}}{P_{dc}} \times 100$$

$$= \frac{4.8}{10} \times 100 = 48\%$$

12.9 A single-ended class A power amplifier is coupled to an $8 \ \Omega$ load, using a transformer with a turn ratio of 5:1. With a 50 V supply the transistor is biased to have a quiscent collector current of 250 mA. When a sinusoidal signal is applied to the base, the collector voltage varies between a minimum of 5 V and maximum of 90 V. Estimate the efficiency, power output and second-harmonic distortion of this stage.

Solution

$$I_{CQ} = 250 \text{ mA}$$
$$V_{CC} = 50 \text{ V}$$

For a transformer-coupled amplifier

$$R'_L = n^2 R_L = (5)^2 \times 8 = 200\ \Omega$$

$$I_{min} = 250 - \frac{90-50}{200\ \Omega} \times 10^3$$

$$= 250 - 200 = 50 \text{ mA}$$

$$I_{max} = 250 + \frac{(50-5)\times 10^3}{200}$$

$$= 475 \text{ mA}$$

Signal power delivered to load is

$$P_{ac} = \frac{(V_{max} - V_{min})(I_{max} - I_{min})}{8}$$

$$= \frac{(90-5)(475-50)}{8} \text{ mW}$$

$$= 4.516 \text{ W}$$

$$\% \text{ efficiency } \eta = \frac{P_{ac}}{P_{dc}} \times 100$$

$$= \frac{4.156}{50 \times 250 \times 10^{-3}} \times 100$$

$$= 36.1\%$$

$$\% D_2 = \left| \frac{A_2}{A_1} \right| \times 100$$

$$A_2 = \frac{I_{max} + I_{min} - 2I_C}{4}$$

$$A_1 = \frac{I_{max} - I_{min}}{2}$$

$$\therefore \quad \% D_2 = \frac{I_{max} + I_{min} - 2I_C}{2(I_{max} - I_{min})} \times 100$$

$$= \frac{475 + 50 - 2 \times 250}{2(475-50)} \times 100$$

$$= 2.9\%$$

12.10 Calculate the efficiency of a transformer-coupled class *A* amplifier for a supply of 12 V and outputs of (a) $V_o = 12$ V, (b) $V_o = 6$ V, (c) $V_o = 2$ V.

Solution

$$V_{CEQ} = V_{CC} = 12 \text{ V}$$

(a)

$$V_{CE\,\max} = V_{CEQ} + V_0 = 12 + 12 = 24 \text{ V}$$

$$V_{CE\,\min} = V_{CEQ} - V_0 = 12 - 12 = 0$$

$$\therefore \quad \%\eta = 50\left(\frac{V_{CE\,\max} - V_{CE\,\min}}{V_{CE\,\max} + V_{CE\,\min}}\right)^2 \times 100$$

$$= 50 \times \left(\frac{24-0}{24+0}\right)^2 \times 100 = 50\%$$

(b)

$$V_{CE\,\max} = 12 + 6 = 18 \text{ V}$$

$$V_{CE\,\min} = 12 - 6 = 6 \text{ V}$$

$$\%\eta = 50 \times \left(\frac{18-6}{18+6}\right)^2 \times 100$$

$$= 50 \times \left(\frac{12}{24}\right)^2 \times 100$$

$$= 12.5\%$$

(c)

$$V_{CE\,\max} = 12 + 2 = 14 \text{ V}$$

$$V_{CE\,\min} = 12 - 2 = 10 \text{ V}$$

$$\%\eta = 50\left(\frac{14-10}{14+10}\right)^2 \times 100$$

$$= 50\left(\frac{4}{24}\right)^2 \times 100$$

$$= 1.39\%$$

12.11 A single transistor is operating as an ideal class *B* amplifier with a 5 K load. A dc meter in the collector circuit reads 20 mA. How much signal power is delivered to the load?

Solution We have $I_{dc} = \frac{I_m}{\pi}$ $I_{rms} = I_m/2$ for a class *B* power amplifer.

$$\text{Power} = I_{rms}^2 R_L$$

$$= \frac{I_m^2}{4} R_L = I_{dc}^2 \cdot \frac{\pi^2}{4} \cdot R_L$$

$$= \frac{\pi^2}{4} (20 \times 10^{-3})^2 \times 5 \times 10^3$$

$$= 4.9 \text{ Watts.}$$

12.12 For a class B amplifier providing a 22 V peak signal to an 8 Ω load, and a power supply of 25 V determine,
(a) input power, (b) output power and (c) circuit efficiency.

Solution (a) Load peak current $I_m = \frac{V_m}{R_L} = \frac{22}{8} = 2.75$ A

$$I_{dc} = \frac{2}{\pi} I_m = \frac{2}{\pi} 2.75 = 1.751 \text{ A}$$

$$P_{dc} = V_{CC} \cdot I_{dc} = 25(1.751) = 43.79$$

(b) $$P_{ac} = \frac{V_m^2}{2R_L} = \frac{(22)^2}{2 \times 8} = 30.25 \text{ W}$$

(c) $$\%\eta = \frac{P_{ac}}{P_{dc}} \times 100 = 69.08\%$$

12.13 For a class B amplifier with V_{CC} = 25 V, driving an 8 Ω load, determine: (a) maximum output power, (b) maximum input power and (c) maximum circuit efficiency.

Solution (a) Maximum output power is $P_{ac\,max} = \frac{V_{CC}^2}{2R_L}$

$$P_{ac,\,max} = \frac{(25)^2}{2 \times 8} = 39.0625 \text{ W}$$

(b) Maximum input power is $P_{dc} = \frac{2V_{CC}^2}{\pi R_L}$

$$= \frac{2 \times (25)^2}{\pi \times 8} = 49.76 \text{ W}$$

(c) Maximum circuit efficiency $\%\eta = \frac{P_{ac}}{P_{dc}} \times 100$

$$= 78.5\%$$

12.14 Calculate the efficiency of a class B amplifier for a supply voltage of V_{CC} = 22 V, driving a 4 Ω load with a peak output voltage of a (a) V_m = 20 V. (b) V_m = 4.0 V.

Solution (a) We have $\%\eta = 78.54 \dfrac{V_m}{V_{CC}} \times 100$ for class B amplifier

$$= 78.54 \times \frac{20}{22} \times 100 = 71.4\%$$

(b) $$\%\eta = 78.54 \times \frac{4}{22} \times 100 = 14.28\%$$

12.15 An input sinusoid signal in fed an ideal clan B push pull amplifier with $R_2 = 0$, V_{CC} = 20 V, $N_2 = 2N_1$, R_L = 20 Ω. The dc current gain h_{FE} is 20. For maximum output signal at $V_m = V_{CC}$, determine (a) output signal power, and (b) dissipation in each transistor collector.

Solution (a) $P_{ac} = \dfrac{1}{2} V_m \cdot I_m$

$$I_m = \frac{V_{CC}}{R'_L} \quad \text{where } R'_L = n^2 \cdot R_L$$

$$n = \text{turns ratio} = \frac{N_1}{N_2} = \frac{1}{2}$$

$$\therefore \quad R'_L = \frac{1}{4} \times 20 = 5\ \Omega$$

$$I_m = \frac{20}{5} = 4\text{ A}$$

$$V_m = V_{CC} = 20$$

$$P_{ac} = \frac{1}{2} \times 20 \times 4 = 40\text{ W}$$

(b) $$P_{dc} = \frac{2I_m}{\pi} \cdot V_{CC} = \frac{2 \times 4}{\pi} \times 20$$

$$= 50.8\text{ W}$$

Collector dissipation in single transistor is half that of the power dissipation in two transistors.

∴ collector dissipation is each transistor is

$$\frac{1}{2}(P_{dc} - P_{ac}) = \frac{1}{2}(50.8 - 40) = 5.4\text{ W}$$

12.16 Given an ideal class B push-pull amplifier whose collector supply voltage is Vcc, and $R_L^1 = n^2 R_L$ are fixed as base-current excitation is varied, show that the collector dissipation P_c is zero at no signal, rises as V_m increases and passes through a maximum at $V_m = 2V_{CC}/\pi$.

Solution We have

$$P_c = \frac{2}{\pi}\frac{V_{CC}\cdot V_m}{R_L'} - \frac{V_m^2}{2R_L'} \quad ...(1)$$

From (1), at $V_m = 0, P_c = 0$

when $V_m = \frac{4}{\pi}V_{CC}, P_c = 0$

P_c will be maximum when $\frac{dP_c}{dV_m} = 0$

$$\Rightarrow \frac{2}{\pi}\cdot\frac{V_{CC}}{R_L'} - \frac{2V_m}{2R_L'} = 0$$

$$\therefore \quad V_m = \frac{2V_{CC}}{\pi}$$

at this value of V_m, $P_c = P_{c\,max}$

$$P_{c\,max} = \frac{2}{\pi}\frac{V_{CC}}{R_L'}\cdot\frac{2V_{CC}}{\pi} - \frac{\left(\frac{2V_{CC}}{\pi}\right)^2}{2R_L'}$$

$$= \frac{4}{\pi^2}\cdot\frac{V_{CC}^2}{R_L'} - \frac{2V_{CC}^2}{\pi^2 R_L'}$$

$$= \frac{2}{\pi^2}\cdot\frac{V_{CC}^2}{R_L'}$$

From Eq. (1) P_c increases as V_m increases, and it will become maximum when $V_m = \frac{2V_{CC}}{\pi}$ and again becomes zero at $V_m = \frac{4V_{CC}}{\pi}$

12.17 (a) A transformer-coupled class A large signal amplifier has maximum and minimum values of collector-to-emitter voltage of 25 V and 2.5 V. Determine its collector efficiency.

(b) A transformer-coupled class A power amplifier supplies the power to an 80 Ω load connected across the secondary of a transformer having turns ratio of 5 : 1. If I_C = 120 mA, find maximum output power.

Solution (a)

$$\% \eta = 50\left(\frac{V_{CE\,max} - V_{CE\,min}}{V_{CE\,max} + V_{CE\,min}}\right) \times 100$$

$$= 50\left(\frac{25 - 2.5}{25 + 2.5}\right) \times 100$$

$$= 40.9\%$$

(b) We have $R_L' = n^2 R_L$

$$= (5)^2 \times 80 = 2000\ \Omega$$

$$I_{rms} = \frac{I_{max} - I_{min}}{2} \cdot \frac{1}{\sqrt{2}}$$

$$= I_c/\sqrt{2} \text{ since for maximum output power}$$

$$I_{max} = 2I_c,\ I_{min} = 0$$

$$\therefore \quad P_{ac} = I_{rms}^2 R_L' = \frac{I_c^2}{2} \cdot R_L'$$

$$= \frac{(120 \times 10^{-3})^2}{2} \times 2000$$

$$= 14.4\ \text{W}$$

EXERCISE PROBLEMS

12.1 Calculate the input power, output power, efficiency of the amplifier for an input voltage that results in a base current of 10 mA peak.

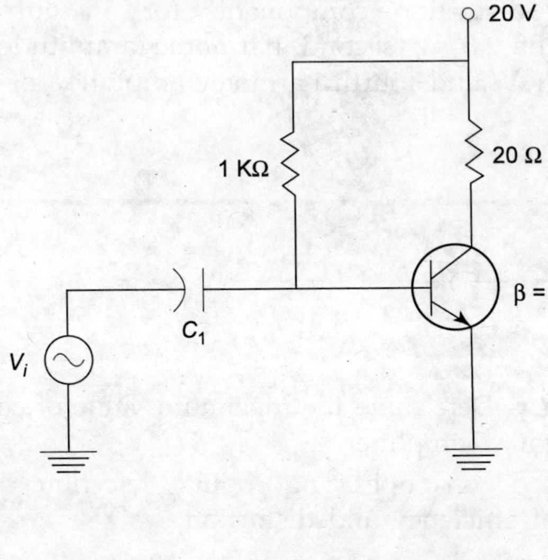

12.2 Calculate the ac power delivered to the 8 Ω speaker for the circuit shown below. The circuit component values result in a dc base current of 6 mA, and the input signal results in a peak base current swing of 4 mA.

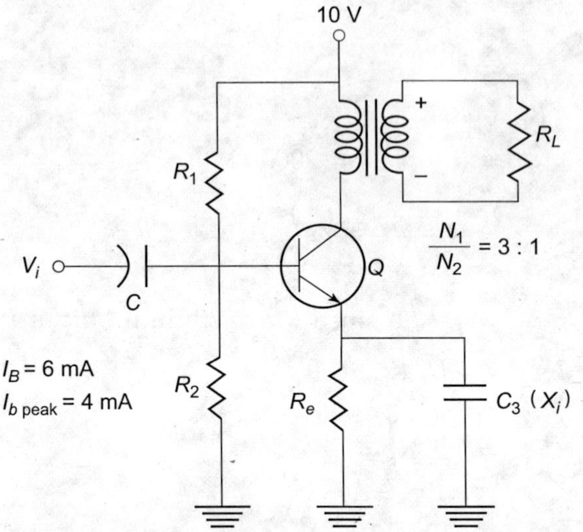

12.3 For a class *B* amplifier providing a 20 V peak signal to a 16 Ω load and a power supply of V_{CC} = 30 V determine the input power, output power, and circuit efficiency.

12.4 For a class *B* amplifier using a supply of V_{CC} = 30 V and driving a load of 16 Ω, determine the maximum input power, output power, and transistor dissipation.

12.5 Calculate the efficiency of a class *B* amplifier for a supply voltage of V_{CC} = 24 V, with peak output voltage of (a) V_m = 22 V (b) V_m = 6 V.

12.6 Calculate the harmonic distortion components for an output signal having fundamental amplitude of 2.5 V, second harmonic amplitude of 0.25 V, third harmonic amplitude of 0.1 V, and fourth harmonic amplitude of 0.05 V.

REVIEW QUESTIONS

1. Define conversion efficiency. Determine the maximum value of conversion efficiency for a series-fed class *A* power amplifier.
2. Classify amplifiers based on operating point selection. Compare various configurations in terms of efficiency and distortion.

3. Discuss how the efficiency of the class *A* power amplifier is improved by the transformer coupled amplifier configuration.
4. What is push-pull configuration and how does this circuit reduce the harmonic distortion? Discuss.
5. What is class *B* amplifier? Why is it employed? Give its circuit, design equations, characteristics and limitations.
6. Give a circuit of class *B* amplifier that eliminates transformer at both input and output. Discuss its operation.
7. How is crossover distortion eliminated in class *AB* amplifier? Discuss in detail.
8. What is class *C* amplifier? How are harmonics avoided in the output of such an amplifier?
9. Classify tuned amplifiers. What is single-tuned amplifier? Give its advantages and limitations.
10. Why is double-tuning employed in a tuned amplifier? Give its characteristics.
11. How is the bandwidth of a tuned amplifier improved? Give a circuit and discuss its operation in detail.
12. What are high-frequency limitations of a tuned amplifier? How are they neutralised?
13. Discuss about cascaded tuned amplifiers and give the effect of gain and bandwidth.
14. Discuss how rectification takes place in power amplifiers.
15. Derive an equation of output power of a class *B* power amplifier.
16. Discuss how the distortion due to harmonics is determined in a power amplifier. Give the equation for total power in the output of power amplifier.
17. Give the design procedure of a transformer-coupled amplifier in detail.
18. Draw a circuit that does not employ output transformer in a class *B* amplifier, and discuss its operation.
19. What is coefficient of coupling in a double-tuned amplifier? Discuss its effect on the frequency response.
20. Discuss the applications of tuned amplifiers.

CHAPTER 13

Operational Amplifiers

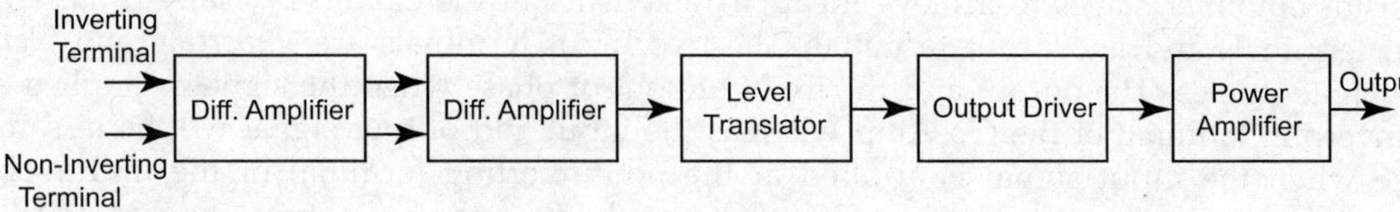

Block Diagram of Op-Amp

Chapter Outline

The concepts introduced in this chapter are:

- *Operational Amplifiers*
- *Block diagram of the Op-Amp*
- *Principle of Differential Amplifier*
- *Offset voltages and currents of an Op-Amp*
- *Various applications of Op-Amp*

13.1 INTRODUCTION

All the types of amplifiers discussed are designed for a specific application or use. They involve either transistor or FET circuits either in single or multi stage configurations. But, Operational Amplifier, known as *Op-Amp,* is an amplifier, which has a number of applications. Op-Amp is a cascade of large number of amplifier stages designed such a way that any electronic circuit whether linear or non-linear, digital or analog, amplifier or oscillator, comparator or converter, ac or dc signal processing, mathematical or logic computations is possible. Operational Amplifiers are the most popular electronic devices in use these days. They are versatile and easy to use. Its popularity is due to the fact that nearly ideal characteristics are achievable in practice. The feedback circuit obtains the performance of an Op-Amp circuit.

An Operational amplifier is an Integrated Circuit (IC) consisting of several stages of transistor amplifier stages to achieve ideal characteristics of the circuit. Basically it has two input terminals and an output terminal. The two input terminals are: *Inverting* and *Non-inverting* terminals. The output and the input are out of phase when the signal is applied at the inverting terminal of the Op-Amp whereas the input and output signal will be in same phase when the input signal is applied at the non-inverting terminal of the input. The Op-Amp has two types of amplifications: differential gain and common mode gain. Ideally the differential gain is infinity while the common mode gain is zero. That is, the Op-Amp amplifies the difference of the two input signals while the common mode signal is suppressed. In other words, if the input to the two terminals of the Op-Amp is same, the output will have to be zero.

Ideally, the Op-Amp has infinite gain. Thus, the Op-Amp circuit can be designed with feedback to the gain magnitude as required. The feedback controls the circuit parameters. The parameters of the circuit designed depend on the external circuit to the Op-Amp but not on the Op-Amp. Thus controlling and achieving required parameters is very easily obtained. The same Op-Amp can be used to achieve very high gain (several lakhs) or very low gain (less than unity) over the entire range of frequencies including zero (dc). Another advantage of the Op-Amp is its stability. Very high stability of the circuit is obtained. In addition to the stability, nonlinearity can be reduced, bandwidth can be improved, input and output impedances can be increased or decreased as per the circuit requirements. Another big advantage with Op-Amp is that replacing an Op-Amp does not affect circuit performance. The gain of the Op-Amp can either be positive or negative. That is either phase inversion between the output and input can be introduced or not.

Nonlinear amplifier design is also possible with Op-Amp. Proper selection of diode and capacitors can lead to the design of any type of nonlinear amplifiers. Precision rectifiers, peak detectors, peak-to-peak detectors, average value detectors or full wave rectifiers and so on can be easily be designed with Op-Amps. A high gain achievable with Op-Amp has another very important application of Op-Amp—comparators. The output of the Op-Amp changes its polarity whenever the input signal crosses zero voltage is the *Zero Cross Detector*. It can be converted to *Level Detector* when the reference voltage is changed to some other value, which is achieved by small change in bias. *Comparators with Hysteresis* are also to be designed with Op-Amp.

Any electronic signal processing can also be achieved with Op-Amp circuit. All types of filters viz. *Low Pass, High Pass, Band Pass, and Band Stop* can be designed with Op-Amps. The greatest advantage of these filters compared to the filters designed with passive devices is that these have gain and also precise frequency response can be obtained easily. They are very stable and use of inductors is eliminated. The ripple, phase, roll off of all the filters can be exactly controlled by selection of filter types—Chebyshev, Butterworth or Bessel etc. Any combinations of these characteristics are also possible with these active filters. Impedance matching either at input or output is easy.

All mathematical operations like summing, difference, multiplication, division, differentiation, integration, squaring, square root etc can be achieved with Op-Amp circuits. When nonlinear devices like diodes or transistors are used in feedback circuit of an Op-Amp circuit, logarithmic characteristics can be devised. Op-Amp circuits can solve differential equations also. In other words, Op-Amp circuit can be treated as *'analog computer'*. All types of multivibrators can be realized with Op-Amp. Flip flop or bistable multi, monostable and astable multivibrators designed with Op-Amp have superior performance when compared to transistor circuits. Power levels can be increased to any desired value with Op-Amp circuits.

Any form of signal can be generated with Op-Amp circuits. Sinusoidal, ramp, saw tooth, square, rectangular or triangular of any frequency, duty cycle, or amplitude can be designed with these Op-Amp Circuits. The step size, step period, and reset time can be independently controlled for stair case generators. Also the rise time and other parameters can be easily be controlled in case of square wave generators. All these adjustments can be done with voltage control. The Op-Amp makes voltage control of parameters much easier than is possible using discrete parts.

Op-Amp finds very wide application as regulator. Voltage regulators designed with Op-Amps can control the voltage, current or temperature very accurately. Special Op-Amp circuits designed with Op-Amps for regulator applications are referred to as *Monolithic Voltage Regulators.* Series pass, shunt, positive output, negative output, switching, foldback, current limited, floating, high voltage and precision regulators are few examples of regulators realized with Op-Amp circuits. In special applications, current regulators are possible. Sampling, sample & hold, analog to digital, digital to analog conversions required in digital circuit applications are few more examples of Op-Amp circuits. Modulators, demodulators, discriminators, detectors, frequency multipliers, limiters, simulators, time and phase circuits are other applications of the Op-Amp in communication engineering and allied fields.

An Op-Amp consists of several inbuilt transistors. Each transistor gain is limited to a value which is of very low of order of 100 and depends on the input and output circuit configuration. Whereas for the Op-Amp, the gain can extend to very large value tending towards infinity. The gain of Op-Amp depends on the feedback circuit involved in the circuit. Dc signals also can be amplified with Op-Amp. The bandwidth of a transistor depends on the gain bandwidth product value of the transistor amplifier circuit. Bandwidth of an Op-Amp is ideally infinity that is, signal of any frequency is amplified with same gain throughout the frequency range. The input impedance of the Op-Amp can be very large extending beyond 1 Mega ohms compared to transistor input impedance, which is of order of few kilo ohms. Similarly, the output impedance of Op-Amp can be very low even up to the order of 1 ohm whereas that of transistor is of order of hundreds or kilo ohms.

13.2 CHARACTERISTICS OF AN IDEAL OP-AMP

Operational Amplifier has very good performance with practical parameters tending towards ideal characteristics. Even though ideal parameters are not achieved in practice, it is interesting to note that nearly very satisfactory results are achieved. Thus, it is very important to understand the characteristics of an ideal Op-Amp. The following are the important characteristics of *Ideal Operational Amplifier*.

- Differential voltage gain, A_d is *infinity*.
- Common mode voltage gain, A_c is *zero*.
- Bandwidth is *infinity*.
- Input impedance is *infinity*.
- Output impedance is *zero*.
- Balanced output that is, output is *zero* when the same signal is applied to two input terminals.
- Parameters *do not* change with temperature.
- Equivalent noise is *zero*.
- Distortion is *zero*.

All the listed characteristics of the Op-Amp are for ideal case. In practical Op-Amp, all these characteristics may not be available at a time, but one can optimize the performance for given set of parameters by comprising on the remaining parameters. In a given application of an Op-Amp, all the parameters are not usually required at a time. Thus, the performance can be tuned for a particular application at the stake of the other unimportant parameters. For example, in a low noise amplifier, noise content can be made to be very low with the use of Op-Amp but the gain of the amplifier is to be compromised. Or, an amplifier designed for very large bandwidth may not give very high input impedance. Thus, optimizing all the parameters may not be possible, but it is very easy to tune the Op-Amp to perform for a set of optimum parameters.

With the characteristics of ideal Op-Amp as listed here, there are two basic rules, which can be derived, and which simplify the design of Op-Amp circuits. They are:

- Input terminals of the Op-Amp draw *no current*.
- Voltage across the input terminals is *zero*.

These two rules are by and large sufficient for design of any Op-Amp circuits. These two rules lead to concept of *virtual ground*. Virtual ground is a condition of short circuit between the two terminals of the Op-Amp. Fig. 13.1 shows an inverting amplifier. The input is fed at inverting terminal through an impedance Z and Z_f is the feedback impedance between inverting terminal and output of the Op-Amp. The non-inverting terminal of the input is at ground potential as shown. By the first rule, the device does not draw any current and so the whole of the input current will flow through the feedback impedance to the output. Also, by the second rule, since the potential drop between the two input terminals is zero, the inverting terminal is also at ground potential. That is a short circuit between the two terminals of the input virtually. This is referred to as virtual ground. This concept eases the design of Op-Amp to very large extent. The input voltage is entirely across the impedance Z and the output voltage is entirely across the feedback Z_f and the current in the two impedances is the same. Thus, calculating gain of the amplifier is simply the ratio of Z_f to Z with a negative sign since the input is fed at the inverting terminal. The impedance between the two terminals at the input is infinity and the voltage between them is zero! This is the concept of virtual ground.

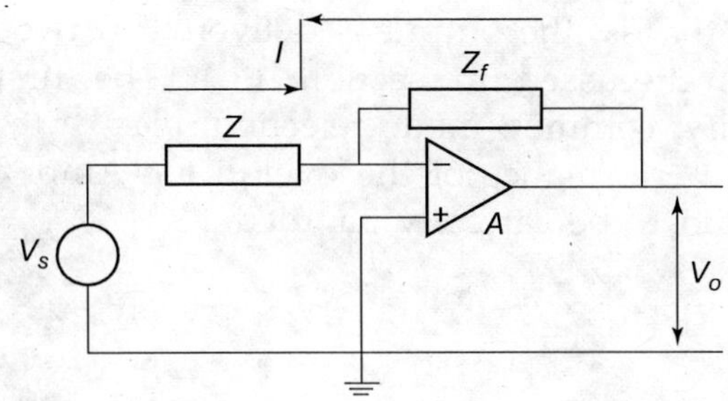

Fig. 13.1 *Concept of Virtual Ground*

In practice the Op-Amp exhibits some offset voltages and currents. For example, the output may not be zero when same input is applied at the two inverting and non-inverting terminals of the Op-Amp. In order to improve the performance of the Op-Amp, one has to adjust these offset parameters and use the Op-Amp. To be more specific, care should be taken to reduce these offset voltages or currents to zero before use of Op-Amp in a circuit. Appropriate voltages are to be applied at the offset adjustment terminals to make the errors to be zero. Also frequency compensation is to be taken care to get the optimum performance of the Op-Amp. The IC package of Op-Amp also brings out two terminals for the power supply biasing requirements.

13.3 BLOCK DIAGRAM OF OP-AMP

The Op-Amp has various transistor stages embedded in the IC. The basic block diagram of such an Op-Amp is as given in Fig. 13.2. The first and second stages of the Op-Amp are differential amplifiers as shown in the figure. The next stage is a level translator followed by a output driving circuit. The last stage of the Op-Amp is a power amplifier.

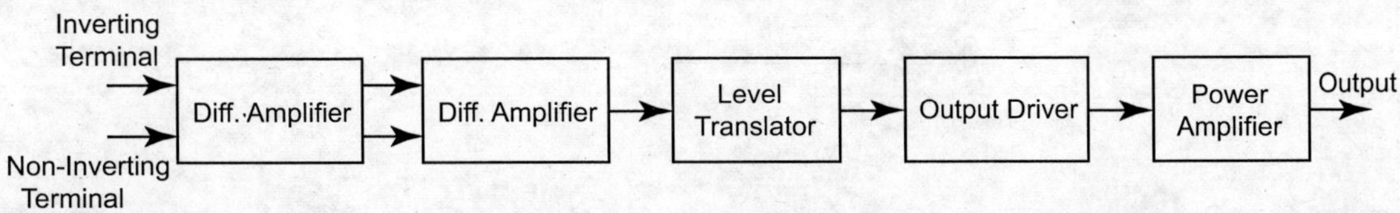

Fig. 13.2 *Block Diagram of Op-Amp*

The *Differential Amplifier* has two inputs and wither one or two outputs. The differential amplifier amplifies the difference of the two input signals and has very high differential gain. Two transistors are employed in this circuit of differential amplifier. The input differential impedance is large but in order to increase this value further, each transistor in the circuit can be replaced by Darlington pair. The performance of the amplifier depends on the common emitter resistance. The higher the resistance value, the higher the common mode rejection ratio achieved. Common mode rejection ratio is the ability of the circuit to eliminate common mode signals in the output and to amplify the difference signal. Thus, this biasing emitter resistance is replaced with a constant current source in order to improve gain and common mode rejection ratio. The first differential amplifier is double ended, that is, the output is delivered in two terminals. The detailed analysis of this circuit is discussed in section 13.4. The first differential amplifier determines the gain, stability, common mode rejection, bias drift, input impedance, slew rate, bandwidth and noise performance of the overall Op-Amp. Thus the design of this first stage is very critical and to be carefully handled.

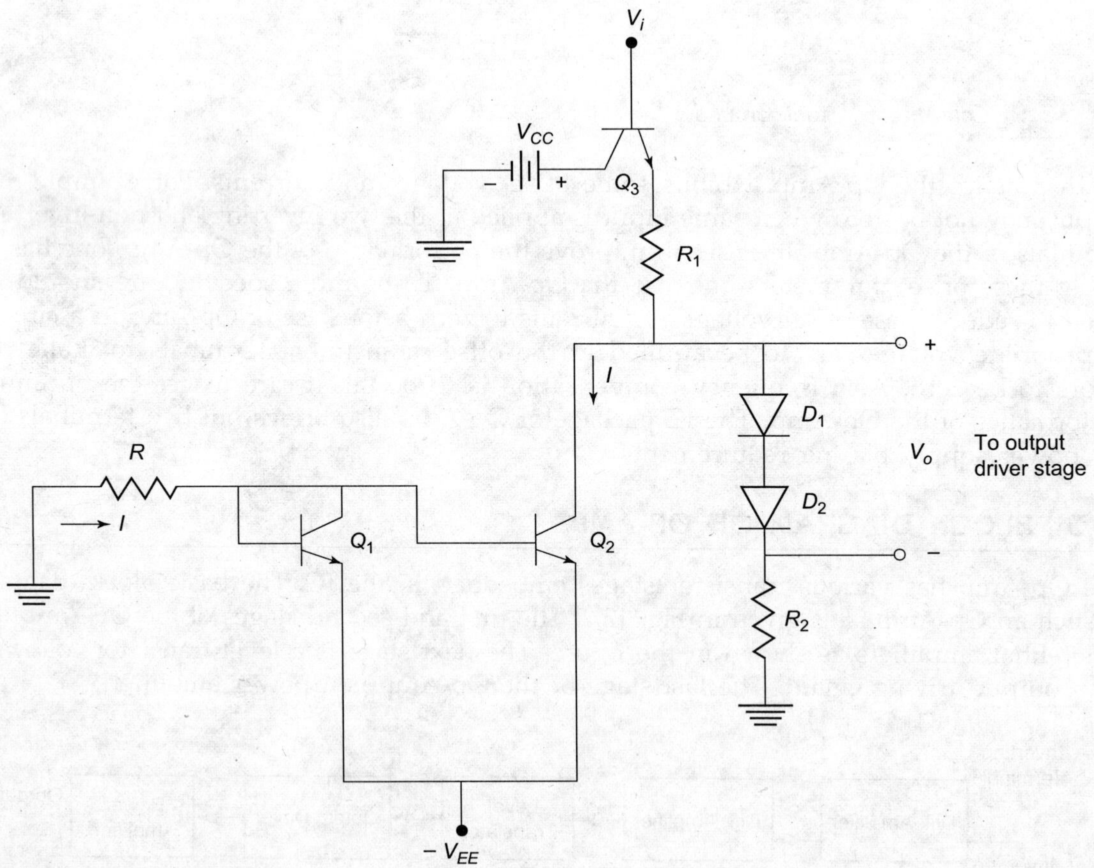

Fig. 13.3 *Level Translator*

The second differential amplifier of the Op-Amp is also similar to that of the first stage except that this is single ended, that is, only one terminal is available at the output. Also the design of this stage is not as critical as that of the first. Since common mode rejection is not of prime concern here, the emitter resistance can be retained while designing the amplifier for very large gain values.

The Op-Amp is to be designed for perfect balance. That is, when the same dc signal is applied to the two input terminals of the amplifier, the output should be zero and also the output should be zero when the two input terminals are grounded. This is to be ensured in the circuit of Op-Amp. Any residue of the voltages after the first two stages of amplifications is to be nullified and made zero. This is achieved by the *Level Translator* circuit. Since no coupling capacitor is used in the IC of Op-Amp, direct coupling is implemented. This increases the dc level of the output of each stage. In order to achieve good performance of the circuit, this dc level is to be adjusted back to zero after each stage output. If dc level is not checked, it will limit the output signal since the quiescent operating points are shifted. A simple level shifter circuit is as shown in Fig. 13.3. It is basically an emitter follower circuit. Thus it also acts as a very good buffer circuit to isolate the high differential amplifier stages with output power amplifier stages. Current mirror circuit is employed in the circuit in order to avoid attenuation of the signal. The level shifting can be given as

$$V_o - V_i = -\,(V_{BB} + I\,R_1) \qquad \text{...(13.1)}$$

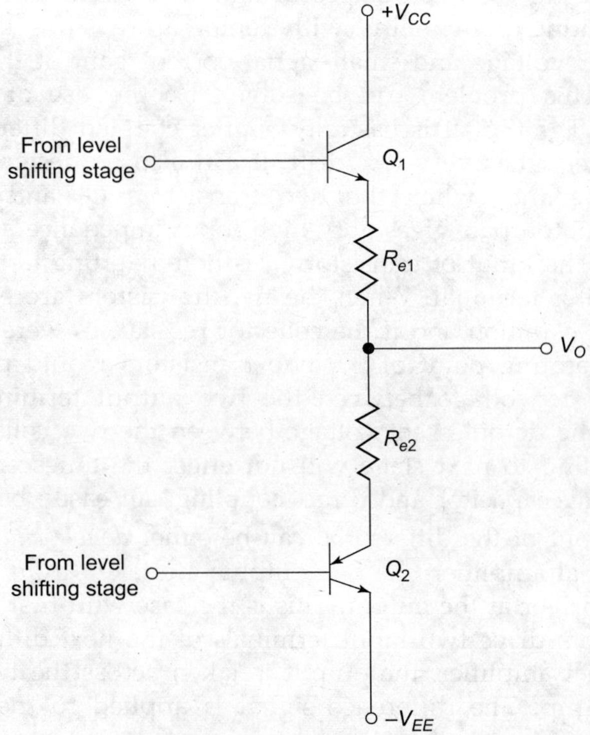

Fig. 13.4 *Output Driver*

The next stage of the Op-Amp is output driver circuit. This circuit is to provide sufficient driving current, very low output impedance and wide bandwidth. This normally employs an emitter with complementary transistors as shown in Fig. 13.4. The npn transistor handles the positive output signal and the pnp transistor handles the negative output signal. This stage drives the output of the Op-Amp and the current driving capacity of this circuit should be high so that loading of the device can be avoided what the current drawn by the output circuit.

Power Amplifier stage is an optional stage at the last stage of the Op-Amp. This is provided whenever very high power at the output is required.

13.4 DIFFERENTIAL AMPLIFIER

The purpose of a *differential amplifier* is to provide very large amplification to the difference signal and to suppress or eliminate the common mode signal (average of the two input signals) at the output. This increases the value of *Common Mode Rejection Ratio* (CMRR), which is defined as the ratio of differential gain to common mode gain. Also the input impedance of the circuit should be very high tending towards several mega ohms. The basic circuit diagram of a differential amplifier employing two similar transistors is as shown in Fig. 13.5. As no coupling capacitors are employed, the gain can be achieved even at dc signals.

The drawback of using cascaded stages of amplifier circuit for achieving very good gain is that the operating point shifts with temperature since the reverse saturation current, base to emitter voltage and small signal current gain of the transistors depends on the temperature. This problem can be solved by the use of balanced differential amplifier as shown in Fig. 13.5. It is basically emitter coupled differential amplifier. Since the circuit is symmetric, it has very low drift. It can also be designed to have very high differential input impedance when the two transistors Q_1 and Q_2 are replaced by Darlington pairs. Darlington pair gives very high input impedance as already discussed in the previous chapters. The input of transistor Q_1 is non-inverting terminal and the input to transistor Q_2 is inverting terminal. When the two transistors are exactly identical, since emitter resistance R_E is common and if the collector resistances were chosen same for both the transistors, the operating point of the two transistors would precisely be the same. This ensures that the dc voltage between the two output terminals of the transistors would be zero. Thus, the dc quiescent voltage between the two collector terminals will be zero and so, the coupling to next stage will not effect the quiescent point of that stage. This avoids the coupling capacitor and direct coupling can easily be incorporated without any difficulty. The output of the differential can be either *double ended* or *single ended.* In a double ended differential amplifier, the two outputs are taken from the two collectors and two more circuits are driver at the output. This is the case with first stage of Op-Amp. The outputs of two collectors drive two input terminals of the next differential amplifier. In a single ended differential amplifier, the output is taken across the two collectors as in the second stage of Op-Amp. The difference signal is applied to the load, which is level translator in this case.

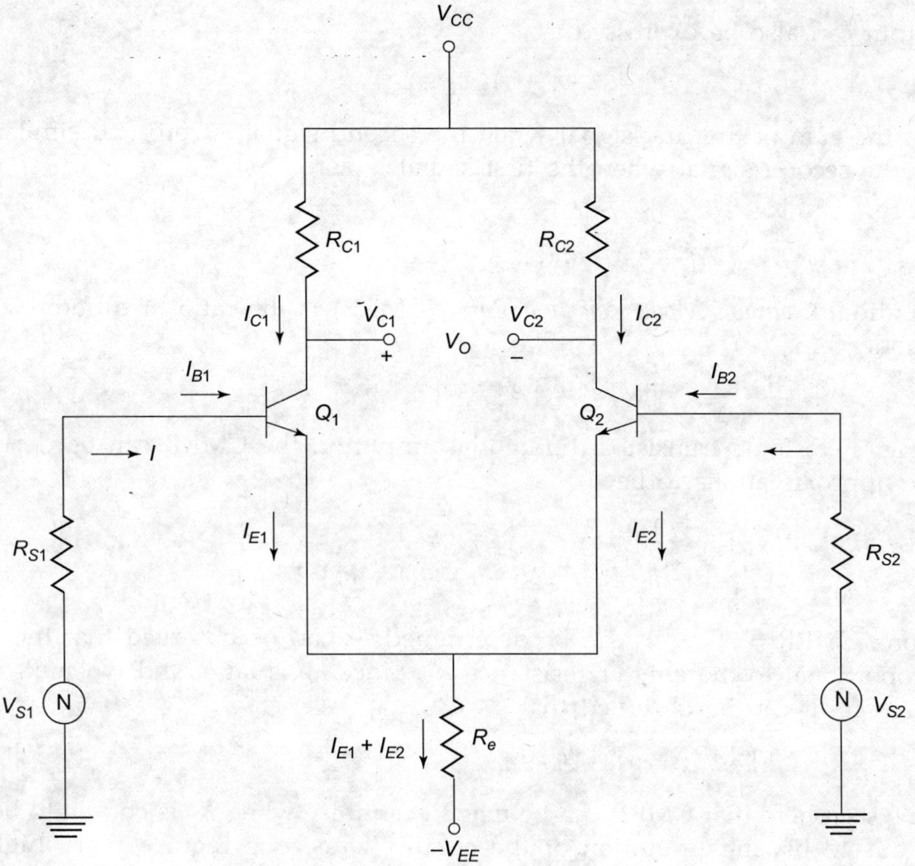

Fig. 13.5 *Differential Amplifier*

If the differential amplifier is ideal, the output would be amplified version of the difference signal without any other signal appearing in the output. Thus,

$$v_o = A_d\, v_d = A_d\,(v_1 - v_2) \qquad ...(13.2)$$

where A_d is the differential gain of the amplifier, v_d is the difference signal and v_1, v_2 are the two input signals. But in practice no differential gives only the difference signal at the output but a common mode signal is also present in the output. The common mode signal can be given as

$$v_c = \tfrac{1}{2}\,(v_1 + v_2) \qquad ...(13.3)$$

Thus the output signal can be rewritten as

$$v_o = A_d\, v_d + A_c\, v_c \qquad ...(13.4)$$

where A_c is the common mode gain of the amplifier.

The output can also be expressed as

$$v_o = A_1 v_1 + A_2 v_2 \quad ...(13.5)$$

where A_1 is the gain of the first signal when the second signal is zero and similarly, A_2 is the gain of the second signal when the first signal is zero.

Therefore,

$$A_d = ½ (A_1 - A_2) \quad \text{and} \quad A_c = A_1 - A_2 \quad ...(13.6)$$

By definition, *Common Mode Rejection Ratio* (CMRR) is the ratio of differential gain to common mode gain.

$$\text{CMRR} = |A_d / A_c| \quad ...(13.7)$$

From Fig. 13.5, for a transistor differential amplifier, the CMRR can be derived with appropriate approximations to be

$$A_d = ½ [h_{fe} R_c / (R_s + h_{ie})] \quad ...(13.8)$$

and

$$A_c = [- h_{fe} R_c / (R_s + h_{ie} + \{1 + h_{fe}\} 2 R_E)] \quad ...(13.9)$$

Therefore, CMRR = A_d / A_c can be determined. It can be observed that the CMRR is directly proportional to the emitter resistance R_E. Since h_{fe} is large and when R_E is chosen to be very large, it can be observed that

$$\text{CMRR} \approx \{1 + h_{fe}\} R_E \quad ...(13.10)$$

It can be seen here that CMRR can be made to infinity when R_E is chosen to be infinity. But this leads to biasing limitation of the circuit. Thus, one requires very high emitter resistance with minimum biasing current for emitter. Thus, the solution to this problem is to replace emitter resistance with a constant current source (common base transistor configuration) which is adjusted to maintain the same biasing current to the emitter circuit. The current gain of a common base circuit is approximately unity and it has low input impedance and very high output impedance. Thus, it is the best choice in these type of circuits. Such a circuit incorporating constant current source is as shown in Fig. 13.6. The CMRR of this circuit is very high and is best suited as the first stage of Op-Amp.

13.5 ERROR VOLTAGES AND CURRENTS

The Op-Amp performance can tend towards ideal characteristics but achieving ideal situation is bleak. In practice, there are some error voltages and currents attached with the Op-Amp. These errors are to be compensated before the use of the device in a given circuit for optimum performance of the Op-Amp in the circuit. The dc parameters of importance in this context are:

- *Input Bias Current* is the average current of the input terminals to balance the OP-Amp. In other words it is half of the sum of input currents when the output voltage is zero. Thus,

$$I_b = ½ (I_1 + I_2)|_{vo = 0} \quad ...(13.11)$$

Ideally this current should be zero so that it will not the feedback incorporated into the circuit. Its value is of order of nA or pA in practical amplifiers.

- *Input Offset Current* is difference in the input currents of a balanced Op-Amp. There is difference in the currents drawn by the Op-Amp between the two terminals. Thus, the difference of the two input currents when output voltage is zero is defined as the input offset current.

$$I_{io} = (I_1 - I_2)\,\big|_{v_0 = 0} \qquad \text{...(13.12)}$$

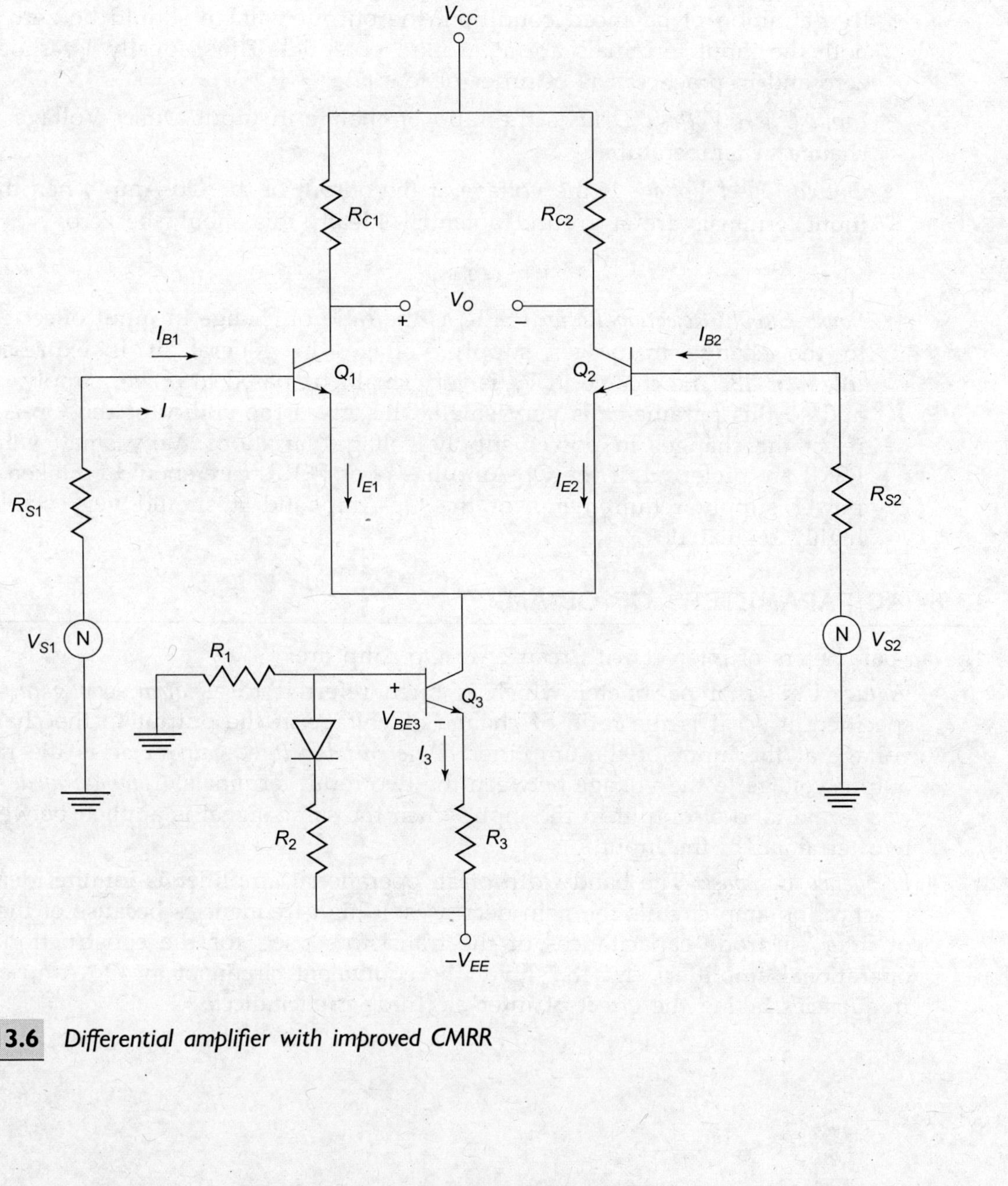

Fig. 13.6 *Differential amplifier with improved CMRR*

This value also should be ideally zero and is of order of few to several hundreds of nA.

- *Input Offset Current Drift* is the ratio of change in Input Offset Current to the change in temperature.
- *Input Offset Voltage* is the voltage that has to applied between the two terminals of the Op-Amp in order to make the Op-Amp balanced. Thus,

$$V_{io} = (V_1 - V_2)\,|_{v_0 = 0} \qquad \text{...(13.13)}$$

By definition of balanced condition, the output voltage should be zero when both the input terminals are at ground potential. Thus, ideally V_{io} should be zero and in practice it is of order of few mV.

- *Input Offset Voltage Drift* is the ratio of change in Input Offset Voltage to the change in temperature.
- *Output Offset Voltage* is the voltage at the output of the Op-Amp when the two input terminals are at ground potential. Ideally this should be zero.

$$V_{oo} = V_o\,|_{v_1 = 0 \text{ and } v_2 = 0} \qquad \text{...(13.14)}$$

- *Power Supply Rejection Ratio* (PSRR) is the ratio of change in input offset voltage to the change in power supply voltage. In general, it is expressed as $\mu V/V$ or dB. The change in V_{io} is very small compared to power supply voltage and so this parameter is very small. This gives the ability of the Op-Amp to reject the changes in power supply voltage variations. Very small values of PSRR are preferred. If the Op-Amp has poor PSRR, care should be taken in the power supply requirements of the Op-Amp and it should very stable and highly regulated.

13.6 AC PARAMETERS OF OP-AMP

The ac parameters of interest with respect to Op-Amp are:

- *Voltage Gain* is a parameter which is often referred to as *open loop gain* and is specified at *dc*. It is the ratio of change of voltage at the output to the change of voltage at the input of the amplifier. *Differential voltage gain* at *dc* is the ratio of output voltage to the voltage between the two input terminals. *Common mode voltage gain* is the ratio of output to the input when the same signal is applied between the two terminals of the input.
- *Frequency Response:* The bandwidth of an operational amplifier is infinite ideally. In practical op-amp circuits the gain decreases at high frequencies because of the effect of inter electrode capacitances of the transistors used for the construction of an operational amplifier. Fig. 13.7 gives the equivalent circuit of an Op-Amp at high frequencies taking the effect of inter electrode capacitances.

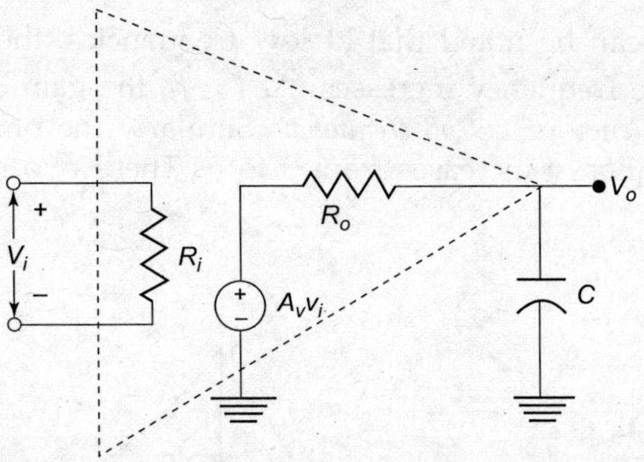

Fig. 13.7 *Equivalent circuit*

C in the figure represents the effect of all capacitances. From the circuit by voltage division,

$$V_o = \frac{1/j\omega C}{R_0 + 1/j\omega C} A_v V_I \qquad ...(13.15)$$

The gain of the amplifier considering the load can be derived to be

$$A = \frac{V_o}{V_I} = \frac{A_v}{1 + j\omega R_0 C} = \frac{A_v}{1 + j\omega/\omega_H} = \frac{A_v}{1 + jf/f_H} \qquad ...(13.16)$$

Where f_H = ½p$R_o C$ is the *corner frequency* or *cut-off frequency* of the operational amplifier. The magnitude and phase angle of the transfer function can be written as

$$|A| = \frac{A_v}{\sqrt{1 + (f/f_H)^2}} \qquad ...(13.17)$$

$$\theta = -\tan^{-1}(f/f_H) \qquad ...(13.18)$$

The values of the normalized magnitude and phase angles for different values of f/f_H are shown in the following Table 13 .1.

Table 13.1 ***Normalized magnitude and phase of the voltage gain***

S.No	f/f_H	Magnitude	Phase angle in degrees
1.	<<1	1	0
2.	≈ 1	1/√2	–45
3.	>>1	≈ f_H/f	–90

From Table 13.1, it can be noted that at low frequencies, the gain remains constant and decreases as the frequency increases. At $f = f_H$ the gain falls by $3dB$. Thus, f_H is called as *corner frequency* or *cut-off frequency*. Similarly the phase angle changes from 0 to –90 as the frequency increases from 0 to ∞. These two responses are plotted in Fig. 13.8.

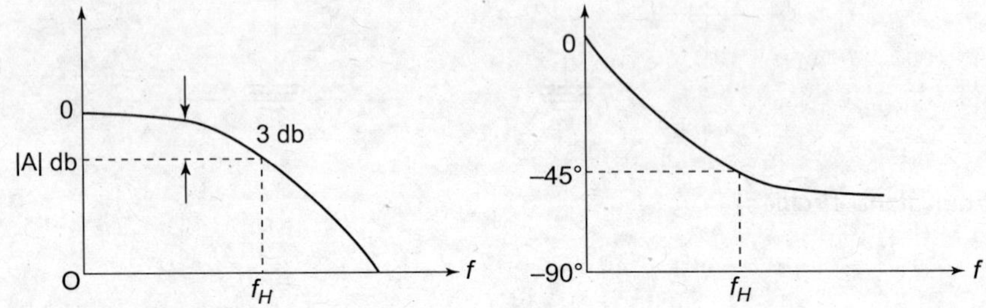

Fig. 13.8 *Frequency of Op Amp*

If the bandwidth required is of high value and closed loop gain is low, compensation techniques are used to achieve these two requirements. Compensation can be provided externally or internally to an Op-Amp. External compensation is of two types named as *dominant pole compensation* and *pole-zero compensation* technique. In dominant pole compensation, a pole is introduced into the transfer function of the Op-Amp. This newly introduced pole will decide the bandwidth of Op-Amp and hence reduces the value drastically. This compensation is used in low-bandwidth applications. In case of pole-zero compensation technique, a new pole and zero are introduced into the transfer function. The new pole decides the bandwidth of Op-Amp. Fig. 13.9 gives the basic circuits that are used to provide external compensation.

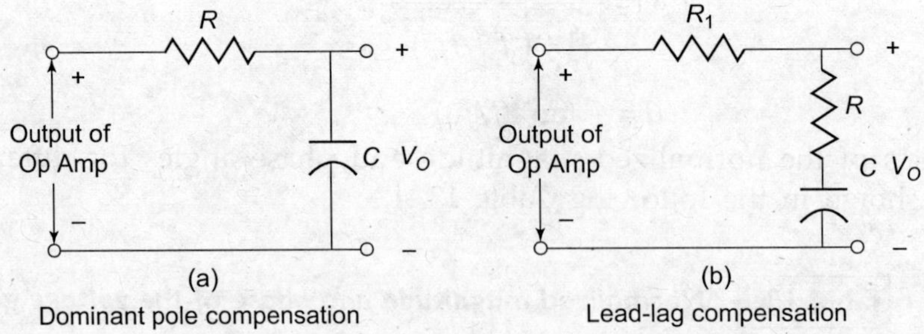

Fig. 13.9 *External compensation circuits*

- *Bandwidth* of Op-Amp can be defined in different ways. They are:
  - * *3dB Bandwidth* is the frequency at which closed loop voltage gain of the amplifier is *3dB* down. In general, closed loop gain is of interest rather than the open loop gain. The bandwidth in an Op-Amp is the upper cutoff frequency since the lower cutoff frequency of an Op-Amp is zero as the Op-Amp operates from *dc* voltages (zero frequency).
  - * *Unity Gain Cross Over Frequency* is the frequency at which the gain of the amplifier is unity (0*dB*). Since the lower cutoff frequency of Op-Amp is *dc* (zero), this also gives the bandwidth at which the gain falls to unity.
  - * *Unity Gain Rise Time* is similar to unity gain cross over gain except that the gain is not measured directly but the small signal rise time calculated for the non-inverting unity gain and bandwidth is calculated (0.35/rise time).
  - * *Full Power Bandwidth* is defined as the maximum frequency at which an undistorted sine wave can be available at the output of Op-Amp. This is usually 10 to 100 times lower than unity gain cross over frequency.
- *Slew Rate* is the slope of the triangular waveform at the output of Op-Amp when the Op-Amp is overdriven by input high frequency sinusoidal signal. When frequency at the input is increased to very large extent, the output signal appears as triangular form. The slope of this waveform is slew rate and is measured as volts per second. Very high slew rates are required in design of Op-Amp circuits since this ensures large bandwidth of operation. Slew rate limiting is characterized by definite flattening on some portion of the sine wave. This flat portion is due to the constant current source charging a capacitance. Slew rate is higher for higher gain circuits. Thus, input signal can be reduced to decrease this effect. In some cases, the compensation capacitor may be too large and some other compensation technique may be used.
- *Input impedance* is either differential or common mode. Differential input impedance is the between two terminals of Op-Amp while the common mode input impedance is between either of the input terminals to the negative power supply terminal. Differential input impedance is an important parameter but the common mode input impedance is not specified in general. The differential input impedance is of order of several mega ohms. The effect of common mode input impedance on the circuit is so small that it is usually not considered. The effect of differential input impedance can also be neglected unless its value is very small of order of 10K or less.
- *Input Capacitance* is of two types—Common mode and differential capacitances. The effect of these parameters on the inverting amplifier is very small such that they are never considered. The typical values of both the capacitances are of the order of 1 to 2pF and a maximum of 3pF. At high frequencies, the common mode capacitance may have some effect of attenuation. The only way out is to choose Op-Amp with low common mode input capacitance values.
- *Output Impedance* of Op-Amp is often very small. In inverting amplifier configuration when feedback is employed, the effective output impedance is further reduced since it is divided by feedback factor times the voltage gain. So, there is no

effect in the mid range of frequencies, but when the frequency of operation tends towards the cutoff value, the voltage gain falls and the output impedance may seem to be increasing.

- *Equivalent Input Noise* affects the dc as well as ac performance of the Op-Amp. Op-Amp manufacturer specifies the noise content up to range of 1 MHz of frequency. The noise generation in the Op-Amp is not at a specified point but to simplify noise calculations, it is specified as equivalent input noise voltage or current.
- *Output Voltage Swing* indicates the value of positive and negative saturation voltages of Op-Amp. These will never exceed the supply voltage.
- *Transient Response* characteristics are rise time and overshoot. These parameters are important in selection of Op-Amp when used for ac applications. Rise time is of order of 0.3*ms* and overshoot is 5 percent.

13.7 MATHEMATICAL COMPUTATIONS USING OP-AMP

There are numerous applications of Op-Amp in various fields. The most important is mathematical computations. Almost all mathematical operations can be performed using an Op-Amp. Any mathematical problem can be solved using Op-Amp. In other words, Op-Amp circuits can be referred to as *Analog Computer*. In this section, few of the mathematical operations that are possible with an Op-Amp are discussed.

1. *Inverting Amplifier*: amplifiers are the basic fundamental application of Op-Amp. In most analog circuits, the basic building block is the inverting amplifier. Inverting amplifier is where the input signal is amplified to the extent required with a 180° phase shift between the input and output signals. The input signal is applied at the inverting terminal of the Op-Amp and the non--inverting terminal is at ground potential as shown in Fig. 13.10. The various names of such a circuit are *Inverting mode amplifier, phase inverter, and inverter.*

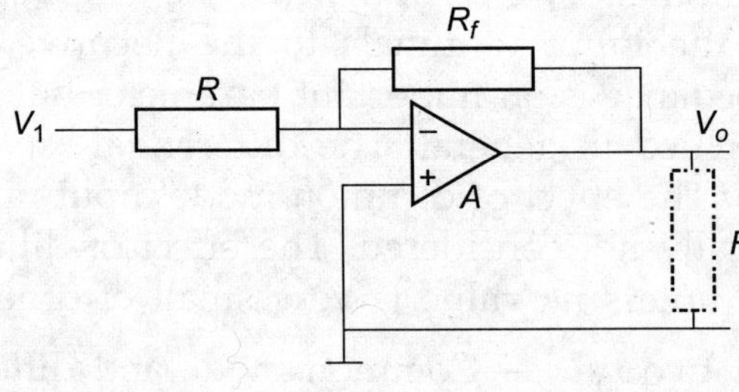

Fig. 13.10 *Inverting Amplifier*

Assuming ideal Op-Amp, the output of the above circuit can be given as

$$V_o = [\,-R_f/R\,]\,V_i \qquad ...(13.19)$$

Thus the gain offered by the amplifier is $-R_f/R$, independent of Op-Amp. The negative sign indicates the phase difference between the input and output signals. The gain offered by the circuit depends on the two resistances and so by choosing these two resistances properly, any gain can be easily achieved. A series resistance R_c may be added between the non-inverting terminal and ground which would compensate any offset errors present in the Op-Amp.

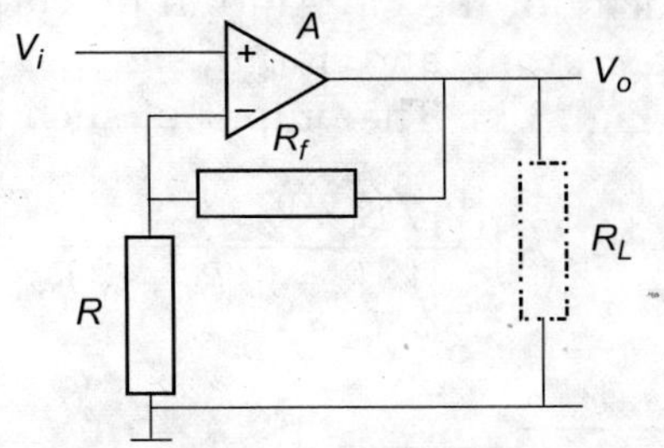

Fig. 13.11 *Non-inverting amplifier*

2. *Non-inverting Amplifier*: Non-inverting amplifier amplifies the input signal applied to its non-inverting terminal. There is no phase difference between the output and input signal. The gain can once again be controlled by the feedback and input resistances. The circuit diagram of such an amplifier is as shown in Fig. 13.11. The output voltage of the circuit may be derived to be

$$V_o = [1 + (R_f/R)]V_i \qquad ...(13.20)$$

Thus the gain offered by the circuit is $1 + (R_f/R)$. A resistor R_c may be added between the inverting terminal and the node of R and R_f in order to compensate for any offset errors, if any.

3. *Summing Amplifier*: The input of the ideal Op-Amp is infinity and so it draws no current and all the current is diverted into the feedback circuit. Fig. 13.12 is an example of summing amplifier. In this configuration, the output is proportional to the sum of all the signals at the input. The current in the feedback resistor R_f is sum of all the currents delivered by various input signals. Thus, the output voltage is R_f times the sum of all these currents. The output voltage of the circuit can be derived to be

$$V_o = -R_f [V_1/R_1 + V_2/R_2 + V_3/R_3] \qquad ...(13.21)$$

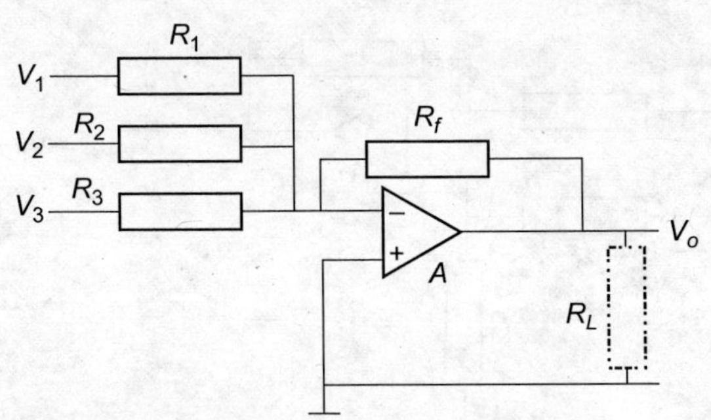

Fig. 13.12 *Inverting summing amplifier*

In the above circuit, it can be observed that output is the inverted version of input. If $R_1 = R_2 = R_3 = R_f$, then the output voltage will be $V_o = -(V_1 + V_2 + V_3)$. This summing can be extended to any number of inputs for an ideal Op-Amp but is limited by the fan-in capacity of the Op-Amp in practice. If one does not require the phase shift between the output and input signals, the non-inverting summing amplifier is as shown in Fig.13.13. The output of such a circuit is

$$V_o = \left[1 + R_f / R\right] \frac{\left[V_1 / R_1 + V_2 / R_2 + V_3 / R_3\right]}{\left[1/R_1 + 1/R_2 + 1/R_3\right]} \quad ...(13.22)$$

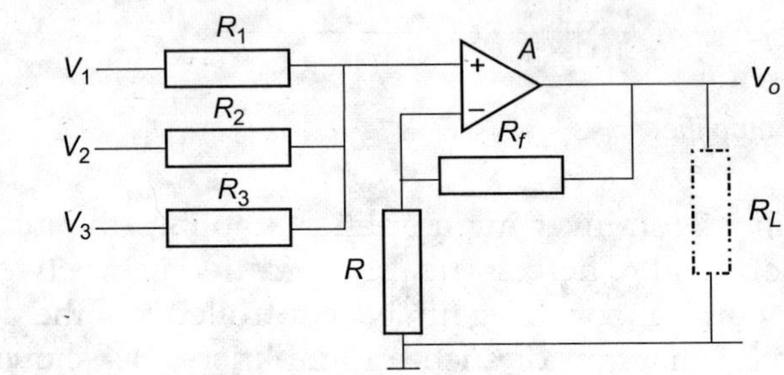

Fig. 13.13 *Non-inverting summing amplifier*

4. *Difference Amplifier*: Difference amplifier amplifies the difference of the input signals. The two signals whose difference is applied to the non-inverting and inverting terminals of the Op-Amp is shown in Fig. 13.14. This figure has the wieghtage to be equal and so the output is simply the subtraction of the two signals. So, it can be referred to as *subtractor*. The wieghtage to the signals can be adjusted with the help of proper selection of the input and feedback resistances. Even *Adder-Subtractor* can be realized when all the signals to be added are fed in at non-inverting terminal and all the signals to be subtracted are fed at the inverting terminal. By choosing proper input resistance of each signal and the feedback resistance, even a simple mathematical equation employing addition and subtraction can be easily realized.

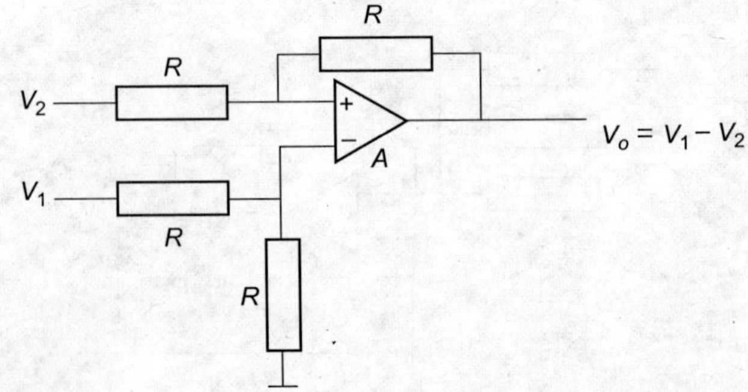

Fig. 13.14 *Subtracting amplifier*

5. *Scale Changer*: If the inverting amplifier, feedback and the input resistors R_f and R are chosen to be pure precision resistors in Fig. 3.10, the output is multiplied by a factor, $-k = -R_f / R$. Thus the input signal is changed by scale equal to the constant $-k$.
6. *Phase Shifter*: Whenever one requires a phase shift of required phase angle, the input and the feedback resistors can be replaced by impedances in the circuit of Fig. 13.10. If the magnitude of the two impedances is chosen to be equal but differing in phase angle, the required phase shift can be achieved. Phase angles of any desired value between 0° and 360° is possible.
7. *Voltage Follower*: Common collector configuration of transistor acts as a buffer circuit providing high input impedance and low output impedance. In order to improve these impedances further, Darlington pair is at rescue. But, in some applications further improvement in these two impedances are required. That is, the input impedance should tend towards infinity (of order of hundreds of mega ohms) and output impedance towards zero (of order of one or two ohms) with a voltage gain of unity. This can be best achieved with Op-Amp circuit. Non--inverting amplifier is with R_f zero and R infinity. Such a circuit is known as *Voltage Follower* and is as shown in Fig. 13.15. This gives unity gain with very high input impedance and very low output impedance.

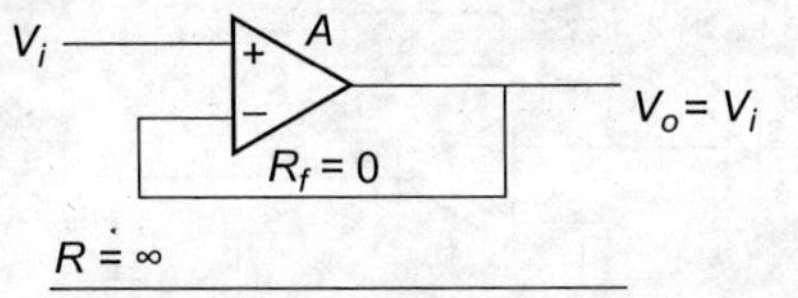

Fig. 13.15 *Voltage follower*

8. *Integrator*: When the feedback resistor of the inverting amplifier is replaced with ideal capacitor, the circuit works as a good integrating circuit. The circuit is as shown in Fig. 13.16.

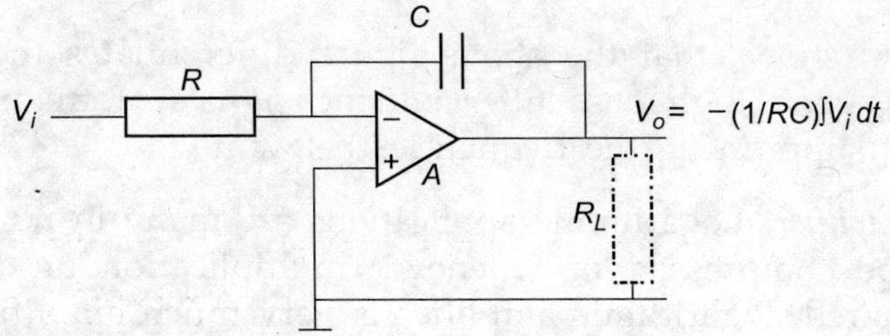

Fig. 13.16 *Integrator*

The current in the input circuit is

$$i = V_i/R \qquad \text{...(13.23)}$$

The same current flows in the feedback capacitor C and thus

$$V_i/R = -\,C(dV_0/dt\,) \qquad \text{...(13.24)}$$

Or,

$$V_o = -\,(1/RC) \int V_i\, dt \qquad \text{...(13.25)}$$

So, the output voltage is the time integral of the input signal. Thus, this circuit acts as very good integrator. This circuit is also a basic *Low Pass Filter,* which allows signals at low frequency to pass through with minimum attenuation and suppresses the high frequency signals above cutoff frequency with very high attenuation. If the circuit is used with variable frequency input, it acts a low pass filter and if the circuit is used at a given frequency, it is a very good integrating amplifier.

9. *Differentiator*: When the positions of resistor and capacitor in Fig. 13.16 are interchanged as shown in Fig. 13.17, a *differentiator* or *High Pass Filter* results.

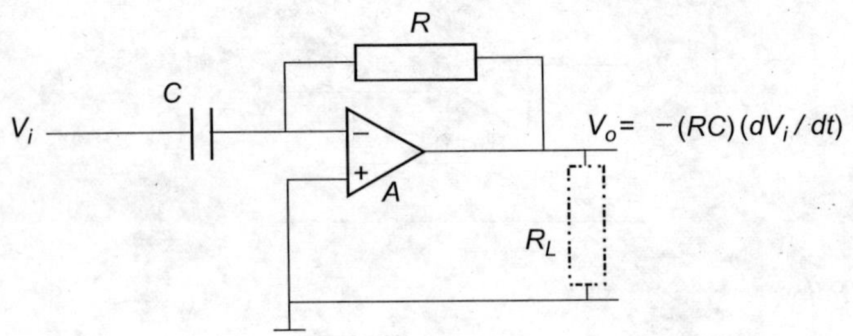

Fig. 13.17 *Differentiator*

The current in the capacitor is $C(dV_i/dt)$ and the same current is flowing the feedback resistor and thus equating the two,

$$V_o = -\,(RC)(dV_i/dt) \qquad \text{...(13.26)}$$

The Eq. (13.26) shows that the above figure differentiates the input signal. This circuit is very good high pass filter passing signals of frequencies above cutoff frequency and suppressing the frequencies below it.

10. *Logarithmic Amplifier*: Logarithmic applications are many in number. Whenever the signal is to be compressed or whenever multiplication or division of different signals is required, logarithmic amplifier is very much amplifier. These amplifiers are basically non-linear. Anti logarithmic amplifiers are inverse of logarithmic amplifiers and they are expanders of the input signal. In mathematical operations like multiplication, division, square root and so on, these are very widely used. The voltage-current relationship of a diode is taken into consideration while designing the logarithmic amplifier. The basic *V–I* equation of a diode is given as

$$I = I_0\,(e^{V/\eta V_T} - 1) \qquad \text{...(13.27)}$$

Or,

$$V = \eta V_T (\ln I - \ln I_0) \qquad ...(13.28)$$

$$\approx \eta V_T \ln I \quad \text{since } I >> I_0 \qquad ...(13.29)$$

The basic logarithmic amplifier is as shown in Fig. 13.18. The feedback resistor R_f of the basic inverting amplifier is replaced by the diode. Thus the voltage across the diode is given by the Eq. (13.29) and thus from the principle of the Op-Amp, the output voltage can be derived as

$$V_o = -\eta V_T \ln (V_i/R) \qquad ...(13.30)$$

Thus the output of the circuit is proportional to the natural logarithm of the input signal.

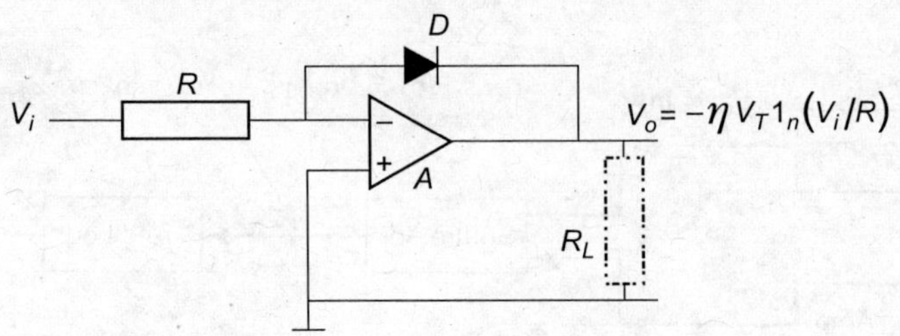

Fig. 13.18 *Logarithmic Amplifier*

The above Circuit is temperature dependent since both η and V_T are temperature dependent variables. The circuit can be made temperature independent by use of matched diodes, two Op-Amps, a thermistor and a temperature independent current source.

When the positions of the resistor and diode are interchanged in the circuit of Fig. 13.18, an anti logarithmic amplifier results as shown in Fig. 13.19.

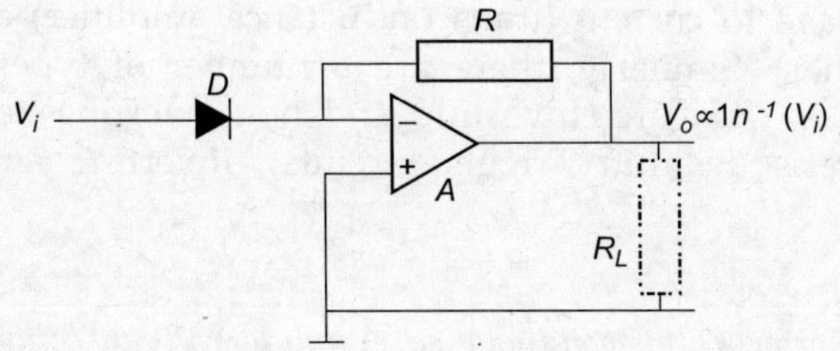

Fig. 13.19 *Anti Logarithmic amplifier*

11. *Multiplier*: Multiplying two signals can be easily achieved with logarithmic and anti logarithmic amplifiers. The two signals to be multiplied are applied to log amplifiers. These two signals are added in a summing amplifier circuit. The output of this circuit is applied to anti logarithmic amplifier. The output of this anti log amplifier is the multiplication of the basic signals. The block diagram of the multiplier circuit is as shown in Fig.13.20.

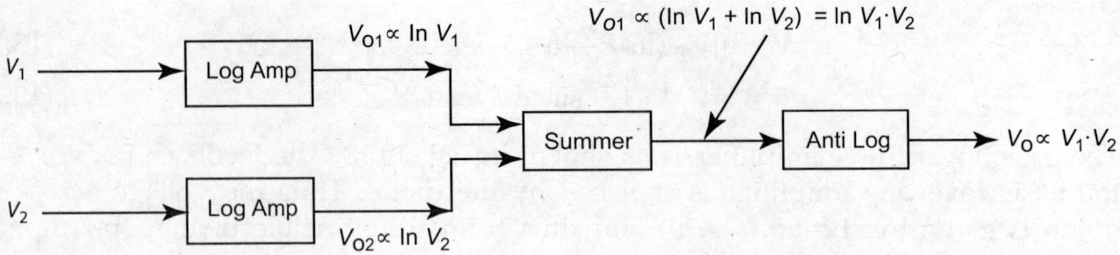

Fig. 13.20 *Multiplier*

12. *Divider*: When summing amplifier of the multiplier circuit is replaced with a subtracting circuit, a good divider circuit results as shown in Fig. 13.21.

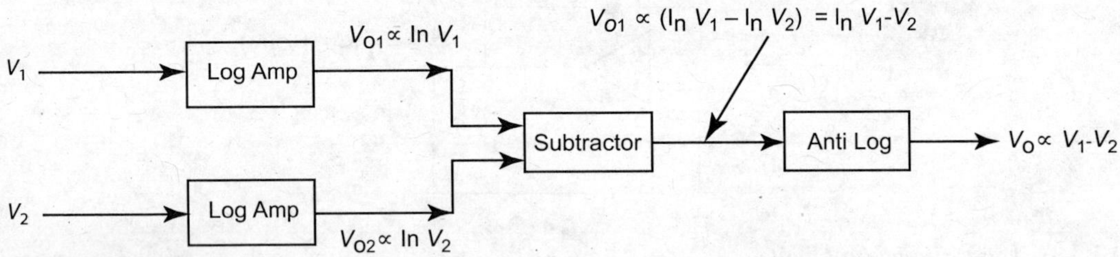

Fig. 13.21 *Divider*

13.8 CONVERTERS AND COMPARATORS

There are lot many applications of Op-Amp. Of these, converters and comparators are very popularly used in many electronic circuits. Some of these are discussed in brief in this section of the chapter. There are large *Converters* in many electronic circuit applications. Some of them are voltage to current (transconductance amplifier) and current to voltage (transresistance amplifier). Similarly, there are a number of types of *Comparators* (also referred to as *Detectors)*. They are: Inverting and non-inverting zero crossing detectors, detectors with Hysteresis (Schmitt Trigger circuits), inverting and non-inverting level detectors and so on.

CONVERTERS

Voltage to current converter (*Transconductance Amplifier*): *Transconductance amplifier* is an amplifier with input parameter as voltage and output parameter as current. In other words, it is a device, which converts voltage to current. The basic circuit configuration of such a converter is as shown in Fig. 13.22.

In Fig. 13.22 (a), the load is floating and the current through this load can be derived as

$$V_i = I_L R \quad \text{or} \quad I_L = V_i / R \qquad \text{...(13.31)}$$

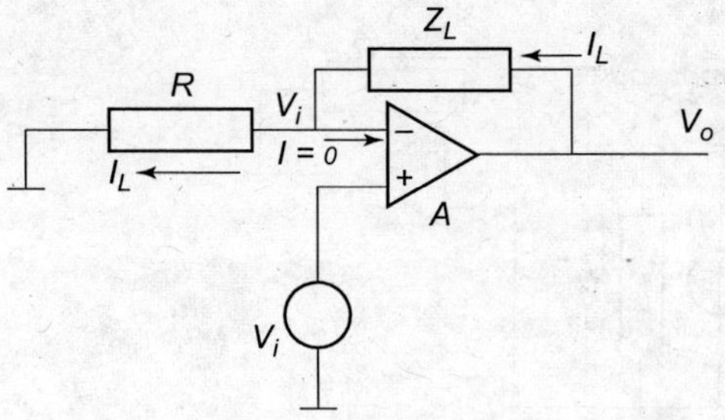

(a) With floating ground

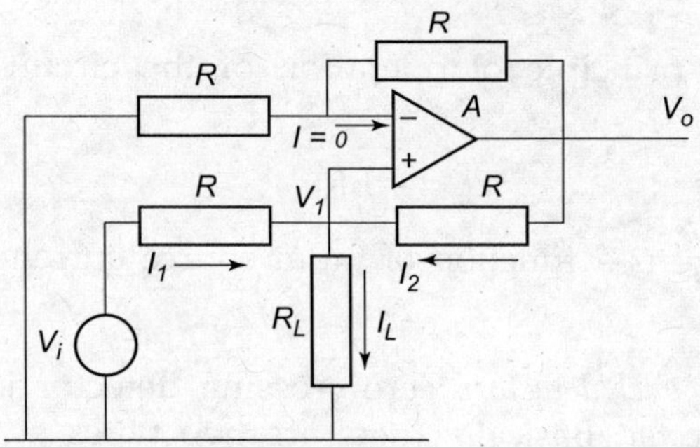

(b) With floating load

Fig. 13.22 *Voltage to current converter*

That is, the output current is the function of input voltage. Since there exists a virtual ground between the two input terminals of the Op-Amp, the inverting terminal voltage is the same as that of the non-inverting terminal, V_i. Also from the figure, it can be seen that the current in the load impedance and the resistor R is the same. Since the load is not grounded in this circuit, it is not very useful in many applications. Fig. 13.22 (b) shows a circuit with one end of the load at ground potential.

$$I_L = I_1 + I_2 = (V_1 - V_i/R + (V_0 - V_1)/R \qquad ...(13.32)$$

Or,
$$V_1 = (V_i + V_0 - I_L R)/2 \qquad ...(13.33)$$

Since the gain of the non-inverting amplifier is $1 + R/R = 2$, the output voltage is given as

$$V_0 = 2\,V_1 = V_i + V_0 - I_L R$$

Thus,
$$I_L = V_i/R \qquad ...(13.34)$$

Current to voltage converter (*Transresistance Amplifier*): A *Transresistance amplifier* is reverse to the transconductance amplifier. Here, the input parameter is current and the output parameter is voltage. The basic circuit diagram of such a current to voltage converter is as

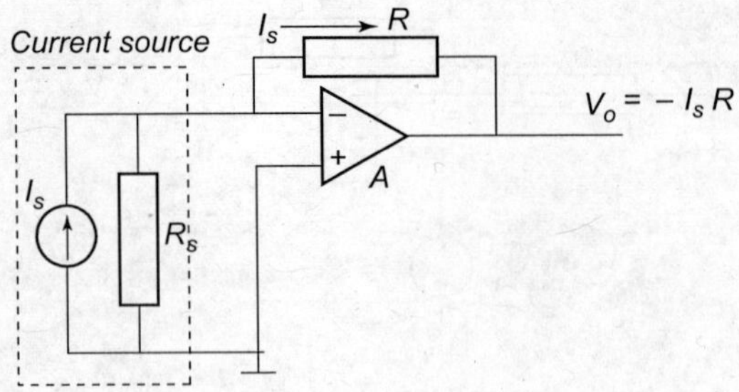

Fig. 13.23 *Current to voltage converter*

shown in Fig. 13.23. The principle of operation of the circuit is self-explanatory. The output voltage is

$$V_o = -I_s R \quad \text{...(13.35)}$$

Thus the output voltage is a function of input source current.

COMPARATORS

Zero crossing detector and *Level detector*: Zero crossing detector is also called *zero crossing comparator* or *zero level detector*. Basically there are two types of these detectors. They are *Inverting* and *Non-inverting* types. A zero crossing detector compares the input voltage to zero voltage and the response of this circuit is the output can be only two voltages (states). In inverting mode, the output voltage will be positive saturation voltage ($+V_{cc}$) when input voltage is less than zero and the output voltage is negative saturation voltage ($-V_{cc}$) when the input voltage is more than zero voltage. Thus, the output voltage changes between positive saturation and negative saturation whenever the input voltage crosses zero voltage level. Whatever be the shape of input voltage, the output voltage is always rectangular switching between $+V_{CC}$ and $-V_{CC}$. Thus, one can convert a sinusoidal voltage to square with such a circuit. In inverting mode, there would be a phase shift between the input sinusoid and output rectangular but both the signals can be made to be in phase when the zero crossing detector is in non-inverting mode. In this non-inverting mode, the output is $+V_{CC}$ whenever the input signal voltage is more than zero and the output voltage will be $-V_{cc}$ whenever the input voltage is less than zero. The circuit diagram for the two types of the zero crossing detectors is as shown in Fig. 13.24.

The explanation of these circuits can be easily understood without any expression for the output. The output is always at saturation voltage. This polarity of this saturation voltage is determined by the voltage difference between the two input terminals of the Op-Amp. If the non-inverting terminal potential is at more potential compared to inverting terminal voltage, the output is positive and if the non-inverting terminal is at less potential compared to inverting terminal, the output is negative. The above two circuits can be converted to *Level Detector* when a reference voltage source is connected in series with the other terminal and the ground. That is, in case of inverting type, the reference voltage is applied at the non-inverting terminal and in case of non-inverting type, the reference voltage is applied at the inverting terminal. In this case, the switchover

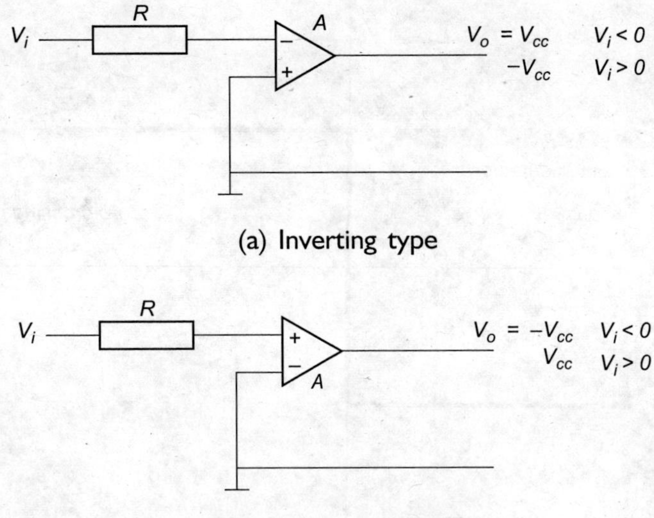

(a) Inverting type

(b) Non-inverting type

Fig. 13.24 *Zero crossing detector*

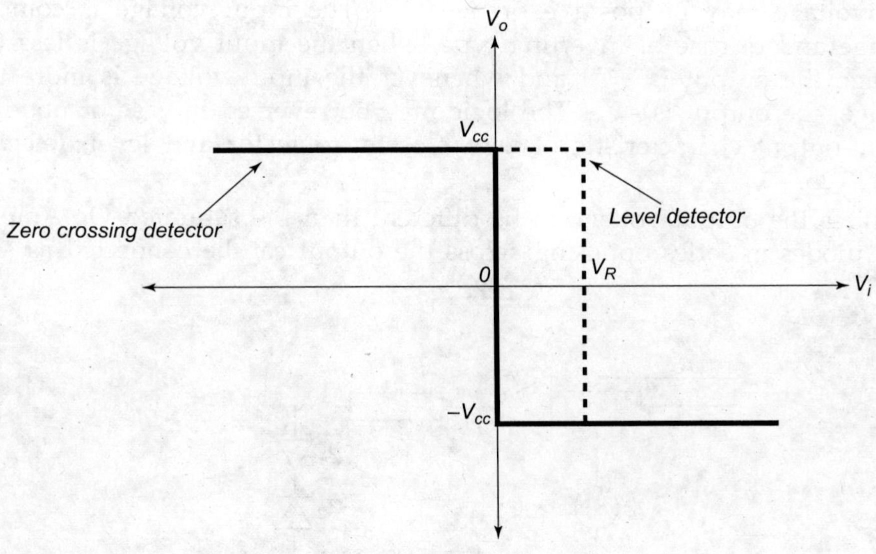

(a) Inverting type

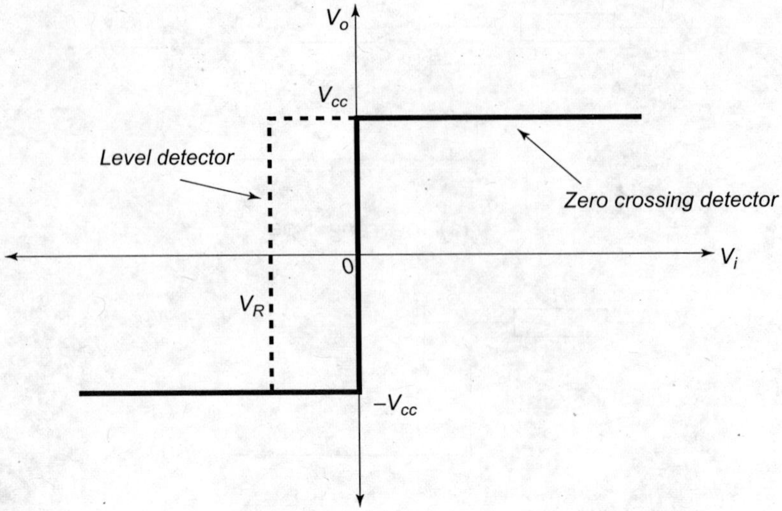

(b) Non-inverting type

Fig. 13.25 *Input-output characteristic of comparator*

of the output voltage is not on the zero voltage but on the reference voltage magnitude. The reference voltage may be positive or negative. The input voltage is compared to reference voltage and in case of inverting type, when the input voltage is less than the reference voltage the output is $+V_{CC}$ and whenever the input voltage is more than the reference voltage, the output is $-V_{CC}$. The logic may be reversed in case of non-inverting type. The input-output characteristic of zero crossing detector and level detector is as shown in Fig. 13.25.

If one requires the output voltage to be different than the saturated Op-Amp voltage V_{CC}, two zener diodes in series opposing across the output can be connected as shown in Fig. 13.26.

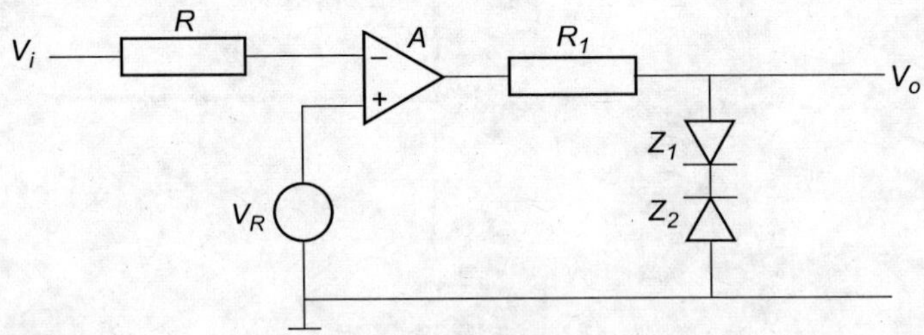

Fig. 13.26 *Inverting Level detector with Zener diodes*

In the inverting level detector in Fig. 13.26, the output depends on the zener voltages of diodes Z_1 and Z_2. The resistance R_1 is so chosen that the current does not exceed the rating of the zener diodes. The output voltage can be given as

$$V_o = V_{z_2} + V_d \qquad V_i < V_R$$
$$= -(V_{z_1} + V_d) \quad V_i > V_R \qquad \text{...(13.36)}$$

where V_d is the cutin voltage of the diode which is forward biased.

Comparators with Hysteresis: In the above discussion, the Op-Amps are assumed to be ideal. But, in practice, they are not so and there are several errors and noise which are to be compensated to achieve good performance of the comparators. The first and foremost error is the input current. The bias and input offset currents cause errors in the performance of the circuit and so these errors have to compensated. This can be done by adding a compensation resistor in the other terminal of the input to reduce the input currents to large extent.

Another problem is the switching time from one state to the other. This is not instantaneous and depends on the slew rate of the Op-Amp. The switching speed cannot be made faster than the slew rate of the Op-Amp.

The Op-Amp comparator has a drawback of *Chatter*. If the input voltage has noise of significant magnitude, the output may switch between the two states several times before the final decision is made. This can be avoided by introducing *Hysteresis* into the circuit. Hysteresis will also help to improve the switching speed of the circuit. Such comparator with hysteresis is referred to as *Schmitt Trigger Circuit*. All types of comparators can be realized with hysteresis of which only inverting zero crossing detector with hysteresis is considered in this discussion. The same principle can be extended to other circuits also.

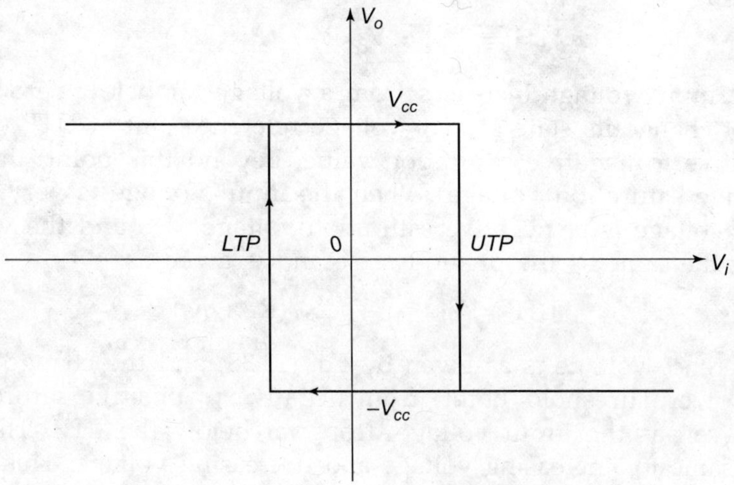

Fig. 13.27 *Input-output characteristic of comparator with Hysteresis*

Comparators with Hysteresis work on the same principle as circuit without hysteresis except that the change between the two states do not occur at the reference voltage but at voltage slightly above and below the reference voltage as defined by the circuit. The change of the states does not occur at the same reference level but it depends on the input voltage change whether is increasing or decreasing. The change of the state occurs at a voltage known as *Upper Threshold Point* when the input voltage is small and increasing to cross the reference voltage. The change of state occurs at a different voltage known as *Lower Threshold Point* when the input voltage is more and decreasing to cross the reference voltage. Thus the input-output characteristic of the comparator exhibits hysteresis and is as shown in Fig. 13.27. Fig. 13.28 shows the circuit of inverting zero crossing detector with hysteresis.

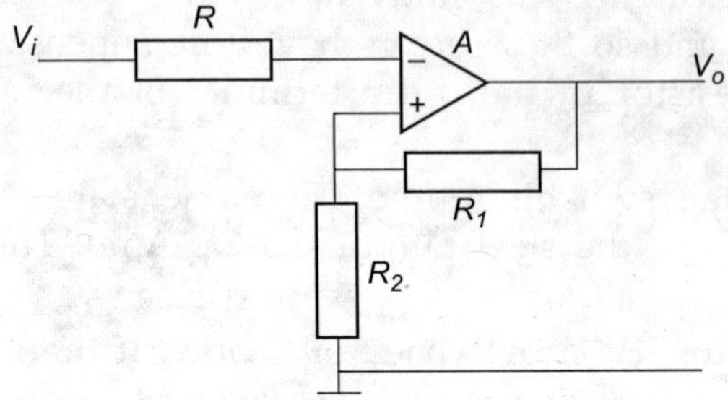

Fig. 13.28 *Inverting zero crossing detector with Hysteresis*

When the input voltage is very small, below zero, the output of the circuit will be at positive saturation voltage V_{cc}. Now, the voltage across the resistor R_2 and the reference voltage is not zero, but

$$\text{UTP} = V_{R2} = V_{cc}\, R_2 \;/\; (R_1 + R_2) \qquad \text{...(13.37)}$$

So, when the input voltage increases from a voltage far below zero towards positive value, it does not change its state at zero voltage reference, but at UTP (Upper Threshold Point) from positive to negative saturation value. Beyond this point, the output voltage remains at positive saturation voltage. When the input voltage is very high well above zero, the output voltage is at negative saturation voltage $-V_{cc}$ and the voltage across the resistor R_2 is not the same as the previous case and is given as

$$\text{LTP} = V_{R2} = -V_{cc}R_2/(R_1 + R_2) \qquad \text{...(13.38)}$$

Thus the reference voltage is as given by Eq. (13.38) and the change of the state does occur at LTP (Lower Threshold Point) from negative to positive saturation value. This state continues even as the input voltage drops far below the LTP. Thus, the reference point is not the same for increasing voltage and decreasing voltage. This is the concept of hysteresis.

The noise present around the reference voltage (zero in this case) does not affect the performance of the detector since the circuit does not respond at zero voltage but at either above or below this reference. The delta change of voltage can be designed based on the noise level present in the input voltage by selecting the values of the resistors R_1 and R_2 appropriately.

SUMMARY

- Operational Amplifier is an amplifier, which is really operational. That is it can perform many operations and its characteristics tend toward ideal values.
- Op-Amps can be used over a very wide frequencies starting from *dc* signals.
- Even though all the characteristics listed for ideal Op-Amp cannot be achieved simultaneously, optimum performance of few parameters at the cost of the others can be achieved easily.
- Virtual ground is nothing but zero potential across infinite impedance.
- Concept of virtual ground between the two input terminals of Op-Amp makes the design of Op-Amp circuit independent of the parameters of Op-Amp. Thus the circuit parameters like gain do only depend on the external components and feedback but not on the Op-Amp.
- When signal is fed at inverting terminal of the Op-Amp, the output signal is out of phase to whereas when signal is fed at non-inverting terminal, the output signal is in phase with the input.
- The Op-Amp amplifies the difference signal between two terminals of the input while it rejects the common mode signal. Measure of ability of Op-Amp to reject common mode signals is CMRR.
- The Op-Amp has two stages of differential amplifiers, in cascade as first stage, a level translator to suppress dc present in the signal and to balance the device, followed by an output driver stage.
- The upper cutoff frequency is the bandwidth of Op-Amp since the lower cutoff for Op-Amp is zero.
- Slew rate is the measure of ability of Op-Amp to retain the shape of sinusoidal signal at the output without distortion.
- PSRR gives the ability of Op-Amp to reject any power supply variations and maintain the performance of the device optimum.
- In spite of all care taken by the manufacturer to reduce error voltages and currents, there are some errors available in the practical Op-Amp. These errors are to be compensated before the use of Op-Amp in a circuit.
- Nearly all mathematical operations like summing, difference, multiplication, division, integration, differentiation, log value and so on, can be performed using Op-Amp circuits.
- Op-Amp circuits can solve mathematical equations and thus, it can be referred to as Analog Computer.
- Op-Amp is a very good comparator and detector. It can compare and detect any signal to any reference per specified.
- Schmitt trigger circuits exhibit hysteresis, which avoids the noise present in the input signal.
- Op-Amp can convert voltage to current or current to voltage signals efficiently.

SOLVED PROBLEMS

13.1 Find the CMRR of the differential amplifier assuming operational amplifier as an ideal one.

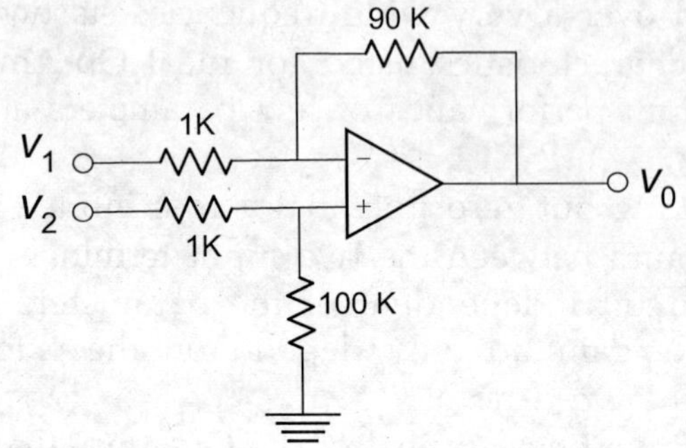

Solution Let V_a and V_b be the voltage at nodes *a* and *b* respectively.

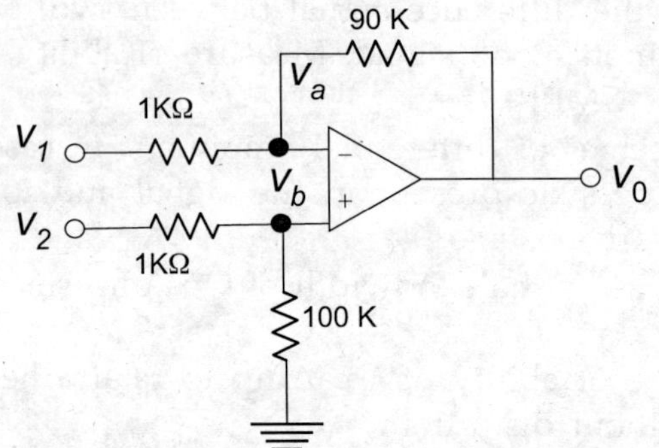

KCL Node '*a*'

$$\frac{V_a - V_1}{1\text{K}} + \frac{V_a - V_o}{90\text{K}} = 0$$

$$91\ V_a - V_o - 90\ V_1 = 0$$

$$\Rightarrow \quad 91\ V_a = V_o + 90\ V_1 \tag{1}$$

KCL at Node '*b*'

$$\frac{V_b - V_2}{1\text{K}} + \frac{V_b}{100\text{K}} = 0$$

$$\Rightarrow \quad 101\ V_b - 100\ V_2 = 0$$

$$\therefore \quad V_b = \frac{100}{101} V_2 = 0.99\ V_2 \tag{2}$$

Because of virtual ground concept, $V_a = V_b$.

$\therefore$ From (1) and (2),

$$91(0.99\ V_2) = V_o + 90\ V_1.$$
$$90.09\ V_2 = V_o + 90\ V_1$$
$$V_o = 90\ V_1 + 90.09\ V_2 \qquad (3)$$

If V_c and V_d are the common mode voltage and differential mode voltages, then

$$V_1 = \frac{V_c + V_d}{2} \text{ and } V_2 = \frac{V_c - V_d}{2}$$

equation (3), becomes,

$$V_o = -90\left(\frac{V_c + V_d}{2}\right) + 90.09\left(\frac{V_c - V_d}{2}\right)$$
$$= 45.00 \times 10^{-3}\ V_c + (-90.045)\ V_d$$
$$= A_c\ V_c + A_d\ V_d$$

where A_c = common Mode gain = 45.00×10^{-3}.

A_d = differential mode gain = -90.045

$$\text{CMRR} = \left|\frac{A_d}{A_c}\right| = \left|\frac{-90.045}{45.00 \times 10^{-3}}\right| = 2001$$

CMRR in decibels = 20 log (2001) = 66.024

13.2 For a differential amplifier gain = 1000, under differential mode of operation, CMRR = 10^6. Calculate output voltage V_{out} when v_1 = 1.1 mV and v_2 = 1.0 mV.

Solution Differential mode signal, $v_d = v_1 - v_2$ = 0.1 mV.

Common mode signal, $v_c = \dfrac{v_1 + v_2}{2}$ = 1.05 mV.

Differential mode gain A_d = 1000

CMRR, $\rho = 10^6$.

$$\text{Output voltage } V_o = A_d\ V_d \left(1 + \frac{1}{\rho}\frac{v_c}{v_d}\right)$$
$$= 1000 \times 0.1 \times 10^{-3}\left(1 + \frac{1}{10^6} \times \frac{1.05}{0.1}\right)$$
$$= 0.1(1 + 10.5 \times 10^{-6})$$
$$= 100.00 \text{ mV}.$$

13.3 A differential amplifier has inputs v_1 = 100 mV and v_2 = 90 mV. It has a differential mode gain of 40 db and a CMRR of 70 db. Find the percentage error in the output voltage and the error voltage.

Solution Given that $v_1 = 100$ mV, $v_2 = 90$ mV.

$$A_d = 40 \text{ db} \Rightarrow A_d = 100$$

$$\text{CMRR db} = 70 \text{ db} \Rightarrow \text{CMRR} = 3162.27$$

For an ideal, differential amplifier

$$V_o = A_d(v_1 - v_2)$$
$$= 100\ (10 \text{ mV})$$
$$= 1 \text{ V.}$$

$$\text{Practically } v_{op} = A_d\, v_d \left(1 + \frac{1}{\rho} \times \frac{v_c}{v_d}\right)$$

$$= 100 \times 10 \times 10^{-3}\left(1 + \frac{1}{3162.27} \times \frac{95}{10}\right)$$

$$= 1.003004 \text{ volts.}$$

$$\text{Error voltage} = V_{op} - V_o = 3.004 \text{ mV}$$

$$\% \text{ error} = \frac{\text{Error voltage}}{V_{op}} \times 100$$

$$= \frac{3.004 \times 10^{-3}}{1.003004} \times 100$$

$$= 0.2995\%$$

13.4 Find the closed loop gain v_o/v_s of the circuit shown in the figure below.

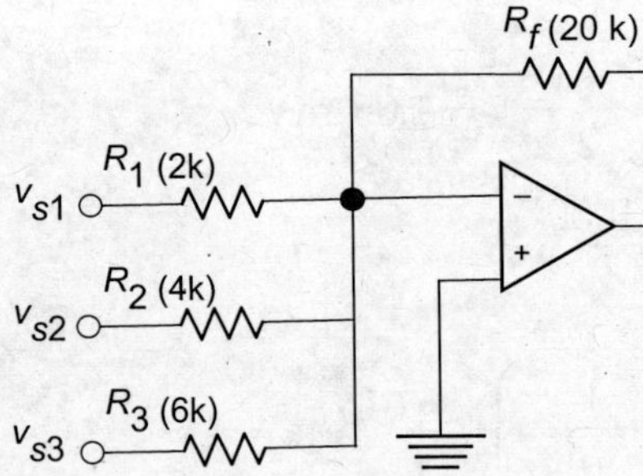

Solution Writing KCL at node 'a', making v_{s1} as active one.

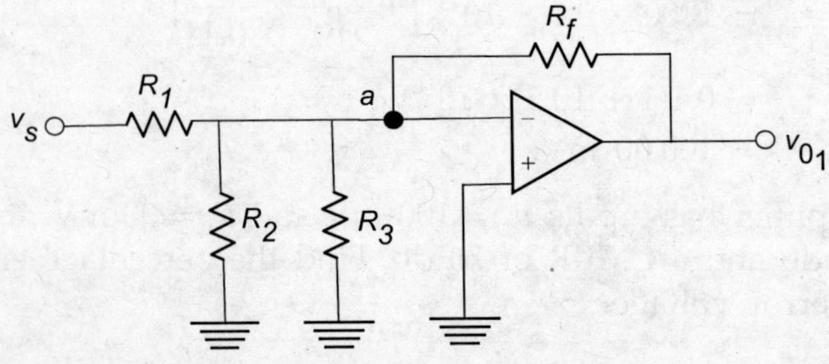

$$\frac{V_a}{R_2}+\frac{V_a}{R_3}+\frac{V_a-V_s}{R_1}+\frac{V_a-V_{o1}}{R_f} = 0.$$

Since $v_a = 0$, because of virtual ground concept,

$$\frac{-v_s}{R_1}+\frac{-v_{o1}}{R_f} = 0$$

$$\Rightarrow \quad \frac{v_{o1}}{v_s} = \frac{-R_f}{R_1} \qquad (1)$$

Making V_{s2} as active one, and killing rest of the sources, the circuit will be KCL at node 'a'

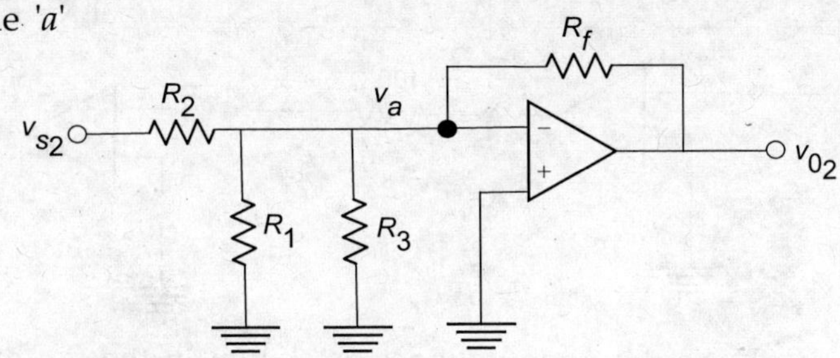

$$\frac{V_a}{R_1}+\frac{V_a}{R_3}+\frac{V_a-V_{s2}}{R_2}+\frac{V_a-V_{o2}}{R_f} = 0$$

Since $v_a = 0$, $v_{o2} = -\dfrac{R_b}{R_2}\, v_{s2}$ (2)

Similarly, making v_{s3} as active one, and killing rest of the sources

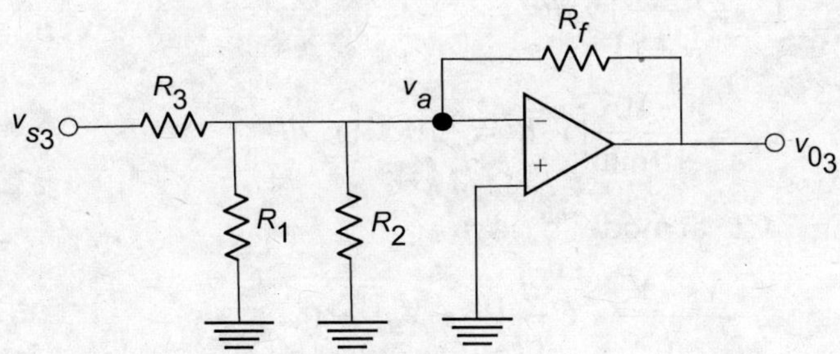

$$\frac{v_a-v_{s3}}{R_3}+\frac{v_a}{R_1}+\frac{v_a}{R_2}+\frac{v_a-v_{o3}}{R_f} = 0$$

$$\therefore \quad v_{o3} = \frac{-R_f}{R_3}\cdot v_{s3} \qquad (3)$$

According to super position theorem

$$v_o = v_{o1} + v_{o2} + v_{o3}$$

$$= \frac{-R_f}{R_1} v_{s1} - \frac{-R_f}{R_2} v_{s2} - \frac{-R_f}{R_3} v_{s3}$$

$$= 10\, v_{s1} - 5\, v_{s2} - \frac{-10}{3}\, v_{s3}$$

13.5 Find the output v_o of the circuit assuming op-amp as an ideal one.

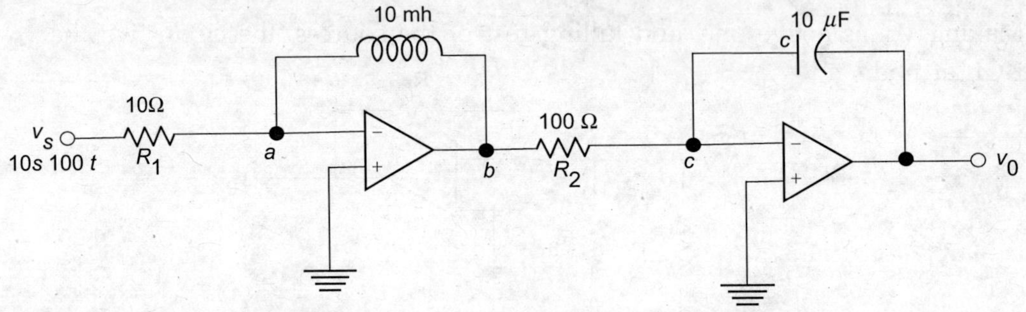

Solution Let V_a, V_b, V_c be the node voltages as shown in the figure. Writing KCL at node '*a*'

$$\frac{V_a - V_s}{R_1} + \frac{1}{L}\int (V_a - V_b)\, dt = 0$$

Since $V_a = 0$.

$$\frac{-V_s}{R_1} - \frac{1}{L}\int v_b dt = 0$$

$$\therefore \quad V_s = \frac{-R_1}{L}\int v_b dt$$

$$\Rightarrow \quad V_s = \frac{-10}{10\,\text{mh}}\int v_b\, dt = -1000\int v_b\, dt \qquad (1)$$

Writing KCL at node '*c*'

$$\frac{V_c - V_b}{R_2} + C\frac{d}{dt}(V_c - V_o) = 0$$

Since $V_c = 0$, because of virtual ground

$$\frac{-V_b}{R_2} + C\frac{d}{dt}(-V_o) = 0.$$

$$V_b = -R_2 C\frac{dV_o}{dt}$$

$$= -100 \times 10 \times 10^{-6} \frac{dV_o}{dt}$$

$$= -10^{-3} \frac{dV_o}{dt} \quad (2)$$

$\therefore$ From (1) and (2),

$$V_s = -1000 \int \left(-10^{-3}\right) \frac{dV_o}{dt} \cdot dt$$

$$= V_o = 10 \cos 1000\ t.$$

13.6 Find the expression for load current i_L in the circuit shown below.

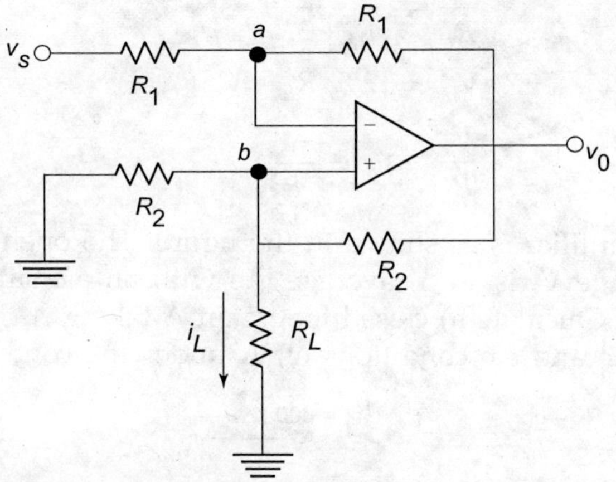

Solution Writing KCL at node 'a'

$$\frac{v_a - v_s}{R_1} + \frac{v_a - v_o}{R_1} = 0$$

$$\Rightarrow v_a = \frac{v_s + v_o}{2} \quad (1)$$

Similarly from KCL at node 'b'

$$\frac{v_b}{R_2} + \frac{v_b}{R_L} + \frac{v_b - v_o}{R_2} = 0$$

$$V_b \left\{ \frac{1}{R_L} + \frac{2}{R_2} \right\} = \frac{V_o}{R_2}$$

$$V_b \left\{ \frac{R_2 + 2R_L}{R_L} \right\} = V_o$$

$$\therefore \quad V_o = V_b \left\{ \frac{R_2 + 2R_L}{R_L} \right\} \quad (2)$$

But load current $I_L = \dfrac{V_a}{R_L}$ or $\dfrac{V_b}{R_L}$

$$\therefore \quad V_o = V_b\left\{\frac{R_2 + 2R_L}{R_L}\right\} \qquad (2)$$

$$\therefore \quad I_L = \frac{V_b}{R_L} = \frac{V_o}{R_L + 2R_L}$$

from (1), $$V_o = 2V_a - V_s = 2\,V_b - V_s.$$

$$\therefore \quad \frac{V_b}{R_L} = \frac{2V_b - V_s}{R_2 + 2R_L}$$

$$\therefore \quad R_2\,V_b + 2R_L\,V_b = 2V_b\,R_L - V_s\,R_L$$

$$\therefore \quad \frac{V_b}{R_L} = I_L = \frac{-V_s}{R_2}$$

13.7 A non-inverting amplifier is as shown in the figure. The open loop gain A_v = 40000, applied input voltage V_s is +1 V average d.c with an a.c sine wave component of 0.4 V peak-to-peak. Calculate a) closed loop gain A taking A_v into account (b) output voltage v_o under d.c and a.c conditions (c) A under the condition that $A_v = \infty$.

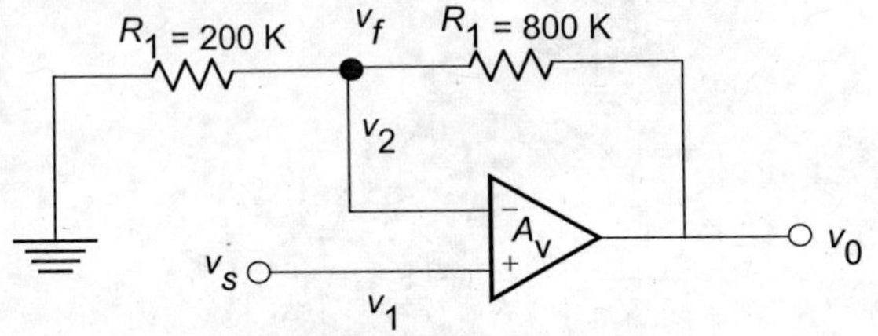

Solution The circuit can be rearranged as

a) For a non-inverting op-amp, there exists voltage-series feed back connection. Applying voltage-division rule

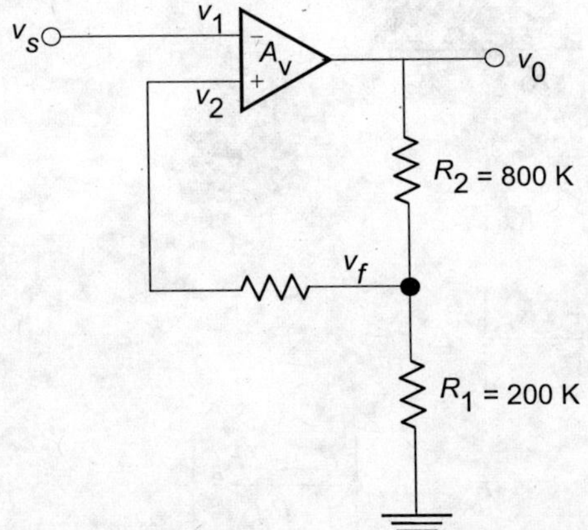

$$V_f = \frac{R_1}{R_1 + R_2}\, v_o = \frac{v_o}{5} \tag{1}$$

and $v_o = A_v\,(V_1 - V_2)$

$$= A_v\,(v_1 - v_f)$$

$$v_o = A_v\left(v_s - \frac{R_1 v_o}{R_1 + R_2}\right)$$

$$v_o\left\{1 + \frac{R_1 A_v}{R_1 + R_2}\right\} = A_v\, V_s.$$

$$\therefore \text{ Closed loop gain} = \frac{V_o}{V_s} = A = \frac{A_v}{1 + \dfrac{A_v R_1}{R_1 + R_2}} \tag{2}$$

$$= \frac{40000}{1 + (40000)\left(\dfrac{1}{5}\right)}$$

$$= 4.9993.$$

From (2), $A = \dfrac{1}{\dfrac{1}{A_v} + \dfrac{R_1}{R_1 + R_2}}$

(b) Substituting $A_v = \infty$,

$$A = \frac{1}{\dfrac{R_1}{R_1 + R_2}} = 1 + \frac{R_2}{R_1}$$

$$= 1 + \frac{800}{200}$$

$$= 5.$$

(c) We have $v_o = \dfrac{A_v}{1 + \dfrac{A_v R_1}{R_1 + R_2}}\, v_s$

given v_s = + 1 V d.c

= 0.4 V a.c peak-to-peak.

Gain $A = \dfrac{v_o}{v_s} \Rightarrow v_o = AV_s$ = 4.993 V d.c = 1.9997 V a.c

13.8 For a particular op-amp the following are the specifications. $I_{CQ} = 10.5\ \mu A$, $C_c = 40$ pF. Amplitude of the input signal is $V_m = 15$ V. Find the slewrate, maximum bandwidth.

Solution

$$\text{Slewrate SR} = \frac{2I_{CQ}}{C_C} = \frac{2\times 10.5\times 10^{-6}}{40\times 10^{-12}}$$

$$= 0.525\ \text{V}/\mu\text{sec.}$$

$$f_{\text{max}} = \frac{\text{slewrate}}{2\pi V_m} = \frac{0.525v/\mu\sec}{2\times 3.14\times 15} = 5.57\ \text{kHz.}$$

13.9 An op-amp has a slew rate of 0.6 V/μs. Determine the highest possible operating frequency for each of the following peak values of o/p voltage (a) 0.2 V (b) 0.6 V (c) 1 V (d) 2 V.

Solution

$$v_m = \frac{\text{slewrate}}{2\pi f_{\text{max}}} \Rightarrow f_{\text{max}} = \frac{\text{slewrate}}{2\pi v_m}$$

a) $f_{\text{max}} = \dfrac{0.6\,\text{v}/\mu\text{s}}{2\pi\times 0.2} = 477.70$ kHz

b) $f_{\text{max}} = \dfrac{0.6\,\text{v}/\mu\text{s}}{2\pi\times 0.6} = 159.23$ kHz

c) $f_{\text{max}} = \dfrac{0.6\,\text{v}/\mu\text{s}}{2\pi\times 1} = 95.54$ kHz

d) $f_{\text{max}} = \dfrac{0.6\,\text{v}/\mu\text{s}}{2\pi\times 2} = 47.77$ kHz

13.10 For an a.c integrator $R_1 = 20$ KΩ, $R_f = 47$ KΩ and $C = 0.1$ μF. Find the 3db cutoff frequency.

Solution An a.c integrator circuit is shown below.

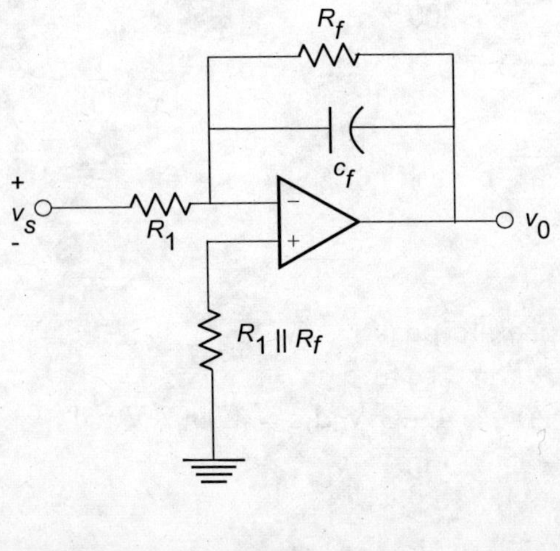

3db Frequency is

$$f_o = \frac{1}{2\pi R_1 C_f}$$

$$= \frac{1}{2 \times \pi \times 20 \times 10^3 \times 0.1 \times 10^{-6}}$$

$$= 79.6 \text{ Hz}$$

13.11 A step signal is given as an input to an integrator. Find the output voltage assuming op-amp as an ideal one if $R_1 C_f = 1$

Solution Given that $V_s = 5 \qquad 0 < t < 4$

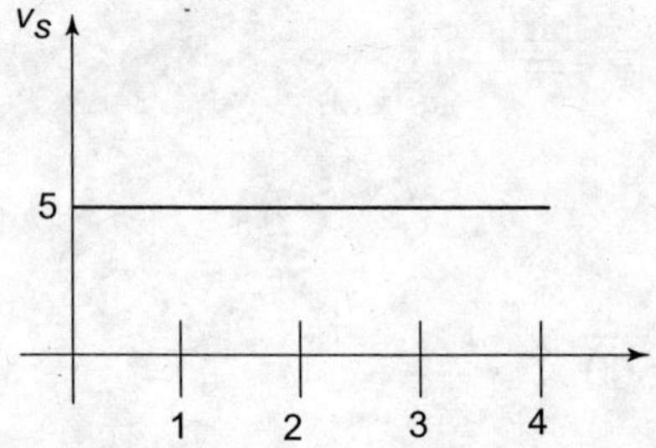

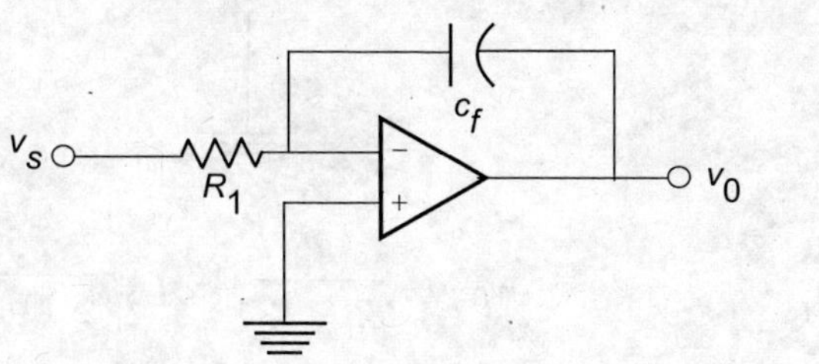

$= 0 \qquad$ otherwise

$$v_o = \frac{1}{R_1 C_f}\int_0^4 v_s\,dt$$

$$\therefore V_o = -\int_0^4 v_s\,dt$$

$$= -\left[\int_o^4 5\,dt\right] = -5 \times 4 = -20 \text{ V.}$$

The output voltage is shown below.

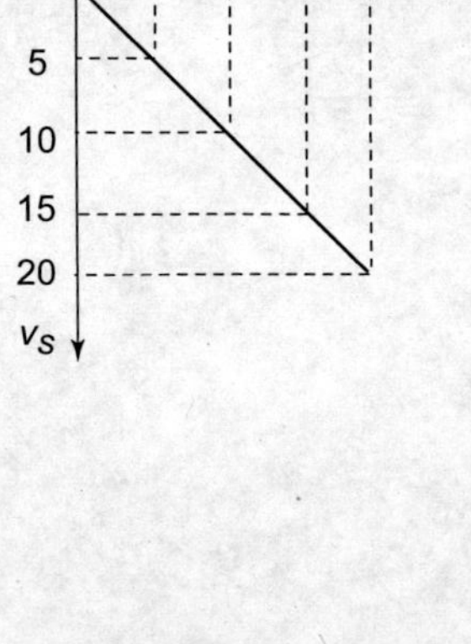

13.12 For the Schmitt trigger circuit shown in the figure below, $R_1 = 500\ \Omega$. $R_2 = 1500\ \Omega$. Saturation voltages are at ± 14 V. Find upper and lower threshold voltages and hysterisis voltage.

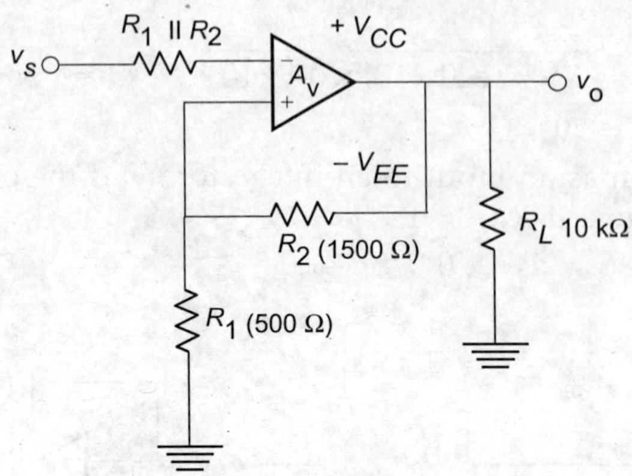

$$V_{UTP} = \frac{R_1}{R_1 + R_2} V_{sat} = \frac{500}{500+1500} \times 14$$

$$= +3.5 \text{ V}$$

$$V_{LTP} = \frac{R_1}{R_1 + R_2} (-V_{sat}) = -3.5 \text{ V}.$$

$$V_H = V_{UTP} - V_{LTP}$$

$$= 3.5 - (-3.5) = 7 \text{ V}$$

13.13 Find the value of R in the circuit shown below. If v_1 and v_2 are amplified by the same amount. Assume op-amp as an ideal one.

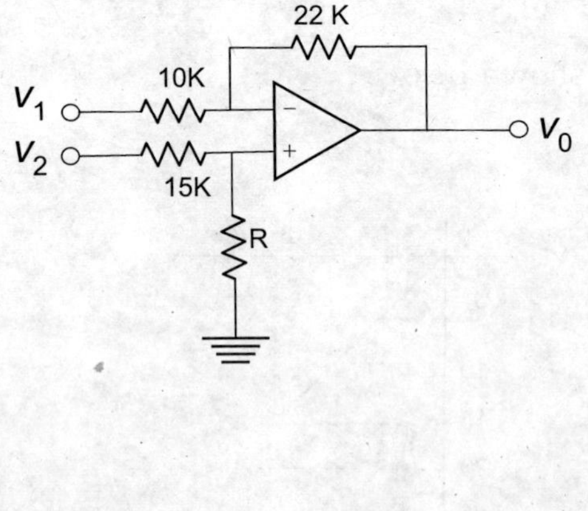

Solution When V_1 is acting, V_2 is made inactive. The circuit acts like an inverting amplifier.

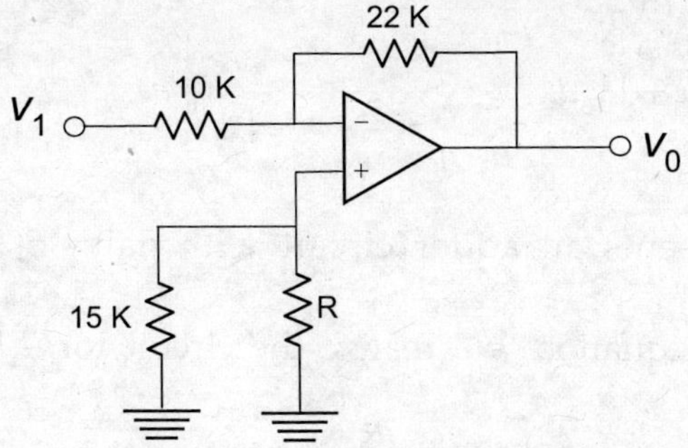

$$\therefore \quad V_{o1} = \frac{-22}{10} V_1 = -2.2 \ V_1 \qquad (1)$$

when V_2 is acting, V_1 is made inactive, the circuit acts like a non-inverting amplifier.

$$V_o = \left(1 + \frac{1+22\text{K}}{10}\right) V_a = 3.2 \ V_a.$$

$$V_a = \frac{R}{15\text{K}+R} \cdot V_2$$

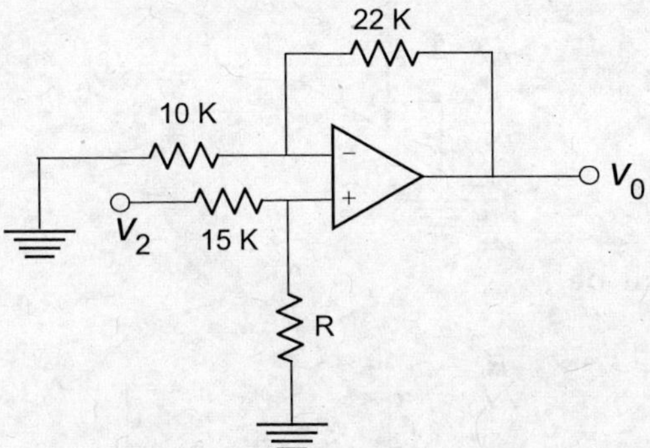

$$\therefore \ V_o = \frac{3.2R}{R+15000} V_2 \qquad (2)$$

From (1) and (2), gain values to be equal.

$$\frac{3.2R}{R+15000} = +2.2$$

$\therefore$ $3.2 \ R - 2.2 \ R = 15000 \times 2.2$

$\therefore$ $R = 33 \ \text{K}\Omega.$

13.14 Realize the following mathematical equations using operational amplifier

(a) $V_o = 5V_1 + 4V_2 + 6V_3$

(b) $V_o = \int v_1 dt + \int v_2 dt + \int v_3 dt$

Solution

(a) First equation represents an adder circuit with gains 5, 4, 6, taking v_1, v_2, v_3 as inputs.

To realize the given equation, we realize the circuit for $-5V_1 - 4V_2 - 6V_3$ and then

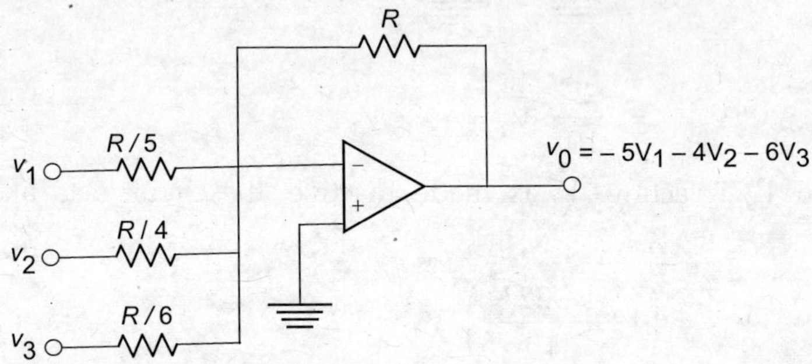

use an inverter.

To invert V_{o1}, we design an inverting amplifier with a gain unity.

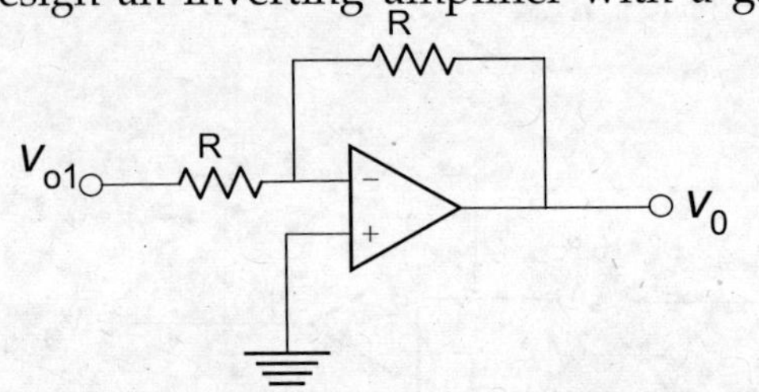

∴ The final circuit will be,

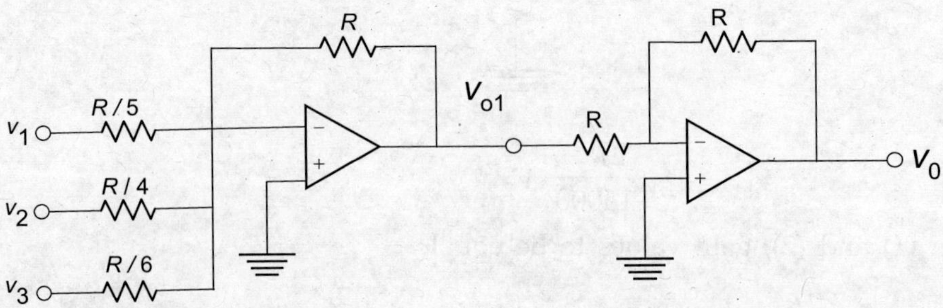

(b) $V_o = \int V_1 dt + \int V_2 dt + \int V_3 dt$

The above expression represents assuming Integrator

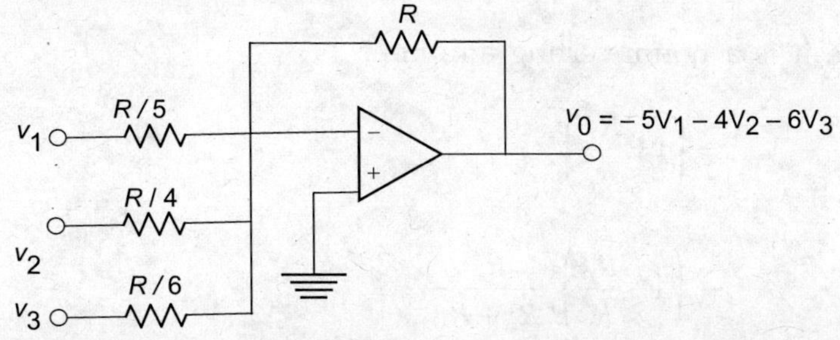

Assume $RC = 1$. It can be shown that $V_{o1} = -\int V_1 dt + \int V_2 dt + \int V_3 dt$. Giving V_{o1} as an input to inverting amplifier.

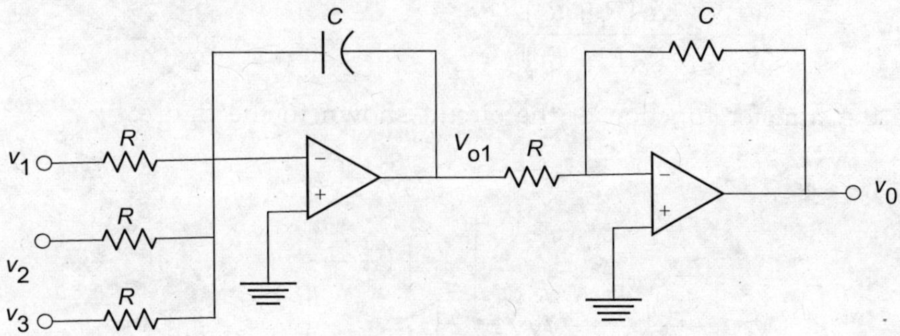

13.15 Derive the expression for gain v_o/v_i of the circuit shown below.

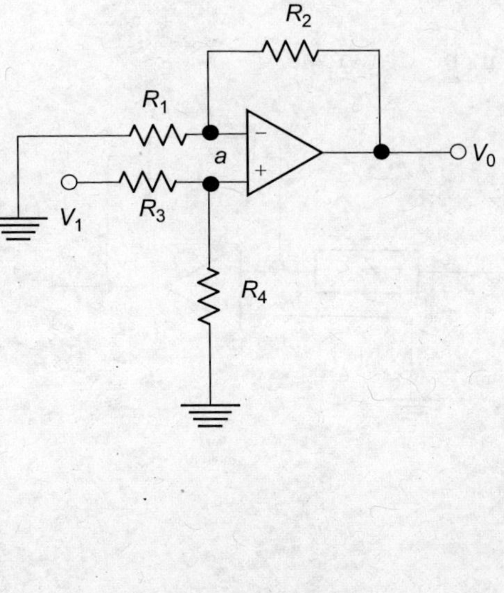

Solution From the circuit,

$$V_a = \frac{R_4}{R_4 + R_3} \cdot V_1$$

Since it is a non-inverting amplifier

$$V_o = \left(1 + \frac{R_2}{R_1}\right) V_a$$

$$= \left(1 + \frac{R_2}{R_1}\right) \cdot \frac{R_4}{R_4 + R_3} V_1$$

$$= \frac{R_1 + R_2}{R_4 + R_3} \cdot \frac{R_4}{R_1} \cdot V_1$$

$$\therefore \quad \frac{V_o}{V_1} = \frac{R_4 (R_1 + R_2)}{R_1 (R_4 + R_3)}$$

13.16 Derive the transfer function of the circuit shown in the figure.

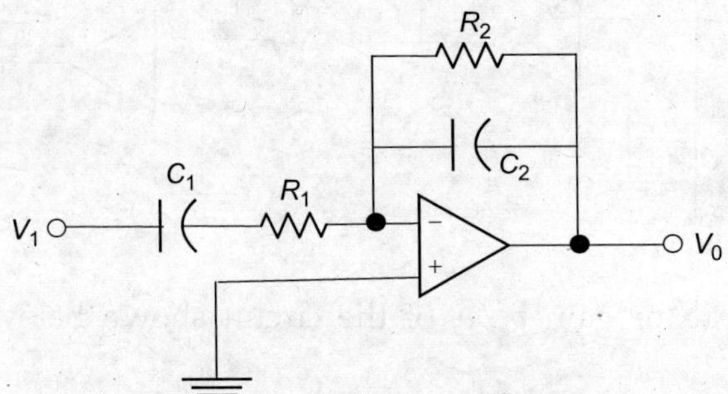

Solution The circuit can be drawn as

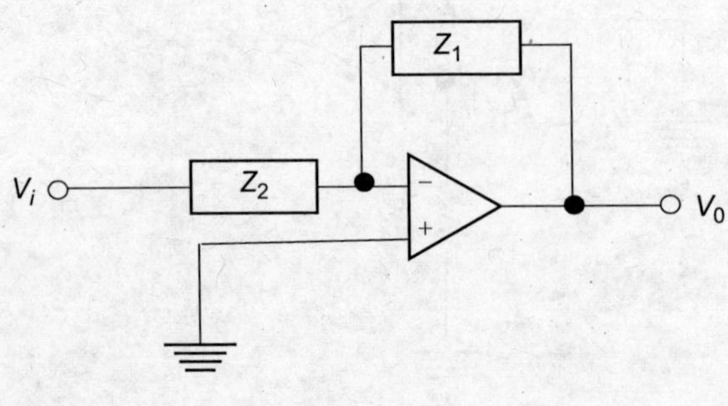

We have, $\dfrac{V_o}{V_i} = \dfrac{-Z_2}{Z_1}$

$$Z_1 = R_1 + \frac{1}{SC_1} = \frac{1+SC_1R_1}{SC_1}$$

$$Z_2 = \frac{R_2 \cdot \dfrac{1}{SC_2}}{R_2 + \dfrac{1}{SC_2}} = \frac{R_2}{1+SC_2R_2}$$

$$\frac{V_o}{V_i} = \frac{-Z_2}{Z_1} = \frac{-R_2}{1+SC_2R_2} \cdot \frac{SC_1}{1+SC_1R_1} = \frac{-R_2C_1S}{(1+SC_1R_1)(1+SC_2R_2)}$$

13.17 Find the input impedance V_s/I_s of the circuit shown

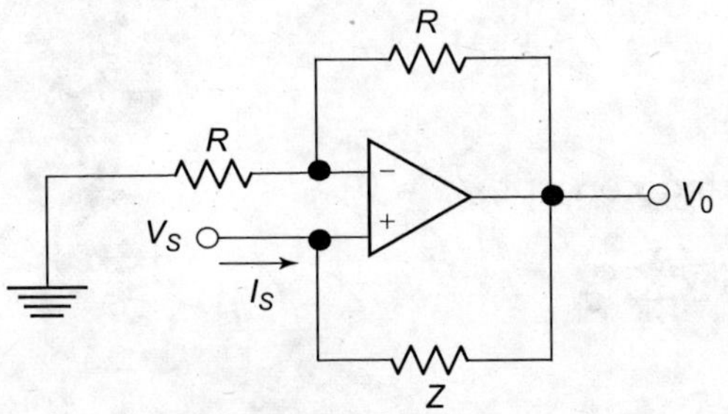

From the circuit $V_o = \left(1+\dfrac{R}{R}\right) V_s = 2\ V_s.$ (1)

Writing KCL at non-inverting terminal,

$$I_S = \frac{V_s - V_o}{Z} = \frac{V_s - (2V_s)}{Z} = \frac{-V_s}{Z}$$

$$\therefore \quad \frac{V_s}{I_s} = -Z.$$

13.18 Find the output waveform if a sine wave of 2 V peak and 200 hZ frequency is applied to a differentiator circuit. Assume C = 10 μF and R = KΩ.

Solution Given input signal $v_s = 2 \sin 2\pi(200)\ t$.

$$V_o = -RC\frac{dv_s}{dt} = -(10 \times 10^3 \times 10 \times 10^{-6})\frac{d}{dt}(v_s)$$

$$= -0.1\frac{d}{dt}(2 \sin 400\pi t)$$

$$= -0.2 \ (\cos 400\pi t) \ 400\pi$$
$$= -80\pi \cos 400 \ \pi t \text{ volts}$$
$$= -251.2 \cos 400 \ \pi t \text{ volts}$$

The input and output waveforms are shown in figure.

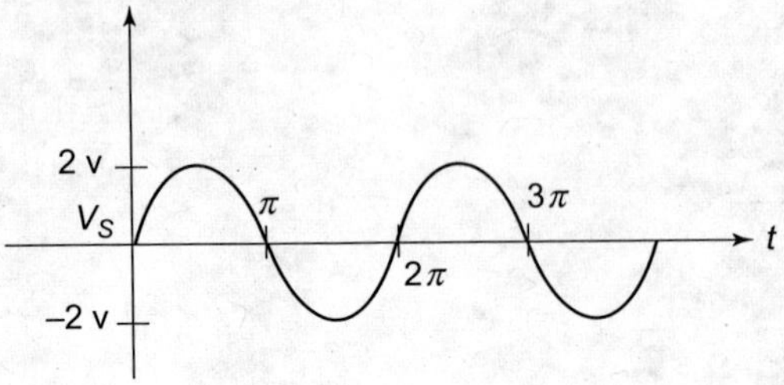

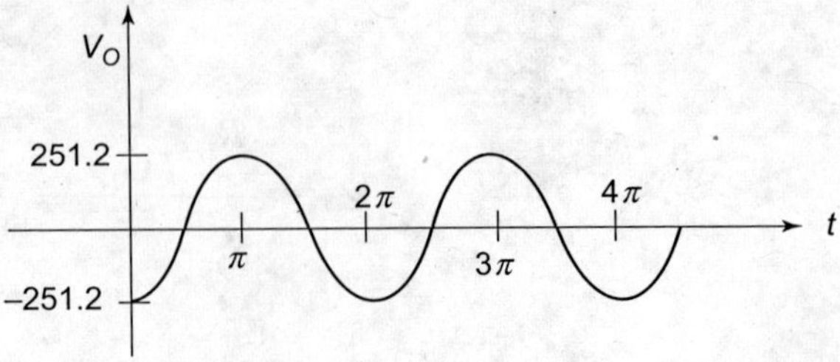

13.19 Find the expression for output voltage V_o if the applied input is a sweep voltage $v_i = At$.

Solution Writing KCL at node '*a*'

$$\frac{V_a - V_i}{Z} + \frac{V_a - V_o}{R_f} = 0$$

Because of virtual ground concept, $v_a = 0$.

$$\therefore \ v_o = \frac{-R_f}{Z} v_i \quad \text{where} \quad Z = R \ || \ \frac{1}{SC}$$

In S-domain, $V_o(s) = \dfrac{-R_f}{Z} \cdot V_i(s)$

$$Z = R \ || \ \frac{1}{SC} = \frac{R \cdot \frac{1}{SC}}{R + \frac{1}{SC}} = \frac{R}{1 + SCR}$$

$$\therefore \ V_o(s) = \frac{-R_f(1 + SCR)}{R} \cdot \frac{A}{S}$$

$$= \frac{-AR_f}{R}\left\{\frac{1}{S} + CR\right\}$$

$$= -AR_f C - \frac{AR_f}{R} \cdot \frac{1}{S}$$

Taking Inverse Laplace transform for $V_o(s)$,

$$v_o(t) = -AR_f \, C u(t) - \frac{AR_f}{R} u(t)$$

EXERCISE PROBLEMS

13.1 A difference amplifier has inputs V_{s1} = 10 mV and V_{s2} = 0.9 mV. It has a difference mode gain of 60 dB. If the CMRR of the amplifier is 80 dB, find the percentage error in the output and error voltage. Derive the expression that is used in this problem.

13.2 The circuit shown is a differential amplifier with ideal Op-Amp. Find the output voltage and show that the output corresponding to common mode voltage is zero when $R'/R = R_1/R_2$.

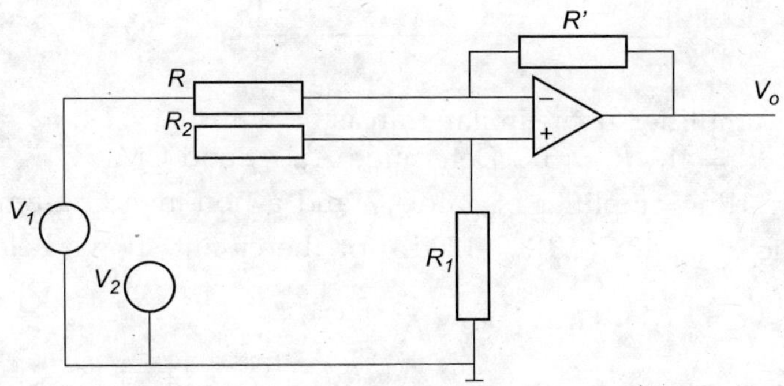

13.3 A differential amplifier has inputs V_{i1} = 10 mV and V_{i2} = 12 mV. It has differential gain of 60 dB and CMRR of 80 dB. Find the error in the output and output error voltage. If the CMRR is made to be 100dB, what is the change in the error?

13.4 Design an Op-Amp with gain of –10 and input resistance of 10K.

13.5 Derive the expression of ideal Op-Amp for the closed loop gain as shown in the figure.

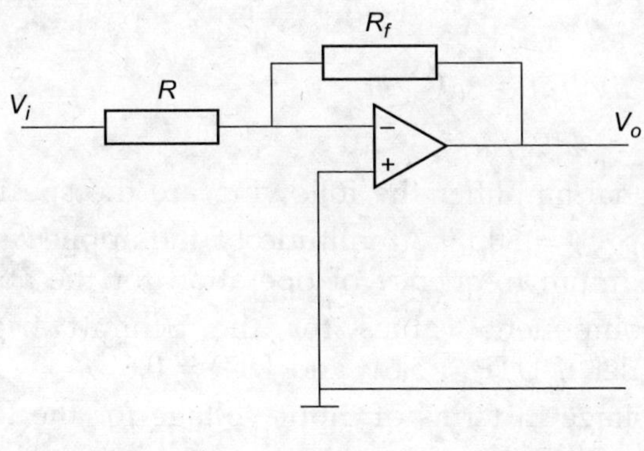

13.6 Find the output voltage of the following circuit.

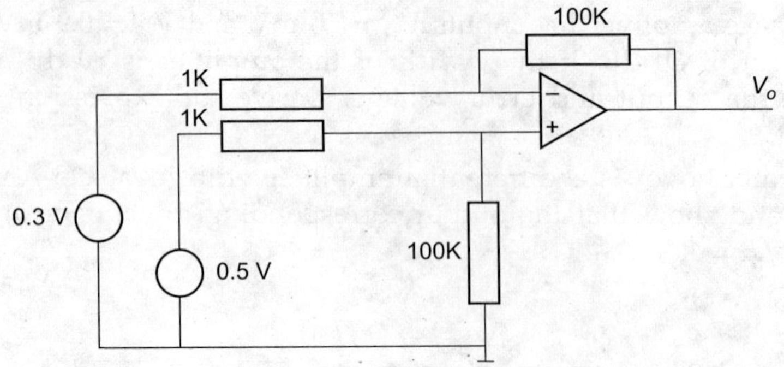

13.7 A differential amplifier uses similar transistor with h_{ie} = 1k, h_{fe} = 100, V_{cc} = V_{EE} = 10V, R_s = lK, R_c = 4K, R_e = 5K. Determine A_d, A_c and CMRR.

13.8 Design an inverting amplifier of gain - 7 and a non-inverting amplifier of gain 11.

13.9 Find the value of V_0 if CMRR = 120dB for the circuit shown below.

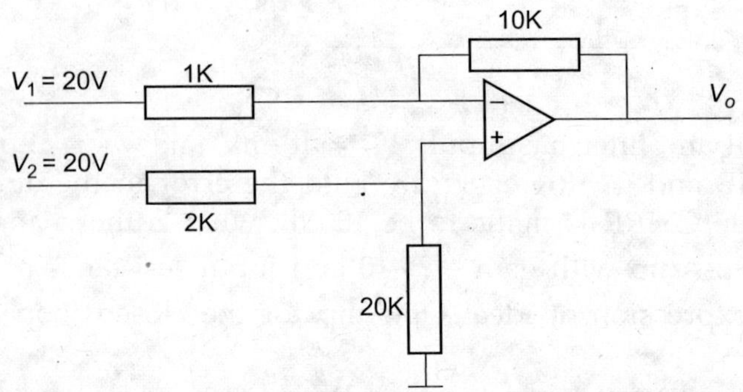

13.10 Realize the following mathematical equations using operational amplifiers

(a) $V_o = V_1 + 4V_2 + 7V_3$

(b) $V_o = dV_1/dt + dV_2/dt + 3dV_3/dt$

(c) $V_o = V_1 + 3V_2 + 4dV_3/dt$

13.11 For a 741 operational amplifier the following are the specifications. I_{CQ} = 7.5microamp C_c = 30pF. Amplitude of the applied signal is 10V. Determine the slew rate, maximum frequency of operation for the above slew rate.

13.12 Determine the component values for the Schmitt trigger circuit using 741 Operational amplifier if UTP = 5.6V and LTP = 0.6V.

13.13 Express output voltage in terms of input voltage for the following circuit

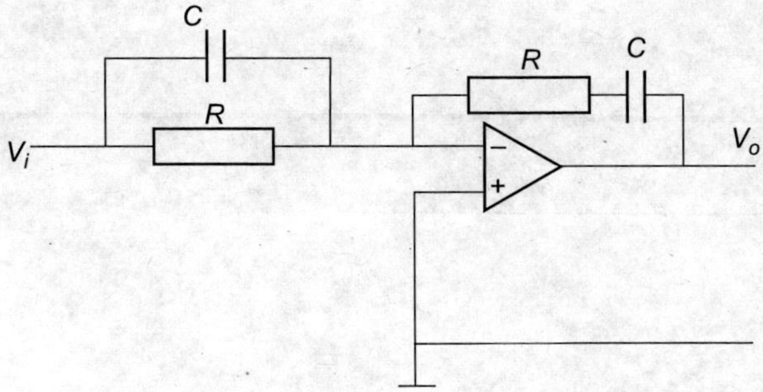

13.14 Find the expression for gain transfer function of the circuit shown below.

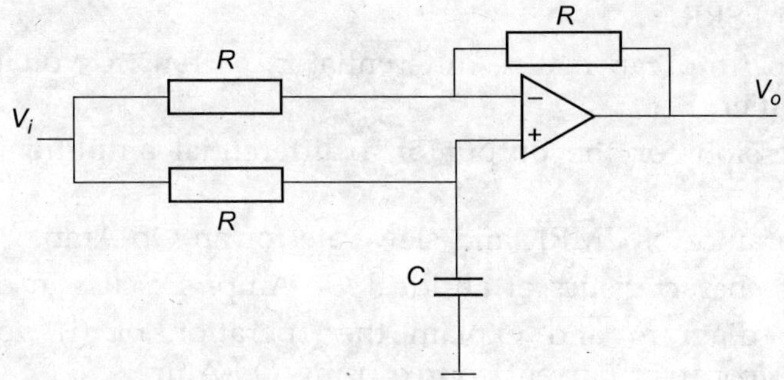

13.15 Find the input impedance of the circuit shown below.

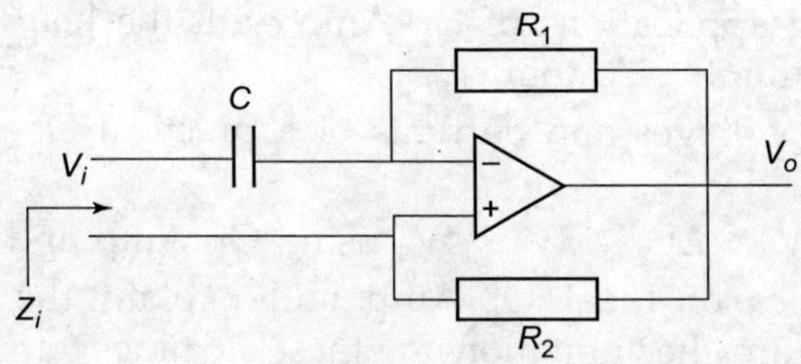

REVIEW QUESTIONS

1. Draw the circuit diagram of a practical logarithmic using Operational amplifier and derive the expression for its output voltage.
2. Explain the terms (i) Input offset current, (ii) CMRR, (iii) Output offset voltage, (iv) Slew rate and (v) PSRR.
3. Illustrate how Op-Amp can realize differentiator, current to voltage converter and logarithmic amplifier.
4. Derive the expression for the output of a differential amplifier as a function of CMRR.
5. Explain the importance of CMRR and slew rate for an Op-Amp.
6. What are various characteristics of an ideal Op-Amp? Discuss in detail.
7. Give the circuit diagram and explain the operation of (i) voltage to current converter and (ii) constant current source using Op-Amp.
8. Draw the circuit diagram of a voltage follower and explain its operation and its application.
9. Draw the circuit diagram of differential amplifier using transistors and determine its CMRR.
10. Explain the following applications of Op-Amp with the help of circuit diagram: (i) Integrator and (ii) Summing Amplifier.
11. Define several offset voltages and currents of a practical Op-Amp. Give the source of each parameter.
12. Realize the function $V_o = 2V_1 + 3V_2 - 4V_3$ using Op-Amp and resistors.
13. List the characteristics of Ideal Op-Amp and explain the significance of these characteristics. What are the limitations of these in practical Op-Amp devices?
14. Draw the basic internal block diagram of the Op-Amp and explain each block in detail.
15. Derive the equation for CMRR of a transistor differential amplifier.
16. Why is a constant current source used in a differential amplifier? Explain.
17. What are various error voltages and currents possible in a practical Op-Amp? Explain each in detail.
18. What is significance of PSRR? Give its details.
19. What is unity crossover frequency? What are the other bandwidth definitions with respect to Op-Amp?
20. What is slew rate? What is its significance?

21. How is multiplication of two signals possible with Op-Amp? Draw the block diagram and explain in detail each block.
22. Draw a circuit using Op-Amp, which can differentiate the input signal with respect to time and explain its operation.
23. Draw the basic circuit diagram of a low pass filter using Op-Amp. It can also perform one mathematical operation. What is that? Explain.
24. Draw a circuit of zero crossing detector and explain why Hysteresis is to be introduced in the circuit.
25. What is Transresistance amplifier? Draw its circuit and explain its operation.

CHAPTER

14

Cathode-ray Oscilloscope

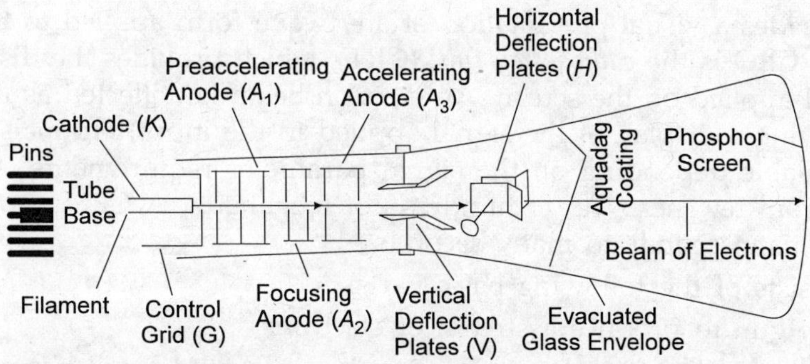

Basic construction of a CRT

Chapter Outline

The concepts introduced in this chapter are:

- *The force on motion of electron in magnetic field*
- *The electron path when subjected to magnetic fields*
- *The behaviour of electron when subjected to both electric and magnetic fields*
- *Types of deflections of an electron beam*
- *Different methods of focusing electrons to a fine spot in an electron beam*
- *Constructional features of a CRT*
- *Applications of CRO*

14.1 INTRODUCTION

The signals employed in electronic circuits may be audio, video, data, timing signals and so on. Generation and manipulation are the main functions of electronic circuits. To manipulate the generated signal, one requires measurement of various parameters of the signal. The electric signal may be dc or ac voltage or current. There are various meters that can measure the parameters of the signals. These meters give the values of the parameters under some assumed conditions. For example, when measuring the rms value of ac signal, the assumption is that the signal is pure sinusoid. The measurement would be wrong if there is any deviation in the shape of the signal under measurement and also no other form of ac signal can be measured with this device. These problems of measurement can be solved when the signal is displayed in its exact form. This is achieved with the help of *cathode-ray oscilloscope* (CRO).

The CRO provides a visual presentation of any wave form applied to the input terminals. The heart of CRO is the *cathode-ray tube* (CRT), which provides the display giving the form of the signal applied on the screen. An electron beam is deflected as it sweeps across the tube face, leaving a display of the signal applied to the input terminals. A wide range of CROs are available depending on the signal parameter requirements. A CRO may be built to operate from few hertz up to hundreds of megahertz. A CRO also measures time from fractions of nanoseconds to many seconds.

The CRT consists of the following basic parts:

- An electron gun to produce a stream of electrons
- Focusing and accelerating elements to produce a well-defined beam of electrons
- Horizontal and vertical deflecting plates to control the path of the electron beam
- An evacuated glass chamber with a phosphorescent screen, which glows visibly when stuck by the electron beam.

The following sections of this chapter present the analysis of electron motion in electrostatic and electromagnetic deflection as well as focusing, and finally the applications of the CRO are discussed.

14.2 TWO-DIMENSIONAL MOTION OF ELECTRON

The force on a unit positive charge at any point in an electric field is defined as electric-field intensity ε *at that point.* Thus, the resulting force on the positive charge q is in the direction of field intensity and is given by

$$f = q\,\varepsilon \qquad \text{...(14.1)}$$

To determine the path of the charged particle in the electric field, by Newton's second law,

$$f = q\,\varepsilon = m a = m\,dv/dt \qquad \text{...(14.2)}$$

where a is acceleration, v is the velocity and m is the mass. When the charged particle is electron, $q = -e$, where e is the charge in coulombs on an electron.

Consider a parallel-plate capacitor as shown in Fig. 14.1. When distance between the plates is much small compared to the area of the plates, the field between the plates can be assumed to be uniform and lines of force pointing in the negative x direction. That is, there is only one field in the $-x$ axis. Since there is no force in the y and z directions, acceleration along these directions must be zero. Thus, the only possible direction of electron movement is along x axis. From Eq. (14.2),

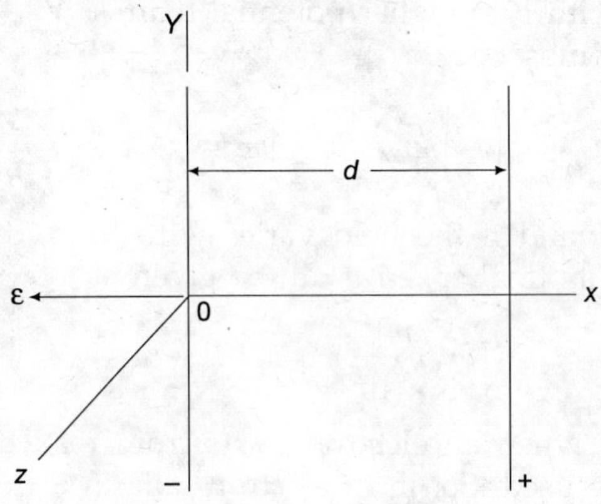

Fig. 14.1 *Parallel-plate capacitor*

$$e\varepsilon = ma_x$$

or,

$$a_x = e\varepsilon/m = \text{constant} \qquad \text{...(14.3)}$$

The velocity and displacement can be given as follows, provided the acceleration in x direction a_x is constant and independent of time as

$$v_x = v_{ox} + a_x t, \quad x = x_o + v_{ox}t + \left(\frac{1}{2}\right)a_x t^2 \qquad \text{...(14.4)}$$

and,

$$a_x = dv_x/dt, \quad v_x = dx/dt \qquad \text{...(14.5)}$$

are acceleration and velocity, respectively.

Now, consider a uniform field; ε_x may be a function of distance but is independent of time, from Eq. (14.2),

$$-e\varepsilon_x = m\,dv_x/dt \qquad \text{...(14.6)}$$

Multiplying the above equation with $dx = v_x\,dt$ and integrating,

$$-\int_{x_o}^{x} \varepsilon_x dx = \int_{v_{ox}}^{v_x} v_x dv_x \qquad \text{...(14.7)}$$

The *potential* V of a point x with respect to point x_o is work done against the field in taking a unit positive charge from x_o to x. Thus,

$$-\int_{xo}^{x} \mathrm{E}_x dx = V \qquad \text{...(14.8)}$$

Equation (14.7) can be written as

$$eV = \left(\frac{1}{2}\right) m\,(v_x^2 - v_{ox}^2) \qquad \text{...(14.9)}$$

If A and B are two points with B at higher potential than A, V_{BA} is the potential difference and q is the charge in coulombs, then,

$$qV_{BA} = \left(\frac{1}{2}\right) mv_A^2 - \left(\frac{1}{2}\right) mv_B^2 \qquad \text{...(14.10)}$$

The above equation is not valid if the field varies with time. If an electron starts at rest and $V_{BA} = V$, $v_A = 0$, $v_B = v$, its final speed v can be given as

$$v = [2eV/m]^{1/2} \qquad \text{...(14.11)}$$

$$= 5.93 \times 10^5\ V^{1/2} \qquad \text{...(14.12)}$$

This equation is valid only when an electron starts from rest. If it possesses some initial velocity, Eq. (14.10) is to be used. So, if an electron falls in a field of 1 V starting at rest, the final velocity achieved by it is 593 km/s. Even though the velocity is very large, the associated kinetic energy is very small compared with that of an electron due to minute mass associated with it.

Two-dimensional motion: The following assumptions are made in the following analysis:

- The charge density is low enough so that the mutual repulsive force may be neglected.
- The movement of the charged particle is in high vacuum so that no collision with gas atoms or ions occurs.
- The mass of the charged particle is extremely small so that the gravitational force may be neglected as compared to the forces exerted by the fields.

Consider a situation as depicted in Fig. 14.2 where an electron enters the region between two parallel plates of a parallel-plate capacitor with an initial velocity v_{ox} in the positive x direction. Assuming the electric field to be uniform and directed in negative y direction, the initial conditions are

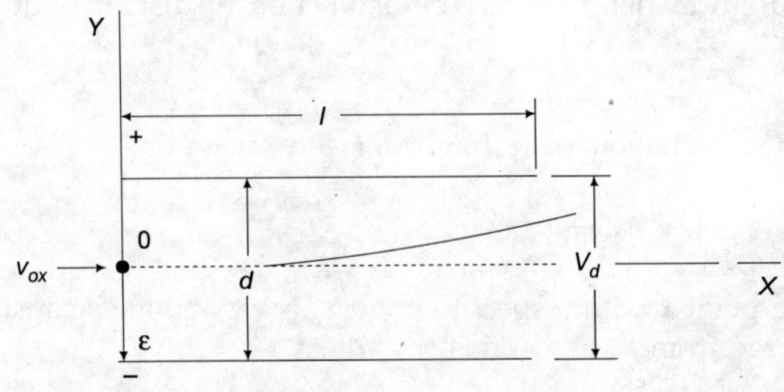

Fig. 14.2 *Two-dimensional motion in a uniform electric field*

$$v_x = v_{ox} \text{ at } x = 0, t = 0$$
$$v_y = 0 \text{ at } y = 0, t = 0 \qquad \text{...(14.13)}$$
$$v_z = 0 \text{ at } z = 0, t = 0$$

Thus, the component of velocity in the z direction remains constant since there is no force in the z direction and thus acceleration is zero. Since the initial velocity in this direction is assumed to be zero, the motion must take place entirely in one plane, the plane of the paper.

Similarly, the velocity along x axis remains constant and equal to v_{ox}. Thus,

$$v_x = v_{ox} \qquad \text{...(14.14)}$$

Then,

$$x = v_{ox}t \qquad \text{...(14.15)}$$

A constant acceleration exists along the y direction and the motion is given by Eq. (14.4) with x replaced by y,

$$v_y = a_y t, \quad y = \left(\frac{1}{2}\right) a_y t^2 \qquad \text{...(14.16)}$$

where,

$$a_y = -(e\varepsilon_y/m) = eV_d/(md) \qquad \text{...(14.17)}$$

where $V = V_d$. These equations indicate that in the region between the plates the electron is accelerated upward, the velocity component v_y varying from point to point, whereas the component v_x remains constant in the passage of electron between the plates. Eliminating t from Eqs. (14.15 and 14.16),

$$y = \left(\frac{1}{2}\right)(a_y/v_{ox}^2)x^2 \qquad \text{...(14.18)}$$

Thus, the particle moves in a parabolic path in the region between the plates.

14.3 FORCE AND MOTION IN MAGNETIC FIELD

The *force* acting on a conductor of length L carrying a current I situated in a magnetic field B is given by

$$f_m = BIL \qquad \text{...(14.19)}$$

The direction of force is perpendicular to the plane consisting of the components of B and I that are mutually perpendicular and is directed along the advance of a right-hand screw as shown in Fig. 14.3.

If an electron takes T seconds to travel a distance of L metres in the conductor, the total number of electrons passing through any cross-section of conductor in unit time is N/T, where N is the total number of electrons contained in the conductor. Thus, the total charge per second crossing any point is

$$I = Nq/T \qquad \text{...(14.20)}$$
$$f_m = BIL = BNqL/T \qquad \text{...(14.21)}$$

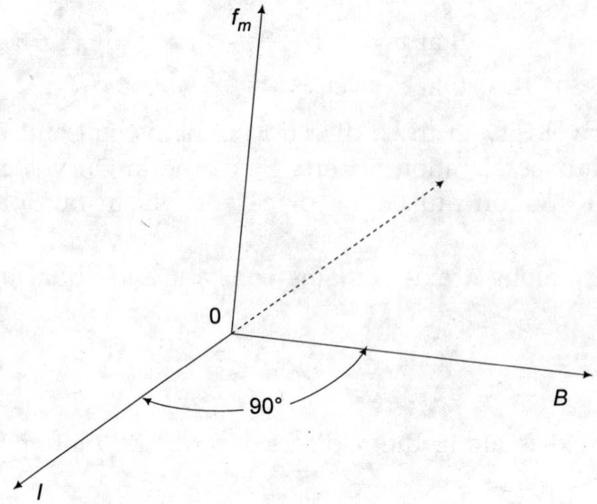

Fig. 14.3 *Direction of force on a particle in magnetic field*

$$= qNBv$$

where v is the drift velocity and the force per electron is

$$f_m = qBv \quad ...(14.22)$$

If the particle is a positive ion, the directions of current and velocity are same. If the particle is an electron, the direction of the current is opposite to that of the drift velocity.

Motion in a magnetic field: The path of a charged particle that is moving in a magnetic field is now investigated. Consider an electron to be placed in the region of the magnetic field. If the particle is at rest, $f_m = 0$ and the particle remains at rest. If the initial velocity of the particle is along the lines of the magnetic flux, there is no force on the particle. Hence, a particle whose initial velocity has no component normal to a uniform magnetic field will continue to move with constant speed along the lines of flux.

Now, assume that an electron moving with a speed v_o enters a constant uniform magnetic field normally as shown in Fig. 14.4. Since the force f_m is perpendicular to v and so to the motion at every instant, no work is done on the electron. That is, the kinetic energy is not increased. Since v and B are both constant in magnitude, f_m is constant in magnitude and perpendicular to the direction of motion of the particle. This type of force results in motion in a circular path with constant speed. The force remains constant in magnitude and is always directed toward the centre of the circle and so is normal to the motion.

A particle moving in a circular path with constant speed v has an acceleration toward the centre of the circle of magnitude v^2/R, where R is the radius of the path. Thus,

$$mv^2/R = eBv \quad ...(14.23)$$

or

$$R = mv/(eB) \quad ...(14.24)$$

Angular velocity can be given as

$$\omega = v/R = eB/m \quad ...(14.25)$$

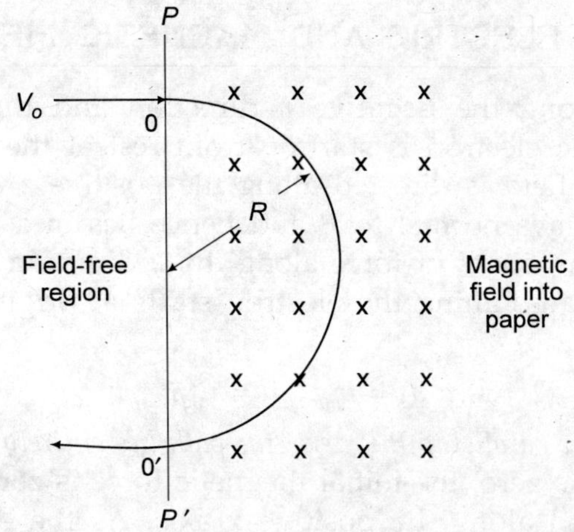

Fig. 14.4 *Circular motion of an electron in a magnetic field*

The *period,* time in seconds, for one complete revolution can be derived to be

$$T = 2\pi/\omega = 2\pi m/(eB) \quad \text{...(14.26)}$$

For an electron, it is

$$T = 3.57 \times 10^{-11}/B \quad \text{...(14.27)}$$

It can be observed that the radius of the path is directly proportional to the speed of the particle. Further, *the period and the angular velocity are independent of speed or radius.* That is, the faster moving particles will traverse larger circles in the same time that a slower particle moves in its smaller circle.

14.4 PARALLEL ELECTRIC AND MAGNETIC FIELDS

When electric and magnetic fields act simultaneously on an electron, whether they are in same or opposite directions and whether the initial velocity is zero or not, no force is exerted due to magnetic field and the motion of the electron is only due to the electric field. The electron will move in a direction parallel to the fields with constant acceleration. If the electric field is directed in positive y direction whereas the magnetic field in the negative y direction, the complete motion of the electron can be given as

$$v_y = v_{oy}t - at, \quad y = v_{oy}t - \frac{1}{2}at^2 \quad \text{...(14.28)}$$

where $a = e\varepsilon/m$ is the magnitude of acceleration. The negative sign in these equations is due to the fact that the direction of the acceleration of an electron is opposite to the direction of the electric field ε.

If initially a component of velocity v_{ox} perpendicular to the magnetic field exists, this component along with the magnetic field will set the electron in a circular motion. The radius of the circular path is independent of the electric field, but the velocity along the field changes with time. As a result of this, the electron travels in a helical path with the pitch changing with time. The distance travelled along y axis per revolution increases with each revolution.

14.5 PERPENDICULAR ELECTRIC AND MAGNETIC FIELDS

If the magnetic field is along the negative y direction and the electric field is along negative x direction and the electron is starting from rest at the origin, the force on the electron due to the electric field is directed along the positive z direction. Any force due to the magnetic field is always normal to B and hence lies in a plane parallel to the x-z plane. Thus, there is no component of force along the y direction and the y component of acceleration is zero. Hence, assuming the electron starts at origin, the motion along y is given by

$$f_y = 0, \quad v_y = v_{oy}, \quad y = v_{oy} t \qquad \text{...(14.29)}$$

If the initial velocity component parallel to B is zero, the path lies entirely in a plane perpendicular to B. Since the initial velocity is zero, the initial magnetic force is zero and due to the electric field the electron is directed along the positive x axis. As the electron is accelerated in positive x direction, the force due to the magnetic field is no longer zero. There will be a component of this force, which is proportional to the x component of velocity and will be directed along the positive z direction. The path will thus bend away from the positive x direction towards the positive z direction. The electric and magnetic fields interact with each other and the net force will finally make the electron travel in a *cycloidal path.* A *cycloid* is defined as the path generated by a point on the circumference of a circle of given radius, which rolls along a straight line.

When an electron is released perpendicular to both the electric and magnetic fields, so that $v_{ox} = v_{oy} = 0$ and $v_{oz} \neq 0$, the electric force is along positive x direction and the magnetic field is along negative x direction. If the net force on the electron is zero, it will continue to move along z axis with the constant speed v_{oz}. Thus, the electron follows a *straight-line path.* This condition is realised when

$$e\varepsilon = eBv_{oz} \qquad \text{...(14.30)}$$

Thus, when an electron is released with the above velocity into perpendicular electric and magnetic fields, it suffers no deflection and the net force on it is zero. It may be noted that this velocity is independent of charge or mass of the particle.

If the initial velocity component in the direction perpendicular to the magnetic field is not zero, the path of the electron is *trochoid.* Trochoid is the locus of a point on a spoke of a wheel rolling on a straight line.

14.6 ELECTROSTATIC DEFLECTION

The deflection system adopted in a cathode-ray tube can be *electrostatic deflection.* It consists of a pair of deflection plates as shown in Fig. 14.5. The deflection voltages are applied between the two plates. For deflecting the beam in the horizontal and vertical directions, two sets of plates are required. The horizontal deflection plates are placed vertically and the vertical deflection plates are placed horizontally. The plane of the plates is always parallel to the axis of the cathode-ray tube. The electrons are attracted towards the positive plates and when they leave the region below the plates they travel in a straight line, at an angle with the axis. These electrons, which leave the region, strike the screen at a point.

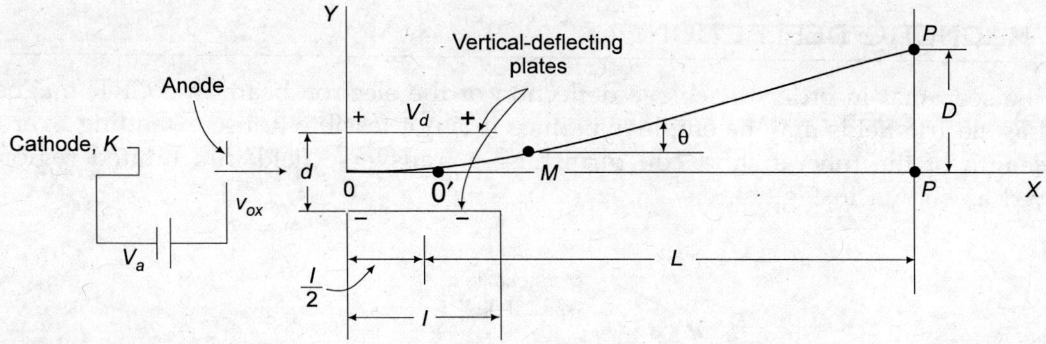

Fig. 14.5 *Electrostatic deflection*

The hot cathode K emits electrons, which are accelerated toward the anode by the applied potential. Those electrons, which are not collected by the anode, pass through the tiny hole in the anode and strike the fluorescent screen. The electrons striking the screen emit energy in form of photons and thus light is emitted from the screen. The displacement D of the electrons is determined by the potential V_d applied between the deflecting plates. The velocity with which the electrons emerge from the anode hole can be given as

$$v_{ox} = \sqrt{[(2eV_a)/m]} \qquad \text{...(14.31)}$$

assuming the velocity of electron emission from cathode to be zero. Since the region between the anode and the edge of the deflection plates is field free, the electrons move in a straight line and with constant velocity. In the region between the plates, the electrons will move in the parabolic path given by Eq. (14.18). The straight-line path in the region from emerging point M of the plates to the point P' on the screen is the tangent to the parabola at the point M.

The deflection D from the axis of the tube can thus be estimated to be

$$D = (lLV_d)/(2dV_a) \qquad \text{...(14.32)}$$

Thus, the deflection on the screen of a CRT is directly proportional to the deflecting voltage applied between the plates. Consequently, a CRT is a linear voltage-indicating device.

The *electrostatic deflection sensitivity* of a CRT is defined as the deflection on the screen per volt of the deflection voltage. Thus,

$$S \equiv D/V_d = (lL)/(2dV_a) \qquad \text{...(14.33)}$$

The sensitivity is independent of both the deflecting voltage and the ratio e/m. Further, he sensitivity varies inversely with accelerating voltage.

The assumptions made in this connection that the deflection field is uniform and the region outside the deflecting plates is field-free may not be possible in practice. This leads to the fringing fields and correction is to be employed up to as much as 40% of the above-achieved result.

14.7 MAGNETIC DEFLECTION

It may be seen that in order to achieve deflection of the electron beam in a CRT, magnetic as well as electric fields may be employed. Since it is not feasible to use extending over the entire length of the tube, a short coil furnishing a transverse field in a limited region is employed as shown in Fig. 14.6.

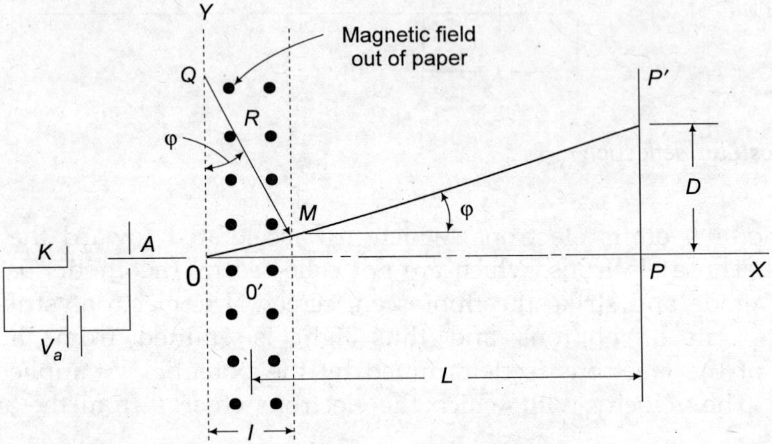

Fig. 14.6 *Magnetic deflection system*

The magnetic field is taken as pointing out of the paper and the beam is deflected upwards. It is also assumed that the magnetic field is uniform in the restricted region and the region outside is field-free. The emerging electron beam experiences a force of magnitude eBv in the region of magnetic field. The path OM will be an arc of a circle whose centre is at Q. The speed of the electrons will remain constant and equal to

$$v = v_{ox} = \sqrt{[(2eV_a)/m]} \qquad ...(14.34)$$

The angle φ is is equal to the length of the arc OM divided by R, the radius of the circle. If φ is small,

$$\varphi \approx l/R \qquad ...(14.35)$$

where

$$R = (mv)/(eB) \qquad ...(14.36)$$

The deflection D on the screen can be derived to be

$$\begin{aligned} D &\approx L \tan \varphi \approx L\varphi \\ &\approx lL/R = (lLeB)/(mv) \\ &= \frac{lL\beta}{\sqrt{V_a}} \sqrt{[e/(2m)]} \end{aligned} \qquad ...(14.37)$$

The *sensitivity of magnetic deflection* can be defined as the ratio of deflection to the magnetic field intensity. Thus,

$$S \equiv D/B = \frac{lL}{\sqrt{V_a}}\sqrt{[e/(2m)]} \qquad ...(14.38)$$

Thus, the sensitivity is independent of B. In electrostatic deflection, the sensitivity is independent of anode voltage but in magnetic deflection, it is inversely proportional to the root of anode voltage. Also, this sensitivity depends on the ratio e/m. Since the sensitivity is proportional to L, the deflecting coils are to be placed as far as possible from the screen. In general, they are placed directly after the accelerating anode.

In a TV, the deflection angle is large and so the assumptions made in the above discussion are not valid. So, the deflection is no longer proportional to B and so the deflection of the beam on the face of the tube will not be linear with time. In a TV, there are two sets of deflection coils mounted on the neck of the tube to achieve horizontal and vertical deflection of the electron beam.

14.8 ELECTROSTATIC FOCUSING

The convergent beam from the accelerating electrode has a tendency to spread because of the mutual repulsion between the electrons. Hence, some focusing device is required to bring the beam to a sharp focus at the screen. There are two methods of focusing. They are *electrostatic* and *magnetic*.

In electrostatic focusing, the electron lens system formed between electrodes makes the electron beam focus into a fine spot on the screen. Both these electrodes are anodes and are at high voltage compared to that of the cathode.

Usually, the electron gun contains two electrostatic focusing systems referred to as electron lenses. The first electron lens comprises the cathode, control grid, and the first accelerating electrode. The dimensions of the electrodes and their voltages are so chosen that the electrons emitted from the cathode converge to a small area located in front of the cathode and on the axis of the tube. This *crossover* surface has an area much smaller than that of cathode and may be treated as the source of electrons for all electrodes beyond this point. Figure 14.7 shows the nature of equipotential contours thus produced. Electrons move normal to these contours.

The second electron lens system is formed in the region between accelerator electrode and the first anode. Figure 14.8 shows the equipotential contours of such a system. The electrons move perpendicular to these contours. The shape of the contours depends on the geometry of the electrode system. The electric-field magnitude depends on the potentials applied to the electrodes.

Usually, the second voltage and accelerating electrode voltage are maintained constant while the first anode voltage is varied to achieve the necessary focusing. The size of the focused spot on the florescent screen depends on the size of the crossover and the magnification of the second lens system. The magnification is generally much less than unity. To reduce the dimension of focused spot, the apertures in the gun may be made small, thus

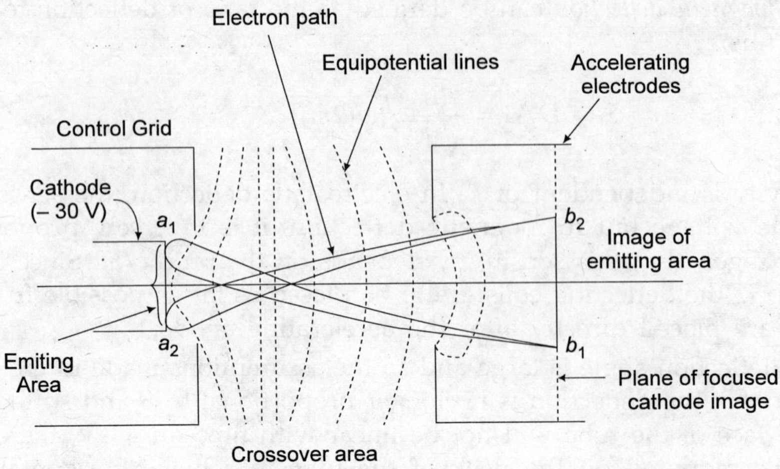

Fig. 14.7 *Electron lens system at the control grid*

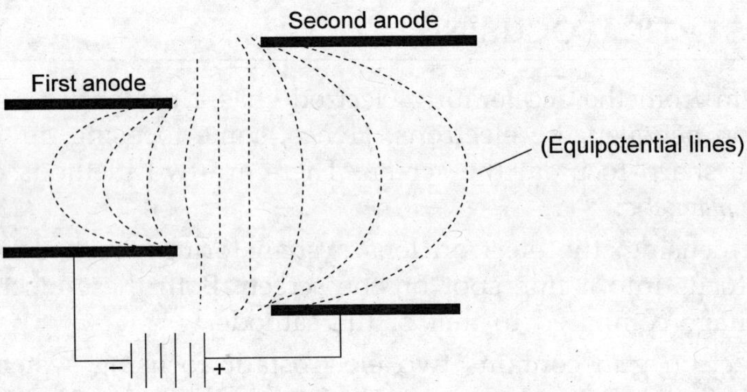

Fig. 14.8 *The second electron lens system*

eliminating divergent electrons. The mutual repulsion of electrons within the cathode-ray beam and unequal velocities of emitted electrons limit the minimum size of spot.

The purpose of the control electrode in the first lens of the electron gun is to vary the current in the beam. However, change in control electrode voltage also changes to some extent the focal length of the first lens, thus changing the size of the focused image on the screen.

14.9 MAGNETIC FOCUSING

Magnetic focusing may be achieved either by a permanent magnet or by an electromagnetic coil, whose axis coincides with the axis of the beam. The coil may extend the whole length of the electron beam or may be limited to a small region. Extended coil is employed in image orthicon camera tube of a TV and short coil type is employed in TV receivers, radar indicators and general purpose CRO.

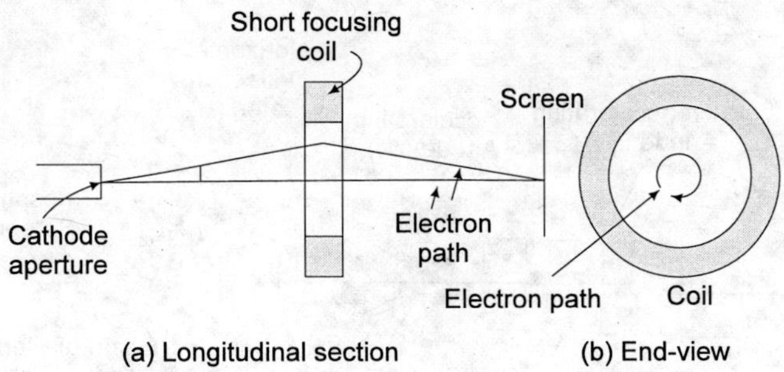

Fig. 14.9 *Magnetic focusing*

The principle of magnetic focusing is as shown in Fig. 14.9. The focusing coil surrounds the tube. The lines of magnetic field are uniformly distributed and are parallel to the direction of the electron motion.

The electron experiences a force from the magnetic field only if its motion is at an angle to the lines of magnetic force. If the line of motion and magnetic lines of force are parallel, the electron moves only by the attractive force of the anode. If the electron leaves the cathode at an angle to the axis of the system, a force is exerted on it by the magnetic field. The direction of this force is at right angle to the direction of motion of electron and to magnetic line of force. The force thus causes the electron to move in a spiral path. If this circle is just completed by the time the electron reaches the screen, the electron must lie on the axis of the tube just opposite the point where it was emitted at the last aperture of the gun. All the electrons emerging from a point are brought to a corresponding point on the screen.

In general, the electrons emitted from the cathode have different initial velocities and varying angles to the axis of the tube. The time required by the electron to complete one circle of the spiral flight depends on the strength of the magnetic field and on the charge to mass ratio of the electron. Since the ratio of the charge to mass of electrons is same for all the electrons and also in a uniform magnetic field, all the electrons complete one circle in same time.

The magnitude of the magnetic field depends on the current flowing in the coil and also number of turns of the coil. The magnitude of the field is adjusted by controlling the current until a perfect focusing on the screen is achieved.

14.10 CATHODE-RAY TUBE

The *cathode-ray tube* (CRT) is the heart of the *cathode-ray oscilloscope* (CRO). CRT provides the visual display of the input signal wave form. Figure 14.10 shows the basic construction of the CRT.

A cathode K containing an oxide coating is heated indirectly by a filament, which results in the release of electrons from the cathode surface. A control grid G provides control of the number of electrons passing through the aperture into the tube. The number of electrons to be emitted to the tube is controlled by the voltage of the control grid.

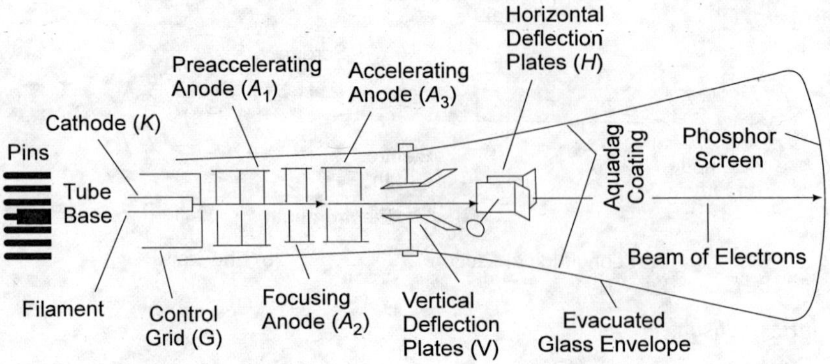

Fig. 14.10 *Basic construction of a CRT*

After the electrons pass the control grid, they are focused into a tight beam and accelerated to a higher velocity by the focusing and accelerating anodes.

The high-velocity well-defined electron beam then passes through the two sets of deflection plates. The first set of plates is oriented to deflect the electron beam vertically up or down. The direction of vertical deflection is determined by the voltage polarity applied to the deflecting plates. The amount of deflection is set by the magnitude of the applied voltage. The beam is also deflected horizontally left to right by a voltage applied to the horizontal deflecting plates. The deflected beam is further accelerated by very high voltages applied to the tube, with the beam finally striking a phosphorescent material on the inside face of the tube. This screen glows when struck by the high-speed electrons. Thus, light is emitted on the other side of the glass screen.

The CRT is a self-contained unit with leads brought out through a base to pins. Various types of CRTs are available in a variety of sizes and with different phosphor screens and deflection electrodes. CRTs find many applications in CRO, TV, radar indicators and so on, of which CRT for CRO is considered in this section.

In a CRO, the electron beam is deflected horizontally by a sweep voltage and vertically by the voltage to be measured. While the electron beam is moved across the face of the CRT by the horizontal sweep signal, the input signal deflects the beam vertically, resulting in a display of the input signal wave form. One sweep of the beam across the face of the tube, followed by a blank period during which the beam is turned off while returned to the starting point across the tube face, constitutes one sweep of the beam.

The display on the screen will be steady when the same image is swept across the tube in each sweep. This requires synchronisation, starting the sweep at the same point in a repetitive wave form cycle. If the signal is properly synchronised, the display will be stationary. If the synchronisation is not proper, the image will drift in horizontal direction.

Figure 14.11(a) shows a basic block diagram of a CRO. To achieve a noticeable deflection on the screen of the CRO, the measured voltage, which is of the order of millivolts to tens of volts, must be increased to order of hundreds of volts for proper operation of the CRO. Amplifiers are adopted for this purpose in both vertical and horizontal deflection

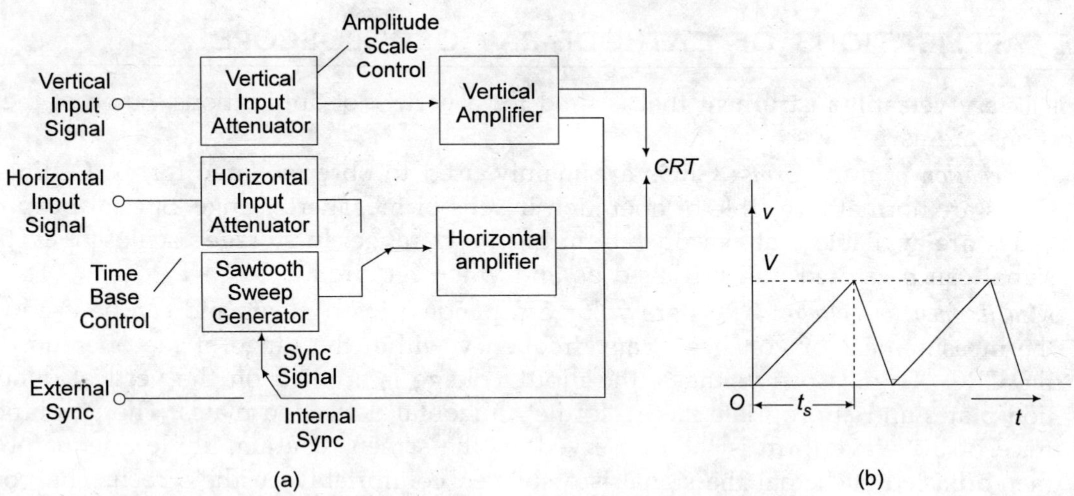

Fig. 14.11 *Basic block diagram of a CRO*

systems. To adjust the amplitude of the signal for comfortable display measurements on the screen, the input is subjected to variable attenuation.

A *sawtooth wave form* is impressed across the horizontal deflection plates. This voltage, as shown in Fig. 14.11(b) is used to sweep the beam across the screen. Thus, it is also known as *sweep voltage*. The electrons are deflected with time in the horizontal direction for a time T. Then, the beam returns to its starting point on the screen very quickly as the sawtooth voltage rapidly falls to its initial value at the end of each period.

If a sinusoidal voltage is impressed across the vertical deflection plates and no sweep is applied to horizontal deflection plates, the display on the screen would be a vertical straight line. Now, when the sweep voltage is applied to horizontal deflection plates, the electron beam sweeps the electron beam across the screen in time and the full sinusoidal voltage is displayed on the CRT screen. The display would be stationary only when the sweep and input signals are properly synchronised. This is achieved when the time period T of the sweep be equal to or is multiple times of the time period T. Thus, it is required to adjust the sweep period of the sawtooth wave form to synchronise with the applied input signal time period.

The input signal can be any wave form. Thus, when a nonrepeating arbitrary wave form is to be displayed on the screen, the transients are taken as the reference and the sweep would start with the start of the transients.

A commercial oscilloscope has many refinements than those discussed above. The sensitivity is greatly increased by means of amplifiers between the input signal and the deflection plates. The electron gun is a complicated structure, which allows for accelerating the electrons through a large potential, for varying the intensity of the beam and for focusing the electrons into a tiny spot. Controls are provided for positioning the beam as desired on the screen.

14.11 APPLICATIONS OF CATHODE-RAY OSCILLOSCOPE

CRO is a very versatile instrument that is used for a variety of applications. Some of these applications are as follows:

- *Observation of wave forms*: CROs are mainly used to observe wave forms. Ordinary CROs are normally capable of operation in tens of megahertz range. Special purpose CROs are available that can operate in gigahertz range. In storage oscilloscopes, the signal can be stored and retrieved as and when required.
- *Measurement of voltage*: CROs are voltage-dependent instruments and can be used for the measurement of voltages at any frequency within the range of the operation of the CRO. To measure voltages, the input voltage is applied on the vertical deflection plates and appropriate sweep to the horizontal deflection plates. The amplitude trace of the wave form is then observed on the screen. The amplitude attenuator is then adjusted such that the signal is displayed comfortably on the screen. The position of the attenuator knob gives the volts/cm position and thus the voltage of the input signal can be measured by multiplying this position value with the number of centimetres the signal is occupying in the vertical direction. Either ac or dc voltages can be measured. The dc voltages are displayed as horizontal straight line whereas the exact wave form is displayed for that of ac signal. From the peak value of the ac signal, the rms and other parameters can be determined. Not only can perfect sinusoidal wave forms be measured, but since the wave form is displayed, the amount of distortion and other parameters of interest can also be determined.
- *Measurement of current*: Current cannot be directly measured with a CRO. To measure the current, a known resistance is taken and the potential drop across the resistance is determined with the help of measurement of potential at both ends of the resistor. This voltage difference divided by the considered resistance value gives the amount of current flowing in the device.
- *Measurement of frequency*: The measurement of frequency is indirect with a CRO. The time period of the input signal is determined with the help of the CRO and the frequency of the input signal is the reciprocal of this time period. The signal whose frequency is to be measured is applied as usual to the vertical deflection plates of the CRT along with the sweep voltage to horizontal deflection plates. The amplitude and the time attenuation knobs are appropriately adjusted such that the signal is displayed comfortably on the screen of the CRO. The time duration between two identical points along the wave form is the time period of the input signal wave form. This is measured by multiplying the time/cm value of time attenuator knob with the distance in cm between two consecutive identical points on the wave form on the screen of the CRO. The reciprocal of this time period is nothing but the frequency of the input signal wave form.
- *Lissajous figures*: Lissajous figures are geometrical figures obtained as a result of the compounding of two simple harmonic wave forms. The two wave forms are applied to the vertical and horizontal deflection plates of the CRO, respectively; it will represent a set of figures called *Lissajous figures*. With the help of these figures, one can determine the ratio of the amplitudes of the signals when the frequency is identical and the two signals are in phase, determine the phase difference between the two

signals of same frequency or determine the unknown frequency of a given signal when the frequency of the other signal is known. If v_x and v_y represent these two wave forms, when $v_x = V_x \sin \omega t$ and $v_y = V_y \sin \omega t$, the trace of the electron beam on the screen of the CRO would be a straight line with slope equal to V_x/V_y as shown in Fig. 14.12.

When $v_x = V_x \sin \omega t$ and $v_y = V_y \cos \omega t$, and when $V_x = V_y$, the trace of the electron beam on the screen would be a circle and when $V_x \neq V_y$, the trace would be an ellipse as shown in Fig. 14.13.

When $v_x = V_x \sin \omega_x t$ and $v_y = V_y \sin (\omega_y t + \theta)$, where θ is the phase angle between the two signals, the trace of the electron beam on the screen would be complex and is as shown in Fig. 14.14. One can determine frequency of an unknown wave form when the other signal frequency is known. The unknown frequency can thus be measured with the following equation:

$$\frac{\omega_x}{\omega_y} = \frac{\text{number of tangent points along the } Y \text{ direction}}{\text{number of tangent points along the } X \text{ direction}} \qquad \text{...(14.39)}$$

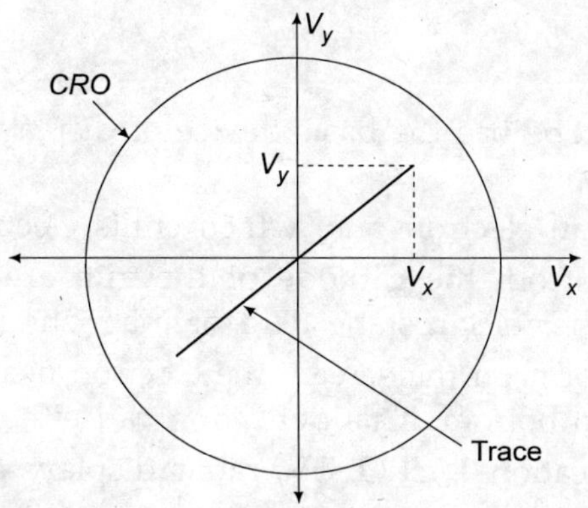

Fig. 14.12 *Lissajous figure for two identical signals*

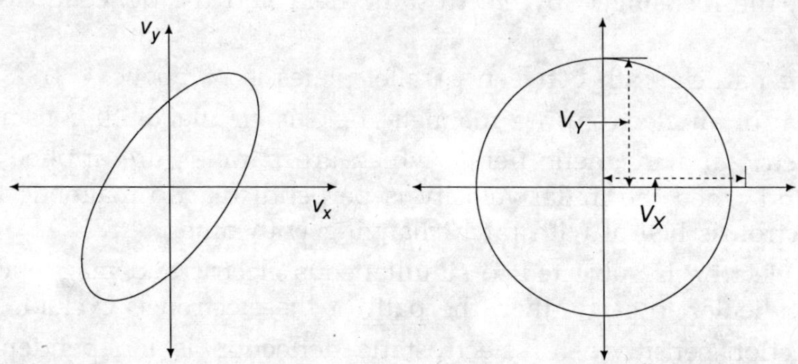

Fig. 14.13 *Lissajous figure for two signals, which are out of phase*

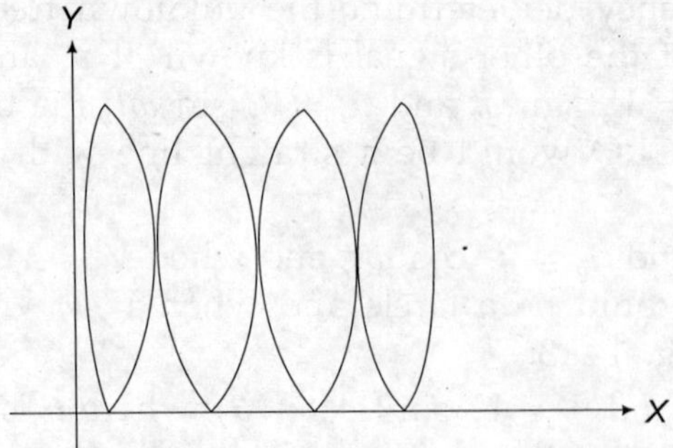

Fig. 14.14 *Lissajous figure of complex wave forms*

SUMMARY

- A cathode-ray oscilloscope has the basic device of displaying the signal wave form as the cathode-ray tube.
- Cathode-ray tube has an electron gun, which emits electrons from a cathode.
- The electrons emitted from the cathode of the gun are accelerated and focused into a fine spot by either electrostatic field or magnetic field.
- To display the wave form on the screen, the electron beam is to be deflected from left to right and top to bottom. This deflection can either be electric or magnetic.
- CRT finds many applications—CRO, TV, radar displays and so on.
- In CRO, the vertical deflection of the electron beam is by the applied input signal whereas the horizontal deflection is by sweep signal wave form, which is a sawtooth wave form.
- In a CRO, the focusing is by electrostatic field and the deflection is by magnetic field.
- The charge particle path between parallel plates is parabolic.
- The motion of an electron in a magnetic field is circular with constant speed.
- When electric and magnetic fields, which are parallel, are applied on a moving electron and when the initial velocity is perpendicular to magnetic field, the path of the electron is helical with pitch changing with time.
- When an electron is subjected to simultaneous electric and magnetic fields, which are perpendicular to each other, the path of the electron is cycloid.
- The deflection sensitivity of electrostatic deflection is independent of both the deflecting voltage and the ratio e/m and varies inversely with accelerating voltage.

- The deflection sensitivity of magnetic deflection is independent of magnetic field, is inversely proportional to the root of anode voltage and depends on the ratio e/m.
- The electron lens system formed between the electrode systems is utilised for focusing the electron beam in electrostatic focusing.
- The deflection coils are mounted at the neck or throughout the length of the CRT for focusing the beam in magnetic focusing.
- The various applications of CRO are measurement of voltage, current, frequency, phase, distortion and so on, whether the input signal is ac or dc.

SOLVED PROBLEMS

14.1 Find the speed and kinetic energy of an electron after it has been accelerated through a potential difference of 1000 V.

Solution

$$\text{Speed of an electron} = \sqrt{\frac{2qV}{m}}$$

$$= \sqrt{\frac{2 \times 1.6 \times 10^{-19} \times 1000}{9.1 \times 10^{-31}}}$$

$$= 5.929 \times 10^{9} \text{ m/sec}$$

$$\text{Kinetic energy K.E.} = q \times V$$

$$= 1.6 \times 10^{-19} \times 1000$$

$$= 1.6 \times 10^{-16} \text{ Joules}$$

$$= 1000 \text{ eV}$$

14.2 A charged particle, having mass equal to 1000 times of an electron and a charge same as that of an electron is accelerated through a potential difference of 1000 V. Calculate the velocity attained and the kinetic energy in eV and Joules.

Solution

$$\text{Mass of charged particle} = 1000 \times 9.1 \times 10^{-31} \text{ kg}$$

$$= 9.1 \times 10^{-28} \text{ kg}$$

$$\text{Charge} = 1.6 \times 10^{-19} \text{ C}$$

$$\text{Velocity} = \sqrt{\frac{2qV}{m}}$$

$$= \sqrt{\frac{2 \times 1.6 \times 1.9 \times 10^{-19} \times 1000}{1000 \times 9.1 \times 10^{-31}}}$$

$$= 0.596 \times 10^{6} \text{ m/sec}$$

$$\text{Kinetic energy} = q \times V$$
$$= 1.6 \times 10^{-19} \times 1000$$
$$= 1.6 \times 10^{-16} \text{ joules}$$
$$= 1000 \text{ eV}$$

14.3 Two plane parallel plates A and B are placed 5 mm apart and potential B is made 300 V positive with respect to plate A. An electron starts from rest from plate A. Calculate (a) the velocity of the electron on reaching plate B (b) time taken for the electron to reach plate B (c) kinetic energy of the electron on reaching plate B.

Solution (a) The electron starts from rest at plate A, so, the initial velocity is zero. The velocity of electron on reacting plate B is

$$v = \sqrt{\frac{2 \times q \times V}{m}} = \sqrt{\frac{2 \times 1.6 \times 10^{-19} \times 300}{9.1 \times 10^{-31}}}$$
$$= 10.27 \times 10^{6} \text{ m/sec.}$$

(b) The average velocity of electron $V_{avg} = \dfrac{\text{initial velocity + final velocity}}{2}$

$$= \frac{0 + 10.27 \times 10^{6}}{2} = 5.135 \times 10^{6} \text{ m/s}$$

Therefore, time taken for travel is

$$\text{time} = \frac{\text{separation between plates}}{\text{average velocity}}$$
$$= \frac{5 \times 10^{-3}}{5.135 \times 10^{6}}$$
$$= 0.973 \times 10^{-9} \text{ s}$$

(c) Kinetic energy of the electron on reaching plate B is

$$\text{KE} = q \cdot V = 1.6 \times 10^{-19} \times 300$$
$$= 4.8 \times 10^{-17} \text{ Joules}$$

14.4 In a vacuum diode, the spacing between the parallel plane plates of cathode and anode is 5 mm, and the potential difference is 250 V. Calculate the time taken by the electron, with an initial velocity of 1×10^{6} m/sec, to travel from cathode to anode.

Solution $V = 250$ V, $d = 5 \times 10^{-3}$ m, $v_{\text{initial}} = 1 \times 10^{6}$ m/s

$\therefore$ The speed acquired by the electron due to the applied voltage is

$$v = \sqrt{V_{\text{initial}}^{2} + \frac{2qV}{m}}$$

$$= \sqrt{(1\times10^{6})^{2} + \frac{2\times1.6\times10^{-19}\times250}{9.1\times10^{-31}}}$$

$$= 9.43 \times 10^{6} \text{ m/s}$$

$$v_{avg} = \text{average velocity} = \frac{v_{initial} + v}{2}$$

$$= \frac{(1+9.43)\times10^{6}}{2}$$

$$= 5.21 \times 10^{6} \text{ m/sec}$$

$$\text{time taken to travel from cathode to anode} = \frac{d}{v_{avg}} = \frac{5\times10^{-3}}{5.2\times10^{6}}$$

$$= 0.958 \times 10^{-9}$$

14.5 An electron enters a uniform magnetic field of flux density 10^{-6} Wb/m^2 with a velocity of 10^8 m/sec, normal to the field. Calculate the radius of the circular path that the electron will trace.

Solution

$$F = ma = Bqv = \frac{mv^2}{r}$$

$$\therefore \quad r = \frac{mv^2}{Bqv} = \frac{mv}{qB}$$

$$= \frac{9.1\times10^{-31}\times10^{8}}{1.6\times10^{-19}\times10^{-6}}$$

$$= \frac{9.1\times10^{-23}}{1.6\times10^{-25}}$$

$$= 568.75 \text{ m}$$

14.6 An electrostatic cathode ray tube has a final anode voltage of 400 V. The deflection plates are 2 cm long and 1 cm apart. The screen is at a distance of 10 cm from the centre of the plates. A voltage of 20 V is applied to the deflection plates. Calculate (a) the velocity of the electron on reaching the field (b) acceleration due to deflection field (c) deflection produced on the screen and (d) deflection sensitivity.

Solution (a) $V = 600$ V, $l = 2$ cm, $d = 1$ cm, $L = 10$ cm $V_d = 20$ V

$$v = \text{Velocity} = \sqrt{\frac{2qV}{m}} = \sqrt{\frac{2\times1.6\times10^{-19}\times400}{9.1\times10^{-31}}}$$

$$= 11.85 \times 10^{6} \text{ m/s}$$

(b) $$ma = q\varepsilon$$

$$a = \frac{q\varepsilon}{m} = \frac{q \cdot \frac{V_d}{d}}{m} = \frac{1.6 \times 10^{-19} \times 20}{10^{-2} \times 9.1 \times 10^{-31}}$$

$$= 3.51 \times 10^{14} \text{ m/s}^2$$

(c) Deflection on the screen, $D = \dfrac{l \cdot L \cdot V_d}{2dV}$

$$D = \frac{2 \times 10^{-2} \times 10 \times 10^{-2} \times 20}{2 \times 10^{-2} \times 400}$$

$$= 0.5 \text{ cm}$$

(d) Deflection sensitivity $= \dfrac{D}{V_d}$

$$= \frac{0.5}{20}$$

$$= 0.025 \text{ cm/V}$$

14.7 In a cathode ray tube having electric deflection system, the deflection plates are 2 cm long and have a uniform spacing of 4 mm between them. The fluorescent screen is 25 cm away from the centre of the deflection plates. Calculate the deflection sensitivity, if the potential of the final anode is (a) 1000 V (b) 2000 V (c) 3500 V.

Solution $d = 4$ mm, $l = 2$ cm, $L = 25$ cm,

Deflection on the screen $D = \dfrac{lLV_d}{2dV}$

$$S = \text{Sensitivity} = \frac{D}{V_d} = \frac{lL}{2dV}$$

(a) $$S = \frac{2 \times 10^{-2} \times 25 \times 10^{-2}}{2 \times 4 \times 10^{-3} \times 1000} \quad \text{if } V = 1000 \text{ V}$$

$$= \frac{50 \times 10^{-4}}{8} = 0.625 \times 10^{-3} \text{ cm/V}$$

(b) When $V = 2000$ V

$$S = 0.3125 \times 10^{-3} \text{ cm/V}$$

(c) When $V = 3500$ V

$$S = \frac{2 \times 10^{-2} \times 25 \times 10^{-2}}{2 \times 4 \times 10^{-3} \times 3500} = 0.1785 \times 10^{-3} \text{ cm/V}$$

14.8 In a CRT, the distance of the screen from the centre of the magnetic field is 22 cm. The deflection magnetic field of flux density 2×10^{-4} Wb/m^2 extends for a length of 2.5 cm along the tube axis. The final anode voltage is 1250 V. Find the deflection of the spot in cm.

Solution $L = 22 \times 10^{-2}$ m, $B = 2 \times 10^{-4}$ Wb/m^2, $l = 2.5 \times 10^{-2}$ m, $V = 1250$ V

The deflection of the spot

$$D = \frac{lBL}{\sqrt{V}} \cdot \sqrt{\frac{q}{2m}}$$

$$= \frac{2.5 \times 10^{-2} \times 2 \times 10^{-4} \times 22 \times 10^{-2}}{\sqrt{1250}} \sqrt{\frac{1.6 \times 10^{-19}}{2 \times 9.1 \times 10^{-31}}}$$

$$= 0.923 \text{ cm}$$

14.9 The electron beam in a CRT is displaced vertically by a magnetic field of flux density 10^{-4} Wb/m^2. The length of the magnetic field along the tube axis is same as that of the electrostatic deflection plates. The final anode voltage is 450 V. Calculate the voltage which should be applied to the Y-deflection plates 0.6 cm apart, to return the spot back to the centre of the screen.

Solution The magnetostatic deflection $D = \frac{lBL}{\sqrt{V}} \cdot \sqrt{\frac{q}{2m}}$

The electrostatic deflection $D = \frac{lLV_d}{2dV}$

For returning the beam to the centre, the electrostatic deflection and the magnetostatic deflection must be equal.

$$\therefore \quad \frac{lBL}{\sqrt{V}} \cdot \sqrt{\frac{q}{2m}} = \frac{lLV_d}{2dV}$$

$$\Rightarrow \quad B\sqrt{\frac{q}{2m}} = \frac{V_d}{2d\sqrt{V}}$$

given $V = 450$, $B = 10^{-4}$ Wb/m^2, $d = 0.6 \times 10^{-2}$ m

$$\therefore \quad V_d = B\sqrt{\frac{q}{2m}} \cdot 2d\sqrt{V}$$

$$= 10^{-4}\sqrt{\frac{1.6 \times 10^{-19}}{2 \times 9.1 \times 10^{-31}}} \times 2 \times 0.6 \times 10^{-2} \times \sqrt{450}$$

$$= 7.547 \text{ V}$$

14.10 (a) If the voltage V_{m1} cos ωt and V_{m2} sin ωt are applied respectively to the horizontal and vertical deflecting plates of an oscilloscope, prove that the electron beam will trace an ellipse on the screen.

(b) A sine wave voltage is displayed on a CRO. Its vertical amplifier sensitivity is set at 5 V/cm and the time base selector swith is set at a sweep speed of 50 μ/cm. The displayed sine wave has peak-to-peak amplitude of 5.4 cm and its 4 complete cycles are accommodated over 8.4 cm of horizontal axis. Find the rms value and the frequency of the input voltage.

Solution (a)

$$X = V_{m1} \cos \omega t \Rightarrow \frac{X}{V_{m1}} = \cos \omega t$$

$$Y = V_{m2} \sin \omega t \Rightarrow \frac{Y}{V_{m2}} = \sin \omega t$$

Squaring and adding, we get

$$\frac{X^2}{V_{m1}^2} + \frac{Y^2}{V_{m2}^2} = 1$$

Hence proved.

(b)

$$V_{PP} = 5\frac{\text{V}}{\text{cm}} \times 5.4 \text{ cm} = 27 \text{ V}$$

$$V_{\text{rms}} = \frac{V_{PP}}{2\sqrt{2}} = \frac{27}{2\sqrt{2}} = 9.54 \text{ V}$$

$$\text{1 cycle takes } T = \frac{8.4 \text{ cm}}{4} \times \text{sweep speed}$$

$$= 2.1 \text{ cm} \times \frac{50\ \mu\text{s}}{\text{cm}} = 105\ \mu\text{s}$$

$$\therefore \quad \text{frequency} = 9.52 \text{ kHz}$$

14.11 Two sinusoidal voltage signals of equal frequency are applied to the vertical and horizontal deflection plates of CRO. Find the phase angle of each trace.

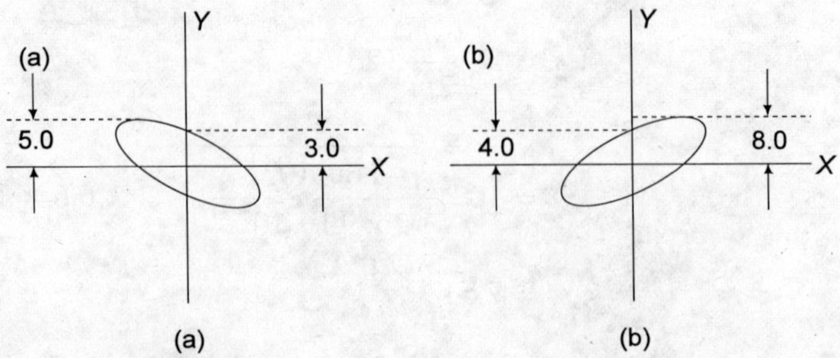

(a) (b)

Solution (a) Phase angle $\phi = \sin^{-1}\dfrac{3.0}{5.0} = \sin^{-1}(0.6) = 36.9°$ or $143.1°$.

(b) Phase angle $\phi = \sin^{-1}\dfrac{4.0}{8.0} = \sin^{-1}(0.5) = 30°$ or $150°$

14.12 (a) Find the ratio of frequencies of vertical and horizontal signals for a Lissajous pattern shown below.

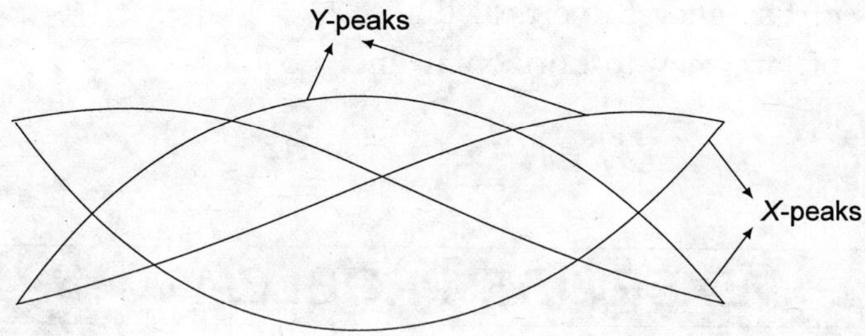

(b) Find the frequency of vertical signal if the frequency of horizontal signal is 3 kHz.

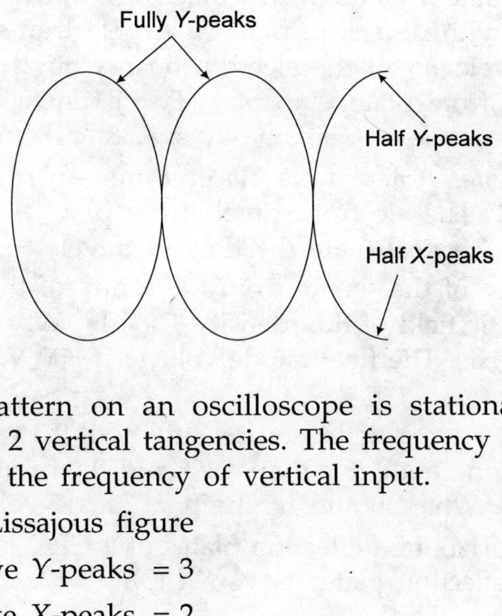

(c) A Lissajous pattern on an oscilloscope is stationary and has 5 horizontal tangencies and 2 vertical tangencies. The frequency of horizontal input is 1000 Hz. Determine the frequency of vertical input.

Solution (a) From the Lissajous figure

Positive Y-peaks = 3

Positive X-peaks = 2

$$\frac{f_y}{f_x} = \frac{\text{Positive } Y\text{ - peaks in pattern}}{\text{Positive } X\text{ - peaks in pattern}}$$

$$= \frac{3}{2}$$

(b) Positive Y-peaks in pattern = 2 + 1/2 = 2½

Positive X-peaks in pattern = 1/2 + 1/2 =1

$$\therefore \quad \frac{f_y}{f_x} = 2\tfrac{1}{2}$$

$$f_y = 3\text{ K} \times 2\tfrac{1}{2} = 7.5\text{ kHz}$$

(c) $f_x = 1000$ Hz

Points of tangency to vertical line = 2

Points of tangency to horizontal line = 5

$$\therefore \quad f_y = f_x \times \frac{5}{2} = 2500\text{ Hz}$$

EXERCISE PROBLEMS

14.1 Find the speed and kinetic energy of an electron after it has moved through a potential difference of 5000 V. (Ans: 4.2×10^7 m/s, 5000 eV)

14.2 Two plane parallel plates *A* and *B* are placed 3 mm apart and potential of B is made 200 V positive with respect to plate *A*. An electron starts from rest from plate *A*. Calculate (a) the velocity of the electron on reaching plate *B*. (b) time taken by the electron to travel from plate *A* to plate *B*. (c) Kinetic energy of the electron on reaching the plate *B*. (Ans: 8.38×10^6 m/s, 0.71 ns, 3.2×10^{-17} J)

14.3 Two plane parallel plates *A* and *B* are placed 8 mm apart and plate *B* is 300 V more positive than plate *A*. The electron travels from plate A to plate *B* with an initial velocity of 1×10^6 m/sec, calculate the time of travel. (Ans: 1.4 ns)

14.4 In a CRT, the distance of the screen from the centre of the magnetic field is 20 cm. The deflecting magnetic field of flux density 1×10^{-4} Wb/m^2 extends for a length of 2 cm along the tube axis. The final anode voltage is 800 V. Find the deflection of the spot. (Ans: 0.42 cm)

14.5 A CRT is designed to have a deflection sensitivity of 0.3 mm per volt. The plates are required to be 3 cm long and 0.6 cm apart. The distance of the screen from centre of plates is to be 20 cm. What should be the final anode voltage? (Ans: 3333 V)

14.6 The voltage across horizontal deflection plates of a CRO is $V_1 \sin(\omega t + \theta_1)$ and that across the vertical deflection plates is $V_2 \sin(\omega t + \theta_2)$. Prove that the trace on the screen is an ellipse.

14.7 Voltage V_1 is applied to the horizontal input and voltage V_2 is applied to the vertical input of the CRO. V_1 and V_2 are having the same frequency. The trace is an ellipse. The slope of the major axis is positive. The maximum vertical value is 2.5 divisions and the point where the ellipse crosses the vertical axis is 1.25 divisions. The ellipse is symmetrical about the horizontal and vertical axis. Determine the possible phase angles of V_2 with respect to V_1.

REVIEW QUESTIONS

1. Give the construction of CRT using electrostatic focusing and deflection systems and describe functions of each in detail.
2. Describe the method of electrostatic focusing in a CRT.
3. Describe the focusing action in a CRT using magnetic focusing.
4. What is the path of an electron entering a uniform magnetic field obliquely?
5. Derive the equation for electrostatic deflection sensitivity.
6. Prove that the electron beam in a CRT appears to emerge from a point at the centre of the electrostatic deflection system.
7. Derive the expression for deflection sensitivity of CRT in a magnetic deflection system.
8. What is the function of time base voltage in a CRO?
9. Draw a basic block diagram of a CRO and explain the features of CRO in detail.
10. What are the applications of a CRO? Explain any two of them in detail.
11. How do you measure frequency of an unknown signal using CRO? Explain.
12. What are Lissajous figures? How do you measure phase between two signals with the help of these figures?
13. Discuss the two-dimensional motion of an electron in a parallel-plate capacitor.
14. Discuss the force experienced by an electron in a magnetic field.
15. Prove that the path of an electron in a magnetic field is circular.
16. Determine the path of the electron when subjected to electric and magnetic fields, which are parallel to each other.
17. Discuss the behaviour of the electron when electric and magnetic fields are applied and when they are perpendicular to each other.
18. How many types of deflection systems are possible in a CRT? Discuss them in detail.
19. Compare the electrostatic and magnetic deflection in terms of deflection sensitivity.
20. How is focusing of electron beam achieved in a CRT? Explain in detail.
21. Explain how focusing of electron beam is achieved using the electrostatic field method.
22. What is magnetic focusing on an electron beam? Explain the principle in detail.
23. How do you measure current with a CRO?
24. Why is synchronisation of sweep signal and input signal required to display a signal on CRO? How is it achieved?
25. How is frequency of the unknown signal determined using Lissajous figures in a CRO?

Bibliography

Bandyopadhyay T.K., *Basic and Applied Electronics*, Books and Allied, Kolkata, 2002.

Bapat Y.N., *Electronic Circuits and Systems – Analog and Digital*, Tata McGraw Hill, New Delhi, 1992.

Boylestad, Robert L., & Nashelsky, Louis, *Electronic Devices and Circuit Theory*, Pearson Education, Delhi, 2002.

Chowdhary, Roy, *Linear Integrated Circuits*, 2nd edition, New Age International Publishers, New Delhi, 2003.

Floyd, Thomas L., *Electronic Devices*, Pearson Education, Delhi, 2003.

Gayakward, Ramakanth A., *Operational Amplifiers & Linear ICs*, 3rd edition, Prentice Hall of India, New Delhi, 2001.

Grob, Bernard, *Basic Electronics*, Tata McGraw Hill, New Delhi, 2000.

Gupta J.B., *Electronic Devices and Circuits*, S.K. Kataria and Sons, Delhi, 2000.

Hassul, Micheal, & Zimmerman, Don, *Electronic Devices and Circuits*, Prentice Hall International, 1997.

Islam S.S., *Semiconductor Physics & Devices*, Oxford University Press, 2006.

Malvino, Albert P., *Electronic Principles*, Tata McGraw Hill, New Delhi, 1984.

Millman, Jacob & Halkias, Christos C., *Integrated Electronics*, Tata McGraw Hill, New Delhi, 1991, reprint 2003.

Millman, Jacob & Grabel, Arvin, *Microelectronics*, McGraw Hill International, NY, 1978.

Mithal G.K., *Electronic Devices and Circuits*, Khanna Publishers, New Delhi, 1994.

Mottershead, Allen, *Electronic Devices and Circuits – An Introduction*, Prentice Hall of India, New Delhi, 1998.

Nair B., Somanathan, *Electronic Devices and Applications*, Prentice Hall of India, New Delhi, 2002.

Paul P., John, *Electronic Devices and Circuits*, New Age International, New Delhi, 2003.

Rao B.V., Narayana, *Principles of Electronics*, Willey Eastern Limited, New Delhi, 1992.

Salivahanan S., *Electronic Devices and Circuits*, Tata McGraw Hill, New Delhi, 1998.

Spencer, Richard R., & Ghausi, Mohammed S., *Introduction to Electronic Circuit Design*, Pearson Education, New Delhi, 2007.

Index

I

Q

R